W0018850

Regional Climate Studies

Series Editors: H.-J. Bolle, M. Menenti, I. Rasool

The BACC Author Team

Assessment of Climate Change
for the Baltic Sea Basin

 Springer

The BACC Author Team

The International BALTEX Secretariat
GKSS-Forschungszentrum
Geesthacht GmbH
Max-Planck-Str. 1
D-21502 Geesthacht
Germany
baltex@gkss.de

ISBN: 978-3-540-72785-9   e-ISBN: 978-3-540-72786-6

Regional Climate Studies ISSN: pending

Library of Congress Control Number: 2007938497

© 2008 Springer-Verlag Berlin Heidelberg

This work is subject to copyright. All rights are reserved, whether the whole or part of the material is concerned, specifically the rights of translation, reprinting, reuse of illustrations, recitation, broadcasting, reproduction on microfilm or in any other way, and storage in data banks. Duplication of this publication or parts thereof is permitted only under the provisions of the German Copyright Law of September 9, 1965, in its current version, and permission for use must always be obtained from Springer. Violations are liable to prosecution under the German Copyright Law.

The use of general descriptive names, registered names, trademarks, etc. in this publication does not imply, even in the absence of a specific statement, that such names are exempt from the relevant protective laws and regulations and therefore free for general use.

Cover design: deblik, Berlin

Printed on acid-free paper

9 8 7 6 5 4 3 2   SPIN 12583621   second corrected printing 2008

springer.com

Preface

Climate change and its impact on our life, our environment and ecosystems in general, are in these days at the forefront of public concern and political attention. The Intergovernmental Panel on Climate Change (IPCC) reports have amply documented that anthropogenic climate change is an ongoing trend, which will continue into the future and may be associated with grave implications. Thus, one conclusion to be drawn is clear – the driver for this climate change should be curtailed to the extent socially responsible and sustainable. We need to strongly reduce the emissions of radiatively active gases into the atmosphere.

However, even the most optimistic emission reduction scenarios envision only a limited success in thwarting climate change. What is possible is to limit this change, but it can no longer be avoided altogether. Even if the challenging goal of a stabilization of global mean temperature at an upper limit of 2 °C above pre-industrial levels at the end of this century will be met, significant pressures on societies and ecosystems for adaptation will be the result. Thus, adaptation to recent, ongoing and possible future climate change is unavoidable.

The BACC initiative has dealt with these pressures for the region of the Baltic Sea Basin, which includes the Baltic Sea and its entire water catchment, covering almost 20% of the European continent. BACC has collected, reviewed and summarized the existing knowledge about recent, ongoing and possible futures regional climate change and its impact on marine and terrestrial ecosystems in this region. The acronym BACC stands for *BALTEX Assessment of Climate Change for the Baltic Sea Basin* and denotes an initiative within BALTEX (Baltic Sea Experiment), which is a Regional Hydrometeorology Project within the Global Energy and Water Cycle Experiment (GEWEX) of the World Climate Research Program (WCRP).

The first chapter of the book places the initiative in context, clarifies a few key concepts and summarizes the key results; Chapters 2 to 5 document the knowledge about recent and ongoing changes in meteorological, oceanographical, hydrological and cryospherical variables, about scenarios of possible future climatic conditions, about changes in terrestrial and freshwater ecosystems, and about changes in marine ecosystems. A series of appendices provide background material relevant in this context.

Two remarkable aspects of the BACC initiative should be mentioned. The first is the acceptance of this report by the Helsinki Commission (HELCOM) as a basis for its intergovernmental management of the Baltic Sea environment. Based on this BACC report, HELCOM has compiled its own conclusions "Climate Change in the Baltic Sea Area – HELCOM Thematic Assessment in 2007". The second aspect is the fact that the BACC report was made possible by the voluntary effort of many individuals and institutions – without dedicated payment from scientific agencies, governments, NGOs, industries or other possibly vested interests. We think this adds significantly to the credibility of this effort, which we expect will be used as a blueprint for assessments of other regions in the world.

The success of BACC has convinced BALTEX that it would be worth to redo the effort in about five years time – assuming that significantly more knowledge has been generated, and that climate change has emerged even more clearly from the "sea of noise" of natural climate variability.

Hans von Storch (Chairman)

The BACC Author Team

The BACC Author Team consists of more than 80 scientists from 13 countries, covering various disciplines related to climate research and related impacts. Each chapter of the book is authored by one to four *lead authors* and several *contributing authors*. While the former established the overall conception, did much of the writing and are largely responsible for the assessment parts of the chapters, the latter contributed pieces of information of various extent to the contents of the book. In order to highlight the teamwork character of this book, both lead and contributing authors of each chapter are mentioned as an author group at the beginning of each chapter, rather than attributing individual section or text contributions to individual contributing authors. Lead authors are mentioned first followed by contributing authors, ordered alphabetically. The authors of annexes, by contrast, are named individually on top of each annex, section or sub-section within the annex part.

The following authors list firstly gives the lead authors of Chaps. 1 to 5, followed by an alphabetically ordered list of all other contributing and annex authors.

Lead Authors

Chapter 1: Introduction and Summary

Hans von Storch
Institute for Coastal Research, GKSS Research Centre Geesthacht, Germany

Anders Omstedt
Earth Sciences Centre – Oceanography, Göteborg University, Sweden

Chapter 2: Past and Current Climate Change

Raino Heino
Finnish Meteorological Institute, Helsinki, Finland

Heikki Tuomenvirta
Finnish Meteorological Institute, Helsinki, Finland

Valery S. Vuglinsky
Russian State Hydrological Institute, St. Petersburg, Russia

Bo G. Gustafsson
Earth Sciences Centre – Oceanography, Göteborg University, Sweden

Chapter 3: Projections of Future Anthropogenic Climate Change

L. Phil Graham
Rossby Centre, Swedish Meteorological and Hydrological Institute, Norrköping, Sweden

Chapter 4: Climate-related Change in Terrestrial and Freshwater Ecosystems

Benjamin Smith
Department of Physical Geography and Ecosystems Analysis, Lund University, Sweden

Chapter 5: Climate-related Marine Ecosystem Change

Joachim W. Dippner
Baltic Sea Research Institute Warnemünde, Germany

Ilppo Vuorinen
Archipelago Research Institute, University of Turku, Finland

Contributing Authors

Anto Aasa
Institute of Geography, University of Tartu, Estonia

Rein Ahas
Institute of Geography, University of Tartu, Estonia

Hans Alexandersson
Swedish Meteorological and Hydrological Institute, Norrköping, Sweden

Philip Axe
Oceanographic Services, Swedish Meteorological and Hydrological Institute, Västra Frolunda, Sweden

Lars Bärring
Rossby Centre, Swedish Meteorological and Hydrological Institute, Norrköping, Sweden; and
Department of Physical Geography and Ecosystems Analysis, Lund University, Sweden

Svante Björck
Department of Geology, GeoBiosphere Science Centre, Lund University, Sweden

Thorsten Blenckner
Department of Ecology and Evolution, Erken Laboratory, Uppsala University, Sweden

Agrita Briede
Department of Geography, University of Latvia, Riga, Latvia

Terry V. Callaghan
Abisko Scientific Research Station, Royal Swedish Academy of Sciences, Abisko, Sweden; and
Department of Animal and Plant Sciences, University of Sheffield, UK

John Cappelen
Data and Climate Division, Danish Meteorological Institute, Copenhagen, Denmark

Deliang Chen
Earth Sciences Centre, Göteborg University, Sweden

Ole Bøssing Christensen
Danish Climate Centre, Danish Meteorological Institute, Copenhagen, Denmark

Gerhard Dahlmann
Federal Maritime and Hydrographic Agency, Hamburg, Germany

Darius Daunys
Coastal Research and Planning Institute, University of Klaipeda, Lithuania

Jacqueline de Chazal
Department of Geography, Catholic University of Louvain, Louvain-la-Neuve, Belgium

Jüri Elken
Marine Systems Institute, Tallinn University of Technology, Estonia

Malgorzata Falarz
Department of Climatology, University of Silesia, Sosnowiec, Poland

Juha Flinkman
Finnish Institute of Marine Research, Helsinki, Finland

Eirik J. Førland
Department of Climatology, Norwegian Meteorological Institute, Oslo, Norway

Jari Haapala
Finnish Institute of Marine Research, Helsinki, Finland

Lars Håkanson
Department of Hydrology, Uppsala University, Sweden

Antti Halkka
Department of Biological and Environmental Sciences, University of Helsinki, Finland

Marianne Holmer
Institute of Biology, University of Southern Denmark, Odense, Denmark

Christoph Humborg
Department of Applied Environmental Science, Stockholm University, Sweden

Hans-Jörg Isemer
International BALTEX Secretariat, GKSS Research Centre Geesthacht, Germany

Jaak Jaagus
Institute of Geography, University of Tartu, Estonia

Anna-Maria Jönsson
Department of Physical Geography and Ecosystems Analysis, Lund University, Sweden

Seppo Kellomäki
Faculty of Forest Sciences, University of Joensuu, Finland

Lev Kitaev
Institute of Geography, Russian Academy of Sciences, Moscow, Russia

Erik Kjellström
Rossby Centre, Swedish Meteorological and Hydrological Institute, Norrköping, Sweden

Friedrich Köster
Danish Institute for Fisheries Research, Technical University of Denmark, Lyngby, Denmark

Are Kont
Estonian Institute of Ecology, University of Tartu, Estonia

Valentina Krysanova
Department of Global Change and Natural Systems, Potsdam Institute for Climate Impact Research, Potsdam, Germany

Ain Kull
Institute of Geography, University of Tartu, Estonia

Esko Kuusisto
Finnish Environment Institute, Helsinki, Finland

Esa Lehikoinen
Department of Biology, University of Turku, Finland

Maiju Lehtiniemi
Finnish Institute of Marine Research, Helsinki, Finland

Göran Lindström
Research and Development – Hydrology, Swedish Meteorological and Hydrological Institute,
Norrköping, Sweden

Brian R. MacKenzie
Danish Institute for Fisheries Research, Technical University of Denmark, Charlottenlund, Denmark

Ülo Mander
Institute of Geography, University of Tartu, Estonia

Wolfgang Matthäus
Baltic Sea Research Institute Warnemünde, Germany

H.E. Markus Meier
Research and Development – Oceanography, Swedish Meteorological and Hydrological Institute,
Norrköping, Sweden; and Department of Meteorology, Stockholm University, Sweden

Mirosław Miętus
Department of Meteorology and Climatology, University of Gdańsk, Poland; and
Institute of Meteorology and Water Management – Maritime Branch, Gdynia, Poland

Anders Moberg
Department of Physical Geography and Quaternary Geology, Stockholm University, Sweden

Flemming Møhlenberg
DHI – Water · Environment · Health, Hørsholm, Denmark

Christian Möllmann
Institute for Hydrobiology and Fisheries Science, University of Hamburg, Germany

Kai Myrberg
Finnish Institute of Marine Research, Helsinki, Finland

Tadeusz Niedźwiedź
Department of Climatology, University of Silesia, Sosnowiec, Poland

Peeter Nõges
Institute for Environment and Sustainability, European Commission, Joint Research Centre, Ispra,
Italy

Tiina Nõges
Institute of Agricultural and Environmental Sciences, Estonian University of Life Sciences, Tartu,
Estonia

Øyvind Nordli
Norwegian Meteorological Institute, Oslo, Norway

Sergej Olenin
Coastal Research and Planning Institute, Klaipeda University, Lithuania

Kaarel Orviku
Merin Ltd, Tallinn, Estonia

Zbigniew Pruszak
Institute of Hydroengineering, Polish Academy of Sciences, Gdansk, Poland

Maciej Radziejewski
Faculty of Mathematics and Computer Science, Adam Mickiewicz University, Poznan, Poland; and
Research Centre of Agriculture and Forest Environment, Polish Academy of Sciences, Poznan, Poland

Jouni Räisänen
Division of Atmospheric Sciences, University of Helsinki, Finland

Egidijus Rimkus
Department of Hydrology and Climatology, Vilnius University, Lithuania

Burkhardt Rockel
Institute for Coastal Research, GKSS Research Centre Geesthacht, Germany

Mark Rounsevell
School of Geosciences, University of Edinburgh, UK

Kimmo Ruosteenoja
Finnish Meteorological Institute, Helsinki, Finland

Viivi Russak
Atmospheric Physics, Tartu Observatory, Estonia

Ralf Scheibe
Geographical Institute, University of Greifswald, Germany

Doris Schiedek
Baltic Sea Research Institute Warnemünde, Germany

Corinna Schrum
Geophysical Institute, University of Bergen, Norway

Henrik Skov
DHI – Water · Environment · Health, Hørsholm, Denmark

Mikhail Sofiev
Finnish Meteorological Institute, Helsinki, Finland

Wilhelm Steingrube
Geographical Institute, University of Greifswald, Germany

Ülo Suursaar
Estonian Marine Institute, University of Tartu, Estonia

Piotr Tryjanowski
Department of Behavioural Ecology, Adam Mickiewicz University, Poznan, Poland

Timo Vihma
Finnish Meteorological Institute, Helsinki, Finland

Norbert Wasmund
Baltic Sea Research Institute Warnemünde, Germany

Ralf Weisse
Institute for Coastal Research, GKSS Research Centre Geesthacht, Germany

Joanna Wibig
Department of Meteorology and Climatology, University of Łódź, Poland

Annett Wolf
Centre for Geobiosphere Science, Lund University, Sweden

Acknowledgements

The BACC Author Team appreciates the work of the BACC Science Steering Committee (SSC) which contributed in various ways to the project by giving advice, providing material and reviewing parts of earlier manuscript versions. BACC SSC members include Sten Bergström, Swedish Meteorological and Hydrological Institute, Norrköping, Sweden; Jens Hesselbjerg Christensen and Eigil Kaas, both at Danish Meteorological Institute, Copenhagen, Denmark; Zbigniew W. Kundzewicz, Research Centre for Agricultural and Forest Environment, Polish Academy of Sciences, Poznan, Poland; Jouni Räisänen, Helsinki University, Finland; Markku Rummukainen, Swedish Meteorological and Hydrological Institute, Norrköping, Sweden; Morten Søndergaard, Copenhagen University, Denmark; and Bodo von Bodungen, Baltic Sea Research Institute Warnemünde, Germany.

An in-depth review of the main book chapters was conducted by several independent evaluators as part of the BACC process which improved the book material significantly. We highly appreciate this valuable contribution to BACC.

The BACC project benefited from countless comments by many scientists on earlier versions of the book's manuscript. The BACC Conference held at Göteborg University in May 2006, where a first version of the material had been presented for discussion to the scientific community and the interested public, was highly instrumental in fostering improvement of the BACC material and the BACC Author Team likes to express its gratitude to all who participated actively at the Conference and for all helpful comments received.

We thank Hartmut Graßl in his capacity as the former BALTEX Scientific Steering Group chairman for initiating both the BACC review process and the cooperation between BACC and the Helsinki Commission (HELCOM).

The cooperation of BACC with HELCOM turned out to be of high mutual benefit and the BACC Author Team appreciates the HELCOM Secretariat staff's continuous interest and cooperation. It is our particular pleasure to thank Janet F. Pawlak at Marine Environmental Consultants, Denmark, and Juha-Markku Leppänen, Professional Secretary at the HELCOM Secretariat in Helsinki, for their enthusiasm and excellent contributions to the progress of the BACC project.

The BACC Author Team would like to thank Hans-Jörg Isemer for keeping up the communication between BALTEX, BACC and HELCOM, and for being the never-stopping engine in the whole preparation process of this book. Without his enthusiasm, the BACC process would not have been as efficient as it was.

The final shaping and editing of the huge BACC material according to the publisher's requirements turned out to be a major effort. The BALTEX Secretariat at GKSS Research Centre Geesthacht undertook to coordinate and perform this editing process and we would like to thank the Secretariat's staff Marcus Reckermann and Hans-Jörg Isemer, painstakingly supported by Silke Köppen, for skilfully coordinating and conducting this editing process. Numerous figures in this book needed to be redrawn or had to be graphically enhanced, which was excellently performed by Beate Gardeike (GKSS, Institute for Coastal Research). Parts of earlier versions of the manuscript needed major language editing and we like to appreciate the meticulous work by Susan Beddig, Hamburg. Last but not least, we thankfully highlight the major contribution by Sönke Rau who accurately established the print-ready book manuscript, thereby eliminating several errors and patiently adding all, even last-minute, changes to the manuscript.

Permissions

Every effort has been made to trace and acknowledge copyright holders. Should any infringements have occurred, apologies are tendered and omissions will be rectified in the event of a reprint of the book.

Contents

The Baltic Sea Basin on 1 April 2004, as seen from the SeaWiFS satellite (NASA/Goddard Space Flight Center, GeoEye)

1 Introduction and Summary

Hans von Storch, Anders Omstedt

In this introductory chapter, we describe the mission and the organisational structure of BACC, the "BALTEX Assessment of Climate Change in the Baltic Sea Basin" (www.baltex-research.eu/ BACC/). Short introductions of the specifics of the Baltic Sea Basin, in terms of geological history, climate, marine and terrestrial ecosystems as well as some aspects of the economic condition, are provided. The different assessments of instationarities in the observational record are reviewed, and the concept of "climate change scenarios" is worked out. Finally, the key findings of the four main Chaps. 2 to 5 are summarised.

1.1 The BACC Approach

1.1.1 General Background – The Global Context

In the last two decades the concept of anthropogenic climate change, mainly related to the release of greenhouse gases, has been firmly established, in particular through the three[1] assessment reports by the Intergovernmental Panel Climate Change IPCC (Houghton et al. 1990, 1992, 1996, 2001). This insight is based on remarkable advances in science and technology related to climate studies. Important progress has been made, for example with respect to the climate archives, correcting data, making data available through data centres, process understanding and modelling. The result of these efforts is an increased understanding of the key aspects of climate dynamics and of climate change on the global scale. These efforts culminated in the famous assertion of the IPCC, according to which the hypothesis that recent climate change is entirely due to natural causes can be rejected with very little risk (global detection), and its conclusion that the elevated greenhouse gas concentrations are the best single explanatory factor. These findings refer mostly to variables and phenomena linked to

the thermal climate regime, e.g. temperature itself, number of frost days, ice and snow.

The situation is less clear when regional scales (less than $10^7 \, \text{km}^2$) are considered. For smaller scales, the weather noise is getting larger, so that the detection of systematic changes becomes difficult or even impossible. In general, very few efforts have been made. For the Baltic Sea Basin no rigorous detection studies have been carried out; however, under the influence of this assessment finding, efforts have now been launched do deal with such questions.

1.1.2 Climate Change Definition

In this book we address the problem of "climate change", which is unfortunately differently understood in different quarters (e.g. Bärring 1993; Pielke 2004). The problem is that "inconstancy" (Mitchell et al. 1966) is an inherent property of the climate system. Some use the term "climate change" to refer to "all forms of climatic inconstancy, regardless of their statistical nature (or physical causes)" (Mitchell et al. 1966). Also, the Intergovernmental Panel on Climate Change (IPCC) defines climate change broadly as "any change in climate over time whether due to natural variability or as a result of human activity." In contrast, the United Nation's Framework Convention on Climate Change (UNFCCC) defines climate change as "a change of climate that is attributed directly or indirectly to human activity that alters the composition of the global atmosphere, and that is in addition to natural climate variability over comparable time periods". Obviously, it is rather important which definition is used, in particular when communicating with the public and the media (Bärring 1993; Pielke 2004).

BACC has decided to essentially follow the IPCC-definition, and to add explicitly "anthropogenic" to the term "climate change" when human causes are attributable, and to refer to "climate variability" when referring to variations not related to anthropogenic influences.

---

[1] The editorial deadline for his book was in 2006 prior to the publication of the 4th IPCC Assessment Report, which is therefore not referenced throughout this book. Statements in this book made with reference to the 3rd IPCC Report are consistent with the assessment of the 4th IPCC report.

1.1.3 The BACC Initiative and the HELCOM link

The purpose of the BACC assessment is to provide the scientific community with an assessment of ongoing climate variations in the Baltic Sea Basin. An important element is the comparison with the historical past, whenever possible, to provide a framework for the severity and unusualness of the variations, whether it may be seen as climate variability or should be seen as anthropogenic climate change. Also, changes in relevant environmental systems due to climate variations are assessed – such as hydrological, oceanographic and biological changes. The latter studies also take account of, and attempt to differentiate, the impacts of changes in other driving factors that co-vary with climate *sensu stricto*, including atmospheric CO_2 concentrations but also acidification, pollution loads, nutrient deposition, land use change and other factors.

The overall format is similar to the IPCC process, with author groups for the individual chapters, an overall summary for policymakers, and a review process. The review process has been organised by the former chair of the BALTEX Science Steering Group, Professor Hartmut Graßl, Hamburg.

Altogether, the BACC team comprises more than 80 scientists from 13 nations, most of them based in the countries around the Baltic Sea, spanning a spectrum of disciplines from meteorology, oceanography and atmospheric chemistry to ecology, limnology and human geography. Each of the Chaps. 1 to 5 has one or more "lead authors", who had the responsibility of organising the work of their assessment groups, consisting of contributors from almost all countries in the Baltic Sea Basin. These groups had the task of considering all relevant published work in their assessment, not only in English but also as far as possible in all of the many languages of the region.

When the BACC initiative was well underway, a contact with the Helsinki Commission, or HELCOM, was established[2]. It turned out that HELCOM was in need for a climatic assessment of the

Baltic Sea area. It was agreed that the BACC report may become a basis for HELCOM's assessment of climate change – which eventually became true: the findings of the BACC report were summarized and put in context in the "HELCOM Thematic Assessment in 2007: Climate Change in the Baltic Sea Area" (Baltic Sea Environment Proceedings No. 111)[3]. This Thematic Assessment was formally adopted at the annual Meeting of the Helsinki Commission by the representatives of the Baltic Sea coastal countries in March 2007. It was announced that this assessment will serve as a background document to the HELCOM Baltic Sea Action Plan to further reduce pollution to the sea and restore its good ecological status, "which is slated to be adopted at the HELCOM Ministerial Meeting in November 2007".

1.2 The Baltic Sea – Geological History and Specifics

In the following a very brief introduction into the specifics of the Baltic Sea Basin is provided; for further details, refer to the Annexes.

1.2.1 Geological History of the Baltic Sea

Since the last deglaciation of the Baltic Sea Basin, which ended 11,000–10,000 cal yr BP, the Baltic Sea has undergone many very different phases. The nature of these phases was determined by a gradually melting Scandinavian Ice Sheet, the glacio-isostatic uplift within the basin, the changing geographic position of the controlling sills, the varying depths and widths of the thresholds between the Baltic Sea and the land surface of the Baltic Sea Basin, and the changing climate. During these phases, salinity varied greatly as did the water exchange with the North Sea.

In the first phase, at the end of the glaciation and during the Younger Dryas, the *Baltic Ice Lake* (BIL) located in front of the last receding ice sheet was formed. It was repeatedly blocked from the ocean, and at least twice, the damming failed at the location of Billingen, with dramatic consequences. The final drainage of the BIL was a turning point in the late geologic development of the Baltic Sea: a warmer climate, a rapidly retreating ice sheet and direct contact with the saline sea in the west characterised the starting point for the *Yoldia Sea* stage, which would last approximately

---

[2]HELCOM is assessing and dealing with environmental conditions of the Baltic Sea from all sources of pollution through intergovernmental co-operation between Denmark, Estonia, the European Community, Finland, Germany, Latvia, Lithuania, Poland, Russia and Sweden. HELCOM is the governing body of the "Convention on the Protection of the Marine Environment of the Baltic Sea Area" – more usually known as the Helsinki Convention.

[3]www.helcom.fi/stc/files/Publications/Proceedings/bsep111.pdf

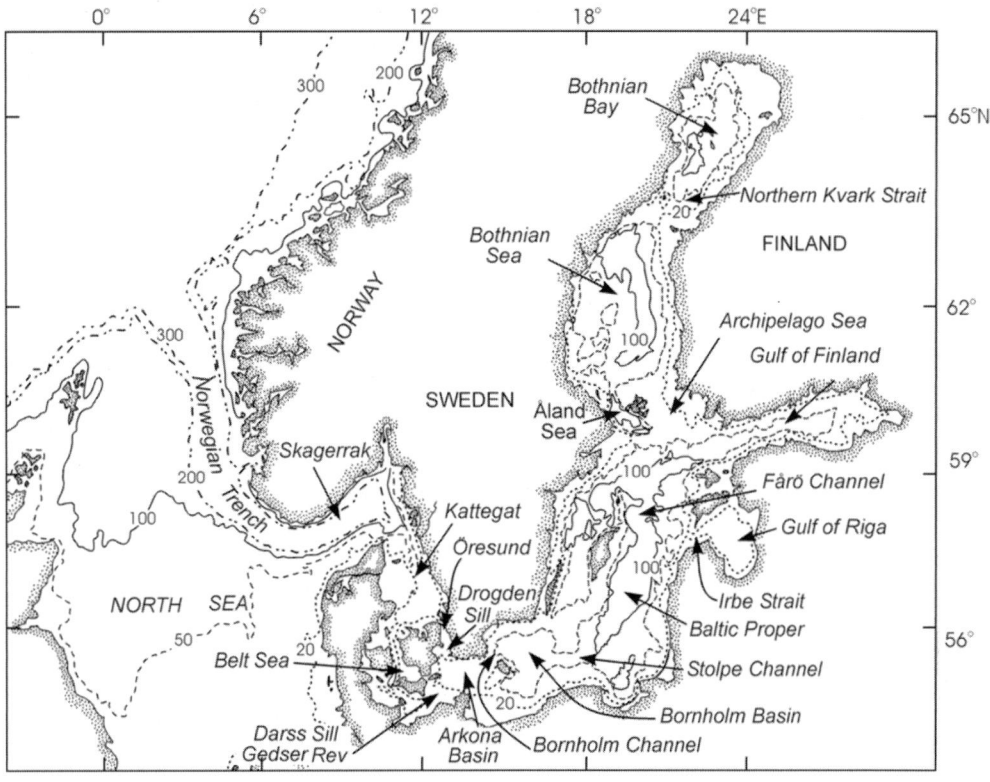

Fig. 1.1. The Baltic Sea-North Sea region with depth contours indicated (from Omstedt et al. 2004a)

900 years, followed by the *Ancylus Lake* transgression, which started around 10,700 cal yr BP. The *Ancylus* transgression ended abruptly with a sudden lowering of the Baltic water level at ca. 10,200 cal yr BP.

A rapid spread of saline influence throughout the Baltic Sea Basin occurred between 9,000–8,500 cal yr BP. The phase of the *Littorina Sea* is reflected in increased organic content in the sediments. With the increased saline influence, aquatic primary productivity clearly increased in the Baltic Sea. During 8,500–7,500 cal yr BP the first and possibly most significant *Littorina* transgression set in. The extent of this and the next two transgressions was of the order of at least 10 m in the inlet areas, with a large increase in water depth at all critical sills. This allowed a significant flux of saline water into the Baltic Sea. The increasing salinity, in combination with the warmer climate of the mid-Holocene, induced a rather different aquatic environment. In terms of richness and diversity of life, and therefore also primary productivity, the biological culmination of the Baltic Sea was possibly reached during the period 7,500–

6,000 cal yr BP. The high productivity, in combination with increased stratification due to high salinities in the bottom water, caused anoxic conditions in the deeper parts of the Baltic Sea.

A last turning point in Baltic Sea development took place after about 6,000 cal yr BP: the transgression came to an end almost everywhere along the Baltic Sea coast line. Due to uplift, a renewed regression occurred, which went along with shallower sills and a reduced flux of marine water into the basin. Baltic sediments suggest that since then salinities in the Baltic Sea have decreased.

1.2.2 Oceanographic Characteristics

The Baltic Sea is one of the largest brackish seas in the world[4]. It is a semi-enclosed basin with a total area of 415,000 km^2 and a volume of 21,700 km^2 (including Kattegat; Fig. 1.1). The Baltic Sea is highly dynamic and strongly influenced by large-scale atmospheric circulation, hydrological processes in the catchment area and by the restricted

---

[4]A detailed description of the Baltic Sea is given in Annex 1

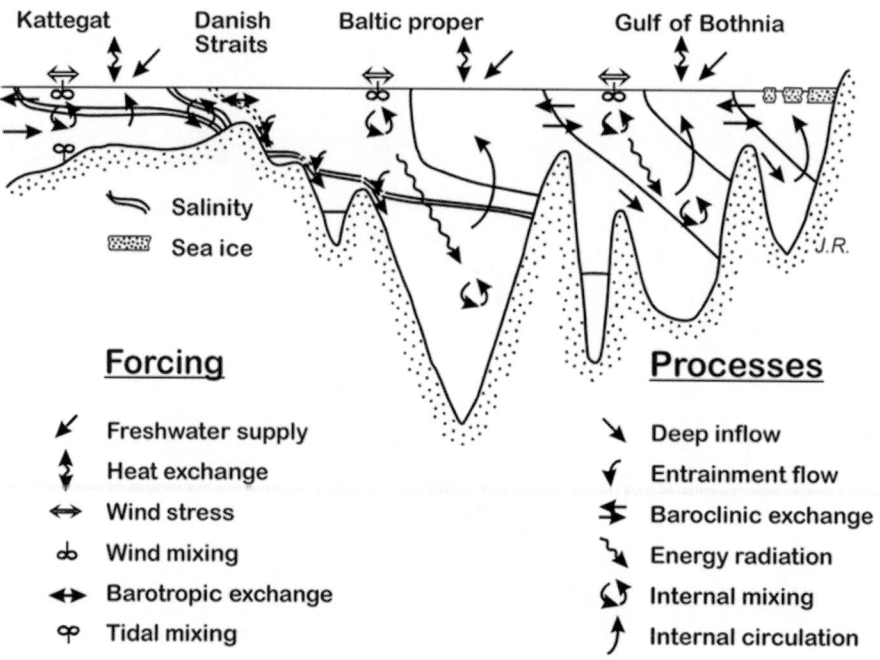

Fig. 1.2. Conceptual model of the Baltic Sea. On the left are processes that force the exchange and mixing and on the right processes that distribute the properties within the Baltic Sea (from Winsor et al. 2001)

water exchange due to its narrow entrance area. It can be divided into a number of different areas; the Kattegat, the Belt Sea, the Öresund, the Baltic Proper, the Bothnian Bay, the Bothnian Sea, the Gulf of Finland and the Gulf of Riga. The Baltic Proper includes the sill areas at its entrance, the shallow Arkona Basin, the Bornholm Basin and the waters up to the Åland and Archipelago Seas.

The complex bathymetry of the Baltic Sea, with its narrow straits connecting the different basins, strongly influences currents and mixing processes (Fig. 1.2). The inflow of freshwater, mainly from rivers into the Baltic Sea, can be described as the engine which drives the large-scale circulation. This inflow generally causes a higher water level in the Baltic Sea than in the Kattegat. The differences in water level force the brackish surface water out of the Baltic Sea. On its way towards the Skagerrak, the brackish water becomes increasingly saline, since the surface water becomes mixed with underlying water and fronts. As compensation for the water entrained into the surface currents, dense bottom water originating from the Skagerrak and Kattegat flows into the Baltic Sea and fills the deeps.

The large scale circulation of the Baltic Sea is due to a non-linear interaction between the estuarine circulation and the exchange with the North Sea. Figure 1.3 represents a conceptual description of the long-term mean circulation (see also Annex 1). Added to the estuarine circulation are large fluctuations, caused by changing winds and water level variations. These influence the water exchange with the North Sea and between the sub-basins, as well as transport and mixing of water within the various sub-regions of the Baltic Sea.

The Baltic Sea has a positive water balance. The major water balance components are inflows and outflows at the entrance area, river runoff and net precipitation. Changes in water storage also need to be considered. Minor terms in the long-term budget are volume change by groundwater inflow, thermal expansion, salt contraction, land uplift and ice export.

The salt balance is maintained by an outflow of low saline water in the surface layer, and a variable inflow of higher saline water at depth. This pattern leads to a permanent stratification of the central Baltic Sea water body, consisting of an upper layer of brackish water with salinities of about

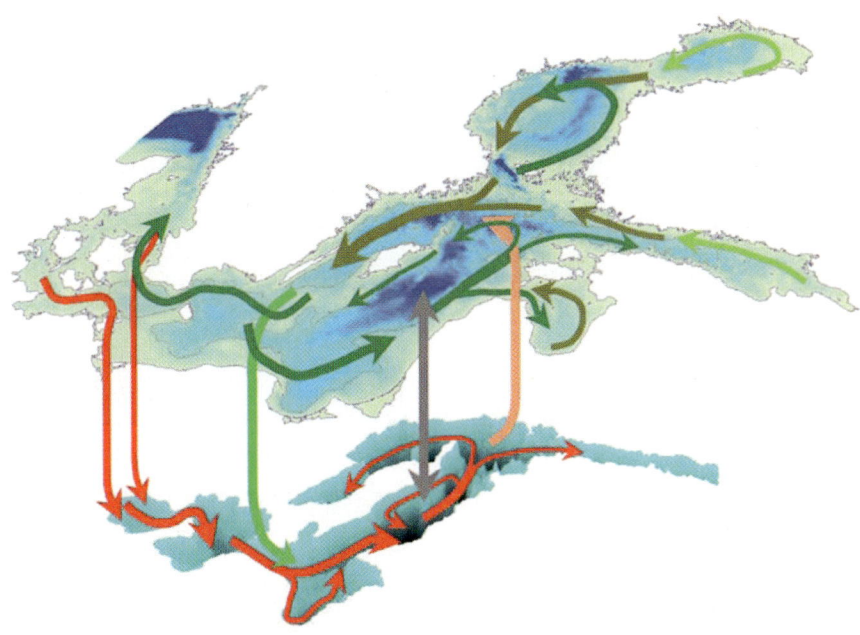

Fig. 1.3. Conceptual model of the Baltic Sea mean circulation. Deep layer circulation below the halocline is given in the lower part of the figure (by courtesy of J. Elken, for details see also Annex 1.1)

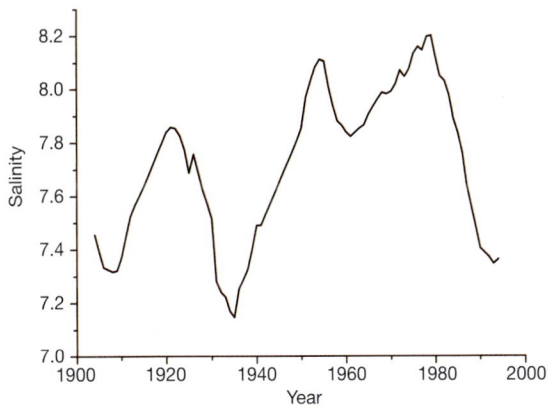

Fig. 1.4. Baltic Sea mean salinity (psu) averaged vertically and horizontally and presented as 5 years running means (from Winsor et al. 2001, 2003)

Fig. 1.5. The maximum ice extent in mild, average, severe, and extremely severe winters is marked analogously with darker colours for the more severe ice winters (redrawn from Seinä and Palosuo 1996, see also Fig. 2.59)

6–8 and a more saline deep water layer of about 10–14[5]. Figure 1.4 illustrates the long-term Baltic Sea mean salinity of 7.7 and its rather slow variation with an amplitude of 0.5 and a time scale of about 30 years.

The temperature undergoes a characteristic annual cycle. During spring, a thermocline develops, separating the warm upper layer from the cold intermediate water. This thermocline restricts vertical exchange within the upper layer until late autumn. Sea ice is formed every year, with a long-

---

[5]The salinity is given according to the Practical Salinity Unit (psu) defined as a pure ratio without dimensions or units. This is standard since 1981 when UNESCO adopted the scale.

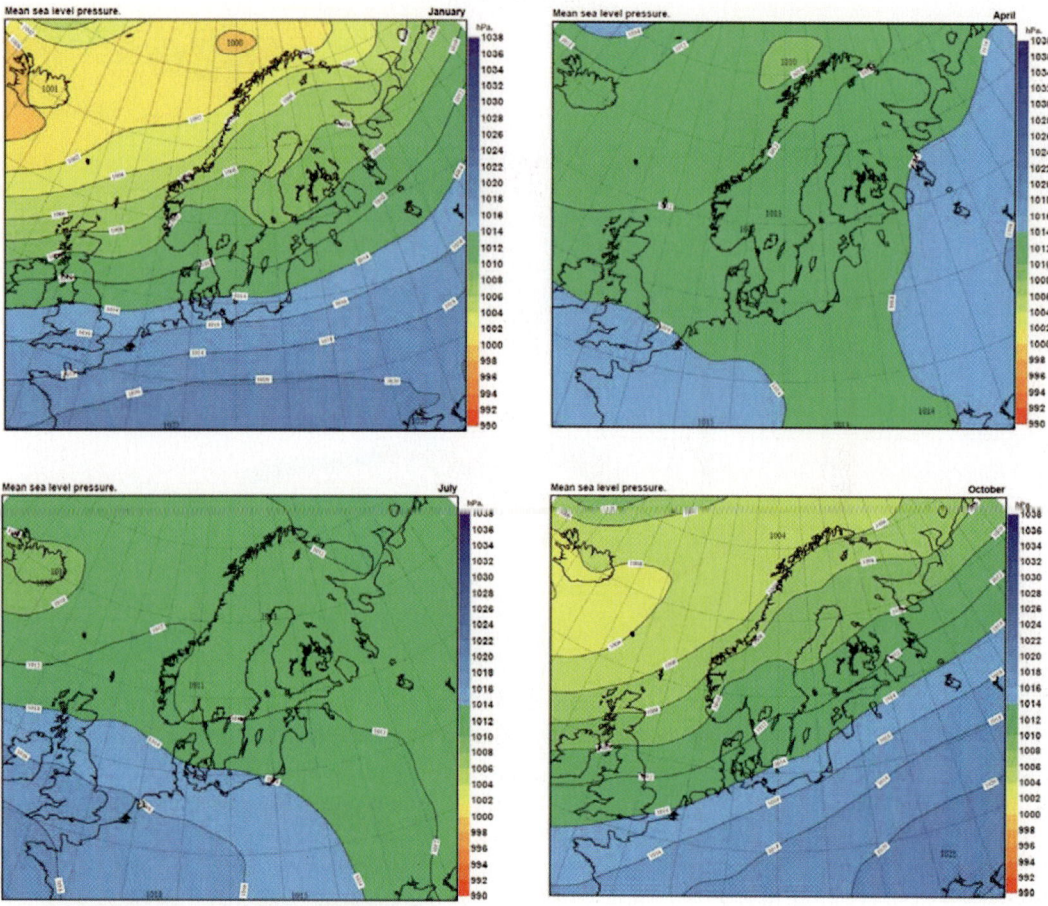

Fig. 1.6. Mean monthly patterns of sea level air pressure during 1979 to 2001 for January, April, July and October (*from top left to bottom right*). The maps were kindly provided by Per W. Kållberg, Swedish Meteorological and Hydrological Institute (SMHI), Norrköping, Sweden, and are extracted from the ERA-40 re-analysis data set (Uppala et al. 2005)

term average maximum coverage of about half of the surface area and with large inter-annual variations (Fig. 1.5).

On average, the Baltic Sea is almost in thermodynamic equilibrium with the atmosphere. The dominating fluxes, with respect to annual means, are the sensible heat, the latent heat, the net long wave radiation, the solar radiation to the open water and the heat flux between water and ice. Minor terms, in relation to long term means, are heat fluxes associated with the differences between inflows and outflows, river runoff and precipitation.

1.2.3 Climate Characteristics

The climate of the Baltic Sea is strongly influenced by the large scale atmospheric pressure sys-

tems that govern the air flow over the region: The Icelandic Low, the Azores High and the winter high/summer low over Russia. The westerly winds bring, despite the shelter provided by the Scandinavian Mountains, humid and mild air into the Baltic Sea Basin. The climate in the southwestern and southern parts of the basin is maritime, and in the eastern and northern parts of the basin it is sub-arctic. The long-term mean circulation patterns are illustrated in Fig. 1.6. Westerly winds dominate the picture, but the circulation pattern shows a distinct annual cycle, with strong westerlies during autumn and winter conditions. The atmospheric circulation is described in more detail in Annexes 1.2 and 7.

The mean near-surface air temperature of the Baltic Sea Basin is, on average, several degrees

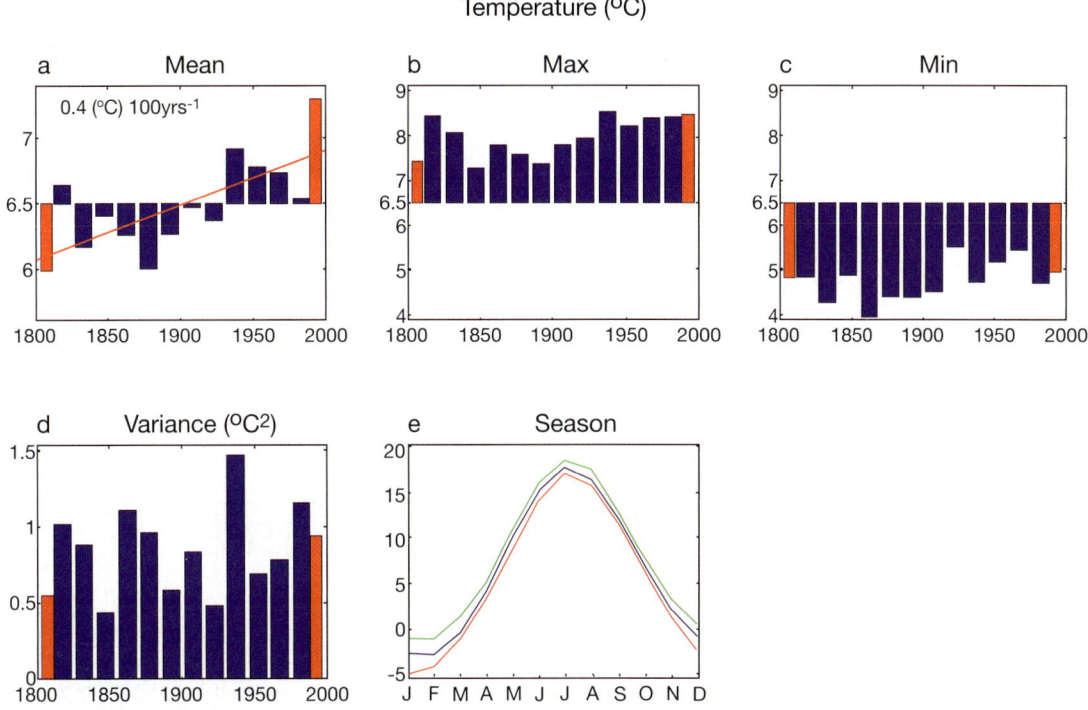

Fig. 1.7. Climate statistics of the Stockholm mean annual air temperature for individual sub-periods during 1800 to 2000. (**a**) Sub-period mean values and trend, (**b**) maximum, (**c**) minimum, and (**d**) variance of individual annual means within each sub-period. The sub-period lengths used are 15 years (*blue bars*) and 10 years (*red bars*). (**e**) shows the mean, maximum and minimum seasonal cycle based on sub-period values (from Omstedt et al. 2004b)

higher than that of other areas located at the same latitudes. The reason is that warm ocean currents bring heat through the Gulf Stream and the North Atlantic Drift to high latitudes along the European coast. The distribution of surface air temperatures is closely linked to the land-sea distribution and the general atmospheric circulation.

The climate variability and trends evident in 200 years of Stockholm temperature records are illustrated in Fig. 1.7. The mean temperature for the whole period is 6.5 °C, and the mean temperatures for sub-periods show a clear positive trend, with the last decade standing out as unusually warm. The increase in the mean sub-period temperature starts at the beginning of the 20th century, as did the increase in maximum air temperatures. However, high maximum temperatures were also recorded at the beginning of the 19th century. The sub-period minimum temperatures were lowest in the mid 19th century. No clear trend can be discerned in either the maximum or minimum temperatures or in their variance. The statistical trend is drawn as a linear trend over the

whole period, but closer inspection indicates that the 19th and 20th centuries behaved differently.

The magnitude of the Stockholm annual air temperature cycle (defined as the difference between the summer and the winter seasonal temperatures, Tsw) appear in Fig. 1.8. For a large index, we would expect a more continental climate influence, while a smaller index would indicate a stronger maritime influence. The mean magnitude of the annual cycle is 18.6 °C, with a decreasing trend over the entire period. The figure indicates that in the 19th century up to 1850, the magnitude of the annual cycle was larger; thereafter, the amplitude was below average with particularly low values at around 1900 and at the end of the 20th century.

The annual temperature cycles for some selected stations in the Baltic Sea Basin are illustrated in Fig. 1.9. The annual mean surface air temperature differs by more than 10 °C in the area. The coldest regions are north-east Finland, and the most maritime region is in the south-western part (Northern Germany and Denmark).

$$T^{SW}_{Index}(^oC)$$

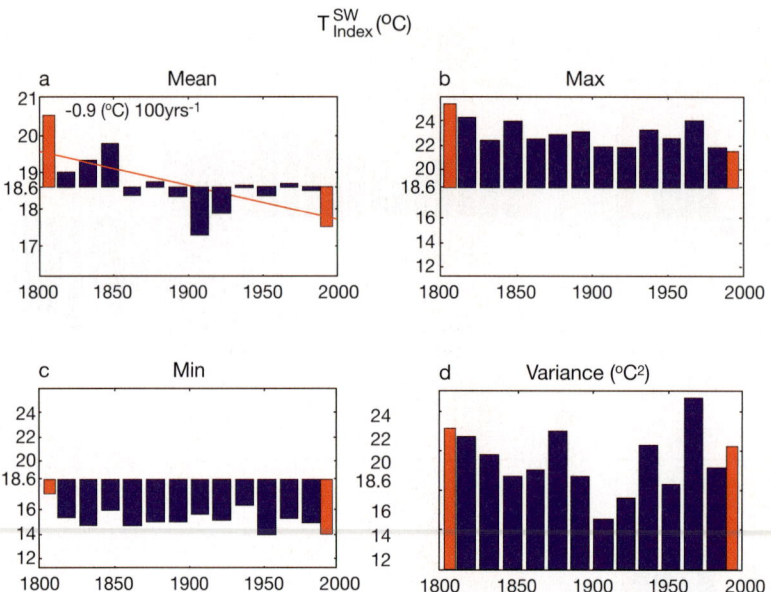

Fig. 1.8. Climate statistics of the magnitude of the seasonal Stockholm air temperature cycle, Tsw index, for individual sub-periods during 1800 to 2000. (**a**) Mean magnitude and trend, (**b**) maximum, (**c**) minimum, and (**d**) variance of the annual cycle within each sub-period. The sub-period lengths used are 15 years (*blue bars*) and 10 years (*red bars*) (from Omstedt et al. 2004b)

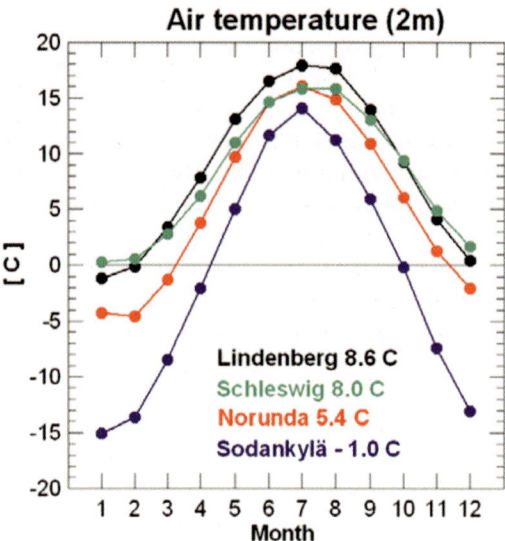

Fig. 1.9. Mean annual cycle of surface air temperature, T_a ($^\circ$C) for the period 1961 to 1990 at Sodankylä (northern Finland, *blue*), Norunda (mid-Sweden, *red*), Lindenberg (eastern Germany, *black*) and Schleswig (northern Germany, *green*). The station name is plotted together with the long-term annual mean value of T_a. Except for Schleswig, data were provided by the respective station managers in the context of CEOP, the Coordinated Enhanced Observing Period of GEWEX, see also www.gewex.com/ceop. Data for Schleswig are taken from Miętus (1998) (by courtesy of Hans-Jörg Isemer, see also Annex 1.2)

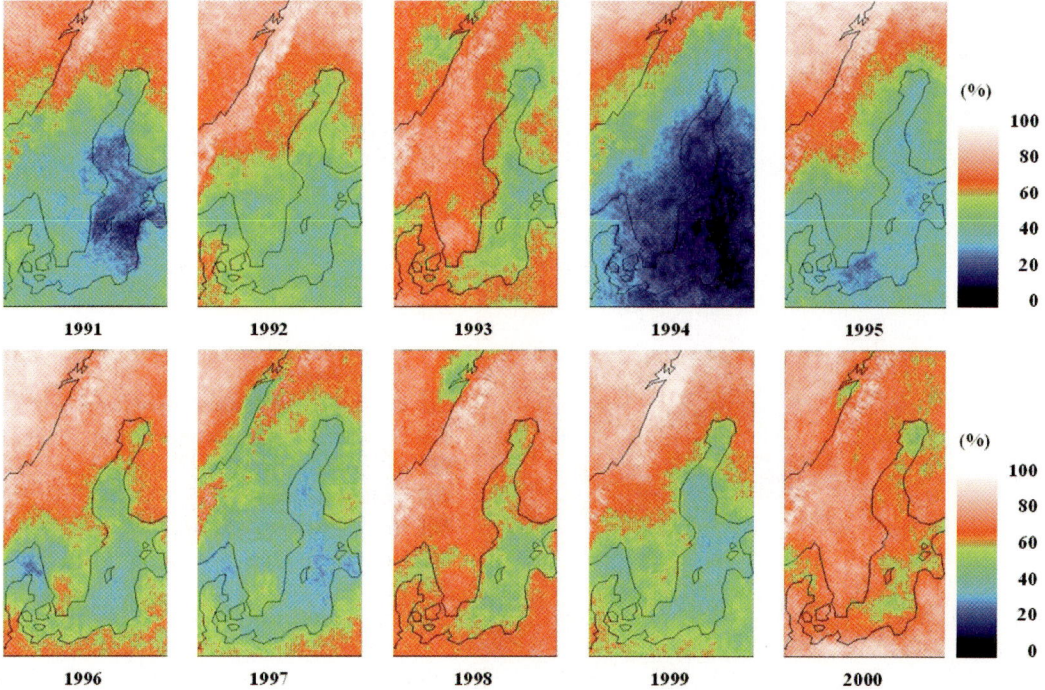

Fig. 1.10. Cloud frequencies over Scandinavia and the Baltic Sea for July during 1991 to 2000 (from Karlsson 2001, see also Annex 1.2 and Karlsson 2003)

Cloudiness, precipitation and humidity all show distinct annual cycles, with large regional variability. This is illustrated by the cloud climate as determined from satellite data, shown in Fig. 1.10. The cloud frequencies in July show large variations from year to year. Also, there is a high degree of regional variability, with high cloud frequencies in the north and low cloud frequencies above the Baltic Sea. The cloud climate reflects the mean atmospheric circulation, the land-sea distribution and the topography.

1.2.4   Terrestrial Ecological Characteristics

The Baltic Sea Basin comprises watersheds draining the Fennoscandian Alps in the west and north, the Erz, Sudetes and western Carpathian Mountains in the south, uplands along the Finnish-Russian border and the central Russian Highlands in the east. The basin spans some 20 degrees of latitude, and climate types range from alpine to maritime to sub-arctic. The Baltic Sea Basin can be divided roughly into a south-eastern temperate part and a northern boreal part, as shown in Fig. 1.11. In the south-eastern part, characterised by a cultivated landscape, the river water

runs into the Gulf of Riga and the Baltic Proper. The northern boreal part is characterised by coniferous forest and peat. The natural vegetation is mainly broadleaved deciduous forest in the lowland areas of the southeast and conifer-dominated boreal forest in northern parts. Cold climate lands and tundra occur in the mountainous areas and in the sub-arctic far north of the catchments region. Wetlands and lakes are a significant feature of the boreal and sub-arctic zones.

Extensive changes in land use have taken place during last centuries, paralleled by a considerable increase in the number of people living in the Baltic Sea Basin. Industrialisation and growing cities have changed the landscape, and much of the forest has been converted to farmland. Only in the northern parts does forest still dominate the landscape. Approximately half of the total catchment area consists today of forest, most of the remainder being agricultural land.

Many plant species are temperature sensitive and cold-limit range boundaries can be correlated to the minimum temperature. The main reason is assumed to be the result of ice formation in plant tissue leading to death. Some other plants are more correlated with growing season heat sums.

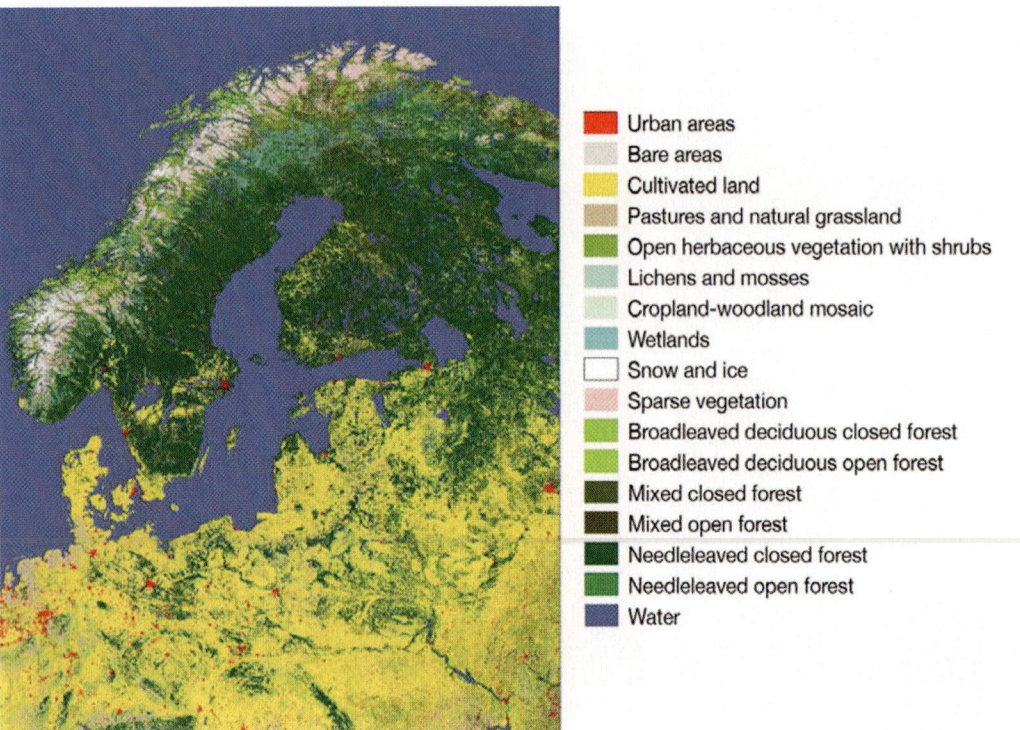

Fig. 1.11. Land cover of the Baltic Sea Basin and surroundings areas (adapted from Ledwith 2002 and the Global Land Cover 2000 database, European Commission, Joint Research Centre 2003; www.gem.jrc.it/glc2000, page visited 26 Feb 2007)

However, at southern or low-altitude range boundaries, temperature relationships are more varied and less well understood.

The most important processes in terms of the overall control of the ecosystem are the physiological processes underlying net primary production, photosynthesis, respiration, stomatal regulation and carbon allocation in plants. Several factors such as temperature, light, soil nutrient concentration, soil water content and atmospheric gases influence these processes; see further discussion in Annex 3.2.

1.2.5   Marine Ecological Characteristics

The Baltic Sea is a large transition area between limnic and marine conditions. Plants and animals are a mix of marine and limnic species together with some genuinely brackish water forms (the Baltic Sea lacks endemics at species level, due to its geological youth). The brackish water affects biodiversity in a profound way due to osmotic stress on plants and animals. Generally, poverty of species number, i.e. low biodiversity, is a common characteristic of brackish waters. The lowest

number of species is not in mid salinities, but displaced close to freshwater (Remane and Schlieper 1971). In the Baltic Sea, the area with lowest number of species is between 5–7 psu, which currently is found north of Gotland (cf. Fig. 1.12). Marine representatives of the Baltic fauna and flora have mostly invaded the area since the end of the last freshwater stage (cf. Annex 2: *The Late Quaternary Development of the Baltic Sea*). Baltic Sea salinity ranges from almost freshwater (the largest single source of freshwater is the River Neva in the Gulf of Finland) to about 25–30 psu. Following this, the number of marine species declines from south to north and east (Fig. 1.12). Marine species also are (submergent according to stratification) more common in deep water due to higher salinity there. Also, due to the stratification of the water and the lack of effective mixing and ventilation, large areas of the bottom are anoxic, covered with hydrogen sulphide and devoid of metazoan life. These zones have been called "the benthic deserts" (Zmudzinski 1978).

Besides distribution, also growth of many marine species, even key species, is influenced negatively by low salinity (rule of size diminution).

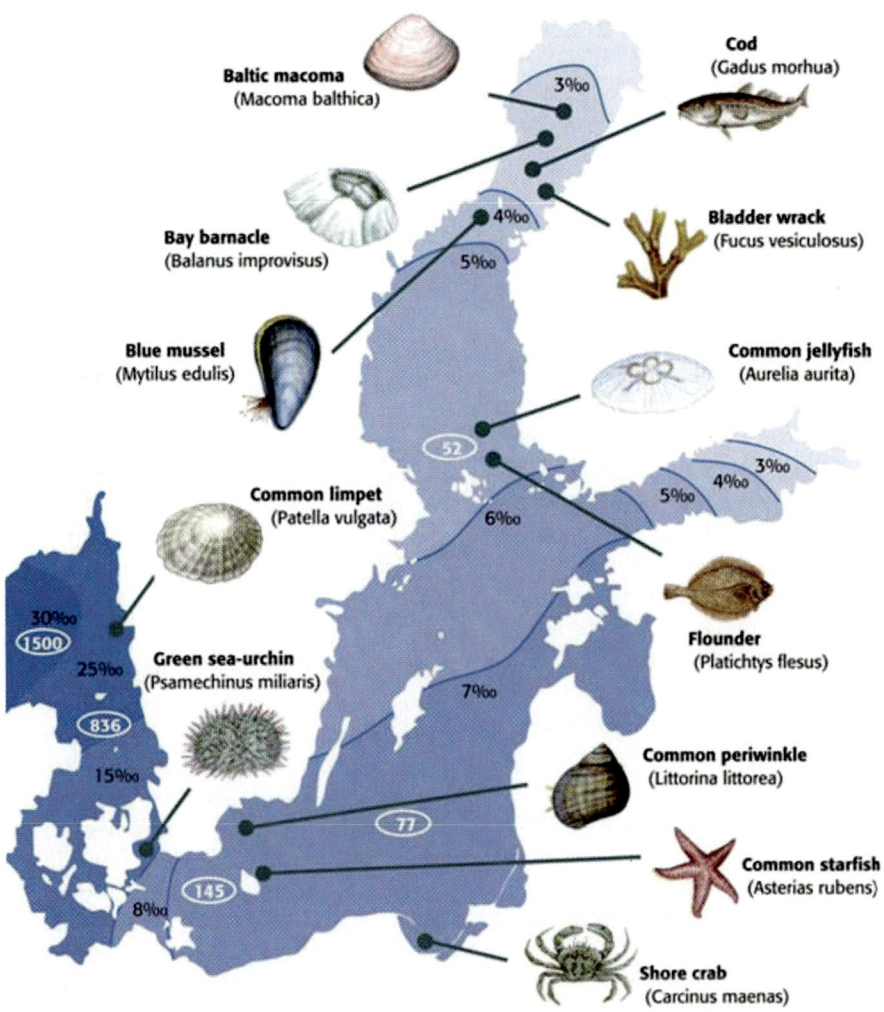

Fig. 1.12. Illustration of how salinity affects biodiversity in the Baltic Sea. The numbers in circles indicate the number of marine macrofauna species found in the area (Figure by Prof. B-O Jansson, Stockholm Marine Research Centre, Stockholm University)

Known examples of reduction in size are the mussels, *Mytilus edulis* and *Mya arenaria*, and the Baltic herring, or streamling, *Clupea harengus membras*.

The Baltic Sea ecosystem is also a transition area between sub-Arctic conditions in the Bothnian Bay to a boreal environment in the central and southern parts of the Baltic Sea. Some of the marine species are Arctic in origin, such as the bivalves *Astarte borealis* and *Macoma calcarea*. They are found in the central and southern half of the Baltic Sea, in the Belt Sea and Kattegat but only sporadically on the Norwegian coasts. In addition to the impoverished fauna of the Arctic and

the North Sea, there is a special group of animals in the Baltic Sea, the glacial relicts (Segerstråle 1966). These arrived in North European waters shortly after the last glaciation. The glacial relicts are often, but not exclusively, closely related to forms living in the Arctic Ocean. Examples of originally Arctic marine species, which are found as glacial relicts in the larger North European lakes and in the Baltic Sea are crustaceans, such as *Mysis relicta*, *Saduria entomon*, and *Monoporeia affinis*, and the fish *Myoxocephalus quadricornis*; freshwater relicts are also found, such as the copepod *Limnocalanus macrurus* and the whitefish *Coregonus lavaretus*. These are found in relatively cool deep water, or in the benthos, which makes them particularly vulnerable to deep water anoxia. Freshwater species, naturally, have entered the Baltic Sea via river mouths. They continue to do so, and many of them are confined to littoral and shallow water, such as the perch, roach and pike, which thrive among the reef beds, but also can be very numerous among truly marine flora, e.g. in the bladder wrack. A description of basic structure and function of this mosaic mixture of species can be found in the Annex 3: *Ecosystem Description*.

Anthropogenic impact is changing the Baltic Sea, not only through contaminants, which have caused reproduction disorders in some key species, such as seals and the white tailed eagle, but also due to eutrophication, which has contributed to anoxia, dead bottoms and disrupted deep water food chains, and has caused local disappearance of bladder wrack, and increased blooms of cyanobacteria in the surface layer of the open water. So far, these influences have not caused any species extinctions, but they have considerably changed species distribution patterns, especially in the coastal areas and deep basins (Leppäkoski and Olenin 2001). On the other hand, the Baltic Sea is a relatively young sea, thus having many vacant niches, and immigrant species arrive with surprising frequency. The first great wave of non-native species came after the building of the inland waterway systems at the end of the 18th to the beginning of the 19th centuries, which connected ultimately Pontocaspian brackish water areas with the Baltic Sea, e.g. bringing in the hydrozoan *Cordylophora caspia* and the zebra mussel *Dreissena polymorpha* (Olenin 2002). The most recent wave of immigration appeared after the collapse of the Soviet Union and the subsequent increase in sea traffic (e.g. the arrival of the Pontocaspian cladoceran *Cercopagis pengoi*). The "original" low biodiversity of the Baltic Sea is thus disappearing rapidly.

Some species occupy several different habitats during their lifetimes; e.g. the Baltic herring spawns in the littoral, often next to freshwater outlets; it feeds in the pelagial in the summer and while young, but on the bottom during winter and when it has grown larger. Finally, it may even turn predator of other fish. A striking example of species utilising widely different habitats during their lives are the Arctic waterfowl, which overwinter in the Southern Baltic Sea Basin and Northern Europe, and migrate through the Baltic Sea twice a year in millions, making spectacular sights. In

environmental management of the Baltic Sea, one must therefore be able to cover a species during the entire life cycle and in every habitat it will use during its lifetime, taking into account the extreme mobility of some species.

1.3   Trends, Jumps and Oscillations – the Problem of Detecting Anthropogenic Climate Change and Attributing it to Causes

People perceive climate not as something stationary – they even think it *should* be stationary – and the deviation from this assumed stationarity is taken as evidence for climate change, in most cases as adverse anthropogenic climate change[6]. This is related to a duality of the term "climate". One meaning is the geophysically defined statistics of weather, which are objectively described by observations and – within limits – by dynamical models. This "geophysical construct" is what this assessment report is about. However, the alternative meaning of a "social construct" refers to what people think about climate, how they perceive it[7]. For the definition of climate policy, it may be questioned whether the key driver is the social construct or the geophysical construct of climate.

This Assessment Report of Climate Change for the Baltic Sea Basin emphasises the need for a rigorous discrimination between long-term systematic changes, related to anthropogenic drivers (foremost greenhouse gas emissions, but also aerosol emissions or land-use change), and natural variations at shorter time-scales, related to internal climate dynamics and to identify the most probable causes of the systematic changes. The

---

[6]The complaints that weather is less predictable than it was in the old days seem to be part of our culture (e.g. Rebetez 1996). An interesting episode in the history of ideas occurred during the 19th and early 20th century, when scientists, and to some extent the public and policymakers, were discussing whether ongoing events were an expression of systematic change or natural variability (e.g. Williamson 1770; Brückner 1890; Stehr et al. 1996; Pfister and Brändli 1999, Kincer 1933, Callendar 1938). In those days, those who claimed that the trends and clustering of extreme events were man-made seem to have been more successful in their arguments.

[7]Thus, there are – even if just emerging – efforts to describe these perceptions scientifically and to understand the social implications of our views of climate and of climate change. Examples of such studies dealing with this social construct are Glacken (1967); Kempton et al. (1995); Kempton and Craig (1993); von Storch and Stehr (2000) and Bray and von Storch (1999).

technical terms are "detection" and "attribution" (Hasselmann 1993; Zwiers 1999; IDAG 2005). For changes on the global and continental scales, a large body of literature has come into being in the last few years; also in the IPCC Assessment Reports this issue plays an important if not dominant role. These efforts led to the famous statement in the Third Assessment Report (Houghton et al. 1996): "The balance of evidence suggests that there is a discernible human influence on global climate." However, for specific regions and impact variables, successful detection has rarely been claimed.

With respect to the Baltic Sea Basin, no formal detection and attribution studies appear to have been carried out so far. It seems that – at least at present – a methodically sound detection of anthropogenic climate change signals in the Baltic Sea Basin due to increased greenhouse gas concentrations has not been achieved (see Chap. 2). This is probably due to the unfavourable signal-to-noise ratio of anthropogenic signals (whose existence on the global scale is well established, at least in terms of air temperature, see IPCC Assessment mentioned above) and natural variability (as for instance related to the NAO). Instead, many studies have attempted to detect trends in recent decades, and sometimes claims are made that such trends are due to global climate change when they are "statistically significant" (see also Sect. 1.3.3).

1.3.1 The Concepts of Formal Detection and Attribution

Detection is formally a statistical test dealing with the null hypothesis that the recent changes of climate are within the limits of natural variations. Usually, the variable considered is the change during the past few, say two or three, decades. This change can take the form of a trend during that time. The test has to be done in a very large multi-dimensional space (spanned by very many locations and variables). Scenarios prepared with climate models help to sort out those dimensions along which a favourable sign-to-noise ratio is expected for changes related to elevated greenhouse gas and aerosol concentrations. We have already mentioned that the regional signal-to-noise ratios at the present level of anthropogenic climate change are seemingly too small for a successful detection. Another requirement is the availability of enough observational data to determine the distribution of natural variations. This is probably not

such a severe limitation for Baltic Sea Basin studies, where several data series extending for more than 100 or even 200 years exist (see Chap. 2). Also, climate model simulations may over or under estimate the level of natural variability.

Attribution is formally a statistical fit based on the assumption that the "signal" of climate change is made up additively of contributions of different influences. Each influence is a function of quantifiable forcings, e.g. the concentration of greenhouse gas concentrations. Then, given the time-dependent forcings, a best mix of contributions to the observed change is determined. The coefficients in this fit are subjected to a statistical test – however having the desired outcome as the null-hypothesis and not as an alternative hypothesis. Thus, a successful attribution, in this formal sense, is a less powerful argument than a successful detection, which would feature the desired outcome as an alternative hypothesis.

Usually, attribution makes sense only after a successful detection. In some cases, however, this is not possible, for instance because of insufficient observational data for determining the level of natural variability. Using attribution in cases where the signal-to-noise ratio is insufficient for a formal detection, one can at least determine to what extent the recent changes are consistent with the hypothesis of anthropogenic forcing. Most claims of ongoing regional anthropogenic climate change are based on such reasoning.

Chapter 2 (see summary in Sect. 1.5.1) reports on a number of such studies. With regard to ecological change (Chaps. 4 and 5, Sects. 1.5.3 and 1.5.4) such efforts have rarely been done; in that academic environment it has been common to relate the emergence of signals to elevated levels of greenhouse gases and anthropogenic aerosols without an adequate statistical analysis.

1.3.2 Homogeneity of Data

An obvious requirement for a meaningful statistical analysis of observational data with respect to their temporal development, the existence of break points, jumps, regime shifts, cycles or stochastic characteristics, is that the data are quality-controlled and *homogeneous* (see also Annex 5). Quality control means ensuring that the data are observed at the ascribed location and time or that they are recorded using prescribed protocols and instruments. It happens, for instance, that log books of ships list incorrect locations, for instance far inland. The quality control of data is a time-consuming task.

The same is true for homogenisation (e.g. Alexandersson 1986; Alexandersson and Moberg 1997; Jones 1995; see Annex 5). A time series is considered homogeneous when it represents the unchanged informational value throughout the observational record. Homogeneity may be compromised by updating old instruments by more accurate ones, by replacing observers, by moving the location of the instrument or the time of the recording, by changing the environment of the instrument (in particular, changing land-uses, for example urbanisation). For instance, when the main weather station of Hamburg was relocated from the harbour area to the airport, the number of reported storms abruptly decreased.

For laymen, it is generally difficult to assess whether meteorological and oceanographic data are of sufficient quality and whether they are homogeneous. There are examples in the literature where significant but false conclusions were drawn from in-homogeneities in data sets. Corrections for in-homogeneities in data sets have started in the meteorological research community but other disciplines such as hydrology, oceanography and ecology have only just recognised the problem. This indicates that many time series in the Baltic Sea Basin may still lack quality and homogeneity controls.

1.3.3 Stationarity of Data – Trends, Oscillations and Jumps

Climate change is expected to emerge in terms of trends or regime shifts, or a blending of jumps and trends (Corti and Palmer 1999). What we subjectively see from data depends much on what we are expecting and on our scientific training. Figure 1.13 illustrates a time series presented as normalised data and as original data. In the figure the same time series is interpreted in three different ways. Obviously, the data is very noisy and the subjective approach doesn't lead anywhere. The original dataset illustrates how the maximum annual ice extent in the Baltic Sea has varied from 1720 up to now. This time series is one of the important data sources for the understanding of the climate in the Baltic Sea and is further discussed in Chap. 2.3.3.

Figures 1.4 and 1.14 provide examples of time series illustrating the methodical problem. The Baltic Sea mean salinity exhibits large variations on long time scales (Fig. 1.4; Winsor et al. 2001, 2003), so that statements of systematic increases or decreases may be derived when limited segments of many years are considered. However, the time series extending across the entire last century indicates that speaking about a long-term trend makes little sense. In particular, the development during the last few decades appears to be inconsistent with the earlier development. Figure 1.14 shows an index of storm frequency for Southern Sweden (Lund; Bärring and von Storch 2004). Again, when limited segments are considered, trends are found, but overall the time series are remarkably stationary. One can certainly not find evidence of a systematically elevated level of storminess in recent decades from this analysis. Both cases tell the same story, namely that short time series are usually insufficient to inform about anthropogenic climate change. They can be used only if other information about the spectrum of natural variability is available (e.g. from extended model simulations).

The above arguments demonstrate that **rigorous statistical analysis is required**. A key concept in the statistical assessment of changing climate is that of stationarity, i.e. the assumption that the statistical parameters such as mean, dispersion, auto-covariance and characteristic patterns are not time-dependent. Detection then refers to rejecting the null hypothesis of stationarity.

In statistical thinking, a time series is a limited random sample of a stochastic process; in principle there may be any number of realisations of this process (von Storch and Zwiers 2002), even if we have only one such realisation available. We may then again, in principle, estimate for each time the statistical parameters across the ensemble of realisations. If they do not depend on time, then the process is stationary. The assumption of ergodicity[8], together with stationarity, allows us to derive estimates from one available time series across time instead of from several samples at the same time across the ensemble.

Examples for non-stationary behaviour in geophysical time series are the diurnal and the annual cycles: The ensemble mean of temperature

---

[8]Ergodicity is a formally difficult concept. It describes the fact that the trajectory (given by the time series) "will eventually visit all parts of the phase space and that sampling in time is equivalent to sampling different paths through phase space. Without this assumption about the operation of our physical system, the study of the climate would be all but impossible" (von Storch and Zwiers 2002).

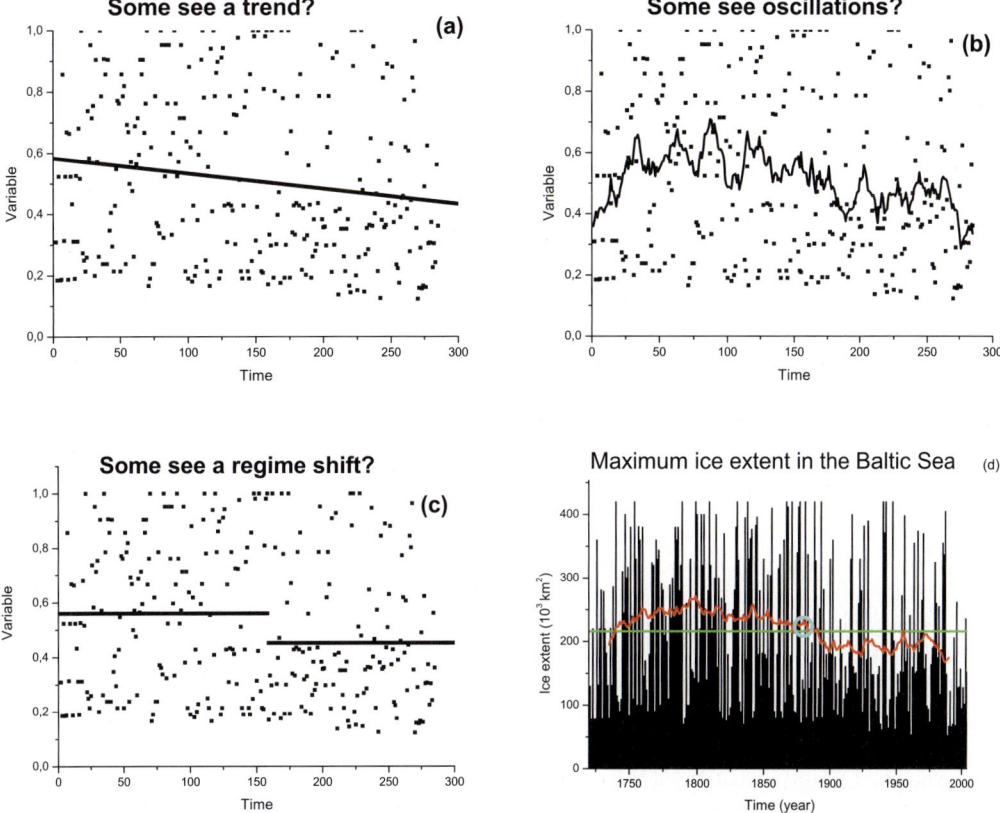

Fig. 1.13. Climate change can be detected by trends, oscillations and jumps or regime shifts. In this figure the same data sets is used and normalised (Figs. **a**–**c**). The original data set is illustrated in Fig. (**d**), for details see Omstedt and Chen (2001)

at 6:00 in the morning is markedly different from the ensemble mean at 18:00 in the evening. Time series with phase-fixed cycles are not stationary; they may represent cyclo-stationary[9] processes, however. But there may be quasi-periodic behaviour, with variable phases, periods and amplitudes, which are not in conflict with the concept of stationarity. Examples are autoregressive processes of the 2nd order with little damping (von Storch and Zwiers 2002). Similarly, jumps and break points do not contradict stationarity, as long as the time between these events is random.

Thus, for a break point, a jump or a regime shift to qualify as evidence for climate change, it needs

to be markedly different from previous such events. If such events have not happened for a very long time, then it would be good evidence – as long as such events are not related to inhomogeneities in the data gathering process.

One may assume that anthropogenic climate change emerges as a mixture of jumps and trends; in scenarios it takes a form something like slowly increasing trends, but in local variables jumps also may be possible. Therefore, much emphasis is on the **detection of trends**. The key question is whether the most recent trend, during the period when we expect the anthropogenic signal to be strongest, is larger than previously recorded trends. If the time evolution has not been monitored for a long enough time to derive an appropriate distribution of past trends, then formal detection studies cannot be done.

Therefore, many studies adopt to the concept of *significance of trends*, i.e., testing the null hypoth-

---

[9]Cyclo-stationary processes are instationary processes, which have parameters, such as the mean or the variance, which depend cyclically on time. The weather is considered cyclo-stationary, with two main cycles, the annual and diurnal cycle. Minor cycles are related to atmospheric tides (see also von Storch and Zwiers 2002).

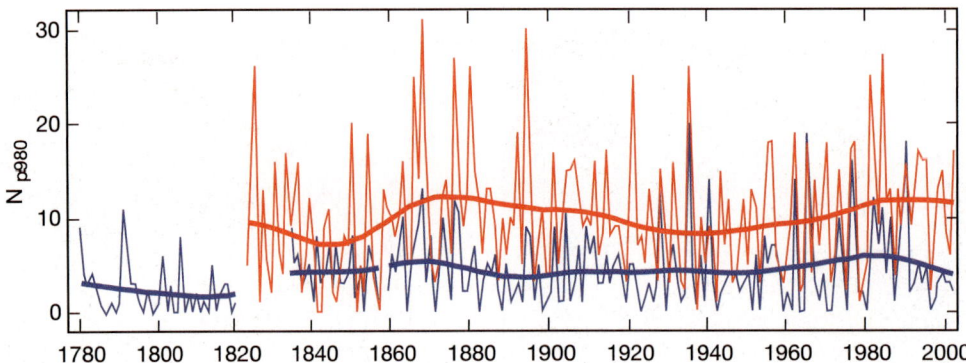

Fig. 1.14. Frequency of daily low pressure readings for Lund (*blue*) and Stockholm (*red*) (Bärring and von Storch 2004). The thin lines show annual variations and the smooth thick lines (Gaussian filter, $\sigma = 3$) show variations at the decadal time-scale

esis of "no trend". However, testing the null hypothesis of no trend is a difficult issue. There are well-established tests in the literature, for instance the Mann-Kendall test, and they are often used in a cookbook recipe like manner. Unfortunately the meaning of the term "trend" used in everyday language is different from the meaning of the same term in statistical analysis. In everyday language it refers to something ongoing for the foreseeable future, whereas the statistical analysis is asking if the trend within a fixed time interval [1, T] could be generated by noise. In the statistical concept, the evolution outside the fixed time interval is not considered.

Instead, as the thought experiment goes, one assumes to have a number of time series over the fixed interval [1, T] generated by the same unknown, these underlying stochastic process with no trend[10]. However, each of these realisations will have a trend, i.e., a linear (or nonlinear) fit to the data from time 1 to time T, simply because of random variations. One determines the distribution of trends within [1, T], associated with realisations of the no-trend random process. If the actual trend, which has to be assessed, is larger than a pre-selected high percentile of this distribution, the null hypothesis of no trend in [1, T] is rejected with a given risk. Obviously, the whole argument does not deal with the question whether we have a trend continuing into the future or not.

A standard method for performing such a test is the Mann-Kendall test. However, this test (and other similar ones) operates with an assumption which is in most cases not fulfilled by geophysical data – that of no serial correlation. If the data are auto correlated, even if only weakly, due to memory, trends or cycles, then the test no longer rejects a correct null hypothesis as rarely as stipulated. The test simply becomes wrong (Kulkarni and von Storch 1995, see Annex 8). It is probable that many claims of the detection of "statistically significant" trends are based on this invalid application of the Mann-Kendall and other similar tests. The same is true for tests of break points.

The proper way to circumvent this problem is to simulate the distribution under a null-hypothesis which explicitly incorporates serial correlation in the data. An alternative is to "prewhiten" the time series before conducting the test (Kulkarni and von Storch 1995). However, these approaches are not commonly used in the present assessment of climate change.

1.4 Scenarios of Future Climate Change

1.4.1 Scenarios – Purpose and Construction

Scenarios are descriptions of possible futures – of different plausible futures. Scenarios are not predictions but "storyboards", a series of alternative visions of futures, which are possible, plausible, internally consistent but not necessarily probable (e.g. Schwartz 1991; Tol 2007). The purpose of scenarios is to confront stakeholders and policymakers with possible future conditions so that they can analyse the availability and usefulness of options to confront the unknown future. Scenar-

---

[10]More precisely, it is the same concept already used to define stationarity: The process is generating a limited series for the times 1 to T. Then, for each time in this interval, one can define statistical parameters such as the mean, standard deviation and so on. If a linear function is fitted to these parameters in the interval [1, T], and the assumption of no trend is valid, then the slope is zero.

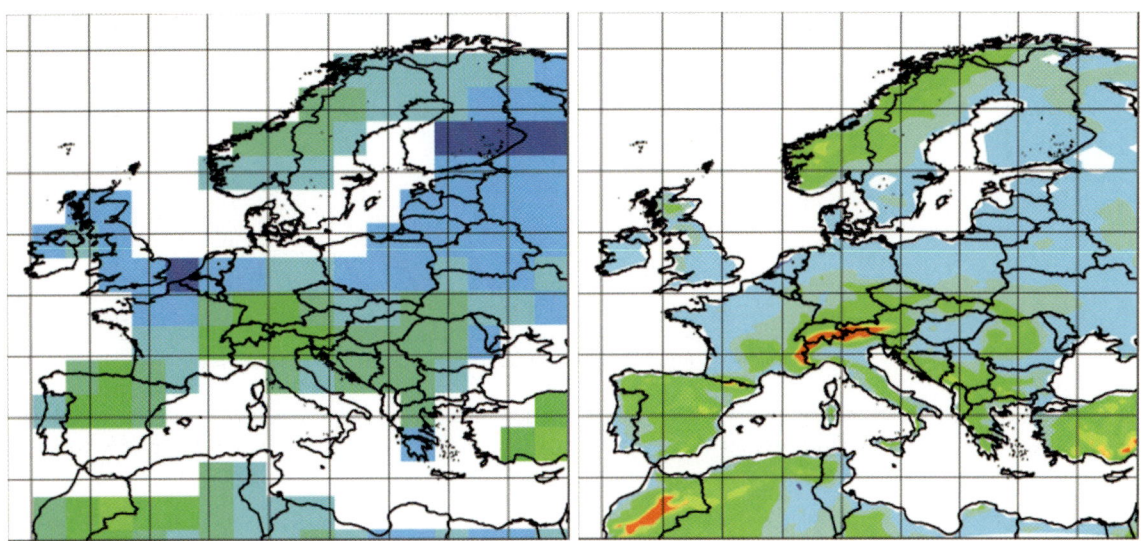

Fig. 1.15. Some typical atmospheric model grid resolutions with corresponding land masks. *Left:* T42 used in global models, *right:* 50 km grid used in regional models (by courtesy of Ole Bøssing-Christensen, Danish Meteorological Institute)

ios allow implementing measures now to avoid unwanted futures; they also may be used to increase chances for the emergence of favourable futures.

In daily life, we frequently operate with scenarios. For instance, when planning in spring for a children's birthday party next summer, we consider the scenarios of an outdoor party on a sunny day or an indoor party on a rainy day. Both scenarios are possible, plausible and internally consistent. Planning for a snowy day, on the other hand, is not considered, as this would be an inconsistent scenario.

In climate research, scenarios have been used widely since the introduction of the IPCC process at the end of the 1980s (Houghton et al. 1990, 1992, 1996, 2001). These scenarios are built in a series of steps. This series begins with *scenarios of emissions* of radiatively active substances, i.e., greenhouse gases, such as carbon dioxide or methane, and aerosols. These scenarios depend on a variety of developments unrelated to climate itself, in particular on population growth, efficiency of energy use and technological development. Many of these factors are unpredictable; therefore, a variety of sometimes ad-hoc assumptions enter these scenarios (Tol 2007).

In the next steps, the construction of scenarios is less ad-hoc, as they essentially process the emissions scenarios. The first step is to transform the emissions into atmospheric concentrations, which

are then fed into global climate models[11]. Thus, possible, plausible and internally consistent future emissions are used to derive estimates of possible, plausible and internally consistent future climate, i.e., seasonal means, ranges of variability,

---

[11]There are different types of **climate models** (cf. McGuffie and Henderson-Sellers 1997; Crowley and North 1991; von Storch and Flöser 2001; Müller and von Storch 2005). In their simplest form they are energy balance models, which describe in a rather schematic way the flux and fate of energy entering the atmosphere as solar radiation and leaving it as long and short wave radiation. These models are meant to be conceptual tools of minimum complexity for describing the fundamental aspects of the thermodynamic engine "climate system". At the other end of the range of complexity are the highly complex tools containing as many processes and details as is can be processed on a contemporary computer. Being limited by the computational resources, these models grow in complexity in time – simply because the computers continually become more powerful. Such models are supposed to approximate the complexity of the real system. They simulate a sequence of hourly, or even more frequently sampled, weather, with very many atmospheric, oceanic and cryospheric variables – such as temperature, salinity, wind speed, cloud water content, upwelling, ice thickness etc. From these multiple time series, the required statistics (= climate) are derived. Thus, working with simulated data is similar to working with observed data. The only, and significant, difference is that one can perform experiments with climate models, which is not possible with the real world. However, present day climate models are coarse, and several processes related to the water and energy cycles are not well understood or described in these models. Direct observations are therefore the main source for the understanding of the climate.

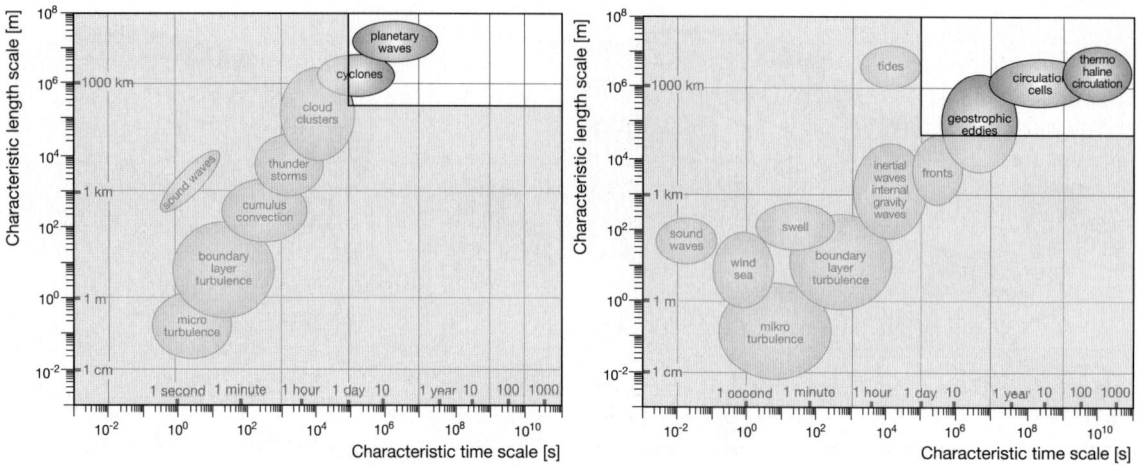

Fig. 1.16. Spatial and temporal scales of some major processes in the atmosphere (*left*) and in the ocean (*right*). In the figures the grey area represents sub-grid scales in common global models, while the visible parts in the upper right corners represent the processes which climate models often aim to resolve (redrawn from Müller and von Storch 2004)

spectra, or spatial patterns. These are the *global climate change scenarios*. Effects of changes in climate forcing factors such as solar radiation, volcano eruptions or land use are often not considered in these scenarios. Instead, the scenarios try to envision what will happen in the future depending only on anthropogenic changes. We may thus call them *anthropogenic climate change scenarios*.

Global climate models are supposed to describe climate dynamics on spatial scales[12] of, say, 1000 km and more. They do not resolve the key geographical features of the Baltic Sea Basin. For instance, in the global models, the Baltic Sea is not connected to the North Sea through narrow sills; instead, the Baltic Sea is something like an extension of the North Sea with a broad link. In addition, the Scandinavian mountain range is shallow (Fig. 1.15). Therefore, in a second step, possible future changes of regional scale climatic features

are derived using regional climate models[13]. These models are based on the concept of "downscaling", according to which smaller scale weather statistics (regional climate) are the result of a dynamical interplay of larger-scale weather (continental and global climate) and specific regional characteristics physiographic detail (von Storch 1995, 1999). There are two approaches for downscaling. One is empirical or statistical downscaling, which employs statistically fitted links between variables representative for the large large-scale weather state or weather statistics, and locally or regionally significant variables (see Chap. 3.4). The alternative is dynamical downscaling, which employs a regional climate model (see Chap. 3.5). Such regional models are subject to the large-scale

---

[12]"**Scales**" is a fundamental concept in climate science. The term refers to typical lengths or typical durations of phenomena or processes. Scales necessarily refer only to orders of magnitude. Global scales refer to several thousand kilometres and more; the continental scale to a few thousand kilometres and more and regional scales to a hundred kilometres and more. The Baltic Sea is a regional feature of the global climate system. When constructing climate models the equations can only be solved within a limited resolution. Dynamic features larger than the grid domain then need to be prescribed, while features below about ten times the grid size need to be parameterised. Typical processes that need to be parameterised in climate models are indicated in Fig. 1.16.

[13]Regional climate models are built in the same way as global climate models - with the only and significant difference that they are set up on a limited domain with time-variable lateral boundary conditions. Mathematically, this is not a well-posed problem, i.e., there is not always one and only one solution satisfying both the ruling differential equations and the boundary constraints. By including a "sponge-zone" along the lateral boundaries, within which the internal solution and the externally given boundary conditions are nudged, it is practically ensured that there is a solution, and that instabilities are avoided. The present day regional climate models only downscale the global models and they thus do not send the information back to the global scale. This implies that they are strongly controlled by the global climate model. The advantage of regional climate models is that they provide an increased horizontal resolution. Therefore, they can simulate the regional details that are often needed in many impact studies.

state simulated by the global models along the lateral boundaries and sometimes in the interior. With horizontal grid sizes of typically 10 to 50 km, such models resolve features with minimum scales of some tens to a few hundred kilometres. They also simulate the emergence of rare events, such as strong rainfall episodes and strong windstorms.

Methodologically, the *anthropogenic* climate change scenarios are *conditional predictions*. After the emissions scenarios are given, no further ad-hoc decisions are required. Significant assumptions are only required for the design of the emission scenarios. These assumptions refer to socio-economic processes, which lead to emissions.

When dealing with regional or local scenarios, one has to keep in mind that not only global climate changes, but also that regional and local climate conditions change, for example due to changing regional and local land use – which may or may not be on a scale comparable to global changes. When assessing impacts of climate change, these have to be compared to the influence of changing usage of the local and regional environment given by another set of scenarios (e.g. Grossmann 2005, 2006; Bray et al. 2003) – e.g. land use, but also the release of anthropogenic substances.

1.4.2 Emission Scenarios

A number of emission scenarios have been published as an "IPCC Special Report on Emissions Scenarios" (SRES; www.grida.no/climate/ipcc/ emission) prepared by economists and other social scientists for the Third Assessment Report of the IPCC. They utilise scenarios of greenhouse gas and aerosol emissions or of changing land use:

(A1) a world of rapid economic growth and rapid introduction of new and more efficient technology,

(A2) a very heterogeneous world with an emphasis on family values and local traditions,

(B1) a world of "dematerialisation" and introduction of clean technologies,

(B2) a world with an emphasis on local solutions to economic and environmental sustainability.

The scenarios do not anticipate any specific mitigation policies for avoiding climate change. The authors emphasise that "no explicit judgments have been made by the SRES team as to their desirability or probability".

The **Scenarios A2 and B2** are widely used. Therefore, we explain the socio-economic background of these scenarios in more detail (for a summary for the other two scenarios, refer to Müller and von Storch 2004): SRES describes the A2-scenario as follows: "... characterised by lower trade flows, relatively slow capital stock turnover, and slower technological change. The world "consolidates" into a series of economic regions. Self-reliance in terms of resources and less emphasis on economic, social, and cultural interactions between regions are characteristic for this future. Economic growth is uneven and the income gap between now-industrialised and developing parts of the world does not narrow.

People, ideas, and capital are less mobile so that technology diffuses more slowly. International disparities in productivity, and hence income per capita, are largely maintained or increased in absolute terms. With the emphasis on family and community life, fertility rates decline relatively slowly, which makes the population the largest among the storylines (15 billion by 2100). Technological change is more heterogeneous. Regions with abundant energy and mineral resources evolve more resource-intensive economies, while those poor in resources place a very high priority on minimizing import dependence through technological innovation to improve resource efficiency and make use of substitute inputs. Energy use per unit of GDP declines with a pace of 0.5 to 0.7% per year.

Social and political structures diversify; some regions move toward stronger welfare systems and reduced income inequality, while others move toward "leaner" government and more heterogeneous income distributions. With substantial food requirements, agricultural productivity is one of the main focus areas for innovation and research, development efforts, and environmental concerns. Global environmental concerns are relatively weak."

In B2, there is "... increased concern for environmental and social sustainability. Increasingly, government policies and business strategies at the national and local levels are influenced by environmentally aware citizens, with a trend toward local self-reliance and stronger communities. Human welfare, equality, and environmental protection all have high priority, and they are addressed through community-based social solutions in addition to technical solutions. Education and welfare programs are pursued widely, which reduces mortality and fertility. The population reaches about

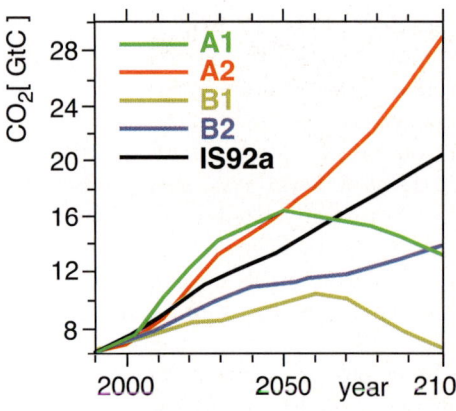 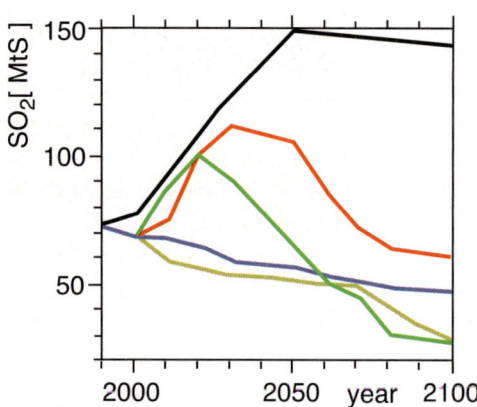

Fig. 1.17. Scenarios of possible, plausible, internally consistent but not necessarily probable future emissions of carbon dioxide (a representative of greenhouse gases; in gigatons) and of sulfur dioxide (a representative of anthropogenic aerosols; in megatons). A1, B1, A2 and B2 are provided by SRES, IS92a is a scenario used in the Second Assessment Report of the IPCC in 1995 (Nakićenović and Swart 2000)

10 billion people by 2100. Income per capita grows at an intermediate rate. The high educational levels promote both development and environmental protection. Environmental protection is one of the few truly international common priorities. However, strategies to address global environmental challenges are not of a central priority and are thus less successful compared to local and regional environmental response strategies. The governments have difficulty designing and implementing agreements that combine global environmental protection. Land-use management becomes better integrated at the local level. Urban and transport infrastructure is a particular focus of community innovation, and contributes to a low level of car dependence and less urban sprawl. An emphasis on food self-reliance contributes to a shift in dietary patterns toward local products, with relatively low meat consumption in countries with high population densities. Energy systems differ from region to region. The need to use energy and other resources more efficiently spurs the development of less carbon-intensive technology in some regions. Although globally the energy system remains predominantly hydrocarbon-based, a gradual transition occurs away from the current share of fossil resources in world energy supply."

Expected emissions of greenhouse gases and aerosols into the atmosphere are derived from these assumptions and descriptions. Figure 1.17 shows the expected SRES scenarios for carbon dioxide (a representative of greenhouse gases; in gigatons per year) and sulfur dioxide (a representative of anthropogenic aerosols; in megatons

sulfur). The SRES scenarios are not unanimously accepted by the economic community. Some researchers find the scenarios internally inconsistent. A documentation of the various points raised is provided by the Select Committee of Economic Affairs of the House of Lords in London (2005). A key critique is that the expectation of economic growth in different parts of the world is based on market exchange ranges (MER) and not on purchasing power parity (PPP). Another aspect is the implicit assumption in the SRES scenarios that the difference in income between developing and developed countries will significantly shrink until the end of this century (Tol 2006, 2007). These assumptions, the argument is, lead to an exaggeration of expected future emissions.

1.4.3 Scenarios of *Anthropogenic* Global Climate Change

The emission scenarios are first transformed into scenarios of atmospheric loadings of greenhouse gases and aerosols. Then, the global climate models derive from these concentrations – without any other externally set specifications – sequences of hourly weather, typically for one hundred years. A large number of relevant variables are calculated for the troposphere, the lower stratosphere and the oceans, but also at the different boundaries of land, air, ocean and sea ice – such as air temperature, soil temperature, sea surface temperature, precipitation, salinity, sea ice coverage or wind speed. Global climate models are subject to some degree of systematic error, so-called

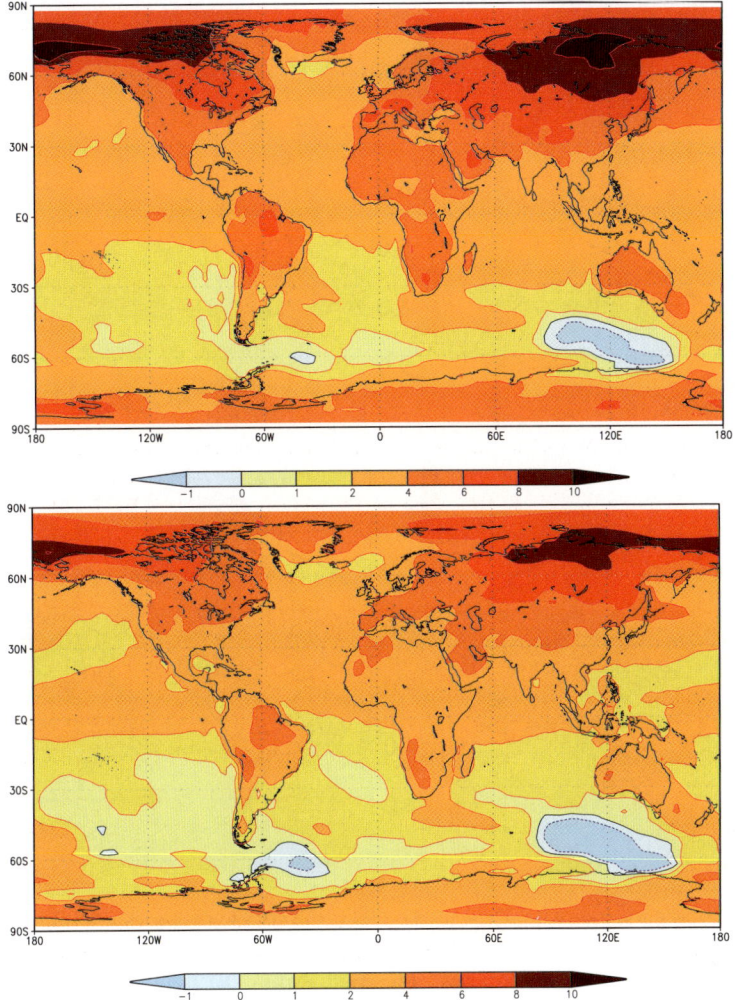

Fig. 1.18. Global scenarios of the winter surface air temperature change (in K) at the end of the 21st century as determined with a global climate model forced with A2 (*top*) and B2 (*bottom*) emissions (by courtesy of Martin Stendel, Danish Meteorological Institute)

biases. This error can be regionally large and is, for example, too large to permit determination of the expected climate change only from a simulation with elevated greenhouse gas concentrations. Instead, the climate change is determined by comparing the statistics of a "scenario simulation" with plausible elevated anthropogenic greenhouse gas and aerosol levels with the statistics of a "control run", which is supposed to represent present conditions with contemporary atmospheric greenhouse gas and aerosol levels. The difference between the control run and present climate conditions provide us with a measure of the quality of the climate simulation. If this difference is large compared to the scenario change, the climate simulation should be interpreted with care.

Figure 1.18 shows as an example of the expected change of winter air temperature for the last 30 years of the 21st century in the scenario A2 and in the scenario B2. This change is given as the difference between the 30 year mean in the scenario run and that of the control run. The air temperature rises almost everywhere; the increase is larger in the higher-concentration scenario A2 than in the lower-concentration B2. Temperatures over land rise faster than over the oceans, which are thermally more inert than land. In Arctic regions, the increase is particularly strong – this is related to the partial melting of permafrost and sea ice.

In Chap. 3.3, global climate change scenarios are discussed in some detail with respect to the Baltic Sea Basin.

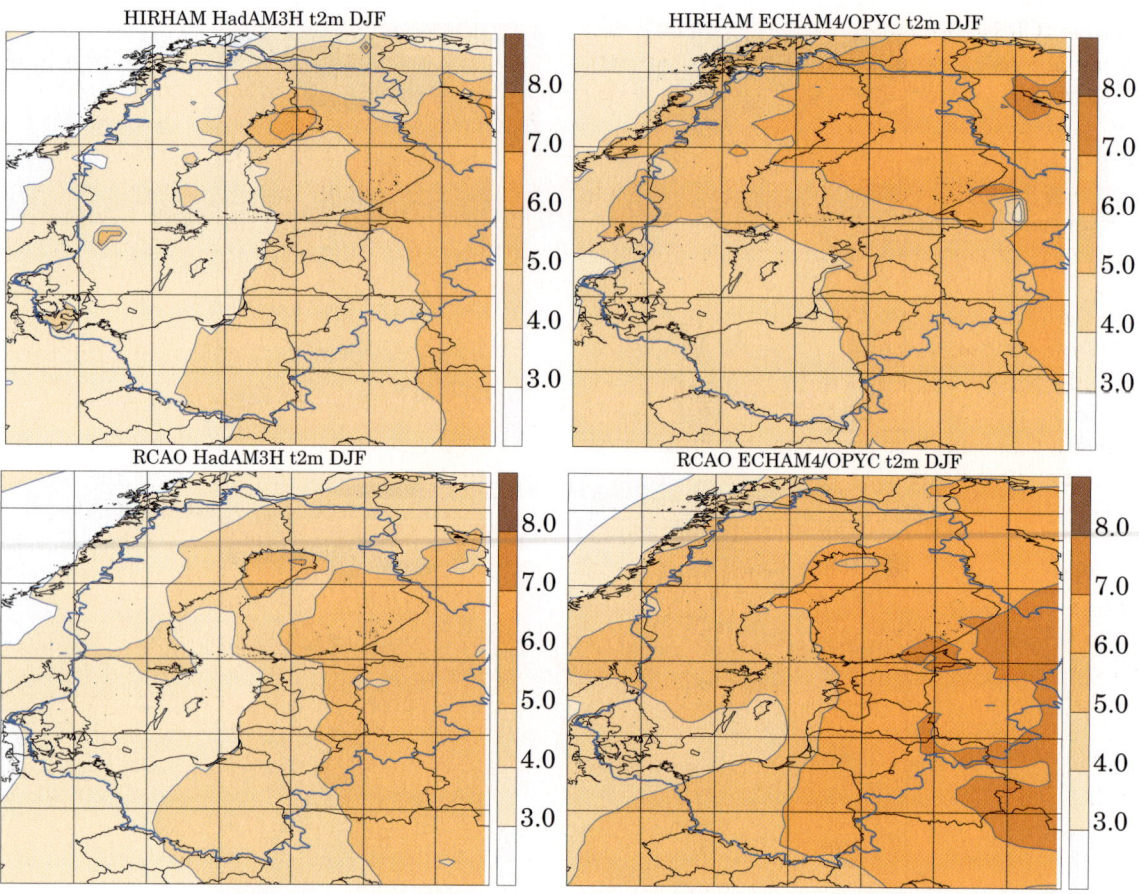

Fig. 1.19. Surface air temperature change (in K) for winter (DJF) between the periods 1961–1990 and 2071–2100 according to the SRES A2 scenario. Plots on the left used HadAM3H as boundary conditions; plots on the right used ECHAM4/OPYC3 as boundaries. For each season the upper row is the DMI regional model HIRHAM and the lower row is the SMHI model RCAO. Note that the ECHAM4/OPYC3 scenario simulations used as boundaries are different for the two downscaling experiments. The Baltic Sea Basin is indicated by the thick pale contour. – For further details see Chap. 3.5

1.4.4 Regional Anthropogenic Climate Change

A number of projects, as e.g. PRUDENCE (Christensen et al. 2002), have used regional climate models (RCM) to derive regional climate change scenarios for Central and Northern Europe. The results have been used to evaluate different sources of uncertainties in the scenarios. Chapter 3.5 (see also Sect. 1.5.2) of this assessment summarises the main and robust results of these studies.

A major result was that the regional models generally return rather similar scenarios for seasonal averages and larger areas when forced by the same global climate change scenario (Déqué et al. 2005). For higher order statistics like daily temperatures, daily rainfall or maximum daily wind speed, the choice of the regional model has an effect (Kjellström et al. 2007; Beniston et al. 2007; Rockel and Woth 2007). Thus, the choice of the regional climate model is of minor relevance when mean changes for larger areas are needed. When different driving global climate change scenarios are used – by using different emission scenarios or different global climate models – the differences become larger (e.g. Woth 2005; Déqué et al. 2005).

If higher levels of anthropogenic forcing are applied, then on average the regional changes become stronger, even if not necessarily in a statistically significant manner. This lack of significance is related to the fact that the signal-to-noise ratio of systematic change and weather noise gets smaller if the considered spatial scales are reduced. In all regional climate change scenarios, a warming is obtained in the entire Baltic Sea Basin and for all seasons (Christensen and Christensen 2003). The warming is most pronounced in the northern and eastern parts of the basin during winter, together with a dramatic reduction in the snow cover, and to the south in summer. Precipitation increases in winter and decreases in southern areas in summer in many scenarios. In general, the projected changes also increase with greenhouse gas emissions.

As an example, the expected changes of winter (DJF) mean temperatures are shown in Fig. 1.19 (see also Chap. 3.5). The different model configurations (two global models, two regional models) indicate that when the snow cover retreats to the north and to the east, the climate in the Baltic Sea Basin undergoes large changes. A common feature in all regional downscaling experiments is the stronger increase in wintertime temperatures compared to summertime temperatures in the northern and eastern part of the Baltic Sea Basin (e.g. Giorgi et al. 1992; Jones et al. 1997; Christensen et al. 2001; Déqué et al. 2005; Räisänen et al. 2004). For further discussion, refer to Chap. 3.5.

Frequency distributions tend to become broader with respect to summer rainfall (Christensen and Christensen 2003), North Sea winter wind speeds (Woth 2005) and continental summer daily temperatures (Kjellström 2004). On the other hand, frequency distributions of daily temperatures become narrower in eastern and northern Europe during winter (Kjellström 2004).

1.5 Ongoing Change and Projections for the Baltic Sea Basin – A Summary

The following provides a summary of the main findings which are presented and discussed in Chaps. 2 to 5 of this book. No specific references are given – they are all listed in the detailed account of the following chapters.

Many of the conclusions depend on the time period studied. A frequent deficiency of contemporary literature is the lack of discrimination between climate variability and anthropogenic climate change. Also, only a few long time series have been quality controlled and are homogeneous. Here, we repeat the main and most robust findings.

When considering the ecological state of the Baltic Sea Basin, assessments of anthropogenic climate change need to distinguish between effects caused by factors such as nutrient and contaminant discharges, eutrophication, over-fishing, land use change, and air pollution. At present no regional studies for the Baltic Sea Basin have been found that can attribute robustly recent trends in climate to increased greenhouse gases. The observed changes in the thermal conditions are, however, consistent with the global signals, which have been attributed reliably to human causes, and consistent with the projections available for the region.

1.5.1 Recent Climate Change in the Baltic Sea Basin

The variability of the large scale atmospheric circulation has a strong influence on the surface climate (temperature, precipitation, wind speed, etc.) of the Baltic Sea Basin. During the 200 years studied, the 19th and 20th centuries' climates differ in several ways. Towards the end of the 19th century, the Little Ice Age ended in the region, and during the period 1871–2004 there were significant positive trends in the mean temperature for the northern and southern Baltic Sea Basin, being 0.10 °C/decade on the average north of 60° N and 0.07 °C/decade on the average south of 60° N. With regard to annual mean temperatures, there was an early 20th century warming that culminated in the 1930s. This was followed by a cooling that ended in the 1960s and then another warming until the present. The warming is characterised by a pattern where mean daily minimum temperatures have increased more than mean daily maximum temperatures. Spring is the season showing the strongest warming. The number of cold nights has decreased and the number of warm days has increased, with the strongest change during the winter season.

The variability in **atmospheric circulation** has a strong influence on the surface climate in northern Europe. From about the 1960s until the 1990s, westerly airflow intensified during wintertime. This increased frequency of maritime air masses contributed to higher wintertime temperatures and enhanced precipitation at regions exposed to westerlies, especially during the 1990s. On a centennial timescale it can be seen that relationships between large scale atmospheric circulation and surface climate elements show strong temporal variability.

Over the latter part of the 20th century, on average, northern Europe has become wetter. The increase in **precipitation** is not spatially uniform. Within the Baltic Sea Basin, the largest increases have occurred in Sweden and on the eastern coast of the Baltic Sea. Seasonally, the largest increases have occurred in winter and spring. Changes in summer are characterised by increases in the northern and decreases in the southern parts of the Baltic Sea Basin. In wintertime, there is an indication that the number of heavy precipitation events has increased.

Characteristics of **cloudiness and solar radiation** show remarkable inter-annual and inter-decadal variability in the Baltic Sea Basin. A decrease in cloudiness and an increase in sunshine duration was observed in the south (Poland) while opposite trends were revealed in the north (Estonia). In the 1990s, all these trends changed their sign. Long-term observations in Estonia show that an improvement in air quality (i.e. a decrease in the aerosol emissions to the atmosphere) reversed the decreasing trend in atmospheric transparency and direct radiation during the 1990s. Presently, atmospheric transparency is at the same level as in the 1930s.

Centennial time series from southern Scandinavia reveal no long-term trend in **storminess** indices. There was a temporary increase in the 1980s–1990s. In the Baltic Sea Basin, different data sources give slightly different results with respect to trends and variations in the extreme wind climate, especially concerning small-scale extreme winds. At the same time, there are indications of increasing impact from extreme wind events. But this increasing impact results from a complex interaction between climate and development trends that increase the exposure to damage and/or the vulnerability of nature and society.

The inter-annual variability in **river water inflow** is considerable, but no statistically significant trend can be found in the annual time series for the period 1921–1998.

The analysis of the long-term dynamics of the **dates of the start and ending of ice events and the duration of ice coverage** for the rivers of the Russian part of the Baltic Sea Basin showed a stable positive trend from the middle of the 20th century to its end.

As to the maximal **ice cover thickness**, a negative tendency has been established for all Polish and Russian lakes studied. In Finland, both decreasing and increasing trends can be found in the maximum ice thickness time series.

A recent decrease in **snow cover duration and water equivalent** has been observed in the southern parts of all the Fennoscandian countries,

while the opposite trend prevails in the north. Changes of **snow depth** are quite similar, i.e. a decrease in the southwestern regions and an increase in the north-eastern regions.

During the 20th century, the Baltic Sea **mean salinity** decreased during the 1980s and 1990s, but similar decreases had also appeared earlier in the century. No long-term trend was found for the 20th century.

There are indications of a sea level rise in the 20th century compared to the 19th century.

A climate warming can be detected from the time series of the maximum **annual extent of sea ice and the length of the ice season** in the Baltic Sea. On the basis of the ice extent, the shift towards a warmer climate took place in the latter half of the 19th century. This gradual shift has been identified as the ending of the Little Ice Age in the Baltic Sea Basin.

Coastal damages appearing in various regions of the southern Baltic Sea generally result from a combination of strong storms, sea-level rise and the decreasing trend in ice cover in the winter, i.e. at times when the most intensive storms occur.

1.5.2 Perspectives for Future Climate Change in the Baltic Sea Basin

Projections of future climate change have been constructed with a series of dynamical models of the regional climate system. Most of these models simulate only the atmosphere and prescribe sea surface temperature and sea ice conditions; in this respect, the model used by the Swedish Meteorological and Hydrological Institute (SMHI) is significantly more sophisticated and an improvement over most other models as it includes a dynamical model of the Baltic Sea itself and regional lakes.

The skill of these models has been demonstrated by their ability to reconstruct the main characteristics of the recent climate. However, the reproduction of recent climate is not perfect, in particular with respect to regional water and energy balances. Also, even a perfect performance with present conditions would be no proof that the models can do a realistic job in describing possible future climate conditions. There is, however, no alternative to the model-based approach (cf. Müller and von Storch 2004), but it is prudent to closely monitor the changing climate conditions closely and to compare these changes with the projections provided by the various models. Also, when the simulated changes are comparable or smaller than the model biases, scenarios should be considered with care. One such example is precipitation over the Baltic Sea Basin, where the differences between present day simulations and observed precipitation are often larger than the simulated regional manifestation of global climate change.

Increasing greenhouse gas concentrations are expected to lead to a substantial warming of the global climate during this century. Cubasch et al. (2001) estimated the annual globally averaged warming from 1990 to 2100 to be in the range of 1.4 to 5.8 °C. This range in temperature change takes into account differences between climate models and a range of anthropogenic emissions scenarios, but it excludes other uncertainties (for example, in natural variability or in the carbon cycle) and should not be interpreted as giving the absolute lowest and highest possible changes in the global mean temperature during the period considered.

Projected future warming in the Baltic Sea Basin generally exceeds the global mean warming in global climate model (GCM) simulations. Looking at the annual mean from an ensemble of 20 GCM simulations, regional warming over the Baltic Sea Basin would be 0.9 °C higher than global mean warming, or some 50% larger in relative terms. In the northern areas of the basin, the largest simulated warming is generally in winter; further south the seasonal cycle of warming is less clear. However, the relative uncertainty in the regional warming is larger than that in the global mean warming. Taking the northern areas of the basin as an example, the warming from the late 20th century to the late 21st century could range from as low as 1 °C in summer (lowest scenario for summer) to as high as 10 °C in winter (highest scenario for winter). The simulated warming would generally be accompanied by an increase in precipitation in the Baltic Sea Basin, except for in the southernmost areas in summer. The uncertainty for precipitation change is, however, larger than that for temperature change, and the coarse resolution of GCMs does not resolve small-scale variations of precipitation change that are induced by the regional topography and land cover.

A more geographically detailed assessment of future anthropogenic climate change in the Baltic Sea Basin requires the use of statistical or dynamical downscaling methods. Yet, as only a limited number of GCM simulations have been downscaled by regional climate models (RCMs) or sta-

tistical downscaling methods, the range of results derived from those downscaling experiments does not fully reflect the range of uncertainties in the GCM projections. Accepting this, the range of results from available downscaling studies is presented below as it gives an indication of plausible future changes. All values refer to changes projected for the late 21st century, represented here as differences in climate between the years 1961–1990 and 2071–2100. All references to "northern" and "southern" areas of the Baltic Sea Basin are defined by the subregions shown in Fig. 3.12 (Chap. 3).

Consistent with GCM studies, all available downscaling studies also indicate increases in **temperature** during all seasons for every subregion of the Baltic Sea Basin. Combined results show a projected warming of the mean annual temperature by some 3 to 5 °C for the total basin. Seasonally, the largest part of this warming would occur in the northern areas of the Baltic Sea Basin during winter months and in the southern areas during summer months. Corresponding changes in temperatures would be 4 to 6 °C in winter and 3 to 5 °C in summer, as estimated from a matrix of regional climate model experiments. As noted above, these ranges most probably underestimate the real uncertainty. The diurnal temperature range – the difference between daily maximum and minimum temperature – would also decrease, most strongly in autumn and winter months. Such levels of warming would lead to a lengthening of the growing season, defined here as the continuous period when daily mean temperature exceeds 5 °C. An example from one RCM indicates that the growing season length could increase by as much as 20 to 50 days for northern areas and 30 to 90 days for southern areas by the late 21st century. The range depends on which of the different emissions scenarios is used.

Projected changes in **precipitation** from downscaling studies also depend both on differences in greenhouse gas emissions scenarios and differences between climate models. Moreover, precipitation results are more sensitive than temperature results to the statistical uncertainty in determining climatological means from a limited number of simulated years, particularly at regional scales. Seasonally, winters are projected to become wetter in most of the Baltic Sea Basin and summers to become drier in southern areas for many scenarios. Northern areas could generally expect winter precipitation increases of some 25 to 75% while the projected summer changes lie between −5 and 35%. Southern areas could expect increases ranging from some 20 to 70% during winter while summer changes would be negative, showing decreases of as much as 45%. Taken together, these changes lead to a projected increase in annual precipitation for the entire basin. In broad terms, these results are consistent with GCM studies of precipitation change, although the projected summer decrease in the southern areas of the basin tends to be larger and extend further north in the available RCM studies than in most reported GCMs. This difference reflects the fact that the few GCM simulations that have been downscaled by RCMs also show this pattern of precipitation change.

Projected changes in **wind** differ widely between various climate models. Differences in the circulation patterns of the driving GCMs are particularly important for the modelled outcome of this variable. From the RCM results presented here, only those driven by the ECHAM4/OPYC3 GCM show statistically significant changes for projected future climate scenarios. For mean daily windspeed over land areas, this would amount to a mean increase of some 8% on an annual basis and a maximum mean seasonal increase of up to 12% during winter. The corresponding mean seasonal increase over the Baltic Sea in winter, when a decrease in ice cover would enhance near-surface winds, would be up to 18%. For RCMs driven by the HadAM3H GCM, the changes are small and not statistically significant. Modelled changes in extreme wind generally follow the same pattern as for the mean wind; however, the spatial resolution of both GCMs and RCMs is far too coarse to accurately represent the fine scales of extreme wind. As the downscaled projections differ widely, there is no robust signal to be seen in the RCM results. Looking at projected changes in large-scale atmospheric circulation from numerous GCMs, and thereby changes in wind, they indicate that an increase in windiness for the Baltic Sea Basin would be somewhat more likely than a decrease. However, the magnitude of such a change is still highly uncertain and it may take a long time before greenhouse gas (GHG) induced changes in windiness, if ever, will emerge from background natural variability. It can be noted, moreover, that ECHAM4/OPYC3 is one of the GCMs that gives higher values for changes in wind.

Hydrological studies show that increases in mean annual **river flow** from the northernmost catchments would occur together with decreases in the southernmost catchments. Seasonally, summer river flows would tend to decrease, while winter flows would tend to increase by as much as 50%. The southernmost catchments would be affected by the combination of both decreased summer precipitation and increased evapotranspiration. Oceanographic studies show that the mean annual sea surface temperatures could increase by some 2 to 4 ˚C by the end of the 21st century. **Ice** extent in the sea would then decrease by some 50 to 80%. The average **salinity** of the Baltic Sea is projected to decrease between 8 and 50%. However, it should be noted that these oceanographic findings are based upon only four regional scenario simulations using two emissions scenarios and two global models.

1.5.3 Changing Terrestrial Ecosystems

The changing climate and other associated environmental and anthropogenic changes may be expected to affect the structure and functioning of ecosystems and threaten the services they provide to society. We assess the potential impacts of the changing environment on terrestrial and freshwater ecosystems of the Baltic Sea Basin, aiming to evaluate the hypotheses:

- that climate change and other associated environmental change over recent decades has affected the ecosystems and their services; and
- that ongoing climate change will cause (further) changes in the ecosystems and their services over the remainder of the 21st century.

In order to highlight the most compelling and societally-relevant aspects of ecosystem change, the analysis focuses on:

- processes and indicators of particular diagnostic value for the attribution of ecosystem changes to identifiable forcing factors; for example, changes in phenology, species distributions and the seasonality of physical, chemical and biological phases in lakes;
- ecosystems and functions of sectorial relevance; for example, productivity and carbon storage in forests; and
- uncertainty associated with ecological complexity and limitations to process understanding; for example, regarding stress responses to changing climatic extremes.

Significant changes in climate, including increasing temperatures and changing precipitation patterns, have occurred over the Baltic Sea Basin in recent decades (Chap. 2). Other associated changes include the continuously rising atmospheric CO_2 concentrations, and increases in deposition loads of atmospheric pollutants, including nitrogen compounds and other acidifying pollutants. A variety of ecosystem impacts of these changes have been identified (hypothesis 1), including the following:

An advancement of spring phenological phases such as budburst and leaf expansion is apparent for many plant species, likely reflecting increasing mean temperatures. Many species also show delayed autumn phases, but trends are less consistent. Phenological trends are stronger in northern Europe than for Europe as a whole, possibly reflecting stronger climate warming.

Species distributional shifts tracking isothermal migration are apparent for both plant and animal species. Possibly related changes include weaker migratory behaviour, for example in some bird species. Treeline advance has been observed in the Fennoscandian mountain range.

Increased growth and vigour of vegetation at high northern latitudes generally is apparent from satellite observations and can be attributed to increased growing season warmth and an extended growing season. Other observations, such as tree ring data, support the existence of a positive growth trend. The magnitude of the trend within the Baltic Sea basin is representative for high latitude areas in Eurasia, and strong compared with similar latitudes in North America.

Physiological stress related to the combined effects of atmospheric pollutants and extreme weather events such as spring frosts and drought are a possible explanation for late 20th-century dieback in boreal and temperate forests.

Degradation of discontinuous permafrost in the subarctic north may be causing a shift towards a greater representation of wet habitats in tundra. Possible consequences include an increased release of methane through (anaerobic) decomposition, which would accelerate greenhouse forcing.

Climate-related changes in lakes including higher water temperatures, advancement of ice break-up, lower water levels and increased influxes of dissolved organic matter from land have consequences for lake ecosystems, including dominance shifts in phytoplankton communities, higher summer algal biomass, and shifts in trophic state.

Climate scenarios described in Chap. 3 consistently point to increased temperatures throughout the Baltic Sea Basin by the end of the 21st century compared with the present. Precipitation scenarios are more variable but generally point to increased precipitation in winter, with southern areas experiencing decreased rainfall in summer. Combined with the effect of higher temperatures on evapotranspiration, this suggests that ecosystems of the temperate zone may face increasingly unfavourable water budgets during the growing season in the future. Potential impacts of these and other associated environmental changes (hypothesis 2) include the following:

Extrapolation of recent phytophenological trends suggests that extension of the vegetation period by 2–6 weeks, depending on the climate scenario, is likely over much of the Baltic Sea Basin.

Further changes in the distributions of some species may be expected, but for many species, lags associated with population and community processes, dispersal limitations etc. are likely. Wholesale biome shifts, such as the northward displacement of the temperate-boreal forest boundary, will be slow compared to the rate of isotherm migration. Natural and semi-natural vegetation of the future may be of a transient character, e.g. aging conifer stands with an increased representation of broadleaved trees in the younger age classes. Changes may be especially marked in subarctic and alpine areas, with forest invading areas that are currently tundra. Increased local richness is likely as species associated with the forest extend their ranges northward and upslope.

Modelling studies generally point to increasing ecosystem production and carbon storage capacity throughout the region of the Baltic Sea Basin in the next 50–100 years, in conjunction with a longer growing season, increased atmospheric CO_2 concentrations and the stimulation of mineralisation processes in warmer soils. However, increased autumn and winter temperatures may be detrimental to hardening processes in trees, increasing susceptibility to spring frost damage. Growing season drought stress may reduce or inhibit production enhancement in temperate parts of the region.

The potential impact of climatic change on the incidence of pest and pathogen outbreaks affecting vegetation is still largely open. It seems reasonable to assume that harmful insects and fungi from central and southern Europe may expand into the Baltic Sea Basin in the warming climate. Warmer water temperatures combined with longer stratified and ice-free periods in lakes may be expected to accelerate eutrophication, increasing phytoplankton production and shifting the phytoplankton community structure towards species with higher temperature optima, including cyanobacteria. Shallow lakes and lake littoral zones may be particularly sensitive to climate warming. Increasing influxes of humic substances in runoff from boreal catchments would steepen light attenuation, with negative impacts on periphyton and benthic communities in lakes. Cold-water fish species may be extirpated from much of their present range while cool- and warm-water species expand northwards.

Uncertainties associated with the assessment of future ecosystem changes are substantial and include uncertainties due to understanding of the biological phenomena being modelled or projected, including system-internal feedbacks and complexity, as well as variation among climate and greenhouse gas emission scenarios on which the assessments are based. The most important source of uncertainty with regard to many impacts is the future development in non-climatic, anthropogenic drivers of ecosystem dynamics including deposition of atmospheric pollutants, land use changes, changes in forest management and agricultural practices, changes in human populations, markets and international trade, and technological development.

1.5.4 Changing Marine Ecosystems

The Baltic Sea is not a steady state system and external drivers acting on different time scales force major changes in the marine ecosystem structure and function. Postglacial isostatic and eustatic processes have shaped the Baltic Sea's coastline, topography, basic chemistry and sedimentary environment on millennium scales (see Annex 2). Climate variability acts on all time scales and, at least over the last 150 years, overlaps with human activities in the drainage basin and the coastal zone, leading to considerable changes in the biogeochemistry of this semi-enclosed sea. Thus, the emerging impacts of anthropogenic climate change (Chap. 2) cannot be separated at this time from natural variability and from other anthropogenic influences.

Studies of past and present ecosystem changes have demonstrated the sensitivity of the marine ecosystem to **temperature** variations. For instance, Northern Baltic annual peaks of the most abundant cladoceran species were found to co-vary with surface water temperature. The higher temperatures during the 1990s were associated with a shift in dominance within the open sea copepod community from *Pseudocalanus* sp. to *Acartia* spp. Increased production and survival rates of sprat and herring populations during the last 5–10 years co-varied with high temperatures and high NAO indices. In the earlier warming period in Fennoscandia between 1870–1940, many range shifts in birds were observed, regarding both the northern and southern borders as well as spring and autumn migration. Furthermore, extreme winter temperatures have long been documented to influence water bird mortality in the Baltic Sea, and winter conditions in the Baltic Sea Basin are known to determine the range of land birds as well as water birds. Spring migration has generally occurred earlier in recent years, although there is a high variation between and within species.

Also, past changes in **salinity** have been associated with marked changes of the ecosystem. An increase in salinity during the first half of the century resulted in a spread of several marine species (e.g. mesozooplankton, barnacles, jellyfish, larvaceans) towards the northern and eastern parts of the Baltic Sea. Correspondingly, the decrease in salinity after the late 1970s in the northern Baltic Sea was reflected in a biomass decline of the large neritic copepod species and an increase of the freshwater cladoceran species. In the deep basins of the open Baltic Sea, the decrease in salinity resulted in reduced standing stocks of *Pseudocalanus elongatus*, an important player in the pelagic food web. In contrast, temperature-sensitive species increased their population sizes. A retreat towards the south has been found in benthic fauna, e.g. *Scoloplos armiger*. The decrease in herring and sprat growth has been related to a salinity-mediated change in the copepod community. The cod, a top predator in the pelagic food chain and a key species in the Baltic Proper which usually regulates the sprat and herring stocks, has seen a decrease. This decrease and the climatically enhanced sprat reproductive success have induced a switch from cod-domination to sprat-domination.

Eutrophication is a phenomenon of the recent past; still it has been documented to change the biota. Several monitoring programmes have been targeted to follow it since the 1970s, mainly because it poses a direct threat to health (toxic algal blooms) and biota (anoxic bottoms develop hydrogen sulphide). Changes of phytoplankton biomass and species composition reflect eutrophication and climatic changes simultaneously. A further twist emerges from the fact that eutrophication itself may be promoted directly by climatic factors, such as runoff and rainfall. There is some evidence that increased primary production has led to an increase of biomass at higher trophic levels (e.g. zooplankton and fish). This trend has been especially clear in benthos. Above the halocline, macrofauna biomass in the 1990s was about five-fold compared with conditions in the 1920s–30s. The deep basins of the Baltic Sea are frequently exposed to hypoxia and anoxia, which results in periodic extinction and recolonisation of bottom fauna.

Anthropogenic climate change **scenarios** for the Baltic Sea Basin (see Chap. 3) describe an increase in temperature, especially during wintertime, and an increase in rainfall in the northern part of the runoff area. The consequence of increasing precipitation is twofold. Increasing precipitation results in a decrease in salinity and in an increase of nutrient leakage and associated eutrophication (see also Sect. 1.5.3).

Projected **increased temperatures**, especially during winter months, will lead to changes in growth and reproduction parameters for fauna and flora, many of which are of boreal origin, i.e., adapted to low temperatures. The following changes are considered possible:

- Increased temperatures stimulate pelagic bacteria growth more than primary production, thus the ratio between bacteria biomass to phytoplankton is expected to increase with increasing temperature in eutrophic waters.
- Diatom spring blooms are subject to species change when winters become milder. Furthermore, it has been suggested that the diatom bloom itself may disappear after milder winters and be replaced by dinoflagellates.
- Increasing summertime temperatures may enhance cyanobacterial blooms.
- Elevated winter temperatures may prevent convection in late winter and early spring with the result that nutrients are not mixed into the upper euphotic zone.

Modelling studies describe the extinction of southern subpopulations of the Baltic ringed seal as a probable effect of expected diminishing **ice** cover suitable for breeding. The grey seal, however, has been shown to have the capability of breeding extensively on land, even in the Baltic Sea Basin.

Expected decreases of **salinity** of the Baltic Sea will modify its ecology in several ways. The most important changes will probably be seen in species distributions (both horizontal and vertical), though growth and reproduction are also likely to be affected. The lower limit of approximate salinity tolerance is 2 psu for *Praunus flexuosus*, *Neomysis vulgaris* and *Gammarus locusta* and 3 psu for *Corophium volutator*; for *Palaemon adspersus* and *Idotea baltica* it is 5.5 psu; for *Pontoporeia femorata* and *Harmothoe sarsi* it is 6 psu; for *Pygospio elegans* and *Laomedea lovéni* it is 7 psu; and for *Terebellides strömii* and *Fabricia sabella* it is 7.5 psu. Thus, along the complete range of Baltic Sea surface salinity we can expect decreases of species number due to changes in species distributions (see also Fig. 1.12).

A decrease of marine fauna is expected to occur first in the northern Baltic Sea surface area, because of the expected intensified rainfall in the northern part of the watershed. In the western Baltic Sea the common starfish (*Asterias rubens*) and common shore crab (*Carcinus maenas*) are among the species expected to decrease if salinity decreases.

We are likely to meet a reversed situation as compared to changes in the 1950s when salinity was rising. Some of this expected trend has already been documented as species like cod, which need a certain level of salinity during a certain life stage, display low reproductive success in the Baltic Sea. Cod eggs need a minimum salinity of 11.5 psu for buoyancy, which they usually find in the halocline regions of the deep Baltic Sea basins. Due to low salinity but also low oxygen concentrations in the deep water, cod eggs are frequently exposed to lethal oxygen conditions in the layer where they are neutrally buoyant.

Finally, decreasing salinity enables all freshwater species to enlarge their distributions in the Baltic Sea. Because of its ecological and evolutionary history, the Baltic Sea predominantly receives species originating from both the adjacent inland waters and the oceanic coasts but also from remote seas. Most of the recent invaders in the Baltic Sea originate from warmer climates. In conditions of increasing water temperature, not only spontaneously spreading European invaders but also exotics from warmer regions of the world can be expected to become established in the Baltic Sea.

Two target species, known to cause severe changes in invaded ecosystems, most likely will spread with climatic warming. The zebra mussel *Dreissena polymorpha* may penetrate to the Gulf of Bothnia. The North American jelly comb *Mnemiopsis leidyi*, which recently invaded the Black and Caspian Seas, may invade the Baltic Sea and cause changes its pelagic system.

In addition, the combination of decreasing salinity and increasing temperature will clearly reduce the general fitness of native benthic species and their adaptability to cope with other stressors, e.g. low oxygen or chemical pollution.

Accelerated **eutrophication** is an expected consequence of anthropogenic climate change in the Baltic Sea due to freshwater runoff determining most of the nutrient load entering it, especially in the near coastal areas.

Eutrophication is expected to enhance the production and biodiversity in the ecosystem up to a certain point, after which a collapse will appear due to several mechanisms such as chemical (anoxia) and biotic interactions (competition, predation, exploitation). After this, a new ecological balance will develop, which will be characterised by low biodiversity and high variability due to episodic outbursts of dominant species. Some effects of eutrophication are clear and predictable, such as the general increase of primary production, but other effects, such as species-specific interactions are extremely hard to predict because of the nonlinearity and complexity of the marine ecosystem. In Chap. 5, a variety of possible effects and ongoing changes are discussed.

1.6 References

Alexandersson H (1986) A homogeneity test applied to precipitation data. J Climatol 6:661–675

Alexandersson H, A Moberg (1997) Homogenization of Swedish temperature data. Part I: A homogeneity test for linear trends. Int J Climatol 17:25–34

Bärring L (1993) Climate – change or variation? Climatic Change 25:1–13

Bärring L, von Storch H (2004) Scandinavian storminess since about 1800. Geophys Res Lett 31 L20202 doi:101029/2004GL020441:1–4

Beniston M, Stephenson DB, Christensen OB, Ferro CAT, Frei C, Goyette S, Halsnæs K, Holt T, Jylhä K, Koffi B, Palutikof J, Schöll R, Semmler T, Woth K (2007) Future extreme events in European climate: An exploration of regional climate model projections. Climatic Change 81:71–95

Bray D, von Storch H (1999) Climate Science. An empirical example of postnormal science. Bull Am Met Soc 80:439–456

Bray D, Hagner C, Grossmann I (2003) Grey, Green, Big Blue: Three regional development scenarios addressing the future of Schleswig–Holstein. GKSS-Report 2003/25, GKSS Research Center, Geesthacht, Germany

Brückner E (1890) Klimaschwankungen seit 1700 nebst Bemerkungen über die Klimaschwankungen der Diluvialzeit (Climate variations since 1700 and remarks on climate variations during diluvial time). Geographische Abhandlungen. Hrsg. von Prof Dr. A. Penck in Wien. Wien and Olmütz, E.D. Hölzel (in German)

Callendar GS (1938) The artificial production of carbon dioxide and its influence on temperature. Q J Roy Met Soc 64:223–239

Christensen JH, Christensen OB (2003) Severe summertime flooding in Europe. Nature 421:805–806

Christensen JH, Räisänen J, Iversen T, Bjørge D, Christensen OB, Rummukainen M (2001) A synthesis of regional climate change simulations – A Scandinavian perspective. Geophys Res Lett 28:1003–1006

Christensen JH, Carter T, Giorgi F (2002) PRUDENCE employs new methods to assess european climate change. EOS 83:147

Corti S, Molteni F, Palmer TN (1999) Signature of recent climate change in frequencies of natural atmospheric circulation regimes. Nature 398:799–802

Crowley TJ, North GR (1991) Paleoclimatology. Oxford University Press, New York

Cubasch U, Meehl GA, Boer GJ, Stouffer RJ, Dix M, Noda A, Senior CA, Raper S, Yap KS (2001) Projections of future climate change. In: IPCC Climate Change 2001: The Scientific Basis. Contribution of Working Group I to the Third Assessment Report of the Intergovernmental Panel on Climate Change. Cambridge University Press, Cambridge New York, pp. 525–582

Déqué M, Jones RG, Wild M, Giorgi F, Christensen JH, Hassell DC, Vidale PL, Rockel B, Jacob D, Kjellström E, de Castro M, Kucharski F, van den Hurk B (2005) Global high resolution versus Limited Area Model climate change projections over Europe: Quantifying confidence level from PRUDENCE results. Clim Dyn 25,6:653–670

Giorgi F, Marinucci M, Visconti G (1992) A $2 \times CO_2$ climate change scenario over Europe generated using a Limited Area Model nested in a general circulation model 2. Climate change scenario. J Geophys Res 97:10011–10028

Glacken CJ (1967) Traces on the Rhodian Shore. University of California Press

Grossmann I (2006) Three scenarios of the greater Hamburg region. Futures 38,1:31–49

Grossmann I (2005) Future Perspectives for the Lower Elbe Region 2000-2030: Climate Trends and Globalisation. PhD thesis, Hamburg University

Hasselmann K (1993) Optimal fingerprints for the detection of time dependent climate change. J Clim 6: 1957–1971

Houghton JT, Jenkins GJ, Ephraums JJ (eds) (1990) Climate Change. The IPCC scientific assessment. Cambridge University Press

Houghton JT, Callander BA, Varney SK (eds) (1992) Climate Change 1992. Cambridge University Press

Houghton JT, Meira Filho LG, Callander BA, Harris N, Kattenberg A, Maskell K (eds) (1996) Climate Change 1995. The Science of Climate Change. Cambridge University Press, Cambridge New York

Houghton JT, Ding Y, Griggs DJ, Noguer M, van der Linden PJ, Dai X, Maskell K, Johnson CA (2001) Climate Change 2001: The Scientific Basis. Cambridge University Press, Cambridge New York

House of Lords, Select Committee on Economic Affairs (2005) The Economics of Climate Change. Volume I: Report 2nd Report of Session 2005–06. Authority of the House of Lords, London, UK; The Stationery Office Limited, HL Paper 12-I (www.publications.parliament.uk/pa/ld/ldeconaf.htm)

IDAG (2005) Detecting and attributing external influences on the climate system. A review of recent advances. J Clim 18:1291–1314

Jones PD (1995) The Instrumental Data Record: Its accuracy and use in attempts to identify the "CO_2 Signal". In: von Storch H, Navarra A (eds) Analysis of Climate Variability: Applications of statistical techniques. Springer, Berlin Heidelberg New York, pp. 53–76

Jones RG, Murphy JM, Noguer M, Keen AB (1997) Simulation of climate change over Europe using a nested regional-climate model 2. Comparison of driving and regional model responses to a doubling of carbon dioxide. Q J Roy Met Soc 123:265-292

Karlsson KG (2001) A NOAA AVHRR cloud climatology over Scandinavia covering the period 1991–2000. SMHI Reports Meteorology and Climatology No 97

Karlsson KG (2003) A 10 year cloud climatology over Scandinavia derived from NOAA advanced very high resolution radiometer imagery. Int J Climatol 23:1023–1044

Kempton W, Craig PP (1993) European perspectives on Global Climate Change. Environment 35: 16–45

Kempton W, Boster JS, Hartley JA (1995) Environmental values in American Culture. MIT Press, Cambridge London

Kincer JB (1933) Is our climate changing? A study of long-term temperature trends. Mon Wea Rev 61:251–259

Kjellström E (2004) Recent and future signatures of climate change in Europe. Ambio 23:193–198

Kjellström E, Bärring L, Jacob D, Jones R, Lenderink G, Schär C (2007) Modelling daily temperature extremes: Recent climate and future changes over Europe. Climatic Change 81:249–265

Kulkarni A, von Storch H (1995) Monte Carlo experiments on the effect of serial correlation on the Mann-Kendall-test of trends. Met Zeitschrift 4 NF:82–85

Ledwith M (2002) Land cover classification using SPOT Vegetation 10-day composite images – Baltic Sea Catchment basin. GLC2000 Meeting, Ispra, Italy, April 18–22

Leppäkoski E, Olenin S (2001) The meltdown of biogeographical peculiarities of the Baltic Sea: The interaction of natural and manmade processes. Ambio 30 4-5:202-209

McGuffie K, Henderson-Sellers A (1997) A climate modelling primer, 2nd ed. Wiley & Sons, Chichester, Great Britain

Mietus M (ed) (1998) The Climate of the Baltic Sea Basin. Marine Meteorology and Related Oceanographic Activities. Report No. 41. WMO/TD-No. 933. World Meteorological Organization, Geneva

Mitchell JM Jr, Dzerdzeevskii B, Flohn H, Hofmeyr WL, Lamb HH, Rao KN, Wallén CC (1966) Climatic Change. WMO Technical Note No 79, World Meteorological Organization, Geneva

Müller P, von Storch H (2004) Computer Modelling in Atmospheric and Oceanic Sciences: Building knowledge. Springer, Berlin Heidelberg New York

Nakićenović N, Swart R (eds) (2000) Emissions Scenarios. A Special Report of Working Group III of the Intergovernmental Panel on Climate Change. Cambridge Univ. Press, Cambridge New York

Olenin S (2002) Black Sea–Baltic Sea invasion corridors. In: Briand F (ed) Alien marine organisms introduced by ships in the Mediterranean and Black Seas. CIESM Workshops Monograph. Comission Internationale pour l'Exploration Scientifique de la mer Mediterranee, Monaco, pp. 29–33

Omstedt A, D Chen (2001) Influence of atmospheric circulation on the maximum ice extent in the Baltic Sea. J Geophys Res 106:4493–4500

Omstedt A, Elken J, Lehmann A, Piechura J (2004a) Knowledge of the Baltic Sea Physics gained during the BALTEX and related programmes. Progr Oceanogr 63:1–28

Omstedt A, Pettersen C, Rodhe J, Winsor P (2004b) Baltic Sea climate: 200 yr of data on air temperature, sea level variations, ice cover, and atmospheric circulation. Clim Res 25:205–216

Pfister C, Brändli D (1999) Rodungen im Gebirge – Überschwemmungen im Vorland: Ein Deutungsmuster macht Karriere (Uprooting in mountain areas – Flooding in the foreland: The career of an interpretation pattern). In: Sieferle RP, Greunigener H (eds) Natur-Bilder. Wahrnehmungen von Natur und Umwelt in der Geschichte. Campus, Frankfurt New York, pp. 9–18 (in German)

Pielke R jr (2004) What is climate change? Issues in Science and Technology Summer 2004, pp. 1–4

Räisänen J, Hansson U, Ullerstig A, Döscher R, Graham LP, Jones C, Meier HEM, Samuelsson P, Willén U (2004) European climate in the late twenty-first century: Regional simulations with two driving global models and two forcing scenarios. Clim Dyn 22:13–31

Rebetez M (1996) Public expectation as an element of human perception of climate change. Climatic Change 32:495–509

Remane A, Schlieper C (1971) Biology of Brackish Water. E. Schweizerbart'sche Verlagsbuchhandlung (Nägele u. Obermiller), Stuttgart

Rockel B, Woth K (2007) Extremes of near-surface wind speed over Europe and their future changes as estimated from an ensemble of RCM simulations. Climatic Change 81:267–280

Schwartz P (1991) The art of the long view. John Wiley & Sons

Segerstråle SG (1966) Adaptational problems involved in the history of the glacial relicts of Eurasian and North America. Rev Roum Biol- Zoologie: 11:59-66

Seinä A, Palosuo E (1996) The classification of the maximum annual extent of ice cover in the Baltic Sea 1720-1995. Meri Report Series of the Finnish Institute of Marine Research 20:79–91

Stehr N, von Storch H, Flügel M (1996) The 19th century discussion of climate variability and climate change: Analogies for present day debate? World Res Rev 7:589–604

Tol RSJ (2006) Exchange rates and climate change: An application of FUND. Climatic Change 75: 59–80

Tol RSJ (2007) Economic scenarios for global change. Proceedings IV, GKSS School on Environmental Research. Springer, Berlin Heidelberg New York (in press)

Uppala SM, Kållberg PW, Simmons AJ, Andrae U, da Costa Bechtold, V, Fiorino M, Gibson, JK, Haseler J, Hernandez, A, Kelly GA, Li X, Onogi K, Saarinen S, Sokka N, Allan RP, Andersson E, Arpe K, Balmaseda MA, Beljaars ACM, van de Berg L, Bidlot J, Bormann N, Caires S, Chevallier F, Dethof A, Dragosavac M, Fisher M, Fuentes M, Hagemann S, Hólm E, Hoskins BJ, Isaksen L, Janssen PAEM, Jenne R, McNally AP, Mahfouf JF, Morcrette JJ, Rayner NA, Saunders RW, Simon P, Sterl A, Trenberth KE, Untch A, Vasiljevic D, Viterbo P, Woollen J (2005) The ERA-40 re-analysis. Q J Roy Met Soc 131:2961-3012.doi:10.1256/qj.04.176

von Storch H (1995) Inconsistencies at the interface of climate impact studies and global climate research. Meteorol Zeitschrift 4 NF, pp. 72–80

von Storch H (1999) The global and regional climate system. In: von Storch H, Flöser G (eds) Anthropogenic Climate Change. Springer, Berlin Heidelberg New York, pp. 3–36

von Storch H, Flöser G (eds) (2001) Models in Environmental Research. Proceedings of the Second GKSS School on Environmental Research. Springer, Berlin Heidelberg New York

von Storch H, Stehr N (2000) Climate change in perspective. Our concerns about global warming have an age-old resonance. Nature 405:615

von Storch H, Zwiers FW (2002) Statistical Analysis in Climate Research. Cambridge University Press, Cambridge New York

Williamson H (1770) An attempt to account for the change of climate which has been observed in the Middle Colonies in North America. Trans Am Phil Soc 1:272

Winsor P, Rodhe J, Omstedt A (2001) Baltic Sea ocean climate: An analysis of 100 yr of hydrographic data with focus on the freshwater budget. Clim Res 18:5–15

Winsor P, Rodhe J, Omstedt A (2003) Erratum: Baltic Sea ocean climate: an analysis of 100 yr of hydrographical data with focus on the freshwater budget. Clim Res 25:183

Woth K (2005) Projections of North Sea storm surge extremes in a warmer climate: How important are the RCM driving GCM and the chosen scenario? Geophys Res Lett:32 L22708 doi: 101029/2005GL023762

Zmudzinski L (1978) The evolution of macrobenthic deserts in the Baltic Sea. 11th Conference of Baltic Oceanographers, vol. 2. Rostock, pp. 780–794

Zwiers FW (1999) The detection of climate change. In: von Storch H, Flöser G (eds) Anthropogenic Climate Change. Springer, Berlin Heidelberg New York, pp. 163–209

2 Past and Current Climate Change

Raino Heino, Heikki Tuomenvirta, Valery S. Vuglinsky, Bo G. Gustafsson, Hans Alexandersson, Lars Bärring, Agrita Briede, John Cappelen, Deliang Chen, Malgorzata Falarz, Eirik J. Førland, Jari Haapala, Jaak Jaagus, Lev Kitaev, Are Kont, Esko Kuusisto, Göran Lindström, H. E. Markus Meier, Miroslaw Miętus, Anders Moberg, Kai Myrberg, Tadeusz Niedźwiedź, Øyvind Nordli, Anders Omstedt, Kaarel Orviku, Zbigniew Pruszak, Egidijus Rimkus, Viivi Russak, Corinna Schrum, Ülo Suursaar, Timo Vihma, Ralf Weisse, Joanna Wibig

2.1 The Atmosphere

This section describes long-term observed climatic changes in atmospheric parameters. The focus is on surface climate conditions, but changes in atmospheric circulation are discussed as they often are behind climatic variability seen on regional and local scales. For a summary introduction on mean atmospheric states and conditions in the Baltic Sea Basin see Annex 1.2 with sections on the general atmospheric circulation (A.1.2.1), surface air temperature (A.1.2.2), precipitation (A.1.2.3), clouds (A.1.2.4), and global radiation (A.1.2.5).

As time periods from several decades to a couple of centuries are described, there are severe limitations to availability (see Annex 4) and quality of data (Annex 5). Figure 2.1 is a map that can be used to locate sites mentioned in Sect. 2.1.

2.1.1 Changes in Atmospheric Circulation

The Baltic Sea Basin is an area of permanent exchange of air masses of different physical features, which results in great variability of weather, from day to day and from year to year. The climate of this region is controlled by large pressure systems that govern the air flow over the continent: The Icelandic Low, the Azores High and the winter high/summer low over Russia.

The influence exerted by the Icelandic Low, in spite of the shelter provided by the Scandinavian Mountains, was predominant in the last WMO normal period 1971–2000 (Miętus 1998b; see also Chap. 1 and Annex 1.2.1). The characteristics of wind over the Baltic Sea and the adjacent coastal region are described in detail by Miętus (1998b). Changes in strong winds and storminess are discussed in Sect. 2.1.5.4.

2.1.1.1 Changes in Large-scale Atmospheric Flow Indices

The North Atlantic Oscillation (NAO) is a circulation factor strongly influencing the climate of Europe. The NAO expresses the intensity of the pressure gradient between the Azores High and the Icelandic Low. The NAO and a hemispheric circulation pattern, the Arctic Oscillation (AO), are described in Annex 7.

The influence of the NAO on Northern European climate is strongest during the winter months. A time series of the winter NAO index is shown in Annex 7 (see also Fig. 2.2). During the last two centuries the variability of the NAO has been rather irregular. However, starting from about the 1960s there has been a positive trend which peaked during the 1990s. This time period of strengthening NAO has influenced the climate of the Baltic Sea Basin by increasing the frequency of mild humid westerlies (see the following Sects. 2.1.2–2.1.5).

There are many indices that can be used for describing circulation changes. Figure 2.2 shows three zonal indices for the December to March (DJFM) seasonal average (Moberg et al. 2005). In addition to the North Atlantic Oscillation Index for the period 1821–2000 (Jones et al. 1997), a Central European zonal index for 1786–1995 (Jones et al. 1999; Slonosky et al. 2000, 2001) and a Fennoscandian zonal index for 1891–1999 by Tuomenvirta et al. (2000) are also shown. All three smoothed series have their largest values in the 1990s, due to many individual years after 1989 having high index values. Hence, the westerly wind flow from the North Atlantic in winter was particularly strong during the last decade of the 20th century.

A review of time series of different circulation indices over Europe (Slonosky et al. 2000, 2001; Niedźwiedź 2000b) shows that atmospheric pro-

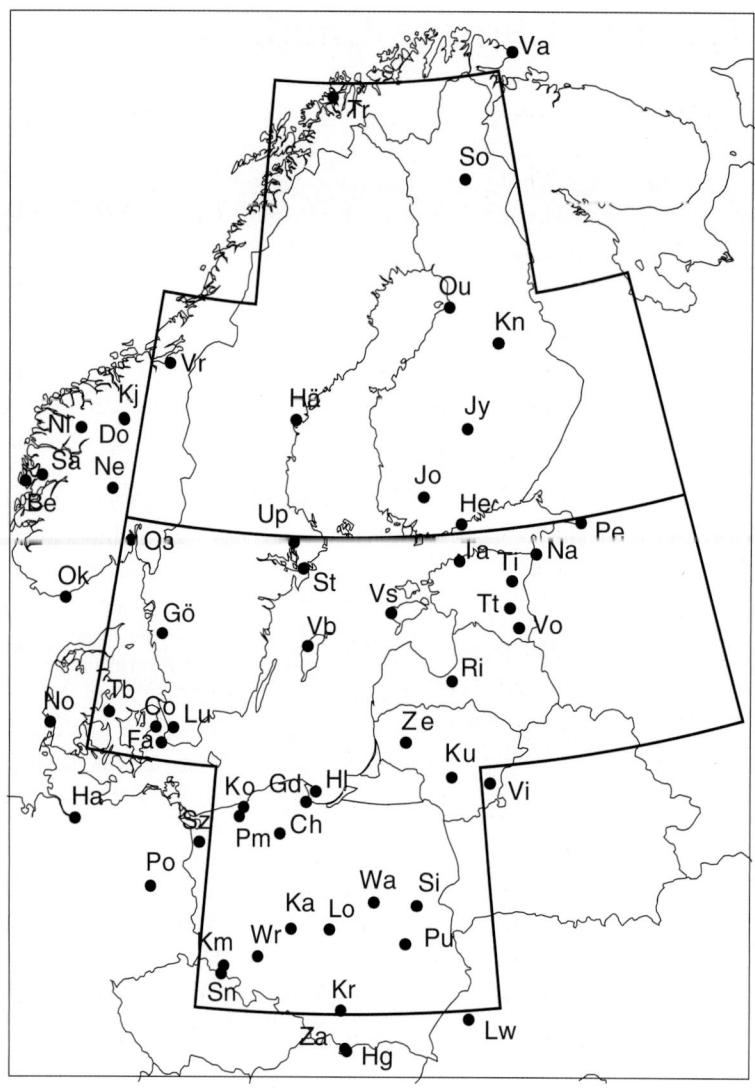

Fig. 2.1. Approximate location of sites (stations, lakes etc.) discussed in Sect. 2.1. See Table 2.1 for a list of sites. The borders of regions used in calculating area-averages for air temperature (southern and northern Baltic Sea Basin land areas) in Sect. 2.1.2 are marked with lines

cesses have varied a lot in the past two centuries. Periods with strong zonal circulation intensity were separated by episodes with more frequent blocking. This variability differed from season to season. During the 1950s and 1960s zonal circulation was weak and had negative values. During the winter (December to February) 1988/1989 the strength of the zonal circulation reached a maximum. Only during winter there is a significant increase through the latter half of 20th century.

Changes in atmospheric circulation covering the Northern Hemisphere have been examined using the macroscale circulation classification of Dzerdzeevskiy over the Baltic Sea Basin and the northern part of the East European plain during the 20th century (Dzerdzeevskiy 1968; Titkova and Kononova 2006).

Whether the observed changes in NAO – or in other circulation patterns – are caused by anthropogenic influence is unknown. Climate models cannot reproduce the observed changes, which, on the other hand, seem to be quite large to have been caused only by natural variability, e.g. see Osborn (2004) and references therein.

Table 2.1. Sites (stations, lakes etc.) shown in Fig. 2.1

Site	Abbrev.	Country	Site	Abbrev.	Country
Copenhagen	Co	Denmark	Tromsø	Tr	Norway
Nordby	No	Denmark	Værnes	Vr	Norway
Tranebjerg	Tb	Denmark	Vardø	Va	Norway
Narva	Na	Estonia	Chojnice	Ch	Poland
Tallinn	Ta	Estonia	Gdánsk	Gd	Poland
Tartu	Tt	Estonia	Hala Gąsienicowa	Hg	Poland
Tiirikoja	Ti	Estonia	Hel	Hl	Poland
Vilsandi	Vs	Estonia	Kalisz	Ka	Poland
Võru	Vo	Estonia	Karkonosze Mountains	Km	Poland
Helsinki	He	Finland	Koszalin	Ko	Poland
Jokioinen	Jo	Finland	Kraków	Kr	Poland
Jyväskylä	Jy	Finland	Łódź	Lo	Poland
Kajaani	Kn	Finland	Pommerania	Pm	Poland
Oulu	Ou	Finland	Pulawy	Pu	Poland
Sodankylä	So	Finland	Siedlce	Si	Poland
Hamburg	Ha	Germany	Snieżka	Sn	Poland
Potsdam	Po	Germany	Szczecin	Sz	Poland
Riga	Ri	Latvia	Warszawa	Wa	Poland
Kaunas	Ku	Lithuania	Wroclaw	Wr	Poland
Vilnius	Vi	Lithuania	Zakopane	Za	Poland
Zemaiciai Upland	Ze	Lithuania	St. Petersburg	Pe	Russia
Bergen	Be	Norway	Falsterbo	Fa	Sweden
Dombås	Do	Norway	Göteborg	Gö	Sweden
Kjøremsgrende	Kj	Norway	Härnösand	Hä	Sweden
Nesbyen	Ne	Norway	Lund	Lu	Sweden
Nigardsbreen	Ni	Norway	Stockholm	St	Sweden
Oksøy	Ok	Norway	Uppsala	Up	Sweden
Oslo	Os	Norway	Visby	Vb	Sweden
Samnanger	Sa	Norway	Lwów	Lw	Ukraine

2.1.1.2 Description of Regional Circulation Variations

A lot of different circulation descriptions on the regional scale exist for many countries or regions within the Baltic Sea Basin. Good examples are studies concerning monthly circulation climatology for Sweden (Chen 2000), with an explanation of its influence on climate and ecosystems (Chen and Hellström 1999; Blenckner and Chen 2003).

Another view on circulation changes are provided by studying winds at the 500 hPa isobaric level (corresponds roughly to a height level of 5 kilometers) over Estonia during 1953–1998, using data registered at the Tallinn Aerological Station (Keevallik and Rajasalu 2001). Wind velocity at the 500 hPa level showed significant inter-annual variability, while variations in the meridional component were somewhat larger than those in the zonal component. A significant change in circula-

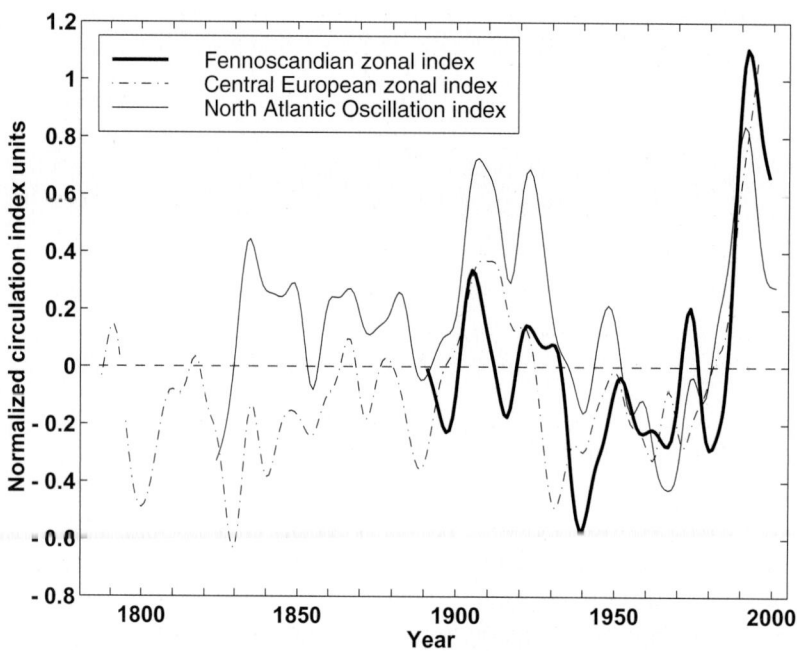

Fig. 2.2. Three zonal circulation index time series based on average DJFM (December to March) pressure gradient data. The line style associated with each index is given in the figure legend. The data are smoothed to highlight variability on timescales longer than 10 years. Each series is normalized by subtracting the mean and dividing by the standard deviation for the 1961–1990 period. Data sources used include: Fennoscandian zonal index from Moberg et al. (2005), Central European zonal index from Jones et al. (1999) and Slonosky et al. (2000, 2001), North Atlantic Oscillation index from Jones et al. (1997)

tion was found in winter (December to February), when the zonal component of the wind velocity increased. The end of winter and the beginning of spring was a key period in climate warming in Estonia which was analysed in more detail using the wind velocity components at two isobaric levels – 850 and 500 hPa (Keevallik 2003). A significant change of +1.04 m/s was found for the zonal component at the 850 hPa level over Tallinn in February for the period 1955–1995. Using circulation indices based on daily sea-level pressure data for the period 1946–1997 over Estonia, Tomingas (2002) also detected an intensification of westerly winds during winter, and a decreasing trend in April, June, September and the summer season (June to August).

Information about circulation variability exists for the Odra and Vistula basins in Poland for the period 1873–2004 (Niedźwiedź 1981, 1996, 2000b, 2005). Among the circulation indices based on the frequency of circulation patterns, Niedźwiedź (1993) calculated one simply on the basis of the circulation types' frequency by using the modified method proposed by Murray and Lewis (1966).

The following types are most important: W – zonal, S – meridional, and C – cyclonicity index (see Fig. 2.3). Some regional indices for Central Europe are also calculated on the basis of meridional and zonal pressure differences, using gridded or station data (Ustrnul 1997; Miętus et al. 2004). They were also compared with circulation indices for other regions in the Baltic Sea Basin and with European Circulation Patterns "Grosswetterlagen" (Hess and Brezowsky 1977; Kaszewski and Filipiuk 2003).

A deep minimum in the frequency of the western sector air flow above Poland (Niedźwiedź 2003a) was observed in the period 1962–1982 (Fig. 2.3). Since 1983 western circulation has been more and more frequent with a maximum around 1990. Ustrnul (1997) also found a positive trend in the zonal (westerly) index over the Baltic Sea Basin during the period 1901–1995. The S index describes the South–North transport of air masses that can provide large contrasts in temperature – best known as a heat wave or cold outbreak. The S index displays a rather steady behaviour (Fig. 2.3).

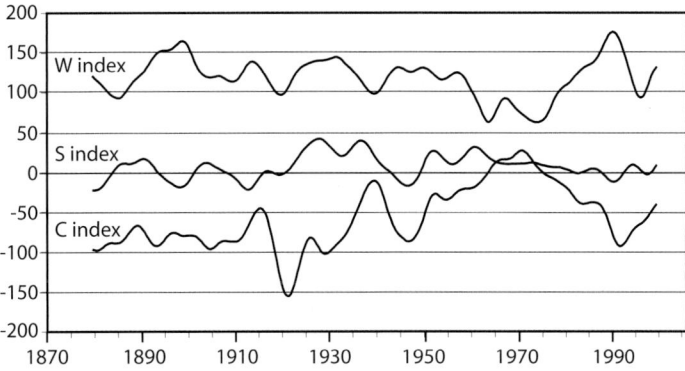

Fig. 2.3. Variability of circulation indices in Poland based on frequency of circulation types (Niedźwiedź 2003a). The curves are smoothed by the 11-years' moving averages

The cyclonicity index (C index) based on the frequency of cyclonic and anticyclonic circulation types was also elaborated for Poland (Fig. 2.3). Anticyclonic situations were generally more frequent than cyclonic ones (negative values of C). A maximum for cyclonic activity was observed in the period 1958–1981 that coincided with a relatively wet period with several floods in Poland (1958, 1960, 1970, 1977, and 1980). The greatest anticyclonicity (C = −230) was connected with the strong droughts in 1920 and 1921.

The highest cyclonicity in the southern part of the Baltic Sea Basin was observed in spring and winter. Anticyclones were most frequent in autumn and sometimes in summer. Taking into account all three indices for Poland (Fig. 2.3), the largest variability after smoothing can be observed in cyclonicity, while the smallest is typical for the meridional S index.

The variability of atmospheric circulation has a strong influence on the surface climate over the whole Baltic Sea Basin. There is a number of papers discussing relationships between atmospheric circulation and local surface climate characteristics (Bartkeviciene 2002; Chen and Li 2004; Degirmendžić et al. 2002; Jaagus 2006a,b; Jönsson and Holmquist 1995; Kożuchowski et al. 1992, 1994, 2000; Kożuchowski and Marciniak 1988, 1990; Kożuchowski and Żmudzka 2002; Marsz 1999; Miętus 1999; Moberg et al. 2005; Niedźwiedź et al. 1994; Omstedt and Chen 2001; Stankunavicius and Bartkevicienė 2002; Twardosz and Niedźwiedź 2001; Ustrnul and Czekierda 2001; Ustrnul 1998; Werner and von Storch 1993; Wibig 1999, 2000, 2001).

Utilising long-term observational series, it has been shown that the relationship between atmo-

spheric circulation and surface climate characteristics show strong temporal variability, indicating that relationships are not stable (Chen and Hellström 1999; Chen 2000; Jacobeit et al. 2001; Slonosky et al. 2001). For example, there is a strong correlation between the strength of the zonal circulation and winter temperatures over the Baltic Sea Basin, but the strength of the correlation has varied with time. The influence of westerly winds in the 1990s certainly contributed to the mild winters in the same period. Mild winters of the 1930s, on the contrary, occurred during a period with low values in zonal indices (Central European and Fennoscandian indices in Fig. 2.2 and Polish W index in Fig. 2.3), exemplifying the non-stationary nature of the relations between circulation and temperature.

2.1.2 Changes in Surface Air Temperature

The Baltic Sea Basin is a region with considerable variations of surface air temperature, both regionally and in time, see Annex 1.2.2 for a brief overview. Due to anthropogenic emissions of greenhouse gases, the interest in variations of air temperature near the surface has increased in recent decades. However, when local or regional air temperatures near the surface are measured, the anthropogenic effect cannot be identified directly. What is measured is the combined effect of an anthropogenic warming trend superimposed upon a large natural variability of the climate system. It is this combined temperature variability and trend that is analysed in this section. Strongly temperature related phenomena are also discussed in Sect. 2.1.5.2 (extremes) as well as in Sects. 2.2.1.5 and 2.2.2 (inland waters).

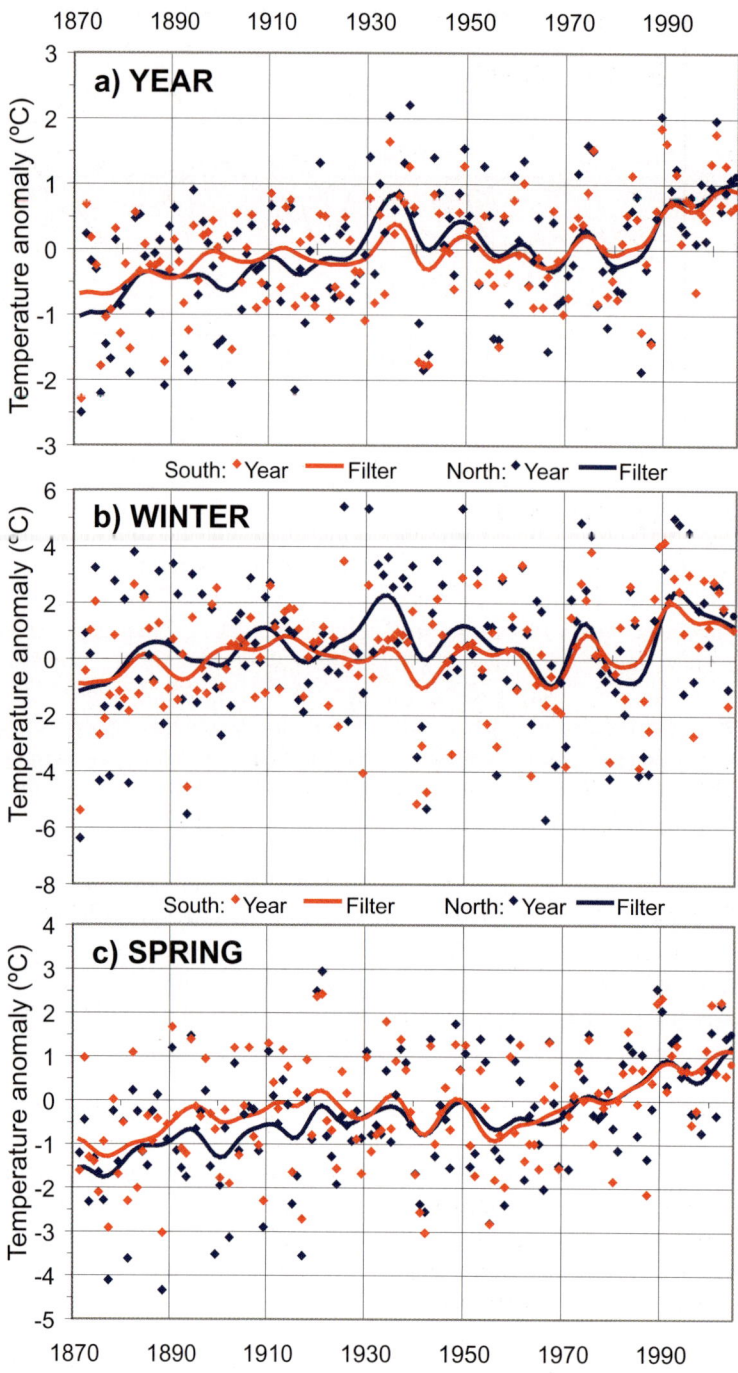

Fig. 2.4. Annual and seasonal mean surface air temperature for the Baltic Sea Basin 1871–2004, calculated from 5° by 5° latitude, longitude box averages taken from the CRU dataset (Jones and Moberg, 2003, and Annex 4) based on land stations (*from top to bottom:* (**a**) = annual, (**b**) = winter (DJF), (**c**) = spring (MAM), (**d**) = summer (JJA), (**e**) = autumn (SON)). Blue colour comprises the Baltic Sea Basin to the north of 60° N, and red colour to the south of that latitude. The dots represent individual years, and the smoothed curves (Gaussian filter, $\sigma = 3$) highlight variability on timescales longer than 10 years

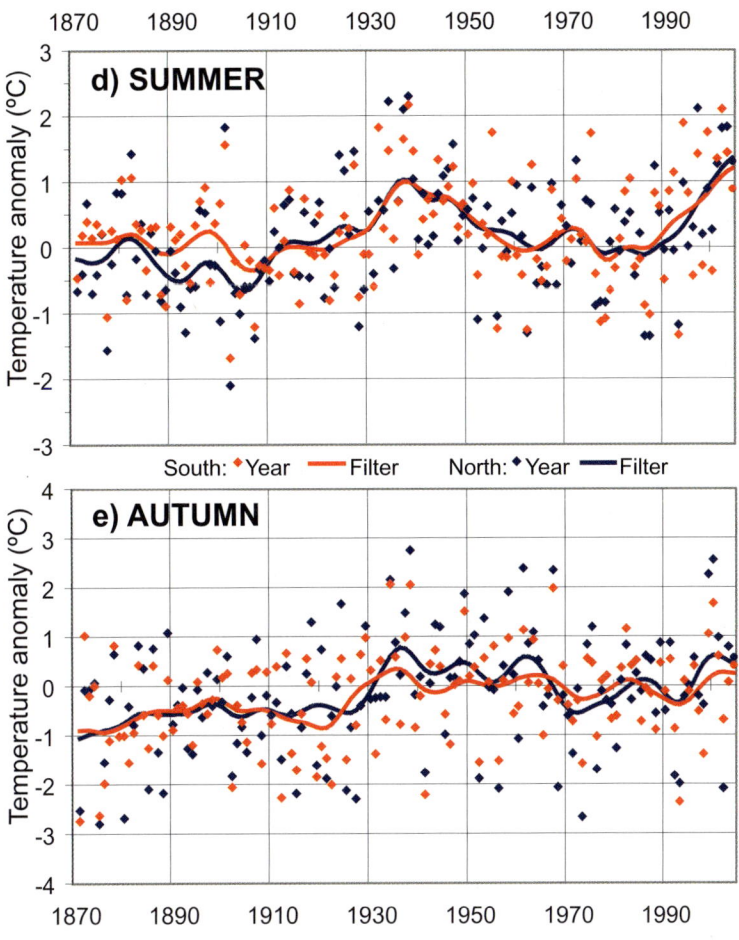

Fig. 2.4. cont.

2.1.2.1 Annual and Seasonal Area Averaged Time Series

The changes of the Baltic Sea Basin mean temperatures during 1871–2004 are inferred here using a gridded data set with a 5° by 5° (latitude, longitude) resolution (Jones and Moberg 2003, see Annex 4). For this analysis, all 5° grid boxes with at least 50% of the box area belonging to the Baltic Sea Basin were considered. Two area-averaged series have been calculated (Fig. 2.4), one including all selected Baltic Sea Basin grid boxes north of 60° N and one including those south of 60° N, thus establishing a northern and southern Baltic Sea Basin surface air temperature record, respectively. All data are anomalies from the 1961–1990 standard normal.

In terms of temperature variations, the Baltic Sea Basin has been a relative uniform area during the period 1871–2004 (Fig. 2.4a). A striking feature in the annual mean series is the early 20th century warming, which is a marked increase in temperature that culminated in the 1930s. This early 20th century warming was followed by a smaller cooling that culminated in the 1960s and then another distinct warming lasting to the end of the series.

Spring is the only season for which the end of the series is clearly warmer than the maximum of the 1930s when analysed on a decadal time scale (Fig. 2.4c). The early 20th century warming is more pronounced in the northern area than in the southern, especially during winter (Fig. 2.4b).

For the period 1871–2004, the largest seasonal trends in the Baltic Sea Basin are observed for the spring. Trends in all seasons are positive and most of them are significant at the 0.05 level (Table 2.2). For the northern area the trend in winter is as

Table 2.2. Linear surface air temperature trends (K per decade) in the period 1871–2004 for the Baltic Sea Basin, its northern (latitude > 60° N) and southern area (latitude < 60° N). Trends written in bold are significant at the 0.05 level. The trends were also tested by the non-parametric Mann-Kendall test. The results were consistent with the linear trend test

Data sets	Year	Winter	Spring	Summer	Autumn
Northern area	**0.10**	0.09	**0.15**	**0.06**	**0.08**
Southern area	**0.07**	**0.10**	**0.11**	0.03	**0.06**

high as 0.09 °C/decade, but as the variability is particularly large in this season, the trend is non-significant.

Annual warming trends (1871–2004) for the northern and southern Baltic Sea Basin, with values of 0.10 and 0.07 °C/decade, respectively, are larger than the trend for the entire globe (1861–2000), which is about 0.05 °C/decade (IPCC 2001). Neither the Baltic Sea Basin temperature series nor the global series show a monotonous temperature increase on the decadal time scale.

The main differences between the Baltic Sea Basin and global series are: (i) The early 20th century maximum occurs in the 1930s for the Baltic, whereas for the globe it occurs in the 1940s. (ii) For the Baltic Sea Basin, the maximum at the end of the series is of about the same magnitude as the one earlier in the 20th century, whereas for the globe the maximum at the end of the series is significantly higher (IPCC 2001) than that in the 1940s. (iii) The variability of annual mean temperature of the Baltic Sea Basin series is about five times larger than the variability of the global mean temperatures. The warming trend characteristic in the Baltic Sea Basin has been observed to extend further to the Arctic (Førland et al. 2002; Overland et al. 2004).

2.1.2.2 *Information Drawn from the Longest Instrumental Time Series*

The Little Ice Age, sometimes defined to be the period 1500–1850 (Ogilvie and Jónsson 2001), manifested itself in a dramatic way in the early to mid 18th century. Well known is the advance of the glacier *Nigardsbreen* in western Norway. The glacier destroyed much farmland and many farmhouses before its advance stopped in 1748 (Nordli 2001).

According to a reconstruction by Nordli et al. (2003) spring-to-summer air temperatures were very low in the early 1740s, and this is also confirmed by a dip in the Uppsala summer air temperature series (Moberg and Bergström 1997). However, the main reason for the glacier advance seems rather to be mild winters in the 1720s and 1730s (Moberg and Bergström 1997; Tarand and Nordli 2001) and a predominance of westerly winds (Bergström 1990), which led to accumulation of snow on the western Norwegian glaciers. This scenario is in line with the conclusion drawn by Nesje and Dahl (2003).

The Stockholm air temperature series shows warm intervals in the 1770s and in the 1790s (Moberg et al. 2002). In the 1770s summer temperatures contributed heavily to the high annual values, whereas in the 1790s it was mainly the winter temperatures that contributed (Fig. 2.5). The warm decade of the 1770s is also seen in the St. Petersburg series (Jones and Lister 2002; Moberg et al. 2000). At the end of the 18th century a change towards colder climate is seen, with low temperatures in the first decade of the 19th century for all seasons. The temperature then increased in the following decades, but in the late 1830s and early 1840s another severe cold period started.

The second half of the 19th century started with increased annual air temperatures, and, fortunately for agriculture in the northern areas, summer temperatures contributed much to this increase. However, the 1850s was the last decade with summers above normal before the onset of the cold period 1860–1920. The most severe decade within this period was the 1860s. The year 1867 was the coldest year in the entire Stockholm temperature series (Fig. 2.5). In Finland, the spring of 1867 was exceptionally cold, e.g. the break-up of lake ice was delayed by more than a month to mid-June (Simojoki 1940), and harvests were lost, leading to the death of about 8% of the Finnish population (Jantunen and Ruosteenoja 2000). The cold period after the 1860s was broken by several mild winters around 1910, followed by the early 20th century warming (e.g. Alexandersson 2002).

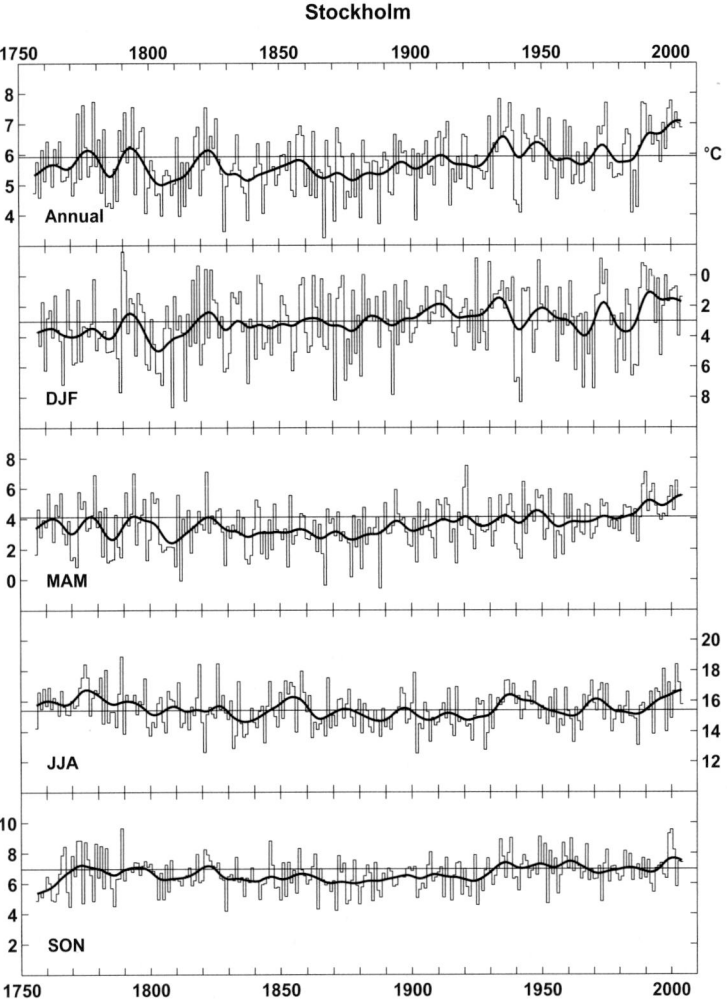

Fig. 2.5. Annual and seasonal mean surface air temperatures (℃) in Stockholm 1756–2004, calculated from the homogenised daily mean temperature series by Moberg et al. (2002) after a correction for a suspected positive bias in summer temperatures before 1859 (Moberg et al. 2003). The correction is the same as used by Moberg et al. (2005). Smoothed curves (Gaussian filter, $\sigma = 3$) highlight variability on timescales longer than 10 years

The St. Petersburg (Jones and Lister 2002) and the Vilnius series (Bukantis and Rimkus 2005) also have decadal air temperature maxima and minima (not shown here) roughly in the same years as for the Stockholm series. But there is a notable difference: The annual maximum of the 1930s at St. Petersburg and Vilnius is not higher than the maxima in the 1890s and 1910s, and, for the winter season, local maxima occur in the 1920s. This differs from the situation in the western and northern parts of the Baltic Sea Basin.

The temperature trends since the start of the Vilnius series in 1777 have been examined by Bukantis et al. (2001) and Bukantis and Rimkus

(2005). A significant trend in the annual values, 0.6 ℃ for the whole period of observation, arises mostly from warming in the cold season (November–March), amounting to 1.0 ℃ (Fig. 2.6).

In Poland, long-term homogeneous time series of air temperature have been reconstructed for Warsaw, 1779–1998 (Lorenc 2000), Szczecin, 1838–1992 (Kożuchowski and Miętus 1996), Gdańsk and Hel, both 1851–1995 (Miętus 1998a) and Koszalin, 1848–1990 (Miętus 2002). There is a statistically significant increase of air temperature in the range of 0.54–0.56 ℃/100 years in Warsaw, Gdańsk and Hel. Szczecin temper-

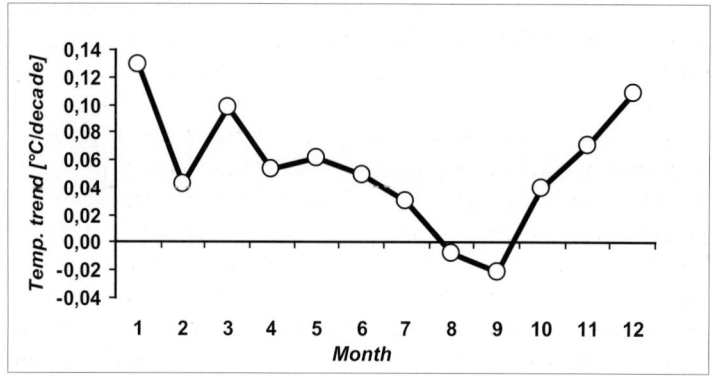

Fig. 2.6. Trends in monthly mean surface air temperature of the Vilnius series during the period 1777–2000

ature series show a larger increase of 1.1 °C/100 years. Warming is strongest in winter, in the range of 1.1 °C/100 years in Gdańsk and Hel, 1.2 °C/100 years in Warsaw and 1.3 °C/100 years in Szczecin.

2.1.2.3 *Temperature Trends in Sub-regions and Relationships with Atmospheric Circulation*

Many analyses of temperature trends and variations have been based on data from the modern instrumental period, from about 1870 to present. Moberg et al. (2005) studied the temperature difference between the two 30 year periods, 1971–2000 and 1891–1920, for 6 individual time series in Fennoscandia (Vardø and Kjøremsgrende in Norway, Oulu and Helsinki in Finland, Uppsala and Stockholm in Sweden). Annual mean changes varied from +0.49 to +0.82 °C. These changes were significant for all stations except for Kjøremsgrende. Also for spring, all differences were significant except for Kjøremsgrende. For the other seasons, most of the station trends were not significant.

The annual temperature increase from the middle of the 19th century to the 1930s or 1940s (see Sect. 2.1.2.1) is characteristic for most of the Baltic Sea Basin. The average warming amounts to almost 1.5 °C in Denmark since 1870 (Cappelen and Christensen 2005) and to 0.7–1.0 °C in Estonia since about 1850 (Jaagus 1996, 1998) and in Finland (Tuomenvirta 2004).

The last decades of the 20th century and the first years of the 21st century are also characterised by a pronounced temperature increase. For example, in Denmark and Estonia the warming started around 1990 (Cappelen and Christensen

2005, Jaagus 1996, 1998), whereas the warming started around 1980 in Sweden and Norway (Alexandersson 2002; Hanssen-Bauer and Nordli 1998). For Poland the mean surface air temperature in the decade 1991–2000 was 0.7 °C higher than in the previous normal period 1931–1960 (Degirmendžić et al. 2004). Ševkunova et al. (1999) found that the region with the strongest warming in the 1980s and the 1990s is located in the eastern Baltic Sea Basin south and east of Tallinn and St. Petersburg.

Temperature trends during the second half of the 20th century have been examined for Poland by Degirmendžić et al. (2004), and for Estonia (10 stations) by Jaagus (2006a). Significant positive linear trends of 0.2 °C/decade and 0.2–0.3 °C/decade for annual temperatures are observed in Poland and Estonia, respectively. For individual months the trends vary considerably, being significant only for March (0.6 °C/decade) and May (0.3 °C/decade) for Poland. In Estonia, March also experiences the largest trend, amounting to 1 °C/decade.

Temperature variability in the Baltic Sea Basin is strongly affected by changes in the atmospheric circulation patterns. For southwestern Sweden and western Denmark, Chen (2000) showed that as much as 70% of the total variance in the monthly mean January surface air temperatures between 1873 and 1994 can be accounted for by a linear combination of three circulation indices; southerly wind, vector wind speed and westerly shear vorticity. Moberg et al. (2005) found, for temperatures averaged over Fennoscandia, correlations with a zonal circulation index in the range +0.44 and +0.75 in the October to March half-year and negative correlations of between −0.23 and −0.36

in the summer months during June to August. Stronger westerly winds in winter thus generally bring milder air towards Fennoscandia, whereas in summer they bring cooler air.

Polish, Lithuanian, and Estonian studies also reveal that the strongest correlations between wind flow indices and surface air temperature occur during winter when zonal indices are used (Jaagus 2006a; Bartkevicicnc 2002). For the transitional seasons, spring and autumn, correlations with temperature are stronger with meridional circulation indices than with zonal indices (Degirmendžić et al. 2004; Kożuchowski et al. 1992, 1994, 2000; Kożuchowski and Marciniak 1988, 1990; Kożuchowski and Żmudzka 2002; Marsz 1999; Miętus 1999; Niedźwiedź and Ustrnul 1994; Ustrnul 1998; Wibig 2000, 2001). Miętus and Filipiak (2002) demonstrated the influence of North Atlantic sea surface temperatures (SST) on thermal conditions in Poland. The strongest connection between SST and air temperature in Poland exists during winter season, the weakest one during spring.

However, when comparing the influence of North Atlantic SST and atmospheric circulation on the temperature in Poland Miętus and Filipiak (2002) conclude that the influence of circulation (described by the pressure field) is dominant over SST forcing. Degirmendžić et al. (2004) also present evidence from Poland supporting the hypothesis that "the real warming observed in the last decade of the 20th century has exceeded circulation-induced changes of temperature".

2.1.2.4 *Mean Daily Maximum and Minimum Temperature Ranges*

Trends and changes occurring at the high and low ends of the air temperature range can have a great impact on society. It is well known that, after 1950, the minimum air temperature on a global scale has increased more than the maximum temperature (e.g. Karl et al. 1993; Easterling et al. 1997). Consequently, this has led to a decrease in the Daily Temperature Range (DTR), defined as DTR = Tx − Tn, where Tx and Tn are the mean daily maximum and minimum air temperatures, respectively. For extreme temperatures see Sect. 2.1.5.2.

By examining the period 1951–1998 for 9 homogeneous series from Poland, Wibig and Głowicki (2002) found that annual mean values of Tn increased significantly during this period for all sta-

tions, with trends ranging from 0.14 °C (Zakopane) to 0.31 °C (Kalisz) per decade, whereas trends of Tx did not increase significantly. This change has led to a decrease in DTR, although significant only with 5 of the 9 series. Seasonally, both Tn and Tx warmed significantly during winter (December to February) and spring (March to May), but with only minor changes in DTR. However, a decrease of DTR is observed for the autumn (September to November), although this is not due to an increased Tn but rather a decreased Tx.

The time evolution of Tx and Tn for Fennoscandia has been studied by Tuomenvirta et al. (2000) over the period 1910–1995. The largest variations occurred in winter (December to February) with the warm 1988–1995 period surpassing even the warm 1930s for both Tx and Tn. For the annual means of Tx and Tn, however, the 1988–1995 period was at about equal as warm as the 1930s. Both the spring Tx and Tn have warmed markedly throughout the 20th century, although only the Tn warming is significant.

The Fennoscandian DTR (Fig. 2.7) decreased during the 20th century, although not steadily. For example, the 1940s was a decade with large positive anomalies for spring, summer and the annual mean. Overall, there was a decrease in DTR from 1910–1939 to 1966–1998 in spring, summer and autumn, resulting in a statistically significant decrease of the annual mean, albeit without any significant decreases in the seasonal values. The overall decrease of the Fennoscandian DTR during the 20th century has been related to an increase in cloudiness and to an intensification of westerly circulation.

Over most of the 20th century, four stations in Poland (Łódź, Puławy, Zakopane and Śnieżka) showed a linear warming of annual Tn that was larger than that of Tx, resulting in a decrease in DTR for all four seasons (Wibig and Głowicki 2002). The decrease in DTR is significant in spring and autumn for all four stations. The strongest warming was found during mid- and late winter for both Tx and Tn. Broadly consistent results were obtained by Miętus and Filipak (2004), who analysed the variability of thermal conditions in the coastal region of the Gulf of Gdańsk during the period 1951–1998. They proved that the increase in mean daily minimum air temperature (close to or higher than 0.2 °C/decade) was definitely higher than that of the mean daily maximum temperature (between 0.1–0.2 °C/decade) at three of the four stations.

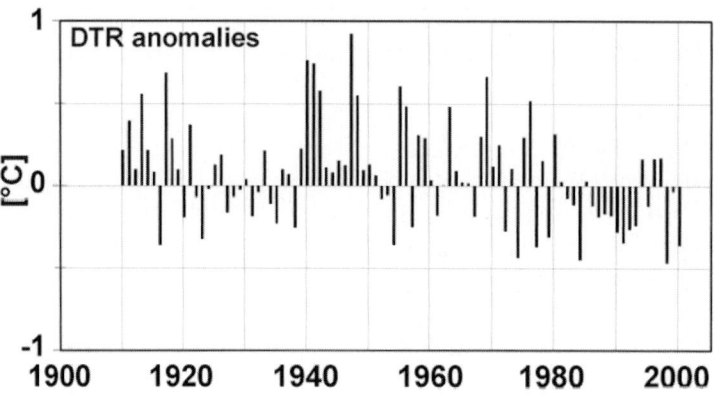

Fig. 2.7. Annual anomalies of Fennoscandian area-averaged daily temperature range (DTR) for the period 1910–2000 (baseline is the 1961–1990 average) (updated from Tuomenvirta et al. 2000)

2.1.2.5 *Changes in Seasonality*

Information on changing seasonality can be drawn from long-term temperature series. Here, we describe results for a cross-section from the west (Norway) to the east (East European Plain) across the Baltic Sea Basin.

For Norway, using a Fourier transformation of seven homogenised monthly temperature series, Grimenes and Nissen (2004) modelled the annual air temperature cycle during the period 1871–1990. These calculations were used to define indices for the length of the growing season (period when daily mean temperature, Tm, is equal to or exceeds 5 °C), the frost-free season (period when Tm > 0 °C), the summer 'half-year' (period when Tm is larger than the annual mean temperature), and the hot season $\frac{d^2 Tm}{dt^2} \leq 0$ as well as the heat sum (sum of Tm during the growing season).

Growing season length and frost-free season length show positive trends for all stations in the period analysed (Table 2.3). This implies that the average level of the annual temperature cycle has been shifted upwards towards higher temperatures. Also, the length of the summer half-year and the hot season show positive trends, which indicates that the shape of the annual temperature cycle has changed. The inflection points of the curve have shifted towards earlier dates in the spring and later dates in the autumn. The summer and winter 'halves' of the curves have become more asymmetric, with the summer side becoming more U-shaped and the winter curve more V-shaped.

A substantial increase in the heat sum (Table 2.3) illustrates the combined effect of the curve's change in shape and level. The varying, and in part smaller changes, in winter strength show that the positive trend described by the former quantities relates less to the changes in winter temperatures than to changes in summer temperatures.

Jones et al. (2002) studied seasonal indices back to the 18th century, using homogenised daily air temperature series from Stockholm, Uppsala and St. Petersburg. For the sub-period after 1870, their results are broadly consistent with those of Grimenes and Nissen (2004): The length of the growing season and the heat sum have increased, whereas the length of the cold season and the frost degrees have decreased. The lengthening of the growing season in Northern Europe is also reported by Chen et al. (2006) and Linderholm et al. (2005). According to Jones et al. (2002), there is little relationship between the temperature of the growing (and also frost) season and its duration. However, day degree counts were found to be very strongly correlated ($r > 0.9$) to average extended summer (May to September) and extended winter (October to April) temperatures.

Jaagus and Ahas (2000) studied long-term trends in climatic seasons during the period 1891–1998 for Tartu, Estonia. In addition to the four main seasons – spring, summer, autumn and winter – four intermediate seasons have also been defined; between autumn and winter (late autumn, early winter), and between winter and spring (late winter, early spring). A general tendency is that the start of the climatic seasons in the spring half-year (late winter, early spring, spring and summer) start earlier, whereas the climatic seasons in the autumn half-year (autumn, late autumn, early winter and winter) start later. But among those,

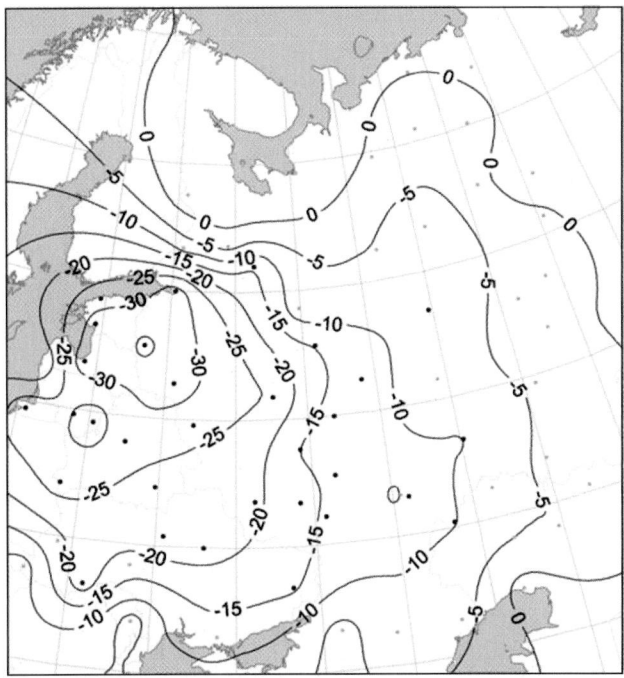

Fig. 2.8. Linear trend of the start date of early spring during the period 1946–1995, expressed in days. Stations with a trend significant at the $p < 0.05$ level are marked with black dots (adapted from Jaagus et al. 2003)

Table 2.3. Means and trends of various indices (Grimenes and Nissen 2004, see text for definitions) for 1871 to 1990. Trends are shown as days/100 years or day-degrees/100 years

	Growth season length (days)		Frost free season length (days)		Summer half-year length (days)		hot season length (days)		Heat sum (day degrees)	
	Mean	Trend	Mean	Trend	Mean	Trend	Mean	Trend	Mean	Trend
Oksøy	211	14.6			175	8.6	154	26.0	2416	188
Oslo	184	12.8	243	14.8	177	5.2	162	18.3	2271	152
Bergen	218	20.7			174	8.8	153	27.4	2362	258
Dombås	138	12.7	193	14.7	177	8.1	162	28.9	1343	118
Værnes	176	10.8	251	8.9	176	6.4	158	21.7	1895	119
Tromsø	129	17.4	202	14.8	163	6.4	130	10.1	1170	258
Vardø	112	16.3	189	18.9	170	4.0	142	7.3	868	178

only the start of late autumn (+8 days) and winter (+17 days) have shifted significantly. Significant are also rather large changes of duration for summer (+11 days), early winter (+18 days), and winter (−30 days) (Jaagus and Ahas 2000).

In the East European Plain, using a similar methodology, the climatic seasons have been analysed for two periods, a long period (1881–1995) and a short period (1946–1995), using 12 and 73 stations, respectively (Jaagus et al. 2003). The most remarkable changes during the shorter pe-

riod were detected in the Baltic Sea Basin part of the Plain (Fig. 2.8), where the start of early spring (when daily mean air temperature permanently rises above zero) has moved towards an earlier start by up to one month.

As a consequence, the duration of winter (when daily mean air temperature permanently drops below zero) has dramatically decreased, and duration of early spring has increased. Similar results are also reported by Kożuchowski et al. (2000) for 8 stations in Poland during the period 1931–1998.

For some of the stations winter is shortened by more than 30 days during this period. Also Markevičiene (1998) reports a reduction of the length of the winter by about 30 days in Vilnius, although the period analysed is much longer (> 200 years).

2.1.3 Changes in Precipitation

Atmospheric circulation and characteristics of air masses (humidity, stability) largely determine the occurrence and rate of precipitation. Although atmospheric processes have a governing role, orography greatly influences spatial distribution and intensity of precipitation, e.g. enhancing rainfall on the windward side of mountains and hills (e.g. Tveito et al. 1997 and Annex 1.2.3). Furthermore, the effect of the underlying surface, e.g. a contrast between the Baltic Sea and adjacent land, can be seen in the regional distribution of precipitation (Annex 1.2.3).

Precipitation measured with gauges is usually an underestimation of true precipitation. There is a substantial undercatch, especially for solid precipitation, frequent light rainfall events or windy conditions (Førland and Hanssen-Bauer 2000). Different mean climatological estimates for the Baltic Sea Basin (for the whole basin as well as separate estimates for the land surface and sea part) are available in the published literature; see Annex 1.2.3 for a brief discussion.

It is thus possible that also changes in other factors, such as gauge design, type of precipitation, wind conditions, etc. contribute to trends in the measured precipitation (see Annex 5). Studies putting recent precipitation changes in the perspective of multi-decadal variability are complicated by homogeneity problems of the time series. Many studies have addressed this question (e.g. Alexandersson 1986; Hanssen-Bauer and Førland 1994; Heino 1994; Peterson et al. 1998; Tuomenvirta 2001; Groisman et al. 2001; Wijngaard et al. 2003). Despite efforts at homogenisation, uncertainties are related to the long-term records.

Modern measurement technologies like satellites and radar have so far been able to produce only short time series that are not free from uncertainties, either (e.g. Koistinen and Michelson 2002; Michelson et. al 2005). In this section we will present variability and trends based on precipitation measurements with gauges. Precipitation extremes are discussed in Sect. 2.1.5.3 and some other precipitation-related variables (e.g. runoff, snow cover) in Sect. 2.2.

2.1.3.1 *Observed Variability and Trends During the Recent 50 Years*

Folland et al. (2001) report in the Third Assessment Report of the Intergovernmental Panel on Climate Change (IPCC TAR) that precipitation has increased by about 12% in all land areas between latitudes $55°$ N and $85°$ N during the 20th century. A closer look reveals that the general high-latitude increase is broken into a complex pattern of changes in northern Europe (New et al. 2001).

In the Baltic Sea Basin, there has been an increase of precipitation in the period 1976–2000 compared to the period 1951–1975 (Beck et al. 2005, see Fig. 2.9). However, the pattern of change is far from uniform. The largest increases occurred in Sweden and on the eastern coast of the Baltic Sea, while southern Poland has received somewhat less precipitation on average. As precipitation varies greatly both spatially and temporally (Annex 1.2.3), it is difficult to establish long-term trends. Because of poor data-coverage in western parts of Norway, the map does not reflect the considerable precipitation increase which was observed in this area during the second half of the 20th century (Hanssen-Bauer 2005). On the other hand, over some areas there is a small artificial enhancement in precipitation increase due to improved techniques for measuring precipitation; e.g. in Finland this amounts to a few percent.

There are distinct patterns of seasonal changes, with spring precipitation increasing over large areas around the Baltic Sea. For example, an increase of more than 15% in spring has occurred in central Sweden, but in some regions in southern Poland spring precipitation has decreased. Rainfall amounts during summer have shown a small decrease over areas mainly in western and southern parts of the Baltic Sea Basin, but some increase has occurred in regions around the northern Baltic Proper as well as in southern Finland and northern Sweden. In autumn there has been an increase almost in the entire Baltic Sea Basin, yet there are regions showing a small decrease, particularly in Germany and Poland. The season with the largest increases is winter. The main area of increase stretches from Norway, Denmark and Germany in the west to the Baltic States and Russia in the east.

Førland et al. (1996) performed an analysis of precipitation changes based on well over a thousand stations during the standard normal periods

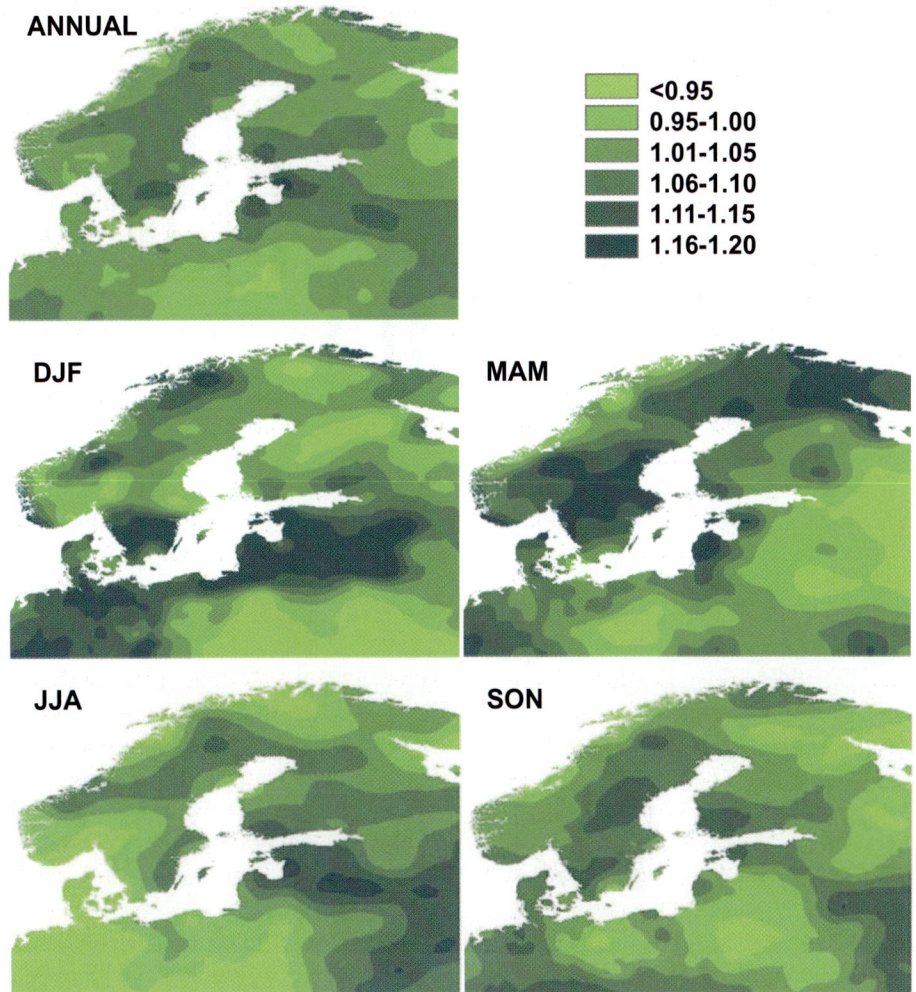

Fig. 2.9. Annual and seasonal precipitation ratios between the periods 1976–2000 and 1951–1975 based on VASClimO (Variability Analysis of Surface Climate Observations) data (Beck et al. 2005)

1931–1960 and 1961–1990, covering a large part of the Baltic Sea Basin. In Førland et al. (1996) the general features are similar to Fig. 2.9, but there are more details over the Scandinavian countries due to better station coverage. The spatial pattern of differences suggests that atmospheric circulation changes have been the main cause. An increase in humid westerlies during the colder half of the year produces an orographic signal where regions exposed to westerlies and which receive a large part of the precipitation in autumn and winter receive more precipitation. In accordance with this, some leeward sides of mountains and hills show a decrease.

The most recent changes are reported in national studies. In Sweden, the 1990s were on average 8% wetter than during the 1961–1990 normal period (Alexandersson and Eggertsson Karlström 2001). All seasons except autumn became rainier. Northern and south-western Sweden show local maxima of positive anomalies. In Denmark as a whole the period 1990–2004 was on average 5% wetter than the 1961–1990 normal period (Cappelen and Christensen 2005). In Finland Hyvärinen and Korhonen (2003) compared measured precipitation amounts at drainage level between the periods 1961–1990 and 1991–2000.

A majority of the studied areas showed an increase of measured precipitation with only a few basins showing no change or a decrease. The increase occurred mainly during the cold season, with a small increase also during summer. In northern Finland during the period 1991–2000, the largest positive anomalies of drainage basin

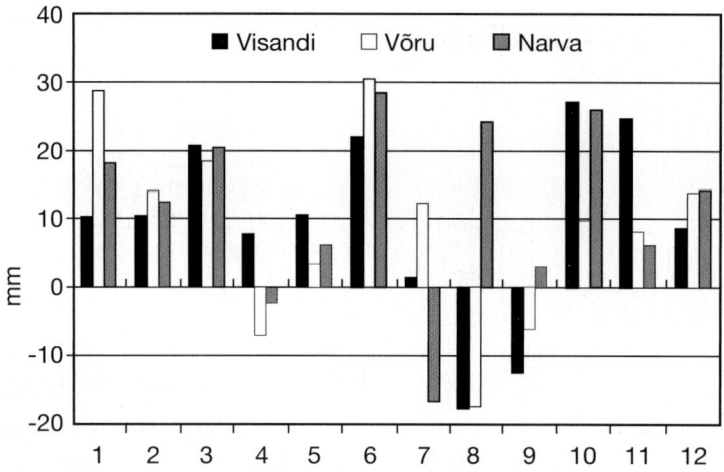

Fig. 2.10. Trend of monthly precipitation (mm/month) at three Estonian stations (Vilsandi, Võru and Narva) during 1951–2000 (Jaagus 2006a)

annual precipitation were up to 50–80 mm ($\sim$ 10–17%) above the standard normal. However, in some cases even half of the difference between the periods 1961–1990 and 1991–2000 can be due to improvements in measurements (Solantie and Junila 1995; Tuomenvirta 2004).

Franke et al. (2004) analysed precipitation trends in Saxony, eastern Germany. The most notable features of the trend analysis for the period 1951–2000 were: A marked decrease in summer rainfall (−10 to −30%) and a significant increase in winter precipitation (10 to 20%). In Poland, during the period 1951–2000 a positve trend (+2.85 mm/decade) in annual area averaged precipitation remains insignificant (Degirmendžić et al. 2004), but there is a significant increase in March and also a significant decrease in the ratio of summer and annual precipitation totals.

In Estonia, the second half of the 20th century is characterised by a statistically significant increase in annual precipitation by 10–25% (80–180 mm) at 8 out of 9 studied stations (Jaagus 2006a). The increase in annual precipitation results partly from the introduction of wetting corrections. Therefore, the real increase may be up to 5–15%. Among the individual months, a statistically significant increase in precipitation is revealed during the period from October to March and in June, whereby trends vary much between stations and months, especially in summer (Fig. 2.10). In Vilnius, Lithuania, cold season precipitation has increased and warm season precipitation has decreased during the last fifty years (Bukantis and Rimkus 2005).

2.1.3.2 *Long-term Changes in Precipitation*

This section discusses evidence concerning precipitation changes based on Baltic Sea Basin records covering time periods of more than a century.

In Norway Hanssen-Bauer and Førland (1998) identified 13 regions having uniform long-term trends and decadal variability. Data for the period 1896–2004 show increasing trends in annual precipitation of between 3 to 18% per 100 years (Hanssen-Bauer 2005). Positive trends were statistically significant in 9 of the 13 regions. The increase has been somewhat larger during the latter half of the record. In southern Norway, the long-term increase is significant only during autumn, while in northern parts of the country spring as well as summer and/or winter precipitation have increased.

In Sweden, Busuioc et al. (2001) identified four regions with similar long-term variability. Figure 2.11 shows annual and seasonal precipitation trends for Sweden (Alexandersson 2004). The increasing trend is significant at the annual level and in all seasons except summer. Alexandersson (2002) divided the country into four regions according to the drainage basins and compared annual totals between the periods 1860–1920 and 1921–2001.

The analysed increase ranges from 23% in the Gulf of Bothnia drainage basin to 15% in the Bothnian Sea basin and 7% in basins in southern Sweden (south of 60° N). Seasonally, the largest increase has taken place in northern Sweden during winter.

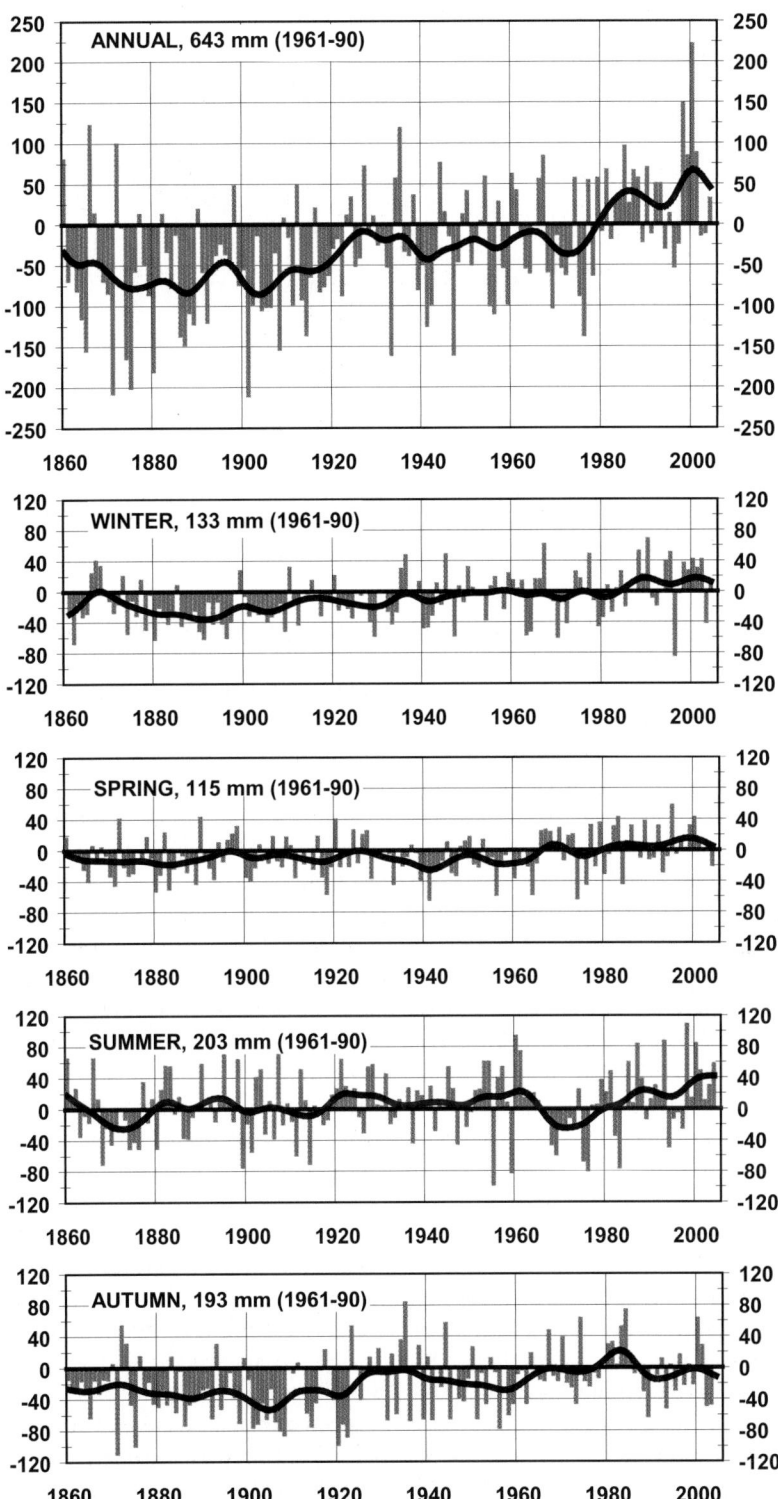

Fig. 2.11. Anomaly time series of annual and seasonal precipitation over Sweden, 1860–2004 (reference period 1961–1990). Curves represent variations in the time scale of about ten years (updated from Alexandersson 2004)

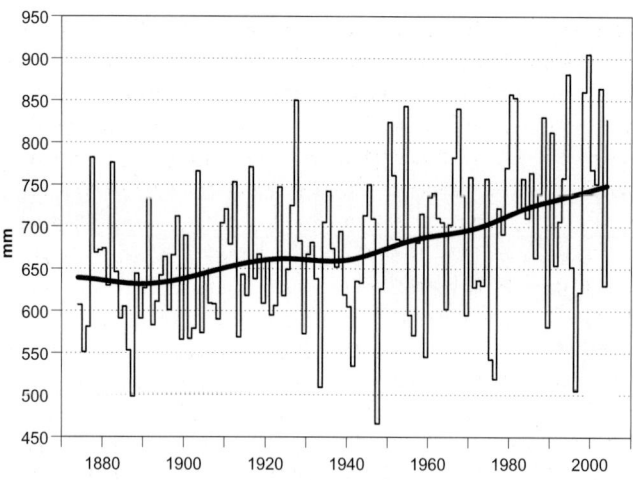

Fig. 2.12. Variation of annual precipitation amount over Denmark, 1874–2004. The curve represents variations in the time scale of about thirty years (Cappelen and Christensen 2005)

Somewhat contrary to its western neighbours, annual precipitation totals over Finland do not show any significant long-term trend during the period 1894–2002 (Tuomenvirta 2004). There are decadal fluctuations of about ±30 mm in annual precipitation, one of which shows a rise of about 10 mm (∼ 2%) from the 1940s to the 1990s. It may well be that averaging over the whole country (an area of more than 330,000 km²) smoothes out some regional trends that are visible e.g. in Finnish discharge records (Hyvärinen 2003). Figure 2.12 shows the annual accumulated precipitation over Denmark (Cappelen and Christensen 2005). There has been a statistically significant increase of approximately 100 mm (≈ 15%) since 1870.

Analysis of spatial mean annual precipitation in Estonia during the period 1866–1995 indicates a weak and insignificant increasing trend. The increase has been larger in autumn and winter, and is absent in spring and summer. Attention has been paid to the analysis of periodical fluctuations of precipitation in Estonia. Using spectral analysis, rather regular periodicity with cycles of 50–60, 25–33 and 5–7 years was detected (Jaagus 1992, 1998; Jaagus and Tarand 1998).

The long-term precipitation series from Latvia show increases during both the cold (Nov–Mar) and the warm (Apr–Oct) seasons, with a more pronounced increase during the cold season (Lizuma 2000; Briede and Lizuma 2002). The significance of precipitation trends shows spatial variability.

A long time series of precipitation from Vilnius, Lithuania (Bukantis and Rimkus 2005) shows a large multi-decadal fluctuation from dry to wet during the first decades of the 20th century. The slight long-term increasing trend in annual precipitation results from an increase during the cold season (Nov–Mar) and a slight decrease during the warm season (Apr–Oct).

Kozuchowski (1985) examined the variation in precipitation during the period 1881–1980 in Poland. A slight increase in annual totals was discovered. The increase was accompanied by a growth of dispersion of precipitation totals. Niedźwiedź and Twardosz (2004) combined three time series from Poland (Kraków, Warszawa, Wroclaw) and one from the Ukraine (Lwów) to represent temporal variability of precipitation to the north of the Carpathian and Sudeten mountains. Over this very long time period (1855–2002), there is no clear long-term trend in annual totals. The second half of the 20th century has a somewhat higher mean level than the earlier time periods, mainly due to a recent increase during winter.

2.1.3.3 *Changes in Number of Precipitation Days*

Besides total amount, also frequency and intensity of precipitation need to be considered (see also Sect. 2.1.5.3 on precipitation extremes). A simple parameter describing frequency is the number of precipitation days, i.e. days when daily precipitation is equal to or exceeds e.g. 0.1 mm or 1.0 mm. The limit of 0.1 mm is quite sensitive to factors such as observer accuracy, observation interval and evaporation characteristics of the precipitation gauge (Heino 1994; Tveito et al. 2001).

Heino (1994) did not find any significant trends in the Finnish data. Bukantis and Valiuskeviciene (2005) found an increasing trend in the number of annual precipitation days in the entire Lithuanian territory for the period 1925–2003. Kożuchowski and Żmudzka (2003) reported an increase in the number of precipitation days in Poland during the last hundred years. In general, autumn and winter showed increasing trends, continuing until March. Førland (2000) studied changes in the distribution of daily precipitation in southern and western Norway. He found a high covariance between probability of rain, total precipitation, and probability of heavy precipitation. During the 20th century at Samnanger (60° 28′ N, 5° 54′ E), where the annual precipitation has increased by 18%, the number of days with precipitation $\geq 0.1\,\text{mm}$ ($\geq 1.0\,\text{mm}$) has increased by 19% (13%). There are many studies on changes of total precipitation in Northern Europe, but there is a lack of information concerning systematic changes in intensity of daily precipitation.

2.1.3.4 *Interpretation of the Observed Changes in Precipitation*

Some of the century-scale precipitation time series from the Baltic Sea Basin display increasing precipitation. Over the whole northern hemisphere latitude band 55° N–85° N, the increase was somewhat steeper during the first half of the century than after the 1950s, whereas in the Baltic Sea Basin it appears that the increasing trend has been larger during the latter half of the 20th century. Winter is usually the season with the largest increase. Homogeneity problems combined with the fact that precipitation varies to a high degree spatially and temporally, make accurate estimation of long-term precipitation trends difficult.

The latter half of the 20th century is characterised by decadal scale fluctuations, but there is evidence of an increase in annual precipitation totals over most of Northern Europe. The observed increase is not uniform and there are also regions lacking statistically significant trends (e.g. in Finland and southeastern Norway). The increase is mostly concentrated in the cold season from October to March. As precipitation amounts display large temporal and spatial variability, there is also variability in trends from month to month and between regions. During summer, convective precipitation processes dominate and there is less spatial coherence of trends between regions during the lat-

ter half of the 20th century. Both increasing (e.g. Sweden) and decreasing (e.g. Poland) trends in large area averages can be found in summertime.

There are many studies relating precipitation to atmospheric circulation (e.g. Kożuchowski and Marciniak 1988; Wibig 1999; Busuioc et al. 2001; Ustrnul and Czekierda 2001; Twardosz and Niedźwiedź 2001; Johansson and Chen 2003), and weak links with remote sea surface temperatures also have been found (Feddersen 2003). During wintertime, part of the increase in precipitation can be attributed to more frequent westerly and/or cyclonic circulation related to large-scale changes in atmospheric circulation (Busuioc et al. 2001; Keevallik 2003; Degirmendžić et al. 2004). Similarly, some of the differences between summer months have been attributed to changes in circulation patterns (Busuioc et al. 2001). Because of large natural variability and somewhat inconsistent signals from future climate simulations (see Chap. 3), it is difficult to attribute the observed increase of precipitation in northern Europe to factors other than changes in atmospheric circulation (Schmith 2001). Although the causes of precipitation fluctuations cannot be identified, these fluctuations can still have remarkable local and regional impacts.

2.1.4 Changes in Cloudiness and Solar Radiation

Nearly all physical processes and phenomena in the atmosphere have originated and are driven by solar energy. Net radiation describes the amount of solar and atmospheric radiation energy absorbed by the Earth's surface. Spatial distribution of solar radiation is, first of all, determined by latitude, but the role of clouds is very significant, especially in the ground-level fluctuations of solar radiation. Cloudiness is, in turn, influenced by atmospheric circulation. See Annexes 1.2.4 and 1.2.5 on mean cloudiness and global radiation, respectively, in the Baltic Sea Basin.

Long-term changes observed in cloudiness and solar radiation have so far hardly been studied. The IPCC Third Assessment Report (IPCC 2001) refers only to a few articles dealing with changes in cloudiness in Europe. They indicate an increase in cloud cover during different periods during the 20th century. Little information can be found about changes in global radiation and other components of the radiation budget.

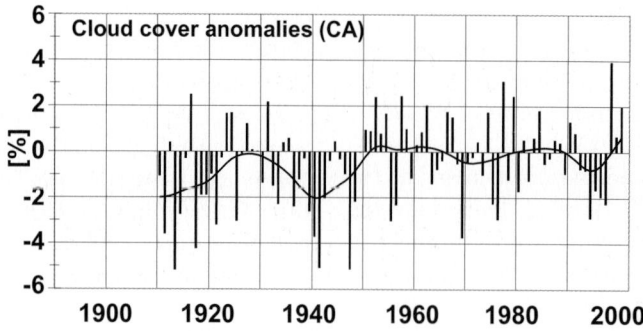

Fig. 2.13. Annual time series of cloud cover anomalies for Fennoscandia, 1910–2000. Low-frequency variability on timescales longer than 10 years is highlighted by the smooth curve (from Moberg et al. 2005)

2.1.4.1 *Cloudiness and Sunshine Duration*

Climatic changes observed in cloudiness over Europe during the last century were studied by Henderson-Sellers (1986, 1992). She detected a generally increasing tendency in total cloudiness, although she found some remarkable spatial differences as well. Later, changes in cloudiness were studied in relation to decreasing trends in diurnal air temperature range (Dai et al. 1997; see also Sect. 2.1.2.4). General research was conducted on cloudiness variations over the former Soviet Union (Sun and Groisman 2000). Low-level cloud cover significantly decreased during the period 1936–1990, as opposed to changes in total cloud cover. The decrease in low-level clouds was mostly due to a decrease in stratiform clouds. A significant increase in cumulus clouds has been observed over the past decades (Sun and Groisman 2000).

A generally increasing trend in cloud cover in Fennoscandia during the period 1910–1995 was documented on the basis of 36 stations by Tuomenvirta et al. (2000). A long-term increasing trend in its annual mean values is 0.17% per decade (Fig. 2.13). This change is characteristic for spring, summer and autumn. Wintertime cloudiness had positive anomalies in the 1920s and 1930s. Since then, it has been decreasing.

Moberg et al. (2003) studied long–term cloud cover time series from Stockholm and Uppsala in Sweden back to 1780 and found large problems in data homogeneity. Both series show strong correlations (in the 20th century) with the Fennoscandian average cloud anomaly series of Tuomenvirta et al. (2000) for high–frequency variability (less than 10-year time scales), but the low-frequency correlations (longer than 10-year time scales) were nearly zero. The authors concluded that it was not possible to construct homogeneous cloud amount series for the whole study period.

Cloudiness variability and trends for Poland have widely been studied. The longest time series is available for Kraków, since 1861. A positive tendency of total cloudiness was revealed up to the 1950s followed by a negative trend (Morawska-Horawska 1985; Matuszko 2003) that has turned again into a weak increase in the 1990s (Matuszko 2003). Two other long-term series from Poland, Wrocław, 1885–1995 (Dubicka and Pyka 2001) and the Karkonosze Mountains, 1885–1995 (Dubicka 2000), show some increase in cloudiness.

Generally, trends in cloudiness in Poland during 1951–2000 at 48 stations show that a decreasing tendency of total cloudiness is prevailing in Poland (Żmudzka 2003), notably in southern, central and northeast parts, and e.g. in the coastal region of Pommerania (Miętus et al. 2004). The decrease is not uniform. There are also some regions with an increase of cloud cover, e.g. in western Poland (Żmudzka 2003) and at stations further from the Baltic Sea in Pomerania (Miętus et al. 2004).

Wibig (2004) analysed a half century long decreasing trend of total cloudiness in Łódź (Fig. 2.14) with respect to cloud types and seasonality. There is an increasing trend in convective clouds, which is statistically significant in winter, spring and summer. Simultaneously, the frequency of stratiform clouds has decreased in all seasons but summer and during the whole year.

For Estonia, an increasing trend is detected in lower, but not in total cloudiness. Changes in the amount of low clouds at 16 stations in Estonia were analysed for the period 1955–1995 by Keevallik and Russak (2001). Trends in annual values were not similar. Significant, increasing trends were detected at 10 stations, but negative trends

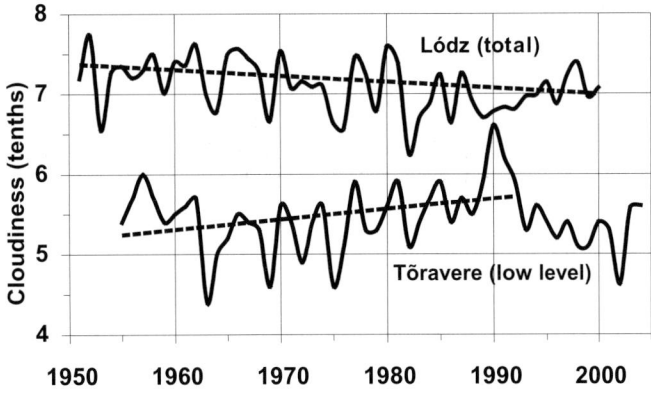

Fig. 2.14. Long-term variability of the mean annual total cloudiness in Łódź at noon during 1951–2000 (Wibig 2004) and annual mean low cloudiness at Tartu-Tõravere in 1955–2004

were found for two stations, possibly due to inaccuracies in observations. The amount of low clouds has increased in March, June and September, while it decreased in some stations in May and October. At Tartu-Tõravere, the monthly mean amount of low clouds has significantly increased in March, June and September, as has the annual mean (Russak 1998). During the earlier time period, 1964–1986, the amount of low clouds from April to September increased by as much as 11% (Russak 1990).

The increase in the amount of low clouds in Estonia was primarily a result of an increase in the occurrence of stratus and stratocumulus clouds. This increase was especially notable in March (Russak 1999; Keevallik and Russak 2001). Keevallik (2003) demonstrated that March has been the month with most significant climate changes in Estonia. The amount of low clouds at the Tiirikoja and Tartu-Tõravere station increased by 0.55 and 0.60 tenths per decade in March during 1955 to 1995, respectively. Since the beginning of the 1990s a decreasing amount of low clouds has been observed at the Tartu-Tõravere station (Fig. 2.14).

Long-term changes in cloudiness in Lithuania were studied by Stankūnavičius (1998) for the period 1925–1995. There was a decrease in total cloud cover in the eastern and central parts in the 1960s and the beginning of the 1970s and an increase in the Zemaiciai Upland. The majority of these trends is not statistically significant. Low cloudiness shows a more pronounced negative trend from the 1940s up to the beginning of the 21st century. Some stations in the hilly eastern and flat northern parts of Lithuania indicate a de-

crease in cloudiness by up to 1–1.2 tenths per 25 years. Seasonal differences are also significant. A negative trend in total and low cloudiness prevails during the cold season (October–February), while a positive trend is evident during the rest of a year, especially in March.

Sunshine duration depends very much on cloudiness. It can be assumed that an increase in cloudiness is closely related to a decrease in sunshine duration. A very long series for sunshine duration in Copenhagen is available since 1876 (Laursen and Cappelen, 1998; Cappelen, 2005). The lowest level was found at the end of the 19th century, rising to the highest level in the middle of the last century, followed by a slight decrease in the 1960s and 1970s and an overall increase from the 1980s to present (Fig. 2.15). Now the mean level of annual sunshine duration is nearly of the same magnitude as in the middle of the last century. All seasons show more or less the same trends.

The longest time series of sunshine duration in Estonia is available from Tartu, dating back to 1901. Substantial fluctuations have been observed during the 20th century, although no general trend exists (Jaagus 1998). The sunniest period was in the 1940s. After that, sunshine duration decreased significantly up to the 1980s. The latest increase in sunshine duration started in the 1990s. Corresponding time series for Helsinki, Finland (Heino 1994) as well as for Riga, Latvia (Lizuma 2000) show much of the same features as in Tartu (Fig. 2.15).

There are long-term low frequency fluctuations causing trends of different signs at different time intervals. For example, annual sunshine dura-

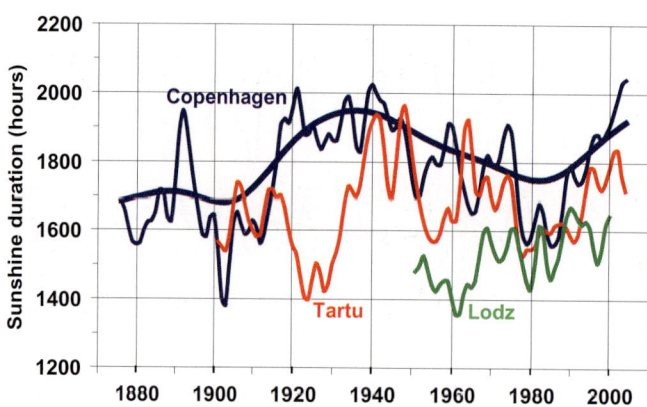

Fig. 2.15. Annual totals of sunshine duration in Copenhagen, 1876-2004 (*black line represents filtered values using a 9-year gauss filter*); Tartu, 1901–2004 and Łódź, 1951–2000. The Łódź series was obtained from Podstawczyńska (2003)

tion in Riga during 1924–1999 decreased by about 200 hours. The decline was most substantial in April, March and September (Lizuma 2000). A significant decrease during the whole 20th century was observed also in Wrocław, Poland (Dubicka and Pyka 2001). A positive trend was found in Łódź, Poland, for annual sunshine duration during 1951–2000 (Fig. 2.15, Podstawczyńska 2003). At the same station, a statistically significant positive trend was observed for May, August, winter (December to February) and spring (March to May), but a decreasing trend was found in September.

Comparing time series of cloudiness and sunshine duration in the second half of the 20th century between the southern and the northern part of the Baltic Sea Basin, a dipole structure can be found (Figs. 2.14 and 2.15). A decrease in cloudiness and an increase in sunshine duration were observed in Poland, while opposite trends were revealed in Estonia. In the 1990s, all of these trends changed their sign. Lithuania seems to be situated between these two regions of opposite fluctuations. Detailed regional studies are lacking from Fennoscandia, where the area average shows decadal fluctuations. The differences in cloudiness and the sunshine duration changes are largely explained by opposite air pressure systems and by changes in cyclone trajectories.

2.1.4.2 Components of the Radiation Budget

Annual totals of global radiation at Tartu-Tõravere show a statistically significant decreasing trend from the 1950s up to the beginning of the 1990s (Russak 1994). Later global radia-

tion started to increase (Fig. 2.16). The decrease was 6.4%. The greatest decrease was observed in March and September (Russak and Kallis 2003). A reversal from decrease to increase in global radiation is found also in Potsdam, Warsaw, Payerne (Switzerland) and Ny Alesund, Spitsbergen (Wild et al. 2005)

Significant decreasing trends in annual totals of global radiation were also found at some other stations during the period 1964–1986 (Russak 1990). At the Kaunas actionometric station, the decrease was 12.5% and in Stockholm (1965–1986) 11.5%. Heino (1994) found a slightly smaller decrease for stations in Finland during the period 1957–1990. Grabbe and Grassl (1994) noted a significant decrease in global radiation in Hamburg, Germany, where the decrease in annual totals of global radiation amounted to $4\,\mathrm{MJm}^{-2}$ per decade during the period 1949–1990 (Liepert 1996).

At Tartu-Tõravere, the decrease in global radiation was caused by a decrease in direct radiation incident on a horizontal surface. These curves are rather similar (Fig. 2.17) and the absolute values of the decrease were close, 230 (global) and 248 (direct) MJm^{-2} during 1955–1992. During 1957–1990 there was a clear maximum in direct radiation in central and northern Finland in the 1970s, but southern Finland behaved quite similar to Tartu (Heino 1994). No significant changes in direct radiation were measured in Hamburg (Liepert 1996).

Two factors have an effect on direct radiation – cloudiness and atmospheric transparency. As described above, low cloudiness increased in Estonia up to the 1990s. Another factor, atmo-

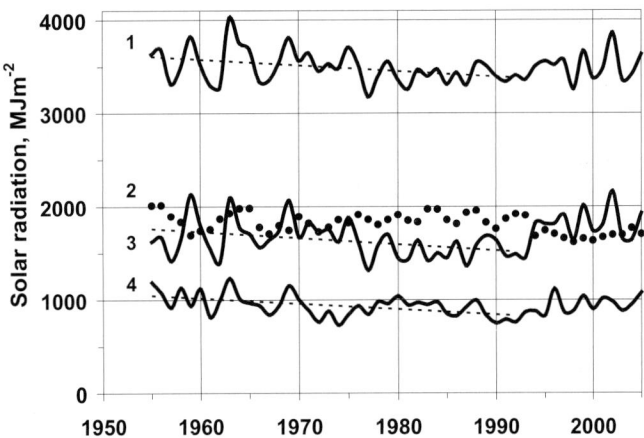

Fig. 2.16. Annual totals of global (**1**), diffuse (**2**), direct (**3**) and reflected (**4**) radiation at Tartu-Tõravere in 1955–2005 (MJm^{-2})

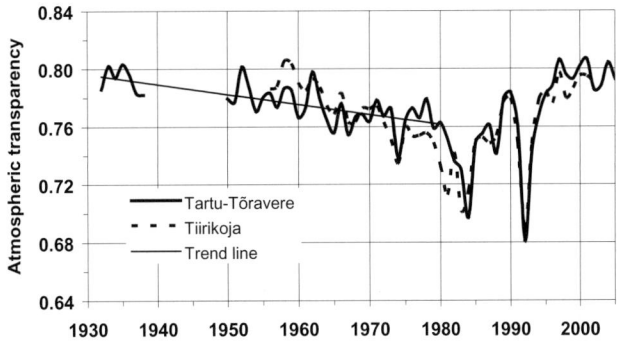

Fig. 2.17. Annual mean atmospheric transparency at Tartu-Tõravere (1932–1938, 1950–2005) and at Tiirikoja (1956–2001)

spheric transparency in cloudless conditions, P_2, can be described by the Bouguer atmospheric transparency coefficient, reduced to optical air mass $m = 2$ and to the mean distance between the Earth and the sun. The scale of P_2 ranges from zero to one.

Besides relatively short-time variations, usually lasting for 1–2 years and often corresponding to major volcanic eruptions, a distinct decreasing trend was found in the time series of annual mean P_2 at Tartu-Tõravere since the 1930s up to the mid-1980s (Russak 1990). The annual values of P_2 in the 1930s were calculated from episodic solar radiation measurements at the University of Tartu (Ohvril et al. 1998).

The similarity of long-term runs of P_2 at Tartu-Tõravere and at another meteorological station in Estonia, Tiirikoja (Okulov 2003), points to the es-

sential role of the aerosol originating from distant sources (Fig. 2.17).

Heikinheimo et al. (1996) studied another measure of atmospheric transparency at three sites in Finland, Tiirikoja, Estonia and Bergen, Norway. During summers from around the 1960s to 1994 the results agreed quite well with those presented in Fig. 2.17.

The improvement of atmospheric transparency at Tartu-Tõravere during the last 10–15 years can be explained by the establishment of strict requirements against air pollution, as well as by an economic decline in the former socialist countries. During the period 1980–2003, the emission of SO_2 decreased in Estonia by 70%, NO_x by 36% and emission of volatile organic compounds was halved (Ministry of the Environment of the Estonian Republic 2004). In the Czech Republic, an impor-

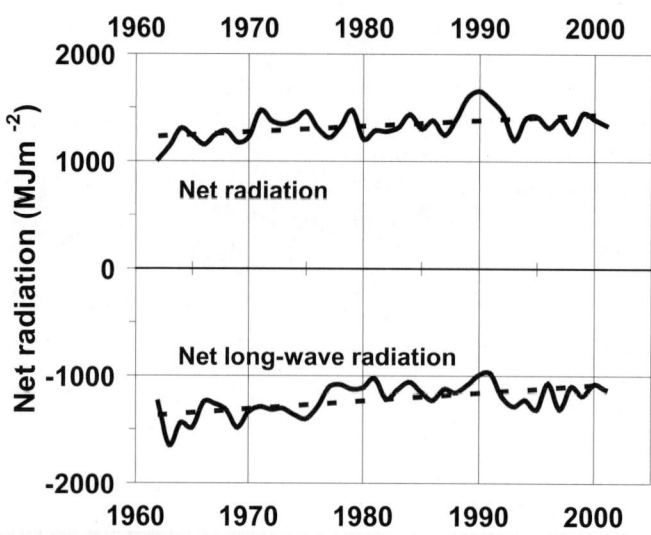

Fig. 2.18. Annual totals of net and net long-wave radiation (MJm^{-2}) at Tartu-Tõravere during 1962–2001 (Russak and Kallis 2003)

tant source of anthropogenic aerosol in Central Europe, the emission of SO_2 was reduced by 86% and NO_x by 55% in 1989–2000 (Hejkrlik 2002). At present, the atmospheric transparency in Estonia has reached the level it had in the 1930s.

No significant trend was found in time series of diffuse radiation during the same period from the 1950s up to the beginning of the 1990s (Russak and Kallis 2003). But a decreasing trend has become evident during the latest 10–15 years (Fig. 2.16). It can be explained by a decrease in both cloudiness and atmospheric turbidity. A significant decrease of diffuse radiation by $4\,\mathrm{MJm}^{-2}$ per decade was observed in Hamburg during the period 1964–1990 (Liepert 1996).

As a result of a decrease in global radiation, the radiation reflected from the surface has also decreased at Tartu-Tõravere (Fig. 2.16) (Russak and Kallis 2003). Time series from central and southern Finland during the period 1960–90 also show a small decline or no change, whereas in Sodankylä, northern Finland, there is a clear drop of about $200\,\mathrm{MJm}^{-2}$ (Heino 1994).

The main reason for the decrease in reflected radiation at Tartu-Tõravere was a decrease in ground surface albedo. In the Baltic Sea Basin, annual mean albedo is highly correlated with the duration of snow cover, especially with snow conditions in March and April (Tooming 1995). According to the analysis of snow data from nearly a hundred observation points in Estonia, the duration of permanent snow cover shortened on av-

erage by 33 days during 1962–1995 (Tooming and Kadaja 1995, 2000a,b; see also Sect. 2.2.3). At Tartu-Tõravere, the annual mean albedo decreased from 0.29 to 0.25 in 1955–1992 (Russak and Kallis 2003).

During the years 1955–1992, the decrease in annual totals of reflected radiation was $216\,\mathrm{MJm}^{-2}$, which is rather close to the decrease in global radiation. In the budget of solar radiation (net short-wave radiation), these changes compensate each other and, as a result, there is no significant trend in the time series of solar radiation absorbed by the surface, also called net short-wave radiation, and the solar radiation budget on the ground (Russak and Kallis 2003).

A breakpoint in the described trends has been found at Tartu-Tõravere for the early 1990s: Solar radiation and transparency of the cloud-free atmosphere began to increase while low cloudiness started to decrease. This kind of change may be related to the diminishing atmospheric aerosol content (air pollution) during recent decades. Changes in the aerosol content lead to the changes in absorption and scattering of radiation, as well as in the amount of possible condensation nuclei and thus in genesis and development of cloudiness. Another plausible explanation for the decrease in low cloudiness could be related to changes in atmospheric circulation.

The radiation budget (net radiation) at the surface can be expressed as an algebraic sum of incoming radiation fluxes and outgoing radiation fluxes. In addition to solar radiation, infrared radiation of the atmosphere and the ground surface contribute to net radiation. Time series of annual totals of net radiation demonstrate a highly significant increasing trend at Tartu-Tõravere during 1962–2001 (Fig. 2.18) (Russak and Kallis 2003). As described above, no trend was found in the short-wave part of the radiation budget. The increase in net long-wave radiation was very similar to the increase in net radiation (Fig. 2.18). Also, the absolute values of the increase were close: $-267\,MJm^{-2}$ in net radiation and $+303\,MJm^{-2}$ in net long-wave radiation.

Unlike solar radiation, no breakpoint was found neither in the time series of net radiation nor for that of net long-wave radiation. Significant trends in net long-wave radiation were found in December, January and in the summer months. In March and April, when the changes in solar radiation were largest, no significant trends could be detected in long-wave radiation. This indicates that possibly different factors are causing the changes in solar and infrared radiation.

The increase in long-wave radiation was more rapid during the first three decades (1960 to 1990) and slower during the last decade. The observed increase in the infrared radiation budget could have been caused by the decrease in the infrared radiation emitted by the ground surface or by the increase in atmospheric radiation (or both). There are no causes leading to a possible decrease in the outgoing infrared radiation. No decreasing trends were noted either in the time series of the surface or in air temperature. The annual mean temperature increased by 1.8 °C at Tartu-Tõravere during that period. Therefore, it can be surmised that the increase occurred in atmospheric counter-radiation. An increase in cloudiness, as well as in the concentration of both the greenhouse gases and the optically active aerosol in the infrared spectral area, should be considered as possible factors causing the increase in the counter-radiation.

2.1.5 Changes in Extreme Events

As the climate changes, the characteristics of extremes may also change (Trenberth 1999; Easterling et al. 2000; Beniston and Stephenson 2004). For many impact applications and decision support systems extreme events are much more important than the mean climate (Mearns et al. 1984; Wise 1999). Changes in extremes may be due to the mean effect (e.g. Wigley 1985), the variance effect (e.g. Katz and Brown 1992), a combination of the mean and variance effects (e.g. Brown and Katz 1995) or the structural change in shape of distribution (e.g. Beniston 2004).

2.1.5.1 *Definition of Extremes*

Extreme climate events can be defined as events that occur with extraordinary low frequency during a certain period of time (rarity), events with high magnitude (intensity) or duration, and events causing sizeable impacts such as losses (severity). According to the IPCC (2001, p. 790), an *extreme climate event* is an average of a number of weather events over a certain period of time, an average which itself is extreme (e.g. rainfall over a season). Some indices for maxima/minima during a certain period of time, magnitude (a threshold of a variable) and rarity have been developed and used to study the changes in climate extremes (e.g. Jones et al. 1999). A recent project funded by the European Union studied 64 climate indices during 1801–2000 in Europe, focusing on the trend of climate extremes (Chen et al. 2006).

While some of the climatic extremes are well described by meteorological variables/indices, others may not be easily defined with data of only a single meteorological variable. For example, freezing rain is a special combination of low temperature and rain that produces major damages through ice loading on wires and structures. Snow damage to forests is another example (Solantie 1994). Landslides are known to be affected by heavy rains (e.g. Schuster 1996). However, their occurrence is also determined by geological and geomorphologic conditions (Brunsden 1999). Yet another type of climate extreme occurs when meteorological conditions that may seem more or less normal from a purely anthropocentric perspective nevertheless could induce a strong impact on other species. Such ecological climate extremes may not be easy to identify as climate extremes. The dependence on factors other than meteorological data makes it difficult to disentangle the specific contribution of weather/climate in producing the impacts and to describe the combinations of extremes. Often, there are no long term homogeneous data to use in climate change studies. Therefore, the next section will focus on the indices which are well defined by meteorological data.

2.1.5.2 *Temperature Extremes*

Data sets enabling studies on changes in extreme
temperature events have been collected by the
projects ECA[1] (Klein Tank et al. 2002b), and
STARDEX[2] (Haylock and Goodess 2004). Recently, in a series of papers (Klein Tank and Können 2003; Frich et al. 2002; Haylock and Goodess
2004; Moberg and Jones 2005) trends in several
cold and warm temperature extremes in Europe
were described. Earlier studies on mean minimum
and maximum temperatures are available from
Brázdil et al. (1996) and Heino et al. (1999) for
central Europe, and Miętus and Owczarek (1994)
for the Polish coast. A more extensive study, taking into consideration seasonal changes of several
indices of cold and warm temperature extremes
in Poland, was presented by Wibig and Głowicki (2002). Seasonal changes in mean maximum
and minimum temperatures in Fennoscandia were
analysed by Tuomenvirta et al. (1998). There
are also papers describing extremes at individual
stations in the Baltic Sea Basin (Piotrowicz and
Domonokos 2002; Jones and Lister 2002; Moberg
et al. 2002; Bergström and Moberg 2002).

There are about 100 stations taken into consideration in papers analysing temperature extremes in the Baltic Sea Basin: the ECA dataset
(Klein Tank et al. 2002b); Tuomenvirta et al.
(1998); Miętus and Owczarek (1994); Brázdil et
al. (1996); Wibig and Głowicki (2002); Heino et
al. (1999); Cracow – Piotrowicz and Domonokos
(2002); Stockholm – Moberg et al. (2002); Uppsala
– Bergström and Moberg (2002); St. Petersburg –
Jones and Lister (2002).

In the second half of the 20th century, the number of days with minimum temperature below 0 °C,
i.e. frost days, strongly decreased in the Baltic Sea
Basin. The decrease was the strongest in southern Norway and Denmark, at some places exceeding 8 days per decade (Nesbye and Tranebjerg).
Figure 2.19 shows the annual trends calculated
for part of the ECA dataset (Klein Tank et al.
2002a) including all stations from 10 countries in
the Baltic Sea Basin (Norway, Sweden, Finland,
Denmark, Germany, Poland, Latvia, Lithuania,
Estonia and Russian Federation), with five additional Polish stations (Łódź, Siedlce, Chojnice,

---

[1] European Climate Assessment & Dataset, see
http://eca.knmi.nl
[2] Statistical and Regional Dynamical Downscaling of
Extremes for European Regions, see www.cru.uea.ac.uk/
projects/stardex/

Śnieżka and Zakopane) considered. The strongest
decrease is observed in the south-western part of
the Baltic Sea Basin and is getting weaker towards
the northeast. The average annual trends for the
countries considered here are listed in Table 2.4.
The strongest decrease in number of frost days was
observed in winter and spring. During autumn, a
slight increasing trend was observed (Klein Tank
and Können 2003; Wibig and Głowicki 2002).

Warming can also be seen in the falling percentage of days with a minimum temperature above
the site and day-specific 90th percentile and below the site and day-specific 10th percentile (warm
and cold nights, Figs. 2.20 and 2.21). The increase in frequency of days with minimum temperature above the threshold is the strongest in
the western part of Baltic Sea Basin, It is weakening towards the east and north during winter
and spring (Fig. 2.20). Seasonally, the weakest
increase is observed in summer, when there are
also regions experiencing a decrease, and similar
behaviour occurs in autumn, too. Comparable results were obtained by Moberg and Jones (2005).

Together with increasing frequency of warm
nights, the frequency of cold nights is decreasing
(Fig. 2.21). The falling tendencies are strongest
in the western part and somewhat weaker or even
positive in the east. In autumn, an increase in
the number of cold nights is observed in eastern
Poland, Latvia, Lithuania, Estonia, Finland and
Russia (map not shown).

There are no statistically significant changes in
the duration of cold waves (i.e. the number of days
in intervals of at least 6 days with mean daily temperature at least 5 °C lower than the mean value
for this calendar day on the basis of the 1961–1990
period). In spring, some decreasing trends can be
observed in Fennoscandia, and in autumn increasing trends occurred in the former Soviet Union
countries (Klein-Tank and Können 2003). The
changes in cold extremes generally follow changes
in mean temperature, but the warming in spring
is usually stronger than in winter, and cooling tendencies in autumn are more evident.

The annual number of summer days (days with
maximum temperature above 25 °C) has undergone a weak increase in the western part of the
Baltic Sea Basin. The increase is significant only
in southern Germany (Fig. 2.19). In the east, a decrease is observed. Similar results were presented
by Moberg and Jones (2005).

Warming can be seen also in changes in the
numbers of days with a maximum temperature

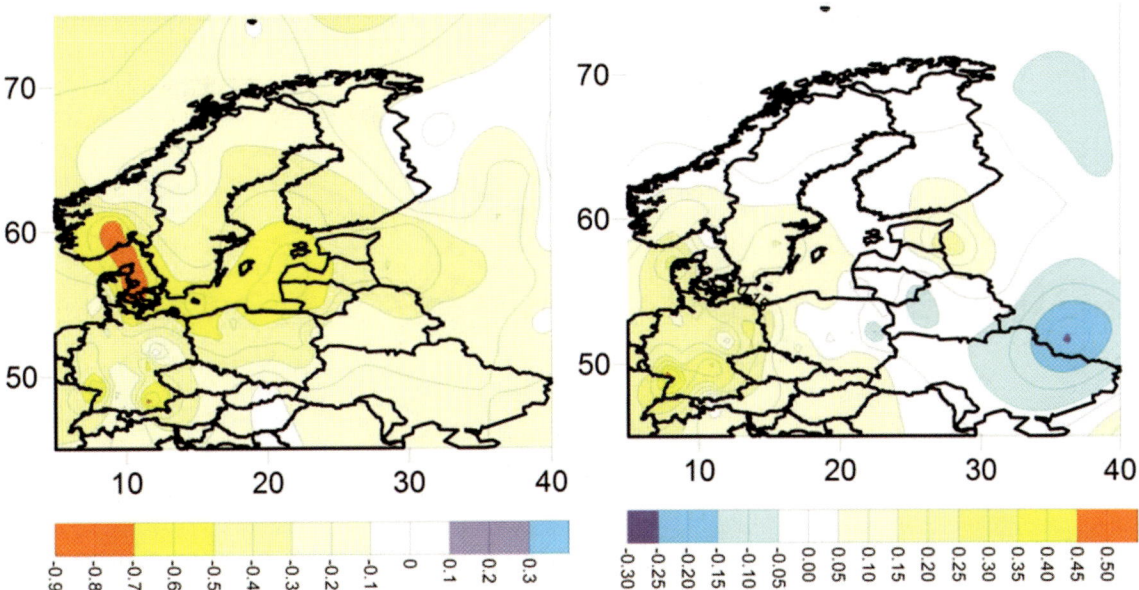

Fig. 2.19. Annual trends of number of frost days (*on the left*) and hot days (*on the right*), both expressed in days/year. The ECA dataset for the period 1951–2000 (Klein Tank et al. 2002a) and five Polish stations (Wibig and Głowicki 2002) were used

Table 2.4. The country average trend values (days/year) for the annual number of frost days in the second half of the 20th century

Country	Change per decade (days/year)	Country	Change per decade (days/year)
Norway	−3.9	Sweden	−3.3
Finland	−2.8	Denmark	−5.7
Germany	−3.9	Poland	−3.8
Latvia and Lithuania	−5.0	Estonia	−5.9

above the site and day-specific 90th percentile and below the site and day-specific 10th percentile (warm and cold days, Figs. 2.22 and 2.23). The strongest increase in warm days has occurred in winter and spring in the southern part of the Baltic Sea Basin region, exceeding 2% per decade, with a maximum above 4% at Nordby (Fig. 2.22). Decreasing trends can be observed in the eastern part of the analysed region in summer and autumn.

The frequency of cold days is mostly decreasing during winter (Fig. 2.23) and even more strongly in spring (map not shown). These falling trends are strongest in the western part of the Baltic Sea Basin, whereas some increase in cold days occurs in the northeast. Summer trends are generally positive. During autumn, the increasing trends dominate more clearly than during summer.

In the STARDEX project, the heat wave duration index was calculated as the number of days in intervals of at least 6 days with a maximum temperature at least 5 °C higher than the mean value for this calendar day during the 1961–1990 period. For the period 1958–2000, an increase in winter is evident in the whole Baltic Sea Basin, whereas during summer the opposite trend can be observed. The frequency of warm spells is increasing in Denmark and Germany and the southernmost parts of Norway and Sweden. In the rest of Scandinavia, the frequency of warm spells were decreasing during the study period. In the south-eastern part, variations are weak and positive trends are mixed with negative ones.

In the ECA project the thresholds for extreme events were defined by similar methods as used in

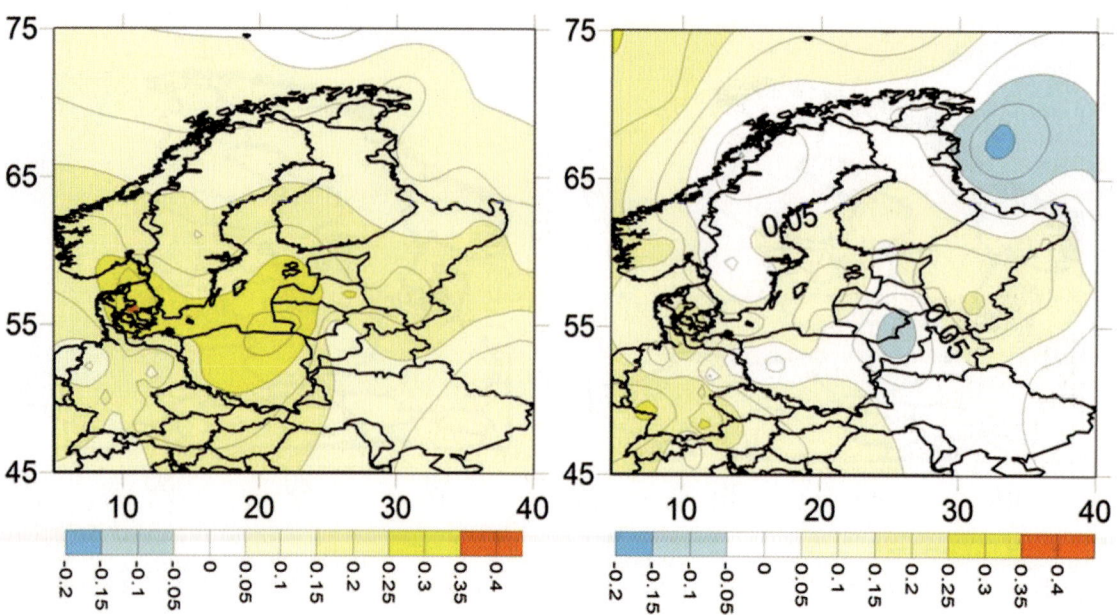

Fig. 2.20. Trends of percentage (fractions of unity) of days with minimum temperature above site and day-specific 90th percentile (*winter on left, summer on right*), days/year. The ECA dataset for the period 1951–2000 was used (Klein Tank et al. 2002a)

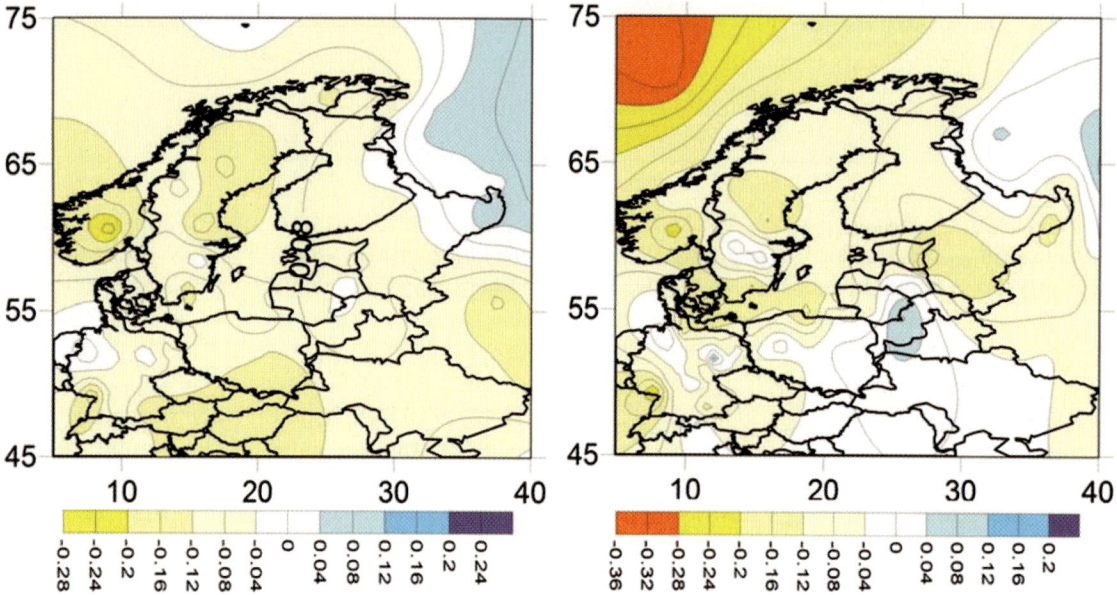

Fig. 2.21. As Fig. 2.20 but for days with minimum temperature below site and day-specific 10th percentile

Figs. 2.19–2.23 but the period analysed was 1946–2004. The calculated trends are a little different from the ones described above. They are weaker (probably because at the end of the forties the maximum temperatures were relatively high), and the summer increase in heat waves in Germany is not noticeable.

Changes in warm extremes are strongest in winter and spring, following the observed changes in mean temperature. In summer, opposite trends are observed within Baltic Sea Basin. Most of increasing trends are observed in western and southern parts of Baltic Sea Basin, whereas cooling trends characterise eastern and northern parts.

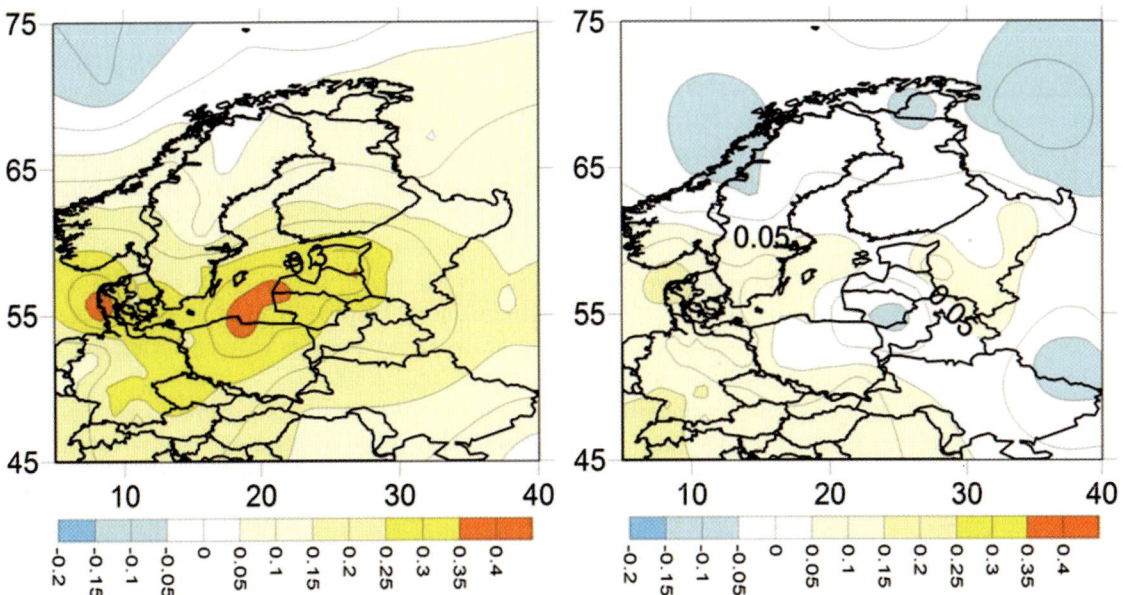

Fig. 2.22. As in Fig. 2.20 but for number of days with maximum temperature above site and day-specific 90th percentile (*winter on left, summer on right*)

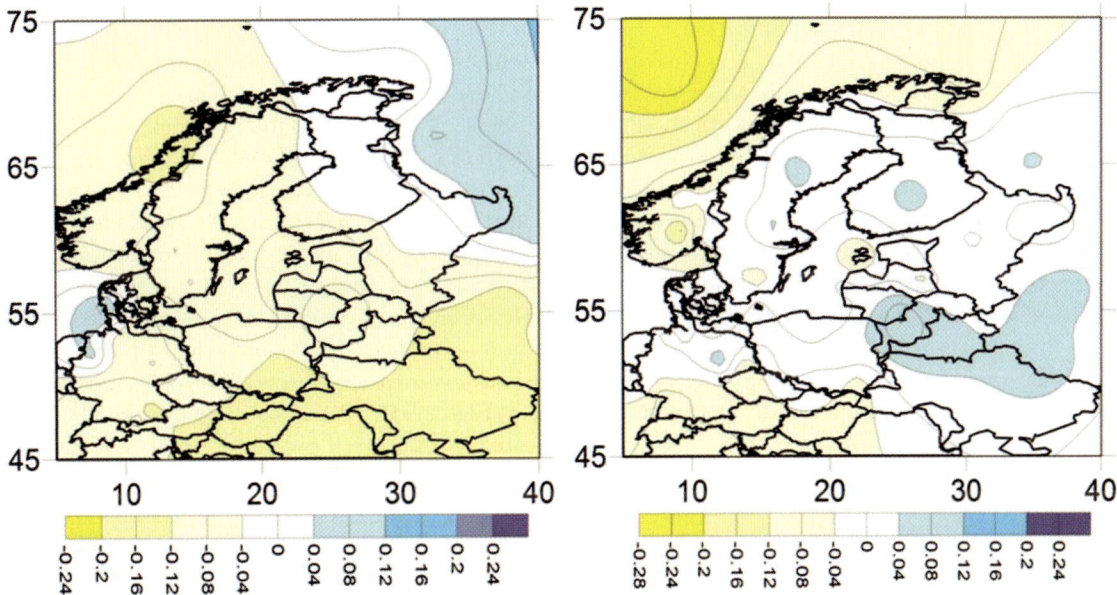

Fig. 2.23. As in Fig. 2.20 but for number of days with maximum temperature below site and day-specific 10th percentile

These trends are broadly consistent with changes in mean temperature observed e.g. in Poland, Estonia and Latvia (see Sect. 2.1.2).

As the main reasons for temporal and spatial differences in trends, the following three are most often mentioned. The first one is contemporary warming. Jones et al. (1999) have shown that winter and summer temperature extremes correlate well with mean winter and summer temperatures. The second one can be related to changes in atmospheric circulation, e.g. NAO variability.

Tuomenvirta et al. (1998) have shown that seasonal mean minimum and maximum daily temperatures are positively correlated with the NAO

index. Thirdly, the intensity of temperature extremes is also related to cloud cover (Tuomenvirta
et al. 2000). In Fennoscandia during winter and
autumn cloud cover prevents cooling. During summer cloud cover strongly diminishes the probability of warm extremes, but is somewhat favourable
to cold extremes.

2.1.5.3   *Precipitation Extremes – Droughts and Wet Spells*

On the global scale, trends in extremes of precipitation are not as clear as those concerning
temperature extremes. Some authors are convinced that the frequency of extreme precipitation
events shows an increasing trend over large areas
(IPCC 2001; Frich et al. 2002; Kundzewicz et al.
2004; Groisman et al. 2005). Extremes related to
precipitation can manifest themselves as extreme
drought or prolonged wet spells and heavy rainfall
events.

The intensity of (atmospheric) drought is usually measured by the annual maximum of consecutive dry days i.e. days with precipitation $< 1\,\mathrm{mm}$
(CDD) (see also Sect. 2.2.1.4 for a hydrological
description of droughts). A significant decreasing
trend in global series of CDD was noticed (Frich
et al. 2002), but over the Baltic Sea Basin trend
maps show a mixed pattern with positive and negative changes (see also Sect. 2.1.3.3). In the second
half of the 20th century the CDD trends over the
Baltic Sea Basin were close to zero and mainly not
significant, as was shown by Haylock and Goodess
(2004) for winters during the period 1958–1999.
ECA results indicate some positive trends of CDD
in Germany during summer for the period 1946–
2004 (Klein Tank and Können 2003). A slightly
decreasing non-significant trend was observed in
Sweden (Alexandersson 2002, 2004) and Norway
(Førland 2000). In Finland and the eastern side
of the Baltic Sea Basin, the number of such days
shows no significant trend (Heino 1994).

In Poland the number of days without precipitation averaged for the whole country was analysed by Niedźwiedź (2000a). A significant negative trend was found, the trend analysis revealed
a decline from on average 208 days in the period
1951–1955 to about 190 days in 1995–1999. The
"least rainy" years with 238 days without precipitation were 1951, 1953, 1959 and 1982. Atmospheric droughts in Poland in the period 1891–
1995 were investigated by Mager et al. (2000).
During this period 75 events occurred, when atmospheric water balance (precipitation minus potential evaporation) was severely negative indicating
drought (Mager et al. 2000).

The frequency of droughts during the second
half of the 20th century was slightly lower than
during the previous period. The longest droughts
lasted 13 months from March 1953 to March 1954,
and from November 1958 to November 1959. Two
periods with a high frequency of droughts can be
distinguished: 1950–1954 (5 years) and 1988–1993
(6 years) and two periods without droughts: 1936–
1939 and 1965–1968.

Most of the droughts covered only part of
Poland. But some of them were associated with
the large blocking anticyclones and covered more
than 85% of the area of Poland and extended to
neighbouring countries. The greatest extent of
coverage was during the droughts of 1992 (92%)
and 1951, 1959 and 1982 (91%). More than 87%
of the territory of Poland was covered during the
droughts in 1921 and 1947 (Mager et al. 2000).

Among different indices of heavy rainfall events,
the most popular is the number of days with precipitation exceeding some threshold, e.g. the frequency of days with precipitation equal to or larger
than 10 mm (R10). For this index an increasing trend is observed globally (Frich et al. 2002).
Heino et al. (1999) analysed the long-term variability of such an index for selected stations in
central and Eastern Europe. An increasing trend
was observed only in German mountain stations
and in western Norway.

However, the situation in the Baltic Sea Basin
is more complex. In Poland, the opposite trend
was observed in the second half of the 20th century (Fig. 2.24). An above-average (> 25 days)
number of days with precipitation 10 mm was observed during two periods: 1958–1981 and 1994–
1998. An absolut maximum (34 days) was found
in 1966 and a secondary maximum in 1997 (30
days), when a large flood occurred in southern
Poland. A similar trend is revealed in the frequency of days with precipitation $\geq 30\,\mathrm{mm}$. This
is contrary to variations in the northern part of
the Baltic Sea Basin and in southern Poland. The
wettest place in Poland (Hala Gąsienicowa, 1520 m
a.s.l.), on the northern slope of the Tatra Mountains, has experienced an increasing trend of 0.5
days/decade in the frequency of days with precipitation $\geq 10\,\mathrm{mm}$. This increase in orographic
precipitation is connected with the positive trend
of frequency of air-flow from the northern sector
(Niedźwiedź 2003a,b,c).

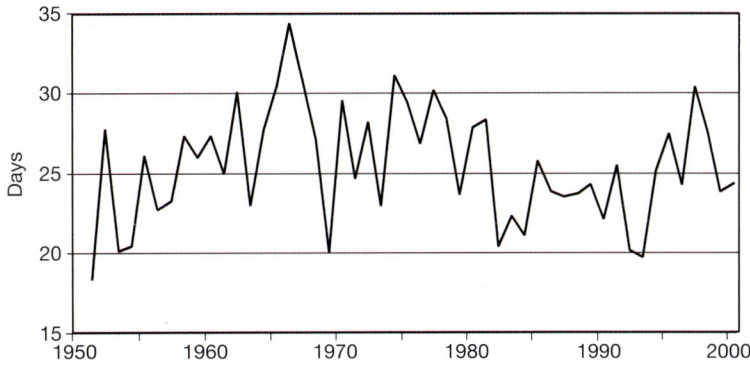

Fig. 2.24. Frequency of days with precipitation $\geq 10\,\mathrm{mm}$ (averaged for Poland) during the second half of the 20th century

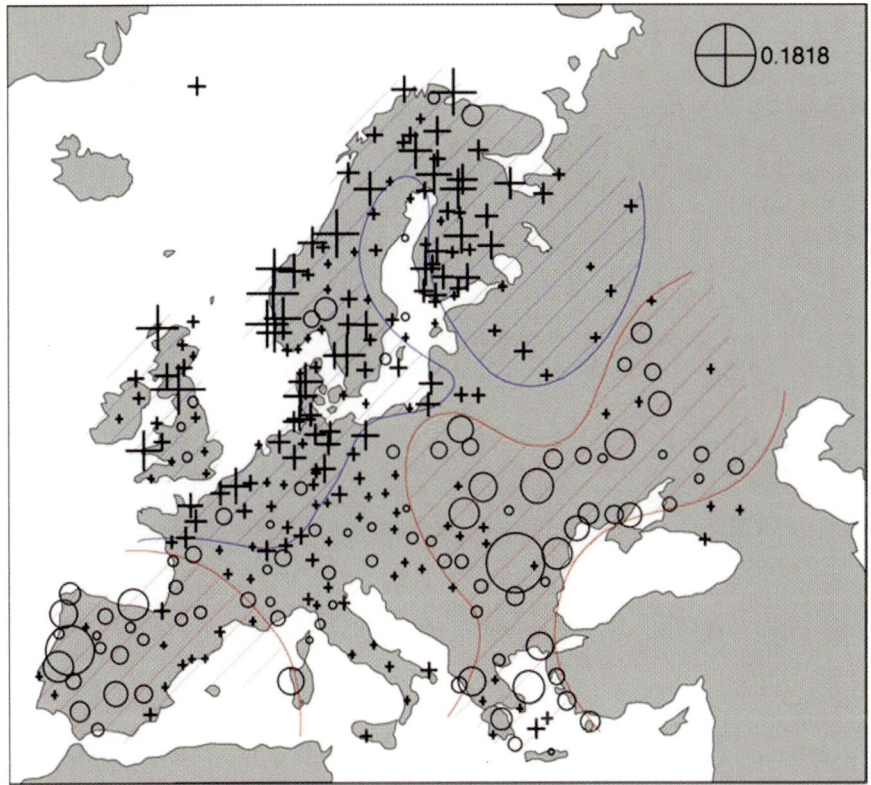

Fig. 2.25. The linear trend in number of precipitation events above the 90th percentile (R90N) during winter (December to February) for 1958–2000. A '+' signifies an increase and a '$\bigcirc$' shows a decrease. The size of the symbol is linearly proportional to the magnitude of the trend. Units are days/year and the maximum trend magnitude is shown in the top right (from Haylock and Goodess 2004, modified)

Heavy precipitation events can also be described using the number of events above the 90th percentile (R90N), using the period 1961–1990 as a baseline (Haylock and Goodess 2004). The number of such days increased significantly in winter in Scandinavia and in Germany (Fig. 2.25). During summer, increasing trends are observed in Scandinavia and the northern part of European Russia while decreasing trends are found in the southern part of the Baltic Sea Basin. The proportion of total rainfall from events above the 90th and 95th percentiles was analysed by Groisman et

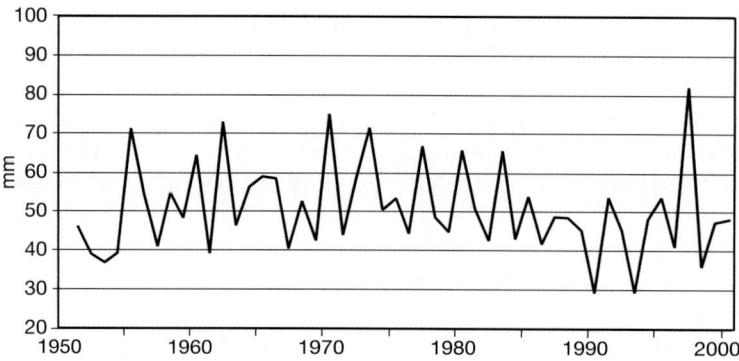

Fig. 2.26. The annual maximum 1-day precipitation averaged for the Polish area during the second half of the 20th century

al. (2005). A significant increase in this proportion was found for the period 1936–1997 in European Russia and the former USSR countries, based on more than 700 long-term records, and in Fennoscandia based on 88 ACIA (Arctic Climate Impact Assessment) data records, but previous papers describing the total precipitation increase in northern Europe (Heino et al. 1999; Førland et al. 1998) show a considerably smaller change in heavy precipitation frequency and proportion.

Wet spells can be characterised by maximum 5-day rainfall total (R5d). This index was analysed within the STARDEX project. Trends in R5d were not uniform in the Baltic Sea Basin. During winter, an increase in Germany and the Scandinavian countries was observed for the period 1958–1999. In summer, a decrease was observed in Germany, whereas Scandinavia and the European part of Russia show an increase.

Achberger and Chen (2006) analysed the trend of another wet spell index, i.e. maximum number of consecutive wet days over 471 stations in Sweden and Norway during the period 1961–2004. They found that there is a considerable spatial variability; however, 72.1% of all stations show an increasing trend over this period, although only 7% have statistically significant trends.

Spatial patterns of maximum 1-day precipitation (R1d) trend are very complicated, with differing values in very close locations (see Førland et al. (1998) for the Scandinavian countries and Heino et al. (1999) for central and Eastern Europe). The R1d averaged for the Polish area during the second half of the 20th century does not show a clear trend (Fig. 2.26). In the Tatra Mountains, R1d shows a slightly increasing trend (1.5 mm/decade). A similar trend was observed for the whole Carpathian

Mountains (Cebulak 1007). The highest precipitation totals often are orographically enhanced. For the Carpathian and Sudetic Mountains such events are connected with the northern or northeastern cyclonic situations during the summer months (Cebulak 1992; Niedźwiedź 2003a).

Long-term variability of thunderstorm frequency in Poland (Bielec-Bąkowska 2002, 2003) was analysed for 56 stations for the period 1951–2000 and for 7 stations for the years 1885–2000. There was no significant trend in the course of the annual number of days with thunderstorms. However, an increasing trend was evident during the winter season, especially for the Baltic Sea coast. Solantie and Tuomi (2000) studied the spatial and temporal distribution of thunder in Finland. A decrease from the 1930s to the period 1987–1996 was detected in the number of thunder days in the middle boreal zone ($\sim$ central Finland). This decrease was related to a decrease in the mean daily maximum temperatures over this short time period.

2.1.5.4 *Extreme Winds, Storm Indices, and Impacts of Strong Winds*

Strong winds can cause destruction to society and natural environments, e.g. forest damage (Nilsson et al. 2004), wind erosion (Ekström et al. 2002; Bärring et al. 2003), snow drift and avalanches resulting from snow being redistributed by prevailing strong winds in the northern mountainous regions (Tushinskiy 1970), and damage to coastal areas (see Sect. 2.3.5) by intensive wave action and/or coastal flooding caused by storm surges. The large scale forcing of synoptic-scale strong wind situations are related to the baroclinic activity concentrated in the North Atlantic storm track and to the phase of the North Atlantic Oscillation (Marshall et al. 2001; Rogers 1997) that influences the overall intensity of the westerly winds. In this section it is thus relevant to look at a larger region extending to the west of the Baltic Sea Basin. While there are comparatively few studies focusing only on the Baltic Sea Basin, there are numerous studies discussing the general wind climate and storminess of this larger region.

The most straightforward way to investigate wind extremes would seemingly be to analyse records of wind observations. However, anemometers were only relatively recently introduced as the standard instrument. Before that, stations equipped with anemometers were less common and instrument problems were more frequent. Instead, wind strength was recorded using the Beaufort scale where the data is based on semi-quantitative observations of the surrounding environment according to set rules. Observations at land stations are very sensitive to the surrounding terrain and vegetation, and any changes are likely to result in inhomogeneities in the observation record. Hence, coastal stations at very exposed locations are most likely to show long-term homogeneity. The WASA (Waves and Storms in the North Atlantic) project (Carretero et al. 1998) carried out comprehensive analyses of several records of wind observations in the North Atlantic and North Sea and stressed the problems of finding long-term homogeneous wind data. Instead, they suggest that geostrophic wind calculated from three sea-level pressure records could be used, see Alexandersson et al. (1998) for details about the calculations. Alternatively, for longer time-series when three suitable station records may not be available, they suggest the use of storminess indices (Schmith et al. 1998) based on the frequency of low pressure or a strong pressure tendency between consecutive observations of sea-level pressure at a single station.

Other approaches towards indirectly getting a handle on long-term variations in synoptic scale wind extremes in the North Atlantic/North Sea region are based on analysing wave heights (e.g. Vikebø et al. 2003) and storm surges (Bijl et al. 1999) or the micro-seismic activity resulting from large storm waves hitting the coastline (Grevemeyer et al. 2000).

During the early 1990s, there were signs of a strengthening of the North Sea wind climate. Early suggestions towards this end were put forth,

for example, by Schmidt and von Storch (1993) based on an analysis of the geostrophic wind over the German Bight derived from three sea-level pressure stations' records. This coincided with an increase of the North Atlantic Oscillation (NAO, see Annex 7) index (Hurrell 1995), beginning in the mid 1960s. This increased NAO, however, came to an end around 1995 (Jones et al. 1997). Alexandersson et al. (2000) came to the same conclusion; they furthermore noted that the wind climate peaked around 1990 and they continued by pointing out that this peak does in fact resemble the situation in the early part of the 20th century. Using four different storminess indices to analyse long sea-level pressure records from Lund and Stockholm, Bärring and von Storch (2004) concluded that there is no long-term trend in storminess indices over southern Scandinavia. They conclude that there was a temporary increase in the 1980s–1990s. A similar increase during the 1980s–1990s can be seen in several other analyses of relatively short time series, such as microseisms (Grevemeyer et al. 2000), wave height off mid-Norway (Vikebø et al. 2003) or storm surge data (Bijl et al. 1999).

Figure 2.27 shows storminess indices, as defined by Bärring and von Storch (2004), for seven stations in the Baltic Sea Basin. The large-scale spatial gradient towards lower surface air pressure in the north is evident in the top panel (counts of low pressure situations below 980 hPa). The increased windiness during the 1980s–1990s is evident at all stations and most pronounced at the northern stations. In addition, the northern stations, especially Härnösand and Kajaani, show a period of enhanced storminess during the 1950s.

Alexandersson and Vedin (2002) suggest that over the period 1880–2001 there is a weak decreasing trend in the incidence of geostrophic storm situations (geostrophic wind speed exceeding $25\,\mathrm{m\,s^{-1}}$) over southern Sweden. They comment that this decreasing trend may not be robust because trustworthy wind observations at Swedish coastal stations do not show this decreasing trend in wind measurements. However, this decreasing trend comes mainly from a period of lower incidence during the 1980s and around 2000, combined with increased incidence during 1910s–1920s. Extending this analysis to include more geostrophic wind triangles shows (Fig. 2.28) that there is a difference between the southern and northern (central) Baltic Sea Basin. In southern Sweden (the Göteborg–Visby–Falsterbo triangle)

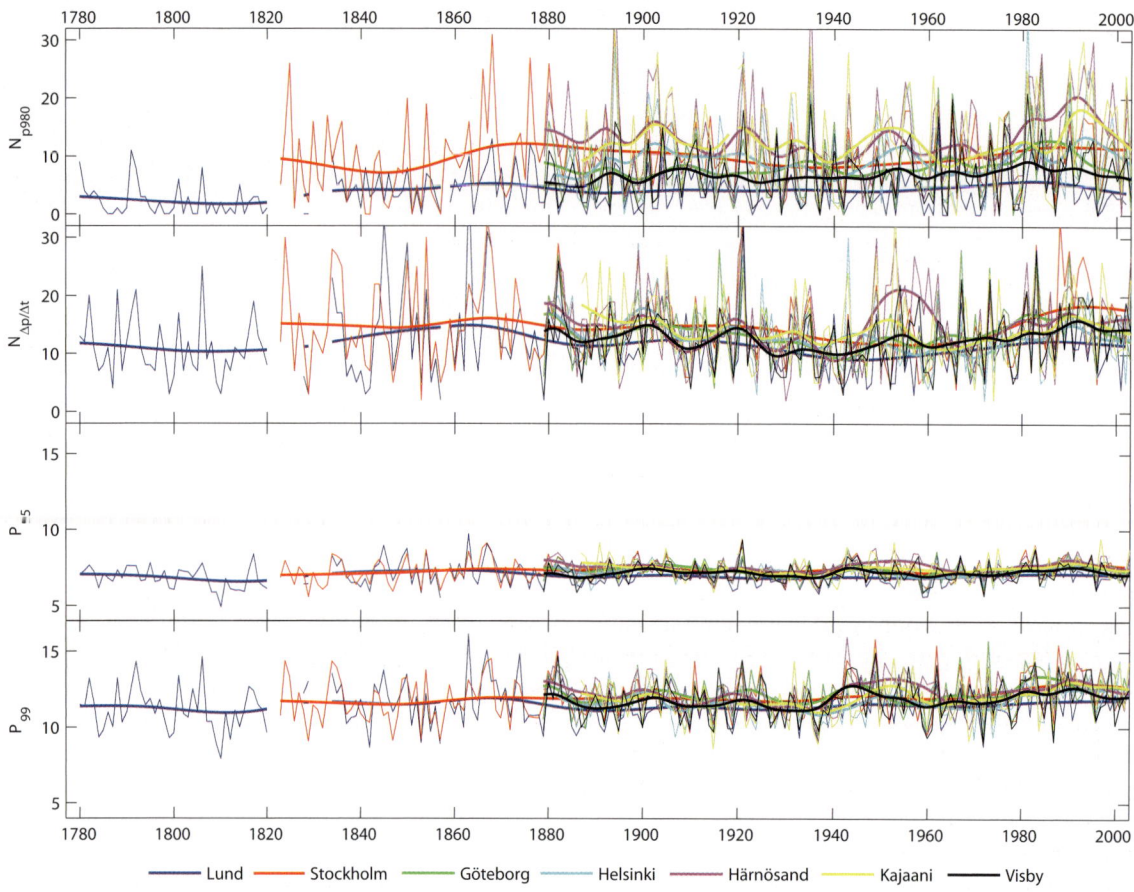

Fig. 2.27. Four different annual storminess indices derived from station pressure records in the Baltic Sea Basin. *From top to bottom:* N_{p980} – number of low pressure observations below 980 hPa, $N_{\Delta p/\Delta t}$ – number of events when the absolute pressure tendency exceeds 16 hPa/12 h, P_{95} and P_{99} – the 95 and 99 percentiles of pressure differences (hPa) between two observations. The three observation hours (morning, midday and evening) vary over the period, in modern times it is typically 05–06, 11–12 and 17–19 UTC (cf. Bärring and von Storch 2004; and Schmith et al. 1997). The thin lines show annual variations and the smooth thick lines (Gaussian filter, $\sigma = 3$) show variations at the decadal time-scale

there are obvious long-term trends toward fewer exceedances above the selected thresholds, but no such trends are to be found in the more northerly Bay of Bothnia region (the Härnösand-Kajaani-Helsinki triangle). The decreasing trend is less pronounced or non-existent during the autumn (September to November). The notion of a more pronounced trend towards weaker winds in the southern part is further supported by Smits et al. (2005), who used high quality surface wind observations from the Netherlands and found a decreasing trend in moderate and strong wind events during the period 1962–2002.

In a similar study of the variability of annual relative wind energy over Sweden, Johansson and Bergström (2004) used geostrophic winds derived from pressure observations. Their overall conclusion is that there are decadal scale variations of the magnitude of ±10% and no long-term trend in the overall relative energy of the wind over Scandinavia. However, their plots suggest that there are some trends in some areas (as defined by the triangles) similar to those presented in Fig. 2.28.

As the region of maximum winds is typically located south of an eastward moving low, the two more windy periods found in the Härnösand and Kajaani storminess indices (Fig. 2.27) are in agreement with wind measurements (Fig. 2.29) analysed by Orviku et al. (2003). They analysed homogenised time series of annual and monthly fre-

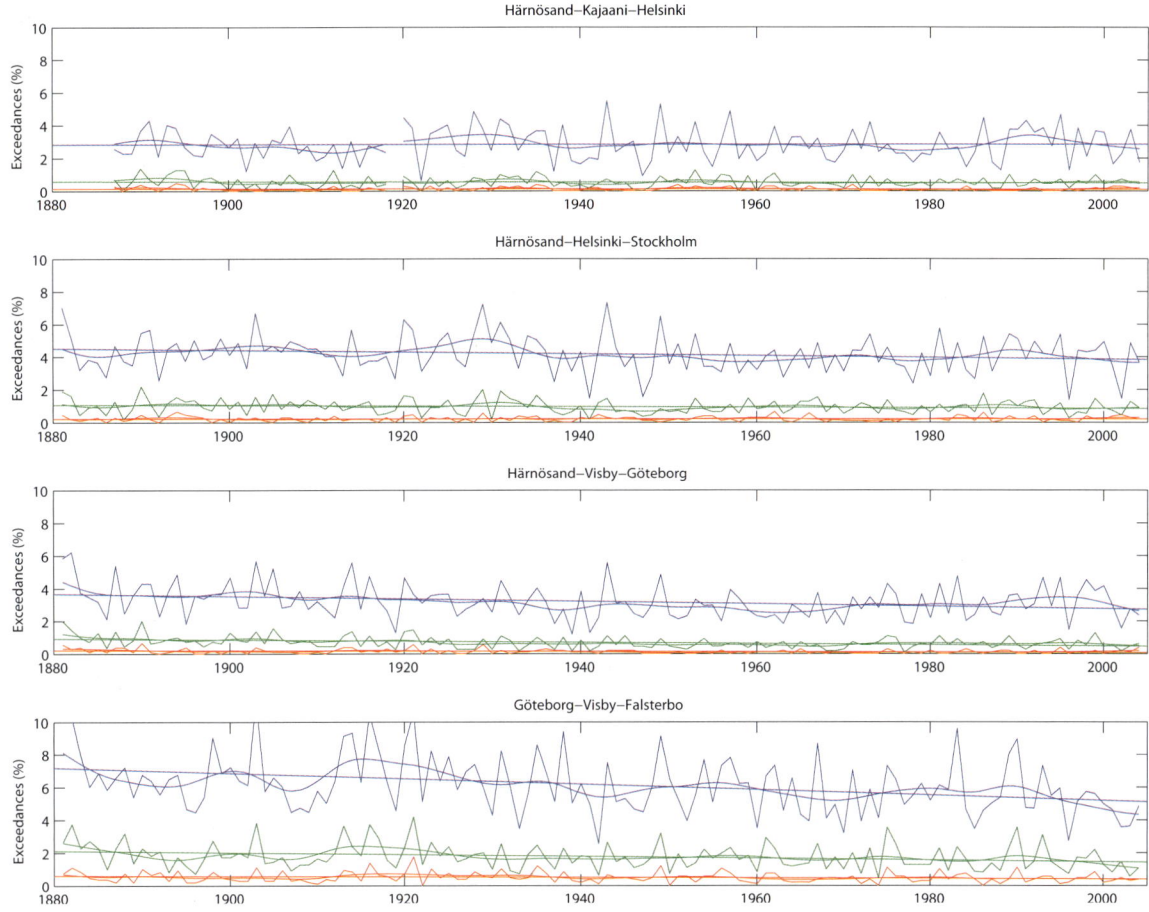

Fig. 2.28. The relative frequency (%) of days where the maximum geostrophic winds exceed selected thresholds: *blue:* $20\,\mathrm{m\,s^{-1}}$; *green:* $25\,\mathrm{m\,s^{-1}}$; *red:* $30\,\mathrm{m\,s^{-1}}$. The maximum is based on pressure readings in the morning, midday and evening. The smooth lines (Gaussian filter, $\sigma = 9$) show variations at the 30-year time-scale

quency of strong winds observed for the period 1948–2003 at three Estonian stations located on the coast of the Baltic Proper and the Gulf of Riga. The main feature of these time series is the two periods of increased frequency of strong winds, during the 1950s and from the late 1970s to the mid 1990s. However, their conclusion of an overall increased frequency of strong winds along the Estonian coast during 1948–2003 supports the notion of different overall trends in the pressure based storminess indices and trustworthy time-series of wind measurements also noted by Alexandersson and Vedin (2002).

The NCEP–NCAR (National Centres for Environmental Predictions – National Centre for Atmospheric Research) Reanalysis dataset shows, in contrast to the direct observational evidence discussed above, a clear trend towards an intensification of both the Pacific and the Atlantic storm tracks (Chang and Fu 2002). This increase may, however, be overestimated (Harnik and Chang 2003). Such an overestimation may influence studies using this data set. Although the NCEP–NCAR Reanalysis dataset may overestimate an increasing trend, the sign of the trend is in agreement with the ECMWF (European Centre for Medium-Range Weather Forecasts) ERA-15 Reanalysis dataset, especially with respect to strong wind cyclones (7 Beaufort and above) (Hanson et al. 2004), even though they concluded that the ERA-15 data provide more comprehensive representation of the northern cyclone climatology.

Nevertheless, the increased wave height found by Bauer (2001) when forcing a wave model with

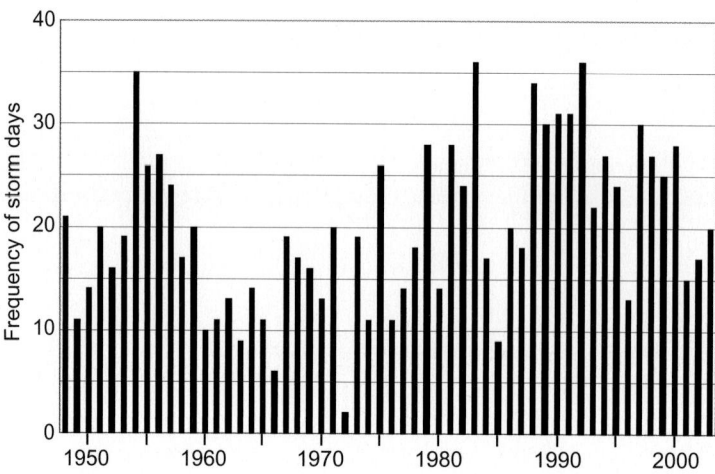

Fig. 2.29. Annual frequency of 'storm days' at Vilsandi, Saaremaa Island, Estonia, derived from homogenised wind measurements. A day was designated as a 'storm day' when mean wind speed during a single observation (10 minutes) was $15\,\mathrm{m\,s^{-1}}$ or higher (adapted from Orviku et al. 2003)

ERA-15 data is consistent with the increased wind speed over the North Sea (Siegismund and Schrum 2001). Weisse et al. (2005) used NCEP–NCAR Reanalysis to force a regional climate model and concluded that the model is capable of reproducing impact-relevant storm indices. Their results are consistent with previous analyses based on reanalysis data in that they find an increase in the number of storms over the North Sea, although this increase has levelled off. For the Baltic Sea they find a weak but statistically significant trend towards increasing frequency of strong wind events (wind speed exceeding $17.2\,\mathrm{m\,s^{-1}}$, Beaufort 8). This trend corresponds to one additional exceedance every 10 years.

Pryor and Barthelmie (2003) report on positive trends in winds at the 850 hPa level over the Baltic Sea Basin region. Their results show that the trend in annual mean wind speed is weak, in excess of $0.25\,\mathrm{m\,s^{-1}}$ per decade, but for strong winds the trend becomes more pronounced. This conclusion is further strengthened in their recent analysis where Pryor et al. (2005) found a similar positive trend in windspeed measured at 10 meter height level from the ECMWF ERA-40 reanalysis over the Baltic Sea.

However, a comparison of storminess indices derived from station pressure readings with indices calculated from the corresponding gridbox of a RCM (Regional Climate Model) simulation using NCEP–NCAR reanalysis data as boundary conditions shows a very good agreement in the high-frequency variability, i.e. variability at the intra-

annual and few years time-scale). However, there seems to be some discrepancy in overall linear trends (Fig. 2.30), mainly induced by the simulated values being too low for the first 10 to 15 years.

On a much smaller spatial scale than the synoptic scale cyclonic storms, strong winds may arise in connection with severe local storms of different origins. For these systems the problems of direct measurements become even more obvious because there is little chance that local wind maxima pass over a meteorological station. Even though recent radar and satellite techniques have improved the situation, indirect measurements and compilations of documentary information on the impact are still the main data sources for studying variations of over time. To a large extent, information on these phenomena are scattered and focus on the description of the impact of individual events and/or an analysis of the meteorological conditions that lead to the event. Tornadoes (over land) or waterspouts (over water) are associated with advection of warm tropical air to high latitudes mostly caused by southern cyclones during summer. They form in a frontal zone where the air temperature gradient is very high and thunderstorms frequently occur. Downdraft is a different phenomenon associated with thunderstorms; it is formed by vigorous downward flow of cold air hitting the ground ahead of the thunderstorm.

During recent years the incidence in Europe of locally strong winds has gained increasing interest (e.g. Brooks et al. 2003; Dotzec 2003; Teitti-

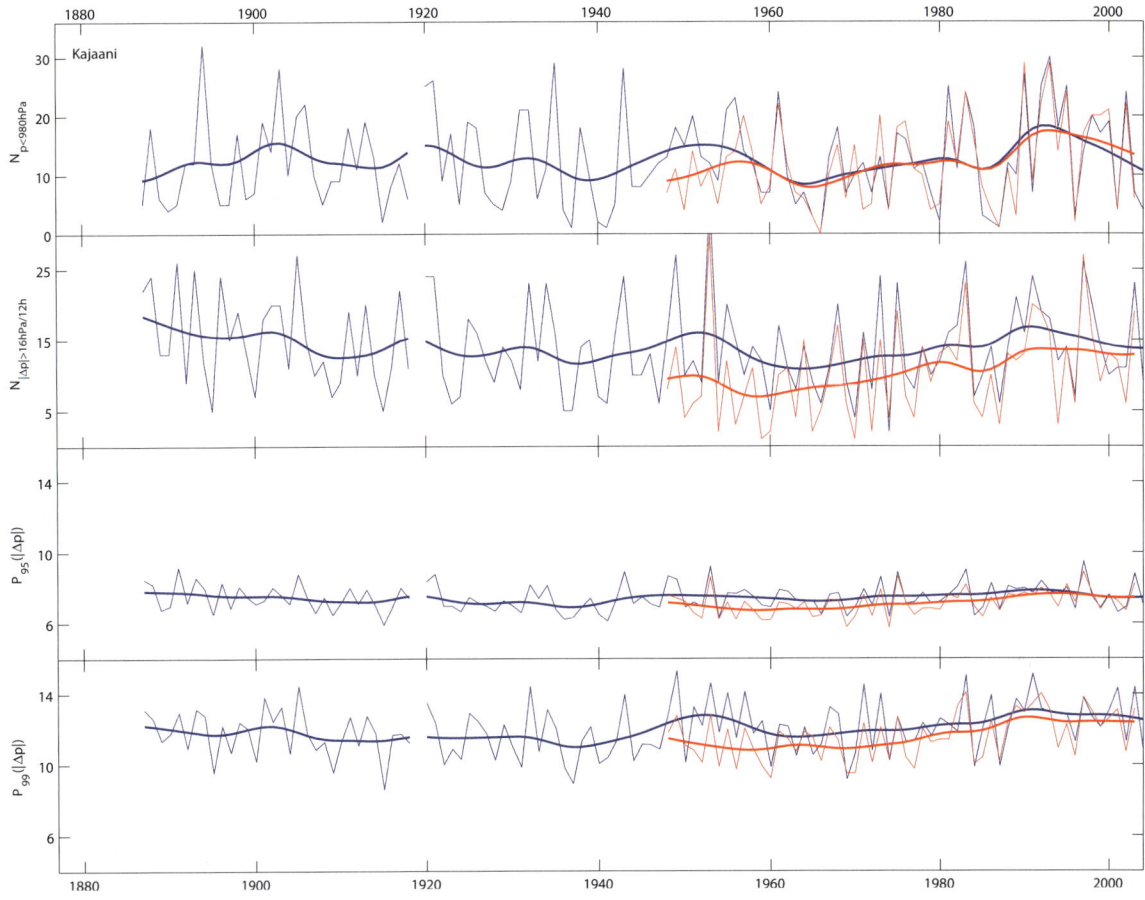

Fig. 2.30. The same four storminess indices as in Fig. 2.27 for the station Kajaani, Finland. Blue curves are based on pressure observations and red curves are calculated from the nearest gridbox of a RCM simulation forced by NCEP reanalysis data (Weisse et al. 2005)

nen 2000; Groenemeijer 2005). Dobrovolný and Brázdil (2003) provide a time-series of strong local winds associated with convective storms in the Czech Republic. The time-series is largely based on historic documents and extends all the way back to AD 500, and the authors conclude that the variations are consistent with the temperature variations during the same period.

Lithuania was hit by a very strong tornado in 1981, and Marcinoniene (2003) analyses the synoptic situation associated with this event and uses a 52 year record of tornado incidence as a reference series. In Estonia, tornadoes occur 2–3 times per year as an average (Merilain and Tooming 2003; Tooming and Merilain 2004). They also present a list of tornadoes observed in Estonia since 1795 and provide a tentative map of tornado tracks (Tooming and Merilain 2004). Although torna-

does in Estonia have been observed much more frequently during recent decades, it doesn't mean that this is a real trend. The main reason for higher reported frequency of tornadoes is probably better communication.

To sum up, different data sources give slightly different results with respect to trends and variations in the extreme wind climate. There is nonetheless a consistent picture across most data sources of an intensification of the wind climate during the 1980s–mid 1990s. While analyses based on the NCEP–NCAR reanalysis dataset tend so show an increasing trend for the period since 1948, geostrophic wind series based on pressure observations suggest a downward trend from the early 20th century, at least in southern Scandinavia. Even though the reanalysis system itself is constant in the NCEP–NCAR reanalysis, the long-term increasing trend in data may be exaggerated because incomplete observational data during the early part of the reanalysis period and the introduction of new observational systems, together with the positive wind climate anomaly during the 1980s–1990s, may partly explain this positive trend. The downward trend in geostrophic winds is strongest in the southern part of the region and hardly discernible in the north. This suggests that it may be due to shifts in the storm tracks; some support of this may be found in wind observations. The single station pressure based storminess indices show clear decadal scale variations but no clear long-term trend. Thus, it is not possible to draw any definitive conclusion regarding long-term trends in the extreme wind climate over the Baltic Sea Basin.

Information on extreme local winds is with few exceptions available as scattered and descriptive studies. There is an emerging interest in compiling long-term and consistent observational datasets from various sources. Such datasets will become instrumental for analysing trends and variations in local extreme winds. As yet, they are not extensive enough for allowing any firm general conclusion regarding trends and variations of small-scale extreme winds in the Baltic Sea Basin.

At the same time there are clear indications of increasing impact from extreme wind events. But this increasing impact results from a complex interaction between climate and development trends that increase the exposure to damage and/or the vulnerability of nature and society.

2.2 The Hydrological Regime

Several rivers, lakes and some observational stations for different hydrological parameters are mentioned in Sect. 2.2, the locations of which are indicated in Fig. 2.31. See Table 2.5 for a related list.

2.2.1 The Water Regime

2.2.1.1 *Annual and Seasonal Variation of Total Runoff*

The river inflow into the Baltic Sea has attracted the attention of many scientists. Mikulski (1970) made the earliest rather comprehensive study, which, however, extended only over the period 1961–1970. Bergström and Carlsson (1994) analysed the period 1950–1990, based on discharge data from around 200 rivers, covering 86% of the total Baltic Sea Basin. They arrived at an average inflow of $14{,}151\,\mathrm{m^3\,s^{-1}}$ or $446\,\mathrm{km^3\,y^{-1}}$. This is equivalent to a water layer of 118 cm over the surface of the Baltic Sea, and is roughly three times the amount of annual evaporation.

Within the BALTEX programme, the runoff to the Baltic Sea during the period 1921–1998 was analysed (Fig. 2.32). The mean for the whole period was $14{,}119\,\mathrm{m^3\,s^{-1}}$ or $445\,\mathrm{km^3\,y^{-1}}$. Unlike the studies referred to above, the runoff to the Danish Belts and Sounds are not included. This explains most of the difference in the mean value for the period 1950–1990, because the runoff to the Belts and Sounds is only about $37\mathrm{km^3\,y^{-1}}$. See also Annex 1.3 for more details on runoff to the Baltic Sea.

The interannual variability in runoff is considerable. The wettest year, 1924, had a mean annual runoff of $18{,}167\,\mathrm{m^3\,s^{-1}}$, while the corresponding value for the driest year, 1976, was only $10{,}553\,\mathrm{m^3\,s^{-1}}$. Even the decadal means vary quite a lot, between $14{,}582\,\mathrm{m^3\,s^{-1}}$ for the 1990s and $12{,}735\,\mathrm{m^3\,s^{-1}}$ for the 1940s. It is noteworthy, that no statistically significant trend can be found in the series.

2.2.1.2 *Regional Variations and Trends*

Most detailed analyses of runoff trends have been made in Sweden (Lindström and Bergström 2004). The study included a total of 61 discharge series with the longest one (at the outlet of Lake Vänern) starting in 1807, but most of them being confined to the 20th century. The last two decades have been very wet in Sweden, with runoff anomalies of about +8% compared with the period 1901–2002. However, they are matched by similar levels in the 1920s. The 1970s was the driest decade, with an anomaly of −9%. A linear trend for annual runoff from 1901–2002 shows an increase by about 5% over the whole period, but the trend is not significant. Records from the 19th century indicate that the runoff was even higher than in the 20th century, although temperatures were generally rather low. On the contrary, recent decades have been characterised by a combination of high temperature and high runoff.

Hisdal et al. (2003) analysed a Nordic data set, encompassing 152 stations in the period 1961–2002, 140 stations in 1941–1960 and 90 stations in 1920–1940. Over one third of these stations are outside the Baltic Sea Basin, e.g. in Norway and western parts of Denmark. The stations were di-

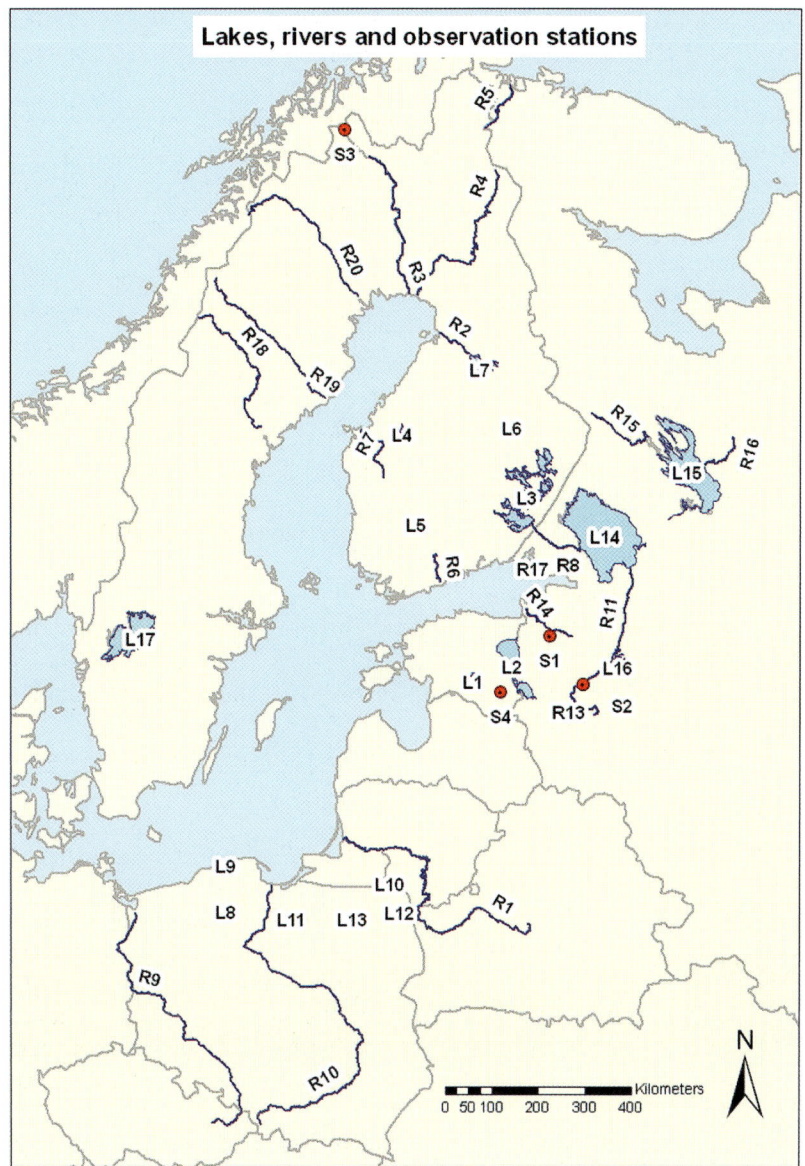

Fig. 2.31. Location of rivers (**R**), lakes (**L**) and stations discussed in Sect. 2.2. See Table 2.5 for a list of sites (by courtesy of Jonas Sjögren, SMHI)

vided into those suitable for the analysis of i) only annual values, ii) annual and monthly values, and iii) annual, monthly and daily values. As to the annual values, this Nordic data set gave positive trends in 1920–2002 at several stations in Denmark, southern Sweden and Lapland.

However, trends at only two stations, both in the south, were significant. Negative trends were rather seldom and regionally scattered. For the period 1941–2002, the number of significant positive trends was 17, in 1961–2002 it was 19. Most of these trends were at stations located in northern

Sweden. There were even fewer negative trends in these periods than in 1920–2002.

As to seasonal runoff, significant positive trends were rather common in winter (DJF) and in spring (MAM) during the period 1941–2002 in the whole northern part of the Baltic Sea Basin In summer and autumn significant positive trends were rather rare, as were negative trends in all seasons.

Annual and monthly runoff from Finland into the Baltic Sea in 1912–2003 were recently analysed (Fig. 2.33). The mean annual flow was $2413\,\mathrm{m^3\,s^{-1}}$, with a maximum of $3589\,\mathrm{m^3\,s^{-1}}$ in

Table 2.5. Rivers, lakes and stations shown in Fig. 2.31

	Rivers		Lakes		Stations
R1	Neman	L1	Võrtsjärv	S1	Kingisepp
R2	Oulujoki	L2	Peipsi	S2	Zapolye
R3	Tornionjoki	L3	Saimaa	S3	Kilpisjärvi
R4	Kemijoki	L4	Lappajärvi	S4	Tiirikoja
R5	Paatsjoki	L5	Näsijärvi		
R6	Vantaanjoki	L6	Kallavesi		
R7	Kyrönjoki	L7	Oulujärvi		
R8	Vuoksa	L8	Jeziorak		
R9	Odra	L9	Lebsko		
R10	Vistula	L10	Hancza		
R11	Volkhov	L11	Charzykowskie		
R13	Shelon	L12	Studzieniczne		
R14	Luga	L13	Mikolajskie		
R15	Suna	L14	Lagoda		
R16	Vodla	L15	Onega		
R17	Perovka	L16	Ilmen		
R18	Ångermanälven	L17	Vänern		
R19	Umeälven				
R20	Luleälven				

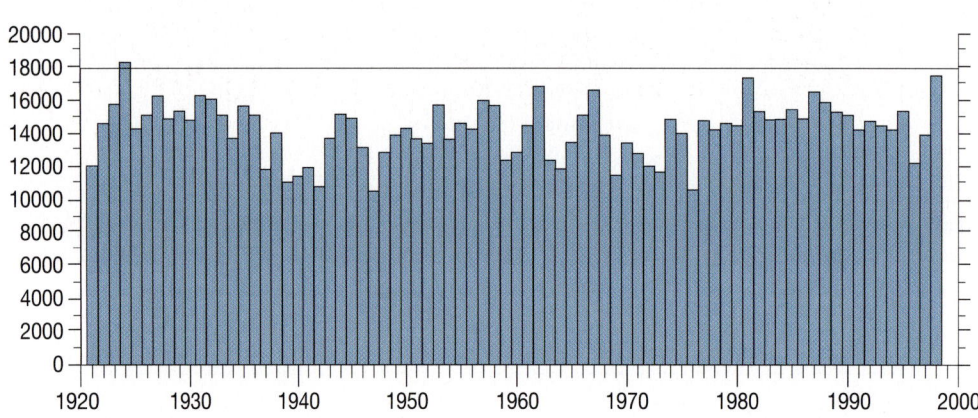

Fig. 2.32. Annual average runoff to the Baltic Sea (in $m^3 s^{-1}$). Data for 1921–1949 after Mikulski, for 1950–1998 from the BALTEX data base (see www.baltex-research.eu). The Danish Belts and Sounds are not included (Bergström 2001)

1981 and a minimum of $1121 m^3 s^{-1}$ in 1941. There is no trend in the annual series. Winter runoff has increased; the linear trend is significant at the level of 99.9%. According to the trend line, the increase has been as high as $785 m^3 s^{-1}$ during the observation period.

This phenomenon was already observed by Hyvärinen and Vehviläinen (1988). It is, however, difficult to judge how much of the increase is due to the regulation of watercourses and what

part of it is caused by climate change and variations. Typical regulation schemes increase wintertime flows considerably. Large northern rivers entering the Bay of Bothnia have been regulated since the 1950s and the rivers of Ostrobothnia since the 1960s.

For the Russian part of the Baltic Sea Basin, mean annual, seasonal and monthly runoff were analysed for the period 1978–2002 (Vuglinsky and Zhuravin 1998, 2001; Shiklomanov and

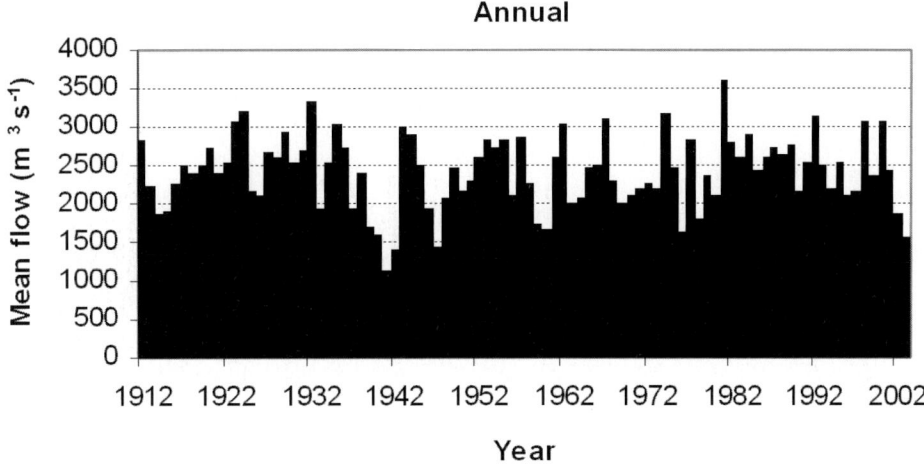

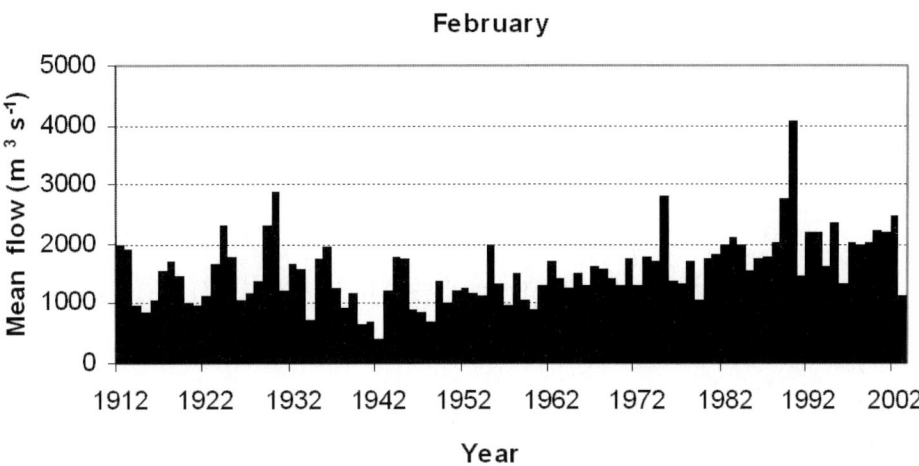

Fig. 2.33. Annual (*top*) and February (*bottom*) runoff ($m^3 s^{-1}$) from the Finnish territory to the Baltic Sea during 1912 to 2002

Georgievsky 2002). Compared to the long-term mean values, the annual flows have increased in some basins south and southwest of the Gulf of Finland by about one third. In the rivers of the Karelian Isthmus, this increase has been 3–11%.

The increase in winter runoff has been remarkable: 40–140% in the basins south and southwest of the Gulf of Finland, and 6–44% in the Karelian Isthmus, in Karelia and in the territory adjacent to Lake Ladoga, the biggest lake in Europe. Summer and autumn runoff have also increased in the southern and south-western river basins (by 6–70%), whereas in some rivers northward and eastward of the Gulf of Finland they have decreased by 2–14%. The spring runoff decreased in the former region by 2–25%, while it

increased by 4–16% in some Karelian rivers and in the Volkhov River.

Monthly data analysis shows that runoff during winter months (December–March) has greatly increased in all the study rivers except the Vuoksa River. Figure 2.34 gives an example of mean annual and winter runoff values from observation sites in the Russian territory.

An increase of wintertime runoff has also been observed in Estonia (Velner and al. 2004). The same trend has occurred in all major rivers in Latvia, but the annual runoff of several Latvian rivers decreased during 1960–1999 (Briede and Lizuma 2004). In Belarus, the winter runoff increased markedly and spring flows decreased in the period 1988–2002 (Nekrasova 2004).

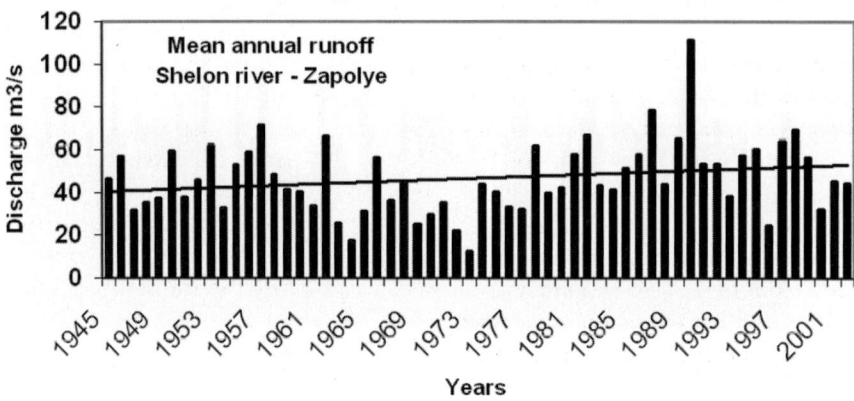

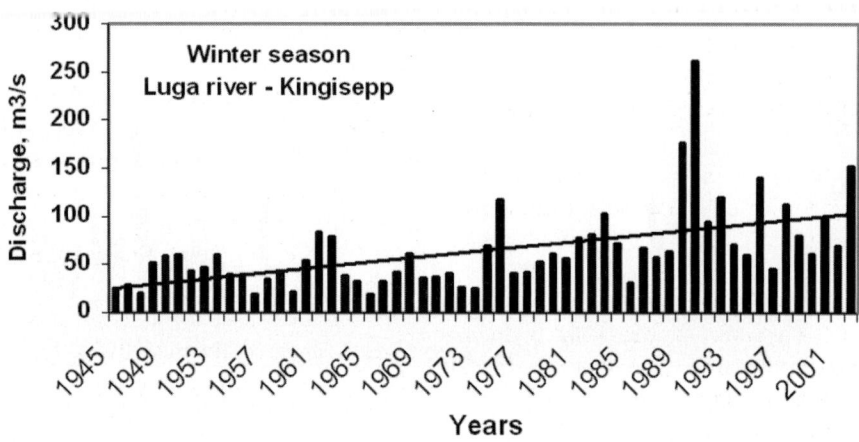

Fig. 2.34. An example of a data series on mean annual runoff and mean winter runoff (Dec–March) in the Russian territory of the Baltic Sea Basin. The observation site of Zapolye on the river Shelon is located at 58° 03′ N, 30° 06′ E, while Kingisepp on the river Luga is located at 59° 23′ N, 28° 36′ E

In the Neman River, shared by Belarus, Lithuania and the Kaliningrad area, an annual runoff series covering 193 years was analysed by Volchak (2004). He found long-term periodicities: Decreasing runoff during 1870 to 1885 was followed by rising flows until 1930, and a decline again is observed from the mid-1960s until the early 1980s.

2.2.1.3 *Floods*

Floods can cause extensive economic damage in the Baltic Sea Basin, although life-threatening events are rare. Extensive literature is available from most countries, but systematic studies covering the whole area in detail have not been made.

Almost half of the Baltic Sea Basin was covered by Hisdal et al. (2003) using the same data set described above. They found an increase in floods caused by rain and a decreasing trend in snowmelt floods in some locations in Sweden, Finland and Denmark. However, the majority of the stations analysed did not show these features and decreasing trends were also found. A more widespread pattern was an earlier occurrence of snowmelt floods in spring, which was likely caused by increased temperatures.

In Sweden, several major floods have occurred in recent decades. During 2000, heavy rains caused extensive damage in Norrland in the summer and later, in the autumn, in south-western Sweden. The lower reaches of the rivers Ångermanälven, Umeälven and Luleälven suffered severe flooding in August 1993, as did Dalarna and Helsingland in September 1985 (SMHI 2001).

In general, flood peaks seem to have increased more in Sweden than the average annual values of runoff. The clearest increase has taken place in the central parts of the country, and the same tendency can be found in the north. In south-eastern Sweden, there is a decreasing rather than an increasing trend (Lindström and Bergström 2004).

In Finland, the worst well-documented flood took place in the Lake District in 1899. The year 1898 had already been wet, and in the winter of 1899 an unusual amount of snow accumulated. Water levels in the central lakes of the area rose to 2 to 2.5 m above the mean (Hyvärinen and Kajander 2005). Another major flood occurred in the Lake District in 1924, as did smaller ones in 1974 and 1981. In southern Finland, the largest flood of the 20th century occurred in the spring of 1966, while in northern Finland, extremes occurred in 1943 (River Oulujoki), in 1968 (River Tornionjoki), 1973 (River Kemijoki) and in 2000 (River Paatsjoki).

In large river basins in northern Finland, the increase in spring flood peaks has been estimated to be about 0.5% per one per cent of drained area (Hyvärinen and Vehviläinen 1988). The effects of drainage attenuate in the course of years. This is partly due to the growth of forest on drained lands, and the water bearing capacity of the ditch network also tends to weaken.

By far the most devastating floods in the Baltic Sea Basin have been the deluges in the Odra Basin in 1997 (with estimated damages of 6 billion Euro) and in the Vistula Basin in 2001 (damages of around 3 billion Euro). The latter two events were followed by the Elbe flood in 2002 (estimated damage of 15 billion Euro), the Elbe river being a neighbouring catchment to the Baltic Sea Basin.

These events have stimulated an intensive discussion about flood causes and new strategies for flood risk management. The proposed strategies clearly show the need for an integrated flood risk management including structural and, even more important, non-structural measures. In addition, these events further stimulated research concerning the possible role of climate change and of land use changes. De Roo et al. (2003) studied historical land use in the Odra Basin. They found that the total area of forested land increased between 1780 and 2001 – contrary to common assumptions. The area of arable land decreased, while the extent of urban areas grew considerably. Overall, the land use changes between 1780 and 2001 have

increased peak discharges, but their impact is not considered very large.

The possible impact of the North Atlantic Oscillation (NAO) on floods in the Baltic Sea Basin was also discussed. Wrzesinski (2004) analysed 17 data series, widely distributed over the Baltic Sea Basin. He found that in the northern part of the area a positive stage of the NAO rather often indicates an early start of spring high water and above-average flood peaks. The opposite is true in the southern part of the Baltic Sea Basin.

2.2.1.4 Droughts

Droughts do not usually cause major damage in the Baltic Sea Basin, although small flows and low soil moisture storage may be harmful to agriculture, particularly in the southern parts of the area.

Hisdal et al. (2003) found a slight trend towards prolonged droughts in southern Scandinavia after the 1960s. In Denmark, a trend analysis of the time series of annual minimum values in 1874–1998 gave decreasing trends at 16 stations (of which three were significant at the 95% level) and increasing trends at 22 stations, six of these being significant.

In Finland the most serious 20th century drought occurred during 1940–42. In 1941 and 1942, the mean annual countrywide discharge was only 49% and 57%, respectively, of the long-term average. These were the two driest years of the whole century. The drought in 2002 and 2003 hit most of Sweden and Finland with a considerable reduction of hydropower production.

Farat et al. (1998) analysed droughts in Poland in the period 1951–90. Fourteen droughts were isolated and described; the highest frequency of drought occurrence was noted in the decades 1951–60 and 1981–90. None of the events was described as an outstanding one.

2.2.1.5 Lakes

The seasonal changes of the total water volume in lakes and reservoirs is dominated by the strong and rather regular variation in Swedish and Finnish water bodies, with a considerable human effect by regulation. In Lakes Ladoga and Onega, the seasonal variation is rather small. In contrast, the water volume of Lake Ilmen (mean area 1350 km^2) varies strongly, because the annual amplitude of water level can be up to six meters.

An analysis of runoff generation in the Baltic Sea Basin should, in fact, take into account the

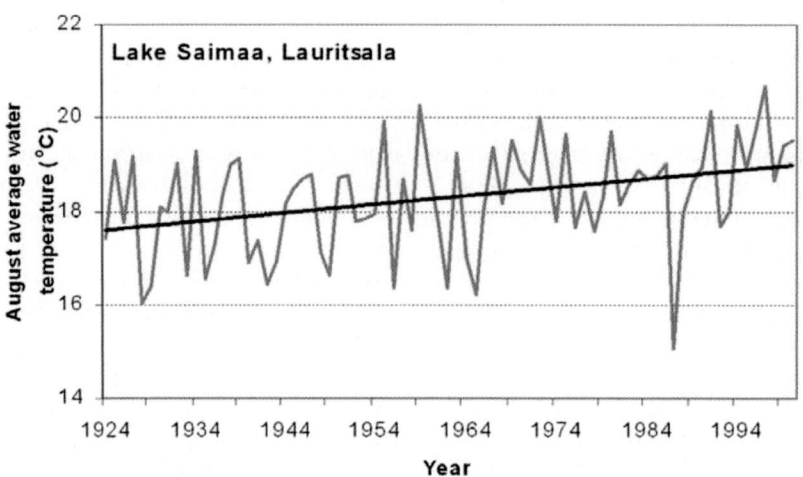

Fig. 2.35. The average water temperature in August in Lake Saimaa (Finland) in the period 1924–2000 (Korhonen 2002)

change in the water storage of lakes. In a dry year, lakes increase the annual runoff into the Baltic Sea, while in a wet year, they decrease it. If this 'lake correction' is included, the true runoff generation in the Finnish territory in 1981 was about $3880 \, \mathrm{m}^3 \, \mathrm{s}^{-1}$ and in 1941 only $850 \, \mathrm{m}^3 \, \mathrm{s}^{-1}$, while the flows into the Baltic Sea were $3589 \, \mathrm{m}^3 \, \mathrm{s}^{-1}$ and $1121 \, \mathrm{m}^3 \, \mathrm{s}^{-1}$, as cited above.

Long-term variations of lake levels in the Baltic Sea Basin have not been analysed as widely as the variations of river discharges. One of the reasons is that many lakes are regulated, and the changes of their levels do not correctly reflect variations in climatic or physiographic factors. The levels of the largest lakes have, however, been of considerable interest, mainly because there are a lot of economic interests along their shorelines. In Lake Vänern (Sweden), the most extreme water level since regulation started in 1938 occurred in the beginning of 2001 (Weyhenmeyer and Sonesten 2002). It exceeded the previous record by 40 cm. In Lake Saimaa (Finland), the flood of 1899 has held the record since the observations started in 1847. In Lake Ladoga, the highest level was measured in 1924, the lowest in 1940.

Mild winters have increased the levels of some lakes in recent decades. Regulation for hydro power production typically has the same effect, which makes the analysis of the real causes of this trend difficult. Long-term periodicities have been found in several lakes; e.g. the levels of Lake Peipsi have been found to follow cycles of 20–30 years

(Jaani and Beljazo 2003) and the levels of Saimaa, Ilmen and Onega to fluctuate with the periodicity of 28–32 years (Masanova and Filatova 1985). On the other hand, Kuusisto (1982) found no evidence of long-term regularities in the annual water levels of Ladoga and Saimaa in the period of 1859–1956. The first autocorrelation coefficients for the annual levels were, not surprisingly, significant (99.9%): Ladoga 0.66 and Saimaa 0.47, but e.g. the seven-year periodicity, generally believed to exist by people living on the shores of Lake Ladoga, did not occur.

Thermal regimes of the lakes in the Baltic Sea Basin have been extensively studied, and long observation series on water temperatures are available. In Finland, the first systematic surface water observations began in Lake Saimaa at Lauritsala in 1916. These observations are still continuing and thus they represent the longest continuous water temperature observation series in Finland. A significant increase in the number of surface water temperature stations took place in 1961, when many water temperature stations were established.

The analysis of the surface water observations of eight sites in Finland did not show many significant long-term changes in the surface water temperature until the year 2000 (Korhonen 2002). For Lake Saimaa many variables (such as average surface water temperature of June, July, August, the number of days exceeding 10 °C 15 °C and 18 °C the temperature sum of June–September) show significant trends of warming (Fig. 2.35). Also, Lake Lappajärvi, in the western part of the country, showed one significant warming trend for the maximum water temperatures of the summer season. None of the other observation sites had statistically significant warming trends, though in southern and central Finland water temperatures in 1991–2000 were slightly higher than during the period 1961–1990. In northern Finland, no changes were detected.

2.2.2 Ice Regime

Ice regimes in the water bodies of the Baltic Sea Basin are formed predominantly by the impact of Atlantic air masses producing a warming effect on the study area during the cold season. The Baltic Sea itself stores much heat in wintertime and also warms the adjacent areas. Therefore, the closer the water body to the sea coast, the later ice cover is formed and the earlier the ice break-up occurs.

The analysis of the dynamics of ice events and ice thickness on rivers and lakes within the Baltic Sea Basin made in this section is based on water bodies in Poland, Russia and Finland as case studies. Sections 2.2.2.1 and 2.2.2.2 deal with rivers and lakes in Poland and Russia, while results for Finnish water bodies are given in Sect. 2.2.2.3.

Changes in ice regimes in lakes within the study area are similar to those in the rivers. But ice on lakes is usually formed later than that on the rivers; ice break-up in lakes also occurs later.

It should be noted that ice regimes in the water bodies are closely connected with energy (heat) resources of the territory. In this sense, any uni-directed change in heat resources will cause changes in the dates of beginning and ending of ice events as well as in the duration of the ice cover and its thickness. Global warming may have caused considerable changes both in the ice regimes of rivers and lakes and their maximum thickness during the last twenty years, which will be described in the following sections.

2.2.2.1 *Ice Events and Ice Thickness in Rivers*

An analysis of the ice event dynamics for rivers in the Russian part of the Baltic Sea Basin has been made by Vuglinsky (2002), this study is summarised in this section.

Mean long-term dates of the start of the ice events in autumn in the rivers vary widely depending on the longitude of the hydrometric station and on the river size (the river size is an indirect index of heat storage in the water mass before ice cover formation). Mean dates of the start of ice events in the rivers in the north of the study area are fixed at the middle of November, whereas in the rivers flowing in the southern part of the study area ice cover is formed 10–15 days later.

The same difference is observed in the dates of ice break-up in spring. In the north, the rivers are ice-free late in April (Suna at Porosozero, Vodla at Kharlovskaya); rivers in the southern territory are ice-free 10–15 days earlier (Luga at Tolmachevo, Shelon at Zapolye). The duration of the complete ice coverage on the rivers is variable, respectively; in the northern rivers it lasts 150–170 days, whereas in the southern rivers it is 110–130 days long. In the Volkhov River at station Volkhov, the mean ice cover duration is 127 days.

The analysis of the long-term dynamics of the dates of the start and ending of ice events and duration of the ice coverage shows that from the middle of the 20th century to its end a stable positive trend was observed in all the study rivers towards changes in the above characteristics of ice regime. During the second half of the 20th century, mean dates of the start of ice events in the rivers of the Russian territory of the Baltic Sea Basin became 10–15 days later; the complete ice melt in rivers occurred earlier by 15–20 days at the end of the 20th century compared with the 1950s). The duration of the complete ice coverage became much shorter. Moreover, in the rivers in the north of the study area the period of complete ice cover duration shortened by 25–30 days, whereas in the southern rivers, this period reduced by as much as 35–40 days (Fig. 2.36).

The maximum ice cover thickness in the rivers studied by Vuglinsky (2000) showed an evident negative trend during the last 30–40 years of the 20th century. The maximum ice cover thickness by the end of the 20th century was 15–20% less in all the study rivers within the Russian territory of the Baltic Sea Basin. A statistically significant trend has been analysed for the Perovka River (Fig. 2.37).

2.2.2.2 *Ice Events and Ice Thickness in Lakes*

Few studies on the dynamics of ice events on lakes within the Baltic Sea Basin have been made, case studies for water bodies in Poland and Russia are summarized in this section.

For all the Polish lakes under consideration, except Lake Hancza (which is the deepest lake in

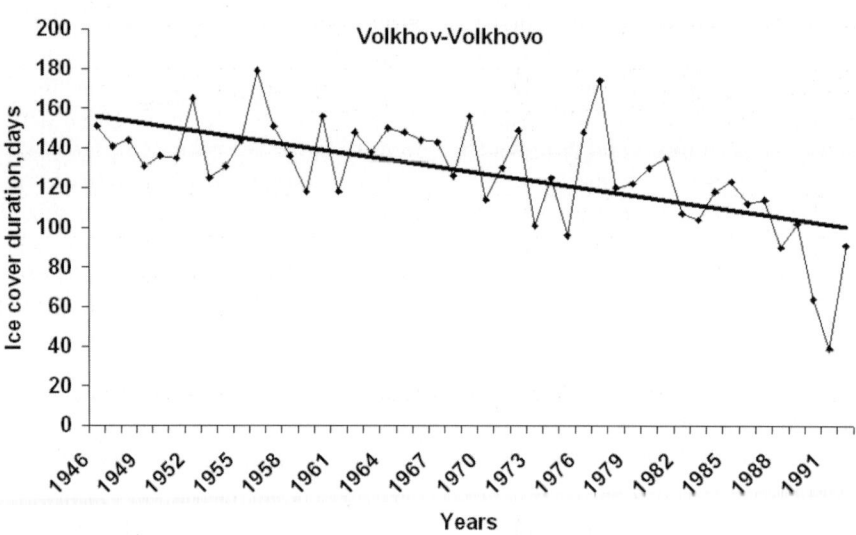

Fig. 2.36. Long-term changes in the ice cover duration of the river Volkhov at station Volkhov during 1946 to 2000 (Vuglinsky 2000)

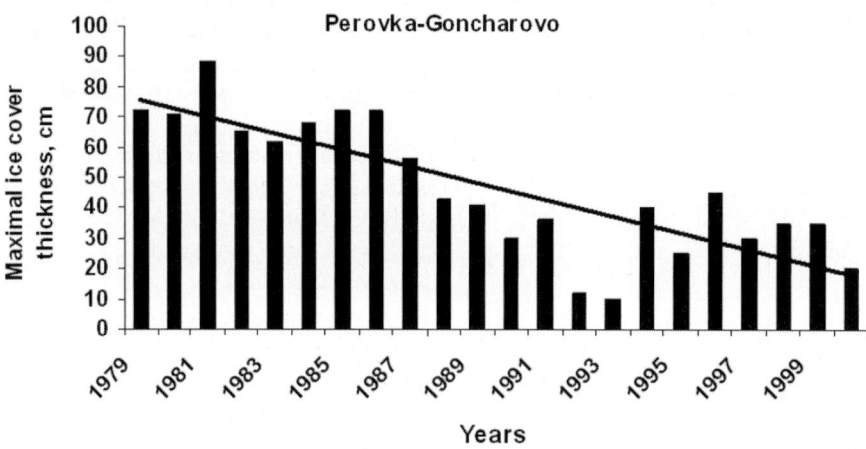

Fig. 2.37. Changes in maximum ice cover thickness in the river Perovka at station Goncharovo during 1979 to 2001 (Vuglinsky 2000)

the European Lowland), a strong negative trend of the ice cover duration (from 0.8 to 0.9 days per year on average) was ascertained for the period 1961–2000. The respective trend of the duration of the freeze-up period (which on average is shorter compared to the other Polish lakes) of Lake Hancza was less and amounted 0.4 day/year on average, mainly due to the set-back of this process at the end of the 20th century (Marszelewski and Skowron 2006).

For the Russian lakes with observation periods of 80–106 years (Lakes Ilmen and Onega, respectively), trends have been discovered towards changes in characteristics of ice events and maximal ice thickness during the years 1980–2000 on the background of a long-term variability of the mentioned quantities. Linear trends indicating later date of ice formation (positive trend) before 1980 and earlier date of ice cover break-up (negative trend) after 1980 have been discovered. But

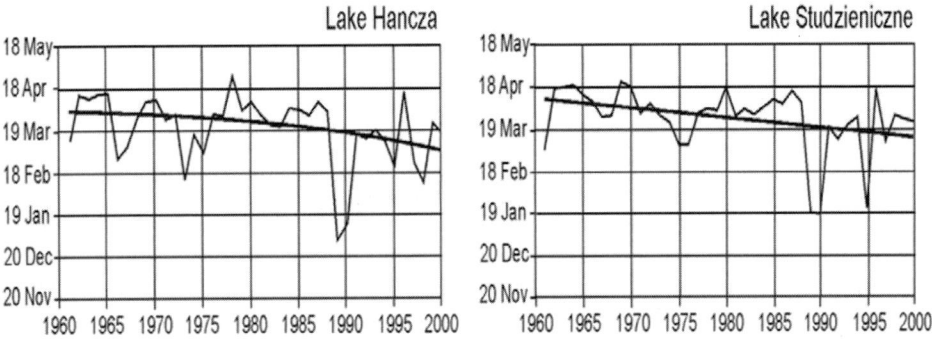

Fig. 2.38. Change of ice cover break-up dates of lakes Hancza and Studzieniczne in Poland during 1961 to 2000 (Marszelewski and Skowron 2006)

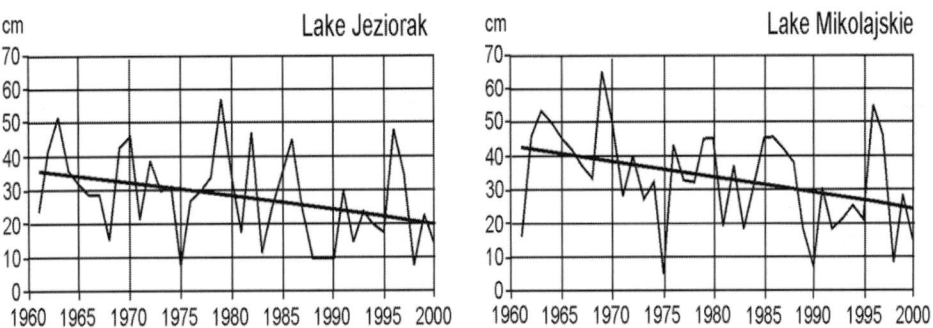

Fig. 2.39. Change of maximal ice cover thickness of lakes Jeziorek and Mikolajskie in Poland during 1961 to 2000 (Marszelewski and Skowron 2006)

all these trends are not statistically significant. The trend towards earlier ice cover break-up is observed for both periods (before and after 1980) across all the study lakes, except Lake Ladoga. For the latter lake, a slightly positive trend was observed for the period before 1980.

Dates of the start and ending of ice events, the duration of ice cover period and maximal ice thickness vary over a large range, depending on geographic location of lakes and their morphometric characteristics.

A trend towards earlier formation of the ice cover for Polish lakes for the period 1961–2000 was found in 5 out of 6 analysed lakes (Marszelewski and Skowron 2006). It ranged from 6–7 days (Lake Mikolajskie) to 16–17 days (Lake Jeziorak). Only for Lake Charzykowskie did the ice cover form later (by 5 days on average).

All Polish lakes investigated (Marszelewski and Skowron 2006) revealed a similar negative trend

(earlier ice break-up in spring) of the ice cover break-up (from 0.6 to 0.8 days year^{-1}, see Fig. 2.38. This trend is in accordance with the general course of the air temperature and surface water temperature in the lakes of this part of Europe (Dabrowski et al. 2004), as well as with the pattern of the ice cover on the lakes of northern Europe (Magnuson et al. 2000).

For all Polish and Russian lakes studied, a negative trend of maximal ice cover thickness was analysed. The mean maximal ice cover thickness for Polish lakes systematically decreased during the period 1961–2000 (Fig. 2.39). On average, this negative trend is estimated to be from 0.26 cm year^{-1} (Lake Charzykowskie) to 0.60 cm year^{-1} (Lake Hancza).

In Russia, only lakes Ladoga and Ilmen show a statistically significant negative trend of the maximal ice cover thickness after 1980, while all other lakes investigated exhibit a negative insignificant

Table 2.6. Trends in time series of ice phenomena and ice thickness for three Russian lakes P1: pre 1980 period, P2: post 1980 period. + positive non-significant trend, − negative non-significant trend, ++ statistically significant positive trend (at $p < 0.05$), −− statistically significant negative trend (at $p < 0.05$) (Vuglinsky et al. 2002)

Lake	Ice-on		Ice break-up		Ice cover duration		Maximal ice cover thickness	
	P1	P2	P1	P2	P1	P2	P1	P2
Ladoga (central part)	+	−	+	−	−	−	−	−−
Onega at station Petrozavodsk	+	−	−−	−	−	−	+	−
Ilmen at station Voitsy	+	−	−	−	−	−	+	−−

trend. Maximal ice cover thickness of the Russian lakes had the most pronounced response to climate warming in wintertime during the last decades of the 20th century (Gronskaya 2000). Trends in time series of ice phenomena and maximal ice cover thickness for the Russian lakes are shown in Table 2.6 for the periods before and after 1980.

2.2.2.3 *Ice Conditions in Rivers and Lakes in Finland*

The ice break-up date has become earlier and the freezing date later at many observation sites in Finland during the last few decades. The trends are statistically significant mainly in those sites which have records at least since the late 19th century (Kuusisto and Elo 2000). Trends of series that started later are in most cases not statistically significant. In the longest series, which started in the late 17th century, the ice break-up has moved 6–9 days earlier per hundred years (Fig. 2.40).

Freezing has been delayed since the late 19th century, in most cases by 0 to 8 days per century. Also, the duration of ice cover has significantly shortened at the sites with the longest records (Fig. 2.41).

Not all of the longest freezing and duration of ice cover trends are statistically significant. The trends of break-up are stronger, because the variation of break-up date is smaller than that of the freezing date. The large variation of freezing dates hides the trends of freezing. There are no statistically significant trends found in northern Lapland but the time series from this part of the country are still relatively short.

Both decreasing and increasing trends can be found in the maximum ice thickness time series. At about half of the sites the trends are statistically significant. The maximum thicknesses have mostly increased in eastern and northern Finland and decreased in southern Finland. The series with records since the 1960s had a significant trend of 2–3 cm per ten years; for longer series the trend was mainly 1–2 cm per ten years. The time series of ice and snow thickness are rather short, about twenty years, thus no long-term analysis can be made. In the 1980s there was much snow and snow ice in southern and central Finland. Thus, the snow ice thickness has in most cases decreased from the 1980s to the year 2000 in southern and central Finland. In northern Finland, the snow ice thickness has increased in some places during the twenty year period.

2.2.3 Snow Cover

Snowfalls occur every winter in the Baltic Sea Basin and seasonal snow cover is formed except in the south-western regions. Typical durations of snow cover over most areas are between four and six months (except for regions on the southern coast of the Baltic Sea). Snow cover is a regularly varying feature of the land areas. It affects the winter and spring climate in several ways. Mean snow conditions as well as regional variations and extremes are described in more detail in Annex 1.3.5.

Annual Northern Hemisphere (NH) average continental snow cover extent has decreased by about 10% from 1966 to 2000 (see Folland et al.

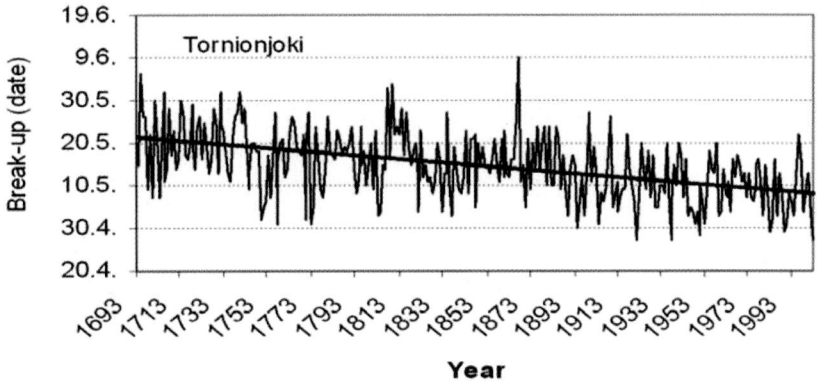

Fig. 2.40. Ice break-up dates of the Tornionjoki river since 1693 (from Korhonen 2004)

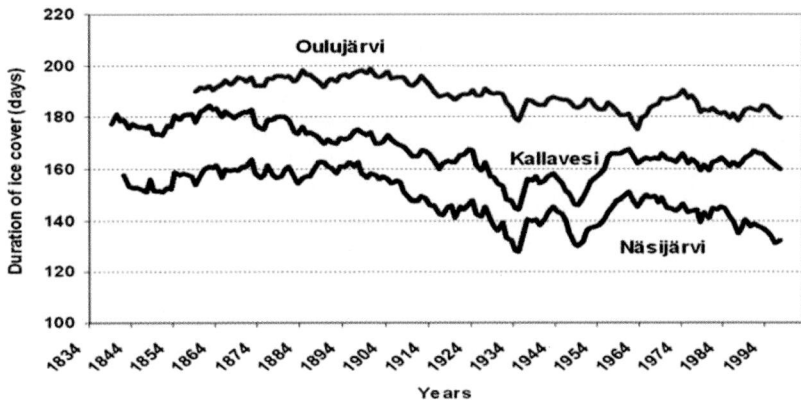

Fig. 2.41. Duration of the ice cover (11-year running mean) in lakes Näsijärvi, Kallavesi and Oulujärvi (Finland) during 1834 to 2000 (adapted from Kuusisto and Elo 2000)

2001). A subset of the NH data was extracted for an area ($\sim 1.2 \times 10^6$ km^2) roughly covering Norway, Sweden and Finland. The time series for this region represents the changes in monthly Fennoscandian snow cover extent anomalies from January 1967 to December 2000.

The Fennoscandian subset has features that are similar to the NH average: mainly positive anomalies before 1979, a period dominated by negative anomalies from May 1987 to March 1995, and a period of small variability from 1995–2000. Due to the small regional extent of snow cover in summer, the months from July to August have the smallest variability, and the corresponding monthly anomalies are therefore close to zero.

A recent decrease of snow cover duration and water equivalent has been observed in southern parts of all the Fennoscandian countries, while the opposite trend prevails in the north. In the Scandic mountains the enhancement of precipitation has overshadowed increases in temperature in the past two decades, and snow cover has become thicker.

The increase of duration of snow cover ($+0.192$ day year^{-1}) and winter air temperature ($+0.010\,°C\,$year^{-1}) is typical for northern Eurasia in total for 1936–2000. In both cases the quality of conformity of calculated regression to empirical data is satisfactory. The revealed situation is expected to be caused by global warming, when an increase of precipitation and, accordingly, snowfall has taken place. And this situation is characteristic both of Nordic countries and the north of the eastern European plain in the corresponding proportion of values of parameters.

In Finland, increasing temperatures have intensified wintertime snowmelt in western and southern parts of the country towards the end of the

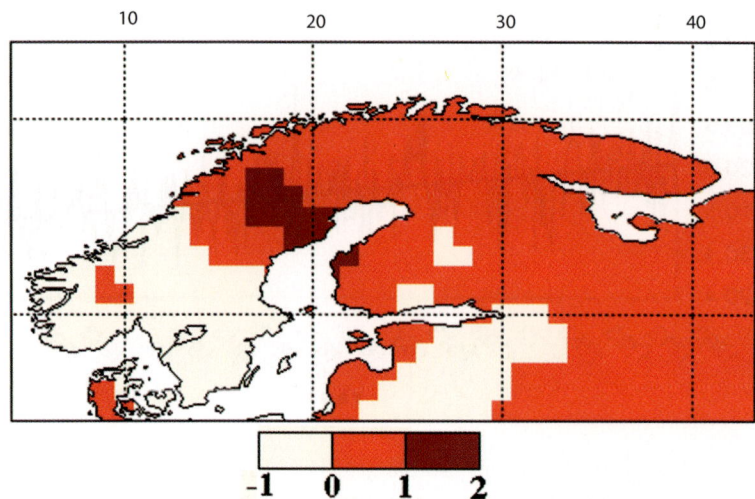

Fig. 2.42. Spatial variability of trends of the durations of snow cover (days year^{-1}) during 1936–2000 (adapted from Kitaev et al. 2006)

Table 2.7. Mean maximum water equivalents of snow (L_{max}) in six river drainage basins in Finland

Drainage basin	L_{max} (mm) 1961–1990	L_{max} (mm) 1991–2005	Difference (%)
Vantaanjoki	109	77	−29
Kyrönjoki	92	70	−24
Vuoksi	146	154	+5
Oulujoki	162	186	+15
Kemijoki	175	195	+11
Paatsjoki	149	178	+19

period 1946–2001, while on the contrary, in eastern and northern Finland the maximum snow storage has increased (Heino 1994; Hyvärinen 2002). This can also be seen from the snow observation series of the Finnish Environment Institute. The Vantaanjoki basin on the southern coast and the Kyrönjoki basin in southern Ostrobothnia had much less snow in 1991–2005 as compared to 1961–1990. The Vuoksi basin, covering a large part of eastern Finland, had a slight increase of snow storage.

The three other basins located in northern Finland show a clear increase of mean maximum water equivalents. In the Finnish snow data, there are also some indications that the variability of snow conditions has increased (Table 2.7).

However, it is too early to claim that the meager snow cover in southern Finland is already an indicator of climate change. The mild winters are connected with a marked predominance of south-westerly winds due to a high NAO index (Annex 7). With a low NAO index, southeasterly winds will dominate and bring abundant orographic precipitation particularly to the Uusimaa region. Even with a high NAO value, interruptions are highly probable (Solantie 2000).

In Sweden, the snow conditions have polarized in a similar way as in Finland: there is more snow in the north, while snow cover has become thinner in the southern part of the country (Sten Bergström, SMHI, pers. comm.). Larsson (2004) analysed over 40 stations in Sweden in 1900–2003 and could not find large changes in snow conditions. In 1961–2003 snow cover days, however, decreased in southern Sweden by 20–40%. In Norway, snow accumulation has increased in some areas at high altitudes.

Changes of snow parameters have, of course, local features. Examining a parity of duration of snow cover and air temperature (Kitaev

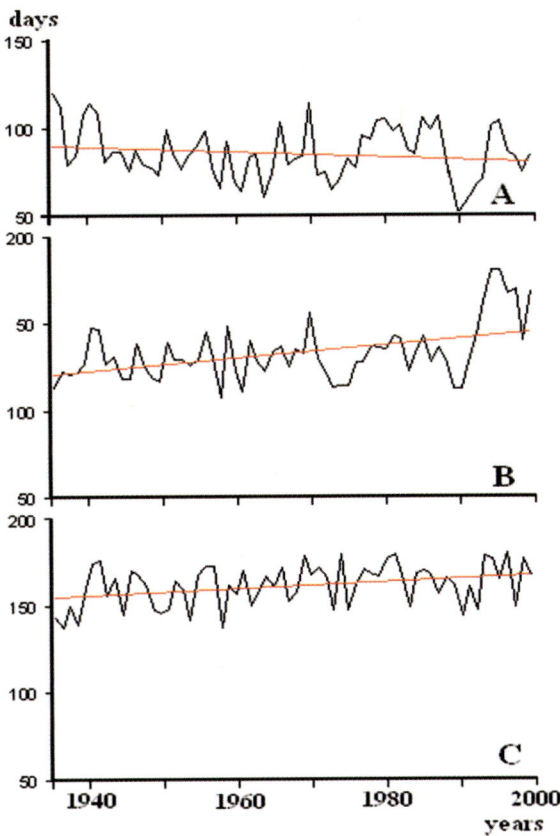

Fig. 2.43. Time series of spatial mean duration of snow cover in western 5–15° E (**A**), central 15–25° E (**B**), eastern 25–45° E (**C**) parts of Northern Europe during 1891/92–2000/01, and its linear trends

et al. 2006), it is possible to allocate three areas (Fig. 2.42). The first is north of 63–65° N, where the long-term increase of duration of snow cover occurs on a background of a reduction of air temperature. The second is the Scandinavian Peninsula to the west of 15–20° E, where there is a gradual increase of air temperature and the duration of the snow period decreases accordingly. The third area is a part of the eastern European plain south of 60–63° N, where the increase of both air temperature and duration of snow cover simultaneously take place.

This situation corresponds to features of trends for western (5–15° E), central (15–25° E) and eastern (25–45° E) parts of the northern area of the Baltic Sea Basin (Fig. 2.43). Values of regional trends are different (−0.146, 0.382 and 0.198 day year^{-1}) with however similar multi-year variation of snow cover duration.

A longer time perspective of changes in snow cover is provided by station observations of the number of days when the ground is more than half-covered by snow. Figure 2.44 shows six long station records of annual anomalies of the number of days with snow cover. The time series show that the trends can be very different in different parts of Fennoscandia.

In Estonia the decrease of the duration of snow cover has been more intense in western and central parts, exceeding in some regions one day per year. The trends are lower in the northeastern part of Estonia. In the period 1961/62–2000/01, negative trends were observed also in snow cover depth and in water equivalent (Tooming and Kadaja 2006). The decrease of spatial mean snow depth was highest in February, 0.30–0.33 cm per year varying in some extent in different types of observations, followed by March. The decrease of spatial mean maximum water equivalent was 0.68 mm per year. The highest decrease was in February (Fig. 2.45).

Snow cover duration showed a negative trend for a 50-year period all over the territory of Latvia. The snow cover duration has decreased on average by 12 days during 1945–1996, but it was

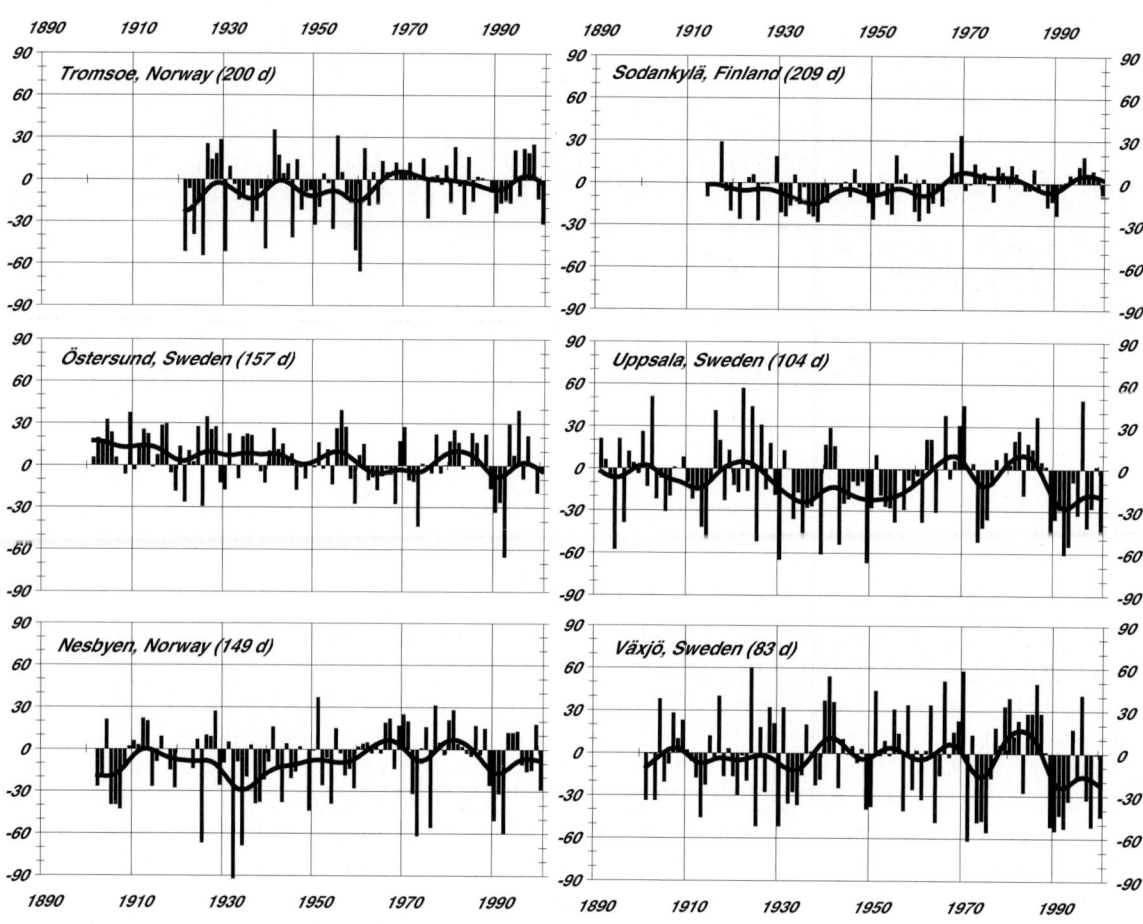

Fig. 2.44. Anomaly time series of the number of days with snow cover (more than half of the ground covered by snow) per year for six stations in Fennoscandia (*Left from top to bottom:* Tromsö, Norway; Östersund, Sweden; Nesbyen, Norway; *Right from top to bottom:* Sodankylä, Finland; Uppsala, Sweden; Växjö, Sweden). The anomalies are with reference to the period 1961–90. Station averages for this period are given in parentheses. The number of snow covered days are calculated for the period from August 1 to July 31 in the following year and plotted against the latter calendar year (data source: the national meteorological services of Sweden, Norway and Finland)

statistically significant only for three stations (Draveniece 1998). No regularities were found in regard to the spatial distribution of the decrease of snow cover duration. Further, the snow cover duration for March was analysed for the station Priekuli, and it was found that the monthly decrease during 1945–1996 comprised almost half of the seasonal value. If time series of the duration of snow cover were extended to around 70 years, these were smaller and statistically insignificant.

When analysing the trends of climatic snow cover indices in Lithuania for 1925–1996, a general tendency of reduced quantity of snow in winter throughout the Lithuanian territory was observed.

This may be associated with the change of atmospheric circulation peculiarities, i.e., an increase in the number of days with cyclonic circulation. In the last decades of the 20th century a permanent snow cover in Lithuanian territory tends to occur earlier and disappear earlier than in the middle of the century.

In Poland, snow cover trends in the 20th century and their relations to the changes of main climatic factors were studied by Falarz (2004a,b). There was a slight negative trend in the duration of the snow cover (up to −4 days/10 years) and its depth (up to −13 cm/10 years) during the 50 investigated winter seasons in the majority of the

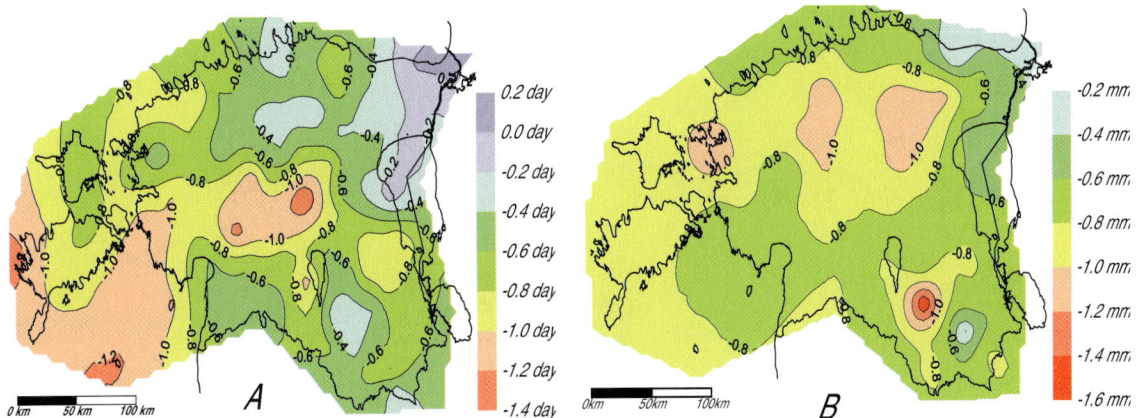

Fig. 2.45. Slopes of linear trends of snow cover duration in days per year (**A**) and snow cover water equivalent in February in mm per year (**B**) in Estonia for the period 1961/62 to 2001/02 (from Tooming and Kadaja 2006)

Polish areas, which is statistically significant (at the 0.05 error level) at some points. A positive trend was observed in areas with a typically thick snow cover. Over the longer periods (60–108 winter seasons) all trends were close to zero. An increasing trend in the variability (expressed by the variation coefficient) of snow cover depth and duration has been observed in the lowland area of Poland since 1950s or 1960s. In the mountainous area the snow cover variability diminishes towards the end of the 20th century.

Thus, it is possible to allocate three areas with special conditions of long-term changes in snow cover. In the first area – the western part of Scandinavia, Poland, Lithuania, Latvia and Estonia – increasing regional winter temperatures lead to an increase in the liquid proportion of precipitation during wintertime and, as a result, in a decrease in snow storage. In the second area, in the north of Scandinavia, the snow storage increases in accordance with a long-term decrease of winter air temperature. In the third area, in the northwest of the eastern European plain, snow storage increases in accordance with the increase in winter temperatures and precipitation (Kitaev et al. 2006).

2.3 The Baltic Sea

2.3.1 Hydrographic Characteristics

Hydrographic measurements in the Baltic Sea started in the 18th century (Fonselius and Valderrama 2003). In the 19th century, titration of salinity was introduced, and in the last decade of the 1800s, hydrographic measurements were initiated in several Baltic Sea countries at coastal stations and lightships. Early deep-sea expeditions were carried out in the summer of 1871 by Germany (Meyer et al. 1873) and in 1877 by Sweden (Pettersson 1893). Regular offshore observations of temperature and salinity at some important deep stations (Fig. 2.46) were established in the beginning of the 20th century after an initiative of Otto Pettersson (Fonselius and Valderrama 2003). In the beginning, only one to two cruises were conducted per year, and the measurements were interrupted during the two world wars. From the 1950s onwards the number of stations increased, and sampling was continuously intensified (Fig. 2.47), with observations made irregularly, both in time and space (Janssen et al. 1999).

Lenz (1971) and Bock (1971) created maps of the temporally averaged vertical and horizontal distributions of temperature and salinity in the Baltic Sea, using monitoring data from 1902–1956. Recently, Janssen et al. (1999) provided an improved climatology of the Baltic Sea hydrography from an extended data base from 1900–1996. This climatology consists of monthly 3-d gridded data with a horizontal resolution of approximately

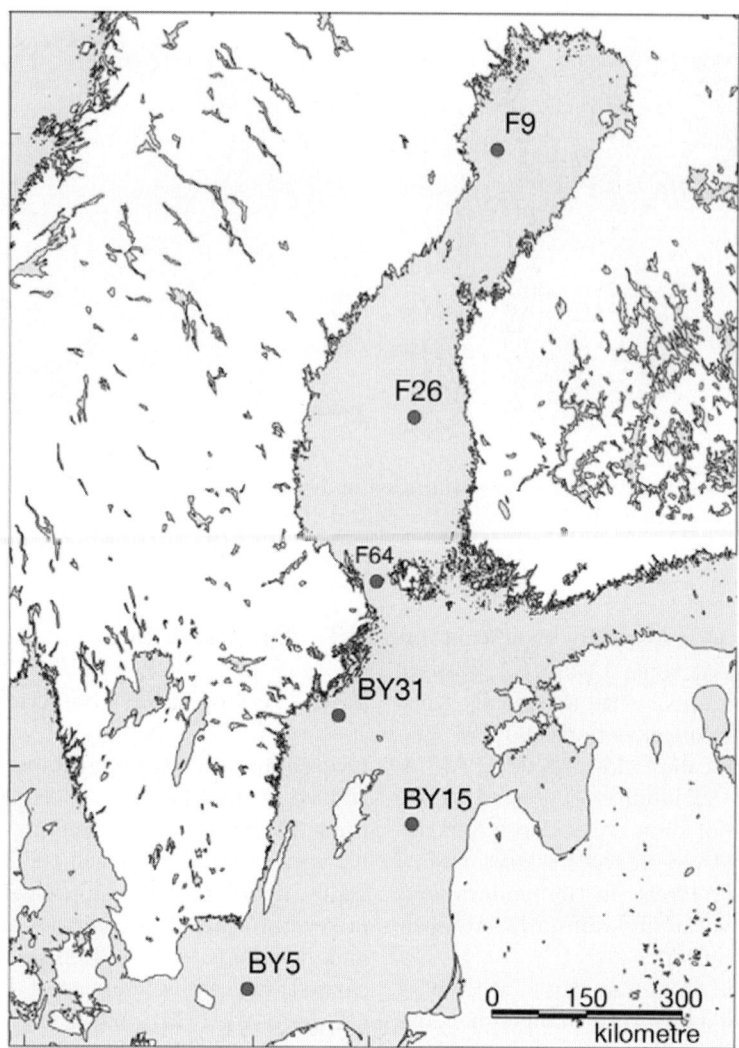

Fig. 2.46. Station map for long-term observations in the Baltic Sea deep basins (from Fonselius and Valderrama 2003)

10 km. About 3.1 million measurements of temperature and 2.9 million measurements of salinity were used. Further information on the hydrography of the Baltic Sea is given in Annex 1.1.

2.3.1.1   *Temperature*

The most detailed knowledge of temperature from observations is derived for the sea surface (SST), since sampling at the surface is easiest and thus most intense (Fig. 2.47). In addition to observations made in the framework of research and monitoring experiments, routine observations of SST are carried out on merchant ships. These data have been used to map the SST (and ice condition) with high temporal resolution since 1960s. The

Swedish Meteorological and Hydrological Institute (SMHI) and the Finnish Institute of Marine Research (FIMR) published weekly and half weekly charts already in early 1970s. The German Hydrographic Services (BSH, IOW) have published weekly temperature charts since 1996, based on remote sensing data (www.bsh.de).

Furthermore, coarser data sets like COADS (Comprehensive Ocean Atmosphere Data Set, Woodruff et al. 1998; Woodruff 2001) or Reynolds SSTs (Reynolds and Smith 1994; since 1981) are available for the Baltic Sea, with $1°$–$2°$ resolution, and have been used as indicators for Baltic Sea surface climate and its variability (Janssen 2002; Dippner et al. 2000). Especially in wintertime, high correlations between SST and air tempera-

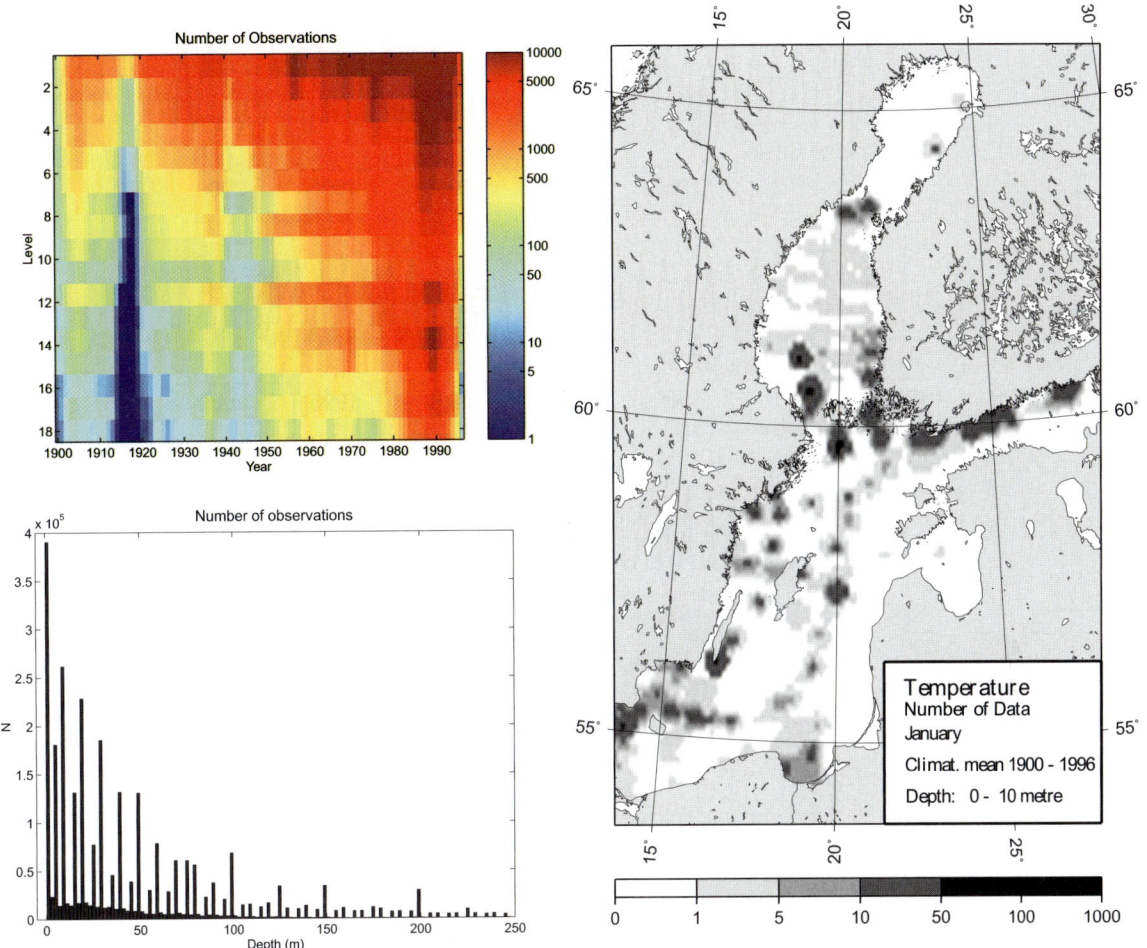

Fig. 2.47. Number of observations since 1900 per year and vertical level (from Janssen et al. 1999)

ture anomalies are found (Tinz 1996). Similarly to ice conditions, strong relations between winter SST anomalies and climatic indicators like the North Atlantic Oscillation (NAO) Index or the Arctic Oscillation (AO) Index were detected by a number of authors (e.g. Tinz 1996; Janssen 2002).

In light of the strong air–sea heat exchange and slow circulation, this is not surprising (Stigebrandt and Gustafsson 2003). The correlations between SST anomalies and air pressure pattern are weaker during the summer months (Janssen 2002).

High resolution monitoring of sea surface temperature variability from remote sensing enables the establishment of multi-year data sets that allow for resolving meso-scale structures, inter-annual variability and shift of seasonal cycles (Siegel et al. 1999; recently extended to cover 1990–2004, see Siegel et al. 2005). Their analy-

sis demonstrated that the late 1990s were characterised by warmer summers and colder winters compared to the years before and after. The more detailed analysis was able to detect anomalies in seasonal extremes as well as an increase in annual mean sea surface temperature of 0.8 °C, and in some areas of the Baltic Sea even higher increases, during most of the period analysed (1990–2004).

Omstedt and Nohr (2004) found an increase in air temperature and wind speed and a decrease in cloudiness and relative humidity over the Baltic Sea during the period 1970–2002. Despite an atmospheric warming of 1 °C, no trend in the modelled mean (averaged over the whole Baltic Sea and all depths) Baltic Sea water temperature was identified (Fig. 2.48). Omstedt and Nohr explained this by the heat balance, which indicated no trend in the net heat loss. However, the period

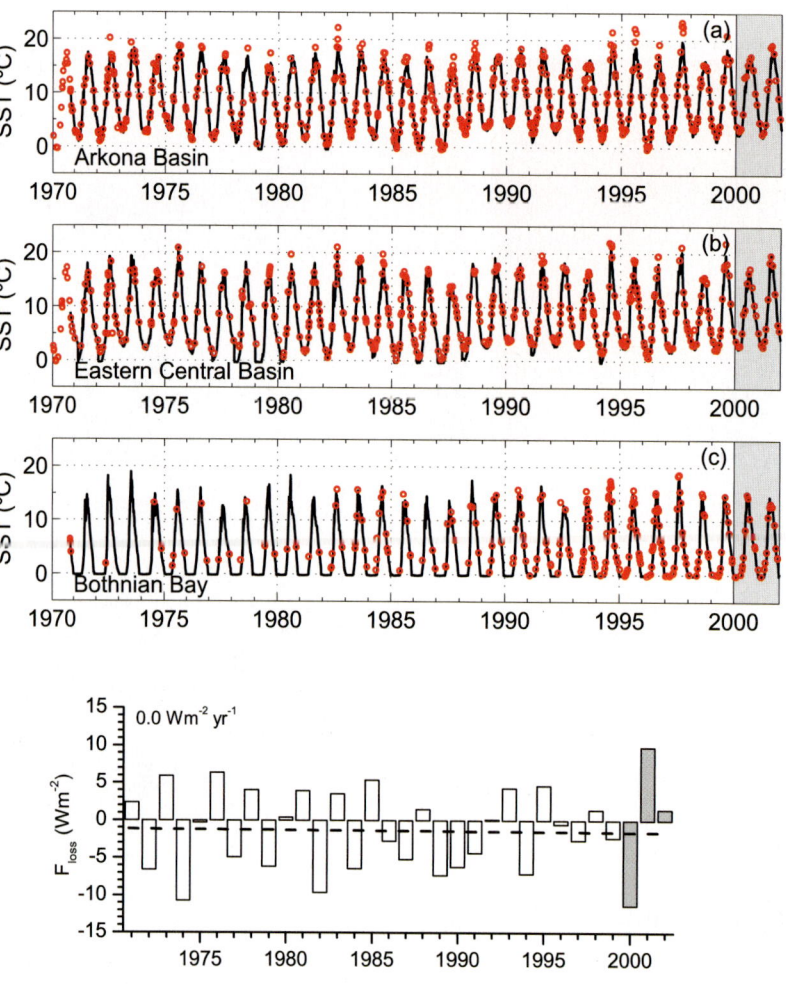

Fig. 2.48. Time series of modelled sea surface temperature in three different Baltic Sea basins (*top*) and annual mean Baltic Sea heat loss (W/m²) during 1970 to 2000 (from Omstedt and Nohr 2004)

1970–2002 is too short to identify statistically significant trends due to the large interannual variability.

By applying optimum analysis to the available historical data, total amounts of heat and salt in the Baltic Sea were calculated for the period 1959–1993 (Toompuu and Wulff 1996; Elken 1998; Toompuu 1998). Heat content maxima were identified around 1975 and 1990. A minimum around 1987 was found both in the Baltic Proper and in the Gulfs of Finland and Riga as a consequence of severe winters between 1985 and 1987. By using a 3-d model, Schrum et al. (2002) computed daily time series of heat content for the whole Baltic Sea (Fig. 2.49). They found similar extremes with maximum heat content periods around 1975 and 1990. These periods had both higher winter and as well as significantly higher summer heat contents.

Similarly to the studies based on observational data, a heat content minimum was found for the Baltic Sea in 1987, when both, winter heat content as well as summer heat content, show minimum values. Rönkkönen et al. (2004) analysed temperature observations taken between 1950 and 1999 at three stations in the outer Gulf of Finland. They computed a vertical mean temperature by taking the average of measurements at 0, 5, 10, 20 and 30 m depth, and then the average of the three stations. They did not find any trend in the annual averaged time series, but they found a significant increase in May–July temperatures. In accordance with the outcome of the investigations by Omstedt and Nohr (2004), neither the results from 3-d modelling (Schrum et al. 2002) nor the observational data analysis (Toompuu and Wulff 1996; Elken 1998; Toompuu 1998;

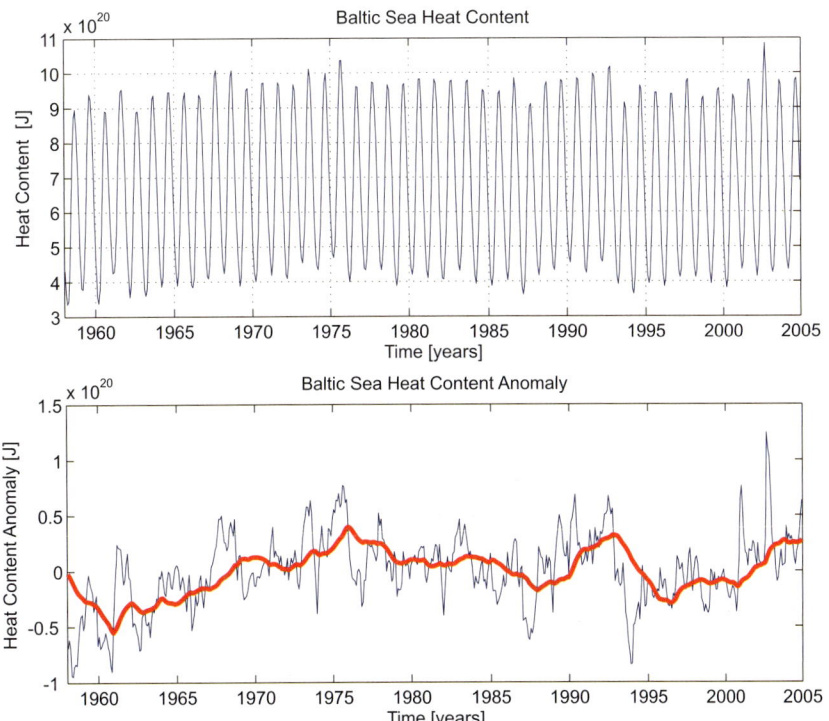

Fig. 2.49. Calculated heat content of the Baltic Sea from a North Sea-Baltic Sea model for the period 1958–2005 based on Schrum et al. (2003). Monthly time-series of heat content in the upper panel, and heat content anomalies (seasonal cycle subtracted) with monthly resolution (*blue*) and as 3-years moving average (*red*) in the lower panel

Rönkkönen et al. 2004) revealed any trends in annual average Baltic Sea heat content and average temperature from 1958 onwards. It seems, however, that changes of the annual cycle have occurred (Rönkkönen et al. 2004).

Fonselius and Valderrama (2003) analysed data from deep monitoring stations for 100 years of observations (1900–2000) and estimated annual temperature time series for different stations (Fig. 2.50). For the second half of the century they found strong variability in annual temperatures, but trends were not identified. However, taking into account the full time series from 1900 onwards, there is evidence that the temperature regime in the Baltic Sea has changed since the beginning of the 1950s. Fonselius and Valderrama detected warming trends with significantly higher temperatures in the second half of the century. Since the temperature increase coincides with the change in sampling frequency (after 1950 the sampling frequency was increased from one to two annual observations to at least four observations annually), it is presently unclear whether the regime as seen from the observations is significantly influ-

enced or even caused by the change in sampling frequency and changing seasonal representation in the data set. However, from Fig. 2.50 and Fig. A.4 (Annex 1.1) it is evident that lower deep water temperatures were observed in the first half of the century than in the second half, which might be an indicator for a trend in annual temperatures.

Recently, aspects of inflow driven temperature variations in the intermediate waters of the Baltic Sea have been discussed in several publications (Feistel et al. 2003, 2005; Hagen and Feistel 2004; Mohrholz et al. 2005). Since 1988, the seasonal temperature minimum of the intermediate winter water has increased by 1.5 °C (Fig. 2.51). This trend was related to an increased frequency and intensity of summer inflow events since the last decade of the 20th century, which were found to contribute significantly to Baltic Sea ventilation.

Although summer inflows such as the exceptional event in 2002 have only little impact on the net salt budget, their impact on stratification and deep water circulation in the following seasons was found to be significant (Feistel et al. 2003; Mohrholz et al. 2005).

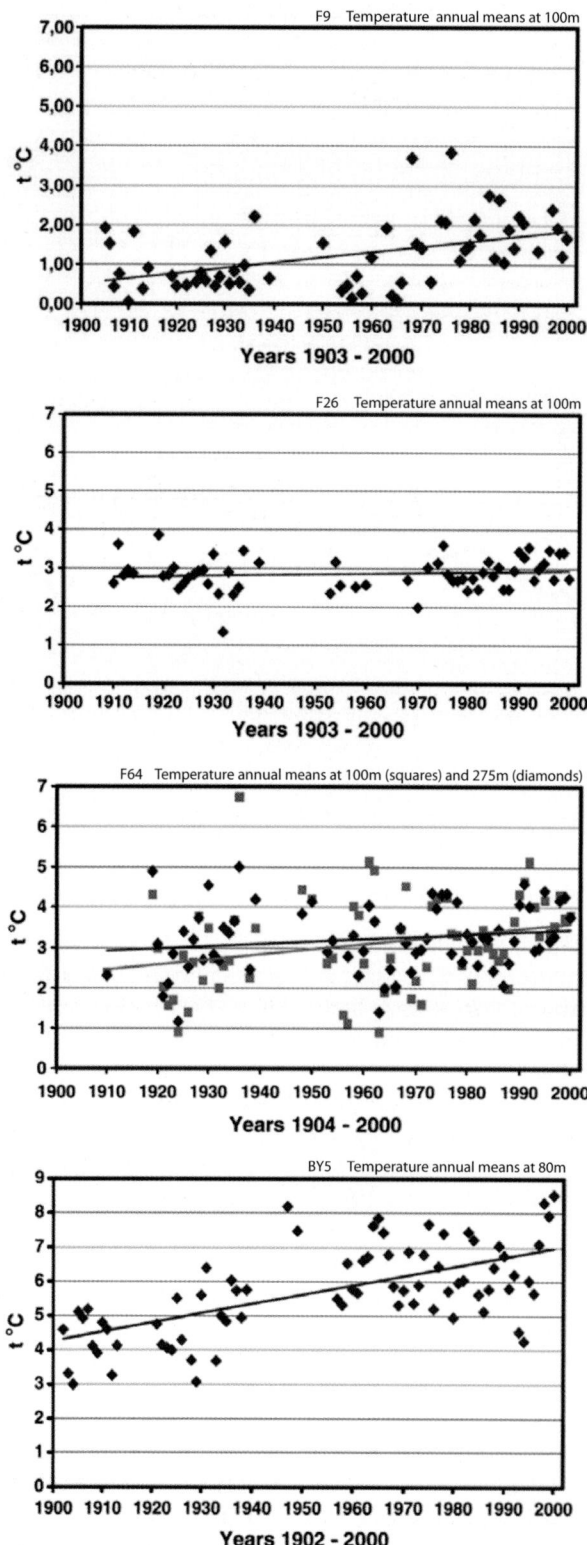

Fig. 2.50. Annual temperatures estimated from long-term measurements at different stations in the Baltic Sea, see Fig. 2.46 (from Fonselius and Valderrama 2003)

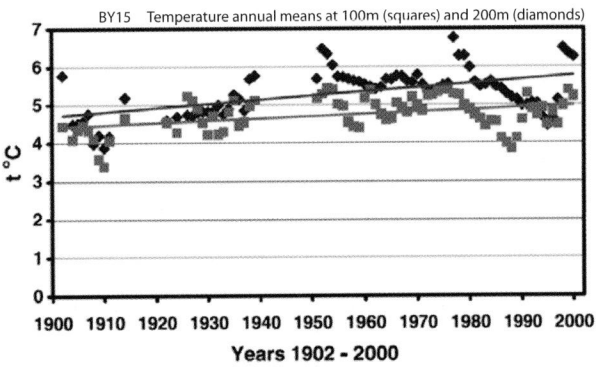

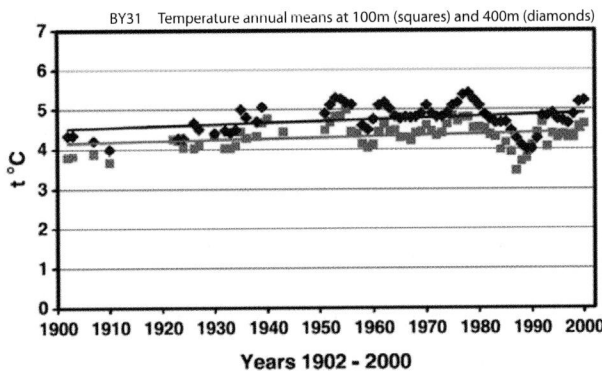

Fig. 2.50. cont.

2.3.1.2 Salinity and Saltwater Inflows

During the 20th century, salinity showed only weak and statistically non-significant trends at all monitoring stations (Fonselius and Valderrama 2003; see also Fig. A.4 in Annex 1.1). The Baltic Sea mean salinity (integrated vertically as well as horizontally) during the 20th century has been reconstructed and analyzed by Toompuu and Wulff (1996), Winsor et al. (2001, 2003), and Meier and Kauker (2003).

In the study by Winsor et al. (2001), the estimated mean salinity was calculated by first using salinity profiles from all major sub-basins of the Baltic Sea for the period 1977–1987, a period with good coverage in all sub-basins. Then the mean was compared with data from the Gotland Deep (BY15), a station that has observations over more than 100 years. The comparison illustrated that the station BY15 represents the Baltic Sea well and, based on data from this station, the mean salinity was reconstructed for the 20th century (Fig. 2.52). In the study by Meier and Kauker (2003) salinities of the 20th century were

calculated utilising a three-dimensional circulation model forced with reconstructed atmospheric surface data (Kauker and Meier 2003).

The mean salinity averaged for the entire Baltic was estimated to be 7.7 (Winsor et al. 2003) or 7.4 psu (Meier and Kauker 2003). The mean salinities calculated from climatological data (Janssen et al. 1999) amount to 7.2–7.4 psu for different months. Daily and decadal variations have maximum amplitudes of 0.1 and 0.5 psu, respectively. No long-term trend was found during the 20th century. On longer time scales salinity variations were much larger. The maximum salinity during the Littorina Sea stage (6,000–5,000 BP) was very likely between 10 and 15 psu due to increased cross-section areas of the inlets (Öresund and Darss) and due to a 15–60% lower freshwater supply than at present (Gustafsson and Westman 2002).

Long records of water exchange variations between the North Sea and Baltic Sea have been calculated from sea level data by Stigebrandt (1984) and Winsor et al. (2001). Andersson (2002) and Ekman (1999) showed that substantial variations

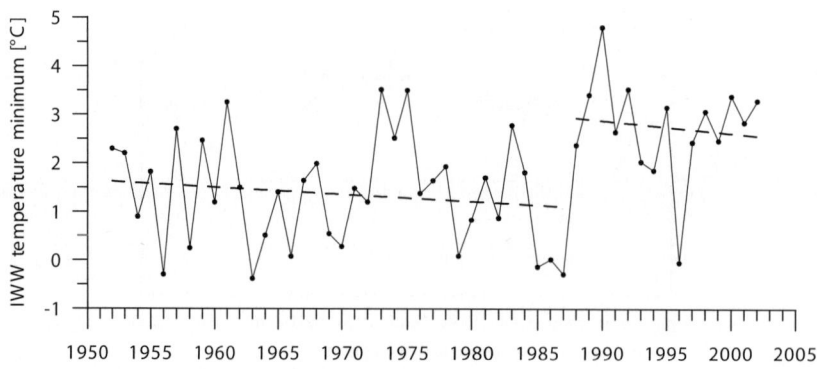

Fig. 2.51. Time series of annual temperature minimum (°C) in the intermediate winter water in the Baltic Sea (from Mohrholz et al. 2006)

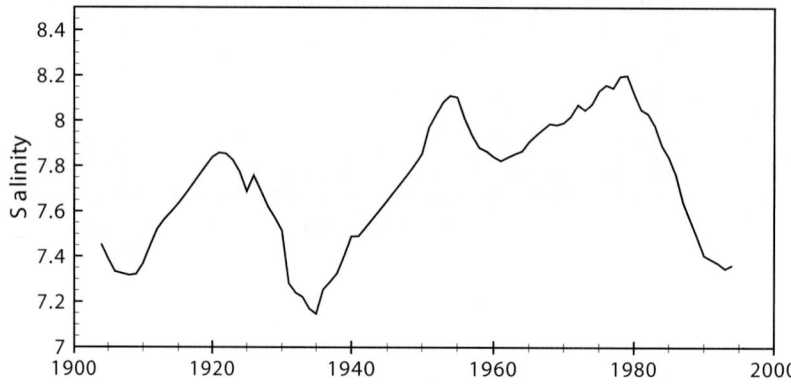

Fig. 2.52. Reconstructed spatially integrated mean salinity (psu) of the Baltic Sea of the 20th century, five-year running mean. Major data gaps occur during the World wars (redrawn from Winsor et al. 2001, 2003)

in the seasonal sea levels have occurred, indicating a substantial variability of the water exchange. However, the mechanisms of baroclinic inflows are more complex (Stigebrandt 1983).

The bottom water in the deep sub-basins is renewed mainly by large perturbations, so-called major Baltic saltwater inflows (Matthäus and Franck 1992; Fischer and Matthäus 1996; see also Annex 1.1). Matthäus and Franck (1992) quantified major inflows of the 20th century from a combination of sea level and salinity observations. A slightly different approach was used by Gustafsson and Andersson (2001). Major saltwater inflows are typically forced by a sequence of easterly winds lasting for about 20 days followed by strong to very strong westerly winds of similar duration (Lass and Matthäus 1996).

Since the mid-1970s, the frequency and intensity of major inflows have decreased. They were completely absent between February 1983 and January 1993. During this low-salinity phase the deep water in the eastern Gotland Basin was poorly ventilated, with oxygen depletion as a consequence. However, this stagnation period was not exceptional. During the last century at least one other long-lasting stagnation period during the 1920/1930s and perhaps a third smaller one during the 1950/1960s were identified (Fonselius et al. 1984; Fonselius 1969; Meier 2005). The latter might have been caused by the strong inflow in 1951, which filled the Baltic Sea depths with highly saline water. Subsequent saltwater inflows might not have been saline enough to replace these earlier water masses.

The two low-salinity phases during the 1920/30s and during the 1980/90s are explained by stronger than normal freshwater inflow and zonal wind velocity. Meier and Kauker (2003) found that about

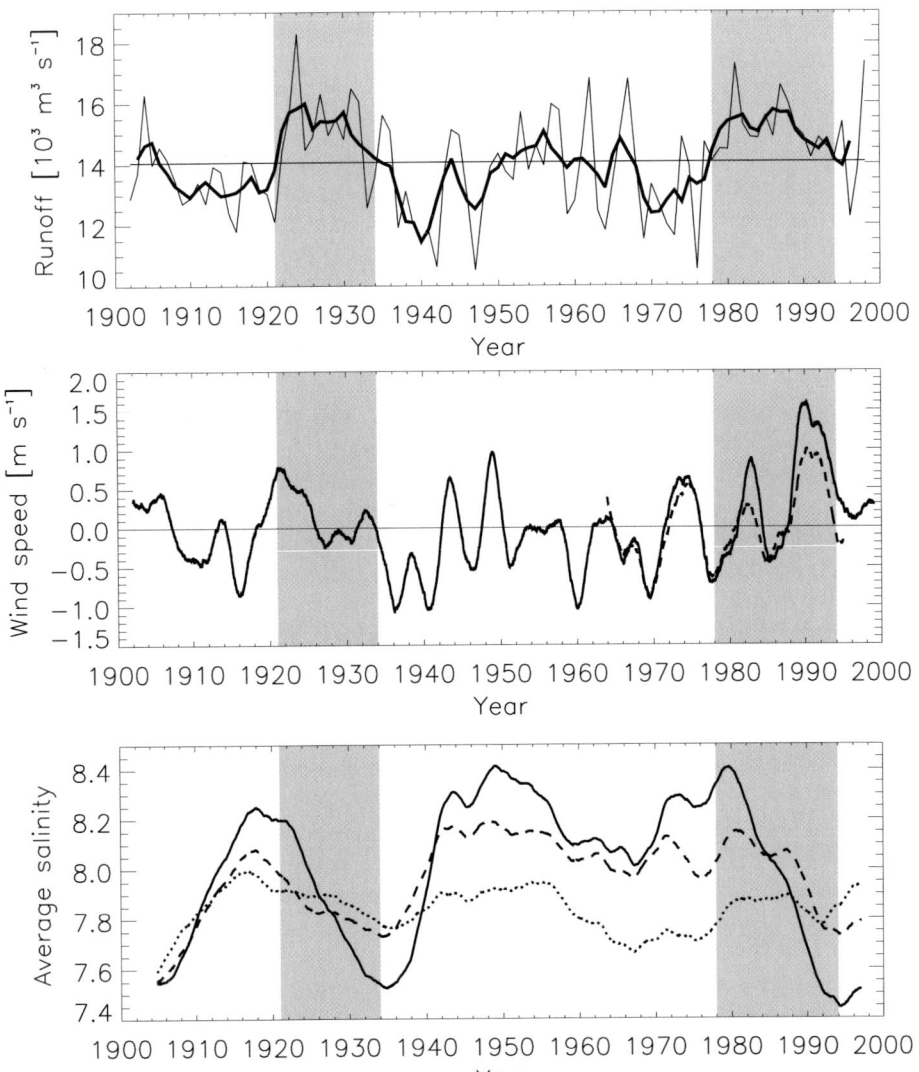

Fig. 2.53. *Upper panel:* Annual mean river runoff to the Baltic Sea (*thin line*). In addition, the 4-year running mean (*thick line*) and the total mean for the period 1902–1998 (*horizontal line*) are shown. *Middle panel:* 4-year running mean reconstructed (*solid*) and observed (*dashed*) zonal wind speed anomalies at Landsort. *Lower panel:* 4-year running mean uncorrected salinity in a reference simulation (*solid*), in a simulation with climatological monthly mean river runoff and precipitation of 1902–1970 (*dashed*), and in a simulation with climatological monthly mean river runoff and precipitation of 1902–1970 and with 4-year high-pass-filtered sea level pressure and associated surface wind (*dotted*). The shaded ranges indicate periods with positively anomalous 4-year running mean runoff, which are related to stagnation phases (from Meier and Kauker 2003)

half of the decadal variability of the average salinity of the Baltic is related to the accumulated freshwater inflow (Fig. 2.53). Thus, the freshwater inflow plays a dominant role for the average salinity in the Baltic. This model result is supported by the analysis of long records of observations (Samuelsson 1996; Schinke and Matthäus 1998; Matthäus and Schinke 1999; Winsor et al. 2001).

Another significant part of the decadal variability of salinity is caused by the low-frequency variability of the zonal wind (Meier and Kauker 2003). The wind stress anomaly is balanced by a sea-level slope anomaly between the Kattegat and the central Baltic Sea. Consequently, an anomalous barotropic pressure gradient hampers saltwater inflows through the Danish straits (Schrum 2001).

Again, the model results are supported by observations (Zorita and Laine 2000). The analysis by Zorita and Laine (2000) of the link between the large-scale atmospheric circulation and annual salinity revealed that roughly one half of the salinity variability is correlated to the meridional atmospheric pressure gradient over the North Atlantic, and thus to the strength of the westerly zonal winds. Lass and Matthäus (1996) found an anomalous west wind component at the station Kap Arkona between August and October for seasons without major Baltic Sea inflow compared to the corresponding seasons with major Baltic Sea inflow during 1951–1990. They suggested that in years without major Baltic Sea inflow the prevailing easterly winds empying the Baltic Sea prior to the main inflow event are reduced It is assumed that the remaining decadal variability shown in Fig. 2.53 is caused by such high-frequency wind fluctuations (Meier and Kauker 2003).

The impact of other external factors on the variability of salinity has been studied as well. A change of the seasonality of the freshwater inflow caused by river regulation has no significant impact on the variability of the average salinity (Meier and Kauker 2003), because the typical response time scale of about 20–30 years (Winsor et al. 2001; Stigebrandt and Gustafsson 2003; Döös et al. 2004; Meier 2005, 2006) is much larger than one year. A rigid ice lid covering the entire Baltic Sea surface area has only a small impact on saltwater inflows according to Meier and Kauker (2003). Neither do decadal variations of the sea level in the Kattegat have an impact.

However, the high-frequency variability of the sea level in the Kattegat is very important for the long-term behaviour of the average salinity. For instance, Stigebrandt (1983) found that approximately one half of the salt transport into the Baltic Sea is carried out by the dispersive mode associated with barotropic fluctuations.

Another important mechanism determining the length of low-salinity phases might be vertical diffusion (Stigebrandt 2003). Axell (1998) showed that variations in the energy supply to mixing are closely related to wind speed variations and found substantial variations in the vertical diffusion. However, long records are not yet available.

2.3.2 Sea Level

2.3.2.1 *Main Factors Affecting the Mean Sea Level*

Investigations of sea level variations have a long tradition in all countries at the Baltic Sea. The findings of various authors (e.g. Stenij and Hela 1947; Hela 1948, 1950; Lisitzin 1957; Lazarenko 1961; Lisitzin 1966a,b, 1974; Krauss 1974; Kalas 1993; Stigge 1993; Ekman 1994; Samuelsson and Stigebrandt 1996; Carlsson 1998a,b; Ekman 1999; Plag and Tsimplis 1999; Johansson et al. 2001; Andersson 2002; Baerens 2003; Ekman 2003; Fenger et al. 2001; Johansson et al. 2004; Dailidien et al. 2005; Raudsepp 1998; Suursaar et al. 2002, 2006) help us to understand the reasons for similarities as well as differences in the sea level fluctuations between various coastal regions of the Baltic Sea.

Sea level is dependent on a series of factors. The long-term trend in sea level observation records from most locations in the Baltic Sea is dominated by the composite of isostatic change due to post-glacial rebound and eustatic (and steric) global, or at least North Atlantic, sea level change. It is rather difficult to separate between the two effects, but using satellite based positioning, GPS (Global Positioning System), yields promising results for the isostatic uplift by direct measurements (Johansson et al. 2002). However, the time series are still rather short and the accuracy is still of the same order of magnitude as the eustatic sea level change, i.e. 1–$2\,mm\,yr^{-1}$. Other methods use various geological shoreline displacement data, for example, comparing the difference in shoreline displacement around lakes (Påsse 1998). The isostatic uplift varies from approximately zero in the southern Baltic Sea to a maximum uplift in Bothnian Bay of about $10\,mm\,yr^{-1}$ (Johansson et al. 2002).

Thus, the spatial variations of long-term sea level change are mainly due to post-glacial rebound of the Scandinavian land plate and the eustatic sea level rise. The latter is larger only on the southern Baltic Sea coast, which sinks more slowly (Lisitzin 1957; Lisitzin 1966b; Ekman 1996; Johansson et al. 2004). Consequently, the net sea level rise was estimated to be about $1.7\,mm$ per year in the southeastern Baltic Sea while it reverses to $-9.4\,mm$ per year in the northwestern Gulf of Bothnia (Vermeer et al. 1988). One of the most significant parts of the above-mentioned studies was the elimination of meteorological ef-

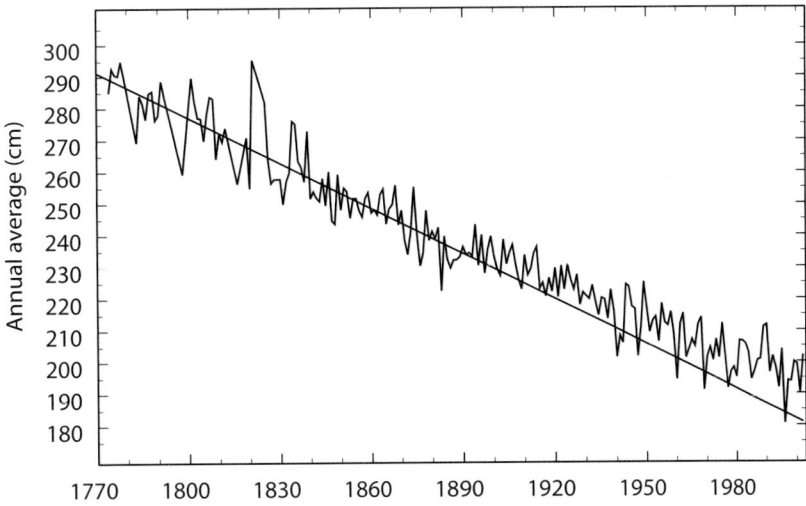

Fig. 2.54. The sea level in Stockholm 1774–2002. The linear trend is computed for 1774–1884 and extrapolated to 2002 (reproduced from Ekman 1999, recomputed and extended)

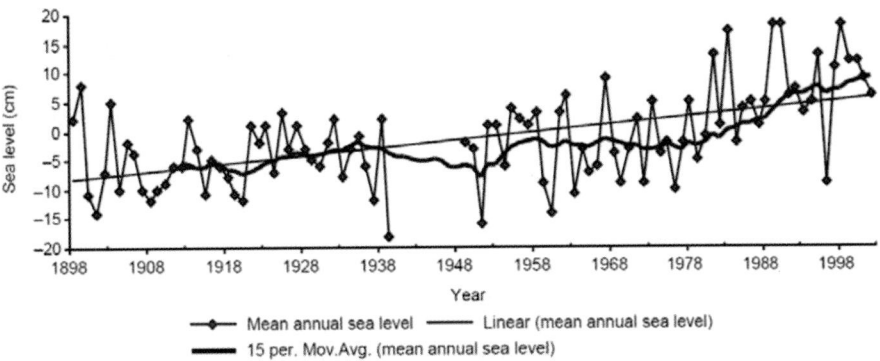

Fig. 2.55. The annual mean water level and the 15-year moving averages in Klaipeda (Dailidiene et al. 2004)

fects on sea level. The effect of meteorological factors on mean sea level was studied by Lisitzin (1966b) and Carlsson (1998a,b).

2.3.2.2 *Changes in the Sea-level from the 1800s to Today*

The eustatic increase in sea level was found to accelerate at many Baltic Sea tide gauge stations at the end of the 20th century (Kalas 1993; Stigge 1993; Kont et al. 1997; Johansson et al. 2001; Fenger et al. 2001; Ekman 2003; Johansson et al. 2004; Suursaar et al. 2006). An illustrative example is the sea level time series from Stockholm shown in Fig. 2.54. There is a long-term decrease in sea level as discussed above, but it is also evident that the sinking rate is lower during

the past 100 years than before. As postglacial rebound does not change on these short time-scales, the reason must be of oceanographic nature. This was noted by Ekman (1999), who attributed the change to an increased eustatic sea level rise of about 1 mm/yr in the 20th century compared to the 19th century. It seems from the figure that an even greater eustatic sea level rise has occurred during the past 20–30 years.

This recent acceleration has been remarked upon in several recent investigations. Dailidiene et al. (2004) showed that sea level rose in the Klaipeda Strait region in toto by about 13.6 cm in 100 years (1898–2001, see Fig. 2.55). However, the rate of sea level rise (1.36 mm year^{-1}) in the Lithuanian region during this period is similar to the rate of global mean sea level rise, which is

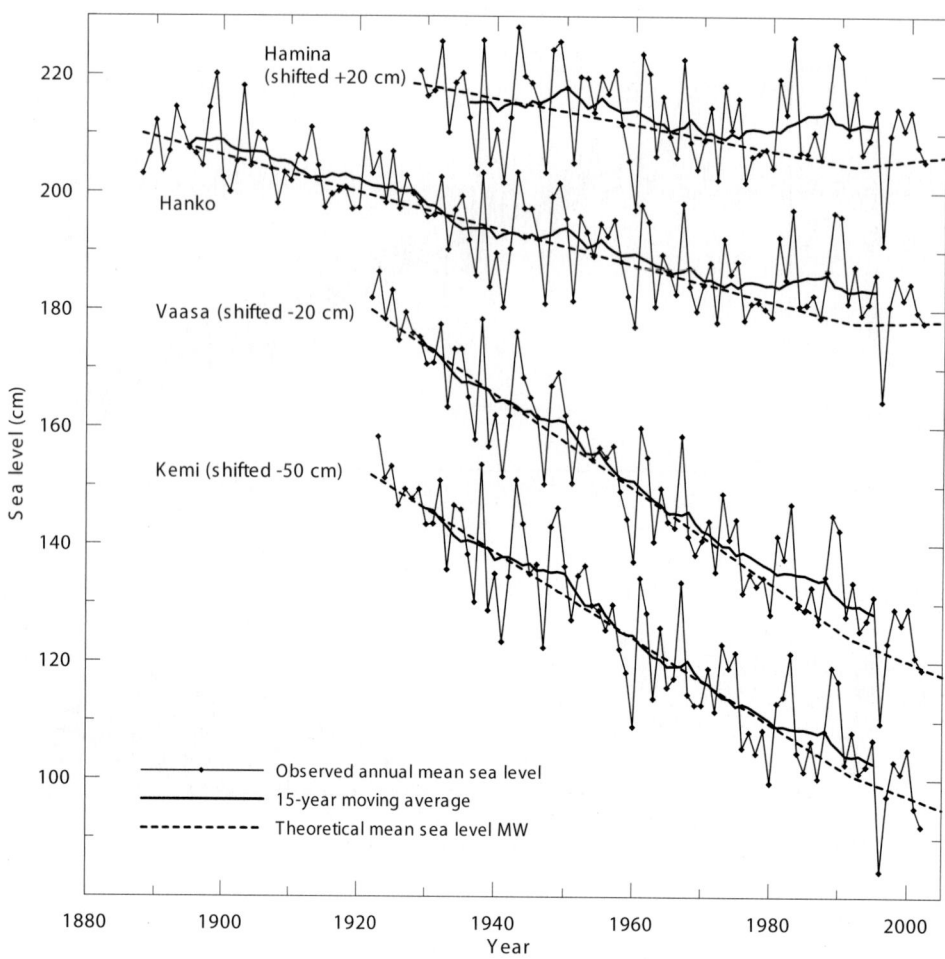

Fig. 2.56. Observed annual mean sea level and the 15-year moving averages at selected Finnish tide gauges (from Johansson et al. 2004). Hamina and Hanko are located at the eastern and western, respectively, Gulf of Finland, Vaasa and Kemi at the southern and northern Gulf of Bothnia, respectively

+1 mm year^{-1} (Basalykas 1985; Raudsepp et al. 1999) and corresponds to the eustatic sea level rise (1.3 mm year^{-1}) around the North Sea during the last century proposed by Christiansen et al. (2001). Suursaar et al. (2006) have studied sea level changes at the Estonian coast for the period 1924–2003. They showed that a faster increase in sea level has taken place during the last decades. They found an eustatic sea level rise varying between 9.5 and 15.4 cm depending on location during the period 1950–2002; i.e. 1.8–3.0 mm year^{-1}, which is higher than global sea level rise, including certainly a local sea-level rise component. The entire increase has occurred during the winter half of the year, from November to March. No changes have been observed during the other months.

The increase in eustatic sea level rise make net sea level rise in parts of the eastern and southern

Baltic Sea (Johansson et al., 2004). For example, land uplift and eustatic sea-level rise more or less balance each other in the Gulf of Finland (see stations Hamina and Hanko in Fig. 2.56) in contrast to the Gulf of Bothnia in the Northern Baltic Sea (stations Vaasa and Kemi in Fig. 2.56).

Long time series have been analysed and various aspects of the changes in sea level have been found and listed, e.g. by Johansson et al. 2001. Ekman and Stigebrandt (1990) studied the time series of Stockholm for the period 1825–1984 and found a statistically significant increase in the amplitude of the annual variation. Ekman (1996, 1997) found that the normal seasonal variation, with a minimum in the spring and a maximum in the autumn, was replaced by a variation with a pronounced maximum for high water years and minimum for low water years, both occurring dur-

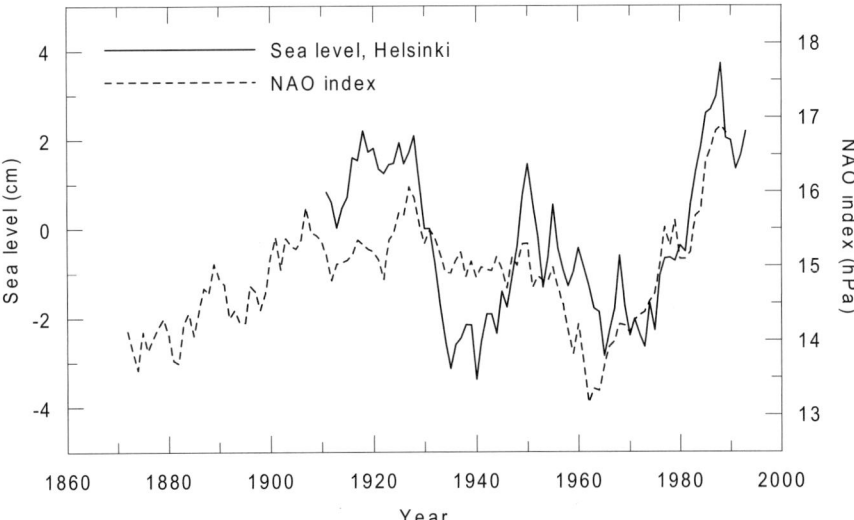

Fig. 2.57. De-trended annual mean sea level at Helsinki and the annual mean NAO index, 15-year running averages (from Johansson et al. 2001)

ing the winter. Ekman (1998) considered the seasonal sea level variation in wintertime using Stockholm monthly sea level data, and found some statistically significant secular changes, these being due to the changes in wind conditions over the transition area between the Baltic Sea and the North Sea. Johansson et al. (2001) have shown that the maxima of sea levels have increased significantly. The increase is proportionally largest in the nodal area of the specific uninodal oscillations of the Baltic Sea. It is possible that the changes in maxima are due to changes in large-scale meteorological and hydrological factors. However, no essential trends were found for minima.

2.3.2.3 *Influence of Atmospheric Circulation*

Andersson (2002) and Ekman (2003) used the historical sea level time series from Stockholm to demonstrate the key role played by winter climate, especially that of wind forcing. Andersson (2002) pointed out that there exists a relation between the Stockholm sea level and the North Atlantic Oscillation (NAO) winter values. Johansson et al. (2001, 2004) studied the Baltic Sea water balance and its correlation with the NAO index and confirmed that the sea level variability along the Finnish coast correlated with the NAO index (Fig. 2.57).

A close correlation between atmospheric circulation and sea level fluctuations was also found at the Estonian coast (Suursaar et al. 2006). Higher

intensity of westerly wind indices (NAO and AO indices, frequency of the zonal circulation) is related to higher sea level. The closest correlation – above 0.6 – was detected during the winter months. A similar correlation was revealed between monthly mean sea level and number of storm days (Suursaar et al. 2006).

Andersson (2002) and Janssen (2002) analysed sea level variations in the Baltic Sea with respect to correlations to NAOWI (NAO Winter Index). Andersson (2002) considered the period 1825–1997 while Janssen (2002) confined his investigation to 1890–1993. They both found variable correlations with time with a maximum at the end of the series, which corresponds well to an increased potential of NAOWI for the regional winter situation in recent decades.

2.3.3 Sea Ice

The ice climate in the Baltic Sea can be characterised by several variables, including the extent and thickness of the ice cover and the duration of the ice season. All these are important for winter navigation and travel on the ice, and for centuries there has been interest in monitoring these variables (e.g. Sass 1866). The historical record of the ice break-up dates has also been used to reconstruct the climate for the last 500 years (Tarand and Nordli 2001). The advance in sea ice research during the last ten years is summarised in Vihma and Haapala (2005) and Omstedt et al. (2004a).

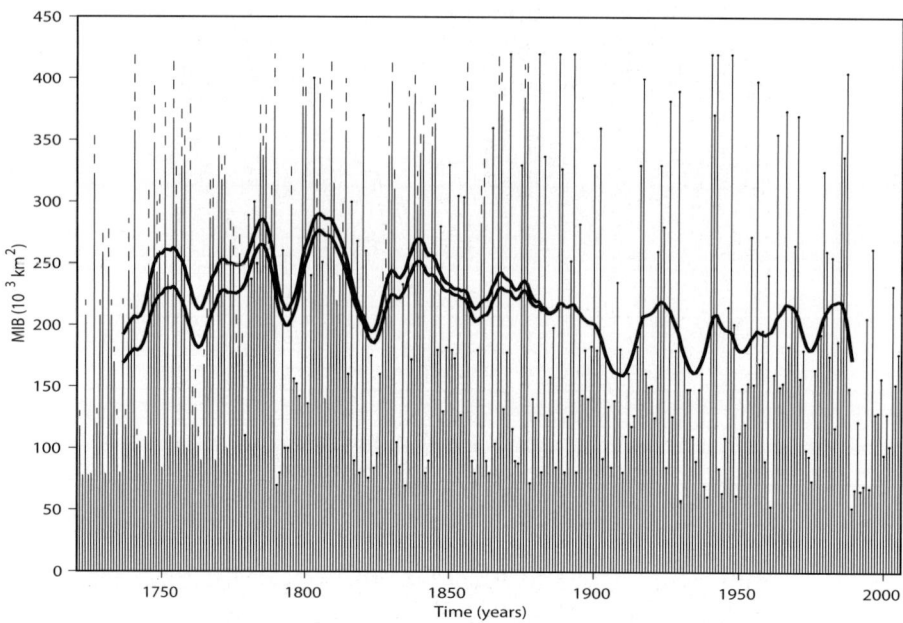

Fig. 2.58. The maximum extent of ice cover in the Baltic Sea 1720–1995. The dashed bars represent the error range of the early estimates. The 30 year moving average is indicated by two lines representing the error range early in the series, converging into one line when high quality data is available (from Seinä and Palosuo 1996)

2.3.3.1 *Ice Extent*

Seinä (1994) and Seinä and Palosuo (1996) have summarised the annual maximum ice extent in the Baltic Sea utilising the material of the Finnish operational ice service from the winters of 1941–1995 and information collected by Prof. Jurva from the winters of 1720–1940. The latter originated from various sources, including observations at lighthouses, old newspapers, records on travel on the ice, scientific articles (Speerschneider 1915, 1927), and air temperature data from Stockholm and Helsinki. Jurva himself never published the whole time series. In his last paper, Jurva (1952) showed the estimated ice extent from only 1830 onwards and commented on the accuracy of the data as follows: "I have tried to determine and estimate the general course of freezing and its different phases, e.g. in cold and correspondingly ice-rich winters from the winter 1829/30 to the eighties of the 19th century, from which period the knowledge of ice conditions in the outer sea is generally lacking. From about the year 1880 onwards we know the extension of the ice cover on the basis of notes made on board ships navigating the middle parts of the Baltic during many winters, or it may be rather easily and sufficiently accurately estimated on the basis of the time analysis of ice winters in

the Archipelago." The time series was published as figures in Palosuo (1953), from where the ice extent has been later digitised by various authors (Lamb 1977; Alenius and Makkonen 1981; Leppäranta and Seinä 1985). In the original figures, the ice extent is illustrated with bar diagrams, and an estimate of the uncertainty of the ice extent is denoted by dashed lines. The uncertainty is largest for severe ice winters, such as 1739/40, when the estimated range was as large as from 350,000 to 420,000 km². In the most commonly used time series (Seinä and Palosuo 1996; their appendices 1 and 2), only the maximum estimates are given. In any case, due to the high correlation between the air temperature and ice extent, even the early data are probably free of drastic errors (Seinä and Palosuo 1996).

The extent of the sea ice cover varies a lot from year to year. Seinä and Palosuo (1996) classified the ice winters so that mild, average and severe winters contain the same percentage ($\sim 33\%$) of the winters in the period of 1720–1995. Mild and severe winters were further classified in extremely mild, mild, severe, and extremely severe ones (Fig. 2.58). The extreme categories both contain $\sim 10\%$ of the winters. In extremely mild winters, only the Bothnian Bay, parts of the Gulf of Finland and Bothnian Sea, and shallow coastal re-

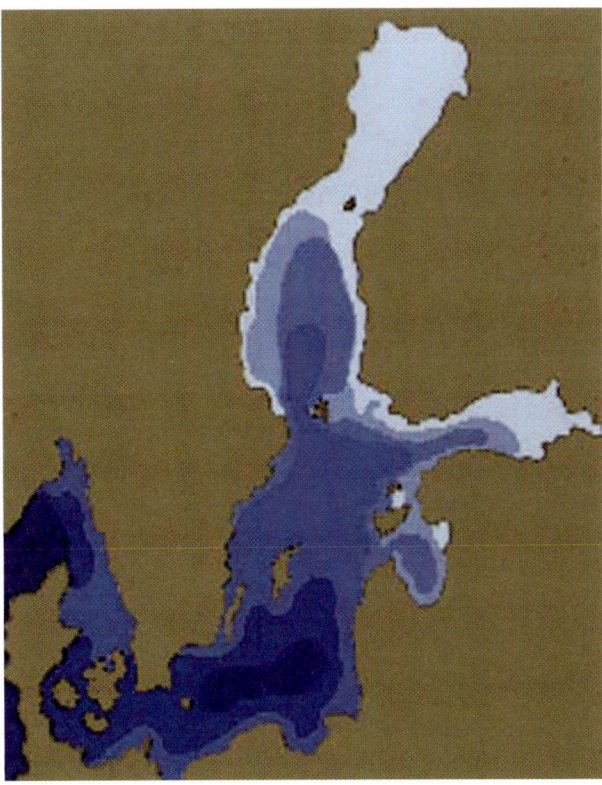

Fig. 2.59. Annual maximum ice extent in the Baltic Sea: in extremely mild winters the maximum ice extent is at least the area marked by the lightest blue and at most that plus the area marked by the next-lightest blue. The maximum ice extent in mild, average, severe, and extremely severe winters is marked analogously with darker colours for the more severe ice winters (redrawn from Seinä and Palosuo 1996, see also Fig. 1.5)

gions in the Gulf of Riga are covered by ice, and the maximum ice coverage is only approximately 12% of the total area of the Baltic Sea (Fig. 2.59).

In average winters, the ice-covered region in March consists of the Gulf of Bothnia, Gulf of Finland, Gulf of Riga, northern parts of the Baltic Proper, and shallow coastal areas further south. In extremely severe ice winters, almost all the Baltic Sea freezes (Fig. 2.59). According to the classification of Seinä and Palosuo (1996), during the last ten years all ice winters have been average, mild, or extremely mild. The latest extremely severe ice winter occurred in 1986–1987, and the latest winters with the Baltic Sea totally frozen have been in 1941–1942 (certainly) and 1946–1947 (most probably; Simojoki 1952). According to Haapala and Leppäranta (1997), the maximum annual ice extent in the Baltic Sea (MIB) did not show clear trends during the 20th century. During 1720 to 2005 we see, however, a decreasing trend (Fig. 2.58), which is further discussed in Sect. 2.3.3.4.

2.3.3.2 Length of the Ice Season

In the Baltic Sea, the first sea ice typically forms in November (at earliest in the beginning of October) in the shallow coastal areas in the northernmost Bothnian Bay. The maximum ice coverage is usually reached in February or March, but sometimes already in January, and sea ice remains in the Bothnian Bay typically until mid-May.

Most analyses on the length of the ice season have been based on local coastal observations. In the southern Baltic Sea, Sztobryn and Stanislawczyk (2002) have found large spatial differences in the sea ice climate along the Polish coast. In general for this region, according to Sztobryn (1994), the length of the ice season has decreased by 1–3 days per decade in the period 1896–1993, see also Girjatowicz and Kożuchowski (1995). Girjatowicz and Kożuchowski (1999) analysed the ice conditions in the region of the Szczecin Lagoon in the period from 1888 to 1995 and found a statistically significant decreasing trend in the duration of the ice season.

In the northern Baltic Sea, Haapala and Leppäranta (1997) analysed ice time series from the Finnish coast and concluded that the length of the ice season shows a decreasing trend during 1889–1994, as also does the probability of annual ice occurrence in Utö (Northern Baltic Proper). Tarand (1993) and Tarand and Nordli (2001) have addressed the break-up dates of sea ice in the port of Tallinn. On the basis of time series over the last 500 years, they concluded that the break-up dates have become earlier since about the mid 19th century and that the changes have been particularly large during the latest decades. This was found also by Jaagus (2006b), who analysed data from nine Estonian stations in the period 1949/50–2003/04. The largest decrease, by more than a month, in the duration of the ice season was observed in the West Estonian Archipelago, while on the southern coast of the Gulf of Finland the decrease has been insignificant. A statistically significant change towards later dates of freezing was detected only at four stations at the west coast of Estonia (Jaagus 2006b).

Jevrejeva (2000) analysed the sea ice and air temperature time series along the Estonian coast in the period of 1900–1990. The results indicated that, at the end of the study period, the date of a stabilised transition of the air temperature to sub-zero values was some 8–14 days later than in the early 1900s, while the onset date of melting air temperatures has become 10–15 days earlier. The number of days with sea ice has decreased by 5–7 days per century in the Gulf of Finland, and by 5–10 days in the Gulf of Riga. These changes have been associated with a climatic warming of 0.5–1.0 °C in Estonia in November–April in the period of 1900–1990; the warming has, however, been statistically significant at 99.9% confidence level only at one of the eight stations analysed by Jevrejeva (2000). Analysing the historical record of ice break-up at the port of Riga in 1529–1990, Jevrejeva (2001) detected a decreasing trend of about 2.0 days per century for the break-up dates for severe winters (statistically significant at the 99.9% level). For mild and average winters, no statistically significant trends were detected.

Considering the whole Baltic Sea, Jevrejeva et al. (2004) did a comprehensive analysis of 20th century time series at 37 coastal stations around the sea. In general, the observations show a tendency towards milder ice conditions, where the largest change is in the length of the ice season, which has decreased by 14–44 days in a century, which, in turn, is largely due to the earlier ice break-up.

2.3.3.3 *Ice Thickness*

Accurate data on the ice thickness is almost entirely restricted to the zone of land-fast ice. In the Bothnian Bay, the level ice thickness is typically 65–80 cm (Alenius et al. 2003), and it reaches 30–50 cm even in mild winters. In the Skagerrak and the coastal areas of Germany and Poland the annual maximum ice thickness varies from 10 to 50 cm (BSH 1994).

In their analysis of 37 time series from the coastal stations around the Baltic Sea, Jevrejeva et al. (2004) did not find any consistent change in the annual maximum ice thickness. According to Haapala and Leppäranta (1997), the level-ice thickness in the Baltic Sea did not show clear trends during the 20th century. Seinä (1993) and Launiainen et al. (2002) reported an increasing trend in the maximum annual ice thickness off Kemi (northernmost Gulf of Bothnia) during the 20th century until the 1980s; in more southerly locations in the Gulf of Bothnia no clear trends were observed for the same period. In all stations, decreasing trends have been observed since the 1980s. In the Gulf of Finland, the maximum annual ice thickness has had a decreasing trend off Helsinki and Loviisa (Alenius et al. 2003). From the point of view of winter navigation, an important climatological parameter is the distance that ships have to cruise in sea ice thicker than some threshold value. Launiainen et al. (2002) calculated the annual maximum distance from the harbour of Hamina (eastern Gulf of Finland) to a zone of sea ice less than 10 cm thick. The results for the period from 1951 to 2000 strongly depended on the air temperature, with short distances in 1990s.

In the drift ice regions, where most of the sea-ice mass locates, we do not have accurate data on the ice thickness. During the last ten years, many field studies have concentrated on the mapping of the ice thickness, but no systematic long-term measurements of the ice thickness have been carried out.

2.3.3.4 *Large-scale Atmospheric Forcing on the Ice Conditions*

During the last decade, increasing attention has been paid to the relationship between the inter-annual variations in the Baltic Sea ice conditions

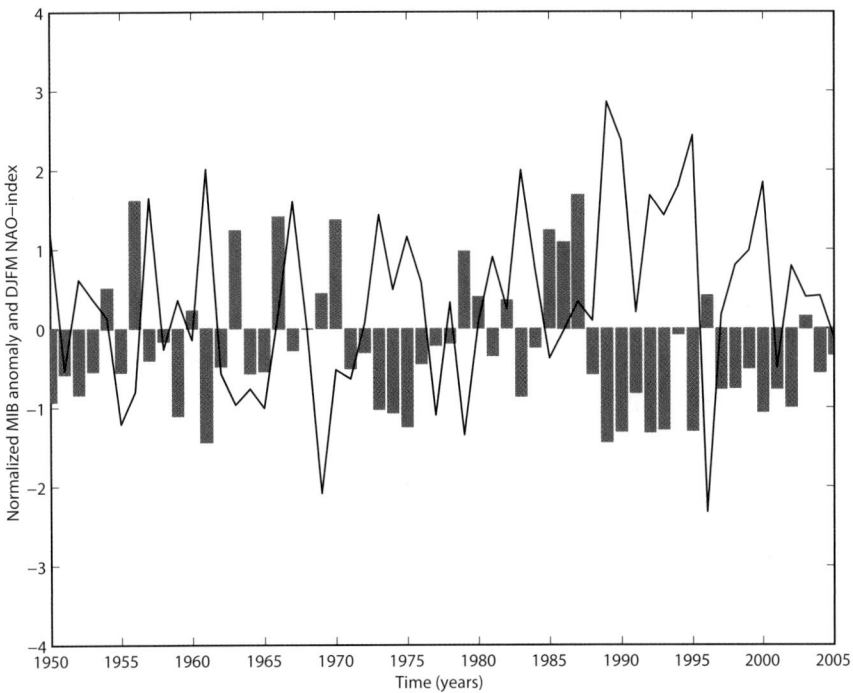

Fig. 2.60. Time series of North Atlantic Oscillation index (Rogers 1984) (*line*) and the annual maximum ice extent of the Baltic Sea, MIB (Seinä and Palosuo 1996) (*bars*). The MIB is presented as an anomaly of the normalised time series

and the indices of the North Atlantic Oscillation (NAO) and Arctic Oscillation (AO). Read more on NAO and AO in Annex 7. The relationship between the large-scale atmospheric circulation and the Baltic Sea ice conditions can be illustrated by comparing the NAO winter index and the MIB (Fig. 2.60). Although the MIB is fairly well correlated with the winter NAO index (Fig. 2.60), moving correlation analyses have demonstrated that the relationship is not constant in time (Omstedt and Chen 2001; Janssen 2002; Meier and Kauker 2004; Schrum and Janssen 2002; Chen and Li 2004). The 31-year moving correlation is high during the 1940s and 1980s but quite low during the 1920s (Meier and Kauker 2002). Changes in the NAO-MIB relationship are explained by changes in the location of the atmospheric pressure patterns (Koslowski and Loewe 1994; Kauker and Meier 2003; Chen and Li 2004). Kauker and Meier (2003) found that 87% of the observed air temperature variance at Stockholm (and consequently most of the MIB variance) is explained by NAO and a sea level pressure pattern similar to the Barents Sea Oscillation, which describes shifts of the Icelandic low pressure center toward the Barents Sea (Skeie 2000).

According to Yoo and D'Odorico (2002), NAO seems to affect mostly the late-winter temperature (January–March), with a significant impact also on the mid-spring (April–May) period, when the air temperature is strongly correlated to the ice break-up dates. The results of Jevrejeva and Moore (2001) suggest that, from the point of view of ice break-up in the Baltic Sea, AO is an index somewhat more essential than NAO: time series of ice break-up date reflect variations in the winter AO index in the 13.9-year period, but not in the NAO 7.8-year period. Calculating cross-wavelet power for the time series, Jevrejeva et al. (2003) found out that the times of largest variance in the Baltic Sea ice conditions were in excellent agreement with significant power in the AO at 2.2–3.5, 5.7–7.8, and 12–20 year periods (previously Alenius and Makkonen (1981) had detected the most distinct cycles in the MIB at the periods of 3.5, 5.2, 8, and 13 years).

It is noteworthy that Jevrejeva et al. (2003) found similar patterns also with the Southern Oscillation Index and El Niño sea surface temperature series (Nino3). Also, according to Omstedt et al. (2004b), 90% of the variance of the time series is for the time scales shorter than 15 years. A concern of these studies is, however, that they have assumed the long-term time series of the MIB as homogeneous in accuracy (see Sect. 2.3.3.1).

Jaagus (2006b) analysed the freezing and break-up dates near the Estonian coast in relation to large-scale atmospheric circulation. NAO indices, the AO index, several teleconnection indices and frequencies of the circulation forms were used to describe circulation conditions. Generally, no correlation was found between the circulation and the date of the first appearance of sea ice. The circulation has the strongest relationship with the date of ice break-up and the length of the ice season. They have a high negative correlation with the characteristics of the intensity of the zonal circulation. The highest value is present in the case of the sea ice break-up and the AO index for the total winter period (December–March): -0.73 as a mean of nine stations.

According to Jaagus (2006b), February is the key month when the circulation plays a main role in determining sea ice conditions for the spring. High intensity of westerlies in winter (February) causes earlier ice melting in the Baltic Sea. Using the conditional Mann-Kendall test, Jaagus (2006b) demonstrated that the significant trends in sea ice near the Estonian coast during 1949/50–2003/04 are caused by the increasing intensity of westerlies in winter, especially in February, and by the corresponding decrease in frequency of meridional circulation types during the same time interval.

From the point of view of climatology, the most relevant parameter for describing the ice conditions is the total mass of ice, but we lack good data on it over large regions. Over a small region, such as the coastal areas of Schleswig-Holstein (Germany), the ice thickness and concentration can be observed sufficiently accurately. Koslowski and Loewe (1994) calculated the areal ice volume, which can be visualised as an equivalent ice thickness in a partly ice-covered sea. Analogously to degree days (accumulated degrees of frost), they further calculated the accumulated areal ice volume, and showed that in the period from 1879 to 1992 it was negatively correlated with the NAO winter index. Weak ice winters were related to a NAO winter index exceeding 1, and strong ice winters were related to a NAO winter index less than -1. Koslowski and Loewe (1994) also noteworthily pointed out that, depending on the exact location and size of the high and low pressure areas, in rare cases a strong ice winter in the south-western Baltic Sea can develop even in spite of a high NAO winter index.

On the basis of the data on the accumulated areal ice volume from the 1878–1993 period, Koslowski and Glaser (1995) reconstructed the ice winter severity since 1701 for the southwestern Baltic Sea, and Koslowski and Glaser (1999) extended the calculations for the period from 1501 to 1995. The present-day ice winter regime has lasted since about 1860, while from about 1760 to about 1860 the ice winters were much more severe. Around 1800 the ice production in the southwestern Baltic Sea was three times larger than it is today (Koslowski and Glaser 1999). According to Omstedt and Chen (2001), the shift towards a warmer climate took place in 1877, associated with a period of an increased low-pressure activity (Omstedt et al. 2004b). This shift in climate was identified as the ending of the Little Ice Age in the Baltic Sea region. Omstedt and Chen (2001) also found that a colder climate is associated with higher variability in the ice extent and with a higher sensitivity of the ice extent to changes in winter air temperature. They further developed a statistical model that links the ice extent and a set of atmospheric circulation indices. Considering the possibilities to make climate-scale predictions for the future sea ice conditions, Omstedt and Chen (2001) argue that such a statistical model could be a useful tool in estimating the mean conditions of the ice extent on the basis of atmospheric pressure fields, which are produced applying a climate model. Tinz (1996) found an exponential relation between the ice extent and the air temperature; applying it and climate model predictions, he forecasted a drastic decrease in the Baltic Sea ice cover in the next 100 years.

2.3.3.5 Summary

A climate warming can be detected from the time series of the maximum annual extent of sea ice and the length of the ice season in the Baltic Sea. On the basis of the ice extent, the shift towards a warmer climate took place in the latter half of the 19th century. During the last ten years, all ice winters have been average, mild, or extremely mild. The record of the length of the ice season shows a decreasing trend by 14–44 days in the latest century, the exact number depending on the location around the Baltic Sea. The indices of AO and NAO correlate with the ice extent, the date of ice break-up, and the length of the ice season.

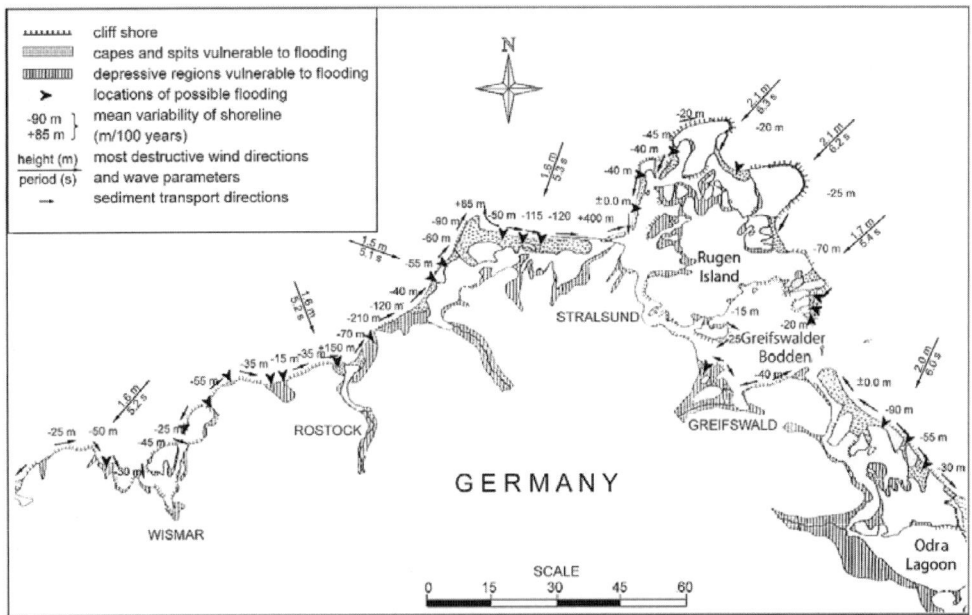

Fig. 2.61. Dynamics of the coast of the southwestern part of the Baltic Sea (after Lampe 1996)

Data on the ice thickness mostly originate from the land-fast ice zone, and basically do not show clear trends during the 20th century, except that during the last 20 years the ice thicknesses have decreased. In the northernmost Bothnian Bay, the ice thickness showed an increasing trend until the 1980s.

2.3.4 Coastal Erosion

The Baltic Sea and its coasts have been subject to continuous, sometimes dramatic evolution due to periodic changes of climate induced by numerous phenomena (Schwarzer et al. 2003), these often on a century scale. Due to specific geological structure and climate conditions, the south Baltic Sea coasts are exposed to much more intensive variations, mainly coastal erosion, than occur in the northern Baltic Sea coastal areas (dominated by rocks). Short segments of the Baltic Sea sandy coast in south Sweden constitute an exception of the above rule. Observations from the last century indicate intensified erosion at many beaches situated on open south Baltic Sea coasts (Lampe 1996; Zawadzka 1999; Eberhards 2003; Orviku et al. 2003; Różyński 2003). Erosive tendencies of coastal change take place from the western part of the South Baltic Sea coast, through the central part, to the easternmost shores. Along the entire, mostly sandy, south Baltic Sea coast, erosive phe-

nomena definitely predominate. Nevertheless, locally, accumulation processes can occur. The latter take place mainly in bays and regions where energy of longshore sediment flux decreases.

2.3.4.1 *Western Part of the South Baltic Sea Coast*

In the western part of the south Baltic Sea, the most long-lasting and visible erosive phenomena occur at the shore segments with cliffs and capes. In the second half of the 20th century, Gudelis and Jemeljanovas (1976) and, earlier, Buelow (1954) pointed out erosion processes along numerous shore segments, mostly at cliffs, in their analyses of the coastal evolution in Denmark and Germany.

In Denmark, cliff erosion amounted on the average to 0.2–0.5 meter per year (m/yr) (locally attaining 2 m/yr). For the German coast, major erosive locations were identified at capes and cliffs in the region of Kiel Bay (on the average 0.3–0.4 m/yr), on the islands of Rügen and Usedom, as well as eastwards of Rostock. Lampe (1996) reported the average value of shoreline retreat for the last 100 years, except for a few places along the German coast, to be smaller than 1 m/yr, see Fig. 2.61. The latest data on the average retreat along Rügen and Usedom Islands show rates of 0.2 m/yr and 0.4 m/yr, respectively, see Pruszak and Okrój (1998) and Schwarzer et al. (2003).

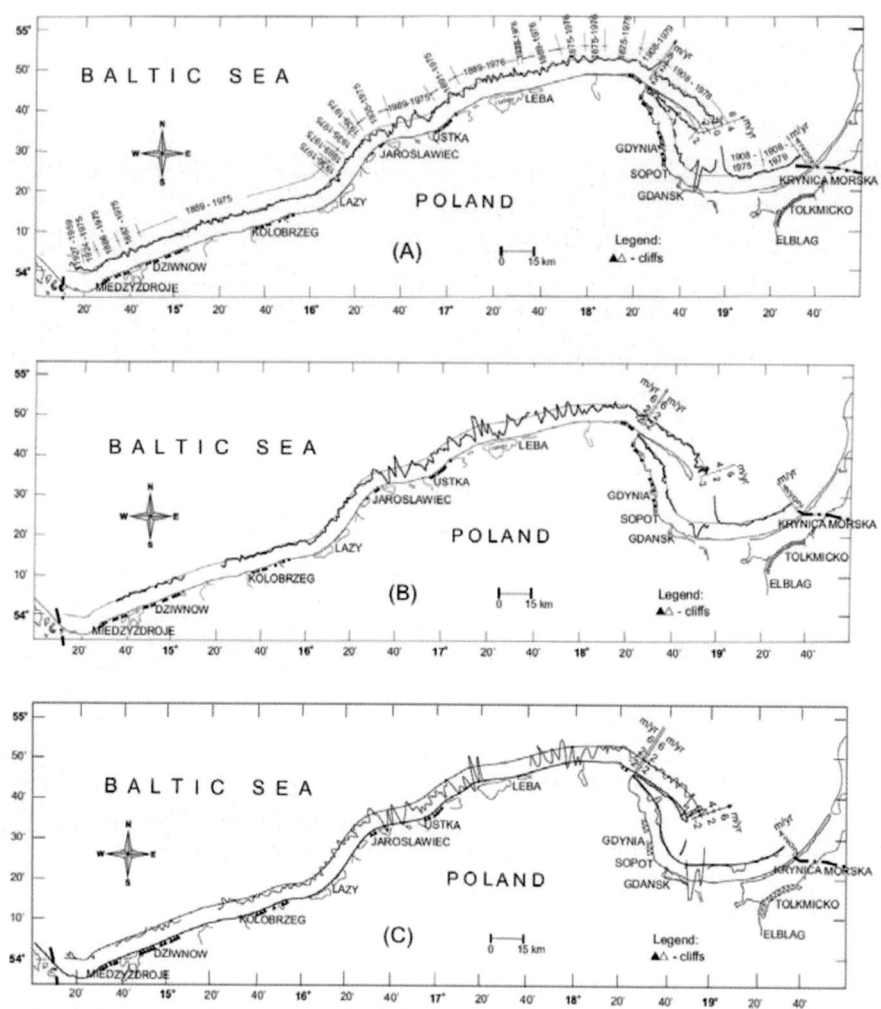

Fig. 2.62. Long-term shoreline changes along the Polish coast (an average rate) during 1875 to 1979 (**A**), 1960 to 1983 (**B**) and 1971 to 1983 (**C**) (after Zawadzka 1999)

2.3.4.2 *Middle Part of the South Baltic Sea Coast*

The behaviour of about 500 km of dunes and cliffs on the Polish coast, comprising a significant central part of the southern Baltic Sea coast, also indicates enhanced and accelerated erosion. The average rate of coastal retreat during the 100 years covering the period between 1875–1979 was 0.12 meter per year (m/yr) (Fig. 2.62A). During the period 1960–1983 (Fig. 2.62B) the average rate of retreat increased to 0.5 m/yr, and between 1971–1983 it reached 0.9 m/yr (Fig. 2.62C). The erosion persisted over an increasing lengths of the coastline: 61% in the 100-year period, 72% in the 24-year period and 74% in the 13-year period. The dune/cliff foot was eroded at a less intensive pace

and in the period 1960–1983, when it retreated at a rate of 0.16 m/yr and in the period 1971–1983, at 0.3 m/yr (Zawadzka 1994). The process of transformation of the coastal zone proceeded with simultaneous erosive and accretive processes. The analysis of changes in the land area showed a decrease in the recovery of eroded beaches and dunes. In the 100-year period, the recovery of eroded land encompassed over 69% of the destroyed area, in the period 1960–1983 it was 20%, and in the period 1971–1983 only 14%. These data clearly document that processes of erosion are intensifying and accelerating, although they are to some extent explained by the difference in the lengths of the time periods in question, where a longer period offers more opportunity for recovery. These

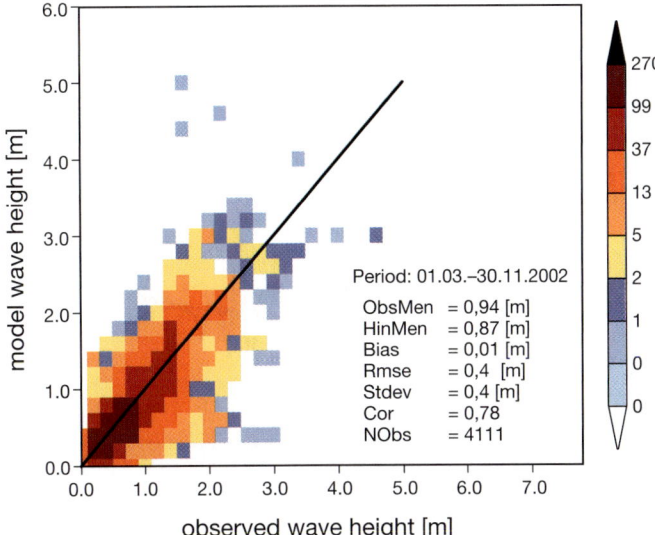

Fig. 2.63. Scatterplot showing observed (*x-axis*) and hindcast (*y-axis*) significant wave height (*in meter*) near Arkona for March to November 2002. Colors indicate the number of observed and model values. The total number of pairs is 4111. Some error statistics are provided in the upper left corner (from Augustin 2005)

erosive processes are now present over about 74% of the length of the Polish part of the Baltic Sea coast.

Since the beginning of the 20th century up to present, coastal defence systems have been continuously installed, which indicates that the beaches are particularly affected by marine erosion. Various types of coastal protection structures are present along 26% of the Polish coast. About 98 km of the coast is under the influence of groynes, while 41 km are protected by light and heavy revetments. The Hel Peninsula is protected along 34% of its length. Coastal protection measures are especially intense in the middle and western part of the open Polish coastline, where coastal defense structures are built along 71 km out of 126 km of the coastline (Pruszak 2001).

2.3.4.3 South-eastern Part of the South Baltic Sea Coast

Similar tendencies of coastal evolution are also observed on other segments of the eastern part of the southern Baltic Sea. In Latvia, practically all cliffs have been subject to various degrees of erosion over the past 50–100 years. In the last 50–60 years, long-term cliff erosion has been 0.5–0.6 m/yr, reaching a maximum of 1–1.5 m/yr along particular stretches of the coast. Since 1980–81, rates of coastal erosion along Latvia's coast have

increased 2–5 fold, reaching 1.5–4 m/yr (Eberhards 2003). Simultaneously, all of Latvia's marine hydrological stations indicate a trend of rising mean sea level during the second half of the last century (see also Sect. 2.3.2). A similar situation is also observed along the Lithuanian coast (Zilinskavas et al. 2001). In recent years a series of very heavy storms completely destroyed sandy beaches in and around Palanga (Lithuania), which were not being eroded previously.

The increasing activity of coastal processes has also been observed along the Estonian coastline during the last decades (Orviku 1992, 1993; Orviku et al. 2003; Raukas et al. 1994). The results of comparing maps from different times as well as results from field measurements in test areas clearly reveal an increased activity of both erosion and accumulation processes (Rivis 2004). Due to such processes, a part of the coastline, aside from migration, is subject to continuous transformation and change of shape. For instance, shore processes during the last century have caused the northwesternmost point of Harilaid Peninsula on Saaremaa (Estonia) to migrate to the northeast and to change its shape.

The decrease in duration or even absence of ice cover in the Baltic Sea in recent decades (Jaagus 2003) has enhanced coastal erosion. The activity of shore processes has increased significantly in Estonia since the 1980s. In the last 20 years,

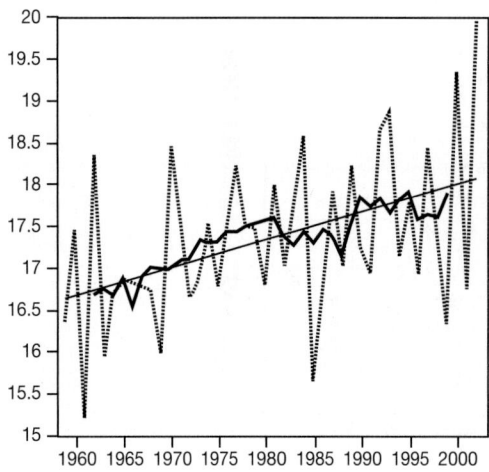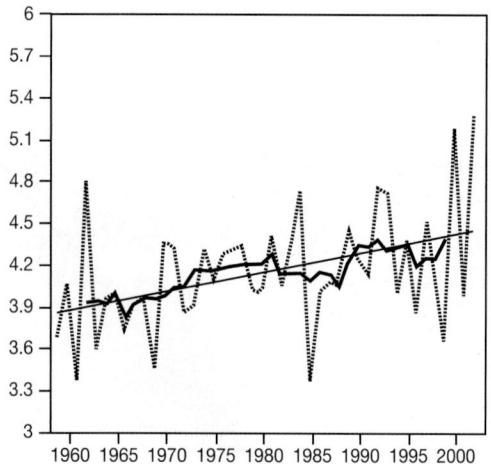

Fig. 2.64. Hindcast annual 99 percentile (*dashed*) wind speed (*left*) and significant wave height (*right*) for a representative model grid point in the Baltic proper (58° N, 20° E). The solid lines represent the 9-year running mean and the liner trend 1958–2002 (from Augustin 2005)

some beach segments eroded with speeds of over 1.5 m/yr (for instance Cape Kiipsaare in Harilaid area, Orviku et al. 2003). The most intensive erosion (Jaerve beach) resulted from extremely strong storms in 1990 and 1999. Intensification of erosion along the eastern coast of the Baltic Sea is associated with global climate change and, particularly, with the increasing trend in annual (winter) storminess over the second half of the last century, simultaneously with less days of ice cover in winter (Orviku et al. 2003).

2.3.4.4 Summary

In various regions of the southern Baltic Sea, coastal damages generally result from a combination of an increased number of strong storms, accelerated sea-level rise, and a decreasing trend of the presence of ice cover in the winter, when the most intensive storms occur. Currently, most of these factors in the Baltic Sea coastal region are strengthened and act simultaneously and jointly, which contributes considerably to the aggravated erosion in the area.

Application of the MSSA method (Multichannel Singular Spectrum Analysis) for long term surveys of shoreline position (evolution) located at the Coastal Research Station at Lubiatowo (Polish coast) show the existence of 8–10 year cycles of changes (Pruszak and Różyński 2001; Różyński 2005). A quasi-rhythmic behavior of shoreline positions with periods of 8–10 years may reflect ef-

fects of large-scale, long-term climatic oscillations generated in the North Atlantic, generally referred to as the North Atlantic Oscillation (NAO). In these periods increased beach erosion is observed. Analogous trends of periodic changes can also encompass longshore bar systems, which was highlighted by Różyński (2003). Analysing long-term field data with a set of data intensive statistical methods, he found periods of intensive shoreline bar dynamics that can match the periodicity of the NAO.

2.3.5 Wind Waves

For any sea area, the climate of wind waves depends primarily on the prevailing wind conditions (speed, fetch and duration) in that area and the given topographic features such as water depth. Any long-term change in the wind climate or the topography (the latter caused for instance by coastal erosion or construction works) will therefore be associated with a corresponding change in the wind wave climate. Away from the coast and with the exception of very long waves, changes in the local wind climate usually represent the most important factor for changes in the wave climate.

While there have been numerous studies on changing wind and wave climate on global and regional scales, to our knowledge the number of such studies for the Baltic Sea remains limited. This may be due to the rather limited and/or relatively short time series of wave measurements

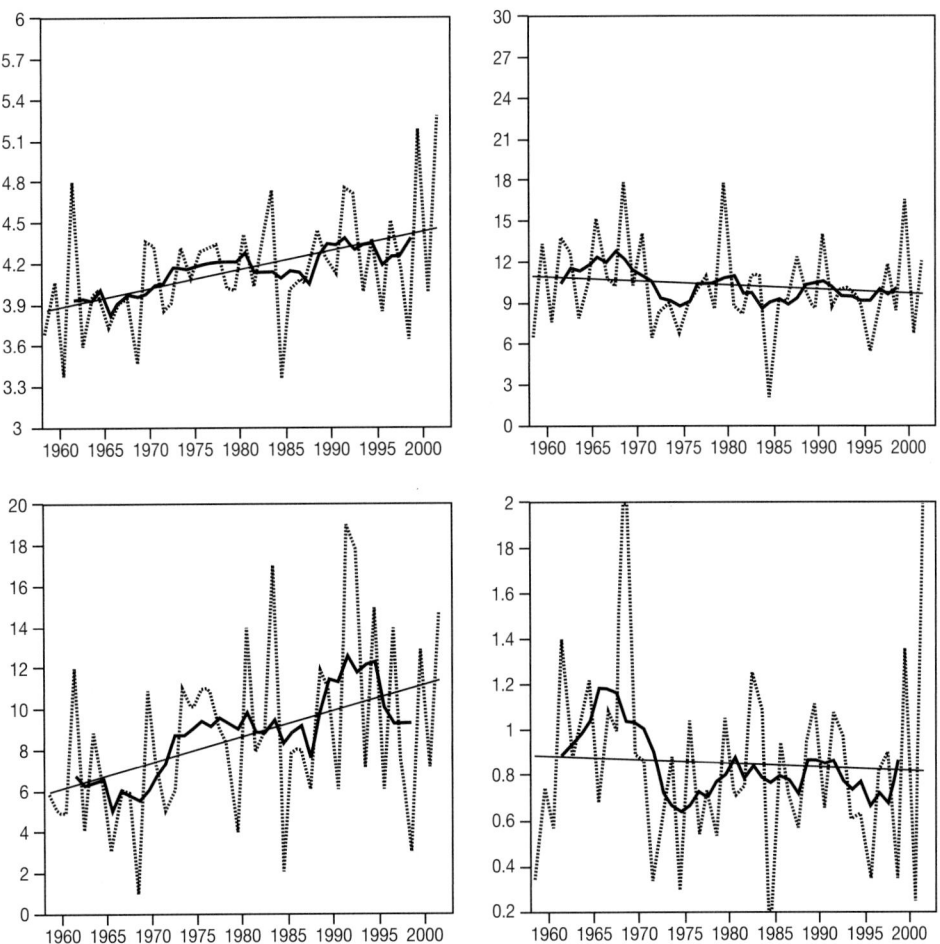

Fig. 2.65. Statistics (*dashed*) of severe wave events at a representative model grid point (58° N, 20° E) in the Baltic proper. *Upper left:* Hindcast annual 99 percentile significant wave height in meters. *Lower left:* The number of events per year that exceed the 45-year average 99 percentile significant wave height. *Upper right:* The yearly average duration of events in hours that exceeds the long-term average 99 percentile significant wave height. *Lower right:* The yearly averaged intensity, i.e. the difference between the maximum wave height of an event exceeding the long-term average 99 percentile significant wave height in meters and the long-term average 99 percentile significant wave height. The solid lines represent the 9-year running mean and the liner trend 1958–2002 (from Augustin 2005)

in the Baltic Sea available so far, which hampers any reliable analysis of long-term changes. While it may be speculated that long-term changes in the Baltic Sea wave climate may have been similar to observed long-term changes in storminess (Sect. 2.1.5.4), there are no long-term measurements that would directly allow for a test of this hypothesis.

An indirect and somewhat limited alternative is provided by the analysis of wave hindcasts. To our knowledge the longest wave hindcast available for the Baltic Sea is a recently completed simulation with the wave model WAM (WAMDI 1988) at

about 5×5 km horizontal resolution for the period 1958–2002. The wave climate of this hindcast and its long-term changes are described in Augustin (2005).

As the quality of any wave model simulation depends to a large extent on the quality of the driving wind fields, some efforts were made to obtain reasonable wind fields. The technique is based on a downscaling of global re-analysis data and is described in von Storch et al. (2000) and Feser et al. (2001). A comparison of observed and downscaled wind fields can be found in Weisse et al. (2005, 2003). A comparison between the observed storm indices from Bärring and von Storch (2004) and those derived from the downscaled wind fields used to drive the wave hindcast for the Baltic Sea can be found in Sect. 2.1.5.4 (Figs. 2.27 and 2.30). In general, a good agreement can be inferred. In particular, the interannual and long-term variability appears to be reasonably captured while there remain some biases for some of the indices. Discrepancies appear to be larger at the beginning of the hindcast period.

Figure 2.63 shows a comparison between observed and hindcast wave height near Arkona, Germany (54° 43.0 N, 013° 44.5 E) for the somewhat brief period March to November 2002. Although there is considerable scatter between model and observations, the hindcast appears to be calibrated, i.e. there is no systematic bias. In particular, the frequency of over- and underestimating severe events seems to be similar, although this cannot be concluded with confidence from the limited material available.

As an example for the hindcast changes in the wind and wave climate in the Baltic proper, Fig. 2.64 shows an analysis of the annual 99 percentiles of wind speed and significant wave height. Both time series show remarkable short-term fluctuations and, considered over the entire 1958–2002 period, display increases of about 1 m/s and 0.3 m in the 99 percentile wind speed and significant wave height, respectively. When compared to the storm indices of Alexandersson et al. (2000) and Bärring and von Storch (2004), this behavior is in agreement with the observed tendency toward more storminess, and the linear increase of significant wave height over the hindcast period may possibly reflect the upward branch of long-term fluctuations in observed storminess.

The increase of hindcast annual 99 percentiles of significant wave height may be caused by more frequent, more severe, and/or longer lasting extreme events. To identify how the characteristics of extreme wave events have changed, Augustin (2005) computed the annual number of extreme events, their average duration and mean intensity relative to a given threshold (Fig. 2.65). It can be inferred that for the Baltic proper the changes are primarily caused by an increase in the frequency of severe wave events while their duration and intensity revealed no significant trend.

Summarising, the analysis of existing long-term atmosphere observations and wave hindcasts for the Baltic Sea suggests that the storm and wave climate in the Baltic Sea has undergone considerable variations in the past. Although wave hindcast results show an increase in the frequency of severe sea states for the period 1958–2002, the analysis of long-term atmospheric data suggests that this increase is within the natural variability of the records and may reflect the upward branch of observed long-term variations in Baltic Sea storminess.

2.4 Summary of Observed Climate Changes

The variability in **atmospheric circulation** has a strong influence on the surface climate in northern Europe (temperature, precipitation, wind speed, etc.). From about the 1960s until the 1990s westerly air flow has intensified during wintertime. This increased frequency of maritime air masses has contributed to higher wintertime temperatures and enhanced precipitation in regions exposed to westerly winds, especially during the 1990s. On a hundred year timescale relationships between large-scale atmospheric circulation and surface climate elements show strong temporal variability.

During the period 1871–2004 there were significant positive trends in the annual **mean temperature** for the northern and southern Baltic Sea Basin, being 0.10 °C/decade on average to the north of 60° N and 0.07 °C/decade to the south of 60° N. These trends are larger than the trend for the entire globe, which amounts to 0.05 °C/decade (1861–2000). The annual mean temperatures show an early 20th century warming that culminated in the 1930s. This was followed by a slighter cooling that finished in the 1960s, followed by another strong warming until present. Warming is characterised by a pattern where mean daily minimum temperatures have increased more than mean daily maximum temperatures.

Spring is the season showing the most linear and strongest warming, whereas wintertime temperature increase is irregular but larger than in summer and autumn. A general tendency is that the start of the climatic seasons in the spring half-year (e.g. spring, growing season, summer) start earlier, whereas the climatic seasons in the autumn half-year (e.g. autumn, frost season, winter) start later. Changes in **extreme temperatures** have broadly followed changes in mean temperatures. The number of cold nights has decreased, while the number of warm days has increased. These changes have been stronger during winter than during summer.

Over the latter part of the 20th century, on average, northern Europe has become wetter. The increase in **precipitation** is not spatially uniform. Within the Baltic Sea Basin the largest increases have occurred in Sweden and on the eastern coast of the Baltic Sea. Seasonally, the largest increases have occurred in winter and spring. Changes in summer are characterised by increases in the northern and decreases in the southern parts of the Baltic Sea Basin. In wintertime, there is an indication that the number of heavy precipitation events has increased.

Characteristics of **cloudiness and solar radiation** have remarkable inter-annual and inter-decadal variations in the Baltic Sea Basin. A decrease in cloudiness and increase in sunshine duration was observed in the south (Poland), while opposite trends were revealed in the north (Estonia). In the 1990s, all these trends changed their sign. Long-term observations in Estonia show that an improvement in air quality (i.e. a decrease in the aerosol emissions to the atmosphere) reversed decreasing trends in atmospheric transparency and direct radiation during the 1990s. At present, the atmospheric transparency is at the same level as in the 1930s.

Hundred-year time series from southern Scandinavia reveal that there is no long-term trend in **storminess** indices. There was a temporary increase of storm activity in the 1980s–1990s. In the Baltic Sea Basin on the whole, different data sources give slightly different results with respect to trends and variations in the extreme wind climate, especially concerning small-scale extreme winds. At the same time, there are indications of an increasing impact of extreme wind events. However, this increasing impact results from a complex interaction between climate and development trends that increase the exposure to damage and/or the vulnerability of nature and society.

The inter-annual variability in **water inflow** (river runoff to the Baltic Sea) is considerable, but no statistically significant trend can be found in the annual time series for the period of 1921–1998.

The analysis of the long-term dynamics of the **dates of the start and ending of ice events and duration of the ice coverage** for rivers of the Russian territory of the Baltic Sea Basin showed a stable positive trend from the middle of the 20th century to its end.

As to the maximal **ice cover thickness**, a negative trend has been analysed for all Polish and Russian study lakes. In Finland, both decreasing and increasing trends can be found in the maximum ice thickness time series.

A recent decrease in **snow cover duration and water equivalent** has been observed in the southern parts of all the Fennoscandian countries, while the opposite trend prevails in the north. Changes of **snow depth** are quite similar, i.e. decrease in south-west regions and increase in the north-east regions.

The Baltic Sea **mean salinity** during the 20th century decreased during the 1980s and 1990s, but similar decreases also appeared earlier in the 20th century. No long-term trend was found during the 20th century.

There are indications of a more rapid eustatic sea level rise in the 20th century than in the 19th century.

A climate warming can be detected from the time series of the maximum **annual extent of sea ice and the length of the ice season** in the Baltic Sea. On the basis of the ice extent, the shift towards a warmer climate took place in the latter half of the 19th century. This gradual shift has been identified as the ending of the Little Ice Age in the Baltic Sea Basin.

Coastal damages appearing in various regions of the southern Baltic Sea generally result from a combination of strong storms, their increased number, accelerated sea-level rise and a decreasing trend of the presence of ice cover in winter, when the most intensive storms occur.

2.5 References

Achberger C, Chen D (2006) Trend of extreme precipitation in Sweden and Norway during 1961–2004. Research Report C72, Earth Sciences Centre, Göteborg University, Sweden

Alenius P, Makkonen L (1981) Variability of the annual maximum ice extent of the Baltic Sea. Meteorol Atmos Phys 29:393–398

Alenius P, Seinä A, Launiainen J, Launiainen S (2003) Sea ice and related data sets from the Baltic Sea. AICSEX – Metadata Report. Meri Report Series of the Finnish Institute of Marine Research 49:3–13

Alexandersson H (1986) A homogeneity test applied to precipitation data. Int J Climatol 6:661–675

Alexandersson H (2002) Temperature and precipitation in Sweden 1860–2001. SMHI Meteorologi No 104

Alexandersson H (2004) Variationer och trender i nederbörden (1860–2003) (Variabilities and trends in precipitation, 1860—2003). Faktablad nr 22 Sveriges Meteorologiska och Hydrologiska Institut, Norrköping, Sverige (in Swedish)

Alexandersson H, Vedin H (2002) Stormar det mera nu? (Do we have more storms today?) SMHI Väder och Vatten, 10:18 (in Swedish)

Alexandersson H, Schmith T, Iden K, Tuomenvirta H (1998) Long-term variations of the storm climate over NW Europe. Global Atmosphere and Ocean Systems 6:97–120

Alexandersson H, Tuomenvirta H, Schmith T, Iden K (2000) Trends of storms in NW Europe derived from an updated pressure data set. Clim Res 14:71–73

Alexandersson H, Eggertsson Karlström C (2001) Temperaturen och nederbörden i Sverige 1961–1990 (Temperature and precipitation in Sweden 1961–1990). Referensnormaler – utgåva 2. SMHI Rapporter 99 Sveriges Meteorologiska och Hydrologiska Institut, Norrköping, Sverige (in Swedish)

Andersson HC (2002) Influence of long-term regional and large-scale atmospheric circulation on the Baltic sea level. Tellus A 54:76–88

Augustin J (2005) Das Seegangsklima der Ostsee zwischen 1958–2002 auf Grundlage numerischer Daten (Sea state climate of the Baltic Sea 1958–2002 based on numerical data). Diploma thesis. Institute for Coastal Research, GKSS Research Center, Geesthacht, Germany (in German)

Axell LB (1998) On the variability of Baltic Sea deepwater mixing. J Geophys Res C: Oceans 103(C10):21667–21682

Baerens C, Baudler H, Beckmann BR, Birr HD, Dick S, Hofstede J, Kleine E, Lampe R, Lemke W, Meinke I, Meyer M, Müller R, Müller-Navarra SH, Schmager G, Schwarzer K, Zenz T (2003) Die Wasserstände an der Ostseeküste. Entwicklung – Stumfluten – Klimawandel (Water levels at the Baltic Sea coast. Trends – storm surges – climate change). In: Hupfer P, Hartt J, Horst S, Stigge HJ (eds) Die Küste (The Coast) Archive for research and technology on the North Sea and Baltic Sea coast. Boyens & Co, Heide in Holstein (in German)

Bärring L, von Storch H (2004) Scandinavian storminess since about 1800. Geophys Res Lett 31 L20202 doi:101029/2004GL020441

Bärring L, Jönsson P, Mattsson JO, Åhman R (2003) Wind erosion on arable land in Scania, Sweden and the relation to the wind climate – A review. Catena 52:173–190

Bartkeviciene G (2002) The impact of North Atlantic oscillation on the Lithuanian climate. Doctoral thesis. Vilnius University (In Lithuanian, summary in English)

Basalykas A (1985) Žemė – žmonijos planeta (The Earth – the habitation of mankind) Mokslas Vilnius (in Lithuanian)

Bauer E (2001) Interannual changes of the ocean wave variability in the North Atlantic and in the North Sea. Clim Res 18:63–69

Beck C, Grieser J, Rudolf B (2005) A New Monthly Precipitation Climatology for the Global Land Areas for the Period 1951 to 2000. (Published in Climate Status Report 2004, pp. 181–190. German Weather Service, Offenbach, Germany)

Beniston M (2004) The 2003 heat wave in Europe. A shape of things to come? Geophys Res Lett 31: 2022–2026

Beniston M, Stephenson DB (2004) Extreme climatic events and their evolution under changing climatic conditions. Glob Plan Change 44:1–9

Bergström H (1990) The early climatological records of Uppsala. Geografiska Annaler 72A,2:143–149

Bergström S (2001) Variability and change in precipitation and runoff to the Baltic Sea. A Workshop on Climate variability and change in the Baltic Sea area. KNMI De Bilt, The Netherlands, 12 November 2001, pp. 43

Bergström S, Carlsson B (1994) River Runoff to the Baltic Sea. Ambio 23:280–287

Bergström H, Moberg A (2002) Daily air temperature and pressure series for Uppsala (1722–1998). Clim Change 53:213–252

Bielec-Bąkowska Z (2002) Zróżnicowanie przestrzenne i zmienność wieloletnia występowania burz w Polsce (1949–1998) (Spatial distribution and the long-term variability of thunderstorm occurrence in Poland (1949–1998)). Wydawnictwa Uniwersytetu Śląskiego Katowice (in Polish)

Bielec-Bąkowska Z (2003) Long-term variability of thunderstorm occurrence in Poland in the 20th century. Atmos Res 67–68:35–52

Bijl W, Flather R, de Ronde JG, Schmith T (1999) Changing storminess? An analysis of long-term sea level data sets. Clim Res 11:161–172

Blenckner T, Chen D (2003) Comparison of the impact of regional and North Atlantic atmospheric circulation on an aquatic ecosystem. Clim Res 23:131–136

Bock KH (1971) Monatskarten des Salzgehaltes der Ostsee dargestellt für verschiedene Tiefenhorizonte (Monthly maps of salinity at different depth levels of the Baltic Sea). Dt Hydrogr Z, Ergänzungsheft Reihe B, 12, 147 pp (in German)

Brázdil R, Budíková M, Auer I, Böhm R, Cegnar T, Faško P, Lapin M, Gajic-Čapka M, Zaninovič K, Koleva E, Niedźwiedź T, Ustrnul Z, Szalai S, Weber RO (1996) Trends of maximum and minimum daily temperatures in central and southeastern Europe. Int J Climatol 16:765–782

Briede A, Lizuma L (2002) Long-term seasonal changes of runoff and its relation to climatic variability in Latvia. XXII Nordic Hydrological Conference NHP, report No 47:623–631

Briede A, Lizuma L (2004) Long-term and seasonal changes of runoff and its relation to climatic variability in Latvia. XXIII Nordic Hydrological Conference, Tallinn, Estonia. NHP Report 48:623–630

Brooks HE, Lee JW, Craven, JP (2003) The spatial distribution of severe thunderstorm and tornado environments from global reanalysis data. Atmos Res 67–68:73–94

Brown BG, Katz RW (1995) Regional analysis of temperature extremes: Spatial analog for climate change? J Clim 8:108–119

Brunsden D (1999) Some geomorphological considerations for the future development of landslide models. Geomorphology 30:13–24

BSH (1994) Eisbeobachtungen an den Hauptfahrwassern der Küste von Mecklenburg-Vorpommern in den Wintern 1956/57 bis 1989/90 (Sea ice observations at the main shipping routes of the Mecklenburg-Vorpommern coast during the winter seasons from 1956/57 until 1989/90). Bundesamt für Seeschifffahrt und Hydrographie, Hamburg und Rostock (in German)

Buelow K (1954) Allgemeine Kuestendynamik und Kuestenrueckgang (General dynamics of coast and retreat of shores). Geologie, Heft 10, Berlin (in German)

Bukantis A, Rimkus E (2005) Climate variability and change in Lithiuania. Acta Zoologica Lituanica 15:100–104

Bukantis A, Valiuskeviciene L (2005) Dynamics of extreme air temperature and precipitation, and determining factors in Lithuania in the 20th century. The Geographical Yearbook 38,1:6–17 (in Lithuanian, summary in English)

Bukantis A, Rimkus E, Stankūnavičius G (2001) Klimato svyravimų poveikis fiziniams geografiniams procesams Lietuvoje (The influence of climatic variations on recent physical geographical processes in Lithuania). In: Lietuvos klimato svyravimai ir kaita, pp. 27–107 (in Lithuanian)

Busuioc A, Chen D, Hellström C (2001) Temporal and spatial variability of precipitation in Sweden and its link with the large scale atmospheric circulation. Tellus 53A 3:348–367

Cappelen J (2005) DMI Monthly Climate Data Collection 1860–2004, Denmark, Faroe Islands and Greenland DMI Technical Report 05–05

Cappelen J, Christensen JH (2005) DMI's bidrag til Danmarks 4 nationale afrapportering til FN's klimakonvention UNFCC (DMI's contribution to Denmarkt's 4th National Report to the UN Climate Convention UNFCC). Danmarks Klimacener rappport 05–05 (In 2005 also to be published as a part of fourth National Communication of Denmark on Climate change. The Danish Environmental Protection Agency) (in Danish)

Carlsson M (1998a) A coupled three-basin sea level model for the Baltic Sea. Cont Shelf Res 18:1015–1038

Carlsson M (1998b) The mean sea-level topography in the Baltic Sea determined by oceanographic methods. Mar Geodes 21:203–217

Carretero JC, Gomez M, Lozano I, de Elvira AR, Serrano O, Iden K, Reistad M, Reichardt H, Kharin V, Stolley M, von Storch H, Gunther H, Pfizenmayer A, Rosethal W, Stawarz M, Schmith T, Kaas E, Li T, Alexandersson H, Beersma J, Bouws E, Komen G, Rider K, Flather R, Smith J, Bijl W, de Ronde J, Mi tus M, Bauer E, Schmidt H, Langenberg H (1998) Changing waves and storms in the northeast Atlantic? Bull Am Met Soc 79:741–760

Cebulak E (1997) Variability of Precipitation in Selected Regions of the Carpathians in the Years 1951–1995. Acta Geophys Pol 70:65–76

Cebulak E (1992) Influence of synoptic situations on maximum daily precipitation in upper Vistula basin. Folia Geographica, Series Geographica Physica 23:81–95 (in Polish with English summary)

Chang EKM, Fu Y (2002) Interdecadal variations in Northern Hemisphere. Winter storm track intensity. J Clim 15:642–658

Chen D (2000) A monthly climatology for Sweden and its application to a winter temperature case study. Int J Climatol 20:1067–1076

Chen D, Hellström C (1999) The influence of the North Atlantic Oscillation on the regional temperature variability in Sweden: Spatial and temporal variations. Tellus 51A:505–516

Chen D, Li X (2004) Scale-dependent relationship between maximum ice extent in the Baltic Sea and atmospheric circulation. Global Planet Change 41(3–4):275–283

Chen D, Moberg A, Jones PD, Jacobeit J, Lister D (2006) Trend atlas of the EMULATE indices Research Report, C73 Earth Sciences Centre Göteborg, University Göteborg Sweden 165 pp

Christiansen J, Bartoldy J, Hansen T, Lillie S, Nielsen J, Nielsen N, Pejrup M (2001) Salt march accretion during sea-level rise and outlook on the future. In: Jørgensen AM, Fenger J, Halsnæs K (eds) Climate change research – Danish Contributions. Danish Climate Centre, Copenhagen, pp. 26–276

Dabrowski M, Marszelewski W, Skoworn R (2004) The trends and dependencies between air and water temperatures in the lakes located in Northern Poland in the years 1961–2000. Hydrol Earth Syst Sci 8,1:79-87

Dai A, DelGenio AD, Fung IY (1997) Clouds precipitation and temperature range. Nature 386: 665–666

Dailidienė I, Stankevičius A, Tilickis B (2004) Peculiarities of the long-term water level fluctuations in the south-eastern part of the Baltic Sea and the Curonian Lagoon. Geodesy and Cartography 30,2:58–64 (in Lithuanian with English summary)

Dailidienė I, Davulienė L, Tilickis B, Stankevičius A, Myrberg K (2006) Sea level variability at the Lithuanian coast of the Baltic Sea. Boreal Env Res 11:109-121

Danilovich I (2004) Influence of climate warming on hydrological regime of lakes and reservoirs in Belarus during 1988–2002. 23th Nordic Hydrological Conference, Tallinn, Estonia. NHP Report 48:691–695

De Roo A, Schmuck G, Perdigao V, Thielen J (2003) The influence of historic land use changes and future planned land use scenarios on floods in the Oder catchment. Physics and Chemistry of the Earth 28:1291–1300

Degirmendžić J, Kożuchowski K, Żmudzka E (2002) Uwarunkowania cyrkulacyjne zmienności temperatury powietrza w Polsce w okresie 1951–2000 (Circulation requirements of the air temperature variations in Poland in the period 1951–2000) Przegląd Geofizyczny 47, 1–2:93–98 (in Polish)

Degirmendžić J, Kożuchowski K, Żmudzka E (2004) Changes of air temperature and precipitation in Poland in the period 1951–2000 and their relationship to atmospheric circulation. Int J Climatol 24:291–310

Dippner J, Kornilovs G, Sidrevics L (2000) Long-term variability of zooplankton in the Central Baltic Sea. J Mar Syst 25:23–31

Dobrovolný P, Brázdil R (2003) Documentary evidence on strong winds related to convective storms in the Czech Republic since ad 1500. Atmos Res 67–8:95–116

Döös K, Meier HEM, Döscher R (2004) The Baltic haline conveyor belt or the overturning circulation and mixing in the Baltic. Ambio 33:261–266

Dotzec N (2003) An updated estimate of tornado occurrence in Europe. Atmos Res 67–8:153–161

Draveniece A (1998) Characteristics of Snow Cover in Latvia. Master thesis, University of Latvia, Riga (in Latvian)

Dubicka M (2000) Variability in cloud cover of the Karkonosze Mountains over the last century. Geographia Polonica 73:117–124

Dubicka M, Pyka JL (2001) Selected problems of climate in Wroc aw during the 20th century. Studies in Geography Warsaw. University Press 29:101–112 (in Polish with English abstract)

Dzerdzeevskiy BL (1968) Circulating mechanisms in the atmosphere of the northern hemisphere in the 20th century. Data of Meteorological Studies, Atmospheric circulation, Moscow

Easterling D, Horton B, Jones P, Peterson T, Karl T, Parker D, Salinger J, Razuvayev V, Plummer N, Jameson P, Folland C (1997) Maximum and minimum temperature trends for the globe. Science 277:364–366

Easterling DR, Meehl GA, Parmesan C, Changnon SA, Karl TR, Mearns LO (2000) Climate extremes: Observations modeling and impacts. Science 289:2068–2074

Eberhards G (2003) The sea coast of Latvia. Morphology Structure. Coastal processes. Risk zone. Forecast. Coastal protection and monitoring. Latvijas Universitate, Riga (in Latvian with extended English summary)

Ekman M (1994) Deviation of mean sea level from the mean geoid in the transition area between the North Sea and the Baltic Sea. Mar Geodes 17:161–168

Ekman M (1996) Extreme annual means in the Baltic Sea level during 200 years. Summer institute for Historical Geophysics, Åland Islands, Small Publications in Historical Geophysics 2

Ekman M (1997) Anomalous winter climate coupled to extreme annual means in the Baltic Sea during the last 200 years. Summer institute for Historical Geophysics, Åland Islands, Small Publications in Historical Geophysics 3

Ekman M (1998) Secular change of the seasonal sea level variation in the Baltic Sea and secular change of the winter climate. Geophysica 34:131–140

Ekman M (1999) Climate changes detected through the world's longest sea level series. Glob Planet Change 21,4:215–224

Ekman M (2003) The World's longest sea level series and winter oscillation index for Northern Europe 1774–2000. Small publications in Historical Geophysics 12:1–31

Ekman M, Stigebrandt A (1990) Secular change of the seasonal variation in sea level and the pole tide in the Baltic Sea. J Geophys Res 95:5379-5383

Ekström M, Jönsson P, Bärring L (2002) Synoptic patterns associated with major wind erosion events in southern Sweden (1973–1991). Clim Res 23:51–66

Elken J (1998) Hydrography and water exchange. In: Tarand A, Kallaste T (eds) Country case study on climate change impacts and adaptation assessments in the Republic of Estonia 6. Baltic Sea and Estonian coast: Potential impacts of climate variability and change on the dynamics, hydrography and related biology of the Baltic Sea. Stockholm Environment Institute, Tallinn Centre, pp. 104–107

Falarz M (2004a) Variability and trends in the duration and depth of snow cover in Poland in the 20th century. Int J Climatol 24:1713–1727

Falarz M (2004b) Changes in the duration and depth of snow cover in Poland in the 20th century and their climatic causes. EMS Annual Meeting Abstracts, vol. 1, 00081

Farat R, Kepinska-Kasprzak M, Kowalczak P, Mager P (1998) Droughts in Poland. A Newsletter of the International Drought Information Center. 10/1:7–10

Feddersen H (2003) Predictability of seasonal precipitation in the Nordic region. Tellus A 55,5:385–400

Feistel R, Nausch G, Mohrholz V, Lysiak-Pastuszak E, Seifert T, Matthäus W, Krüger S, Sehested Hansen I (2003) Warm waters of summer 2002 in deep Baltic Proper. Oceanologica 45:571–592

Feistel R, Nausch G, Hagen E (2006) Unusual inflow activity 2002/3 and varying deep-water properties. Oceanologia 48:21–35

Fenger J, Buch E, Jacobsen PR (2001) Monitoring and impacts of sea level rise at Danish coasts and nears shore infrastructures. In: Jørgensen AM, Fenger J, Halsnæs K (eds) Climate change research – Danish Contributions. Danish Climate Centre, Copenhagen pp. 237–254

Feser F, Weisse R, von Storch H (2001) Multi-decadal atmospheric modeling for Europe yields multi-purpose data. Eos Transact 82:305–310

Fischer H, W Matthäus (1996) The importance of the Drogden Sill in the Sound for major Baltic inflows. J Mar Sys 9:137–157

Folland CK, Karl TR, Christy JR, Clarke RA, Gruza GV, Jouzel J, Mann ME, Oerlemans J, Salinger MJ, Wang SW (2001) Observed Climate Variability and Change. In: Houghton JT et al. (eds) Climate Change (2001) The Scientific Basis Contribution of Working Group I to the Third Assessment Report of the Intergovernmental Panel on Climate Change. Cambridge University Press, pp. 99-181

Fonselius S (1969) Hydrography of the Baltic deep basins III, Fishery Board. Sweden, Ser. Hydrogr. Rep. No 23:1–97

Fonselius S, Valderrama J (2003) One hundred years of hydrographic measurements in the Baltic Sea. J Sea Res 49:229-241

Fonselius S, Szaron J, Öström B (1984) Long-term salinity variations in the Baltic Sea deep water. Rapports et Procès-Verbaux des Réunions, International Council for the Exploration of the Sea 185:140–149

Førland EJ (2000) Trends in precipitation intensity in Norway and the nordic region during the 20th Century. ECAC 2000: 3rd European Conference on Applied Climatology, 16–20 October 2000. Pisa, Italy, pp. 1–6

Førland EJ, Hanssen-Bauer I (2000) Increased precipitation in the Norwegian Arctic: True or false? Clim Change 46:485–509

Førland EJ, van Engelen A, Hanssen-Bauer I, Heino R, Ashcroft J, Dahlström B, Demarée G, Frich P, Jónsson T, Miętus M, Müller-Westermeier G, Palsdottir T, Tuomenvirta H, Vedin H (1996) Changes in "normal" precipitation in the North Atlantic region (second edition). DNMI-Report 7/96 Klima

Førland EJ, Alexandersson J, Dahlström H, Drebs A, Frich P, Hanssen-Bauer I, Heino R, Helminen J, Jónsson T, Nordli PØ, Pálsdóttir T, Schmith T, Tuomenvirta H, Tveito OE, Vedin H (1998) Relating extreme weather to atmospheric circulation using a regionalised dataset. Final Report (1996–1998) DNMI-Report 17/98 Klima

Førland EJ, Hanssen-Bauer I, Jónsson T, Kern-Hansen C, Nordli PØ, Tveito OE, Vaarby Laursen E (2002) Twentieth-century variations in temperature and precipitation in the Nordic Arctic. Polar Record 38:203–210

Franke J, Goldberg V, Eichelmann U, Freydank E, Bernhofer C (2004) Statistical analysis of regional climate trends in Saxony, Germany. Clim Res 27:145–150

Frich P, Alexander LV, Della-Marta P, Gleason B, Haylock M, Klein-Tank A, Peterson T (2002) Observed coherent changes in climatic extremes during the second half of the 20th Century. Clim Res 19:193–212

Girjatowicz KP, Kożuchowski K (1995) Contemporary changes of Baltic Sea ice. Geographia Polonica 65:43–50

Girjatowicz KP, Kożuchowski K (1999) Variations of thermic and ice conditions in the Szczecin Lagoon region. In: Järvet A (ed) Publ 2nd Workshop on the Baltic Sea Ice Climate, Dept Geography, Univ Tartu, 84:69-73

Grabbe GC, Grassl H (1994) Solar radiation in Germany – observed trends and an assessment of the causes. Part II: Detailed trend analysis for Hamburg. Beitr Phys Atmosph 2:31–37

Grevemeyer I, Herber R, Essen HH (2000) Microseismological evidence for a changing wave climate in the northeast Atlantic Ocean. Nature 408:349-352

Grimenes AA, Nissen Ø (2004) Mathematical modelling of the annual temperature wave based on monthly mean temperatures; and comparisons between local climate trends at seven Norwegian stations. Theor Appl Climatol 78:229–246

Groenemeijer P, Teittinen J, Punkka AJ (2005) The meteorological environment of significant tornadoes in northern and central Europe. Poster P15R16, 32nd Conference on Radar Meteorology, 24–29 October 2005. Albuquerque, New Mexico, USA

Groisman PY, Rankova EY (2001) Precipitation trends over the Russian permafrost-free zone: Removing the artefacts of pre-processing. Int J Climatol 21:657–678

Groisman PY, Knight RW, Easterling DR, Karl TR, Hegerl TC, Razuvaev VN (2005) Trends in intense precipitation in the climate record. J Clim 18:1326–1350

Gronskaya TP (2000) Ice thickness in relation to climate forcing in Russia. Verh. Internat. Verein Limnol 27,5:2800–2802

Gudelis WK, Jemeljanovas EM (1976) Bierega Bałtyjskowo Moria (Baltic Sea shores) Geołogija Bałtyjskowo Moria Vilnius Part 6:142–160 (in Lithuanian)

Gustafsson BG, Andersson HC (2001) Modeling the exchange of the Baltic Sea from the meridional atmospheric pressure difference across the North Sea. J Geophys Res-Oceans 106,C9:19731–19744

Gustafsson BG, Westman P (2002) On the causes for salinity variations in the Baltic Sea during the last 8500 years. Paleoceanography 17:1040

Haapala J, Leppäranta M (1997) The Baltic Sea ice season in changing climate. Boreal Env Res 2:93–108

Hagen E, Feistel R (2004) Observations of low-frequency current fluctuations in deep water of the Eastern Gotland Basin/Baltic Sea. JGR 109 C03044 doi:101019/2003JC002017

Hanson CE, Palutikof JP, Davies TD (2004) Objective cyclone climatologies of the north Atlantic – A comparison between the ECMWF and NCEP reanalyses. Clim Dyn 22:757–769

Hanssen-Bauer I (2005) Regional temperature and precipitation series for Norway: Analyses of time-series updated to 2004 met. no Report 15/2005 Climate Norwegian Meteorological Institute, Oslo

Hanssen-Bauer I, Førland EJ (1994) Homogenizing long Norwegian precipitation series. J Clim 7: 1001–1013

Hanssen-Bauer I, Nordli Ø (1998) Annual and seasonal temperature variations in Norway 1876–1997. DNMI-report 25/98

Harnik N, Chang EKM (2003) Storm track variations as seen in radiosonde observations and reanalysis data. J Clim 16:480–495

Haylock MR, Goodess CM (2004) Interannual variability of European extreme winter rainfall and links with large-scale circulation. Int J Climatol 24:759–776

Heikinheimo MJ, Ohvril H, Venäläinen A, Skartvelt A, Olseth JA, Laine V, Teral H, Arak M, Teral K (1996) Recent variations of atmospheric turbidity at selected sites in Finland, Estonia and Norway as revealed by surface solar radiation measurements. Geophysica 32:195–215

Heino R (1994) Climate in Finland during the period of meteorological observations. Finnish Meteorological Institute, Contributions 12

Heino R, Brázdil R, Førland EJ, Tuomenvirta H, Alexandersson H, Beninston M, Pfister C, Rebetez M, Roesner S, Rosenhagen G, Rösner S, Wibig J (1999) Progress in study of climatic extremes in Northern and Central Europe. Climatic Change 42:151–181

Hejkrlik L (2002) Recent changes in air pollution in the Czech Republic. Fourth European Conference on Applied Climatology ECAC 2002, Brussels, 12.11.2002-15.11.2002, Abstract Volume

Hela I (1948) On the stress of the wind on the water surface. Geophysica 3:146–161

Henderson-Sellers A (1986) Cloud changes in a warmer Europe. Climatic Change 8:25–52

Henderson-Sellers A (1992) Continental cloudiness changes this century. GeoJournal 27:255–262

Hess P, Brezowsky H (1977) Katalog der GroSSwetterlagen Europas 1881–1976 (Catalogue of the GroSSwetterlagen of Europe). Berichte des Deutschen Wetterdienstes Nr 113 Bd 15, Selbstverlag des Deutschen Wetterdienstes, Offenbach am Main (in German)

Hisdal H, Holmqvist E, Hyvärinen V, Jónsson P, Kuusisto E, Larsen SE, Lindström G, Ovesen N, Roald LA (2003) Long time series. A review of Nordic studies. Report by the CWE Long Time Series Group, CHIN/Nordic Council of Ministers, Reykjavík

Hisdal H, Holmqvist E, Kuusisto E, Lindström H, Roald LA (2004) Has streamflow changed in the Nordic countries? Järvet A (ed) NHP Report No 48:633–643

Hurrell JW (1995) Decadal trends in the north-Atlantic oscillation – regional temperatures and precipitation. Science 269:676–679

Hyvärinen V (2003) Trends and Characteristics of Hydrological Time Series in Finland. Nordic Hydrology 34:71–90

Hyvärinen V, Kajander J (2005) Rivers and lakes of Fennoscandia. In: Seppälä M (ed) The Physical Geography of Fennoscandia. Oxford University Press, Great Britain

Hyvärinen V, Korhonen J (eds) (2003) Hydrological Yearbook 1996–2000. The Finnish Environment 599, Finnish Environment Institute, Helsinki, Finland

Hyvärinen V, Vehviläinen B (1988) The effects of climatic fluctuations and man on discharge in Finnish river basins. IAHS Publication 130:97–103

IPCC (2001) Climate Change 2001: The Scientific Basis. Contribution of Working Group I to the Third Assessment Report of the Intergovernmental Panel on Climate Change (Houghton JT, Y Ding, DJ Griggs, M Noguer, PJ van der Linden, X Dai, K Maskell, and CA Johnson (eds)). Cambridge University Press, Cambridge New York

Jaagus J (1992) Periodicity of precipitation in Estonia. Man and Nature. In: Kaare T et al (eds) Estonian Geographical Society, Tallinn, pp. 43–53

Jaagus J (1996) Climatic trends in Estonia during the period of instrumental observations and climate change scenarios. Estonia in the system of the global climate change. Institute of Ecology, 4:35–48

Jaagus J (1998) Climatic fluctuations and trends in Estonia in the 20th century and possible climate change scenarios. In: Kallaste T, Kuldna P (eds) Climate change studies in Estonia. Tallinn, Stockholm Environment Institute, Tallinn Centre, pp. 7–12

Jaagus J (2003) Muutused Eesti rannikumere jääoludes 20 sajandi teisel poolel (Changes in sea ice conditions near the Estonian coast during the second half of the 20th century). Publicationes Instituti Geographici Universitatis Tartuensis 93:143–152 (in Estonian)

Jaagus J (2006a) Climatic changes in Estonia during the second half of the 20th century in relationship with changes in large-scale atmospheric circulation. Theor Appl Climatol 83:77–88

Jaagus J (2006b) Trends in sea ice conditions on the Baltic Sea near the Estonian coast during the period 1949/50–2003/04 and their relationships to large-scale atmospheric circulation. Boreal Env Res 11:169–183

Jaagus J, Ahas R (2000) Space-time variations of climatic seasons and their correlation with the phenological development of nature in Estonia. Clim Res 15:207–219

Jaagus J, Tarand A (1998) Precipitation. Periodical fluctuations and a seasonal shift. Country case study on climate change impacts and adaptation assessments in the Republic of Estonia. In: Tarand A, Kallaste T (eds) Report to the UNEP/GEF project No GF/2200-96-45. Tallinn, Stockholm Environment Institute, Tallinn Centre, pp. 21–23

Jaagus J, Truu J, Ahas R, Aasa A (2003) Spatial and temporal variability of climatic seasons on the East European Plain in relation to large-scale atmospheric circulation. Clim Res 23:111–129

Jaani A, Belzano V (2003) Long-term water level fluctuations of Lake Peipsi (Chudskoe) and their extra-terrestrial reasons. Uurimusi Eesti klimaast. Jaagus J (ed) Publicationes Instituti Geographici Universitatis Tartuensis 93:254–266

Jacobeit J, Jönsson P, Bärring L, Beck C, Ekström M (2001) Zonal indices for Europe 1780–1995 and running correlations with temperature. Climatic Change 48:219-241

Janssen F (2002) Statistical Analysis of multi-year variability of the hydrography in North Sea and Baltic Sea. PhD thesis, University of Hamburg (in German)

Janssen F, Schrum C, Backhaus J (1999) A climatological dataset for temperature and salinity in the North Sea and the Baltic Sea. Dt Hydrogr Z, Suppl 9

Jantunen J, Ruosteenoja K (2000) Weather conditions in northern Europe in the exceptionally cold spring season of the famine year 1867. Geophysica 36:69–84

Järvet A (2004) Influence of hydrological factors and human impact on the ecological state of shallow lake Võrtsjärv in Estonia. Dissertationes Geographicae Universitatis Tartuensis 19

Jevrejeva S (2000) Long-term variability of sea ice and air temperature conditions along the Estonian coast. Geophysica 36:17–30

Jevrejeva S (2001) Severity of winter seasons in the northern Baltic Sea between 1529 and 1990: reconstruction and analysis. Clim Res 17:55–62

Jevrejeva S, Moore JC (2001) Singular Spectrum Analysis of Baltic Sea ice conditions and large-scale atmospheric patterns since 1708. J Geophys Res 28:4503–4506

Jevrejeva S, Moore JC, Grinsted A (2003) Influence of the Arctic Oscillation and El Niño–Southern Oscillation (ENSO) on ice conditions in the Baltic Sea: The wavelet approach. J Geophys Res 108 doi 101029/2003JD003417

Jevrejeva S, Drabkin VV, Kostjukov J, Lebedev AA, Leppäranta M, Mironov U Ye, Schmelzer N, Sztobryn M (2004) Baltic Sea ice season in the 20th century. Clim Res 25:217–227

Johansson B, Chen D (2003) The influence of wind and topography on precipitation distribution. A case study in Sweden. Int J Climatol 23:1523–1535

Johansson C, Bergström H (2004) Variabiliteten i vindens energiinnehåll över Skandinavien mellan 1900–2000 (Variability in wind energy over Scandinavia during 1900–2000). Elforsk Rapport No 04:36. Elforsk AB SE-10153 Stockholm, Sweden (in Swedish)

Johansson JM, Davis JL, Scherneck HG, Milne GA, Vermeer M, Mitrovica JX, Bennet RA, Jonsson B, Elgered G, Elósegui P, Koivula H, Poutanen M, Rönnäng BO, Shapiro II (2002) Continuous GPS measurements of postglacial adjustment in Fennoscandia 1 Geodetic results. J Geophys Res 107 B8 2157 101029/2001JB000400

Johansson M, Boman H, Kahma KK, Launiainen J (2001) Trends in sea level variability in the Baltic Sea. Boreal Env Res 6:159–179

Johansson M, Kahma KK, Boman H, Launiainen J (2004) Scenarios for sea level on the Finnish coast. Boreal Env Res 9:153–166

Jones PD, Lister DH (2002) The daily temperature record for St Petersburg (1743–1996). Climatic Change 53:253–267

Jones PD, Moberg A (2003) Hemispheric and large-scale surface air temperature variations: An extensive revision and update to 2001. J Clim 16:206–223

Jones PD, Jónsson T, Wheeler D (1997) Extension to the north Atlantic oscillation using early instrumental pressure observations from Gibraltar and south-west Iceland. Int J Climatol 17: 1433–1450

Jones PD, Davies TD, Lister DH, Slonosky V, Jónsson T, Bärring L, Jönnson P, Maheras P, Kolyva-Machera F, Barriendos M, Martin-Vide J, Alcoforado MJ, Wanner H, Pfister C, Schuepbach E, Kaas E, Schmith T, Jacobeit J, Beck C (1999) Monthly mean pressure reconstructions for Europe for the 1780–1995 period. Int J Climatol 19:347–364

Jones PD, Horton EB, Folland CK, Hulme M, Parker DE, Basnett TA (1999) The use of indices to identify changes in climatic extremes. Climatic Change 42:131–149

Jones PD, Briffa KR, Osborn T, Moberg A, Bergstrøm H (2002) Relations between circulation strength and the variability of growing-season and cold season climate in northern and central Europe. The Holocene 12 No 6:643–656

Jönsson P, Holmquist B (1995) Wind direction in southern Sweden 1740–1992: Variation and correlation with temperature and zonality. Theor Appl Climatol 51:183–198

Jurva R (1952) On the variations and changes of freezing in the Baltic during the last 120 years. Fennia 75:17–24

Kahma KK, Boman H, Johansson MM, Launiainen J (2003) The North Atlantic Oscillation and sea level variations in the Baltic Sea. ICES Marine Science Symposia: Hydrobiological variability in the ICES area 1990–1999. Edinburgh, Scotland, 8–10 August 2001: Actes du Symposium: Posters presented at the Symposium 219:365–366

Kalas M (1993) Characteristic of sea level changes on the Polish Coast of the Baltic Sea in the last forty-five years. International Workshop, Sea Level Changes and Water Management 19–23 April 1993, Noorsdwijerhout Nederlands, pp. 51–61

Karl TR, Jones PD, Knight RW, Kukla G, Plummer N, Razuvayev V, Gallo KP, Lindseay J, Charlson RJ, Peterson TC (1993) A new perspective on recent glogal warming: Asymmetric trends of daily maximum and minimum temperature. Bull Am Met Soc 74:1007–1023

Kaszewski BM, Filipiuk E (2003) Variability of atmospheric circulation in Central Europe in the summer season 1881–1998 (on the basis of the Hess-Brezowski classification) Meteorologische Zeitschrift 12:123–130

Katz RW, Brown BG (1992) Extreme events in a changing climate: Variability is more important than averages. Climatic Change 21:289–302

Kauker F, Meier HEM (2003) Modeling decadal variability of the Baltic Sea: 1 Reconstructing atmospheric surface data for the period 1902–1998. J Geophys Res 108(C8) 3267 doi:101029/ 2003JC001797

Keevallik S (2003) Changes in spring weather conditions and atmospheric circulation in Estonia (1955–95). Int J Climatol 23:263–270

Keevallik S, Rajasalu R (2001) Winds on the 500 hPa isobaric level over Estonia (1953–1998). Phys Chem Earth (B) 26:425–429

Keevallik S, Russak V (2001) Changes in the amount of low clouds in Estonia (1955–1995) Int J Climatol 21:389–397

Kitaev L, Razuvaev V, Heino, R, Forland E (2006) Duration of snow cover over Northern Europen. Russian Meteorology and Hydrology 3:95–100

Klein Tank AMG, Können GP (2003) Trends in indices of daily temperature and precipitation extremes in Europe 1946–99. J Clim 16:3665–3680

Klein Tank AMG, Wijngaard J, van Engelen A (2002a) Climate of Europe: Assessment of observed daily temperature and precipitation extremes. KNMI, De Bilt, The Netherlands (data and metadata available at http://ecaknminl)

Klein Tank AMG, Wijngaard JB, Können GP, Böhm R, Demarée G, Gocheva A, Mileta M, Pashiardis S, Hejkrlik L, Kern-Hansen C, Heino R, Bessemoulin P, Müller-Westermeier G, Tzanakou M, Szalai S, Pálsdóttir T, Fitzgerald D, Rubin S, Capaldo M, Maugeri M, Leitass A, Bukantis A, Aberfeld R, van Engelen AFV, Førland E, Mi tus M, Coelho F, Mares C, Razuvaev V, Nieplova E, Cegnar T, Antonio López J, Dahlström B, Moberg A, Kirchhofer W, Ceylan A, Pachaliuk O, Alexander LLV, Petrovic P (2002b) Daily dataset of 20th century surface air temperature and precipitation series for the European Climate Assessment. Int J Climatol 22:1441–1453

Koistinen J, Michelson DB (2002) BALTEX weather radar-based precipitation products and their accuracies. Bor Env Res 7:253–263

Kont A, Ratas U, Puurmann E (1997) Sea-level rise impact on coastal areas of Estonia. Climatic Change 36:175–184

Korhonen J (2002) Water temperature conditions of lakes and rivers in Finland in the 20th century (Ch 8) In: Suomen vesistojen lampoolot 1900-luvulla. Soumen Ympäristö 566

Korhonen J (2004) Analysis of lake ice cover observations in Finland. 23th Nordic Hydrological Conference, Selected Articles Vol 2 Tartu, pp. 685–690

Koslowski G, Glaser R (1995) Reconstruction of the ice winter severity index since 1701 in the western Baltic. Climatic Change 31:79–98

Koslowski G, Glaser R (1999) Variations in reconstructed ice winter severity in the western Baltic from 1501 to 1995 and their implications for the North Atlantic Oscillation. Climatic Change 41:175–191

Koslowski G, Loewe P (1994) The western Baltic Sea ice seasons in terms of mass-related severity index 1879-1992. Tellus 46A:66–74

Kożuchowski K (1985) Variation in precipitation in the years 1881–1980 in Poland. Acta Geographica Lodziensia No 48 (in Polish)

Kożuchowski K, Marciniak K (1988) Variability of mean monthly temperature and semi-annual precipitation totals in Europe in relation to hemisphere circulation patterns. Int J Climatol 8:191–199

Kożuchowski K, Marciniak K (1990) The influence of global circulation patterns on inter-annual temperature changes in Europe. Meteorol Z 40:237–240

Kożuchowski K, Miętus M (1996) Historia zmian klimatu w Szczecinie (The history of climate variability in Szczecin) In: Kożuchowski K (ed) Współczesne zmiany klimatyczne. Klimat Szczecina i współczesne zmiany klimatyczne w rejonie Morza Bałtyckiego (Contemporary climate changes. Climate of Szczecin and contemporary climate changes in the Baltic Sea region). Uniwersytet Szczeciński, Rozprawy i Studia, T. (CCXCVIII), pp. 45–60 (in Polish)

Kożuchowski K, Żmudzka E (2002) The warming in Poland: The range and seasonality of changes in air temperature during the half of the 20th century. Miscellanea Geographica vol. 10

Kożuchowski K, Żmudzka E (2003) 100-year series of areally averaged temperatures and precipitation totals in Poland. Studia geograficzne 75 Acta Universitatis Wratislaviensis No 2542:116–122

Kożuchowski K, Wibig J, Maheras P (1992) Connections between air temperature and precipitation and the geopotential height of the 500 hPa level in a meridional cross-section in Europe. Int J Climatol 12:343–352

Kożuchowski K, Trepińska J, Wibig J (1994) The air temperature in Cracow from 1826 to 1990: Persistence fluctuations and the urban effect. Int J Climatol 14:1035–1049

Kożuchowski K, Degirmendžić J, Fortuniak K, Wibig J (2000) Trends to changes in seasonal aspects of the climate in Poland. Geographia Polonica 73 No 2:7–24

Krauss W (1974) Two-dimensional seiches and stationary drift currents in the Baltic Sea. ICES Special Meeting on Models of Water Circulation in the Baltic. Paper 10

Kundzewicz Z, Graczyk D, Pińskwar I, Radziejewski M, Szwed M, Bärring L, Giannakopoulos Ch, Holt T, Palutikof J, Leckerbusch GC, Ulbrich U, Schwarb M (2004) Changes in the occurrence of extremes part I. Climatic background. Papers on Global Change IGBP 11:9–20

Kuusisto E (1982) Euroopan suurimman järven hydrologiaa (The hydrology of the largest lake in Europe) Vesitalous 1:10–12 (in Finnish)

Kuusisto E, Elo AR (2000) Lake and river ice variables as climate indicators in Northern Europe. Verh Internat Verein Limnol 27:2761–2764

Lamb HH (1977) Climate: Present, past and future. Vol 2: Climatic history and the future. Methuen & Co Ltd London

Lampe R (1996) Küsten und Küstenschutz in Mecklenburg-Vorpommern (Coast and coastal protection in Mecklenburg-Vorpommern). Erdkundeunterricht 9:364–372 (in German)

Larsson M (2004) Syns den globala uppvärmning i den svenska snöstatistiken? (Can global warming be detected in Swedish snow statistics?) Examen arbete vid Institutionen for geovetenskaper Nr. 90, Uppsala universitetet (in Swedish)

Lass HU, Matthäus W (1996) On temporal wind variations forcing salt water inflows into the Baltic Sea. Tellus 48A:663–671

Launiainen J, Seinä A, Alenius P, Johansson M, Launiainen S (2002) Atmospheric reflections to the Baltic Sea ice climate. In: Omstedt A, Axell L (eds) Fourth Workshop on the Baltic Sea Ice Climate, Norrköping, Sweden. SMHI Oceanografi No 72:19–30

Laursen EV, Cappelen J (1998) Observed hours of bright sunshine in Denmark – with climatological standard normals 1961–90. DMI Technical Report 98-4

Lazarenko NN (1961) (Fluctuations of water level) Gidrometizdat Leningrad (in Russian)

Lenz W (1971) Monatskarten der Temperatur der Ostsee dargestellt für verschiedene Tiefenhorizonte (Monthly temperature maps for various depths levels in the Baltic Sea). Dt Hydrogr Z, Ergänzungsheft, Reihe B 11 (in German)

Leppäranta M, Seinä A (1985) Freezing maximum annual ice thickness and break up of ice on the Finnish coast during 1830–1984. Geophysica 21:87–104

Liepert BG (1996) Regionale Klimadiagnose mittels Messungen der solaren Strahlung (Regional climate diagnosis using solar radiation measurements). PhD dissertation, University of Munich (in German)

Linderholm H, Walther A, Chen D (2005) Trend of growing season length over the great Baltic sea region. Research Report C69, Earth Sciences Centre, Göteborg University, Sweden

Lindström G, Bergström S (2004) Runoff trends in Sweden 1807–2002. Hydrological Sciences Journal 49:69–83

Lisitzin E (1957) The annual variation of the slope of the water surface in the Gulf of Bothnia. Commentationes Physico-Mathematicae, Societas Scientiarum Fennica XX 6:1–20

Lisitzin E (1966a) Land uplift in Finland as a Sea level problem. Ann. Acad. Sci Fennicae A III 90: 237–239

Lisitzin E (1966b) Mean sea level heights and elevation systems in Finland. Societas Scientiarum Fennica, Commentationes physico-Mathematica 32 No 4

Lisitzin E (1974) Sea-level changes. Elsevier Oceanography Serie 8, Elsevier

Lizuma L (2000) An analysis of a long-term meteorological data series in Riga. Geographical articles/
  Folia Geographica VIII. Living with Diversity in Latvia, Riga, Societas Geographica Latviensis, pp.
  53–60

Lorenc H (2000) Studia nad 220-letnią (1779–1998) serią temperatury powietrza w Warszawie oraz
  ocena jej wiekowych tendencji (Analysis of the 220 year (1779–1998) air temperature record from
  Warshaw and assessment of its centennial tendencies). Materiały Badawcze IMGW, S. Meteorologia
  31 (in Polish)

Mager M, Kuźnicka M, Kępińska-Kasprzak M, Farat R (2000) Changes in the intensity and frequency
  of occurrence of droughts in Poland (1891–1995) Geographia Polonica 73:41–47

Magnuson JJ, Robertson DM, Benson BJ, Wynne RH, Livingstone DM, Arai T, Assel RA, Barry RG,
  Card V, Kuusisto E, Granin NG, Prowse TD, Stewart KM, Vuglinsky VS (2000) Historical trends
  in lake and river cover in the Northern Hemisphere. Science 289:1743–1746

Marcinoniene I (2003) Tornadoes in Lithuania in the period of 1950–2002 including analysis of the
  strongest tornado of 29 may 1981. Atmos Res 67:475–484

Markevičiene I (1998) Klimato elementų kintamumas Lietuvos teritorijoje (The variability and
  changes of climatic elements in Lithuanian territory). In: Metų terminiai sezonai, pp. 110–132
  (in Lithuanian)

Marshall J, Kushner Y, Battisti D, Chang P, Czaja A, Dickson R, Hurrell J, McCartney M, Saravanan
  R, Visbeck M (2001) North Atlantic climate variability: Phenomena impacts and mechanisms. Int
  J Climatol 21:1863–1898

Marsz AA (1999) Oscylacja Północnoatlantycka a reżim termiczny zim na obszarze północno-
  zachodniej Polski i na polskim wybrzeżu Bałtyku. (North Atlantic Oscillation and thermic regime
  of the winters on the NW Poland and polish Baltic shore) Przegląd Geograficzny 71: 225–245

Marszelewski W, Skowron R (2006) Ice cover as an indicator of winter air temperature changes: case
  study of the Polish Lowland lakes. Hydrol Sci J 51,2:336—349

Masanova MD, Filatova IV (1985) Probability structure of interannual water level change in north-
  western lakes. In Problemy Issledovaniya Krupnyh Ozer. Leningrad, 81–84 pp

Matuszko D (2003) Cloudiness changes in Cracow in the 20th century. Int J Climatol 23:975–984

Matthäus W, Franck H (1992) Characteristics of major Baltic inflows – A statistical analysis. Continent Shelf Res 12:1375–1400

Matthäus W, Schinke H (1999) The influence of river runoff on deep water conditions of the Baltic
  Sea. Hydrobiologia 393:1–10

Mearns LO, Katz RW, Schneider SH (1984) Extreme high-temperature events: Changes in their probabilities with changes in mean temperature. J Clim Appl Met 23:1601–1613

Meier HEM (2005) Modeling the age of Baltic Sea water masses: Quantification and steady state
  sensitivity experiments. J Geophys Res 110 C02006 doi: 101029/2004JC002607

Meier HEM (2006) Baltic Sea climate in the late 21st century – a dynamical downscaling approach
  using two global models and two emission scenarios. Clim Dyn 27:39-68

Meier HEM, Kauker F (2002) Simulating Baltic Sea climate for the period 1902–1998 with the Rossby
  Centre coupled ice-ocean model. Reports Oceanography 30, Swedish Meteorological and Hydrological Institute, Norrköping, Sweden

Meier HEM, Kauker F (2003) Modeling decadal variability of the Baltic Sea: 2. Role of freshwater inflow and large-scale atmospheric circulation for salinity. J Geophys Res 108(C11) 3368
  doi:101029/2003JC001799

Merilain M, Tooming H (2003) Dramatic days in Estonia. Weather 58:119-125

Meyer HA, Möbius K, Karsten G, Hensen V (1873) Die Expedition zur physikalisch-chemischen und
  biologischen Untersuchung der Ostsee im Sommer 1871 auf SM Avisodampfer Pommerania. Jahres-
  bericht der Commission zur wissenschaftlichen Untersuchung der deutschen Meere. (The expedition
  to investigate the physics, chemistry and biology of the Baltic Sea in summer 1871 on "SM Avisodampfer Pommerania". Annual report of the commission for the scientific exploration of the German
  oceans). Verlag Wiegandt & Hempel, Berlin (in German)

Michelson DB, Jones CG, Landelius T, Collier CG, Haase G, Heen M (2005) 'Down-to-Earth' modelling of equivalent surface precipitation using multisource data and radar. Q J Roy Met Soc 131: 1093–1112

Miętus M (1998a) O rekonstrukcji i homogenizacji wieloletniej serii średniej miesięcznej temperatury ze stacji w Gdańsku-Wrzeszczu, 1851–1995 (On the reconstruction and homogenization of the long term record of mean monthly air temperature at Gdańsku-Wrzeszczu, 1851–1995). Wiadomo ci IMGW, XXI(XLII), 2:41–63 (in Polish)

Miętus M (ed) (1998b) The climate of the Baltic Sea Basin. World Meteorological Organization, Marine Meteorology and Related Oceanographic Activities, Rep No 41 WMO/TD-No933

Miętus M (1999) Rola cyrkulacji atmosferycznej w kształtowaniu warunków klimatycznych i oceanograficznych w polskiej strefie brzegowej Morza Bałtyckiego (The impact of atmospheric circulation on climatic and oceanographic conditions in the Polish coastal zone of the Baltic Sea). Materiały Badawcze IMGW, ser Meteorologia 29 (in Polish)

Miętus M (2002) O ciągłości i jednorodności wieloletnich serii klimatologicznych na przykładzie rezultatów pomiarów w Koszalinie (On the homogeneity of long term climatic records and the example of measurements in Koszalin). In: Wójcik G, Marciniak K (eds) Działalności naukowa Prof Wł Gorczyńskiego i jej kontynuacja (The scientific activity of Prof Wł Gorczyńskiego and its continuation). UMK Toruń, pp. 321–336 (in Polish)

Miętus M, Filipiak J (2002) Struktura czasowo-przestrzennej zmienności warunków opadowych w rejonie Zatoki Gdańskiej (The structure of spatio-temporal variability of pluvial conditions in the Gulf of Gdansk region). Materiały Badawcze IMGW, S Meteorologia 34 (in Polish)

Miętus M, Filipiak J (2004) The temporal and spatial patterns of thermal conditions in the area of the southwestern coast of the Gulf of Gdańsk (Poland) from 1951 to 1998. Int J Climatol 24:499–509

Miętus M, Owczarek M (1994) Variability of mean annual amplitude of air temperature in relation to the mean cloudiness on the Polish coast since 1945. In: Heino R (ed) Climate Variations in Europe. Proceedings of the European Workshop held in Kirkkonummi (Majvik), Finland, 15–18 May 1994. Publ of the Academy of Finland 3/94:128–135

Miętus M, Filipiak J, Owczarek M (2004) Klimat wybrzeża południowego Bałtyku. Stan obecny i perspektywy zmian (The climate of the Baltic Sea coast. Present conditions and future perspectives.) In: Cyberski J (ed) Środowisko Polskiej Strefy Południowego Bałtyku – stan obecny i przewidywane zmiany w przededniu integracji europejskiej (The environment of the Polish zone of the southern Baltic Sea – Present conditions and projected changes in the perspective of European integration). Gdańsk, Wyd. Gdańskie, pp. 11–44 (in Polish, with English summary)

Mikulski Z (1970) Inflow of river waters to the Baltic Sea in 1961–1970. Nordic Hydrology 4:216–227

Ministry of Environment of the Estonian Republic (2004) www.envir.ee

Moberg A, Bergström H (1997) Homogenisation of Swedish Temperature Data. Part III: The long temperature record from Stockholm and Uppsala. Int J Climatol 17:667–699

Moberg M, Jones PD (2005) Trends in indices for extremes in daily temperature and precipitation in central and western Europe 1901–1999. Int J Climatol 25:1149-1172

Moberg A, Jones PD, Barriendos M, Bergstrøm H, Camuffo D, Cocheo C, Davis TD, Demaree G, Martin-Vide J, Maugeri M, Rodriguegues R, Verhoeve T (2000) Day-to-day temperature variability trends in 160– to 275–year-long European instrumental records. J Geophys Res 105:22849–22868

Moberg A, Bergström H, Krigsman JR, Svanered O (2002) Daily air temperature and pressure series for Stockholm (1756–1998). Climatic Change 53:171–212

Moberg A, Alexandersson H, Bergström H, Jones PD (2003) Were the southern Swedish summer temperatures before 1860 as warm as measured? Int J Climatol 23:1495–1521

Moberg A, Tuomenvirta H, Nordli Ø (2005) Recent Climatic Trends. In: Seppälä M (ed) Chapter 7: The Physical Geography of Fennoscandia. Oxford Regional Environments Series, pp. 113–133

Mohrholz V, Dutz J, G Kraus (2006) The impact of exceptionally warm summer inflow events on the environmental conditions in the Bornholm Basin. J Mar Syst 60:285–301

Morawska-Horawska M (1985) Cloudiness and sunshine in Cracow 1861–1980 and its contemporary tendencies. Int J Climatol 5:633–642

Nekrasova L (2004) Influence of climate fluctuation during 1988–2002 on a hydrological regime of the rivers of Belarus. XXIII Nordic Hydrological Conference, Tallinn, Estonia. NHP Report 48:696–703

Nesje A, Dahl SO (2003) The 'Little Ice Age' – only temperature? The Holocene 13:139-145

New M, Todd M, Hulme M, Jones P (2001) Precipitation measurements and trends in the twentieth century. Int J Climatol 21:1899-1922

Niedźwiedź T (1981) Sytuacje synoptyczne i ich wpływ na zróżnicowanie przestrzenne wybranych elementów klimatu w dorzeczu górnej Wisły (Synoptic situations and their influence on spatial differentiation of the selected climatic elements in the Upper Vistula basin), Rozprawy Habilitacyjne Uniwersytetu Jagiellońskiego, nr 58, Kraków (in Polish)

Niedźwiedź T (1993) Changes of atmospheric circulation (using the P, S, C, M indices) in the Winter season and their influence on air temperature in Cracow "Early Meteorological Records in Europe – Methods and Results", Zeszyty Naukowe UJ – Prace Geograficzne 95:107–113

Niedźwiedź T (1996) Long-Term Variability of the Zonal Circulation Index above the Central Europe. In: Obr bska-Starkel B, Niedźwiedź T (eds) Proceedings of the International Conference Clim Dynam and the Global Change Perspective, October 17–20 1995, Kraków, Zeszyty Naukowe UJ – Prace Geograficzne, 102:213–219

Niedźwiedź T (2000a) Dynamics to selected extreme climatic events in Poland. Geographia Polonica 73:25–39

Niedźwiedź T (2000b) Variability of the atmospheric circulation above the Central Europe in the light of selected indices. In: Obrębska-Starkel B (ed) Reconstructions of Climate and its Modelling, Prace Geograficzne, Institute of Geography of the Jagiellonian University Cracow 107:379-389

Niedźwiedź T (2003a) The extreme precipitation in Central Europe and its synoptic background. Papers on Global Change, Warszawa IGBP 10:15–29

Niedźwiedź T (2003b) Extreme precipitation events on the northern side of the Tatra Mountains. Geographia Polonica 76:13–21

Niedźwiedź T (2003c) Variability of Atmospheric Circulation in Southern Poland in the 20th Century. In: Pyka JL, Dubicka M, Szczepankiewicz-Szmyrka A, Sobik M, Błaś M (eds) Man and Climate in the 20th Century. Acta Universitatis Wratislaviensis No 2542 Studia Geograficzne 75:230–240

Niedźwiedź T (2005) Catalogue of synoptic situations in the upper Vistula river basin (1873–2004). Computer file available at: Department of Climatology, Faculty of Earth Sciences, University of Silesia, Bedzinska 60, 41–200 Sosnowiec, Poland; niedzwie@ultra.cto.us.edu.pl

Niedźwiedź T, Twardosz R (2004) Long-term variability of precipitation at selected stations in Central Europe. Papers on Global Change IGBP, 11:73–100

Niedźwiedź T, Ustrnul Z (1994) Maximum and Minimum Temperatures in Poland and the Variability of Atmospheric Circulation. Proceedings of the meeting of the Comission on Climatology of the IGU, 15–20 August 1994, Brno pp. 420–425

Niedźwiedź T, Ustrnul Z, Cebulak E, Limanówka D (1994) Long-Term Climate Variations in Southern Poland Due to Atmospheric Circulation Variability. In: Heino R (ed) Climate Variations in Europe Proceedings of the European Workshop on Climate Variations held in Kirkkonummi (Majvik), 15–18 May 1994, Painatuskeskus Helsinki, pp. 263–277

Nilsson C, Stjernquist I, Bärring L, Schlyter P, Jönsson AM, Samuelsson H (2004) Recorded storm damage in Swedish forests 1901–2000. Forest Ecol Manag 199:165–173

Nordli PØ (2001) Reconstruction of Nineteenth Century summer temperatures in Norway by proxy data from farmers' diaries. Climatic Change 48:201–218

Nordli Ø, Lie Ø, Nesje A, Dahl SO (2003) Spring-Summer Temperature Reconstruction in western Norway 1734–2003: A data-synthesis approach. Int J Climatol 23:1821–1841

Ogilvie A, Jónsson T (2001) "Little Ice Age" Research: A Perspective from Iceland. Climatic Change 48:219-241

Ohvril H, Okulov O, Jaagus J (1998) Atmospheric transparency in Estonia during last 60 years. In: Proceedings of the 2nd International Conference on Climate and Water. Espoo, Finland, 17–20 Aug 1998, 2:682–690

Okulov O (2003) Variability of atmospheric transparency and precipitable water in Estonia during last decades. PhD theses in Geophysics, Dissertationes Geophysicales Universitatis Tartuensis

Omstedt A, Chen D (2001) Influence of atmospheric circulation on the maximum ice extent in the Baltic Sea. J Geophys Res 106:4493–4500

Omstedt A, Nohr C (2004) Calculating the water and heat balances of the Baltic Sea using ocean modelling and available meteorological hydrological and ocean data. Tellus 56:400–414

Omstedt A, Elken J, Lehmann A, Piechura J (2004a) Knowledge of the Baltic Sea Physics gained during the BALTEX and related programmes. Progr Oceanogr 63:1–28

Omstedt A, Pettersen C, Rodhe J, Windsor P (2004b) Baltic Sea climate: 200 yr of data on air temperature sea level variation ice cover and atmospheric circulation. Clim Res 25:205–216

Orviku K (1992) Characterization and evolution of Estonian seashores. Summary of doctoral thesis. University Tartu.

Orviku K (1993) The present state of Estonian coasts and evolution tendencies. Proceedings of the International Coastal Congress ICC-Kiel '92 (Kiel Germany) pp. 621–624

Orviku K, Jaagus J, Kont A, Ratas U, Rivis R (2003) Increasing activity of coastal processes associated with climate change in Estonia. J Coast Res 19:364–375

Osborn TJ (2004) Simulating the winter North Atlantic Oscillation: The roles of internal variability and greenhouse gas forcing. Clim Dyn 22:605–623, data available at: www.cruueaacuk/~timo/proj-pages/nao_update.htm

Overland JE, Spillane MC, Percival DB, Wang M, Mofjeld HO (2004) Seasonal and Regional Variation of Pan-Arctic Surface Air Temperature over the Instrumental Record. J Clim 17:3263–3282

Palosuo E (1953) A treatise on severe ice conditions in the Baltic Sea. Publ Finnish Institute of Marine Research, No 156

Påsse T (1998) Lake-tilting a method for estimation of isostatic uplift. Boreas 27:69–80

Peterson TC, Easterling DR, Karl TR, Groisman P, Nicholls N, Plummer N, Torok S, Auer I, Boehm R, Gullet D, Vincent L, Heino R, Tuomenvirta H, Mestre O, Szentimrey T, Salinger J, Førland EJ, Hanssen-Bauer I, Alexandersson H, Jones P, Parker D (1998) Homogeneity adjustments of in situ atmospheric climate data: A review. Int J Climatol 18:1493–1517

Pettersson O (1893) Resultaten av den svenska hydrografiska expeditionen år 1877 (Results from the Swedish hydrographic expedition in 1877). Kungliga Svenska Vetenskaps-akademiens Handligar. Bandet 25. 1:73–161 (in Swedish)

Piotrowicz K, Domonkos P (2002) Fluctuations in annual sums of winter cold days and summer warm days with selected central European stations as examples and their connection to the large scale circulation. Prace Geograficzne 110:25–45

Plag HP, Tsimplis MN (1999) Temporal variability of the seasonal sea-level cycle in the North Sea and Baltic Sea in relation to climate variability. Glob Planet Change 20:173–203

Podstawczyńska A (2003) Variability of sunshine duration in Łódź in 1951–2000. Man and Climate in the 20th Century, Studia Geograficzne 75, Acta Universitatis Wratislaviensis 2542:282–291

Pruszak Z (2001) Vulnerability and Adaptation of Polish coast to impact of sea-level rise (SLR), Archives of Hydro-Engineering and Environmental Mechanics 48:73–90

Pruszak Z, Okrój T (1998) Sediment transport between land and sea around the Southern Baltic. International Decade for Natural Disaster Reduction (INDR), German INDR-Series No 13 pp. 18

Pruszak Z, Różyński G (2001) Data-driven analysis and modeling of shoreline evolution trends. Proceedings International Conference on Coastal Research through Large Scale Experiments – CD'01 ASCE Lund pp. 741–750

Pryor SC, Barthelmie RJ (2003) Long-term trends in near-surface flow over the Baltic. Int J Climatol 23:271–289

Pryor SC, Barthelmie RJ, Schoof JT (2005) The impact of non-stationarities in the climate system on the definition of 'a normal wind year': A case study from the Baltic. Int J Climatol 25:735–752

Raudsepp U (1998) Climatic aspects of the sea level variations in Estonia. In: Climate change studies in Estonia. Tallinn: Stockholm Environment Institute Tallinn Centre, 1998, pp. 79-84

Raudsepp U, Toompuu A, Kõuts T (1999) A stochastic model for the sea level in the Estonian costal area. J Mar Syst 22:69-87

Raukas A, Bird ED, Orviku K (1994) The provenance of beaches on the Estonian islands of Hiiumaa and Saaremaa. Proceedings Estonian Academy of Sciences, Geology 43:81–92

Reynolds RW, Smith TM (1994) Improved global sea surface temperature analyses. J Clim 7:929-948

Rivis R (2004) Changes in shoreline positions on the Harilaid Peninsula, West Estonia, during the 20th century. Proceedings of the Estonian Academy of Sciences. Biology, Ecology 53:179-193

Rogers JC (1984) The accociation between the North Atlantic Oscillation and the Southern Oscillation in the northern hemisphere. Mon Weather Rev 112:1999-2015

Rogers JC (1997) North Atlantic storm track variability and its association to the North Atlantic Oscillation and climate variability of northern Europe. J Clim 10:1635–1647

Rönkkönen S, Ojaveer E, Raid T, Viitasalo M (2004) Long-term changes in Baltic herring (*Clupea harengus membras*) growth in the Gulf of Finland. Can J Fish Aquat Sci 61:219-229

Różyński G (2003) Coastal nearshore morphology in terms of large data sets, DSc thesis, IBW PAN Publishers, Gdańsk

Różyński G (2005) Long-term shoreline response of a nontidal barred coast. Coast Eng 52:79–91

Russak V (1990) Trends of solar radiation cloudiness and atmospheric transparency during recent decades in Estonia. Tellus 42:206–210

Russak V (1994) Is the radiation climate in the Baltic Sea region changing? Ambio 23:160–163

Russak V (1998) Seasonal peculiarities of long-term changes in radiation regime in Estonia. In: Kallaste T, Kuldna P (eds) Climate change studies in Estonia. Stockholm Environment Institute, Tallinn, pp. 13–20

Russak V (1999) Muutustest kliimaelementide aegridades varakevadises Eestis (Changes in time series of some climate elements in Estonia in early spring). Publicationes Instituti Geographici Universitatis Tartuensis, 85:52–59 (in Estonian)

Russak V, Kallis A (2003) Eesti kiirguskliima teatmik (Handbook of Estonian solar radiation climate). Eesti Meteoroloogia ja Hüdroloogia Instituut, Tallinn, 384 (in Estonian)

Samuelsson M (1996) Interannual salinity variations in the Baltic Sea during the period 1954–1990. Continent Shelf Res 16:1463–1477

Samuelsson M, Stigebrandt A (1996) Main characteristics of long-term sea level variability in the Baltic Sea. Tellus 48:672–683

Sass AF (1866) Untersuchungen über die Eisbedeckung des Meeres an den Küsten der Inseln Ösel und Moon (Investigations on sea ice cover at the coasts of the islands Ösel and Moon). Bulletin de l'Academie Imperiale des Sciences de St-Petersbourg No 9:145–188 (in German)

Schinke H, Matthäus W (1998) On the causes of major Baltic inflows. An analysis of long time series. Continent Shelf Res 18:67–97

Schmidt H, von Storch H (1993) German bight storms analyzed. Nature 365:791–791

Schmith T (2001) Global warming signature in observed winter precipitation in Northwestern Europe? Clim Res 17:263–274

Schmith T, Alexandersson H, Iden K, Tuomenvirta H (1997) North Atlantic pressure observations 1868–1995. (WASA dataset 10; CD-ROM included) Technical Report No 97-3. Danish Meteorological Institute Copenhagen, Denmark

Schmith T, Kaas E, Li TS (1998) Northeast Atlantic winter storminess 1875–1995 re-analysed. Clim Dyn 14:529–536

Schrum C (2001) Regionalization of climate change for the North Sea and Baltic Sea. Clim Res 18:31–37

Schrum C, Janssen F (2002) Decadal variability in Baltic Sea ice development. Analysis of model results and observations. Fourth Workshop on Baltic Sea Ice Climate, Norrköping, Sweden 22–24 May 2002, SMHI, Oceanografi 72:49–58

Schrum C, Martinez-Lopez B, Siegismund F (2002) Modellierte Klimatologie von Wärmeinhalt und Salzgehalt in Nordsee und Ostsee (Modelled climatological heat content and salinity in the North and Baltic Seas). Berichte aus dem Zentrum für Meeres- und Klimaforschung der Universität Hamburg, Reihe B 45, 205 pp. (in German)

Schrum C, Hübner U, Jacob D, Podzun R (2003) A coupled atmosphere/ice/ocean model for the North Sea and the Baltic Sea. Clim Dyn 21:131–151

Schuster RL (1996) Socio-economic significance of landslides: In: Turner AK, Schuster RL (eds) Landslides: Investigation and Mitigation: Sp. Rep. 247, Transportation Research Board, National Research Council. National Academy Press, Washington DC, pp. 12–31

Schwarzer K, Diesing M, Larson M, Niedermeyer R, Schumacher W, Furma czyk K (2003) Coastline evolution at different time scales-examples from the Pomeranian Bight, Southern Baltic Sea. Mar Geol 194:79-101

Seinä A (1993) Ice time series of the Baltic Sea. Proc 1st Workshop of the Baltic Sea Ice Climate, Tvärminne, Finland, 24–26 August 1993. Dept of Geophysics, Univ of Helsinki, Report Series in Geophysics 27:87–90

Seinä A (1994) Extent of ice cover 1961–1990 and restrictions to navigation 1981–1990 along the Finnish Coast. Finnish Marine Research No 262.

Seinä A, Palosuo E (1996) The classification of the maximum annual extent of ice cover in the Baltic Sea 1720–1995. Meri – Report Series of the Finnish Institute of Marine Research 20:79–910

Ševkunova E, Meštšerskaja A, Jaani A (1999) Baltikum kui kliimamuutuste eriline piirkond (The Baltic area – a specific region of climate change). Publicationes Instituti Geographici Universitatis Tartuensis 85:156–163 (in Estonian with summary in English)

Shiklomanov I, Georgievsky V (2002) The influence of anthropogenic climate changes on hydrological regime and water resources. In: Climate changes and their consequences. St. Petersburg Nauka (in Russian)

Siegel H, Gerth M, Tiesel R, Tschersich G (1999) Seasonal and interannual variations in satellite derived sea surface temperature of the Baltic Sea in the 1990s. Dt Hydrogr Z 51:407–422

Siegel H, Gerth M, Tschersich G (2006) Sea surface temperature development of the Baltic Sea in the period 1990–2004. Oceanologia 48:119-131

Siegismund F, Schrum C (2001) Decadal changes in the wind forcing over the North Sea. Clim Res 18:39-45

Simojoki H (1940) Über die Eisverhältnisse der Binnenseen Finnlands (On the ice conditions of Finnish lakes). Mitteilungen des Meteorologischen Institutes der Universität Helsinki 43 (in German)

Simojoki H (1952) Die Eisverhältnisse in den Finnland umgebenden Meeren in den Wintern 1946–50 (Sea ice conditions in Finnish coastal waters in the winters 1946–1950). Publ Finnish Institute of Marine Research No 154 (in German)

Skeie P (2000) Meridional flow variability over the Nordic sea in the Arctic Oscillation framework. Geophys Res Lett 27:2569-2572

Slonosky VC, Jones PD, Davies TD (2000) Variability of the surface atmospheric circulation over Europe 1774–1995. Int J Climatol 20:1875–1897

Slonosky VC, Jones PD, Davies TD (2001) Atmospheric circulation and surface temperature in Europe from the 18th century to 1995. Int J Climatol 21:63–75

SMHI (2001) Erfarenheter av sommarens översvämningar (Experience of the summer flooding). Regeringsrapport 18 (in Swedish)

Smits A, Klein-Tank AMG, Können GP (2005) Trends in storminess over the Netherlands 1962–2002. Int J Climatol 25:1331–1344

Solantie R (1994) Effect of weather and climatological background on snow damage of forests in Southern Finland in November 1991. Silva Fennica 28:203–211

Solantie R (2000) Snow depth on January 15th and March 15th in Finland 1919–1998 and its implications for soil frost and forest ecology. Finnish Meteorological Institute, Meteorological Publications 176

Solantie R, Junila P (1995) The correction of precipitation measurements based on comparisons between Tretyakov and Wild gauges. Finnish Meteorological Institute, Meteorological Publications 33, 142 pp (in Finnish with summary in English)

Solantie R, Tuomi T (2000) On the Areal and Temporal Distribution of Thunder in Finland. Geophysica 36:49-68

Speerschneider CIH (1915) Om isforholdene i Danske farvande i aeldre og nyere tid aarene 690–1860 (On ice conditions in Danish waters in past and recent times, the years 690–1860). Publikationer fra det Danske Meteorologiske Institut Meddelelser Nr 2, Copenhagen (in Danish)

Speerschneider CIH (1927) Om isforholdene i Danske farvande aarene 1861–1906 (On ice conditions in Danish waters for the years 1861–1906). Publikationer fra det Danske Meteorologiske Institut Meddelelser Nr 6, Copenhagen (in Danish)

Stankūnavičius G (1998) Cloudiness. In: The variability of changes of climatic elements in Lithuanian territory 110–132 (in Lithuanian)

Stankūnavičius G, Bartkevicien G (2002) The extreme circulation conditions in North Atlantic. Part I: The centers of action. Geography 38:22–33 (in Lithuanian with English summary)

Stenij SE and Hela I (1947) Suomen merenrannikoiden vedenkorkeuksien lukuisuudet (Frequency of the water heights on the Finnish coasts). Merentutkimuslaitoksen julkaisu (Finnish Marine Research) 138 (in Finnish with English summary)

Stigebrandt A (1983) A model for the exchange of water and salt between the Baltic and the Skagerrak. J Phys Oceanogr 13:411–427

Stigebrandt A (1984) Analysis of an 89-year-long sea level record from the Kattegat with special reference to the barotropically driven water exchange between the Baltic and the sea. Tellus 36:401–408

Stigebrandt A (2003) Regulation of the vertical stratification length of stagnation periods and oxygen conditions in the deeper deepwater of the Baltic proper. In: Fennel W, Hentzsch B (eds) Marine Science Reports, Baltic Sea Research Institute, Warnemünde 54:69–80

Stigebrandt A, Gustafsson BG (2003) Response of the Baltic Sea to climate change – Theory and observations. J Sea Res 49:243–256

Stigge HJ (1993) Sea level changes and high-water probability on the German Baltic Coast. International Workshop, Sea Level Changes and Water Management 19–23 April 1993. Noorsdwijerhout Nederlands, pp. 19–29

Sun B, Groisman PY (2000) Cloudiness variations over the former Soviet Union. Int J Climatol 20:1097–1111

Suursaar Ü, Kullas T, Otsmann M (2002) A model study of the sea level variations in the Gulf of Riga and the Väinameri Sea. Cont Shelf Res 22:2001–2019

Suursaar Ü, Jaagus J, Kullas T (2006) Past and future changes in sea level near the Estonian coast in relation to changes in wind climate. Boreal Env Res 11:123–142

Sztobryn M (1994) Long-term changes in ice conditions at the Polish coast of the Baltic Sea, Proc. IAHR Ice Symposium, Norwegian Inst. Techn, pp. 345–354.

Sztobryn M, Stanislawczyk I (2002) Changes of sea ice climate during the XX century – Polish coastal waters. In: Omstedt A, Axell L (eds) Fourth Workshop on the Baltic Sea Ice Climate. Norrköping, Sweden, SMHI, Oceanografi No 72:69–76

Tarand A (1993) The Tallinn time series of ice break-up as a climate indicator. Report Series in Geophysics University of Helsinki, Finland 27:91–93

Tarand A, Nordli PO (2001) The Tallinn temperature series reconstructed back half a millennium by use of proxy data. Climatic Change 48:189-199

Teittinen JJ (2000) Tornadoes in Finland during the years 1997-1999. Poster P39. 20th Conference on Severe Local Storms. Orlando, Florida, USA, 11–16 September 2000

Tinz B (1996) On the relation between annual maximum extend of ice cover in the Baltic Sea and sea level pressure as well as air temperature field. Geophysica 32:319-341

Titkova TB, Kononova NK (2006) Relationship between snow accumulation anomalies and general atmospheric circulation. Annals of Russian Academy of Sciences, Series Geography 1:35–46

Tomingas O (2002) Relationship between atmospheric circulation indices and climate variability in Estonia. Boreal Env Res 7:463–469

Tooming H (1995) Dependence of surface albedo on snow cover duration and other snow parameters in Estonia. Meteorol Zeitschrift NF 4:62–66

Tooming H, Kadaja J (1995) Changes in snow cover and surface albedo in Estonia during the last 100 years. Meteorol Zeitschrift NF 4:67–71

Tooming H, Kadaja J (2000a) Snow cover and surface albedo in Estonia. Meteorol Zeitschrift NF 9:97–102

Tooming H, Kadaja J (2000b) Eesti lumikatte atlas (Snow cover atlas of Estonia). EMHI, Tallinn, pp. 305 (in Estonian)

Tooming H, Kadaja J (2006) (eds) Eesti Lumikatte Teatmik – Handbook of Estonian snow-cover. Estonian Meteorological and Hydrological Institute, Tallinn-Saku

Tooming H, Merilain M (2004) A tornado map for Estonia. J Meteorol 29:51–57

Toompuu A (1998) Total amount of heat and salt in the Baltic. In: Tarand A, Kallaste T. Tallinn (eds) Country case study on climate change impacts and adaptation assessments in the Republic of Estonia Sea: 6. Baltic Sea and Estonian coast: Potential impacts of climate variability and change on the dynamics hydrography and related biology of the Baltic Sea. Stockholm Environment Institute, Tallinn Centre, pp. 110–115

Toompuu A, Wulff F (1996) Optimum spatial analysis of monitoring data on temperature salinity and nutrient concentrations in the Baltic Proper. Env Monit Assess 43:283–308

Trenberth KE (1999) Conceptual Framework for Changes of Extremes of the Hydrological Cycle with Climate Change. Climatic Change 42:327–339

Tuomenvirta H (2001) Homogeneity adjustments of temperature and precipitation series. Finnish and Nordic data. Int J Climatol 21:495–506

Tuomenvirta H (2004) Reliable estimation of climatic variations in Finland. Finnish Meteorological Institute Contribution No 43

Tuomenvirta H, Alexandersson H, Drebs A, Frich P, Nordli PØ (1998) Trends in Nordic and Arctic extreme temperatures. DNMI-Report 13/98 Klima

Tuomenvirta H, Alexandersson H, Drebs A, Frich P, Nordli PØ (2000) Trends in Nordic and Arctic temperature extremes and ranges. J Clim 13:977–990

Tushinskiy GK (1970) Soviet Union regions with avalanche dangers. Moscow State University, Moscow

Tveito OE, Førland EJ, Dahlström B, Elomaa E, Frich P, Hanssen-Bauer I, Jónsson T, Madsen H, Perälä J, Rissanen P, Vedin H (1997) Nordic Precipitation maps. DNMI-Report 22/97 KLIMA

Tveito OE, Førland EJ, Alexandersson H, Drebs A, Jonsson T, Vaarby-Laursen E (2001) Nordic climate maps. DNMI report 06/01 KLIMA

Twardosz R, Nied wied T (2001) Influence of synoptic situations on the precipitation in Kraków. (Poland) Int J Climatol 21:467–481

Ustrnul Z (1997) Zmienność cyrkulacji atmosfery na półkuli północnej w XX wieku (Variability of atmospheric circulation in Northern Hemisphere during the 20th century). Materia y Badawcze Seria: Meteorologia – 27, IMGW, Warszawa (in Polish)

Ustrnul Z (1998) Zmienność temperatury powietrza na wybranych stacjach Europy Środkowej na tle warunków cyrkulacyjnych (Variability of air temperature at the selected stations in Central Europe presented on the circulation conditions background). Acta Universitatis Lodziensis, Folia Geographica Physica 3, pp. 307—318 (in Polish with summary in English)

Ustrnul Z, Czekierda D (2001) Circulation background of the atmospheric precipitation in Central Europe (based on the Polish example). Meteorologische Zeitschrift 10:103–111

Velner H, Pärnapuu M, Saks A (2004) Water-flow and management of rivers in Estonia. 23th Nordic Hydrological Conference, Tallinn, Estonia. NHP Report 48:466–473

Vermeer M, Kakkuri J, Mälkki P, Boman H, Kahma KK, Leppäranta M (1988) Land uplift and sea level variability sepctrum using fully measured monthly means of tide gauge readings. Finnish Marine Research 256:1–75

Vihma T, Haapala (2005) Sea Ice. In: BALTEX Phase 1, State of the Art Report. International BALTEX Secretariat Publication 31:32–59

Vikebø F, Furevik T, Furnes G, Kvamsto NG, Reistad M (2003) Wave height variations in the North Sea and on the Norwegian continental shelf 1881–1999. Continent Shelf Res 23:251–263

Volchak A (2004) Calculation and forecast of the annual discharge of the Neman River in Byelorussia. Proceedings of the Fourth Study Conference on BALTEX Bornholm Denmark 24–28 May 2004. International BALTEX Secretariat Publication 29:141–142

von Storch H, Langenberg H, Feser F (2000) A spectral nudging technique for dynamical downscaling purposes. Mon Wea Rev 128:3664–3673

Vuglinsky VS (2000) Extremely early and late dates of lake freezing and ice break-up in Russia. Proceedings of the 27. Limnological Congress in Dublin, vol. 27. Stuttgart, pp. 2793–2795

Vuglinsky VS (2002) Peculiarities of ice events in Russian Arctic rivers. Hydrolog Process 16:905–913

Vuglinsky VS, Zhuravin SA (1998) The estimation of river inflow into the Baltic Sea – provision with information peculiarities of forming variability. Proceedings of the Second Study Conference of BALTEX. International BALTEX Secretariat Publication 11:231–232

Vuglinsky VS, Zhuravin SA (2001) Long-term variations of inflow to the Gulf of Finland from the Neva River basin and the Lake Ladoga role in its control Proceedings of the Third Study Conference of BALTEX. International BALTEX Secretariat Publication 20:247–248

Vuglinsky VS, Gronskaya TP, Lemeshko NA (2002) Long-term characteristics of ice events and ice thickness on the largest lakes and reservoirs of Russia. "Ice in the Environment": Proceedings of the 16th International Symposium on Ice, v. 3, pp. 88–91

The WAMDI Group (1988) The WAM model – a third generation ocean wave prediction model. J Phys Oceanogr 18:1776–1810

Weisse R, Feser F (2003) Evaluation of a method to reduce uncertainty in wind hindcasts performed with regional atmosphere models. Coastal Eng 48:211–225

Weisse R, von Storch H, Feser F (2005) Northeast Atlantic and North Sea storminess as simulated by a regional climate model 1958–2001 and comparison with observations. J Clim 18:465–479

Werner PC, von Storch H (1993) Interannual variability of central European mean temperature in January–February and its relation to large-scale circulation. Clim Res 3:195–207

Weyhenmeyer G, Sonesten L (2002) Klimat och vattenstånd under 2001 (Climate and water level in 2001). In: Christensen A, (2002) Vänern – Årsskrift. Vänerns vattenvårdsförbund Report 22 (in Swedish)

Wibig J (1999) Precipitation in Europe in relation to circulation patterns at the 500 hPa level. Int J Climatol 19:253–269

Wibig J (2000) Oscylacja Północnoatlantycka i jej wpływ na kształtowanie pogody i klimatu (North Atlantic oscillation and its impact on weather and climate). Przegląd Geofizyczny R 45:121–137 (in Polish)

Wibig J (2001) Wpływ cyrkulacji atmosferycznej na rozkład przestrzenny anomalii temperatury powietrza i opadów w Europie (Impact of the atmospheric circulation on spatial distributions of air temperature and precipitation anomalies in Europe). Wydawnictwo Uniwersytetu Łódzkiego, Łódź, pp. 208 (in Polish)

Wibig J (2004) Long-term variability of cloudiness and its relation to precipitation on the example of Łódź. Institute of Geophysics, Polish Academy of Science, Monographic Volume E-4 (377) Potential climate changes and sustainable water management, pp. 25–32

Wibig J, Głowicki B (2002) Trends of minimum and maximum temperature in Poland. Clim Res 20: 123–133

Wigley TML (1985) Impact of extreme events. Nature 316:106–107

Wijngaard JB, Klein Tank AMG, Können GP (2003) Homogenity of 20th century European daily temperature and precipitation series. Int J Climatol 23:679–692

Wild M, Gilgen H, Roesch A, Ohmura A, Long CN, Dutton EG, Forgan B, Kallis A, Russak V, Tsvetkov A (2005) From dimming to brightening: Decadal changes in solar radiation at earth's surface. Science 308:847–850

Winsor P, Rodhe J, Omstedt A (2001) Baltic Sea ocean climate: An analysis of 100 yr of hydrographic data with focus on the freshwater budget. Clim Res 18:5–15

Winsor P, Rodhe J, Omstedt A (2003) Erratum: Baltic Sea ocean climate: An analysis of 100 yr of hydrographical data with focus on the freshwater budget. Clim Res 25:183

Wise (1999) Workshop report of the Workshop on Economic and social impacts of climate extremes: Risks and benefits, October 14–16, 1999. Amsterdam, pp. 39

Woodruff SD (2001) Quality control in recent COADS updates. Proceedings of International Workshop on preparation, processing and use of historical marine meteorological data. Tokyo, Japan, 28–29 November 2000. Japan Meteorological Agency and the Ship & Ocean Foundation pp. 49–53

Woodruff SD, Diaz HF, Elms JD, Worley SJ (1998) COADS Release 2 data and metadata enhancements for improvements of marine surface flux fields. Phys Chem Earth 23:517–527, www.cdc.noaa.gov/coads/egs_paper.html

Wrzesinski D (2004) Flow regimes of rivers of northern and central Europe in various circulation periods of the North Atlantic Oscillation (NAO). XXIII Nordic Hydrological Conference, Tallinn, Estonia, NHP Report 48:671–679

Yoo JC, D'Odorico P (2002) Trends and fluctuations in the dates of ice break-up of lakes and rivers in Northern Europe: The effect of the North Atlantic Oscillation. J Hydrol 268:100–112

Zawadzka E (1994) South Baltic coastal changes during last 100 years. Book of Abstracts of ICCE'94 Kobe, pp. 568–569

Zawadzka E (1999) Development tendencies of the Polish south Baltic coast (in Polish). GTN Gdansk, pp. 147

Zilinskavas G, Jarmalavicius D, Minkevicius V (2001) (Aeolian processes on the marine coast). The Institute of Geography, Vilnius (in Lithuanian with English summary)

Żmudzka E (2003) Wielkość zachmurzenia w Polsce w drugiej połowie XX wieku (Cloudiness in Poland in the 2nd half of the 20th century). Prz. Geof. XLVIII 3–4:159–185 (in Polish)

Zorita E, Laine A (2000) Dependence of salinity and oxygen concentrations in the Baltic Sea on large-scale atmospheric circulation. Clim Res 14:25–41

3 Projections of Future Anthropogenic Climate Change

L. Phil Graham, Deliang Chen, Ole Bøssing Christensen, Erik Kjellström, Valentina Krysanova, H. E. Markus Meier, Maciej Radziejewski, Jouni Räisänen, Burkhardt Rockel, Kimmo Ruosteenoja

3.1 Introduction to Future Anthropogenic Climate Change Projections

This chapter focuses on summarising projections of future anthropogenic climate change for the Baltic Sea Basin. This includes the science of climate change and how future projections are made, taking into account anthropogenic influence on greenhouse gases (GHG). Looking forward toward future climates requires using state-of-the-art modelling tools to represent climate processes.

The chapter begins with an overview of the current understanding of global anthropogenic climate change and how this is applied for projections into the 21st century. Important processes for the global climate and their representation in global climate models (GCMs) are introduced. Projected global changes are summarised and then put into specific context for the Baltic Sea Basin. This includes discussion of the performance of such models for the present climate. A range of future climate outcomes is presented, originating from using several GCMs and from using a set of different projected GHG emissions scenarios.

Due to the coarse scales of GCMs, downscaling techniques are used to produce detailed results on regional to local scales. Methods for both statistical downscaling and dynamical downscaling using regional climate models (RCMs) are described. Results for the key climate variables of precipitation and temperature, and others, are summarised for the Baltic Sea Basin. Projections of anthropogenic climate change are further coupled to hydrological and oceanographic processes via models to assess basinwide climate change impacts. Hydrological modelling shows how climate-driven changes impact on the distribution and timing of runoff into the Baltic Sea. Oceanographic modelling shows corresponding changes in water temperature, sea ice, salinity and sea levels.

3.2 Global Anthropogenic Climate Change

Before presenting anthropogenic climate change projections for the Baltic Sea Basin, it is necessary to give a brief overview of the current understanding of global anthropogenic climate change during and after the 21st century. In our discussion, we draw heavily on the IPCC Third Assessment Report (IPCC 2001a), particularly its chapter on projections of future anthropogenic climate change (Cubasch et al. 2001).

Changes in the global climate can occur both as a result of natural variability and as a response to anthropogenic forcing. Part of the natural variability is forced, that is, caused by external factors such as solar variability and volcanic eruptions; part is unforced, that is, associated with the internal dynamics of the climate system. The most important source of anthropogenic climate forcing is changes in the atmospheric composition. Increases in CO_2 and other greenhouse gases make the atmosphere less transparent for thermal radiation and therefore tend to warm up the surface and the troposphere.

However, human activities have also increased the concentrations of several aerosol types. The net effect of anthropogenic aerosols is thought to be to cool the global climate, although this effect is quantitatively much less well known than the impact of increasing greenhouse gases. The relative importance of aerosol-induced cooling, as opposed to greenhouse-gas-induced warming, is likely to decrease in the future.

External factors that may cause changes in the global climate are commonly compared in terms of globally averaged *radiative forcing*. Radiative forcing measures, in approximate terms, the change in the energy balance of the Earth–atmosphere system that a given change in external conditions would induce with no compensating changes in climate (for the exact definition, see IPCC 2001a, p. 795). Positive radiative forcing tends to increase and negative forcing decrease the global mean temperature. The magnitude of the temperature response depends on several feedback processes acting in the climate system and needs to be estimated with climate models. Model simulations suggest that the ratio of the response to the magnitude of the forcing is approximately the

133

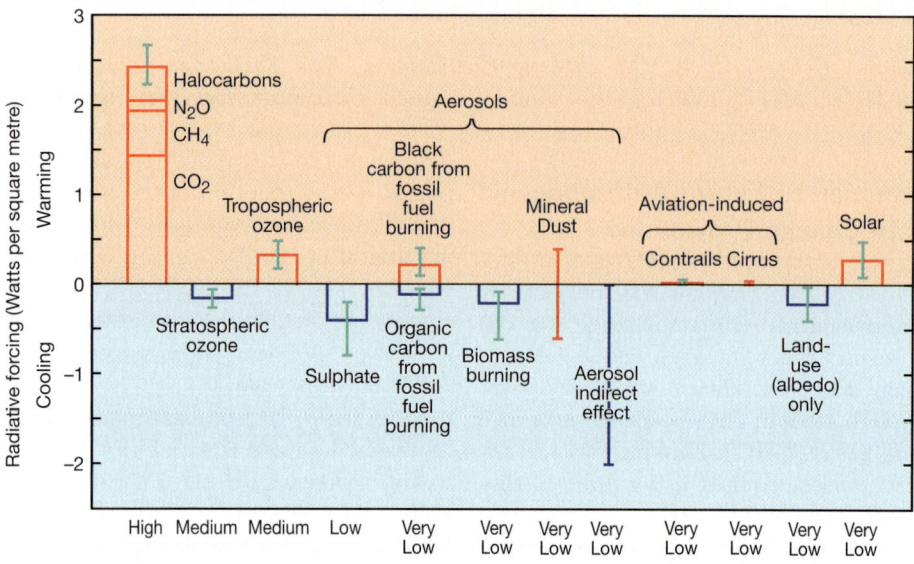

Fig. 3.1. Estimates of globally averaged radiative forcing resulting from various changes in external conditions from the year 1750 to the year 2000 (IPCC 2001a, Summary for Policymakers, Fig. 3). For each forcing agent, the bar shows the best estimate of the forcing and the vertical line the uncertainty range mainly based on the variation among published studies. See IPCC (2001a) for further details

same for different forcing agents (e.g. Forster et al. 2000; Joshi et al. 2003).

Estimates of the present-day radiative forcing from IPCC (2001a) are shown in Fig. 3.1. Increases in CO_2, CH_4, N_2O and other long-lived greenhouse gases since the preindustrial time are estimated to have caused a positive forcing of about $2.5\,Wm^{-2}$. This value is associated with only a relatively small uncertainty, unlike the effects of many other forcing agents. Stratospheric ozone depletion has caused a slight negative and increases in tropospheric ozone probably a slightly larger positive forcing. The direct effect of anthropogenic aerosols, associated with the scattering and absorption of solar radiation by aerosol particles, varies in sign between different aerosol types but the net forcing is probably negative. The indirect aerosol forcing associated with aerosol-induced increases in cloud albedo and lifetime is most probably negative. However, as the estimated uncertainty range 0 to $-2\,Wm^{-2}$ indicates, the magnitude of this effect is poorly known. Among the other forcing agents included in the figure, changes in land use are estimated to have caused a slight negative forcing. The forcing asso-

ciated with changes in solar irradiance is thought to be positive, mainly because of increases in solar irradiance in the first half of the 20th century.

Many of the forcing agents that are thought to have affected the global climate in recent decades and centuries are poorly known in quantitative terms. Nevertheless, the forcing estimates shown in Fig. 3.1 clearly suggest that the largest contribution to the observed global warming in the industrial era has come from increased greenhouse gas concentrations. This is the case especially for the last few decades when the increase in greenhouse gas concentrations has been most rapid. Estimates of greenhouse gas emissions and concentrations for the rest of the 21st century (discussed below) suggest that positive greenhouse gas forcing will become increasingly dominant in the future.

3.2.1 Global Warming in the 21st Century

In their projections of global climate change in the 21st century, Cubasch et al. (2001) focused on climate changes resulting from anthropogenic changes in the atmospheric composition. They estimated that changes in greenhouse gas and aerosol concentrations would raise the global mean temperature[1] by 1.4–5.8 °C between the years 1990 and 2100. The lower limit of this uncertainty interval is approximately twice the global mean warming observed in the 20th century. The upper limit is similar to the difference between present-day and ice-age conditions (e.g. Weaver et al. 1998). The wide uncertainty interval takes into account two sources of uncertainty: that due to the future emissions of greenhouse gases and aerosol precursors, and that associated with the response of the global mean temperature to a given change in the atmospheric composition. These two factors are discussed in more detail below. It is important to note that there are additional sources of uncertainty that may not be adequately accounted for by this range of projected changes, such as the conversion of emissions to atmospheric concentrations of GHGs and aerosols. Such issues are also further discussed below.

Projections of future anthropogenic climate change require estimates of future greenhouse gas and aerosol concentrations. Such estimates are dependent on information about future emissions. Because the emissions depend on factors such as population growth, economic growth, structure of economy, methods for producing energy and so on, precise prediction of them is impossible. Instead, several emissions scenarios are used. Such scenarios are based on alternative but plausible and internally consistent sets of assumptions about the demographic, socioeconomic and technological changes that together determine the evolution of emissions in the future.

A comprehensive set of emissions scenarios, the so-called SRES (Special Report on Emissions Scenarios) scenarios were developed and described by Nakićenović et al. (2000). The SRES scenarios were built around four narrative storylines that describe the evolution of the world in the 21st century. Altogether, 40 different emissions scenarios were constructed, 35 of which were detailed enough to be used in anthropogenic climate change projections. Six of these (A1B, A1T, A1FI, A2, B1 and B2) were chosen by the IPCC as illustrative marker scenarios. The main underlying assumptions behind these scenarios are described in Annex 6.

The various SRES emissions scenarios remain relatively similar during the early parts of the 21st century. In particular, they all indicate an increase in global CO_2 emissions in the next few decades (Fig. 3.2a), as a result of increasing energy consumption required by increasing population and world economy. Towards the late 21st century, the scenarios tend to diverge. Some of them (e.g. A1FI and A2) project a strong increase in CO_2 emissions throughout the century, leading by the year 2100 to emissions several times larger than today. In some other scenarios (e.g. A1T and B1), however, the CO_2 emissions peak by the mid-21st century and fall below the present level by the year 2100. These differences are reflected in the CO_2 concentrations derived from the emissions (Fig. 3.2b). In the year 2100, the B1 emission scenario is calculated to lead to a CO_2 level of about 550 parts per million (ppm), as compared to a present-day level of about 375 ppm. The corresponding value for the highest scenario (A1FI) is about 970 ppm. However, even for the B1 scenario with the largest decrease in CO_2 emissions after 2050, the CO_2 concentration still continues to rise slowly in the end of the 21st century.

The SRES scenarios also describe the emissions of several other greenhouse gases, including methane (CH_4) and nitrous oxide (N_2O). The resulting concentrations are quite variable across the scenarios, particularly for CH_4 which has a relatively short lifetime and therefore responds rapidly to changes in emissions. The A2 scenario is calculated to lead to a CH_4 concentration of about 3700 parts per billion (ppb) in the year 2100 (as compared to 1760 ppb in the year 2000), while the corresponding value for the B1 scenario is below 1600 ppb. For N_2O, all the SRES scenarios indicate an increase in the atmospheric concentration, although the rate of the increase depends on the projected emissions that vary substantially among the scenarios. Many of the scenarios also indicate an increase in average tropospheric ozone (O_3) concentration, as a result of increases in pollutants that participate in the formation of O_3. Most of the SRES scenarios suggest that the net effect from changes in non-CO_2 greenhouse gases will strengthen the global warming during this century; this effect is likely to be smaller than that of increasing CO_2, but it is not negligible.

Anthropogenic increases in atmospheric aerosol concentrations, which are thought to have suppressed the greenhouse-induced warming during the 20th century (Mitchell et al. 2001), will not

---

[1]The temperature discussed here and in the following refers to the two-meter level air temperature, if not otherwise stated.

The global climate of the 21st century

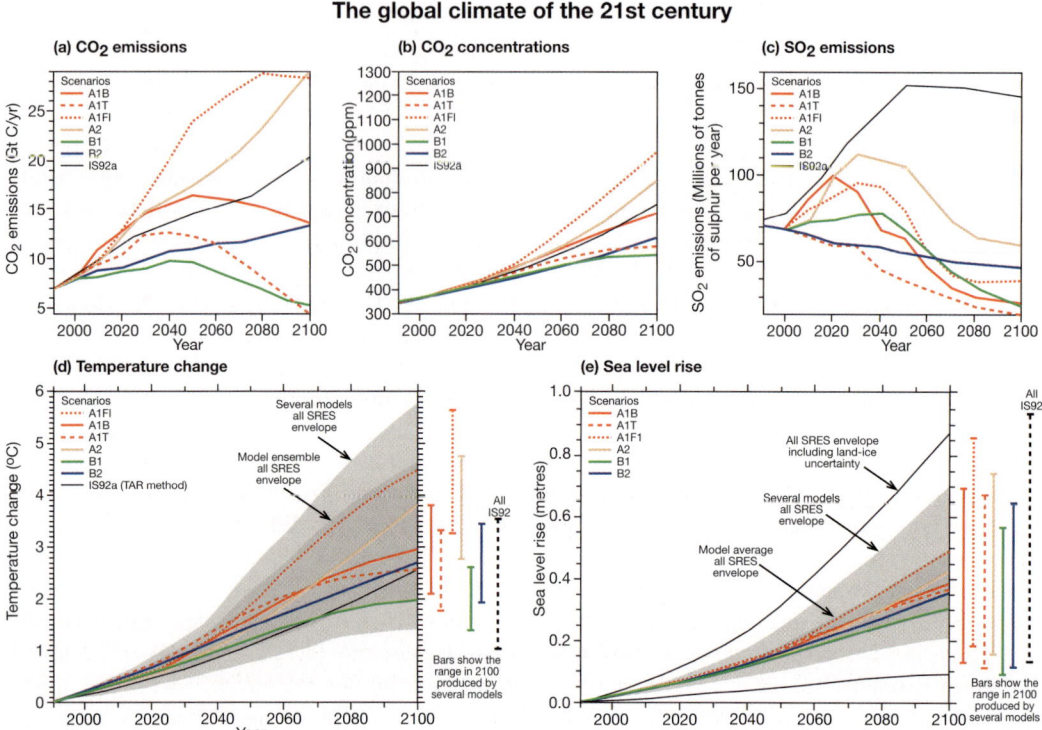

Fig. 3.2. Summary of some key factors related to global anthropogenic climate change in the 21st century, as presented in the IPCC Third Assessment Report (IPCC 2001a, Summary for Policymakers, Fig. 5). (**a**) shows the CO$_2$ emissions of the six illustrative SRES scenarios along with an older scenario (IS92a) used in the IPCC Second Assessment Report. (**b**) shows projected CO$_2$ concentrations. (**c**) shows anthropogenic SO$_2$ emissions. Note that the older IS92a scenario, with very large SO$_2$ emissions in the late 21st century, is now believed to be unrealistic. (**d**) and (**e**) show the projected global mean temperature and sea level responses, respectively. The "several models all SRES envelope" in (**d**) and (**e**) shows the temperature and sea level rise, respectively, for a simple climate model forced with all 35 SRES scenarios and tuned separately to mimic the behaviour of seven complex climate models. The "model average all SRES envelope" shows the average from these models for the range of scenarios. Note that the diagrams do not include all sources of uncertainty

necessarily continue to do so in the future. The main contributor to anthropogenic aerosol-induced cooling, SO$_2$ emissions, are still projected to increase in a global mean sense during the first decades of the 21st century in most SRES scenarios, which would suppress the warming during this period. By the end of the century, however, the world-wide introduction of cleaner technologies is projected to reduce the global SO$_2$ emissions distinctly below present-day levels (Fig. 3.2c). Due to the very short lifetime of tropospheric aerosols, this would result in an immediate decrease in sulphate aerosol concentrations. The large projected increases in greenhouse gas concentrations, in combination with small or negative changes in aerosol concentrations, would imply a large increase in the total anthropogenic radiative forcing (Fig. 3.3).

The second source of uncertainty included in the quoted 1.4–5.8 °C range is the imprecisely known response of the climate system to changes in atmospheric composition. To account for this uncertainty, Cubasch et al. (2001) used the results of seven different general circulation models (GCMs) to calibrate a simple climate model, which was run separately for all 35 SRES scenarios and with parameters corresponding to each of the seven GCMs. The GCMs were not used directly to simulate the climate evolution under all the SRES scenarios, because this would have been extremely demanding in terms of computing resources. However, the differences in simulated global mean warming between the GCMs and the calibrated simple model are likely to be small. Various types of climate models are described in more detail in Annex 6.

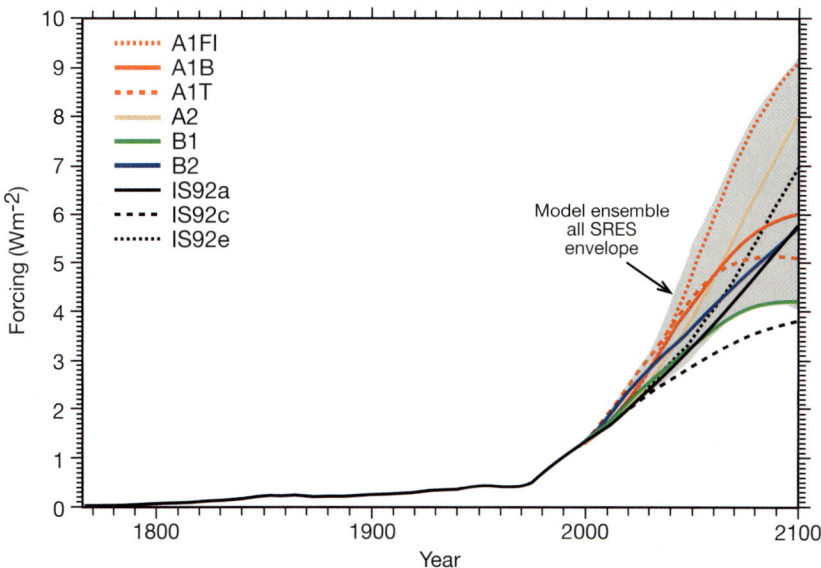

Fig. 3.3. Best-estimate historical anthropogenic radiative forcing up to the year 2000 followed by radiative forcing for the six illustrative SRES scenarios. The shading shows the envelope of the forcing that encompasses the full set of 35 SRES scenarios. Forcing estimates for three earlier scenarios (IS92a, IS92c and IS92e) are also shown (IPCC 2001a, Technical Summary, Fig. 19)

The computed evolution of the global mean temperature, presented as differences from 1990, is shown in Fig. 3.2d. The differences among the SRES scenarios remain relatively modest during the first decades of the century but grow rapidly thereafter, when the different emission and concentration projections diverge. Averaging the results of the seven models, the estimate of warming from the year 1990 to the year 2100 is 2.0 °C for the lowest and 4.6 °C for the highest SRES scenario. A direct comparison of these numbers with the full range of 1.4–5.8 °C would suggest that differences between emission scenarios are a larger source of uncertainty in century-scale global warming than differences among climate models. However, because the set of seven models used for generating these temperature projections did not include the least and most sensitive of all climate models, Cubasch et al. (2001) concluded that the uncertainties related to emission scenarios and climate models are of comparable importance.

The 1.4–5.8 °C range should not be interpreted as giving the absolutely lowest and highest possible global mean temperature changes by the year 2100. In fact, several sources of uncertainty are excluded or included only partially. One factor that is excluded is the uncertainty related to deriving the concentrations of CO_2 and other greenhouse gases from emissions. Due to uncertainties

in modelling the carbon cycle and particularly its response to anthropogenic climate changes, the actual CO_2 concentration resulting from the high A1FI emissions scenario in the year 2100 might be as low as 820 ppm or as high as 1250 ppm, as compared to the estimate used of 970 ppm (Prentice et al. 2001). The full uncertainty in estimating the radiative effects of atmospheric aerosols is also excluded, although the relative importance of this uncertainty will decrease when the greenhouse-gas-induced warming becomes increasingly dominant over the aerosol-induced cooling (Wigley and Raper 2001). Moreover, as noted above, the range does not necessarily capture the total uncertainty associated with climate models. Similarly, although a wide range of different emission scenarios are used, it is not inconceivable that the real emissions would fall outside this range. For example, because the SRES scenarios exclude new policy measures to control greenhouse gas emissions, success in international climate policy could in principle allow the emissions to fall below the SRES interval, or become less for some certain world development alternative.

The 1.4–5.8 °C range also excludes natural climate variability which could either amplify or counteract the anthropogenic warming. Model simulations (e.g. Bertrand et al. 2002) and recent estimates of past climate variability (Moberg et al. 2005) suggest that the natural variations of global mean temperature may amount to a few tenths of °C per century. This is not negligible, but considering the wide uncertainty interval in anthropogenic warming it only represents a relatively small additional uncertainty. On the other hand, natural climate variability increases towards smaller spatial scales. It may therefore have a substantial effect on regional climate changes, particularly in the near future when the anthropogenic forcing is still relatively weak (e.g. Hulme et al. 1999).

Despite the excluded uncertainties, temperature changes that fall somewhere in the middle of the 1.4–5.8 °C range seem to be more likely than changes that fall at the extremes or outside of this range. Wigley and Raper (2001) estimated probability distributions of future global warming by assuming equal likelihood for all 35 SRES scenarios and a log-normal probability distribution for climate sensitivity approximately representing the models used by Cubasch et al. (2001). Unlike Cubasch et al. (2001), they also allowed for uncertainty in the carbon cycle and aerosol forcing, but they found these additional sources of uncertainty to be of only secondary importance. They estimated the 5–95% uncertainty range of global warming from 1990–2100 as 1.7–4.9 °C, with a median of 3.1 °C. The corresponding 5–95% range for the warming in 1990–2070 was 1.3–3.3 °C and that for 1990–2030 0.5–1.2 °C. However, their analysis makes several simplifying assumptions, so that the actual uncertainty ranges may be wider.

3.2.2 Geographical Distribution of Anthropogenic Climate Changes

Global warming is expected to vary both geographically and seasonally (Cubasch et al. 2001). Continents are generally expected to warm more rapidly than the oceans, so that nearly all land areas are likely to warm faster than the global average (Fig. 3.4a). Particularly strong warming is projected for Northern Hemisphere high-latitude areas in winter, not only over land but even more over the Arctic Ocean, where the warming is greatly amplified by reduced sea ice. Most other ocean areas are likely to warm less rapidly than the global average. The simulated warming tends to be particularly modest over the Southern Ocean and in the northern North Atlantic. In these areas, the ocean is well-mixed to great depths, and surface warming is therefore retarded

by the slow warming of the deep ocean. In addition, most models simulate a decrease in the North Atlantic thermohaline circulation. This also acts to reduce the warming in the northern North Atlantic and actually leads in some models to local cooling in this area. Although the large-scale patterns of temperature change are reasonably similar between various models (e.g. Harvey 2004), there are substantial variations at smaller horizontal scales. These variations are caused mostly by differences in the models themselves, but also by unforced variability ("noise") superimposed on the forced anthropogenic climate change signal (e.g. Räisänen 2001a). The variation between different GCM simulations within the Baltic Sea Basin is addressed in more detail in Sect. 3.3.

There will likely be a slight increase in the globally averaged precipitation during this century. Most models suggest a 1–2% increase in global precipitation for each 1 °C increase in global mean temperature (Cubasch et al. 2001; Räisänen 2001a). However, more so than with temperature, precipitation changes will vary geographically (Cubasch et al. 2001; Räisänen 2001a; Harvey 2004; see also Fig. 3.4b). High-latitude areas are expected to experience a general increase in precipitation, particularly in winter, when increases are also likely in many mid-latitude areas. Increases are also expected in most tropical regions, although there is less consistency in precipitation change between different models at low latitudes than at high latitudes (Cubasch et al. 2001; Giorgio et al. 2001; Räisänen 2001a). By contrast, most subtropical regions and many regions in the lower midlatitudes are likely to suffer a decrease in mean precipitation. Simulated precipitation changes have a lower signal-to-noise ratio than temperature changes. Precipitation changes will therefore be more difficult to discern from natural variability than temperature changes, at least during the early stages of anthropogenic climate change. Partly for the same reason, precipitation changes are generally less consistent among the models than temperature changes, although the agreement tends to be better at high latitudes than at low latitudes (Räisänen 2001a).

The patterns of anthropogenic climate change shown in Fig. 3.4b were generated by averaging the results of 20 climate models. These are given in a normalized form as the ratio of the local temperature or precipitation change to the change in the global mean temperature. This way of presentation is justified by the model-based evidence

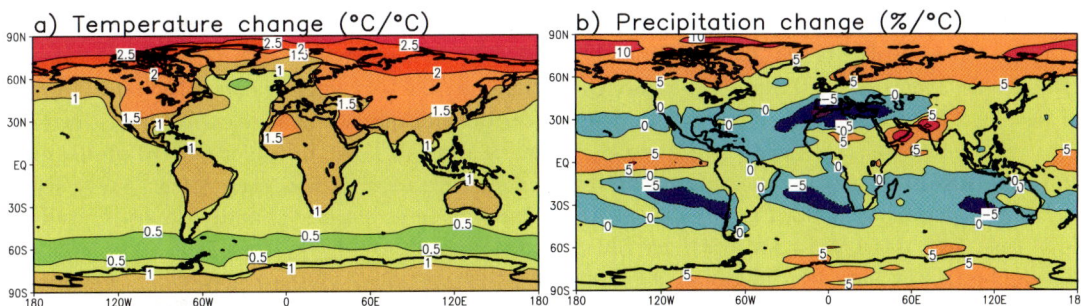

Fig. 3.4. Typical patterns of annual mean temperature and precipitation change in global climate models, both scaled for a 1 °C increase in global mean temperature. These maps are based on the idealised CMIP2 simulations in which CO_2 gradually doubles in 70 years (see Sect. 3.3.1). The temperature and precipitation changes in the 20-year period surrounding the doubling of CO_2 were first evaluated for 20 models and then averaged over all of them. Finally, the changes were divided by the global 20-model mean warming in these simulations (1.7 °C). Note that the geographical distribution of the changes differs between the models

that local anthropogenic climate changes tend to scale approximately linearly with the change in the global mean temperature (Mitchell et al. 1999; Huntingford and Cox 2000; Mitchell 2003; Harvey 2004). However, this pattern scaling principle only holds for the deterministic climate change signal associated with external forcing such as increased greenhouse gas concentrations, not for the noise associated with internally generated natural variability. For some climate variables, such as precipitation and windiness, it might take a long time before the anthropogenic climate change signal grows large enough that it can be easily discerned from the noise.

The detailed geographical patterns of change are model-dependent. They also depend to some extent on the forcing agents included in the simulation. Simulations that include changes in aerosol concentrations may give somewhat different patterns of anthropogenic climate change from simulations that only include increasing greenhouse gas concentrations, even for the same global mean warming.

A warmer atmosphere will be able to contain more water vapour. As a result, the high extreme values of daily precipitation are likely to increase even in many of those areas where the average precipitation remains unchanged or decreases (e.g. Zwiers and Kharin 1998; Hegerl et al. 2004). On the other hand, the number of precipitation days is likely to decrease in many regions of the world, including some regions with a slight increase in the average precipitation (Hennessy et al. 1997; Räisänen and Joelsson 2001; Räisänen et al. 2003).

3.2.3   Global Sea Level Rise

The thermal expansion of sea water, together with melting of glaciers, is computed to increase the globally averaged sea level by 9–88 cm from the year 1990 to the year 2100 (Fig. 3.2e). The uncertainty range is mostly associated with uncertainties in modelling anthropogenic climate change, ocean heat uptake and glacier behaviour. Because globally averaged sea level responds to anthropogenic climate changes with a substantial time lag, differences between emissions scenarios have only a limited effect on the projected sea level changes during this century.

3.2.4   Global Warming and Sea Level Rise After the Year 2100

The changes in global climate are expected to continue after the year 2100. Even if the atmospheric greenhouse gas concentrations were stabilised, the global mean temperature would still continue to rise for several centuries, although at a reduced rate. The additional global warming following the stabilisation of the atmospheric composition might exceed 1 °C (Cubasch et al. 2001, Fig. 9.19). The thermal expansion of sea water will continue for several centuries even after the surface temperature has stabilised, as the warming gradually penetrates deeper into the ocean. The total increase in global sea level due to thermal expansion alone might reach 1–3 m by the year 3000, assuming that atmospheric greenhouse concentrations were stabilised at a level equivalent to a quadrupling of the pre-industrial CO_2 concentration (Church et al. 2001). In addition to this, a gradual melting of the Greenland ice sheet appears likely if the local warming in Greenland exceeds 3 °C (Huybrechts and De Wolde 1999). This would increase the global sea level by about 7 m. For a mid-range scenario with a warming of 5.5 °C in Greenland, about 3 m of this increase would be realized by the year 3000 (Church et al. 2001, Fig. 11.16). A decrease in the mass of the West Antarctic Ice Sheet could also potentially increase the global sea level, by up to 6 m for a complete melting of the ice sheet. Whether changes in this ice sheet are likely to make a significant contribution to global sea level rise during the next millennium is, however, still debated (Church et al. 2001).

3.3 Anthropogenic Climate Change in the Baltic Sea Basin: Projections from Global Climate Models

Global climate models, also known as general circulation models (the acronyme GCM is used in both meanings and the two terms are used interchangeably in this text) are used for numerical simulations of the global climate. These models aim to represent all the physical processes of the atmosphere, land surface and oceans which are thought to be important for determining the evolution of climate on time scales extending up to several centuries. Modelling over climatological time scales on a global basis is computationally demanding, and this constrains the resolution that can be used in the models. The horizontal resolution of GCMs tends to be some 300 km in the atmosphere and often about 150 km in the oceans. This is sufficient for the models to reproduce the major atmospheric and oceanic circulation patterns and trends for climatological variables over continental scales.

3.3.1 Global Climate Model Experiments

This section addresses projections of 21st century anthropogenic climate change in the Baltic Sea Basin using simulations made with global atmosphere–ocean general circulation models (GCMs). The focus is on changes in time mean temperature and precipitation. However, to help the interpretation of the temperature and precipitation changes, changes in the atmospheric circulation and in the North Atlantic thermohaline circulation are also discussed briefly.

Before proceeding to specific aspects for the Baltic Sea Basin, some general issues related to anthropogenic climate change simulations are discussed. Three main questions are addressed in this introductory subsection: (i) what kind of anthropogenic climate change experiments have been made with GCMs, (ii) which sets of model experiments are used in this assessment, and (iii) how anthropogenic climate changes are estimated from the simulations.

Anthropogenic climate change experiments can be broadly divided into two classes: scenario experiments and sensitivity experiments. Scenario experiments are conducted to provide plausible projections of future climate. A prerequisite for this is that the external forcing used in the simulations – changes in the concentrations of different atmospheric greenhouse gases and acrosols – is consistent with a plausible and internally consistent emissions scenario, such as one of the SRES scenarios. By contrast, sensitivity experiments are mainly motivated by the need to study model behaviour. The forcing in these experiments is usually simple, such as an increase in atmospheric CO_2 with no other changes.

In this section, results from both sensitivity experiments and scenario experiments on anthropogenic climate changes in the Baltic Sea Basin are discussed. In addition to results extracted from published literature, some updates of earlier calculations are presented. These updates are based on two sets of model simulations: the so-called CMIP2 experiments (Meehl et al. 2000), and experiments based on the SRES forcing scenarios available from the IPCC Data Distribution Centre (http://ipcc-ddc.cru.uea.ac.uk/).

CMIP2, the second phase of the Coupled Model Intercomparison Project, is an intercomparison of standard idealised anthropogenic climate change experiments made with many climate models. For this assessment, CMIP2 results for 20 models were available. Each model has been used to make two 80-year simulations: a control simulation with constant (approximately present-day) CO_2 concentration and an increased greenhouse gas simulation with gradually (1% per year compound) increasing CO_2.

The increase in CO_2 concentration in the CMIP2 greenhouse runs, with a doubling over 70 years, is faster than that projected to occur under any of the SRES scenarios. On the other hand, the concentrations of other greenhouse gases such as CH_4 and N_2O are kept constant in CMIP2, although increases in these gases are actually likely to amplify the global warming in the real world.

As a result, the rate of global warming in the CMIP2 greenhouse runs compares well with simulations based on mid-range SRES forcing scenarios. The same conclusion holds for the main geographic patterns of anthropogenic climate change. This is both because the increase in CO_2 is projected to be the main cause of anthropogenic climate change during this century and because the geographic patterns of simulated anthropogenic climate change tend to be reasonably insensitive to the exact nature of the forcing (Boer and Yu 2003; Harvey 2004). Anthropogenic climate changes in the CMIP2 simulations are reported in Sect. 3.3.3.

The main limitation of the CMIP2 data set in the context of projecting anthropogenic climate changes is the fact that the experiments are based on a single forcing scenario and lack aerosol effects. As a result, this set of simulations will tend to underestimate the uncertainty of anthropogenic climate changes in the real world.

An analysis of the temperature and precipitation changes in the Baltic Sea Basin in the SRES simulations is given in Sect. 3.3.4. This analysis is based on a smaller number of models than are available in CMIP2, but it allows us to explore how the simulated anthropogenic climate changes depend on the assumed evolution of greenhouse gas and aerosol emissions. Readers interested in quantitative, internally consistent projections of anthropogenic climate change in the Baltic Sea Basin should primarily use the SRES-based information in Sect. 3.3.4, rather than the CMIP2-based results in Sect. 3.3.3. In addition, the SRES-based GCM simulations are more directly comparable with the regional climate model results reported in Sect. 3.5 than the CMIP2 simulations.

The models used in this section are listed in Table 3.1. CMIP2 simulations are available for 20 models and SRES simulations for seven models. One of the SRES models (CCSR/NIES2) is, however, excluded for most of the analysis as explained below. Most of the models in the SRES data set also participated in CMIP2. However, the SRES data set includes a more recent version of the Canadian Centre for Climate Modelling and Analysis model (CGCM2) than CMIP2 (CGCM1).

Model-based estimates of anthropogenic climate change are usually computed from the difference in climate between two simulated periods, a scenario period and a control period. In transient scenario simulations, which typically span the whole time range from some time in the 19th

or 20th century to the late 21st century, the control period is often chosen as 1961–1990, corresponding to the presently used WMO normal period. One or more scenario periods are chosen from later times in the same continuous simulation, as demonstrated in Fig. 3.5. In sensitivity experiments, a separate control run with no anthropogenic forcing is often made to estimate the control period climate, as already mentioned for the CMIP2 simulations. In neither case are anthropogenic climate changes estimated by comparing simulated future climate directly with observed present-day climate, due to systematic errors in the models. Differences between simulated and observed present-day climate are often comparable to, or in some cases even larger than, differences between simulated future and present-day climates. However, if model errors are assumed to have similar effects on the simulated present-day and future climates, their net effect approximately vanishes when taking the difference.

The performance of models in simulating present-day climate provides one means for estimating their reliability. Prior to presenting model-based projections of future anthropogenic climate change, the simulation of present-day climate in the Baltic Sea Basin is therefore discussed (Sect. 3.3.2). It is difficult to compress the evaluation of model-simulated climate into a single figure of merit, and it is even more difficult to convert this information to a quantitative estimate of model credibility in simulating anthropogenic climate changes. This is partly because the models are complex, but also because good performance in simulating the present climate in some area might hide compensating errors between different parts of the model. In fact, models that simulate the present climate with similar skill may simulate widely different changes in climate when forced with increased greenhouse gas concentrations (e.g. Murphy et al. 2004; Stainforth et al. 2005). Thus, when anthropogenic climate change simulations are available for several models, a comparison between the simulated changes probably gives a more direct measure of uncertainty than an evaluation of the control climates. However, it is important to remember the caveat that anthropogenic climate changes in the real world may in principle fall outside the range of model results.

In what follows, the ability of global climate models to simulate the present climate in the Baltic Sea area (Sect. 3.3.2) is first addressed. After this, model-simulated anthropogenic climate

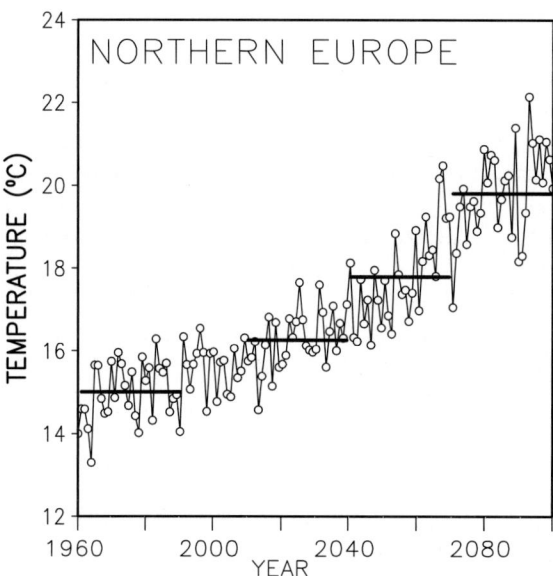

Fig. 3.5. Estimation of anthropogenic climate changes from a model-simulated time series (in this case, the summer mean temperature in northern Europe in the HadCM3 simulation forced by the SRES-A2 scenario). The first horizontal line shows the mean of the time series for years 1961–1990, and the remaining three lines the means for periods 2010–2039, 2040–2069 and 2070–2099. Temperature changes for the latter 30-year periods are calculated by subtracting the mean for 1961–1990 from the mean in the corresponding period

changes in the region are addressed (Sects. 3.3.3–3.3.4). The aims in Sect. 3.3.3 are more qualitative than quantitative. The purpose is to give an overview of the typical features of anthropogenic climate change in the Baltic Sea Basin in greenhouse gas experiments and of the intermodel differences, but the results are not intended to be interpreted as quantitative anthropogenic climate change projections for the real world. Examples of such quantitative projections are given in Sect. 3.3.4 for different SRES emission scenarios.

3.3.2 Simulation of Present-day Climate from Global Climate Models

Although the global geographic distributions of present-day temperature and precipitation climates are generally well simulated by GCMs, biases on the regional scale may be substantial (McAvaney et al. 2001; Giorgi et al. 2001). Model performance in simulating the climate in northern Europe has been the subject of several studies, including multi-model intercomparisons by Räisänen (1994, 2000) and Jylhä et al. (2004). In this subsection, these earlier studies are complemented and updated by presenting some results specifically tailored for the Baltic Sea Basin. Area means

of simulated temperature and precipitation were then calculated over the land area in the Baltic Sea Basin, indicated by shading in Fig. 3.9. The model results were compared with observational estimates derived from the University of East Anglia Climate Research Unit (CRU) climatology (New et al. 1999) representing the period 1961–1990.

Figure 3.6 compares the seasonal cycles of temperature and precipitation, as averaged over the land area in the Baltic Sea Basin, between the CMIP2 control simulations and the CRU analysis. The variation between the models is large. In a few extreme cases, simulated monthly temperatures differ by about 8–10 °C from the observational estimate. The largest cold and warm biases tend to occur in the winter half-year, from November to April, but in one model there is also a very large warm bias in summer. However, because the simulated temperatures are distributed on both sides of the observed values, the 20-model mean temperatures are mostly close to those observed. A slight average cold bias is present in most months of the year, but this bias is generally small compared with the differences among the models. The only exception is spring, when the cold bias is about 2 °C.

Table 3.1. Models used in Sect. 3.3. The columns labeled as CMIP2 and SRES indicate the model runs included in the analysis of the CMIP2 and the SRES simulations in Sects. 3.3.2–3.3.4. The atmospheric resolution includes both the horizontal and the vertical resolution. The former is expressed either as degrees latitude × degrees longitude or as a spectral truncation and the approximate equivalent grid size (in parentheses). In the IPSL–CM2 model, the meridional grid has 50 points evenly distributed in the sine of latitude and the actual resolution is therefore coarser than 3.6° in high latitudes. The vertical resolution is given as "Lmm", where mm is the number of levels. The last column indicates whether flux adjustments are used for the heat (H), freshwater (W) and momentum (M) fluxes, respectively

MODEL NAME	COUNTRY OF ORIGIN	REFERENCE	CMIP2	SRES	ATMOSPHERIC RESOLUTION	FLUX ADJ.
ARPEGE/OPA2	France	Barthelet et al. (1998)	X		T31 (3.9° × 3.9°) L19	–
BMRCb	Australia	Power et al. (1993)	X		R21 (3.2° × 5.6°) L17	HW
CCSR/NIES1	Japan	Emori et al. (1999)	X		T21 (5.6° × 5.6°) L20	HW
CCSR/NIES2	Japan	Nozawa et al. (2000)	X	(A1FI, A2 , B1, B2)	T21 (5.6° × 5.6°) L20	HW
CGCM1	Canada	Flato et al. (2000)	X		T32 (3.8° × 3.8°) L10	HW
CGCM2	Canada	Flato and Boer (2001)		A2, B2	T32 (3.8° × 3.8°) L10	HW
CSIRO Mk2	Australia	Hirst et al. (2000)	X	A2, B1, B2	R21 (3.2° × 5.6°) L9	HWM
CSM 1.0	USA	Boville and Gent (1998)	X		T42 (2.8° × 2.8°) L18	–
ECHAM3/LSG	Germany	Voss et al. (1998)	X		T21 (5.6° × 5.6°) L19	HWM
ECHAM4/OPYC3	Germany	Roeckner et al. (1999)	X	A2, B2	T42 (2.8° × 2.8°) L19	HW
GFDL_R15_a	USA	Manabe et al. (1991)	X		R15 (4.5° × 7.5°) L9	HW
GFDL_R30_a	USA	Knutson et al. (1999)	X	A2, B2	R30 (2.25° × 3.75°) L14	HW
GISS2	USA	Russell and Rind (1999)	X		4.0° × 5.0° L9	–
GOALS	China	Zhang et al. (2000)	X		R15 (4.5° × 7.5°) L9	HWM
HadCM2	UK	Johns et al. (1997)	X		2.5° × 3.75° L19	HW
HadCM3	UK	Gordon et al. (2000)	X	A1FI, A2, B1, B2	2.5° × 3.8° L19	–
INM	Russia	Diansky and Volodin (2002)	X		4.0° × 5.0° L21	–
IPSL–CM2	France	Braconnot et al. (1997)	X		3.6°s × 5.6° L15	–
MRI1	Japan	Tokioka et al. (1995)	X		4.0° × 5.0° L15	HW
MRI2	Japan	Yukimoto et al. (2000)	X		T42 (2.8° × 2.8°) L30	HWM
PCM	USA	Washington et al. (2000)	X	A2, B2	T42 (2.8° × 2.8°) L18	–

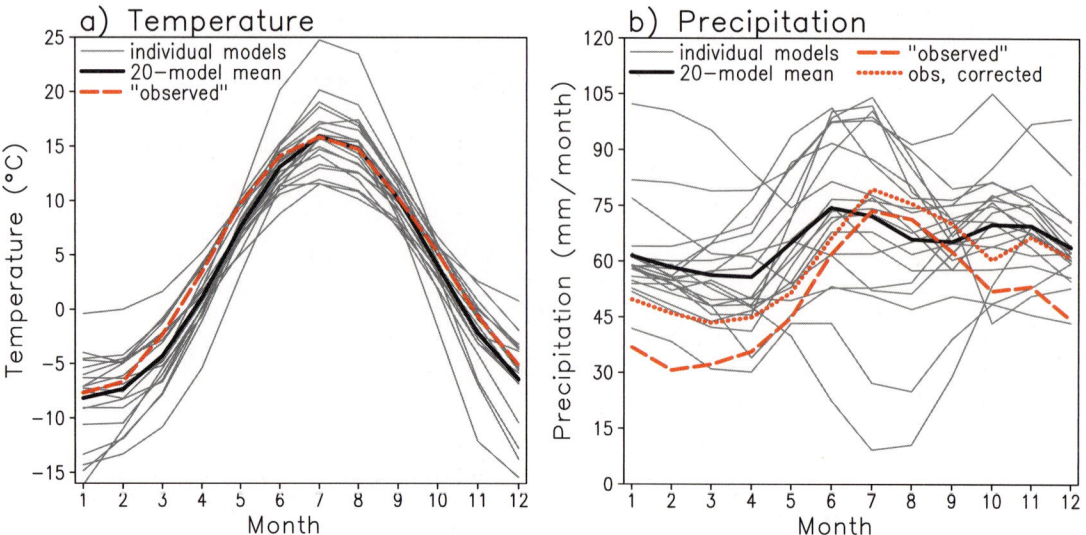

Fig. 3.6. Average seasonal cycles of (**a**) temperature and (**b**) precipitation over the total land area of the Baltic Sea Basin. The thin solid lines represent the control run climates of the 20 individual CMIP2 models and the thick solid line the 20-model mean. The dashed lines give observational estimates derived directly from the CRU climatology (New et al. 1999). For precipitation, a corrected observational estimate is also given (*dotted, see text for details*)

A more detailed look at different parts of the Baltic Sea Basin reveals that temperature biases vary across the region. In winter, in particular, there is a marked contrast between the eastern parts (Finland and western Russia) and the northwestern parts (particularly central and northern Sweden) of the basin. In the former area, average simulated winter temperatures are 1–4 °C below the observed values, in the latter several °C above them (see Fig. 2 in Räisänen 2000). Räisänen (2000) attributed the warm bias in north-western Scandinavia to the relatively coarse resolution of the models, which allows the influence of the Atlantic Ocean to extend further inland in the simulations than in reality. He likewise noted that the smoothening and lowering of the Scandinavian mountains associated with the coarse resolution might also contribute to the warm bias in this area.

The seasonal cycle of precipitation is simulated less well by the models than that of temperature (Fig. 3.6b). The scatter among the models is large in all seasons, but it is particularly pronounced in late summer (July–August). A few of the models simulate an annual minimum of precipitation in this time of the year, in contrast to the maximum shown by observations, but there are also a few models with an over-pronounced summer maximum in precipitation.

Verification of model-simulated precipitation is complicated by the tendency of gauge measurements to underestimate the actual precipitation, particularly in winter when much of the precipitation falls as snow. The CRU analysis is based in most regions on uncorrected measurements (New et al. 1999) and it thus also suffers from this problem. Figure 3.6b therefore also shows a corrected precipitation estimate, which was obtained by multiplying the uncorrected CRU values with coefficients based on the work of Rubel and Hantel (2001), as detailed in Jones and Ullerstig (2002) and Räisänen et al. (2003) (see also Sect. 2.1.3). The correction is relatively small in summer but it amounts to about 40% in winter and 19% in the annual mean (Räisänen et al. 2003). The 20-model mean simulated annual precipitation is 30% above the uncorrected observational estimate, but only 9% above the corrected estimate. If the correction is not too small, a large majority of the models simulate too much precipitation in winter and in spring (Fig. 3.6b).

The ability of GCMs to simulate small-scale regional variations in precipitation is severely limited by the coarse resolution of these models. This is particularly evident in the vicinity of high orography, such as the Scandinavian mountains. Observations show a very steep contrast between abundant precipitation on the western slopes of

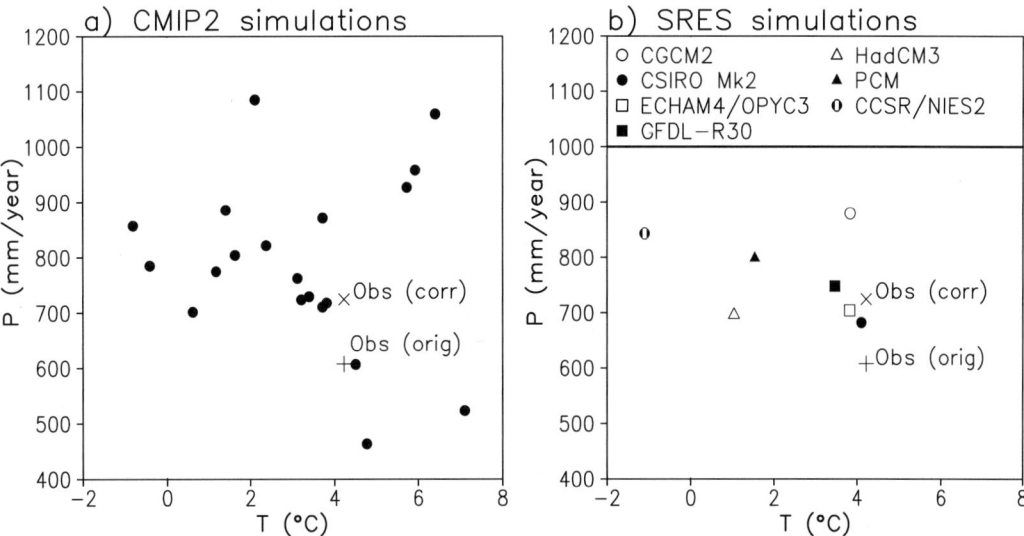

Fig. 3.7. Annual area means of temperature (*horizontal axis*) and precipitation (*vertical axis*) over the Baltic Sea Basin land area (**a**) in the CMIP2 control simulations and (**b**) in the SRES simulations for the years 1961–1990. The plus sign (+) represents the observational estimate derived directly from the CRU climatology and the cross (×) the estimate including the precipitation correction mentioned in the text

the mountain range and much less precipitation on the eastern side, but in GCMs this contrast is much less pronounced (Rummukainen et al. 1998; Räisänen and Döscher 1999; Räisänen 2000). Thus, some of the precipitation that should fall down in Norway spills in the models over to the Baltic Sea Basin, particularly to central and northern Sweden.

Simulated annual area means of temperature in the Baltic Sea Basin land area vary from $-0.8\,°C$ to $7.1\,°C$, 14 of the 20 models being colder than the CRU observational estimate of $4.2\,°C$ (Fig. 3.7a). The 20-model range in annual area mean precipitation is from 460 to 1080 mm, to be compared with uncorrected and corrected observational estimates of 610 and 720 mm. For comparison, Fig. 3.7b shows the control period (1961–1990) annual area means of temperature and precipitation for the set of models that will be used in Sect. 3.3.4 for deriving projections of anthropogenic climate change under the SRES forcing scenarios. Among these models, the annual area mean precipitation is between 680 and 880 mm, being thus reasonably close to the corrected observational estimate. The simulated annual area mean temperature is within $1\,°C$ of the observed value in four of the seven models but clearly too low in the remaining three models.

One of the factors that affect the simulated temperature and precipitation climate is the simu-

lation of atmospheric circulation. The observed (National Centres for Environmental Prediction reanalysis; Kistler et al. 2001) and average CMIP2 control run distributions of winter and summer mean sea level pressure are compared in Fig. 3.8. The general features of the observed and simulated pressure distributions are very similar, although this similarity hides substantial variations among the individual models (Räisänen 2000). However, the wintertime extension of the Icelandic low towards the Barents Sea is not simulated well. The average simulated pressure in winter is slightly too high over the northernmost North Atlantic and the Barents Sea, and slightly too low over Europe approximately at $50°–65°\,N$. A similar pattern of pressure biases occurs in spring (not shown). These biases in the average pressure field suggest that in most models the eastern end of the North Atlantic cyclone track is too zonally oriented, with too little cyclone activity over the European sector of the Arctic Ocean and too much activity over the European mainland. Räisänen (2000) speculated that this may contribute to the mentioned excess of winter and spring precipitation over at least some parts of northern and central Europe.

A few studies have compared model-simulated interannual variability with observations. Jylhä et al. (2004) note that the interannual standard deviation of Finland area mean temperature in 1961–1990 was $1.1\,°C$, whereas the six models stud-

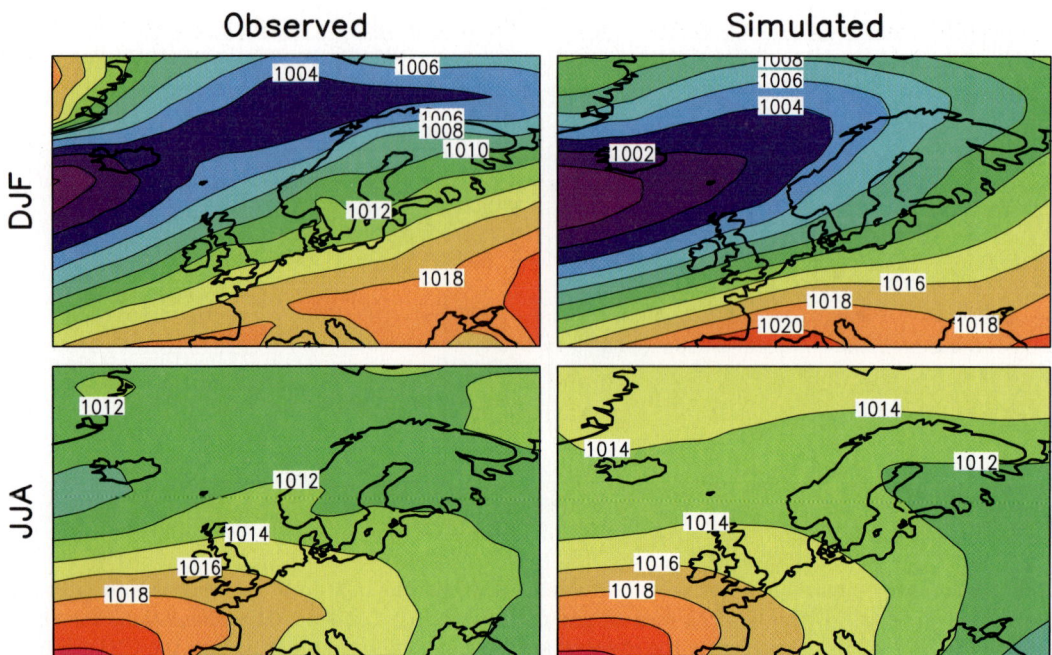

Fig. 3.8. Distribution of mean sea level pressure in winter (*top*) and summer (*bottom*) as observed in 1961–1990 (Kistler et al. 2001) and as averaged over the CMIP2 simulations. Contours are drawn at every 2 hPa

ied by them simulated values between 0.9 °C and 1.3 °C. The coefficient of variation of observed annual area mean precipitation was 12%, whereas model-simulated values ranged from 8% to 13%. Räisänen (2001b) averaged the local interannual standard deviation of monthly temperature and precipitation over 19 models and over all land grid boxes within Finland, Sweden, Norway and Denmark. He found the models to reproduce the observed seasonal cycle of temperature variability, with much larger variability in winter than in summer. The magnitude of the simulated variability was also in agreement with observations except in April and November, when the simulated variability was too large. The average standard deviation of monthly precipitation was close to that observed in winter and spring but below the observed values in summer and autumn. However, because the simulated mean precipitation was too high in winter and spring, the relative variability of monthly precipitation was too small in all months.

This is qualitatively as expected, because the resolution of the models is not sufficient to capture the details of individual precipitating weather systems. However, the simulated relative variability still remained below the observational estimate when the latter was derived from observations first aggregated to model grid boxes.

The atmospheric circulation over Europe (see also Annex 1.2) varies substantially from year to year, which leads to variations in temperature, precipitation, windiness and other aspects of the surface climate. A large part of this variation is associated with the North Atlantic Oscillation (NAO; Hurrell and van Loon 1997; Hurrell et al. 2003; Annex 7), particularly in winter. A positive (negative) NAO index indicates stronger (weaker) than average westerly flow over the North Atlantic at about 50–60° N. Global climate models simulate NAO variability with many properties in agreement with observations, including its spatial signature in the pressure (Osborn et al. 1999) and temperature fields (Stephenson and Pavan 2003), and the temporal autocorrelation structure (Stephenson and Pavan 2003). However, as pointed out by Stephenson and Pavan (2003) and Osborn (2004), different models simulate NAO variability with different skill, the amplitude of the variability also varying between models.

The causes of the observed strong positive trend in the NAO index from the early 1960's to the early 1990's are still poorly understood. In the light of model simulations, the trend appears to be at the outer limits of what could be expected from internal variability alone, although this conclusion may be sensitive to the exact definition of the NAO index and to the magnitude of the simulated internal variability (Osborn 2004; Selten et al. 2004). This suggests that the trend might have been at least in part externally forced. However, it is not clear whether the trend can be attributed to increased greenhouse gas concentrations, since most models simulate only modest changes in NAO in response to greenhouse gas forcing (Räisänen and Alexandersson 2003; Osborn 2004). Stratospheric ozone depletion might also have affected the trend in the NAO index (Volodin and Galin 1999), but recent model results do not support the idea that it would have made a major contribution (Shindell et al. 2001; Gillett et al. 2003). Scaife et al. (2005) show that the observed NAO trend is reproduced well in a climate model simulation with a prescribed increasing trend in stratospheric westerlies resembling that observed from 1965 to 1995. However, their study does not explain why the stratospheric winds changed. Similarly, although studies by Hurrell et al. (2004), Hoerling et al. (2004) and Selten et al. (2004) suggest that the observed trend in the NAO index has been partly triggered by changes in the distribution of sea surface temperature in the tropical Indian Ocean, the cause of these changes (internal variability or anthropogenic forcing) is still unclear.

3.3.3 Projections of Future Climate from Global Climate Models

In this and the following subsection, we assess GCM-simulated anthropogenic climate changes in northern and central Europe in general and within the Baltic Sea Basin in particular. The aim of this subsection is to give an overview of the temperature and precipitation changes typically simulated by GCMs when forced by increasing greenhouse gas concentrations, of the differences in the model projections and of the factors that may affect the simulated temperature and precipitation changes and the intermodel differences. Most of this subsection is based on the idealised CMIP2 simulations, which are available for a significantly larger number of models than simulations based on the more detailed SRES forcing scenarios.

3.3.3.1 *Temperature and Precipitation*

The globally and annually averaged warming at the time of the doubling of CO_2 in the CMIP2 simulations varies from 1.0 to 3.1 °C, with a 20-model mean of 1.7 °C. The 20-model mean in northern Europe is about 2.5 °C, or 50% larger than

the global mean warming, with somewhat greater warming in winter than in summer (left column of Fig. 3.9). The warming in winter typically increases from southwest to northeast, from the Atlantic Ocean towards the inner parts of Eurasia and the Arctic Ocean. The warming in summer has a slight tendency to increase towards southeast, but the gradient in this season is weaker than in winter. The changes in spring and autumn are generally between those in winter and summer (e.g. Räisänen 2000). The simulated annual precipitation increases, on the average, by about 10% in northern Europe (middle column of Fig. 3.9). In Central Europe, including the southernmost parts of the Baltic Sea Basin, the average model results suggest an increase in precipitation in winter but a decrease in summer. In northern Europe, precipitation increases in most models throughout the year, but less in summer than in the other seasons (Räisänen 2000, 2001b).

All the CMIP2 models simulate an increase in both the annual area mean temperature and the annual area mean precipitation in the Baltic Sea Basin land area. However, the magnitude of the changes varies substantially among the models (Fig. 3.10a). The range in annual temperature change is from 0.7 °C to 6.9 °C with a mean of 2.6 °C, and that in precipitation change from 3% to 22% with a mean of 8%. One of the models (CCSR/NIES2) simulates much larger increases in both temperature and precipitation than any of the others. This model also stands out as the one with the largest global mean warming in the CMIP2 data set (3.1 °C, to be compared with the second largest value of 2.1 °C).

Some variation between the anthropogenic climate changes simulated by the various models would be expected simply as a result of the internal variability that accompanies the forced anthropogenic climate change in the simulations (Cubasch et al. 2001; Räisänen 2000, 2001a). However, the differences in change between the models are too large to be explained by this factor alone. An estimation of internal variability in the CMIP2 simulations using the method of Räisänen (2001a) suggests that only about 7% of the intermodel variance in Baltic Sea Basin land area annual mean temperature change is due to internal variability. For annual precipitation change, the corresponding fraction is 17%. This implies that most of the differences in precipitation change and, particularly, temperature change are directly caused by the differences between the models. The

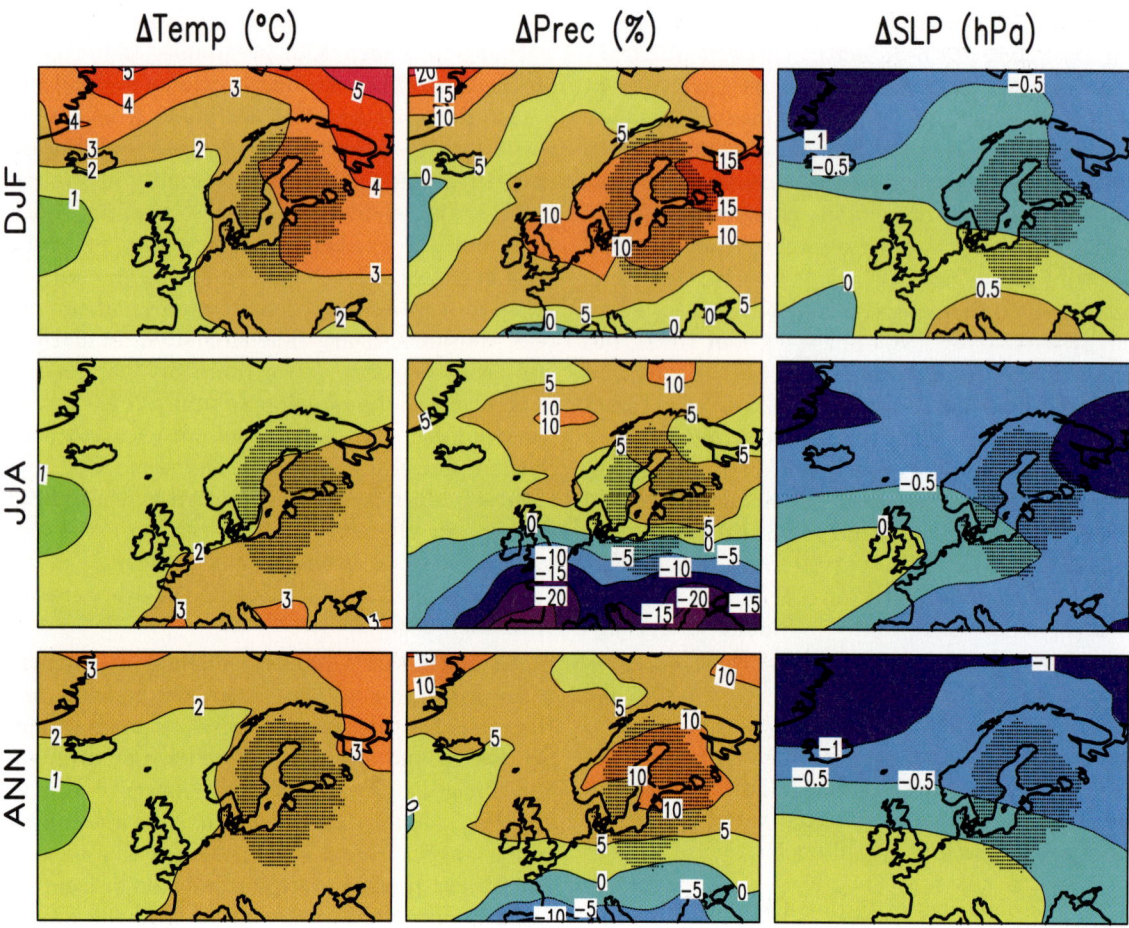

Fig. 3.9. Changes in temperature (*left*), precipitation (*middle*) and sea level pressure (*right*) around the time of the doubling of CO_2 (model years 61–80) as averaged over the CMIP2 models. Results are shown for winter (DJF = December–January–February), summer (JJA = June–July–August) and the annual mean (ANN). The contour interval is 1 °C for changes in temperature, 5% for changes in precipitation and 0.5 hPa for changes in sea level pressure. The shading indicates the Baltic Sea Basin land area

same conclusion also generally holds for seasonal temperature and precipitation changes, although the relative importance of internal variability is larger on the seasonal than on the annual time scale.

The changes in temperature and precipitation tend to be positively correlated (Fig. 3.10a). This is the case in winter, spring and autumn, but not in summer. As argued by Räisänen (2000), the positive correlation in winter and in the transitional seasons is consistent with the Clausius–Clapeyron relationship. The larger the increase in temperature, the larger the increase in the capability of air to bring moisture from lower latitudes and the Atlantic Ocean to northern Europe. The lack of positive correlation in summer is consis-

tent with the smaller relative importance of atmospheric moisture transport, as opposed to the local evaporation, in providing the precipitating water in this season (Numaguti 1999). In addition, feedbacks associated with soil moisture and/or cloudiness may play a role. When evaporation becomes restricted by a drying out of the soil, this tends to induce both a decrease in precipitation and an increase in temperature due to increased sensible heat flux and increased solar radiation allowed by reduced cloudiness (Wetherald and Manabe 1995; Gregory et al. 1997).

The simulated annual mean warming in the Baltic Sea Basin is strongly correlated ($r = 0.8$, although this is reduced to $r = 0.6$ if CCSR/NIES2 is excluded) with the global mean warming as

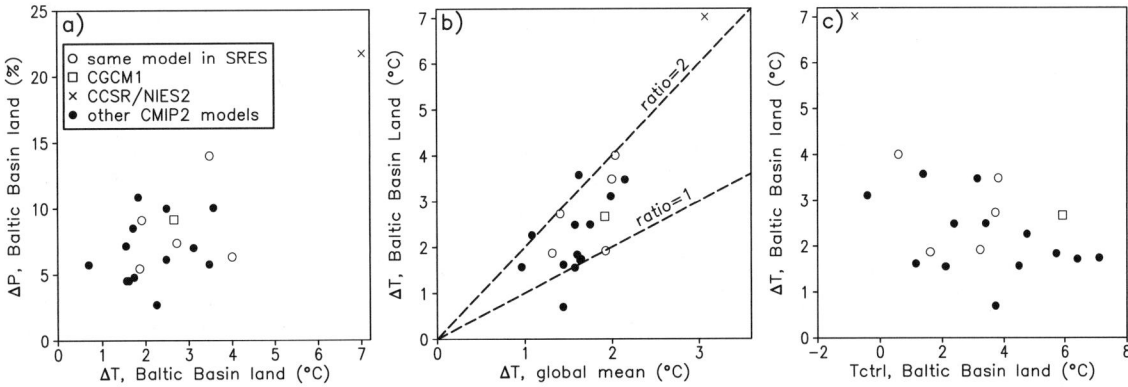

Fig. 3.10. (a) Changes in annual area mean temperature (*horizontal axis*) and precipitation (*vertical axis*) in the Baltic Sea Basin land area in the CMIP2 simulations, (b) Changes in global mean temperature (*horizontal axis*) and Baltic Sea Basin land area mean temperature (*vertical axis*; note the difference in scale), (c) Control run mean temperature (*horizontal axis*) and temperature change in the Baltic Sea Basin land area (*vertical axis*). The models for which SRES simulation data are also available are shown with special symbols, as detailed in the first panel

shown in Fig. 3.10b. Despite this correlation, the ratio between the warming in the Baltic Sea Basin and the global mean warming varies considerably between the models. In three of the 20 models, the warming in the Baltic Sea Basin is more than twice the global mean warming, and in another three models the Baltic Sea Basin warms less than the world on the average. In the remaining 14 models, the warming in the Baltic Sea Basin exceeds the global mean warming but less than by a factor of two.

There is a slight tendency of the Baltic Sea Basin annual mean temperature to increase more in those models in which the simulated control run temperatures are low (Fig. 3.10c). This tendency, which is strongest in winter, is consistent with the idea that feedbacks associated with changes in snow and ice cover should be stronger in a colder climate with more extensive snow and ice cover (Rind et al. 1995). In addition, models with low wintertime near-surface temperatures are more likely to have a stable boundary layer characterized by a surface inversion (e.g. Räisänen 1994). Strong stability suppresses vertical mixing in the atmosphere and may therefore allow strong greenhouse-gas-induced warming in the lowest troposphere even if the warming at higher levels is modest (Mitchell et al. 1990).

3.3.3.2 Atmospheric and Oceanic Circulation

Temperature and precipitation changes in the Baltic Sea Basin may also be affected by changes in atmospheric and oceanic circulation. To characterise the changes in atmospheric circulation in the CMIP2 simulations, the average changes in time mean sea level pressure are shown in the right column of Fig. 3.9 (because of incomplete data availability, only 17 models are included in the average in this case). The maps suggest a slight decrease in sea level pressure over the northernmost North Atlantic and the Arctic Ocean throughout the year, and smaller decreases or slight increases further south over the Atlantic Ocean and in central Europe. This pattern of change implies a slight increase in westerly winds over the northern North Atlantic and northern Europe, qualitatively similar to the recent wintertime changes. However, the amplitude of the 17-model mean pressure changes is very small, although partly as a result of opposing changes in different models (e.g. Räisänen 2000).

Some studies have addressed how simulated circulation changes impact on changes in temperature and precipitation. Regression-based calculations of Rauthe and Paeth (2004) suggest that, in winter, average model-simulated changes in atmospheric circulation would warm northern Europe by 0.2–0.6 °C by the year 2050 from the period 1900–1980, as compared with a total warming of about 3 °C for the forcing scenario used in their calculations. They also found a circulation-induced increase in winter precipitation in northern Europe, particularly Norway, but little contribution to either the changes in temperature or precipitation in central Europe. Their results indicate that, although circulation changes may make some contribution to European temperature and precipitation changes, they are not likely to be the dominant agent of change. Similarly, Stephenson et al. (2006) found that changes in the NAO index only have a minor effect on the changes in winter temperature and precipitation in the CMIP2 simulations. In another study, Jylhä et al. (2004) compared the simulated temperature changes in Finland under the SRES A2 and B2 forcing scenarios with the changes in the westerly component of the geostrophic wind. In all seasons except for summer, they found a positive correlation between the two parameters, with models with a larger increase in westerly flow simulating larger warming. However, they only studied results from five models.

In summary, results from global climate models would suggest that changes in the atmospheric circulation are likely to have only a secondary impact on future changes in temperature and precipitation in the Baltic Sea Basin. On the other hand, simulations with regional climate models (see Sect. 3.5) indicate that circulation changes may nevertheless be very important for some aspects of anthropogenic climate change, such as changes in average and extreme wind speeds and the regional details of precipitation change in mountainous areas.

The relatively modest role that circulation changes appear to play in most anthropogenic climate change simulations is in apparent contrast with recently observed interdecadal variations in climate. In particular, a persistently positive phase of the NAO has been pointed out as the main cause of the series of mild winters that has characterised the climate in northern Europe since the late 1980's (Tuomenvirta and Heino 1996; Räisänen and Alexandersson 2003; Sect. 2.1.2.1). If the model results are realistic, they imply that the recent circulation changes, particularly the strong positive trend in NAO from the 1960's to the 1990's, cannot be necessarily extrapolated to the future. At least in some climate models, NAO trends comparable to that recently observed occur purely as a result of internal climate variability (e.g. Selten et al. 2004). On the other hand, some studies (Shindell et al. 1999; Gillett et al. 2003) have suggested that current climate models may underestimate the sensitivity of the atmospheric circulation to greenhouse gas forcing.

Another significant source of uncertainty for future anthropogenic climate change in Europe is the behaviour of the North Atlantic Ocean circulation.

At present, the Atlantic Ocean transports about 1.2 PW of heat poleward of 25° N, or 20–30% of the total heat flux carried by the atmosphere–ocean system at this latitude (Hall and Bryden 1982). The large northward heat transport in the Atlantic Ocean is one of the reasons for the relatively mild climate in northern and central Europe, as compared with other regions in the same latitude zone. Much of the oceanic heat transport is due to the so-called thermohaline circulation (THC). This circulation includes the Gulf Stream and the North Atlantic Drift at the surface, and a southward return flow in the deep ocean. The surface and deep currents are connected by convective sinking of water that is thought to occur at several locations in the northernmost North Atlantic.

In experiments with increased greenhouse gas concentrations, most GCMs simulate a decrease in the strength of the THC. This is caused by a decrease in the density of the surface water in the northern North Atlantic, which suppresses ocean convection. Both increased water temperature and reduced salinity associated with increased precipitation and river runoff may reduce the water density, but the precise mechanisms seem to be model-dependent (Cubasch et al. 2001). More importantly, the magnitude of the change differs greatly between different models. While most of the models studied by Cubasch et al. (2001) simulated a 30–50% decrease in the THC by the year 2100 when forced with a middle-of-the-range forcing scenario, some of them showed only a small decrease and one (Latif et al. 2000) no decrease at all. None of these models simulated a complete shutdown of the THC in the 21st century, although some simulations (e.g. Stocker and Schmittner 1997; Stouffer and Manabe 2003) suggest that this might be possible later.

Another important question is the sensitivity of the European climate to changes in the THC. In some simulations with increasing greenhouse gas concentrations, weakening of the THC leads to slight local cooling in the northern North Atlantic. However, the cooling is generally limited to a small area. Thus, higher temperatures are simulated in Europe despite reduced THC intensity (as shown by the results presented in this report), although the decrease in THC may act to reduce the magnitude of regional warming.

However, there have been at least two model simulations in which changes in the THC have lead to slight cooling along the north-western coasts of Europe. In the GISS2 CMIP2 simulation (Rus-

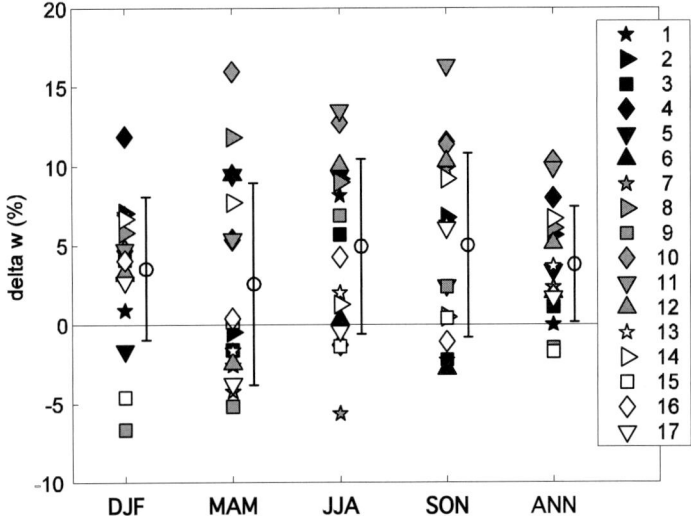

Fig. 3.11. Differences in percent between control and scenario for the seasonal means of the geostrophic wind speed from 17 CMIP2 GCM simulations. The mean and the error bar (± one standard deviation) of the changes in the 17 GCMs are also plotted (from Chen et al. 2006)

sell and Rind 1999), slight cooling is simulated in northern and western Scandinavia. Most of the Baltic Sea Basin warms up even in this model, but the warming is weaker than in the other CMIP2 simulations. A later version of the same model simulates stronger warming in northern Europe (Russell et al. 2000), but results similar to those of Russell and Rind (1999) were recently reported for another model by Schaeffer et al. (2004). In both cases the cooling in northern Scandinavia is an extension of stronger cooling in the sea area between Scandinavia and Svalbard. Schaeffer et al. (2004) show that the cooling is caused by a local expansion of sea ice triggered by a shutdown of ocean convection in this area. Other convection sites remain active in their model, so that the overall intensity of the THC is only moderately reduced. Their analysis suggests that the impact of THC changes on the atmospheric climate depends not only on the overall change in THC intensity but also on the regional details of the change. The atmospheric climate appears to be particularly sensitive to changes in ocean convection near the ice edge.

In summary, there is uncertainty in both the future changes in the Atlantic THC and the sensitivity of the European climate to these changes. However, models give at most very limited support to the idea that changes in the THC would turn the greenhouse-gas-induced warming to cooling. Furthermore, the risk of cooling appears to

decrease with increasing distance from the northwestern coastlines of Europe.

3.3.3.3 *Large-scale Wind*

Analysis of surface wind at regional scale for present climate is rare, as are such studies in the anthropogenic climate change context. This is due to the facts that homogeneous wind measurements hardly exist (Achberger et al. 2006) and analysis of model simulations is usually focused on temperature and precipitation. The few works dealing with change in wind concern wind high up in the atmosphere (Pryor and Barthelmie 2003), surface wind (10 m) from reanalysis (Pryor et al. 2005a; "reanalysis" is described in Sect. 3.5.1.1), or geostrophic wind derived from surface pressure data (Alexandersson et al. 2000; see also Sect. 2.1.5.4). Recently, Chen and Achberger (2006) analysed surface wind measurements at a few Swedish stations in relation to the large scale geostrophic wind over the Baltic Sea. They expressed this in terms of circulation indices (Chen et al. 2006; see also Fig. 3.16) as well as gridded 10 m wind in relation to the geostrophic wind of the NCEP reanalysis for the grid box over Stockholm (57.5–60° N, 15–17.5° E). They concluded that there is a strong relationship between the observed station and the grid surface wind speeds, as well as with the geostrophic wind speeds. This opens the possibility of estimating changes in regional surface wind using information on geostrophic wind derived from surface pressure data. They used simulated sea level pressure from the standardised global climate model simulations of the CMIP2 project (Meehl et al. 2000; see Sect. 3.3.1) to estimate changes in the surface wind over the Baltic region centered around Stockholm.

The GCM-simulated changes in the geostrophic wind are estimated by comparing means from the last 30-year period in the enhanced greenhouse runs with means from the entire 80-year period in the control runs. The last 30 years of the CMIP2 scenario approximately represent the period when doubling of CO_2 from present day levels is reached, and the difference between the future and present climates can be considered as a response to the doubled CO_2.

Since 17 different GCMs were used, the spread of the estimates can be taken as a measure of the uncertainty associated with global climate models (Chen et al. 2006). Figure 3.11 displays the relative change in geostrophic wind. As there exists a close linear relationship between the large scale geostrophic and the grid surface (10 m) winds over the period 1948–2004, the percent change in the geostrophic wind speed would reflect the percent change in the surface wind if the linear relation also holds in the future. Whether this assumption will be valid or not depends on the changes in the vertical stability of the atmosphere. The majority of models indicate an increase in the wind speed, mainly caused by an increased westerly wind. With the help of a t-test at the 5% level, it was determined that none of the negative changes are considered significant, while some of the positive changes in each season are significant (2, 2, 3, 3 and 5 of 17 for winter, spring, summer, autumn and annual means, respectively). The mean changes for winter, spring, summer, autumn and annual means are 3.5%, 2.6%, 4.9%, 5% and 3.8%, respectively.

3.3.4 Probabilistic Projections of Future Climate Using Four Global SRES Scenarios

According to present understanding, it is likely that future radiative forcing will fall within the range defined by the four SRES scenarios B1, B2, A2 and A1FI (for a brief description of the scenarios, see Annex 6). Therefore, it is reasonable to employ this set of SRES scenarios in constructing anthropogenic climate change projections, instead

of a single CMIP2 or SRES scenario. Each SRES scenario is based on internally consistent assumptions about future development, in contrast to the idealised CMIP2 scenario.

In this subsection we concentrate on presenting anthropogenic climate change projections for the time period 2070–2099, relative to the baseline period 1961–1990. This is a commonly studied period as the ratio of climate change signal to noise due to internal variability is higher than for earlier years. Moreover, the various SRES scenarios do not diverge much until the middle of the 21st century (Fig. 3.2d). Scenarios for time periods earlier than 2070–2099 will be discussed briefly at the end of this subsection.

As far as the authors are aware, there exists no published research dealing with SRES-based climate projections just for the Baltic Sea Basin. Therefore, we have carried out the same calculations made in Ruosteenoja et al. (2007), tailoring the analysis for this region.

In the IPCC Data Distribution Centre, SRES-forced simulations were available for seven GCMs (Table 3.1). Of these models, CCSR/NIES2 simulates very large increases in temperature and precipitation compared to the remaining models (Fig. 3.10a), the projections of the other six models being much closer to one another. Furthermore, the control climate of CCSR/NIES2 is far too cold. Giorgi and Mearns (2002) suggested that, in order for a GCM to have good reliability in simulating regional anthropogenic climate change, the model should simulate the present-day climate with a small bias and climate projections should not be too different from those given by other GCMs. CCSR/NIES2 fails to fulfill either of these conditions. In addition, the spatial resolution of that model is lower than in any other GCM for which SRES runs are available (see Table 3.1). Consequently, we omit CCSR/NIES2 in this analysis. The six models incorporated in the analysis have a horizontal grid spacing of $\sim 400\,\mathrm{km}$ or smaller in the Baltic Sea Basin.

The responses to the A2 and B2 forcing scenarios have been simulated by all six of the models (see Table 3.1). The low-forcing B1 response was available for HadCM3 and CSIRO Mk2, while the high-forcing A1FI response was only available for HadCM3. Furthermore, HadCM3 was the only model for which ensemble runs were available, the ensemble size being three for the A2 and two for the B2 scenario. For most of the models, projections for the highest (A1FI) and lowest

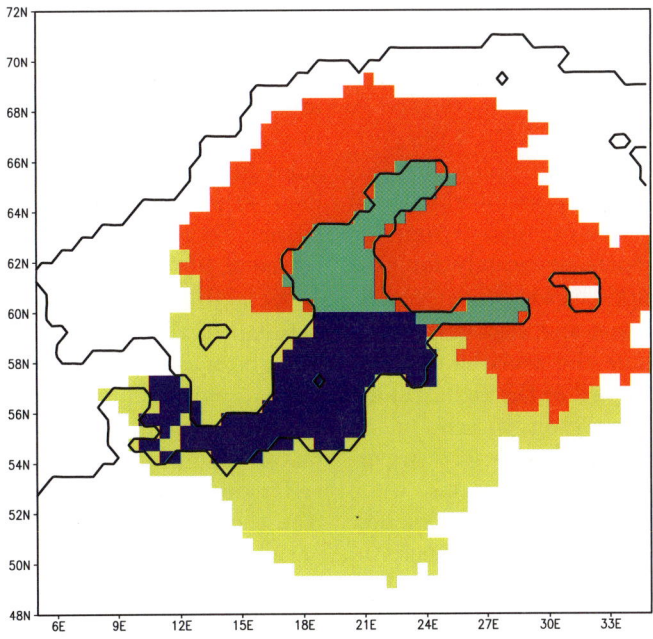

Fig. 3.12. The four subregions employed in representing the probability intervals of temperature and precipitation change

(B1) forcing scenario were composed employing a super-ensemble pattern-scaling technique, developed and assessed in Ruosteenoja et al. (2007). This method uses linear regression to represent the relationship between the local GCM-simulated temperature/precipitation response and the global mean temperature change simulated by a simple climate model. The method has several advantages, for example, the noise caused by natural variability is reduced, and the method utilises the information provided by GCM runs performed with various forcing scenarios effectively. The super-ensemble method proved especially useful in a situation with only one A2 and one B2 simulation available for an individual GCM. In such a case, the conventional time-slice scaling from an individual GCM response would excessively transfer noise to the scaled response.

In anthropogenic climate change impact research, it is generally not adequate only to consider a deterministic estimate of the change, but it is also of interest to know a range inside which the projected anthropogenic climate change is expected to fall. We constructed 95% probability intervals of spatially averaged temperature and precipitation change for two maritime and two continental regions of the Baltic Sea Basin. In winter in particular, there is a distinct south-west to north-

east gradient in the geographical distribution of average model-simulated temperature change.

The mean of 20 CMIP2 simulations is given in Fig. 3.9, and a qualitatively similar pattern was obtained by averaging the six SRES simulations (not shown). Therefore, both the Baltic Sea and the continental runoff area were divided into a south-western and north-eastern subregion (Fig. 3.12). The NE maritime region covers the Gulf of Finland and the Gulf of Bothnia, the SW region the Baltic Proper. Considering hydrological applications, the two continental subregions correspond to the runoff areas of the maritime subregions. In individual GCMs horizontal resolution is much coarser than in Fig. 3.12, and therefore the results are not sensitive to the details of the subdivision.

Since we have analysed six GCMs only, modelled anthropogenic climate change projections applied as such do not give a statistically representative picture of regional anthropogenic climate change. Instead, we have fitted the normal (Gaussian) distribution to the set of model projections. The validity of the normal distribution approximation was assessed in Ruosteenoja et al. (2007). In all, the projections of the various GCMs were found to follow the normal distribution fairly well. Utilising the Gaussian approximation, we can construct 95% probability intervals (i.e. 2.5–97.5%) for the temperature (T) and precipitation (P) change:

$$I_{\Delta T} = \overline{\Delta T} \pm 1.96 s_{\Delta T}; \ \ I_{\Delta P} = \overline{\Delta P} \pm 1.96 s_{\Delta P}$$

where the means (denoted by an overline) and standard deviations (s) are calculated from the responses of the six models, separately for each season, region and forcing scenario.

In calculating the means and standard deviations needed to determine the probability intervals, the total weight assigned to the HadCM3 ensemble runs was twice that given to each of the other models. Probability intervals for temperature change are given in absolute terms, while precipitation changes are expressed here in percentages to facilitate application of the results to impact studies. In transforming the precipitation changes into percentages ($100\% \times DP/P$), the denominator P is the baseline-period precipitation averaged over the six GCMs, the weights being as stated above.

The 95% probability intervals for temperature change for each season, region and scenario are depicted in Fig. 3.13. Even at the lower end of the interval, the inferred temperature change is invariably positive. In winter and spring, the north-eastern part of the Baltic Sea Basin tends to warm more than the south-eastern one, while in the other seasons differences among the regions are smaller. As far as the medians of the intervals are concerned, the warming is a monotonic function of the strength of the radiative forcing, the A1FI forcing producing nearly double warming compared to B1.

However, the probability intervals are quite broad, reflecting the large scatter among the projections of the various models. For example, in both north-eastern regions the extreme estimates for springtime temperature response to the A1FI forcing range from < 3 °C to more than 10 °C. Another striking feature is the fact that the probability intervals representing different forcing scenarios overlap strongly. Even for the B1 forcing, the upper estimates are 4–6 °C, the lower estimates mostly being ~1–2 °C.

The 95% probability intervals for precipitation change are presented in Fig. 3.14. With the exception of some regions in the intermediate seasons, the 95% probability intervals intersect the zero line. Consequently, in most cases even the sign of the future precipitation change cannot be

established firmly. Especially in winter, the projections provided by the various models diverge strongly, and probability intervals of the change are broad. Compared with the large uncertainties in the projections, differences among the various subregions are fairly small. In studying the median estimate, summertime precipitation seems to change little, by ±10%. In other seasons an increase of 10–30% is projected.

As a rule of thumb, other probability intervals can be derived from Figs. 3.13 and 3.14 in a straightforward manner by adjusting the length of the bar while keeping the middle point unchanged. For instance, to obtain the 90% interval (i.e., an interval excluding 5% of probability density at *both* ends) the bar lengths must be multiplied by the factor 1.645/1.960, in accordance with the fitted normal distribution.

The time trends of change in temperature and precipitation for the four emissions scenarios are shown in Fig. 3.15. In the response to the A2 radiative forcing scenario, temperature increases fairly linearly in time (Fig. 3.15a). For the A1FI scenario, warming is more rapid in the second half of the century, while both B scenarios show decreasing rates of warming during this time. The scenarios deviate little from one another before 2030. However, the actual temperatures in individual decades are affected by natural variability as well as by anthropogenic forcing. In the curves depicted in Fig. 3.15a, this influence is suppressed by taking an average of several models and by the application of temporal smoothing. As a consequence of internal natural variability, one can expect that the actual tridecadal mean temperatures may fluctuate up to 1 °C around the general warming trend (the estimate for the magnitude of natural variability is given in Fig. 3.15a). These assessments do not take into account possible future changes in external natural forcing agents, such as the intensity of solar radiation.

In the simulations of future precipitation, as shown in Fig. 3.15b, the influence of natural variability is thought to be much stronger than in the temperature projection. For example, the B2-forced precipitation change shows an apparent rapid increase during the coming decades, whilst later in the century the rate of change appears to be much slower. However, it is difficult to know if this is a reliable feature of the B2 scenario or just a consequence of the fact that the effects of natural variability are not filtered out completely even when averaged over several models.

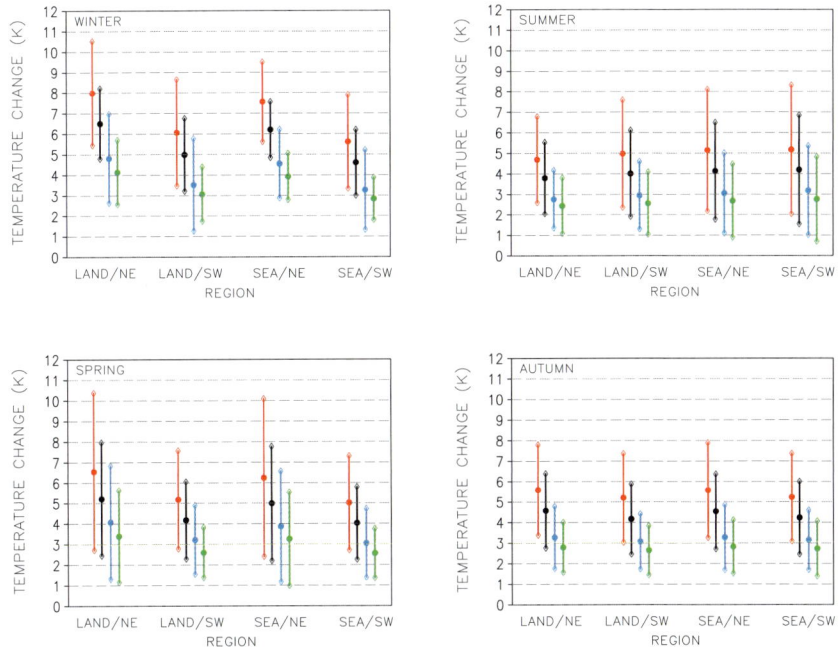

Fig. 3.13. Seasonal GCM-driven 95% probability intervals of absolute temperature change (*vertical bars*) from 1961–1990 to 2070–2099 for four subregions (defined in Fig. 3.12), derived from SRES-forced simulations performed with six GCMs. Intervals are given separately for the A1FI (*red*), A2 (*black*), B2 (*blue*) and B1 (*green*) scenarios. The dot in the centre of the bar denotes the median of the interval

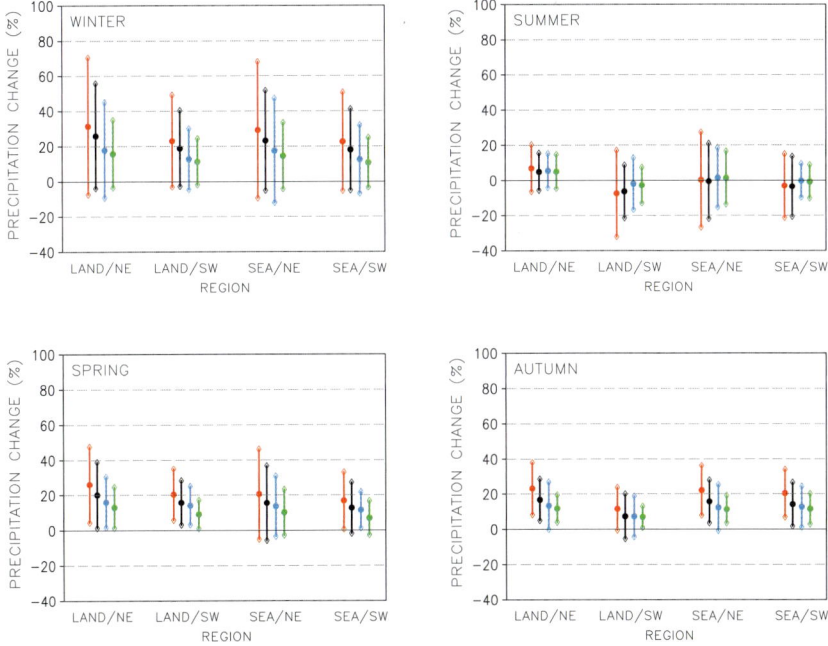

Fig. 3.14. Seasonal GCM-derived 95% probability intervals of precipitation change in percent. Intervals are given separately for the A1FI (*red*), A2 (*black*), B2 (*blue*) and B1 (*green*) scenarios. The dot in the centre of the bar denotes the median of the interval

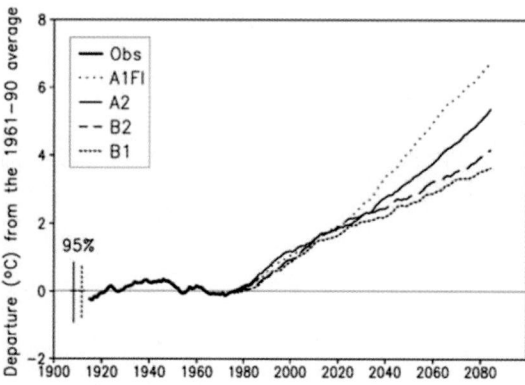
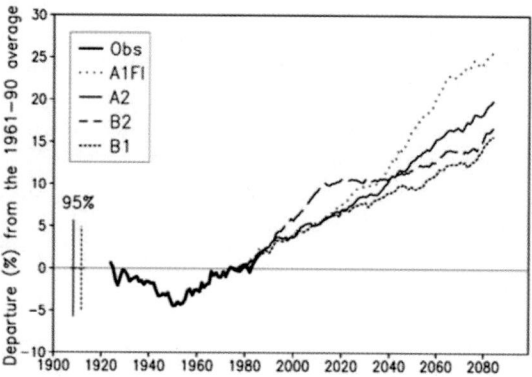

Fig. 3.15. Annual mean temperature (*left*) and precipitation anomalies for Finland relative to the mean of the baseline period 1961–1990. Before the 1980's the anomalies are based on observations, after that on the mean of projections given by four GCMs (HadCM3, ECHAM4, NCAR-PCM and CSIRO-Mk2). Projections are given separately for the A1FI, A2, B2 and B1 SRES radiative forcing scenarios. Curves are smoothed by applying 30-year running means. The vertical bars on the left indicate the 95% probability intervals for differences between two arbitrarily-chosen 30-year mean temperatures/precipitations in a millennial control run with unchanged atmospheric composition, performed with two coupled GCMs (*left bar* – HadCM3; *right bar* – CGCM2); these bars give a measure for internal natural variability (adapted from Jylhä et al. 2004)

Climate models are evolving continuously, and we can anticipate that their ability to simulate the response to anthropogenic climate forcing will improve. The results discussed in this subsection are based on a fairly small number of climate models. In the future, new more accurate models are likely to improve our picture of expected anthropogenic climate change.

3.4 Anthropogenic Climate Change in the Baltic Sea Basin: Projections from Statistical Downscaling

Statistical downscaling (also called empirical downscaling) is a way to infer local information from coarse scale information by constructing empirical statistical links between large scale fields and local conditions (e.g. Zorita and von Storch 1997). Such statistical links can be used to develop detailed local climate scenarios based upon the output from global climate models. A number of statistical downscaling studies have been performed for the Baltic Sea Basin during the last few years (Hanssen-Bauer et al. 2005). Statistical downscaling often involves analysis of observed predictands and predictors, establishing downscaling models between the two, and applying them to global climate model outputs. This section focuses

on the last step for the Baltic region. Special attention is paid to future projections.

3.4.1 Statistical Downscaling Models

Most of the downscaling studies for the Baltic Sea Basin have so far been focused on monthly mean temperature and precipitation (e.g. Murphy 1999, 2000), although other variables have also been used as predictands.

For example, Linderson et al. (2004) tried to develop downscaling models for several monthly based statistics of daily precipitation (e.g. 75 and 95 percentiles and maximum values for daily precipitation) in southern Sweden, but found that skillful models can only be established for monthly mean precipitation and frequency of wet days. Kaas and Frich (1995) developed downscaling models for monthly means of daily temperature range (DTR) and cloud cover (CC) for 10 Nordic stations. Omstedt and Chen (2001), Chen and Li (2004), and Chen and Omstedt (2005) developed downscaling models for annual maximum sea ice cover over the Baltic Sea and sea level near Stockholm. Even the phytoplankton spring bloom in a Swedish lake has been linked to large scale atmospheric circulation and may thus be projected by statistical downscaling (Blenckner and Chen 2003).

Several large scale climate variables have been used as predictors of the statistical models. Due to its strong influences on the local climate (e.g. Chen and Hellström 1999; Busuioc et al. 2001a,b), atmospheric circulation is usually the first candidate as predictor. Among various ways to characterize the circulation, the Sea Level Pressure (SLP) based geostrophic wind and vorticity, u, v and ζ (see Fig. 3.16), are widely used (e.g. Chen 2000; Linderson et al. 2004).

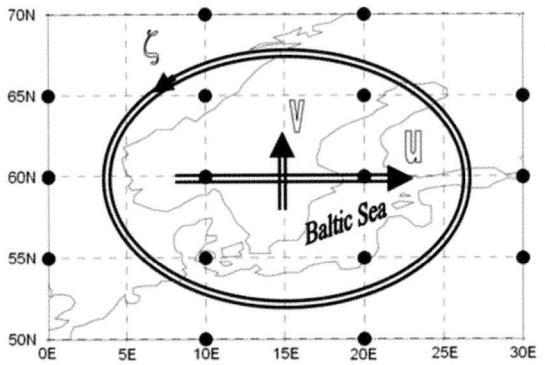

Fig. 3.16. Map showing the predictor domain of the statistical model (adapted from Chen et al. 2006)

In their downscaling study of CC and DTR, Kaas and Frich (1995) stated that the inclusion of tropospheric temperature information among the predictors is of fundamental importance for estimating greenhouse gas induced changes. They thus used both the 500–1000 hPa thickness and the sea level pressure (SLP) fields as predictors. Several potential "signal-bearing" predictors have been tested for downscaling precipitation. Hellström et al. (2001) used large-scale absolute humidity at 850 hPa (q850) as predictor for precipitation, in addition to circulation indices. They conclude that changes in q850 seem to convey much of the information on precipitation changes projected by ECHAM4. Linderson et al. (2004) tested several predictors for monthly mean precipitation and frequency of wet days, including large-scale precipitation, humidity and temperature at 850 hPa and a thermal stability index. They concluded that large-scale precipitation and relative humidity at 850 hPa were the most useful predictors in addition to the SLP based predictors u, v and ζ. Relative humidity was more important than precipitation for downscaling frequency of wet days, while large-scale precipitation was more important for downscaling precipitation.

3.4.2 Projections of Future Climate from Statistical Downscaling

3.4.2.1 *Temperature*

Benestad (2002b, 2004) downscaled temperature scenarios for localities in northern Europe using 17 climate simulations from 10 different global climate models, mainly based on the emission scenario IS92a. A total of 48 downscaled temperature scenarios were produced by using different global simulations, predictors and predictor domains. Though the models show a considerable spread concerning projected warming rates, some results seem to be robust. The projected warming rates are generally larger inland than along the coast. The 48 scenario ensemble mean projected January warming rate during the 21st century increases from slightly below 0.3 °C per decade along the west coast of Norway to more than 0.5 °C per decade in inland areas in Sweden, Finland and Norway (Benestad 2002b). This is shown in terms of probabilities by Benestad (2004), who concludes that under IS92a, the probability of a January warming of 0.5 °C per decade or more is less than 10% along the Norwegian west coast, but more than 70% in some inland areas in Sweden and Finland. Another robust signal is that the projected warming rates in Scandinavia are larger in winter than in summer. Some models also show a tendency for larger warming rates at higher latitudes, though distance to the open sea seems to be more important than latitude.

3.4.2.2 *Precipitation*

Some statistically downscaled scenarios for precipitation were produced applying only SLP-based predictors (Busuioc et al. 2001b, Benestad 2002b). These may be used to evaluate possible consequences of changes in the atmospheric circulation for future precipitation but not for estimating the total effect of increased concentrations of greenhouse-gases on precipitation conditions. The following is thus focused on precipitation studies including additional predictors.

Hellström et al. (2001) used the SLP-based predictors (geostrophic wind and vorticity: u, v, ζ) and large-scale q850 to deduce precipitation scenarios for Sweden from the global models HadCM2 and ECHAM4/OPYC3. Changes in precipitation conditions were projected by studying the differences between 10-year control and scenario timeslices. The downscaled precipitation scenarios for

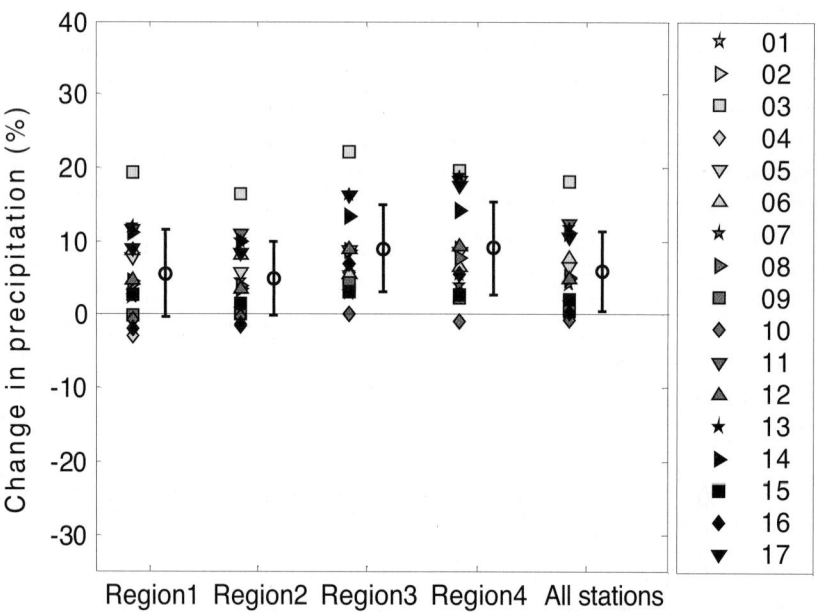

Fig. 3.17. Regional changes in percent for downscaled annual precipitation for four regions in Sweden and for all of Sweden. Region 1 = southernmost, Region 2 = south, Region 3 = north, Region 4 = northernmost. The numbered models are specified in Fig. 3.11. The mean and the error bar (± one standard deviation) of the changes in the 17 GCMs are also plotted (adapted from Chen et al. 2006)

winter and spring show increased precipitation in northern and north-western Sweden (approximately +20%) and reduced precipitation (approximately −20%) in southern Sweden.

During autumn both models project a substantial increase in north-western Sweden, but only minor changes in the southernmost part. During summer HadCM2 projects a substantial increase over the entire country, while ECHAM4/OPYC3 indicates an increase in northern and a reduction in central and southern Sweden. Hellström et al. (2001) conclude that change in vorticity is the greatest contributor to the projected precipitation changes in southern Sweden, while q850 have greater effect in the northern parts of the country. The modelled reduction in spring and winter precipitation in southern Sweden is linked to reduced vorticity, while the projected all-season precipitation increase in northern Sweden is mainly attributed to increased humidity.

Based on the Canadian Global Circulation Model 1 (CGCM1), Linderson et al. (2004) established scenarios of precipitation amount and frequency of wet days for the Scania region in southern Sweden. The CGCM1 simulation applied a greenhouse forcing corresponding to the observed one during the 20th century and thereafter the

IS92a emission scenario up to the year 2100. The downscaled scenario shows a significant increase of the annual mean precipitation (∼ 10%) and a slight decrease (∼ 1.5%) in the frequency of wet days. The downscaled increase is slightly more than the GCM based large scale change in the annual precipitation (8%), which may be interpreted as an enhanced effect of local topography implicitly included in the statistical downscaling model (Achberger 2004).

The results indicate an increase in precipitation intensity almost all year round, but especially during winter. The increase in precipitation during winter and spring is associated with an increase in westerly flow and vorticity, but also with an increase in the large-scale precipitation. The summer decrease is linked to a decrease in vorticity and westerly flow and an increase of northerly flow. The circulation changes, however, may to some extent be specific to CGCM1.

Chen et al. (2006) downscaled precipitation scenarios for Sweden based on the 17 CMIP2 GCMs (Meehl et al. 2000; Sect. 3.3.1 introduces CMIP2), using large scale precipitation as predictor in addition to geostrophic wind and vorticity. They compared the precipitation conditions during years 50–80 of the scenario period (in CMIP2 the doubling of CO_2 occurs in year 70) with an 80-year control run, and concluded that the ensemble of scenarios suggests an overall increase in annual precipitation, as shown in Fig. 3.17. The increase in precipitation is more significant in northern than in southern Sweden. This overall positive trend can be attributed to the increased large-scale precipitation and the westerly wind. The seasonal precipitation in autumn, winter and spring is expected to increase, whereas there is an indication of decreasing summer precipitation in the southern half of the country. The estimated uncertainty is nearly independent of region. However, there is a seasonal dependence; the estimates for winter show the highest level of confidence, and the estimate for summer the least.

3.5 Anthropogenic Climate Change in the Baltic Sea Basin: Projections from Regional Climate Models

Dynamical downscaling describes the process of downscaling from global scales to regional or local scales using dynamical models. For climate studies, this typically consists of applying a coupled atmosphere–land surface model to a limited area of the globe at scales considerably finer than those used for global climate models. For example, horizontal scales are typically some tens of kilometres versus hundreds of kilometres for GCMs. Like statistical downscaling, dynamical downscaling requires driving inputs from a global model. However, dynamical downscaling differs from statistical downscaling in that it includes explicit representation of physical processes for every grid square included in its domain. Critical variables from GCM simulations define driving inputs at the boundaries of the regional climate model (RCM) domain.

3.5.1 Interpreting Regional Anthropogenic Climate Change Projections

Regional climate modelling has been developed and used for dynamical downscaling of GCM results over the past 15 years. The first studies for Europe, including the southern half of the Baltic Sea Basin, were those of Giorgi et al. (1990) and Giorgi and Marinucci (1991). Since their introduction, the spatial and temporal resolution of RCMs has become finer and the level of output detail has increased considerably (e.g. Giorgi et al. 1992;

Christensen et al. 1998; Jones et al. 1995 and 1997; Christensen et al. 2007).

Only a few studies have specifically focused on the Baltic Sea Basin (e.g. Jacob et al. 2001; Räisänen and Joelsson 2001; Kjellström and Ruosteenoja 2007), but a succession of European Union funded research projects with focus on regional downscaling over Europe has produced a host of European-wide studies with results relevant for the Baltic Sea Basin (i.e. Regionalization, RACCS, MERCURE, PRUDENCE, MICE, ENSEMBLES).

There is uncertainty associated with regional anthropogenic climate change projections, which can be summarised as the combination of biases related to the formulation of the regional climate model, lateral boundary conditions and initial conditions from the driving global model, and choice of emissions scenario. This section first addresses the question of the formulation of the RCMs by reviewing experiments in which the RCMs should represent the present climate. This is followed by an inventory of available RCMs and anthropogenic climate change experiments, and then results for projections of the future climate.

3.5.1.1 *Simulation of Present-day Climate from Regional Climate Models*

An important step in climate modelling is to evaluate how well models perform for the present climate (e.g. Achberger et al. 2003). This typically consists of performing model simulations for retrospective observed periods using boundary conditions that best represent observations. In lieu of actual observations at RCM boundaries, "reanalysis" data are often used. These data sets consist of results from numerical weather prediction models that are driven (and constrained) by as many actual observations as possible to produce representation of meteorological variables on a uniform model grid at sub-daily time scales (Uppala et al. 2005). Reanalysis data produced within Europe to date include ERA-15 (1979–1993) and ERA-40 (1957–2001). Other reanalysis data that are commonly used are the NCEP reanalysis data (1948–present, Kalnay et al. 1996; Kistler et al. 2001).

The subject of RCM evaluation was the theme of the European project MERCURE, which used reanalysis data from ERA-15 as reported by Hagemann et al. (2002, 2004) for the Baltic Sea Basin. Five regional models were analysed. In general, temperature and precipitation values averaged over larger areas like the Baltic Sea Basin matched closely with results from observations and the driving reanalysis data. For the Baltic Sea Basin the authors concluded that although average precipitation is generally overestimated to some extent, except during summer, the annual cycle is well described. For temperature, two of the models show a very close match to the observed annual cycle, while two of the others show an exaggerated seasonal cycle with biases of ± 2 °C for individual months. For other quantities, like model generated runoff, evapotranspiration and snowpack, the agreement between individual models and observations is not as good as for temperature and precipitation. Examples include both under- and overestimations for individual months as well as leads and lags of up to 2 months in the timing of the seasonal cycles. A further conclusion was that the higher resolution of RCMs resulted in more realistic smaller-scale variation as compared to GCMs, for example better orographic precipitation and better temporal detail. Additional aspects of the same simulations were analysed by Frei et al. (2003) and by Vidale et al. (2003).

A prominent deficiency of European climate simulations is a tendency toward excessive summer drying that most RCMs show, especially in southeastern Europe, but also in the southern part of the Baltic Sea Basin for some RCMs. This is characterised by temperatures that are too high and both precipitation and evapotranspiration that are too low. This was also noted in the first European multi-RCM project using reanalysis data, as reported in Christensen et al. (1997). Jones et al. (2004), using another RCM, attributed this phenomenon to model deficiencies, in particular for cloud and radiation processes.

In summary, RCMs have been shown to reproduce the mean climate and observed climate variability over Europe for the last decades. This includes not just the large-scale circulation but also other variables. For instance, near-surface temperature is most often simulated to within 1–2 °C from the observations over areas similar in size to the Baltic Sea Basin. Also, the seasonal cycle of precipitation is reproduced to within the uncertainties given by the observational data sets. Despite this agreement between the observed and simulated climate, individual RCMs do show more substantial errors for some variables in different regions and seasons. Furthermore, extreme values are by their nature difficult to validate and an extensive evaluation of model performance in terms of simulating

extreme conditions is lacking. However, it can be noted that errors in simulating extreme conditions often tend to be larger than errors in mean conditions. This is discussed in more detail in Sect. 3.6.

An example of a common deficiency in the RCMs is the inability of RCMs to simulate high wind speeds over land without an additional gust parameterization (Rockel and Woth 2007).

3.5.1.2 Regional Climate Models and Anthropogenic Climate Change Experiments

Regional anthropogenic climate change projections covering the entire Baltic Sea Basin first became available with the work reported by Jones et al. (1995, 1997). A general compilation of RCMs and earlier studies was undertaken by the IPCC (Giorgi et al. 2001). Since then, a number of simulations have been performed, some of which are compared in Christensen et al. (2001) and Rummukainen et al. (2003) for the Nordic region.

In the European project PRUDENCE, ten different regional climate models were used to carry out more than 25 experiments, a majority of which were based on a common global anthropogenic climate change experiment (Christensen et al. 2007). The PRUDENCE matrix of experiments addressed some of the uncertainties mentioned above (Déqué et al. 2007). These most recent studies were based on anthropogenic climate change simulations using a 30-year period as a control to represent the present climate and a 30-year future period based on a documented emissions scenario.

A summary of the RCM future climate simulations referred to in the following discussion is shown in Table 3.2. Here, and in Sect. 3.6, focus is on reporting projections for the key variables of temperature, precipitation, wind and snow. The ability of the RCMs to reproduce the control climate, typically for the period 1961–1990, is first discussed. Future climate scenario experiments for the period 2071–2100 are then presented. In addition to material from published literature, this includes the most recently available results from the PRUDENCE data centre (http://prudence.dmi.dk/). Seasonal averages are highlighted from two specific PRUDENCE RCMs – RCAO and HIRHAM – that were used to downscale simulations from two GCMs – HadCM3/HadAM3H and ECHAM4/OPYC3 – forced by the SRES-A2 emissions scenario (Nakićenović et al. 2000; see also Annex 6). Thus, a range of experiment results reflecting some of the uncertainties originating from both boundary conditions and RCM formulation differences is presented.

In addition, scatter plots for different subregions of the Baltic Sea Basin that include simulations addressing further aspects of uncertainty are presented to illustrate a more comprehensive spread of anthropogenic climate change projections in the region. This includes simulations with numerous RCMs, GCM ensemble simulations downscaled with the same RCM, RCM simulations with different horizontal resolution, and simulations using additional emissions scenarios. It should be noted, however, that the GCMs and emissions scenarios reported here are only a subset of those available (cf. Sect. 3.3).

3.5.2 Projections of Future Climate from Regional Climate Models

3.5.2.1 *Temperature*

The air temperature discussed below refers to the two-meter level air temperature in the models. This corresponds to a typical height common to most observation stations.

Temperature, control climate

Temperature biases in the Baltic Sea Basin for control experiments downscaled with RCMs have been shown to be generally positive for the winter season when compared to observations (Räisänen et al. 2003; Giorgi et al. 2004a). However, these biases are typically less than 2 °C. An exception is larger warm biases in the mountainous interior of northern Scandinavia. These larger biases, which are also seen in the global models (cf. Sect. 3.3.3), may partly be related to the fact that the observations come primarily from cold valley stations while RCM output is averaged from gridboxes covering a range of elevations, as discussed in Räisänen et al. (2003). The milder winter climate shown in these studies can be related to the excessive north–south pressure gradient over the North Atlantic inherited from the global models (cf. Sect. 3.3.3). It should be noted that the size and even sign of the temperature bias is sensitive to the boundary conditions (cf. Fig. 3.6a).

Regarding summer temperature, Räisänen et al. (2003) and Giorgi et al. (2004a) found a relatively large bias (1–2 °C) in summer temperature specific

to the south-eastern part of the Baltic Sea Basin, while biases in spring and autumn were small for the entire region. Vidale et al. (2007) also showed a positive bias in temperature in the south-eastern part of the Baltic Sea Basin for one of the RCMs that they analysed. This bias in the south-eastern part of the Baltic Sea Basin is thought to be related to the summer drying bias mentioned above for the reanalysis boundary experiments in southeastern Europe (cf. Sect. 3.5.1.1).

Seasonal cycle of temperature, future climate

In future scenarios, as snow cover retreats north and east, the climate in the Baltic Sea Basin undergoes large changes, particularly during the winter season. A common feature in all regional downscaling experiments is the stronger increase in wintertime temperatures compared to summertime temperatures in the northern and eastern parts of the Baltic Sea Basin (e.g. Giorgio et al. 1992; Jones et al. 1997; Christensen et al. 2001; Déqué et al. 2007), as shown in Fig. 3.18. This pattern of anthropogenic climate change is also seen from the global climate models (e.g. Fig. 3.9), but here, with higher horizontal resolution, regional features have a more pronounced impact on the results. For instance, the strong reduction in sea ice in the Bothnian Bay (presented in Sect. 3.8) leads to a substantial increase in air temperature over the Bothnian Bay.

The projected temperature change for summer is shown in Fig. 3.19. Warming in this season is most pronounced further south in Europe, and consequently, it is to the south of the Baltic Sea that the warming is strongest in the Baltic Sea Basin. Figures 3.20 and 3.21 show a wider range of RCM model results, whereby additional experiments are included as outlined in the figure caption. These figures summarise change in temperature against change in precipitation for northern and southern subregions of the Baltic Sea Basin. In some models the summertime warming south of the Baltic Sea is as large, or even larger, than that during winter.

As seen in Fig. 3.19, a local maximum over the Baltic Sea stands out in the experiments forced with boundary conditions from HadAM3H. This feature is associated with a strong increase in Baltic Sea SSTs (sea surface temperatures) for these experiments, which is larger than in any of the other GCMs discussed in Sect. 3.3.4 (Kjellström and Ruosteenoja 2007).

Table 3.2. List of studies that include RCM simulations of future climate covering parts of the Baltic Sea Basin. The scenarios SRES-A2 and SRES-B2 refer to the special report on emission scenarios from the IPCC (Nakićenović et al. 2000). [1]Pattern-scaling to a common time frame was performed; the original control and scenario simulations covered different time periods. [2]Pattern-scaling to a common emission scenario was applied; the original simulations used different scenarios (cf. Christensen et al. 2001). [3]PRUDENCE experiments

Reference	RCM	Resolution	GCM	Scenario	Timeperiod
Giorgi et al. (1992)	MM4	70 km	CCM	$2 \times CO_2$	–
Jones et al. (1995, 1997)	HadRM	50 km	HadCM	$2 \times CO_2$	–
Déqué et al. (1998)	Arpège stretched	60–700 km	N/A	$2 \times CO_2$	2×10 yrs
Räisänen et al. (1999)	RCA1	44 km	HadCM2	GHG	Pre-ind. (10 yrs) & 2039–2049
Räisänen et al. (2001)	RCA1	44 km	HadCM2 ECHAM4/OPYC3	GHG	Pre-ind. (10 yrs) & 2039–2049 10 yrs from 1980s & 2070s
Christensen et al. (2001, 2002)	RCA, HIRHAM	18,44,55 km	ECHAM4/OPYC, HadCM2	GHG, GHG +sulf+trop. O_3	10–20 yrs from 1990s & 2050s[1]
Rummukainen et al. (2003)	RCA, HIRHAM	18,44,55 km	ECHAM4/OPYC HadCM2	B2[2]	10–20 yrs from 1990s & 2050s[1]
Räisänen et al. (2003)	RCAO	50 km	HadCM3/HadAM3H ECHAM4/OPYC3	A2, B2	1961–1990, 2071–2100
Christensen and Christensen (2004)	HIRHAM	50 km	HadAM3H ECHAM4/OPYC3	A2	1961–1990, 2071–2100
Räisänen et al. (2004)	RCAO	50 km	HadCM3/HadAM3H ECHAM4/OPYC3	A2, B2	1961–1990, 2071–2100
Moberg and Jones (2004)	HadRM3P	50 km	HadCM3/HadAM3P	–	1961–1990
Kjellström (2004)	RCAO	50 km	HadCM3/HadAM3H ECHAM4/OPYC3	A2, B2	1961–1990, 2071–2100
Giorgi et al. (2004a,b)	RegCM	50 km	HadCM3/HadAM3H	A2, B2	1961–1990, 2071–2100
de Castro et al. (2007)	9 RCMs[3]	≈ 50 km	HadCM3/HadAM3H	A2	1961–1990, 2071–2100
Déqué et al. (2007)	10 RCMs[3]	≈ 50 km	HadCM3/HadAM3H ECHAM4/OPYC3 Arpège/IFS	A2, B2	1961–1990, 2071–2100
Jylhä et al. (2007)	7 RCMs[3]	≈ 50 km	HadCM3/HadAM3H ECHAM4/OPYC3	A2, B2	1961–1990, 2071–2100
Kjellström et al. (2007)	10 RCMs[3]	≈ 50 km	HadCM3/HadAM3H ECHAM4/OPYC3	A2, B2	1961–1990, 2071–2100
Vidale et al. (2007)	10 RCMs[3]	50 km	HadCM3/HadAM3H ECHAM4/OPYC3	A2, B2	1961–1990, 2071–2100
Ferro et al. (2005)	HIRHAM	50 km	HadAM3H	A2	1961–1990, 2071–2100
Beniston et al. (2007)	10 RCMs[3]	≈ 50 km	HadCM3/HadAM3H ECHAM4/OPYC3	A2, B2	1961–1990, 2071–2100
Pryor and Barthelmie (2004)	RCAO	50 km	HadCM3/HadAM3H ECHAM4/OPYC3	A2, B2	1961–1990, 2071–2100
Rockel and Woth (2007)	8 RCMs	≈ 50 km	HadCM3/HadAM3H	A2, B2	1961–1990, 2071–2100
Leckebusch and Ulbrich (2004)	HadRM3H	≈ 50 km	HadCM3/HadAM3H	A2, B2	1960–1989, 2070–2099
Rummukainen et al. (2004)	RCAO	50 km	HadCM3/HadAM3H ECHAM4/OPYC3	A2, B2	1961–1990, 2071–2100
Pryor et al. (2005b)	RCAO	50 km	HadCM3/HadAM3H ECHAM4/OPYC3	A2, B2	1961–1990, 2071–2100

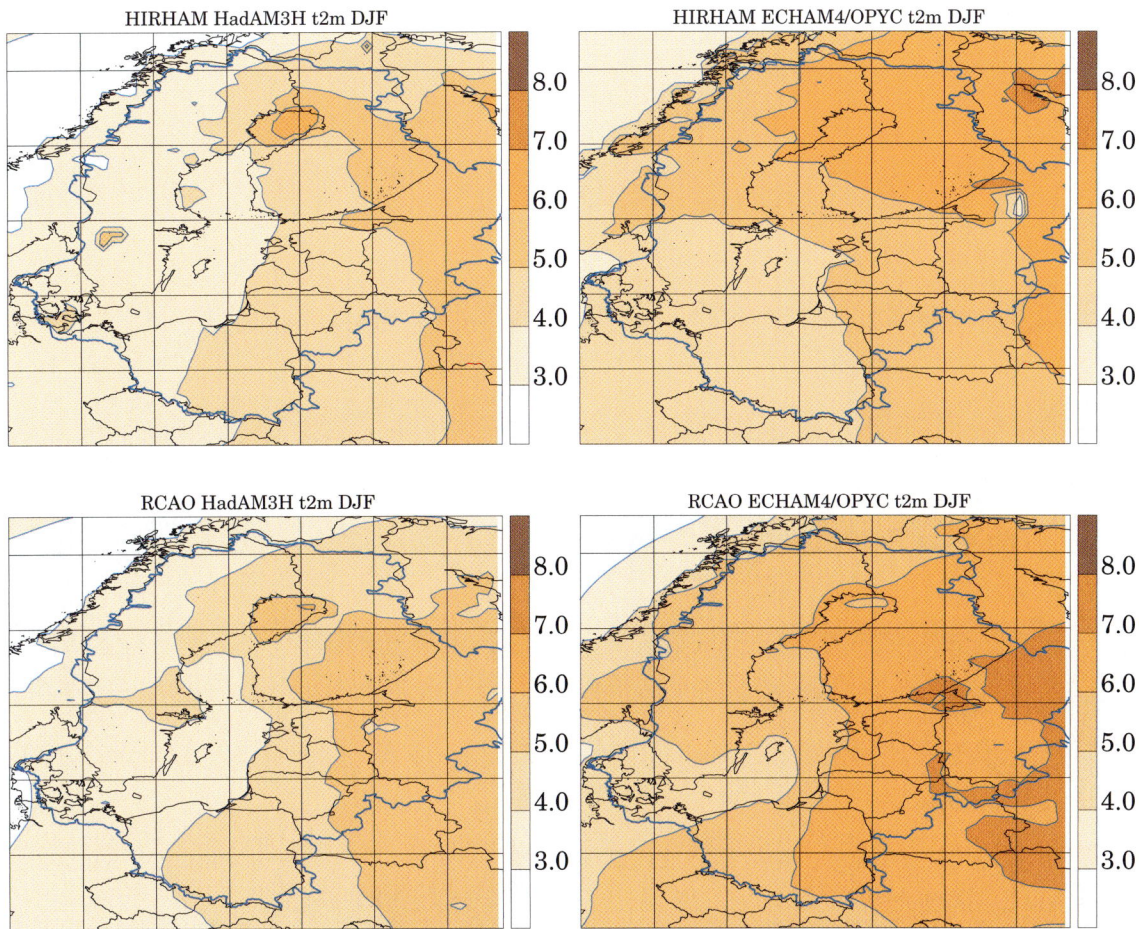

Fig. 3.18. RCM simulated temperature change in °C for winter (DJF) between the periods 1961–1990 and 2071–2100 using the SRES-A2 emissions scenario. The upper plots show results from the HIRHAM Model and the lower plots are from the RCAO Model. Plots on the left used GCM boundary conditions from HadAM3H; plots on the right used ECHAM4/OPYC3. The Baltic Sea Basin is indicated by the thick blue line (*note: ECHAM4/OPYC3 scenario simulations used as boundaries are different for the two RCM downscaling experiments, see Sect. 3.5.2.3*)

Experiments to investigate this further with the help of a regional coupled atmosphere–ocean model showed the excessive SSTs used in HadAM3H to be unrealistic (Kjellström et al. 2005). Such anomalies from GCMs can have consequences for the hydrological cycle over the Baltic Sea and parts of the surrounding land areas.

Figures 3.20 and 3.21 show area mean changes of temperature and precipitation for winter and summer, respectively, for 25 of the PRUDENCE experiments, which included 10 different RCMs, some run for three different GCM ensemble members, three different driving GCMs and two emission scenarios. The figures summarise results for the same four areas shown in Fig. 3.12 (two over

land and two over sea). The overall ranges in temperature in Fig. 3.20 are smaller than the ranges estimated for the SRES-A2 and SRES-B2 experiments based on 6 GCMs, as illustrated in Fig. 3.13.

This implies that the largest uncertainty in these regional climate projections is due to the boundary conditions from the GCMs, as shown by Déqué et al. (2007). Nevertheless, there is additional uncertainty illustrated in Fig. 3.20 that is due to the formulation of the individual RCMs.

The open dots in the plot show the different projections from the common experiment performed in PRUDENCE, in which the RCMs were forced by lateral boundary conditions and SSTs from one GCM (HadAM3H). This experiment illus-

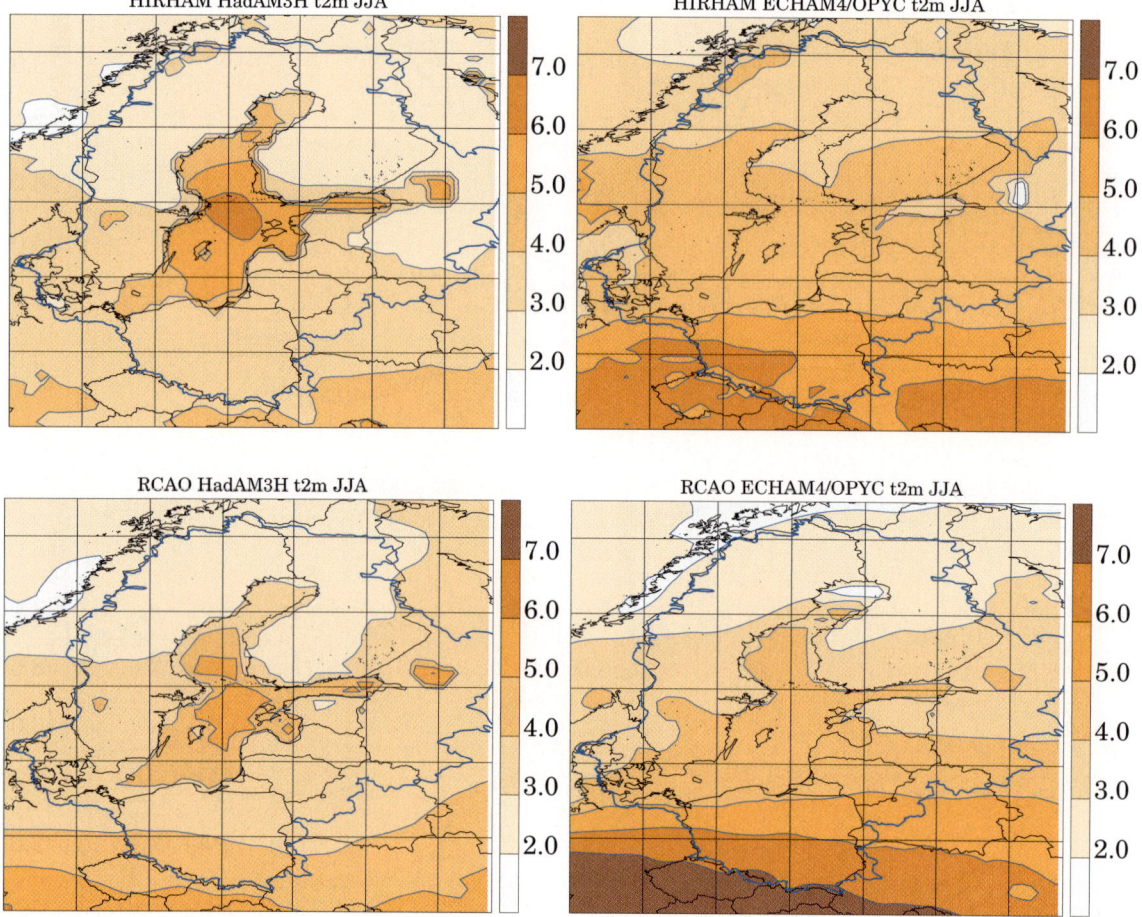

Fig. 3.19. RCM simulated temperature change in ℃ for summer (JJA) between the periods 1961–1990 and 2071–2100 using the SRES-A2 emissions scenario. The upper plots show results from the HIRHAM Model and the lower plots are from the RCAO Model. Plots on the left used GCM boundary conditions from HadAM3H; plots on the right used ECHAM4/OPYC3. The Baltic Sea Basin is indicated by the thick blue line (*note: ECHAM4/OPYC3 scenario simulations used as boundaries are different for the two RCM downscaling experiments, see Sect. 3.5.2.3*)

trates that there is a considerable spread between the projections due to RCM formulation. This spread is even larger during summer, as shown for the land areas in Fig. 3.21 (since the excessive HadAM3H Baltic Sea summer SSTs are unrealistically high, as discussed above, all experiments utilising those as lower boundary conditions have been excluded from the sea regions in Fig. 3.21). Compared to the ranges given by the GCMs in Fig. 3.13 the ranges from Fig. 3.21 are almost as large, indicating the relative importance of RCM uncertainty in climate projections for summer in this region.

Interannual variability of temperature, future climate

Räisänen et al. (2003) and Giorgi et al (2004a) showed that the interannual variability of temperature was simulated well in their respective control simulations as compared to gridded observations (CRU data). In the future anthropogenic climate change simulations analysed by Räisänen et al. (2003), interannual variability decreased in northern Europe during winter. They related this to the reduction in snow and ice in mid and high latitudes, as also seen in global models (Räisänen 2002). They also found large differences between different emission scenarios that they attributed to a relatively low signal-to-noise ratio.

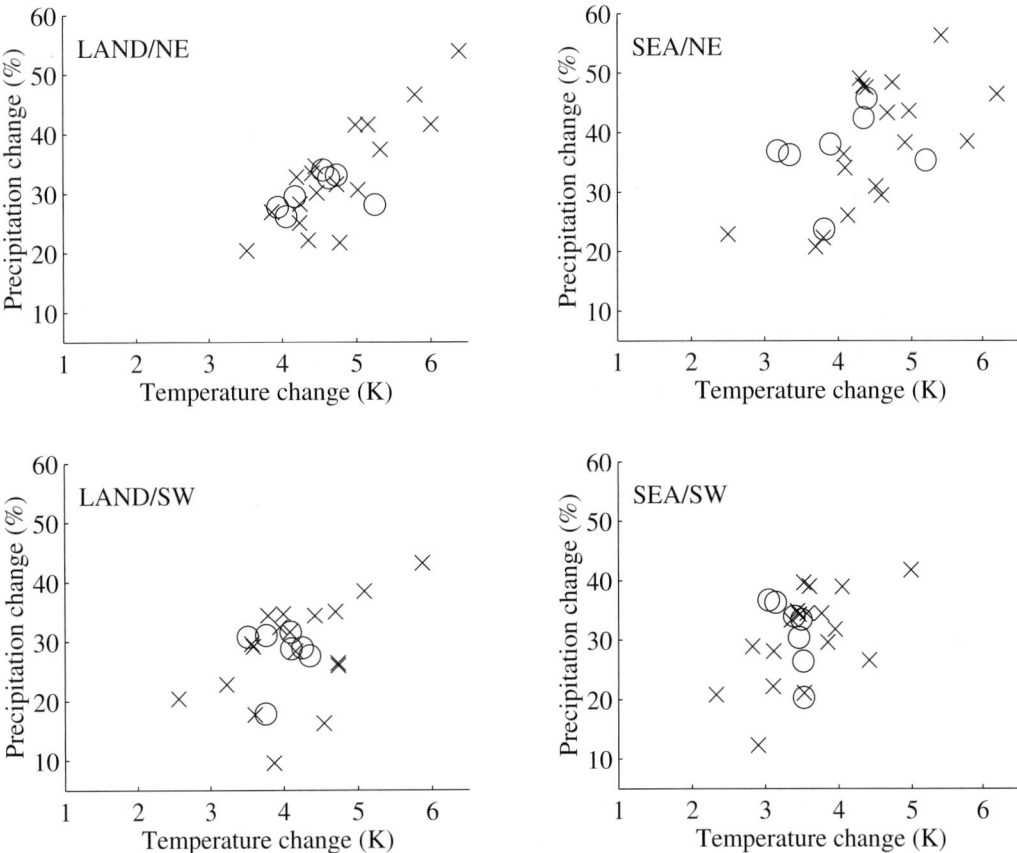

Fig. 3.20. Changes in winter (DJF) area mean temperature and precipitation for the areas defined in Fig. 3.12 The "○" symbol denotes 7 RCMs from the common PRUDENCE experiment based on the same GCM (HadAM3H). The "✕" symbol denotes other regional downscaling experiments from PRUDENCE, which included different driving GCMs, different emission scenarios, higher horizontal resolution and several ensemble members

Giorgi et al. (2004b) found increasing interannual variability south of the Baltic Sea during summer (JJA) and autumn (SON) in two future scenarios, SRES–A2 and SRES–B2. Schär et al. (2004) and Vidale et al. (2003) also found increasing interannual summer (JJA) temperature variability for Central Europe, including the southern part of the Baltic Sea Basin. The standard deviation of the interannual variability increased by 20 to 80% between the different RCMs in their study, all of which used the same emissions scenario. They linked this to the dynamics of soil-moisture storage and the associated feedbacks on the surface energy balance and precipitation.

Diurnal temperature range, future climate

The diurnal temperature range (DTR) is defined as the average difference between daily maximum and minimum temperatures. In future anthropogenic climate change scenarios, both Christensen et al. (2001) and Rummukainen et al. (2003) found a stronger decrease of the DTR in fall and winter than during summer for the Nordic region. The reduction around 2050 amounts to $0.6 \pm 0.4\,°C$ in winter, where the central value is the mean from all experiments and the range is given by the spread within the ensemble.

Räisänen et al. (2003) also found a larger change in DTR in northern Europe during the period late autumn to spring than in summer. The reduction of DTR in their scenarios, analysed over Sweden for the period 2071–2100, was of the order of $1\,°C$ for the winter months. They also noted that the night-to-day temperature variability in northern Europe is small during mid-winter and that much of the simulated changes in DTR may be affected by irregular day-to-day variations.

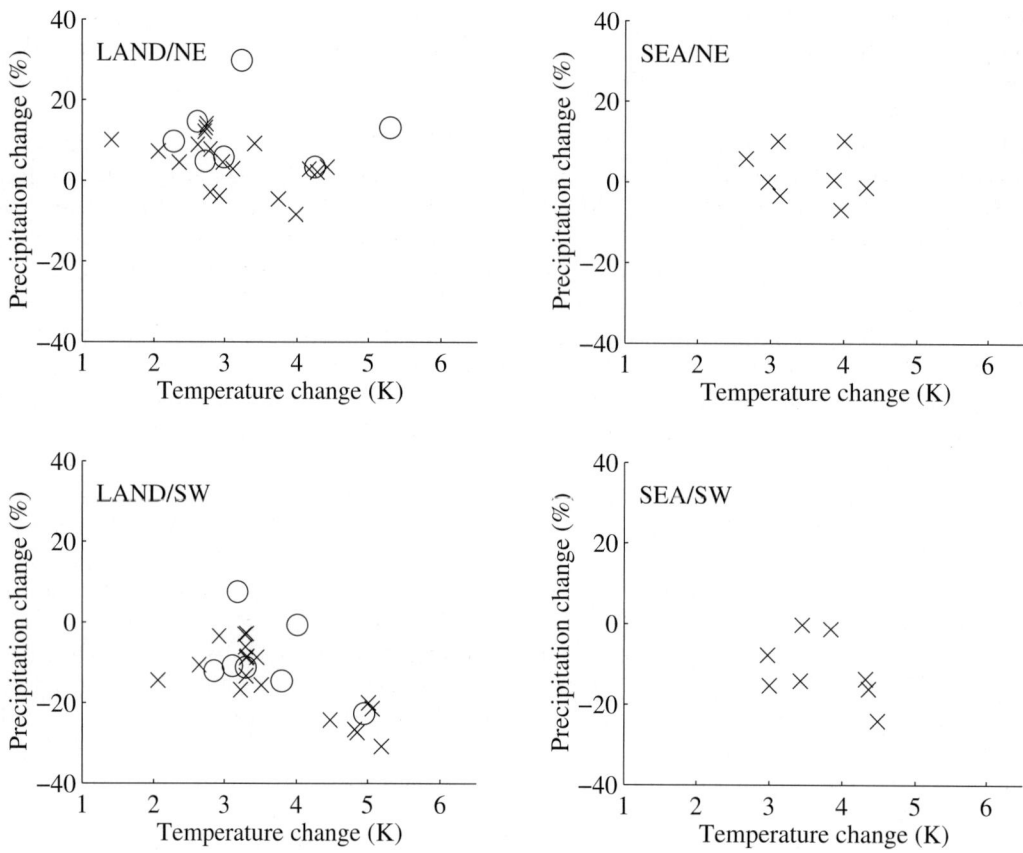

Fig. 3.21. Changes in summer (JJA) area mean temperature and precipitation for the areas defined in Fig. 3.12 The "○" symbol denotes 7 RCMs from the common PRUDENCE experiment based on the same GCM (HadAM3H). The "×" symbol denotes other regional downscaling experiments from PRUDENCE, which included different driving GCMs, different emission scenarios, higher horizontal resolution and several ensemble members (*note that the common experiment is not shown for the sea areas due to unrealistically high Baltic Sea summer SSTs in HadAM3H*)

It should be noted that some models have problems in adequately simulating daily maximum and minimum temperatures for control periods (Moberg and Jones 2004; Kjellström et al. 2007; see also Sect. 3.6.2.1).

3.5.2.2 Precipitation

Precipitation, control climate

Inadequate observations contribute to difficulties in verifying precipitation results from climate models. Jones et al. (1995) compared large-scale precipitation over Europe from an RCM and its driving model to observed precipitation climatology (Legates and Wilmott 1990). They showed the RCM precipitation to be around 30% higher than the driving GCM and higher than observations, except for summer. They concluded that

the large scale precipitation from the RCM could be realistic anyway, as observed precipitation was probably underestimated due to gauge undercatch errors, particularly for winter (see also Annex 5).

Several studies (Christensen et al. 1998; Rummukainen et al. 2001; Rutgersson et al. 2002) have similarly concluded that RCM precipitation in northern Europe is higher than the driving models and generally higher than observed data sets except for summer. Christensen et al. (1998) linked positive precipitation biases to exaggerated cyclone activity and high SSTs in the GCM control simulation, as compared to observations. They also attributed some of the bias to undercatch in precipitation observations.

The credibility of precipitation in climate models does not depend only on mean values. Hellström et al. (2001) studied the annual cycle of precipitation in Sweden and Hanssen-Bauer et al. (2003) focused on regional variability in Norway. Both concluded that RCM simulations improved significantly, according to the driving GCM. Rummukainen et al. (2001) further concluded that although a large part of RCM biases can be attributed to the driving models, the RCMs added value to GCM simulations.

However, division of precipitation into intensity classes reveals a "drizzle problem", which is that the models exhibit too many days with light precipitation (defined as less than $1 \, mm \, day^{-1}$) and too few dry days (defined as less than $0.1 \, mm \, day^{-1}$), compared to observations (Christensen et al. 1998). The problem was shown to be more pronounced at finer RCM resolutions. According to Räisänen et al. (2003), relative interannual variability of precipitation from RCMs seems to be underestimated in the Baltic area, probably connected to the unrealistically high number of rainy days.

In more recent studies, Hagemann and Jacob (2006) compared the PRUDENCE RCM results to two observed precipitation databases that attempt to correct for gauge undercatch (CMAP, Xie and Arkin 1997; and GPCP, Huffman et al. 1997). They showed annual precipitation values for the Baltic Sea Basin to be close to an average of the two databases. They argued that the average of the two databases should exhibit realistic corrections for winter precipitation. Kjellström and Ruosteenoja (2007) also compared Baltic Sea Basin precipitation to observed databases (GPCP as above; and CRU, New et al. 1999). Their analysis of the seasonal cycle showed a general overestimation of winter precipitation. They also suggested that observational databases over the sea may be biased toward high summer precipitation due to erroneous influence of coastal precipitation.

Winter precipitation, future climate

As with results from global model experiments, regional projections for winter precipitation show increases over most of Europe (e.g. Déqué et al. 1998). Details of the geographical distribution for precipitation changes vary with different RCM simulations, however. The main source of disagreement in the RCM results is likely the different large-scale anthropogenic climate change signals from the GCM simulations employed. These differences are not just due to the consequences of different model formulations, but are also due to the fact that slow climate variations like the NAO can have different phases in the global simulations used to drive the RCMs.

The analysis by Déqué et al. (2007) of the multiple PRUDENCE simulations systematically attributes variations in results to different RCMs, emissions scenarios, GCM boundaries and variation between GCM ensemble members. For winter precipitation over the Baltic area, GCM boundary conditions are estimated to account for some 61% of the total variance and the choice of RCM for some 34% (Déqué, pers. comm. 2005). In summer, these roles are reversed with 74% of the variation attributable to choice of RCM for the Baltic area. However, there is an even wider distribution between choice of RCM, GCM and emissions scenario for most other European subregions.

This can be illustrated through an examination of Fig. 3.22, which shows winter precipitation change. The large-scale anthropogenic climate change in the ECHAM4/OPYC3 simulation leads to an intensification of the zonal flow, basically increasing the number and intensity of low-pressure systems from the Atlantic that hit Norway. The HadAM3H model turned the flow in a more south-easterly direction, leading to a decrease in precipitation in mid-Norway.

Simulations from the PRUDENCE common experiment tend to agree about an increase of more than 25% in a southwest to northeast band from England to Finland. In Fig. 3.20 this increase in winter precipitation can also be seen in the four subregions. The ECHAM4/OPYC3 experiments using the RCAO (Räisänen et al. 2003, 2004) and HIRHAM models (Christensen and Christensen 2004) show less agreement, as seen in Fig. 3.22. While both simulations exhibit increased precipitation of over 25% for most of northern Europe, the increase in precipitation is larger and covers an area much farther east in the RCAO simulation.

It is important to note that the boundary fields for these two sets of simulations are the same for the control period but not for the scenario period. The two simulated 2071–2100 time periods have different NAO phases, which is reflected in the downscaling results (see e.g. Ferro 2004). Hence, the larger difference between the two RCM projections using ECHAM4/OPYC3 scenario simulations compared to the two RCM projections using HadAM3H simulations can be attributed to differences in large-scale circulation in two different ECHAM4/OPYC3 scenario simulations.

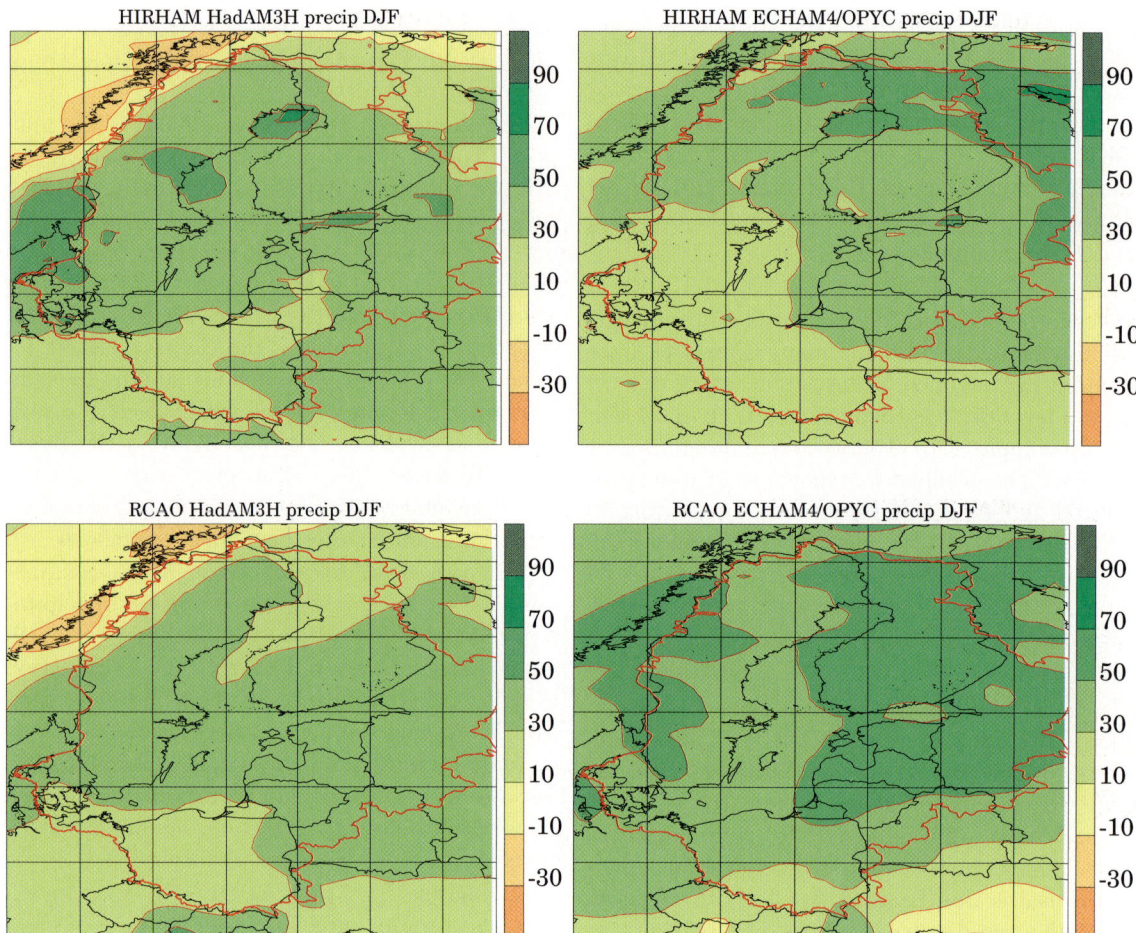

Fig. 3.22. RCM simulated precipitation change in percent for winter (DJF) between the periods 1961–1990 and 2071–2100 using the SRES-A2 emissions scenario. The upper plots show results from the HIRHAM Model and the lower plots are from the RCAO Model. Plots on the left used GCM boundary conditions from HadAM3H; plots on the right used ECHAM4/OPYC3. The Baltic Sea Basin is indicated by the thick red line (*note: ECHAM4/OPYC3 scenario simulations used as boundaries are different for the two RCM downscaling experiments, see Sect. 3.5.2.3*)

Summer precipitation, future climate

For future summers, regional anthropogenic climate change simulations show increases in precipitation for northern parts of the Baltic Sea Basin and decreases to the south, as shown in Fig. 3.23 for four simulations. This results in only a small average change for the basin as a whole. The dividing line between increase and decrease for the full range of PRUDENCE simulations generally goes across the southern half of Norway and Sweden continuing eastward through the Baltic countries (Kjellström and Ruosteenoja 2007).

As discussed above (Sect. 3.5.2.1) SSTs from simulations driven by HadAM3H show a large

anomalous summer heating of the Baltic Sea, which in turn leads to a local increase in precipitation (Kjellström and Ruosteenoja 2007). Representation of summer precipitation over parts of the Baltic Sea Basin is therefore particularly sensitive to how thermal conditions in the Baltic Sea itself are input to the RCMs.

Precipitation and temperature, future scenarios

Figures 3.20 and 3.21 provide a comprehensive comparison of the projected anthropogenic climate change signals in temperature and precipitation between RCMs in the northern and southern sub-

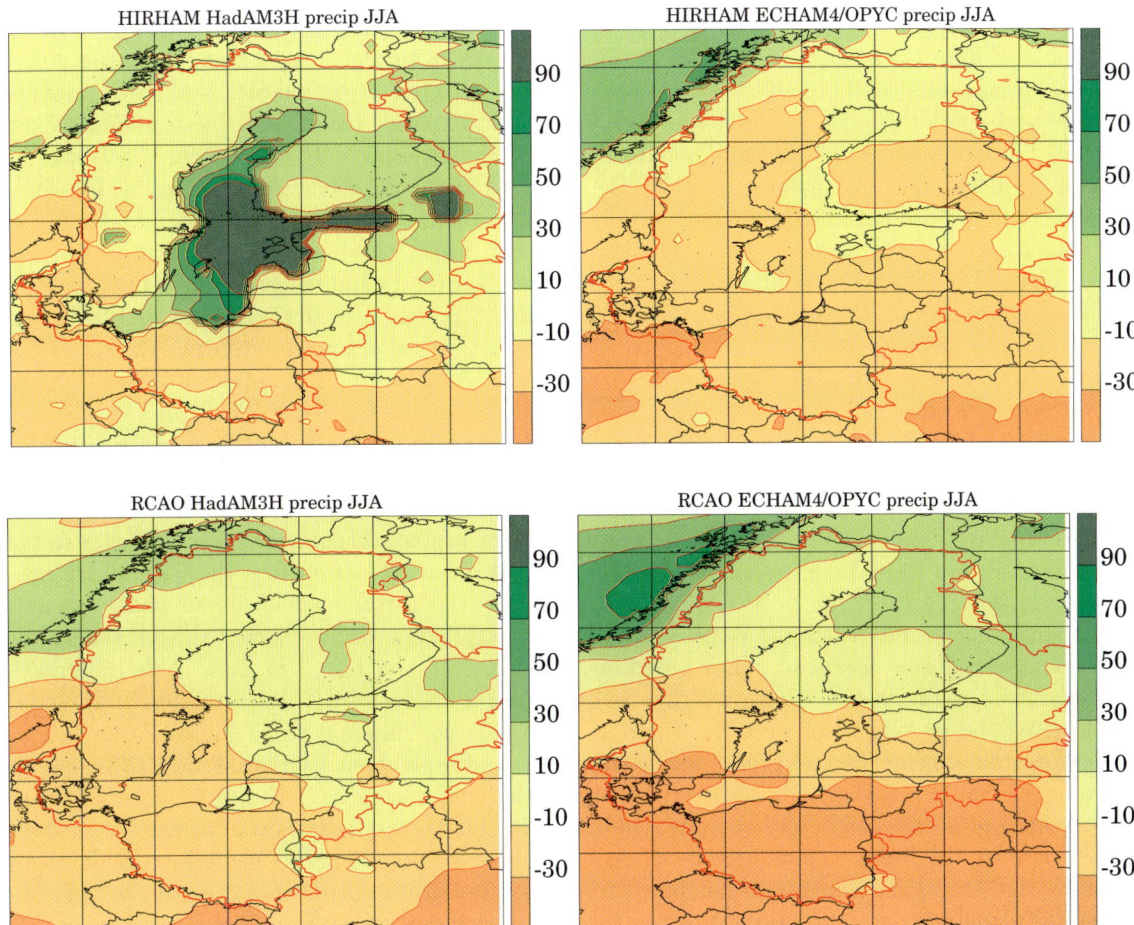

Fig. 3.23. RCM simulated precipitation change in percent for summer (JJA) between the periods 1961–1990 and 2071–2100 using the SRES-A2 emissions scenario. The upper plots show results from the HIRHAM Model and the lower plots are from the RCAO Model. Plots on the left used GCM boundary conditions from HadAM3H; plots on the right used ECHAM4/OPYC3. The Baltic Sea Basin is indicated by the thick red line (*note: ECHAM4/OPYC3 scenario simulations used as boundaries are different for the two RCM downscaling experiments, see Sect. 3.5.2.3*)

regions of the Baltic Sea Basin. The trend is for wintertime climate to become milder and generate more precipitation in both sub-regions. As discussed for the GCMs in Sect. 3.3.3.1, there tends to be a positive correlation between increasing temperature and precipitation, particularly over the north-eastern land area in most RCMs – the warmer the wetter. During summer there is no such correlation in the north, where all models get warmer to a different degree and most get wetter by between 5 to 20%. In the south, there is a tendency of an inverse correlation for summer – the warmer the drier.

3.5.2.3 Wind

To date, studies on future wind changes using RCMs were based mainly on two sets of simulations. One such set was performed within SWECLIM (Swedish Regional Climate Modelling Programme) and is described in detail by Rummukainen et al. (2000; 2004). The other set is more recent and includes eight different RCMs from the PRUDENCE project (Christensen et al. 2002). The results presented below focus mostly on results from the same two RCAO control simulations discussed above for temperature and precipitation. Wind speed in the following discussion refers to the mean velocity of the near surface wind. This is generally the model wind speed at a height of 10 m from the surface.

Several parameters can be used to describe the temporal and spatial changes in wind speed and direction. A summary of the relevant quantities is given by Pryor and Barthelmie (2004). Besides generally applied parameters such as means, autocorrelations and percentiles, they also described quantities especially useful for the wind-power industry. These are the Weibull distribution and energy density, which are described in detail by the Danish Wind Industry Association (www.windpower.org).

Wind, control climate

Pryor and Barthelmie (2004) compared wind speed from the two RCAO control simulations to NCEP reanalysis data for the period 1961–1990. They found that these RCM simulations accurately represented the dominant wind direction from southwest to southeast. However, they underestimated the prevalence of westerly winds in a band oriented southwest to northeast across the centre of the Baltic Sea Basin and overestimated the frequency of northeasterly and easterly winds in the south of the basin. Qualitatively RCAO driven by HadAM3H showed greater similarity to the NCEP reanalysis data than RCAO driven by ECHAM4/OPYC3.

Comparing to gridded observations (CRU data), Räisänen et al. (2003) found that the simulated seasonal cycle of wind speed in the RCAO simulations was in good agreement for the Baltic Sea Basin as a whole. However, summer minimums occurred one month earlier (July) in the control simulations compared to observations. In amplitude, both simulations overestimated the average observed wind speeds in winter and underestimated them in summer. Räisänen et al. (2003) ascribe this to two possible factors: 1) deficiencies in the RCM in simulating the boundary layer near surface conditions, and 2) an uneven distribution of observation stations.

Wind, future climate

Overall, Pryor and Barthelmie (2004) found that the spatial patterns of wind results from RCAO driven by ECHAM4/OPYC3 show larger changes between control and scenario than RCAO driven by HadAM3H. This is particularly true for mean wind speed as shown in Figs. 3.24 and 3.25, energy density, and the upper percentiles of the wind speed distribution. They argued that the differences in the two projected SRES-A2 future climate simulations from RCAO are due to differences in mean sea level pressure and transient activity.

This is in line with the findings by Räisänen et al. (2003), who show future changes in mean annual wind speed to fall mostly between −4 to +4% over Scandinavia for RCAO driven by HadAM3H for both SRES-A2 and SRES-B2 scenarios. Corresponding results from RCAO driven by ECHAM4/OPYC3 are about 8% for SRES-A2 and slightly less for SRES-B2.

A statistical analysis by Pryor and Barthelmie (2004) showed only small similarities in mean wind speed change between the two RCAO simulations, both for winter and summer and even less so in the annual mean. They ascribe the large differences between the anthropogenic climate change signals to the different GCMs used to drive the RCM. Figures 3.24 and Fig. 3.25 show these differences related to choice of GCM for summer and winter.

Again, the different NAO phases in the boundary conditions from ECHAM4/OPYC3 show up in the winter climate. The more positive phase of NAO in the boundaries is indicative of stronger winds on average. This is due both to the stronger pressure gradient in itself and to the warmer climate with less stably stratified conditions on average.

Regarding the seasonal cycle, Räisänen et al. (2003) found that the largest increases in future climate wind speed occur in simulations driven by ECHAM4/OPYC3 in winter and early spring over Sweden and northern Europe, as shown for SRES-A2 in Fig. 3.24 for winter. They occur when the increase in north–south pressure gradient is largest. Over land areas, these increases are up to 12% for the SRES-A2 and some 7% for the SRES-B2 scenarios, as an average over Sweden for DJF. Corresponding simulations driven by HadAM3H show almost no change over land for winter. Figure 3.25 illustrates that there is an opposite trend for summer from these simulations, as they show a decrease in wind speed over most of the Baltic Sea Basin. However, statistical analysis showed that only the ECHAM4/OPYC3 driven results for winter are statistically significant at the 95% level.

Over the Baltic Sea itself, simulated wind changes are modified by stability effects associated with changes in SST and ice cover. The winter increases in the ECHAM4/OPYC3 driven RCM simulations are up to about 18% in SRES-

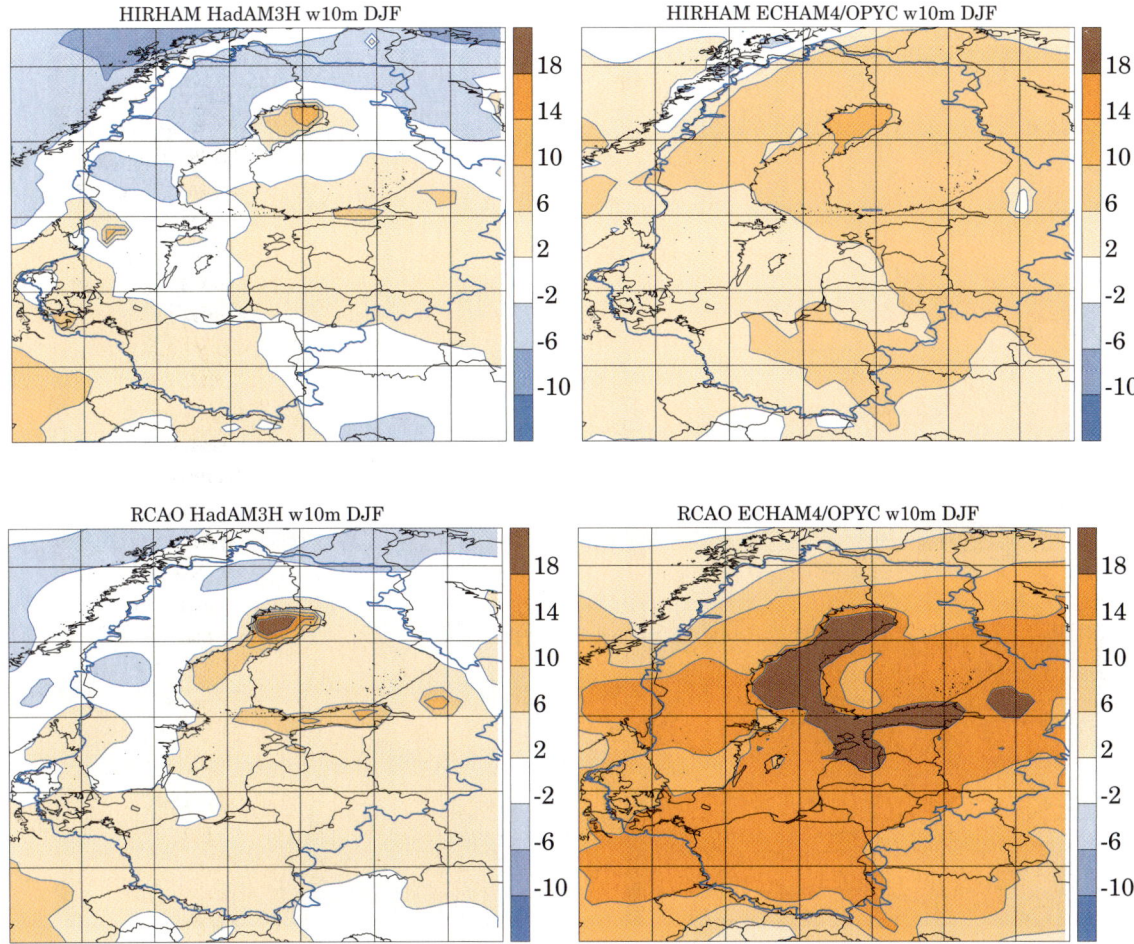

Fig. 3.24. RCM simulated wind speed change in percent for winter (DJF) between the periods 1961–1990 and 2071–2100 using the SRES-A2 emissions scenario. The upper plots show results from the HIRHAM Model and the lower plots are from the RCAO Model. Plots on the left used GCM boundary conditions from HadAM3H; plots on the right used ECHAM4/OPYC3. The Baltic Sea Basin is indicated by the thick blue line (*note: ECHAM4/OPYC3 scenario simulations used as boundaries are different for the two RCM downscaling experiments, see Sect. 3.5.2.3*)

A2 and 13% in SRES-B2, as an average over the entire Baltic Sea for DJF. Corresponding HadAM3H driven simulations show an increase of less than 5%, as an average over the entire Baltic Sea. The largest increases occur over the central and northern part of the sea, where ice cover decreases in the scenario runs. Summer changes over the Baltic Sea show a decrease of up to 7% for ECHAM4/OPYC3 driven simulations and an increase of about 5% for HadAM3H driven simulations, as an average over the entire Baltic Sea for JJA.

Regarding the latter, this reflects the large increase in Baltic Sea SSTs discussed above (Sect. 3.5.2.1), which leads to reduced surface stability and thereby higher wind speed (Räisänen et al. 2003).

Pryor and Barthelmie (2004) interpreted future wind changes with respect to the use of wind as an energy source by defining wind resource classes calculated from energy density in each model grid cell. This is defined as "poor" for energy density less than $70\,\mathrm{Wm}^{-2}$, or "good" for energy density greater than $140\,\mathrm{Wm}^{-2}$. Using these definitions, Pryor and Barthelmie (2004) found that the number of grid cells rated as "poor" decreased and the number of grid cells rated as "good" increased for almost all the simulations they analysed.

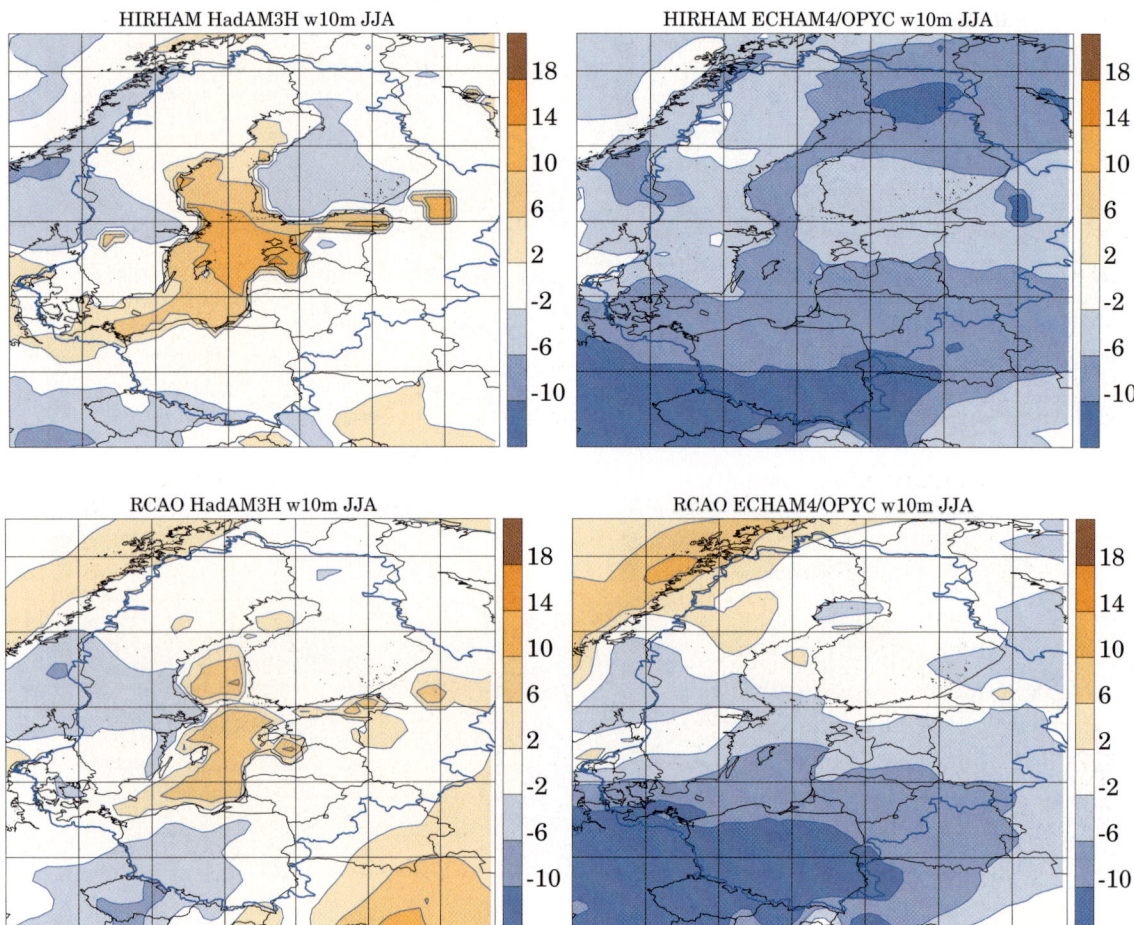

Fig. 3.25. RCM simulated wind speed change in percent for summer (JJA) between the periods 1961–1990 and 2071–2100 using the SRES-A2 emissions scenario. The upper plots show results from the HIRHAM Model and the lower plots are from the RCAO Model. Plots on the left used GCM boundary conditions from HadAM3H; plots on the right used ECHAM4/OPYC3. The Baltic Sea Basin is indicated by the thick blue line (*note: ECHAM4/OPYC3 scenario simulations used as boundaries are different for the two RCM downscaling experiments, see Sect. 3.5.2.3*)

3.5.2.4 Snow

Snow, control climate

Räisänen et al. (2003) found that the mean annual duration of snow season as simulated by RCAO is in good agreement with the observations of Raab and Vedin (1995) for Sweden. The mean annual maximum water content of the RCM snow pack tends to be slightly too low in southern Sweden and slightly too high in northern Sweden. A large positive bias occurs for the inland of northern Sweden (looking specifically at two available observation stations). This was attributed to orographic effects. In reality much of the precipitation in the

north falls on the western side of the Scandinavian mountains. Due to orographic smoothing, the RCMs tend to generate more precipitation on the eastern side of the actual mountain divide.

Christensen et al. (1998) examined the dependence of snow cover on RCM resolution and found that a higher resolution significantly increases snow cover and delays spring snow melt due to the more realistic description of mountain topography. Räisänen et al. (2003) showed that differences due to GCM driving models were found to be modest, except for southern Sweden where ECHAM4/OPYC3 driven simulations show more snow and a longer snow season than for HadAM3H driven simulations. Jylhä et al. (2007) evaluated

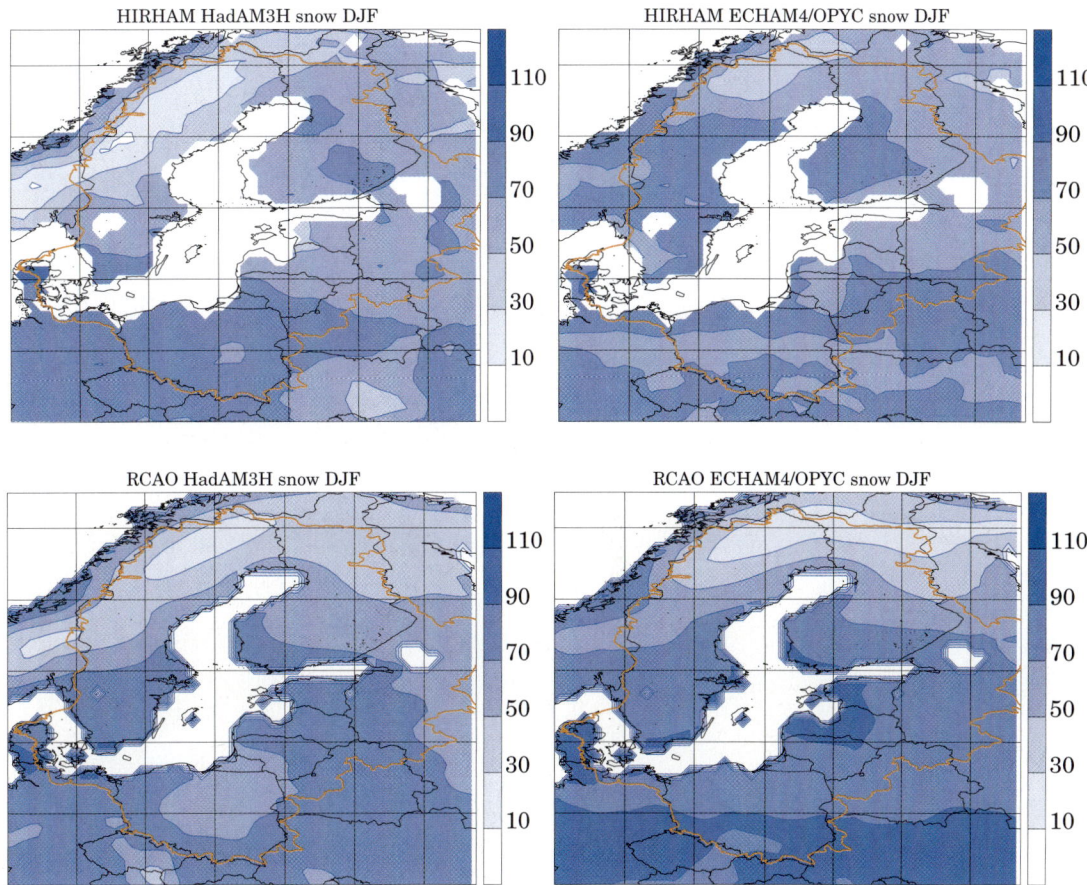

Fig. 3.26. RCM simulated snow depth reduction in percent for winter (DJF) between the periods 1961–1990 and 2071–2100 using the SRES-A2 emissions scenario. The upper plots show results from the HIRHAM Model and the lower plots are from the RCAO Model. Plots on the left used GCM boundary conditions from HadAM3H; plots on the right used ECHAM4/OPYC3. The Baltic Sea Basin is indicated by the thick red line (*note: ECHAM4/OPYC3 scenario simulations used as boundaries are different for the two RCM downscaling experiments, see Sect. 3.5.2.3*)

an ensemble mean of results from seven different PRUDENCE RCM simulations and found good agreement with the observations of Heino and Kitaev (2003) for the simulated number of days with snow cover (see also Annex 1.3.5).

Snow, future climate

Since snow changes follow changes in temperature, there is a general decrease in snow variables due to atmospheric warming in the RCM future scenarios. Results described by Räisänen et al. (2003) show a future decrease in mean annual maximum snow water equivalent everywhere over northern Europe from RCAO simulations. ECHAM4/OPYC3 driven results show a clear north–south gradient in change of snow wa-

ter equivalent for both scenarios with only small deviations over Swedish inland areas. Reduction in snow water equivalent is some 60 to 80% in the southern part of the Baltic Sea Basin up to about latitude 62° N, some 40 to 60% between 62° N and 66° N, and less than 40% north of 66° N. There is also a clear region with relatively small changes down to 20% east of the Scandinavian mountains. For northern areas of the Baltic Sea Basin differences in results due to different driving GCMs are small, as shown in Fig. 3.26. Further south, for example in Denmark, the differences are larger in relative terms (Fig. 3.26) but small in absolute terms (Fig. 3.27).

Giorgi et al. (2004b) also found a general decrease in snow depth in their simulations with another RCM. There is a general decrease over the

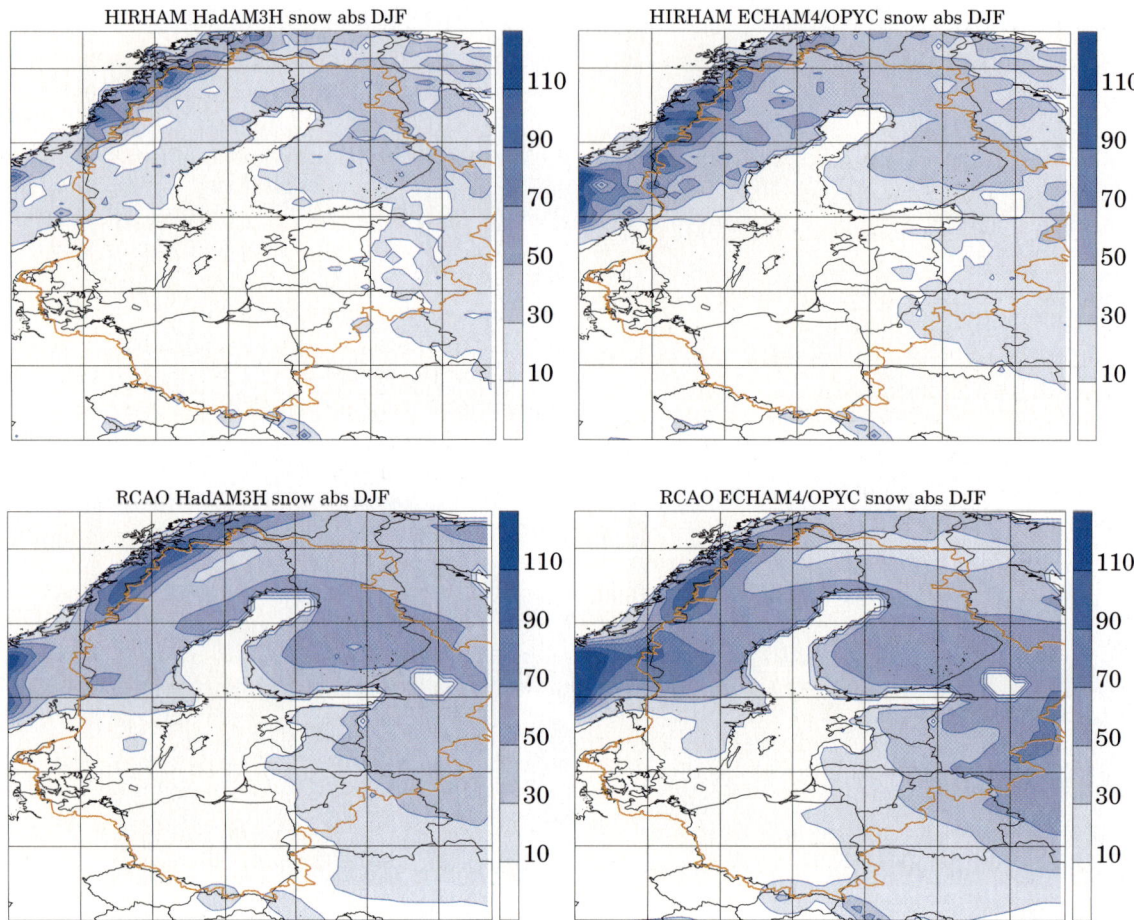

Fig. 3.27. RCM simulated snow depth reduction in mm of equivalent water for winter (DJF) between the periods 1961–1990 and 2071–2100 using the SRES-A2 emissions scenario. The upper plots show results from the HIRHAM Model and the lower plots are from the RCAO Model. Plots on the left used GCM boundary conditions from HadAM3H; plots on the right used ECHAM4/OPYC3. The Baltic Sea Basin is indicated by the thick red line (*note: ECHAM4/OPYC3 scenario simulations used as boundaries are different for the two RCM downscaling experiments, see Sect. 3.5.2.3*)

whole model domain. Over southern Sweden and Norway the decrease is about 50 to 100%. Over Denmark, Germany, and Poland and most parts of the Baltic States, where the present climate snow depth is already small, snow vanishes totally in the scenario simulations. These results were further confirmed by Jylhä et al. (2007). All seven RCMs analysed agreed about substantial decreases in snow depth. The mean annual decrease evaluated from the RCMs was shown to be some 50 to 70% for northern Europe and 75 to 90% for eastern Europe.

Räisänen et al. (2003) also found that the decrease in the duration of the snow season is greater in the SRES-A2 scenario than in SRES-B2. It is also greater in ECHAM4/OPYC3 driven simula-

tions than in HadAM3H simulations. For all cases, the greatest changes (some 45 to 90 days) occur in the same area, a belt extending from central Scandinavia to the Baltic countries. This region showed a reasonable snow season in the control run, but with milder temperatures and is therefore more sensitive to scenario temperature increases than northern Scandinavia. South of this belt, the snow season was generally short even in the control run. Jylhä et al. (2007) found from the multimodel ensemble mean of RCMs driven with the HadAMH3 SRES-A2 scenario that mean changes were largest in areas that also had the largest decline in the number of frost days. This occurred primarily in mountainous areas and around the northern Baltic Sea, with a projected decrease of more than 60 snow cover days. They also found that the portion of days with only a thin snow cover increases.

3.6 Projections of Future Changes in Climate Variability and Extremes for the Baltic Sea Basin

In the previous section, mean changes in atmospheric parameters were described. Such changes do not imply per se that extreme values also change in the same way. The occurrence of extremes is of great interest due to their considerable impact on mankind – e.g. droughts, floods and storms. Reinsurance companies, for instance, undertake their own investigations on extremes, since windstorms and floods are the two natural hazards that have caused the highest economic losses over Europe during the past century (Munich Re Group 1999).

3.6.1 Interpreting Variability and Extremes from Regional Anthropogenic Climate Change Projections

A common way to interpret climate variability and changes in extremes from regional anthropogenic climate change projections is to study the change in the distribution of key variables. Changes in mean and/or variance of the distribution function lead to different future projections. For some quantities, such as temperature, a normal distribution may be applied. More difficult to interpret are those quantities that generally cannot be approximated by a normal distribution, such as precipitation.

Some examples of how to interpret changes to normally distributed temperature are as follows:

- *Increase in mean* shifts the distribution function to a warmer climate. There will be fewer cold and more hot days, and more hot extremes.
- *Decrease in mean* shifts the distribution to a colder climate with fewer hot and more cold days, and more cold extreme events.
- *Increase in variance* broadens the distribution function, resulting in a larger variability of temperatures, but also in both more cold and extremely cold days and more hot and extremely hot days.
- *Decrease in variance* narrows the distribution function, leading to less variability and fewer extreme events, both cold and hot.

For more quantitative assessment of future changes, the percentiles of a distribution function can be calculated. Change in percentile values between 1st to 5th and between 95th to 99th are commonly used to define changes in extremes. The percentiles can be calculated either directly from the empirical distributions or from fitted distribution functions. As another approach for investigating changes in extremes, one can define a threshold value for the chosen variable and then determine the number of events exceeding this threshold for both present and future climates. For example, evaluating the number of times that the wind speed exceeds a certain value can be used to assess changes in storm events.

3.6.2 Projections of Future Climate Variability and Extremes

3.6.2.1 Temperature Variability and Extremes

Temperature variability and extremes, control climate

Overestimation of maximum temperatures in summer dominates in central and southern Europe and is associated with excessive drying of soils in the RCMs, as discussed in Sect. 3.5.1.1 (e.g. Vidale et al. 2007). In northern Europe, including the Baltic Sea Basin, the problem of dry soils and excessively high temperatures was not as large, and most PRUDENCE RCMs instead tended to underestimate the highest daily maximum temperatures (Kjellström et al. 2007). Räisänen et al. (2003) discussed the underestimation of high temperatures during summer in northern Europe, which they related to an over-representation of cloudy and rainy conditions in the RCAO simulations.

In winter, Räisänen et al. (2003) found that 30-year average minimum temperatures were too low for Sweden. Kjellström et al. (2007) showed that RCAO produces a cold bias for the 1st percentile of the daily minimum temperatures over Scandinavia during winter, while the other RCMs exhibit a warm bias for these extreme temperatures. Further, they found that all the RCMs overestimate the 5th percentile of diurnal average temperature for the 1961–1990 period when compared to the long observational records of daily temperatures in Stockholm, Uppsala and St. Petersburg, as shown in Fig. 3.28. Jylhä et al. (2007) found that the PRUDENCE RCMs generally capture the observed spatial patterns in the annual number of frost days (defined as days with a minimum air temperature below 0 °C) and freezing point days (defined as days with a minimum air temperature below 0 °C and a maximum temperature above 0 °C, i.e. days during which the air temperature crosses the 0 °C threshold).

Kjellström et al. (2007) further compared the simulated daily maximum and minimum temperatures from the control climate of ten different RCMs to observations from a large number of European stations. They found considerable biases in some of the models. Taken as regional averages, these biases fall within ±3 °C in most regions for the 95th and 5th percentiles of daily maximum and minimum temperatures in summer and winter, respectively. A general tendency is that the biases are smaller for the more central percentiles than for the more extreme ones, and biases in the extremes are substantially larger than biases in the seasonal averages reported in Sect. 3.5.2.1. This implies that the conclusions regarding extremes are not as robust as those regarding seasonal averages.

Summertime warm temperatures, future climate

Kjellström (2004) investigated how probability distributions of diurnal averaged temperatures change in four different future scenarios. It was found that the asymmetry of these distributions changes differently depending on location and season. For summer, the changes are almost uniform over northern Scandinavia, while there are large differences between different parts of the probability distributions in the southern parts of the Baltic Sea Basin. The differences are manifested as a larger change of, for instance, the 99th percentile than the median, implying larger than average temperature increase on the warmest days. It was noted that the largest differences are found in areas where projected changes to components of the hydrological cycle are large, such as for cloud cover and soil moisture.

For the same set of simulations Räisänen et al. (2004) investigated changes in 30-year averages of yearly maximum and minimum temperatures. They found increases in maximum temperatures that were similar to the increases in summer mean temperatures for the Baltic Sea and Sweden while the maximum temperatures increased more than the average south of the Baltic Sea. The changes related to heat waves investigated by Beniston et al. (2007) showed that while the duration of heat waves (defined as the maximum length of all heat waves) increased only slightly, the number, frequency (defined as the total length of all heat waves) and intensity of heat waves increased substantially (by more than a factor of 5) in the Baltic Sea Basin.

Kjellström et al. (2007) showed that the differences between RCM projections of daily maximum temperatures are larger than differences between mean temperatures. For the Baltic Sea Basin some RCMs project the 95th percentile of daily maximum temperature to increase by 3 to 5 °C, while in others the increase lies from 5 to more than 10 °C. They also found that, although there is a large inter-model variability, the anthropogenic climate change signal is well beyond natural variability, as derived from the long series of daily temperature measurements in Stockholm, Uppsala and St. Petersburg, as seen in Fig. 3.28.

Wintertime cold temperatures, future climate

A snow covered surface is a crucial requirement to attain really low temperatures. The large wintertime temperature increase projected for northeastern Europe in Sect. 3.5.2.1 is to a large extent related to the withdrawal of the snow cover. Temperatures on the coldest days increase dramatically in the future scenarios. For instance, Kjellström (2004) showed individual daily increases of more than 15 °C in parts of eastern Europe and Russia, while the average daily temperatures increase more modestly by 3 to 7 °C. The largest differences between the increase of temperature on the coldest days and the increase in the median were found to be similar to the area where the length of the snow season decreases the most (cf. Sect. 3.5.2.4).

Ferro et al. (2005) showed some of the complex changes in the probability distributions for daily minimum temperatures in the Baltic Sea Basin, including greater changes on the coldest days as compared to the mean. Räisänen et al. (2004) showed a high degree of nonlinearity between the average increase in winter temperature and the increase in the 30-year average of yearly minimum temperatures. They also pointed out the fact that the average minimum temperatures in the control run simulations were too cold over Sweden, a fact that could contribute to the large anthropogenic climate change signal (see above). However, they concluded that this is unlikely to be the only cause of the large increases.

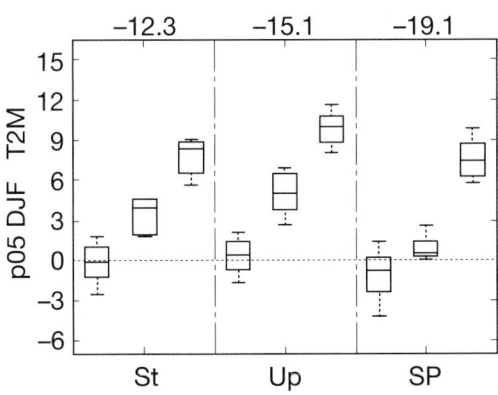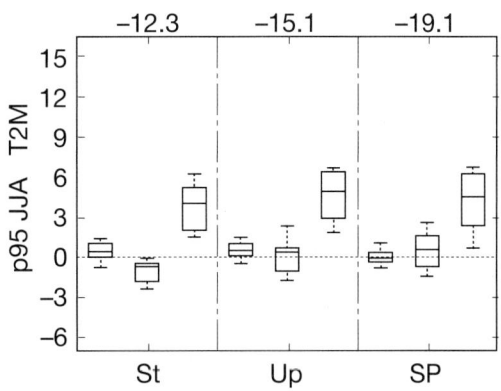

Fig. 3.28. Daily mean temperature deviation from the 1961–1990 observed median in °C, values of which are shown along the top. Shown are the 5th percentile for winter on the left and the 95th percentile for summer on the right. Three stations are given – Stockholm (St), Uppsala (Up) and Saint Petersburg (SP). Three boxplots are shown for each station; the left shows the observed spread between different overlapping 30-year periods from the last 200 years, the middle shows the spread between the different RCMs for the control period, and the right shows RCM spread for the future SRES-A2 simulations. In each boxplot the box extends from the lower to the upper quartile, with the line inside the box denoting the median. The vertical lines extend from the lower (upper) quartile to the minimum (maximum) value. Values of the 1961–1990 observed median are shown along the top (based on data from Kjellström et al. 2007)

Kjellström et al. (2007) found, similarly to the case for high temperatures in summer, a larger inter-model variability for extremely low temperatures than for the mean. Some RCMs project changes in the 5th percentile of daily average temperatures of 4 to 7 °C while in others the changes are 7 to 12 °C in the Baltic Sea Basin. Again, and even more pronounced, it was shown that the projected changes of minimum temperatures are well outside of the observed climate variability during the last 200 years, as seen in Fig. 3.28.

3.6.2.2 Precipitation Extremes

Due to its very nature as infrequent, sporadic events occurring on small spatial scales, simulating extreme precipitation in RCMs is much more difficult than simulating mean precipitation. Hence, care has to be taken to extract meaningful results from RCM studies. To date, only limited evaluation has been performed.

Precipitation extremes, control climate

Christensen et al. (1998) investigated heavy precipitation (defined as exceeding 10 mm day^{-1}) in an RCM at two different resolutions (57 km and 19 km) and its driving GCM. The results were compared to observations over the Nordic countries. The RCM simulations showed a more repre-

sentative number of high-intensity rain days than the GCM, with the finer resolution simulation showing the highest number.

Christensen et al. (2002) compared results from RCM simulations at 22 km resolution to 20 years of gridded observations over Denmark (Scharling 2000). They also showed high-intensity events to be more frequent with higher spatial resolution, but the regional model did not produce a sufficient number of extreme events even though the mean precipitation was realistic (see also Christensen and Christensen 2004). However, in contrast to the driving GCM, the high-resolution RCM simulation showed a realistic annual variation of the decay exponent (i.e. the exponent of an exponential fit to the probability of exceedance as a function of daily precipitation values). Räisänen et al. (2003) concluded that extreme precipitation was underestimated in RCMs. Semmler and Jacob (2004) looked at daily precipitation over the German state of Baden–Württemberg and found realistic magnitudes and regional variation for 10-year return periods compared to gridded observations.

Available validation studies are thus insufficient to provide definite statements about the relative merits of different regional climate models, or at what resolution they could best be applied. However, the studies do show that RCMs provide more realistic descriptions of extreme precipitation than GCMs.

Precipitation extremes, future climate

Christensen et al. (2001) compiled results of several independent regional simulations showing that heavy precipitation (defined as exceeding $10 \, \text{mm day}^{-1}$) increased significantly over Scandinavia following a similar but less significant increase in mean precipitation over the same area. Räisänen and Joelsson (2001) showed that spatial aggregation of a simple statistic, the annual-maximum precipitation event, increased the statistical significance. Their 10-year simulations showed a significant increase of this quantity when averaged over the entire RCM domain, which encompassed central and northern Europe. The method was extended in Räisänen et al. (2004) and applied to 30-year simulations (as presented in Sect. 3.5). Christensen et al. (2002) found that the decay exponent decreased in summer, which indicates more intense extreme rainfall, although the statistical significance of this result was not assessed.

Some results for projected future changes in precipitation extremes are common for several downscaling experiments. The decrease of summer precipitation in southern Europe is seen in several numerical experiments (Cubasch et al. 2001; Giorgi et al. 2001). However, in spite of this reduction the extreme precipitation generally shows an increase (e.g. Christensen and Christensen 2003, 2004; Räisänen and Joelsson 2001; Räisänen et al. 2004; Beniston et al. 2007). Christensen and Christensen (2004) similarly show a larger increase for heavy precipitation than mean precipitation for two river catchments in the Baltic Sea Basin, Oder and Torne, as simulated with two different driving models, HadAM3H and ECHAM4/OPYC3. This result was also found to apply over the entire Baltic Sea Basin and the Baltic Sea itself; however, the anomalous increase in Baltic Sea SSTs from HadAM3H are problematic for this experiment (see Sect. 3.5.2.1).

A further analysis over Europe in Beniston et al. (2007) shows that several models share the tendency to exhibit increasingly positive changes for higher return periods (see also Kjellström 2004). However, some models (e.g. HadRM3H) have such large reductions in precipitation frequency that even the highest extremes have negative changes. Räisänen et al. (2004) found that the projected future reduction in precipitation in Central and Southern Europe is due to a reduction in precipitation frequency and not in intensity. They showed that average intensity changes only slightly, while extreme values tend to increase. Winter precipitation extremes were also analysed by Beniston et al. (2007). For most of Europe, including the Baltic Sea Basin, extreme winter precipitation was shown to increase, roughly proportional to the increase in mean precipitation.

3.6.2.3   Wind Extremes

Characteristics of changes in wind extremes can be expressed by several different parameters, for example as changes in the upper percentiles (e.g. 90th, 95th, 99th) of the daily mean wind or of the daily maximum wind speed. For calculations of higher wind speeds additional techniques are needed, such as gust parameterisation. Using such parameterisations, the number of storm peak events can be determined. Another method for assessing changes in storm events is applying a combination of maximum wind speed and pressure changes (Leckebusch and Ulbrich 2004).

Wind extremes, control climate

Rockel and Woth (2007) studied the 99th percentile of daily mean wind speed for SRES-A2 scenario simulations from eight different RCMs. For north-eastern Europe monthly averages of the 99th percentile from all the control simulations varied between about 13 to $17 \, \text{m s}^{-1}$ for January and 10 to $12 \, \text{m s}^{-1}$ for July (Fig. 3.29). However, over land these quantities can only be used to assess a qualitative change in wind speed, as regional models are hardly producing wind speeds above $17 \, \text{m s}^{-1}$ (cf. Sects. 3.5.1.1 and 3.5.2.3).

Wind extremes, future climate

An increase of strong winds over the whole Baltic Sea Basin was reported by Pryor and Barthelmie (2004). They investigated the change in the 90th percentile and found an increase in the southern part of the Baltic Sea Basin and southern to mid-Sweden of up to $0.7 \, \text{m s}^{-1}$ over land areas. Over the northern part of the basin, the increase was lower, less than $0.4 \, \text{m s}^{-1}$. The largest increase occurs over the Baltic Sea itself, with more than $0.7 \, \text{m s}^{-1}$. These numbers were taken from the RCAO simulation driven by ECHAM4/OPYC3; values from the corresponding simulation driven by HadAM3H are generally lower as discussed below (see also Sect. 3.5.2.3). Following Pryor

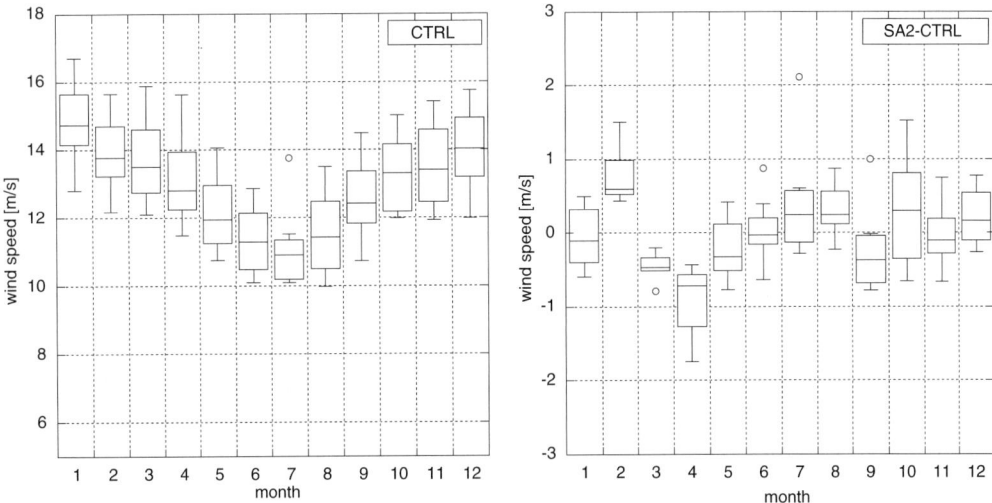

Fig. 3.29. 99th percentile of daily mean wind speed over Scandinavian land area from eight different RCMs driven by HadAM3H boundary conditions. The plot on the left shows results for the present climate (1961–1990). The plot on the right shows change in future climate for the SRES-A2 scenario (scenario 2071–2100 minus present day 1961–1990). Open circles denote outliers (i.e. where the distance from either the lower 25% or the upper 75% quartile is larger than 1.5 times the interquartile distance) (from Rockel and Woth 2007)

and Barthelmie, the differences in future change of extreme wind speed may be indicative of a change in the NAO teleconnection patterns. In agreement with dynamic scales, up to 50% of the interannual variability in the 90th percentile winter wind speeds in the Baltic Sea Basin can be attributable to variations in NAO (Pryor and Barthelmie 2003).

Rockel and Woth (2007) studied daily mean wind speed over land for SRES-A2 scenario simulations from eight different RCMs. Looking at the 90th percentile, half of the models determine a future increase in wind speed of around $1\,m\,s^{-1}$ in February and a decrease of around $1\,m\,s^{-1}$ in April over north-eastern Europe (Fig. 3.29). Individual models show values of up to $2\,m\,s^{-1}$ (July) and nearly $-2\,m\,s^{-1}$ (April). For September the 99th percentile of wind speed decreases between 0 and about $1\,m\,s^{-1}$, with a model mean of $0.5\,m\,s^{-1}$.

Generally, the changes in extreme wind speed follow those in mean wind speed (Räisänen et al. 2003). RCAO driven by ECHAM4/OPYC3 shows an increase of about 8%; in terms of annual maximum wind speed, the projected changes over Sweden are 8% and 6% for the SRES-A2 and SRES-B2 scenarios, respectively. RCAO driven by HadAM3H shows a decrease of about 4%; the corresponding changes in annual maximum wind speed are -3% and -2% over Sweden for the SRES-A2 and SRES-B2 scenarios, respectively.

However, only the results from the RCAO simulations driven with ECHAM4/OPYC3 are statistically significant at the 95% level. The large-scale geographical patterns of change and differences between the different simulations of annual maximum wind speed broadly follow those of the annual mean wind speed (Räisänen et al. 2004). The largest increases occur in northern Europe in the regions of western Norway and Sweden.

The results described above are based directly on model calculated wind. As such, this wind speed does not realistically reflect the occurrence of wind peaks or gusts. Gusts occur on finer temporal and spatial scales than those resolved by RCMs. Thus, a sub-grid parameterisation is necessary to properly account for them. For two of the eight RCMs used in PRUDENCE, maximum wind speed results included gust parameterisation. Rockel and Woth (2007) studied the future change in the number of storm peak events (defined as wind speeds greater than $17.2\,m\,s^{-1}$) for the SRES-A2 scenario with HadAM3H boundary conditions. They found an increase of about 10% over the southern part of the Baltic Sea Basin and a decrease of about 10% in the northern part for both models. In the middle of the Baltic Sea Basin the two models show opposite behaviour. One gives a decrease of about 10%, whereas the other shows storm peak events to increase by about 10%. As the same gust parameterisation is implemented in both models, other differences in the models must be responsible for these discrepancies.

3.7 Projections of Future Changes in Hydrology for the Baltic Sea Basin

Hydrological regimes vary according to how the local and regional climate varies; looking toward future climate, both change and variability in climate will produce changes in hydrological conditions. This section focuses on the hydrological response to projected changes in climate for the Baltic Sea Basin. Hydrological studies focusing on anthropogenic climate change are often associated with analysing impacts to water resources, thus combining the science of anthropogenic climate change with applications for society. The following hydrological assessment strives primarily to summarise responses of the hydrological system and does not attempt to cover the full details of the studies on impacts included in the literature, although some overlap is unavoidable.

Most studies conducted within the Baltic Sea Basin do not cover the entire basin. Many are often of national interest and concentrate only on certain river basins. Therefore to be complete, a short summary of relevant studies is included here, even though they do not address the continental scale of the full Baltic Sea Basin.

3.7.1 Hydrological Models and Anthropogenic Climate Change

Although both global and regional climate models include representation of the hydrological cycle and resolve the overall water balance, they typically do not provide sufficient detail to satisfactorily address how a changing climate can impact on hydrology (Varis et al. 2004). Due to this, hydrological models are used to further investigate hydrological responses to anthropogenic climate change. Many researchers have estimated how hydrological conditions may change with anticipated climate change for a host of different drainage basins around the world (e.g. Arnell 1999; Bergström et al. 2001; Gellens and Roulin 1998; Grabs et al. 1997; Hamlet and Lettenmaier 1999; Kaczmarek et al. 1996; Sælthun et al. 1998; Vehviläinen and Huttunen 1997). The common approach for such studies is to first evaluate representative anthropogenic climate changes from the climate models and then to introduce these changes to a hydrological model for the basin in question.

Many such studies were based on anthropogenic climate change results from global general circulation models (GCMs), some used statistical downscaling methods, and more recent studies included results from regional climate models (RCMs).

3.7.2 Interpreting Anthropogenic Climate Change Projections for Hydrology

Transferring the signal of anthropogenic climate change from climate models to hydrological models is not a straightforward process. In a perfect world one would simply use outputs from climate models as inputs to hydrological models, but meteorological variables from climate models are often subject to systematic biases. For example, in the Alpine region of Europe, many RCMs exhibit a dry summertime precipitation bias on the order of 25% (Frei et al. 2003). For northern Europe, including parts of the Baltic Sea Basin, precipitation biases tend toward overestimation (Hagemann et al. 2004; see also Sect. 3.5.1.1). Hydrological regimes are particularly sensitive to precipitation, and such biases strongly affect the outcome of hydrological model simulations. Uncertainties in observations further complicate the analysis of precipitation biases.

Due to climate model biases, most studies of the hydrological response to anthropogenic climate change to date have resorted to the practice of adding the change in climate to an observational database that is then used as input to hydrological models to represent the future climate (Andréasson et al. 2004; Bergström et al. 2001; Kilsby et al. 1999; Lettenmaier et al. 1999; Middelkoop et al. 2001; Sælthun et al. 1999). This common approach for impacts modelling has been referred to as the *delta change* approach (Hay et al. 2000), and variations of this approach have been the de facto standard in anthropogenic climate change impacts modelling for some time. According to Arnell (1998), this requires two important assumptions. One is that the base condition represents a stable climate both for the present and for a future without anthropogenic climate change. Secondly, the atmospheric model scenarios represent just the signal of anthropogenic climate change, ignoring multi-decadal variability. However, the longer the time period of climate model simulations, the more multi-decadal variability is smoothed out.

A major disadvantage of the delta change approach is that representation of extremes from future climate scenarios effectively gets filtered

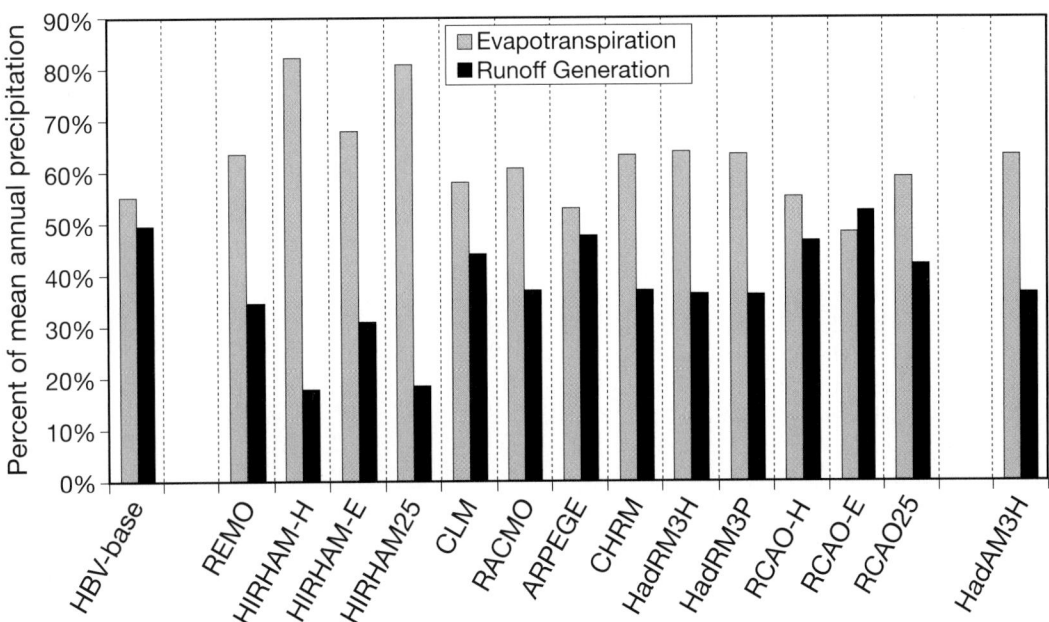

Fig. 3.30. RCM partitioning of precipitation into evapotranspiration and runoff generation over the total Baltic Sea Basin for control simulations representing the period 1961–1990. Also shown are calibrated results from the HBV-Baltic hydrological model (HBV-base; see Sect. 3.7.4), which are thought to give a reasonably accurate representation of the partitioning. All RCM simulations with the exception of 3 were forced by the global HadAM3H (*also shown*); the exceptions are HIRHAM-E and RCAO-E forced by ECHAM4/OPYC3, and HadRM3P forced by HadAM3P (from Graham et al. 2007b)

out in the transfer process. The delta change extremes are simply the extremes from present climate observations that have either been enhanced or dampened according to the delta factors. For this reason, researchers have recently been investigating more direct methods for representing the future climate in assessments of the hydrological response to anthropogenic climate change. This employs applying some form of scaling (modification) to RCM outputs to try to correct for biases before transfer to hydrological models. Such methods also have limitations, which can be severe, but they are more consistent with the RCMs and provide additional answers that are missing in the delta change approach (Arnell et al. 2003; Graham et al. 2007a; Lenderink et al. 2007).

Yet another approach is to use runoff results directly from climate models. This applies primarily to RCMs, where horizontal model scales are becoming finer and are approaching scales more representative of large-scale hydrological processes. RCM model runoff output is in the form of *runoff generation*, which is the instantaneous excess water per model grid square, without any transla-

tion or transformation for groundwater, lake and channel storage, or transport time. As such, this runoff value is difficult to compare to observations and it does not provide flow rates through rivers into the sea. *River routing* schemes can be used to route climate model runoff ($mm\,day^{-1}$) to river discharge ($m^3\,s^{-1}$) (Hagemann and Dümenil 1999; Lohmann et al. 1996); this mainly affects timing and seasonal distribution. However, since this approach makes no corrections to runoff volumes, water balance biases from the climate models greatly influence the results.

The partitioning of precipitation into evapotranspiration and runoff is critical for realistic representation of the hydrological cycle. Graham et al. (2007b) investigated the hydrological performance of 13 RCM control simulations over the Baltic Sea Basin with a simple comparison of the partitioning of annual RCM precipitation into evapotranspiration and total runoff generation, as shown in Fig. 3.30. They found that the majority of RCMs investigated tended to underestimate the partitioning of precipitation into runoff. This is likely due to a general overestimation of evapotranspiration in the basin.

3.7.3 Country Specific Hydrological Assessment Studies

As mentioned above, many hydrological studies do not cover the entire Baltic Sea Basin. This section gives a concise summary of known studies on a country specific basis. Some of these were conducted over the total territory of the country in question, while others concentrated on specific river basins or specific subbasins. Figure 3.31 shows a map giving the approximate locations of documented studies. A short section on each country follows. Note that although the Czech Republic, Slovakia and Ukraine also have areas within the Baltic Sea Basin, they are not included here due to their relatively small contributions of runoff.

Belarus

One study was conducted and reported in the Assessment of Potential Impact of Climatic Changes in the Republic of Belarus (World Bank 2002) and in BALTEX conference proceedings (Kalinin 2004). Three different incremental climate scenarios were used, 1) an increase in temperature by 2 °C, 2) a decrease in precipitation by 10%, and 3) a combination of both changes. A water balance model was used to calculate monthly mean and annual mean river runoff and total evapotraspiration. The entire territory of Belarus was included.

According to the first scenario, river runoff would decrease by 10%, and total evapotranspiration would increase by 4.7% (World Bank 2002). According to the second scenario, river runoff would decrease by 24.5%, and total evaporation would decrease by 5.4%. In this case, the maximum runoff and total evaporation reduction would take place in July with 29.7% and 7%, respectively. In the third scenario runoff would decrease by 29.3% on average, and total evapotranspiration would decrease by 0.7% on average. The maximum runoff and total evapotranspiration reduction would take place in July with 45.2% and 5.1%, respectively. River runoff appeared to be quite sensitive to the simultaneous precipitation reduction and air temperature rise.

A further analysis looked at how a temperature increase of 1.5 °C by the year 2025 would affect groundwater. This showed a groundwater level recession of approximately 0.03–0.04 m relative to the current level (Kalinin 2004).

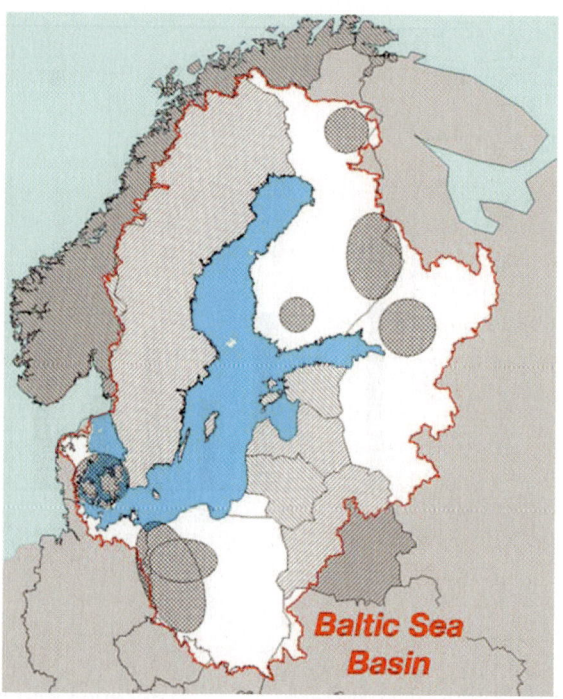

Fig. 3.31. Locations of country specific studies conducted to analyse the hydrological response to projected anthropogenic climate change in the Baltic Sea Basin. Countries that included their total territory in the studies are indicated with diagonal striping. Otherwise, the general location of basins studied is indicated with hatched circles

Denmark

Andersen et al. (2006) used 30-year HIRHAM RCM control and SRES-A2 simulations driven by ECHAM4/OPYC3 in the NAM rainfall-runoff model for the Gjern River basin. They found mean annual runoff to increase by 7.5% in the future climate. Seasonally, their results show considerably higher runoff during winter. They found summer runoff to increase in streams that are predominantly groundwater fed and to decrease in streams with a low base-flow index, typically loamy catchments. Summer runoff reductions of 40–70% were projected for the latter stream type.

Thodsen et al. (2005) and Thodsen (2007) looked at the impact of projected anthropogenic climate change on the Odense River, also using the NAM rainfall-runoff model with a HIRHAM SRES-A2 simulation (GCM unspecified). They found that runoff would increase for the period December to August, with as much as 30% in February. A decrease in runoff was shown for September to November, with as much as 40% in September.

They also found that extremes would be more pronounced, both at high flows and at low flows.

Estonia

A coordinated study using the same set of anthropogenic climate change scenarios generated by two GCMs was reported in the Country Case Study on Anthropogenic Climate Change Impacts and Adaptation Assessments in the Republic of Estonia (Tarand and Kallaste 1998) and other publications (Kallaste and Kuldna 1998; Järvet et al. 2000; Roosaare 1998). Additionally, another study using five incremental scenarios was conducted for the Lake Võrtsjärv (Järvet 1998).

The MAGICC model (Wigley and Raper 1987; 1992a; 1992b) and SCENGEN program were used for anthropogenic climate change scenario generation in the country case study (Keevallik 1998). Three alternative IPCC GHG emissions scenarios (IS92a, IS92c, IS92e) were combined with results of two GCM experiments (HadCM2 and ECHAM3). As a result, six anthropogenic climate change scenarios up to year 2100 were prepared for modelling anthropogenic climate change impact on river runoff. The three IPCC scenarios were qualitatively labeled as MIN (IS92c), MID (IS92a) and MAX (IS92e), as used in the discussion below.

Several different hydrological models and tools were used for the analysis of river runoff, evapotranspiration, groundwater and water supply. The water balance model WATBAL was used for river runoff with a monthly time step. The entire territory of Estonia, subdivided into western Estonia with strong influence from the Baltic Sea, and central and eastern Estonia with a more continental climate. The territory was further subdivided into 36 river basins. In some studies the watershed of the Väike-Emajõgi River was studied in greater detail.

Evapotranspiration was shown to increase under all six scenarios studied. The predicted changes would affect evapotranspiration more in the cold season than in the growing season. However, the change in the magnitude is much smaller on the annual scale, as the cold season evapotranspiration constitutes only 10–13% of the annual evapotranspiration. The most significant increase was simulated using the ECHAM3-MAX scenario. In absolute values, the ECHAM3-MAX scenario predicted evapotranspiration increases of 16 mm in June and about 4 mm in January.

The modelled changes in the mean annual runoff in different basins and scenarios range from −1% to +74%. The largest increases were found for the Emajõgi River, and for a number of small river basins. An increase in total annual runoff by 20–40% (HadCM2-MID) and 30–60% (ECHAM3-MID) was modelled for the year 2100.

Seasonal dynamics of runoff were analysed for several rivers and showed the projected runoff increase in winter to have the largest impact. The maximum increase for the Emajõgi River was projected for April or May (depending on the model and scenario). Runoff maximums in spring were shown to decrease considerably in the central and western parts of Estonia. This is related to the projection that the duration of snow cover would also decrease considerably. Among the single river basins studied, the projection for the small Lõve Rive on the Saaremaa Island stands out; it showed the lowest increase in every scenario.

Groundwater recharge was shown to increase on average by 20–40% according to these simulations with a maximum increase of up to 75% increase. The ratio of the groundwater contribution to river runoff would increase from 30 to 40%. The modelling results also indicated a rise of long-term mean annual groundwater levels by about 0.5–0.8 m in northern Estonia, and 0.2–0.4 m in southern Estonia. Furthermore, considerable changes would occur in the seasonal dynamics of the groundwater regime, with rising water levels in spring and autumn. This would tend toward an earlier onset of flooding.

Finland

As outlined by Bergström et al. (2003), three major studies concerning anthropogenic climate change and hydrology have been conducted in Finland. The Nordic research programme on Climate Change and Energy Production (CCEP; Sælthun et al. 1998) was carried out during 1991–1996. The multidisciplinary Finnish Research Programme on Climate Change (SILMU) was carried out during 1991–1995 (SILMU 1996; Vehviläinen and Huttunen 1997) and more recently, the ILMAVA project (ILMAVA 2002). Additional studies have focused on hydrological impacts and design floods (Tuomenvirta et al. 2000).

The CCEP research programme used an anthropogenic climate change scenario based on statistically downscaled information from four different General Circulation Models. The SILMU sce-

narios were based on an intercomparison study of
GCM simulation results (Räisänen 1994). For the
ILMAVA project (ILMAVA 2002), two SRES scenarios (A2 and B2; Nakićenović et al. 2000) from
the HadCM3 GCM were used.

All of the studies used the operational watershed models of the Finnish Environment Institute (SYKE; Vehviläinen 1994), which are
based on a Finnish version of the HBV-model
(Bergström 1976). All studies thus far have used
the delta change approach for transferring the anthropogenic climate change signal from climate
models to hydrological models and have relied on
GCM models.

The drainage basins used for hydrological studies were selected to represent different regions of
Finland, as shown in Fig. 3.31. Starting with
the CCEP project, subbasins from southern and
northern Finland were used. In the SILMU
project, a similar selection of drainage basins was
chosen, but larger areas in the southeast were included. For the ILMAVA project, basins producing most of Finland's hydropower were chosen.

The trend of results from the more recent
ILMAVA project is similar to results from the earlier CCEP and the SILMU projects. The projected anthropogenic climate change was shown to
strongly affect the seasonal distribution of runoff
and other water balance terms. With increased
temperature, snow cover diminishes or almost vanishes in southern Finland, and its duration becomes shorter. Frequent thawing periods result
in increased occurrence of winter floods and decreased spring floods. Summers become drier due
to the longer summer season, and increases occur
in both evapotranspiration and lake evaporation.

There are differences between the results of
these studies, as seen for changes in runoff. Annual runoff from the Kemihaara subbasin (Kemijoki basin, northern Finland) was found to increase by 2% in the CCEP project, whereas in
the SILMU project essentially no annual change
was found. Results from ILMAVA showed annual runoff in this sub-basin to increase by 5 to
8%. In the Oulujoki drainage basin (mideastern
Finland), the CCEP project reported nearly no
change in annual runoff, but results from the IL-
MAVA project showed an annual increase of 2 to
7%. For the Vuoksi drainage basin (south-eastern
Finland), changes in annual runoff varied between
−1 to +4% (CCEP), −2% (SILMU) and 0 to +8%
(ILMAVA). These differences were due primarily
to differences of climate scenarios, especially re-

garding precipitation. However, it was also found
that results were quite sensitive to how evapotranspiration and lake evaporation are represented.

Results from the SILMU project also showed
that changes in maximum flows for large basins
with a high concentration of lakes depend strongly
on the location of the site within the lake system. For upper subbasins of large basins and
basins without lakes, the maximum discharge decreased by 20 to 60% due to smaller spring floods.
However, maximum inflows to the central lakes
of large basins increased by some 3 to 17%, as
snowmelt and precipitation accumulate into these
lakes during winter, when no evaporation takes
place. Thus, due to increased volume accumulat-
ing in large lakes, the maximum discharge of the
lakes would increase.

Germany

No specific studies were found for German basins
flowing into the Baltic Sea. However, runoff from
German territory is included in projections of the
hydrological response for the Oder River. See related studies under Poland.

Latvia

Related studies were reported in the Third National Communication of the Republic of Latvia
under the United Nations Framework Convention on Climate Change (Ministry of Environmental Protection and Regional Development 2001),
BALTEX conference proceedings (Butina et al.
1998a), and Proceedings of the Second International Conference on Climate and Water (Butina
et al. 1998b; Jansons and Butina 1998).

Anthropogenic climate change scenarios from
the UKMO GCM transient scenario (Murphy and
Mitchell 1994) and the GENESIS GCM (Thompson and Pollard 1995) scenario were used. The
results of the GENESIS GCM were represented in
the form of monthly corrections to meteorological
parameters received from maps of low resolution
(Henderson-Sellers and Hansen 1995). Doubling of
the atmospheric CO_2 concentration was assumed.
The climate scenario predicts a 3–3.5 °C rise in air
temperature and a 20–25% increase in precipitation.

All of Latvia was included in the Third National
Communication. Studies by Butina et al. (1998b)
were done in the Liulupe basin (17,600 km^2),
its subbasin Viesite-Sudrabkalni, and the Berze

basin. Hydrological assessments were made using the HBV hydrological model (Bergström 1995).

According to the assessment in the Third National Communication, groundwater levels in the lowest coastal zone of Riga Bay could rise by 50–70 cm. Risk of floods would increase in the lower reaches of the large rivers Liulupe, Daugava and Gauja. Rising groundwater level could cause serious problems to people living in lowlands in the coastal zone where elevation above sea level is only 0.7 to 2.0 m.

Butina et al. (1998b) reported that, due to higher temperatures, more winter precipitation would fall as rain instead of snow. The spring snowmelt would shift from April to February or even earlier. River flow would be higher during all seasons according to the GENESIS scenario, but not for the UKMO GCM scenario. The UKMO GCM transient scenario showed river flow to increase by 11% on average (ranging from −7 to +36%), while, according to the GENESIS scenario, river flow would increase on average by 83% (ranging from 55 to 120%).

Jansons and Butina (1998) used the same scenarios to investigate changes in runoff and nutrient load for the small agricultural Berze catchment. According to the GENESIS scenario, annual flow would increase by 57%, but the UKMO GCM scenario showed no significant increase in flow. Flood peaks increased by 32% with GCM GENESIS input, while the UKMO GCM scenario predicted a decrease in flood peaks. The GCM GENESIS scenario generated an increase in summer flow, whereas a moderate decrease was projected using the UKMO GCM scenario.

Lithuania

Studies were reported in Lithuania's Second National Communication under the Framework Convention on Climate Change (Ministry of the Environment 2003) and in BALTEX conference proceedings (Rimkus 2001). The National Communication used scenario results from the GFDL GCM. Rimkus used results from five GCMs over Lithuania; these were HadCM2, ECHAM4, CGCM1, GFDL-R15 and CSIRO-Mk2.

According to the GFDL model, the average summer temperature by 2050 would exceed the recent average by 1.7 °C, and the average winter temperature would be 1.2–1.3 °C higher. The summer precipitation would increase slightly until 2020 and then start decreasing in the subse-

quent period. The amount of summer precipitation by 2050 would be 5–6% lower than the present level, and the winter precipitation would be 5–6.5% higher.

No hydrological models were used for the assessment described in the Second National Communication; only analysis of the climate model outputs was carried out. This included assessment of possible effects over all of Lithuania. One outcome from this study was that the extensive wetlands of the country would become dryer with accelerated succession.

Rimkus (2001) used an analysis of regression links between climate variables. Data from the nearest grid points were used to assess changes in snow water equivalent in Lithuania. Two cases were analysed: 1) the mean temperature for the snow accumulation period would rise by 1.5 °C and precipitation would rise by 8 mm (as expected around the year 2040); 2) the mean temperature of the snow accumulation period would rise by 3.0 °C and precipitation would increase by 14 mm (as expected around the year 2065).

Results from Rimkus (2001) showed that maximum snow depth in Lithuania would decrease significantly under anthropogenic climate change scenarios. For the reference period 1961–1990, maximum snow water equivalent was 40 mm on average, ranging from 21 to 60 mm. Average maximum snow water equivalent would decrease to 34 mm, with temperature and precipitation increasing by 1.5 °C and 8 mm, respectively. Average maximum snow water equivalent would decrease to 28 mm, with temperature and precipitation increasing by 3.0 °C and 14 mm, respectively.

Norway

In addition to the Nordic research programme on Climate Change and Energy Production (CCEP; Sælthun et al. 1998) during 1991-1996 (see also Finland and Sweden), two national projects on anthropogenic climate change and hydrological impacts have been performed in Norway (Bergström et al. 2003). These were "Climate change and water resources" (Sælthun et al. 1990) conducted prior to CCEP, and "Climate change and energy production potential" (Roald et al. 2002; Skaugen et al. 2002; Skaugen and Tveito 2002), which was carried out during 2000–2002.

Sælthun et al. (1990) used anthropogenic climate change scenarios based on the NCAR model (Washington and Meehl 1989). No downscaling procedure was applied. Two climate scenarios were used: one based on what was considered the most probable changes of precipitation and temperature, and one based on greater changes. As described under Finland, CCEP used an anthropogenic climate change scenario based on statistically downscaled information from four different General Circulation Models (Sælthun et al. 1998). Local and regional climate scenarios have also been studied in the RegClim project since 1997. This has focused mainly on results originating from ECHAM4/OPYC3 simulations using the IPCC IS92a scenario. Regional downscaling from both statistical techniques (Hanssen-Bauer et al. 2000, 2001; Benestad 2002a) and RCM modelling (Bjørge et al. 2000) was used.

Runoff simulations were performed with the HBV Model (Bergström 1995). Evapotranspiration was estimated according to temperature based methods developed by Sælthun et al. (1990), which were further improved in later studies (Sælthun 1996). In Sælthun et al. (1990) 7 basins were used, in Sælthun et al. (1998) 10 basins, and in Roald et al. (2002) 42 basins. These basins are distributed over all of Norway and represent different hydrological regimes. Future water balance changes for the whole of Norway were also included in Roald et al. (2002), Engen-Skaugen et al. (2005) and Roald et al. (2006).

Sælthun et al. (1990) concluded that correct modelling of evapotranspiration is important when it comes to estimating the future water balance (Fossdal and Sælthun 1993). The annual evapotranspiration increase was 40–55 mm in mountainous areas, 45–100 mm in transitional areas and 50 to 110 mm in lowland areas. The corresponding increase in annual runoff for mountainous areas was more than 750 mm. In lowland and forested inland basins, annual runoff was shown to decrease in response to increased evapotranspiration. The wettest scenario resulted in increased runoff over all of Norway, up to 15% on the west coast.

Sælthun et al. (1998) drew similar conclusions. Evapotranspiration was shown to increase due to increased summer temperature and longer snow free periods. Annual evapotranspiration would increase between 100 and 200 mm over 100 years, and precipitation would increase between 15 and 20% for the same period. An evapotranspiration increase of 100 mm would counterbalance the precipitation increase in areas where present annual precipitation is less than 700 mm. The annual runoff would therefore increase in western areas

and decrease in inland areas. Results from Roald et al. (2002) showed that runoff would increase over almost all of Norway, following the same pattern as the increase in precipitation.

The scenarios in Sælthun et al. (1990) and Sælthun et al. (1998) showed a drastic change in the seasonal distribution of runoff, with increases in winter, reduced spring flood peaks with earlier occurrence and decreases in summer. The changes were mostly controlled by the effect of temperature on snow processes. The seasonal distribution would not change much in coastal regions that do not have stable snow cover under the present climate. The largest changes were shown for the lower elevations of regions that now have a stable snow cover during winter. Roald et al. (2002) showed that summer runoff would decrease and autumn runoff would increase, especially on the west coast.

Sælthun et al. (1990) concluded that melting of Norwegian glaciers would increase and the summer discharge in glacier basins would therefore increase. Most of the glaciers would also experience a negative mass balance, resulting in reduced volume. The long-term effect would be reduced summer runoff in basins that have glaciers today. However, high altitude glaciers in maritime climates with high precipitation might maintain their volume and even grow. In Roald et al. (2002), simulations suggested that glaciers would accumulate and the effect on runoff would be negative, as opposed to a state of equilibrium.

Poland

Kaczmarek et al. (1997) used data from several GCMs with their hydrological model CLIRUN (Kaczmarek 1996; Kaczmarek 1993) for studying three middle-size catchments in Poland. These were followed up by additional studies with updated models (Kaczmarek 2003; Kaczmarek 2004). De Roo and Schmuck (2002) report on studies using different incremental scenarios. The early studies by Kaczmarek focused on the Warta basin, which is the largest tributary to the Oder River. De Roo and Schmuck analysed the entire Oder basin. Later studies by Kaczmarek covered other regions of Poland.

Kaczmarek et al. (1997) noted that current (i.e. 1997) climate models did not offer the degree of watershed specific information required for hydrological modelling. Moreover, the results for river flow showed great differences depending on the different models used (Kaczmarek 1996). Gutry-Korycka (1999) also noted that the information from GCMs was not sufficient for definite projections of the influence of warming on river flow in the second half of the 21st century. Similarly, results on hydrological drought frequency differed considerably between the models, both in magnitude and direction (Kaczmarek and Jurak 2003).

De Roo and Schmuck (2002) used the LIS-FLOOD Model in their studies together with seven different incremental scenarios. They looked at, 1) annual precipitation increases of 15% and 22%, 2) annual precipitation decreases of 10% and 15%, 3) an average annual temperature increase and decrease of 1 °C and 4) a combined 15% increase in precipitation and 1 °C increase in temperature. They reported that a 15% increase in precipitation showed a sharp increase in maximum discharge of 600–$900 \, \mathrm{m^3 \, s^{-1}}$ at all major gauge locations. A decrease in precipitation by 10% resulted in a decrease of peak discharge of 430–$560 \, \mathrm{m^3 \, s^{-1}}$. A precipitation decrease by 15% led to a decrease of peak discharge of 590–$810 \, \mathrm{m^3 \, s^{-1}}$.

Kaczmarek (2003) studied the influence of climate change on the water balance in Poland for the period 2030-2050 using the CLIRUN3 hydrological model and two anthropogenic climate change scenarios: GFDL (warm and dry) and GISS (warm and wet). They projected a decrease in river flow and soil moisture in summer and autumn. The flood season would also shift from March–April to January–February, in accordance with results from the EU project "Impact of Climate Change on Water Resources in Europe" (from the 4th Framework Programme).

Kaczmarek (2004) stated that although it is certain that a warmer climate will accelerate the hydrological cycle, less is known about impacts at river basin levels. In the maritime parts of Europe, he reports a tendency toward increasing streamflow during winter. Furthermore, a reduction during low flow periods is expected, which could lead to increased drought frequency and, in most catchments, increased flood frequency. For Poland itself, projected flow characteristics vary between models and scenarios.

Russia

Two studies were reported in Meteorology and Hydrology (Kondratyev and Bovykin 2003; Meleshko et al. 2004), one study in Water Resources (Grigoryev and Trapeznikov 2002), and one study in an INTAS Report (Kondratyev 2001). The study performed by Meleshko et al. (2004) covers the entire Baltic Sea Basin and is summarised in Sect. 3.7.4. Kondratyev and Bovykin (2003) used climate scenarios from ECHAM4/OPYC3 calculated for the Lake Ladoga drainage basin ($258,000 \, \mathrm{km^2}$) for the period 2001–2100 (Arpe et al. 2000; Golitsyn et al. 2002). They coupled this to a model of hydrological regimes and nutrient fluxes for the system catchment and lake. This was applied for the much smaller Lake Krasnoye and its drainage basin ($168 \, \mathrm{km^2}$) with the help of regression equations (Kondratyev et al. 1998; Kondratyev and Bovykin 2000). For Kondratyev (2001), eight different incremental scenarios were used instead.

According to Kondratyev and Bovykin (2003), a moderate increase in river discharge was projected for Lake Krasnoye. Snow water equivalent was shown to decrease by 25–28%. Soil moisture in the watershed would increase by 7–8% in autumn and winter, and decrease by 10–18% in summer. Spring floods would start earlier. The lake level would be characterised by earlier spring maximums and a 10 cm lower water level in summer, as compared to the 1964–1984 reference period.

Results from Kondratyev (2001) summarised characteristics of annual runoff, maximum water discharge in the tributaries, soil moisture in autumn, and numerous nutrient transport variables for the Lake Krasnoye catchment. Their findings showed mean annual runoff changes from −26% to +35%, maximum runoff changes from −59% to +66% and autumn soil moisture changes from −26% to +14%.

Grigoryev and Trapeznikov (2002) used probabilistic incremental scenarios for Lake Ladoga. They applied a transfer function model with climate characteristics as input and water level in the lake as output. According to their probabilistic climate scenario, the water level in Lake Ladoga would decrease by 50 cm.

Sweden

Sweden also participated in the Nordic research programme on Climate Change and Energy Production (CCEP; Sælthun et al. 1998) during 1991–1996 (see also Finland and Norway). This resulted, among others, in a comprehensive study of evapotranspiration effects (Lindström et al. 1994). Sensitivity studies of climate change effects on hydrology were conducted in an early study of effects on river regulation (Carlsson and

Sanner 1996). From 1997 to 2003, hydrological impacts studies concerning anthropogenic climate change were mainly produced within the Swedish Regional Climate Modelling Programme (SWECLIM; Bergström et al. 2001; Gardelin et al. 2002a; Gardelin et al. 2002b; Graham et al. 2001; Andréasson et al. 2002; Andréasson et al. 2004). Most recently, hydrological impact studies were conducted within the EU PRUDENCE project (Graham et al. 2007a).

As described under Finland, the CCEP research programme used an anthropogenic climate change scenario based on statistically downscaled information. The anthropogenic climate change scenarios from SWECLIM came primarily from RCM modelling. Most simulations were based on three different GCMs, HadCM2, HadAM3H and ECHAM4/OPYC3, downscaled with the Rossby Centre regional climate models RCA (Rummukainen et al. 2001; Jones et al. 2004) and RCAO (Döscher et al. 2002). Some use of statistically downscaled scenarios was also made (Bergström et al. 2003). Earlier SWECLIM simulations were based on business as usual (BaU) emissions scenarios (Houghton et al. 1992), while more recent simulations used the SRES (A2 and B2) scenarios (Nakićenović et al. 2000).

The HBV hydrological model was used in the Swedish studies (Bergström et al. 2001; Graham et al. 2001). Particular effort was placed on developing appropriate representation of evapotranspiration processes and on the interface used for transferring anthropogenic climate change information from climate models to hydrological models. Most studies to date used the delta change approach for transferring the signal of anthropogenic climate change to hydrological models. The PRUDENCE studies also investigated the use of precipitation scaling.

Six test basins representing different climate and hydrological regimes in Sweden were initially chosen for impacts analysis within both CCEP and SWECLIM. They were later supplemented by two additional basins, the Lule River Basin, representing a high degree of river regulation, and Lake Vänern, the basin containing Sweden's largest lake. Limited analyses to address soil frost and groundwater dynamics were also made in the Svartberget experimental forest site in northern Sweden. In the last year of SWECLIM, a hydrological model was developed to conduct hydrological change studies over all of Sweden using some 1000 subbasins.

Results from the HBV model using eight different regional climate simulations from RCA1 and four simulations from RCAO were reported by Andréasson et al. (2004). A general tendency is the shift in the runoff regime towards decreasing spring flood peaks and increasing autumn and winter flows. Mean annual change in runoff from the simulations shows increasing runoff in northern basins and decreasing runoff in southern basins. In the northernmost basin (Suorva) there is little impact on the magnitude of spring runoff as snow accumulation is less affected in this region. However, the timing of snowmelt is affected in all basins. Summer flows are severely reduced in the two southernmost basins, Blankaström and Torsebro.

The more recent RCAO scenario results generally show higher runoff during winter and lower spring runoff than results from the earlier RCA1 scenarios. The relative range of changed runoff between different simulations is larger towards the south of the country. This may be explained partly by increasing rates of projected evapotranspiration further south and associated uncertainty in modelling changes to future evapotranspiration. For a majority of the basins, the RCA1-HadCM2 simulations show greater impact on runoff than for RCA1-ECHAM4/OPYC3 simulations. Regarding RCAO simulations, impacts driven by ECHAM/OPYC3 are generally larger than those driven by HadAM3H.

Attempts to make some assessment of extreme flows were included in Bergström et al. (2001), Gardelin et al. (2002a) and Andréasson et al. (2004). These papers present changes in 100-year flood events obtained from hydrological modelling and frequency analysis. They report a decrease in the frequency of high spring floods and an increase in the frequency of flooding events for autumn and winter in many basins. However, such conclusions are subject to great uncertainty as the delta change approach used for these studies does not provide good representation of changes in variability coming from the climate models.

3.7.4 Baltic Sea Basinwide Hydrological Assessment

Studies addressing the hydrological response to anthropogenic climate change specifically for the Baltic Sea Basin are not numerous. On a global scale, IPCC (2001b) presented hydrological modelling results for all continents. On a European scale, Arnell (1999), Strzepek and Yates (1997), Lehner et al. (2001) and Meleshko et al. (2004) presented hydrological modelling results that include the Baltic Sea Basin. On regional scales, Graham (1999b) presented early results that specifically address the Baltic Sea Basin. These were updated to include improved methods and newer RCM scenarios in Graham (2004) and to include an ensemble of RCM models in Graham et al. (2007b). Finally, both Hagemann and Jacob (2007) and Graham et al. (2007b) present results for the Baltic Sea Basin using runoff routing methods. A summary of results from these studies follows below. All used the delta change approach, unless otherwise noted.

3.7.4.1 Projected Changes in Runoff

Included in the global scale results presented by the IPCC are maps showing change in annual runoff at the end of the 21st century from macroscale hydrological modelling. Results using two different GCM simulations (HadCM2, HadCM3) are shown there in mm yr^{-1} (IPCC 2001b, Fig. TS-3). Although the two simulations show contradictory trends for other regions around the globe, results for the Baltic Sea Basin are qualitatively similar. The trend shown is for increased annual runoff in the north of the basin and decreased runoff in the south. Magnitudes range from up to $+150$ mm yr^{-1} in the north to -150 mm yr^{-1} in the south. For comparison, the observed total mean annual runoff to the Baltic Sea is some 280 mm yr^{-1} (Bergström and Carlsson 1994). Close examination of the IPCC maps, however, reveals that the location where projected runoff change goes from positive to negative varies considerably between the two simulations. For instance, in Finland for simulations using HadCM2 a positive change in runoff is shown, while for simulations using HadCM3 much of the country is shown with a negative change.

More detail is found in Arnell (1999), where the results from macroscale hydrological modelling using four different GCM simulations (based on HadCM2 and CCC) are shown. In this case, the projected future climate is the middle of the 21st century, the 2050s. Here again, an increase in runoff is simulated for the northern areas of the Baltic Sea Basin and a decrease is simulated for the southern areas. Results are expressed in terms of percent change in annual runoff and range from some $+50\%$ to -25% for the Baltic Sea Basin.

Although Arnell (1999) comments that there are large areas of agreement over Europe between the simulations, he specifically notes that discrepancies are high in the eastern Baltic region. Results for this area show a reduction of runoff by up to 20% in one simulation and an increase of over 25% in another simulation. Arnell (1999) also notes that by the 2050s snow cover at the end of March will have disappeared across eastern Poland, Belarus, Ukraine and the Baltic sea coast. This is broadly comparable to more recent findings from RCMs (see Sect. 3.5.2.4).

Meleshko et al. (2004) used results from seven GCMs (CGCM2, CSIRO-Mk2, CSM1.4, ECHAM4/OPYC3, GFDL-R30-c, HadCM3 and PCM). They applied a simple balance equation to define annual discharge for large river basins from GCM inputs of precipitation and evapotranspiration. This was done for the entire Baltic Sea Basin. Projections using the ensemble of seven GCMs showed an overall increase in total river runoff to the Baltic Sea of 1.9% for the period 2041–2060 and 5.7% for the period 2080–2099.

Graham (1999b) presented results using the HBV-Baltic hydrological model (Graham 1999a) together with climate model simulations from the RCA Model (Rummukainen et al. 1998). HBV-Baltic is a large-scale hydrological modelling application that covers the total Baltic Sea Basin up to its outflow into Öresund and the Danish Belts, as shown in Fig. 3.32. Three RCM simulations representing "business as usual" GHG emissions scenarios with forcing from two different GCM simulations (HadCM2, ECHAM4/OPYC3) were presented. Two of these simulations used a horizontal resolution of 88 km and were referred to as RCA88-H and RCA88-E for HadCM2 and ECHAM4/OPYC3 forcing, respectively. The effects from the RCA88-E scenario simulations were quite different to those from RCA88-H, particularly in the southern Baltic river basins. Total river discharge to the Baltic Sea decreased considerably for the RCA88-E simulation but increased somewhat for the RCA88-H simulation. These early simulations were based on 10-year time periods for the present compared to 10-year time periods for the future, which are relatively short in terms of representing interannual variability.

In Graham (2004) and Graham et al. (2007b), a number of additional hydrological response simulations were carried out using HBV-Baltic. These used 30-year time periods to represent both the future climate and the control climate (present

Fig. 3.32. Basin boundaries for HBV-Baltic. The five main Baltic Sea drainage basins are shown with thick lines (adapted from Graham 1999a)

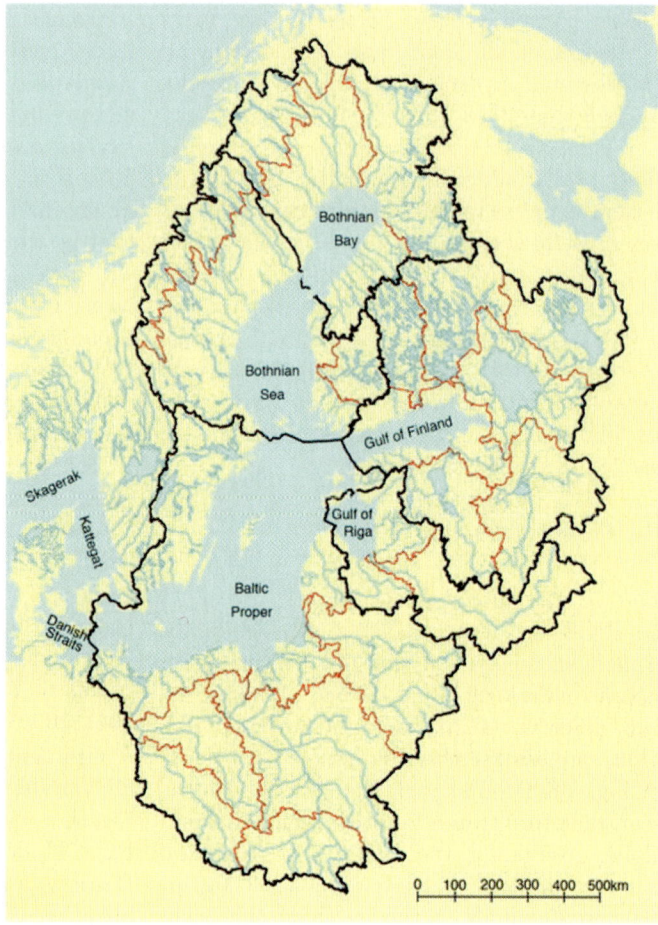

climate). Graham et al. (2007b) used many of the RCM simulations from PRUDENCE (Christensen et al. 2007; see Sect. 3.5.1.2). This showed the ensemble of RCMs driven by the same GCM with the same GHG emissions scenario to result in a range of potential outcomes that follow similar mean trends. Based on the SRES-A2 emissions scenario (see Annex 6), Fig. 3.33 shows results of river discharge summarised as total inflows to the five main subregional Baltic Sea drainage basins and for the total Baltic Sea Basin. Results using nine RCM simulations with global forcing from HadAM3H (referred to as the common PRUDENCE experiment in Sect. 3.5.2) are shown in the left plot of the figure. Results using 2 RCM simulations with global forcing from ECHAM4/OPYC3 are shown in the right plot of the figure.

General trends in the north show increases in wintertime river flow coupled with somewhat lower and earlier springtime peak flows. This reflects the substantial changes that warmer temperatures will inflict on the snow regime in the north. Trends in

the south show more pronounced effects on summertime river flow. River flow to the Gulf of Finland exhibits a combination of these effects, depending on which simulations one examines, even though these flows are highly dictated by the outflow from Lake Ladoga. River flow to the total Baltic Sea Basin is an integration of the combined effects to the five main sub-drainage basins.

The range of outcomes helps to characterise the uncertainty contributed from using different RCMs. As shown in Fig. 3.33 for RCMs driven by HadAM3H, this range is fairly narrow for much of the year. However, during late summer and autumn months larger deviations occur, which is most obvious in the plot for the total Baltic Sea Basin. Although it is not easily seen from the plots, much of this deviation originates from the Gulf of Finland and other eastern drainage basins.

According to Kjellström and Ruosteenoja (2007), the climate change signal for precipitation in this area is affected by different approaches in the RCMs for representing feedback from the

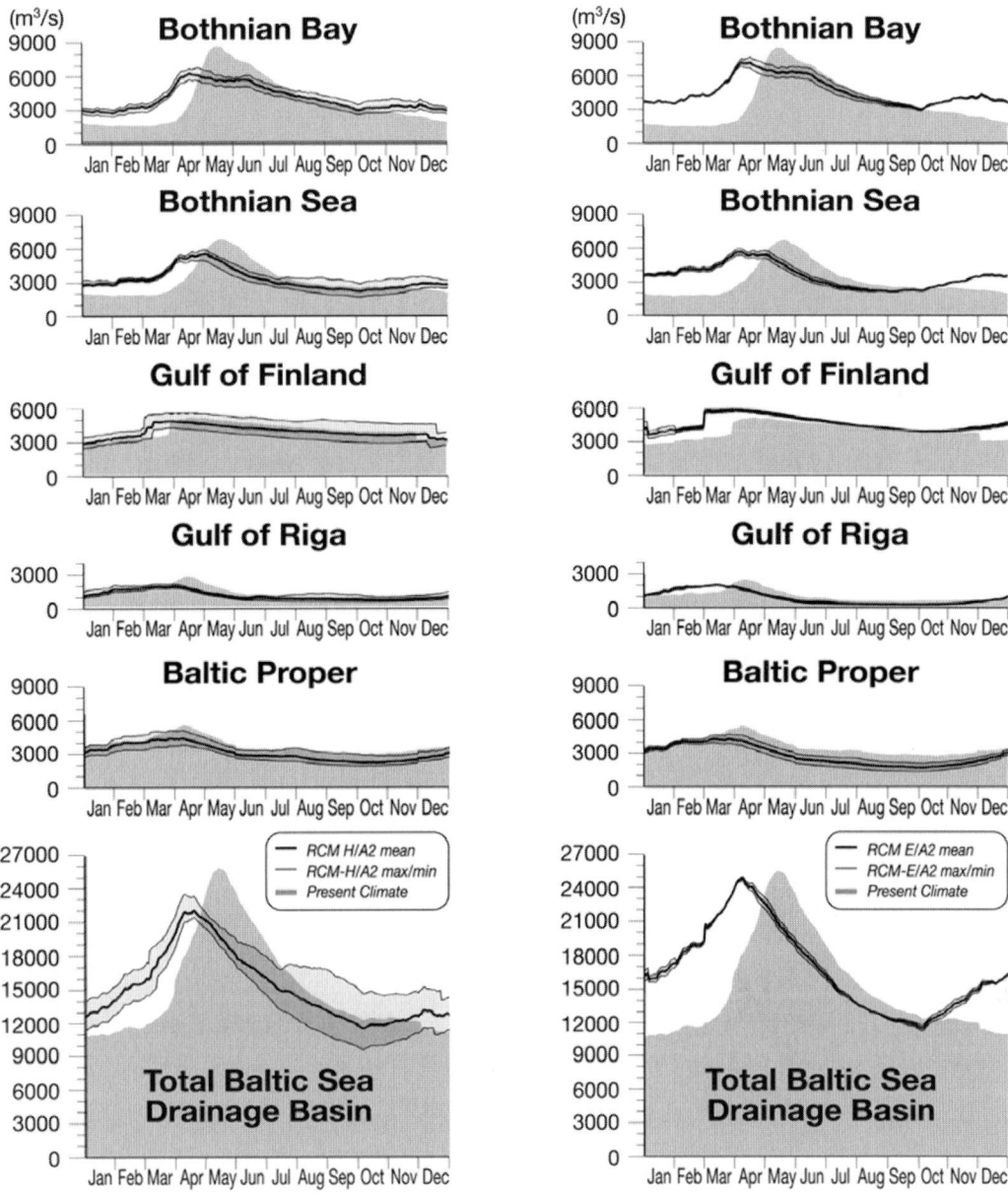

Fig. 3.33. Mean river discharge from HBV-Baltic using the delta change approach for RCM-A2 scenarios at ∼ 50 km resolution, driven by global forcing from HadAM3H (*left*) and ECHAM4/OPYC3 (*right*). The plots summarise results using nine different RCMs with HadAM3H forcing and two RCMs with ECHAM4/OPYC3. The scenarios represent future climate for the period 2071–2100 compared to the control period 1961–1990 (adapted from Graham et al. 2007b)

Baltic Sea itself. In particular, anomalously high sea surface temperatures can have an effect (SSTs; see also Sect. 3.5.2.2). One of the two models that shows the greatest increase in river flow from the eastern side of the Baltic sea Basin is also the model that Kjellström and Ruosteenoja (2007) show to produce the greatest precipitation change in that region due to anomalous SSTs.

Effects on modelled river discharge from using different GCMs to drive the RCMs are seen by comparing the left and right plots in Fig. 3.33, although not as many simulations were available using ECHAM4/OPYC3. According to these results, forcing with ECHAM4/OPYC3 produced a considerably different river discharge response than simulations with forcing from HadAM3H.

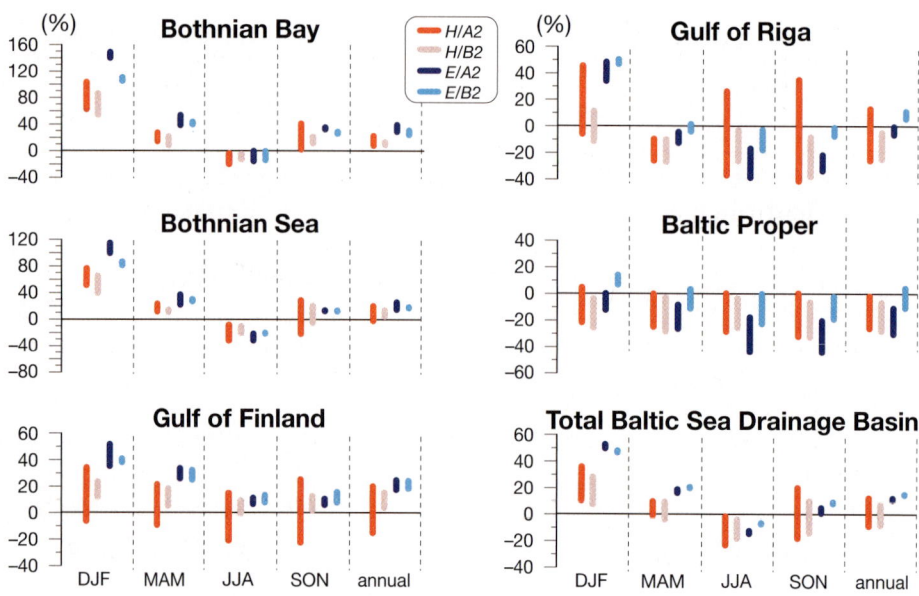

Fig. 3.34. Modelled percent volume change in river discharge from HBV-Baltic simulations using RCM scenarios for the period 2071–2100 compared to the control period 1961–1990. This is summarised by season for the five main Baltic Sea drainage basins (Fig. 3.32); December, January, February (DJF); March, April, May (MAM); June, July, August (JJA), and September, October, November (SON). Each bar represents the range of results between the simulations performed for SRES-A2 forced by HadAM3H (H/A2, 9 simulations), SRES-B2 forced by HadAM3H (H/B2, 3 simulations), SRES-A2 forced by ECHAM4/OPYC3 (E/A2, 2 simulations), and SRES-B2 forced by ECHAM4/OPYC3 (E/B2, 2 simulations) (created with results from Graham 2004 and Graham et al. 2007b)

River discharge in general tends to be higher for the ECHAM4/OPYC3 driven simulations. Such differences were also reported by Graham (2004), where simulations from the SRES-B2 scenarios were presented as well. There it was observed that the GCM model used for boundary conditions has as much impact on total river flow as the emissions scenarios used.

Changes in river flow to the Baltic Sea are further summarised in Fig. 3.34 as percent volume change per season. This figure includes results based on RCMs using two different GHG emissions scenarios (SRES-A2, SRES-B2) and two different GCMs (HadAM3H, ECHAM4/OPYC3). The length of each bar in the figure shows the range of results for each case, although the number of simulations varies between the cases (see figure caption for detail). Qualitatively, there are overall similarities between the different simulations. In many instances, the results tend to fall within the same sign (positive or negative) for the given seasons. However, the degree of similarity varies among the different subregions. The considerable differences obtained by using different GCMs are

also apparent in these plots (e.g. compare H/A2 results to E/A2 results). This figure also shows that the largest range of uncertainty with respect to the relative change in volume occurs in the Gulf of Riga drainage basin, as evidenced by numerous long bars.

Two runoff routing schemes have been applied to the Baltic Sea Basin to date. These are the RCroute scheme (Graham 2002; Graham et al. 2007b) and the HD Model (Hagemann and Dümenil 1999; Graham et al. 2007b; Hagemann and Jacob 2007). Both of these were used to produce routed river flow directly from RCM results. As stated above, hydrological response studies from river routing techniques are highly influenced by RCM biases. Therefore they are best used when converted to percent change in river discharge, as done in the reported literature. It was reported that despite large differences in individual RCM simulations, the overall signal of the ensemble mean response was in agreement between RCroute and the HD Model (Graham et al. 2007b). These were also qualitatively in agreement with the various results using HBV-Baltic as presented above.

However, choosing a single RCM from this group of results as a basis for further impact studies would result in quite different answers depending on the RCM used.

There is a notable difference between the two river routing approaches in that RCroute uses runoff generation directly from the RCMs and the HD Model performs its own re-partitioning of RCM precipitation into runoff and evapotranspiration (Hagemann and Jacob 2007). This is a likely explanation to why results from the HD Model show a narrower range of uncertainty around the mean than those from RCroute, as it effectively filters out some of the biases in precipitation partitioning in the RCMs.

3.7.4.2 *Projected Changes in Evapotranspiration*

Although evapotranspiration is a critical component of the water balance, comprehensive anthropogenic climate change effects on this variable are little reported for the Baltic Sea Basin. One reason for this is that there is a large amount of uncertainty associated with evapotranspiration and projected future climates (Bergström et al. 2001).

Hydrological studies typically include calculation of their own estimates of evapotranspiration, both for the present climate and for future climates. This has shown to produce reasonable estimates for present climates as calibration can be performed against observations of river flow. However, using the same calibrated evapotranspiration parameterisations for the future can be suspect, particularly for temperature based methods.

For this reason, delta change techniques have come into use for estimating evapotranspiration as well (Andréasson et al. 2004; Lenderink et al. 2007). There are various ways to perform such estimates, but a main objective is that the annual percent change in evapotranspiration matches the annual percent change as simulated by climate models, while preserving the water balance in the hydrological simulations. This approach was applied by Graham (2004); estimates of future change in evapotranspiration from four simulations for the Baltic Sea Basin are presented from that work in Table 3.3.

3.7.5 Synthesis of Projected Future Hydrological Changes

Many different studies using numerous models and approaches to evaluating projections of hydrological change within the Baltic Sea Basin are sum-marised above. The studies were conducted on a broad range of scales, using different levels of detail and different future scenario simulations. Although it is difficult to assemble such an array of results into definite conclusions, there are common signals and similarities shown. A fundamental conclusion is that the assumed projected future anthropogenic climate changes provide the greatest source of uncertainty for projected future hydrological changes.

Some robust findings are that snow and cold weather processes were shown to be sensitive to anthropogenic climate change throughout the Baltic Sea Basin. Warmer temperatures will impact greatly on snowpack volumes and duration, resulting in considerable impact to timing of runoff. Simultaneous increases and/or decreases in precipitation will strongly affect corresponding runoff volumes. However, the response of evapotranspiration is a key process in determining how runoff volumes will change and how groundwater levels will in turn be affected.

Further conclusions are that there will be a north–south gradient in how projected future hydrological changes occur over the Baltic Sea Basin, and effects during cold months show larger relative change than for warm months. According to analyses using an ensemble of RCM anthropogenic climate change scenarios, the following concluding remarks were made by Graham et al. (2007b) for scenario simulations for the period 2071–2100 compared to control simulations for the period 1961–1990.

- On average for the total basin, summer river flows show a decrease of as much as 22%, while winter flows show an increase of up to 54%.
- On the large scale, annual river flows show an increase in the northernmost catchments of the Baltic Sea Basin, while the southernmost catchments show a decrease.
- The occurrence of medium to high river flow events shows a higher frequency.
- High flow events show no pronounced increase in magnitude on the large scale.
- The greatest range of variation in flow due to different RCMs occurs during summer to autumn.

The authors point out, however, that there are deficiencies in the methods used for performing hydrological response studies. Although the delta change approach may provide usable estimates of mean changes, representation of changes to ex-

Table 3.3. Mean annual evapotranspiration change in percent estimated from four anthropogenic climate change scenarios simulated with the RCAO Model. This is the difference between the scenario period 2071–2100 and the control period 1961–1990. These are summarised for the five main Baltic Sea drainage basins (Fig. 3.32) and the total Baltic Sea Basin. H/A2, H/B2, E/A2 and E/B2 are simulation descriptors for the HadAM3H and ECHAM4/OPYC3 GCMs with SRES-A2 and SRES-B2 scenarios, respectively (adapted from Graham 2004)

	Bothnian Bay Basin	Bothnian Sea Basin	Gulf of Finland Basin	Gulf of Riga Basin	Baltic Proper Basin	Total Baltic Sea Basin
RCAO-H/A2	23%	19%	19%	15%	11%	16%
RCAO-H/B2	13%	11%	12%	10%	6%	10%
RCAO-E/A2	20%	24%	22%	19%	15%	19%
RCAO-E/B2	15%	17%	17%	16%	13%	15%

treme events is inadequate. Using methods incorporating a precipitation bias correction approach with RCM simulations versus the delta approach resulted in higher peak flows for the projected future climate. However, there are considerable differences in the performance between different RCMs, not only with regard to precipitation, but also temperature (see Sect. 3.5). Such differences can vary regionally as well. It is difficult to establish uniform procedures for scaling such critical variables and the question also arises as to how much scaling is reasonable without adversely affecting the original anthropogenic climate change signal. Furthermore, river flow routing of RCM-generated runoff can be used to analyse both model performance and scenario trends, but regard must be given to the precipitation biases that most RCMs show.

3.8 Projections of Future Changes in the Baltic Sea

3.8.1 Oceanographic Models and Anthropogenic Climate Change

The Baltic Sea is located in the transition zone between continental and maritime climate regimes. Under present climate conditions (see e.g. Annex 1.1 and Sect. 2.3), about half of the Baltic Sea is ice-covered in winter. Baltic Sea salinity is controlled by river runoff, net precipitation, and water exchange with the North Sea. Regional sea surface temperature varies with season but is also affected by the ocean circulation. The region is also characterised by land uplift and subsidence, which exert long-term effects on the coastal geometry. Anthropogenic climate change will likely affect regional sea ice and water temperature, as well as sea level and, possibly, salinity and oxygen conditions in the Baltic Sea deeps.

These aspects have been studied thoroughly using four regional coupled atmosphere–ocean modelling projections (Döscher and Meier 2004; Meier et al. 2004a) based on HadAM3H and ECHAM4/OPYC3 GCM driven simulations from RCAO, each forced by both B2 and A2 emission scenarios (cf. Table 3.1). In addition, so-called delta-change experiments have been performed (Meier 2006). In these, the 30-year monthly mean changes of the forcing functions for the Baltic Sea model RCO were calculated from the time slice experiments. These changes were added to reconstructed atmospheric surface fields and runoff for the period 1902–1998 (Kauker and Meier 2003). The results of both RCAO and RCO are compared with other studies on anthropogenic climate change in the Baltic Sea.

3.8.2 Projected Changes in Sea Ice

Since anthropogenic climate change might affect the ice season in the Baltic Sea considerably, the Baltic Sea ice in changing climate has been investigated in several studies (e.g. Tinz 1996, 1998; Haapala and Leppäranta 1997; Omstedt et al. 2000; Haapala et al. 2001; Meier 2002b, 2006; Meier et al. 2004a). These authors have applied different methods, based upon either statistical or dynamical downscaling of GCM results. The models used vary in complexity.

The main conclusion from these studies is that the projected decrease of ice cover over the next 100 years is dramatic, independent of the applied models or scenarios. For instance, Haapala et al. (2001) applied two different coupled ice–ocean models for the Baltic Sea using the same atmospheric forcing. They found that overall the simulated changes of quantities such as ice extent and ice thickness, as well as the interannual variations of these variables, were fairly similar in both mod-

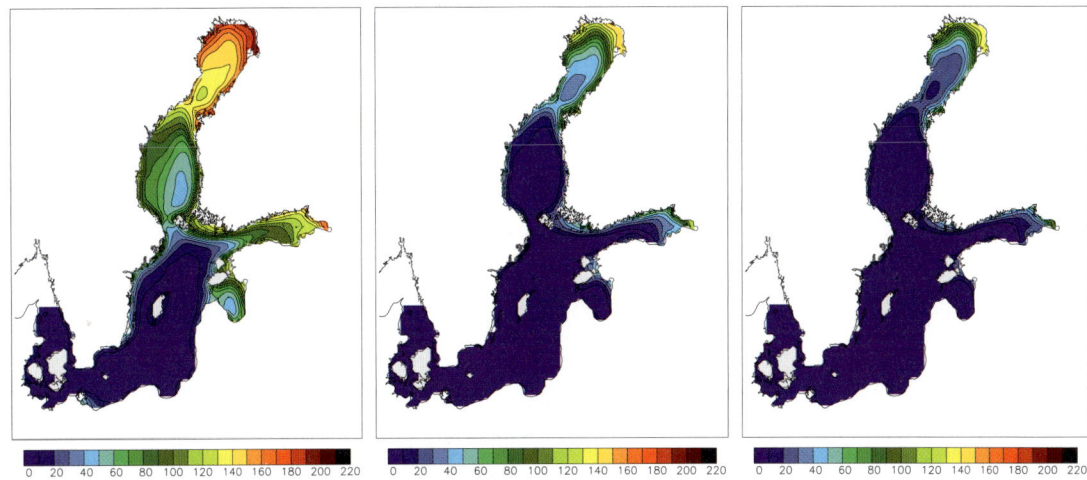

Fig. 3.35. Mean number of ice days averaged for regional downscaling simulations of HadAM3H and ECHAM4/OPYC3: control (*left panel*), B2 scenario (*middle panel*), and A2 scenario (*right panel*) (adapted from Meier et al. 2004a)

els. However, looking in more detail, differences were also reported, such as the spatial distributions of ice thickness.

RCAO results suggest that the Baltic Sea ice extent may decrease by 57 or 71% towards the end of the 21st century in the B2 and A2 scenarios, respectively (Meier et al. 2004a). The Bothnian Sea, large areas of the Gulf of Finland and Gulf of Riga, and the outer parts of the south-western archipelago of Finland would become ice-free in the mean. The length of the ice season would decrease by 1–2 months in the northern parts and 2–3 months in the central parts of the Baltic Sea (Fig. 3.35).

None of the simulated winters in 2071–2100 are completely ice-free due to a non-linear sensitivity of the simulated sea ice cover on the winter mean air temperature (Fig. 3.36). Severe ice winters are projected to be more sensitive to anthropogenic climate change than mild ice winters. These results are in accordance with earlier studies based on uncoupled ice-ocean modelling (Meier 2002b).

However, based upon the variability of the entire 20th century, an ice-free winter was found assuming changes of atmospheric surface variables corresponding to an A2 scenario (Meier 2006). Using the process-oriented PROBE-Baltic model and results from the first simulations with the RCA model, Omstedt et al. (2000) found that the scenario simulation indicates a maximum ice extent close to the observed long-term minimum and that there is no ice during 3 out of 10 winters. Miętus et al. (2004) also report that the ice season is likely

to become shorter due to higher sea water temperature.

In addition to scenarios, sensitivity studies were performed (e.g. Omstedt and Nyberg 1996; Omstedt et al. 1997; Meier 2002b, 2006). These studies show that the summer heat content may affect only the subsequent ice season. The time scale of the upper layer heat content amounts to a few months at the maximum. The sensitivity of ice cover and ice thickness to changes in salinity is relatively small.

3.8.3 Projected Changes in Sea Surface Temperature and Surface Heat Fluxes

The ensemble average annual mean sea surface temperature (SST) increases by 2.9 °C from 1961–1990 to 2071–2100. The ensemble consists of the four RCAO scenario simulations described by Räisänen et al. (2004). The SST increase is strongest in May and June (Fig. 3.37), and in the southern and central Baltic Sea (Döscher and Meier 2004). Details of the spatial SST patterns in the scenarios can partly be explained by sea ice reduction. In the northern basins the future year-to-year variability of mean SST was projected to increase because of melting ice. Results based on coupled and uncoupled ocean simulations are rather similar, as seen for example in Fig. 3.37 (Meier 2006).

All RCAO scenarios showed changes in the seasonal cycle of atmosphere–to–ocean heat transfer (Döscher and Meier 2004). There is a reduced heat

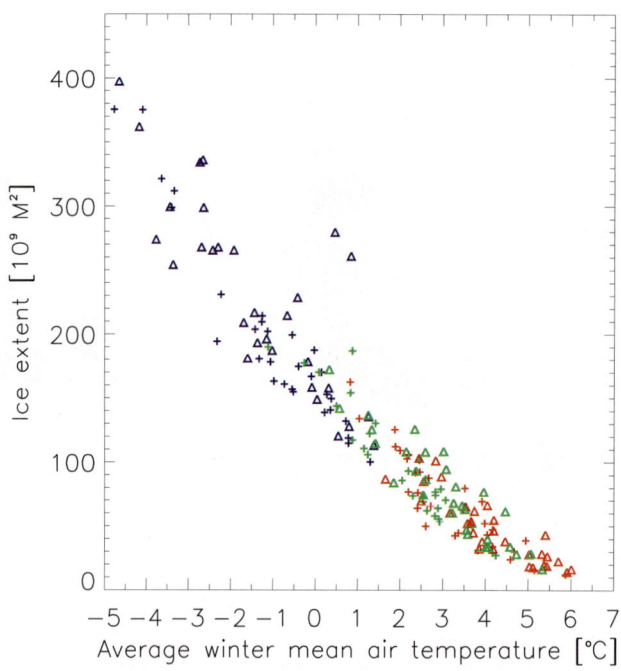

Fig. 3.36. Scatterplot of annual maximum ice extent in the Baltic Sea and winter mean (December through February) air temperature at Stockholm: RCAO-H (*plus signs*), RCAO-E (*triangles*), control (*blue*), B2 (*green*), and A2 (*red*). RCAO-H and RCAO-E denote simulations using RCAO with lateral boundary data from HadAM3H and ECHAM4/OPYC3, respectively (from Meier et al. 2004a)

loss in autumn, increased heat uptake in spring, and reduced heat uptake in summer. The overall heat budget change is characterised by increased solar radiation (due to reduced cloudiness and reduced surface albedo in winter), which is balanced by changes in the remaining heat flux components, i.e. net longwave radiation (out of the ocean) is increased, sensible heat flux (out of the ocean) is reduced, and latent heat flux (out of the ocean) is increased.

To date, dynamical downscaling experiments of Baltic Sea climate were only performed for limited periods. As the time slices were too short to properly spin up initial stratification for future climate, Meier (2002a) investigated the uncertainty of SST scenarios caused by the unknown future salinity, comparing scenarios with and without spin-up. He found that area mean SST changes do not differ much and that horizontal anomaly patterns are similar. However, some local differences were not negligible. The largest positive and negative differences were found in winter and summer, respectively, both in the Bothnian Bay.

Sensitivity studies showed that the heat content of the Baltic Sea is much more sensitive to changes in the wind forcing than the heat content of the

North Sea (Schrum and Backhaus 1999). However, the opposite is true for the heat flux from the water to the atmosphere during autumn, because advective and atmospheric heat flux changes are working in the same direction in the Baltic Sea but in opposite directions in the North Sea.

3.8.4 Projected Changes in Sea Level and Wind Waves

In the following, scenarios of mean sea level and storm surges are discussed separately, because many studies focus only on mean sea level changes.

Sea level change is not expected to be geographically uniform in the Baltic Sea, so information about its distribution is needed for the assessments of the impact on coastal regions. It is therefore important to analyse the long-term trend in changes of sea level, to study the variability of annual mean sea level in regions of interest, and to assess the importance of the corresponding meteorological and oceanographic parameters, especially wind distributions, as well as air and water temperatures.

Since the end of the 19th century a possibly anthropogenic climate change related eustatic sea

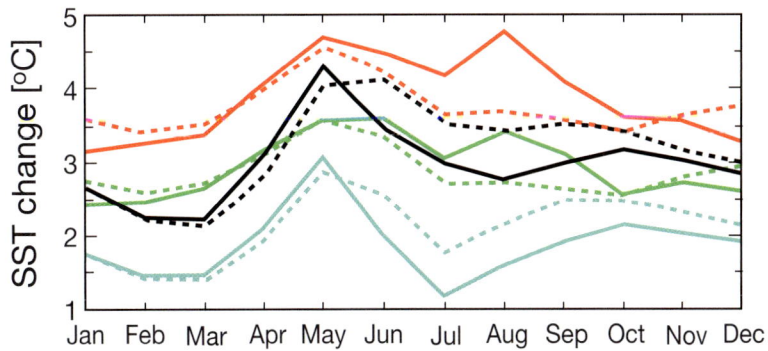

Fig. 3.37. Mean annual cycle of monthly sea surface temperature change: RCAO-H/B2 (*blue solid line*), RCAO-H/A2 (*black solid line*), RCAO-E/B2 (*green solid line*), and RCAO-E/A2 (*red solid line*). Dashed lines denote the corresponding RCO scenarios (adapted from Meier 2006)

level rise of about $1\,\mathrm{mm}\,\mathrm{yr}^{-1}$ has been observed at the Swedish station Stockholm (Ekman 1988). Other long-term sea level records indicate similar trends (Church et al. 2001; Sect. 2.3.2).

Utilising different methods, several studies suggest an accelerated sea level rise by the end of the 21st century. For instance, using a statistical downscaling method Heyen et al. (1996) found a slight increase of Baltic sea level anomalies in winter when air pressure from a GCM greenhouse experiment is downscaled. It was found that a global mean sea level rise of 50 cm from 1990 to 2080 would lead to a sea level rise of 33–46 cm in Danish waters (Fenger et al. 2001).

It can be expected that by the year 2100 many regions currently experiencing a relative fall in sea level will instead have a rising relative sea level (Fenger et al. 2001). Johansson et al. (2004) calculated mean sea level scenarios for the 21st century at the Finnish coast. They considered land uplift, the projected global average sea level rise and the projected trends of the leading sea level pressure component in GCM scenarios. The latter was used to estimate changes of the water balance associated with changes of the NAO.

Johansson et al. (2004) concluded that the past trend of decreasing mean sea level in the Gulf of Finland (Sect. 2.3.2) will not continue in the future because the accelerated global average sea level rise will balance the land uplift. Indeed, land uplift and the global average sea level rise, according to Church et al. (2001), seem to be the dominant contributions to the future changes of mean sea level in the Baltic Sea (Meier et al. 2004b).

Model studies concerning future sea levels have been carried out as well. The local hydrodynam-

ically driven sea level change component in the semi-enclosed sub-basins of the Estonian coastal sea due to changes in wind climate was analysed on the basis of sensitivity and scenario runs of a 2D hydrodynamic model (Suursaar et al. 2006). It was demonstrated that every change in long-term wind regime (e.g. in average wind speed, variability or directional distribution) has an effect on the established sea level regime; the effect is different along the coastline, and it depends on coastline configuration. Following the observed trend towards an increase in atmospheric westerlies, the hydrodynamic model simulations predicted an increase of up to 5–6 cm in annual means in some windward bays of the Gulf of Riga, if the average wind speed increases by $1–2\,\mathrm{m\,s}^{-1}$. This local sea level rise component could be up to 9–11 cm in winter months, while in summer a sea level rise is unlikely. Further enhancement of the seasonal signal in sea level variations in the form of lower return periods for extreme sea level events is anticipated.

In dynamical downscaling experiments for the entire Baltic Sea performed at the Rossby Centre using either HadCM2 or ECHAM4/OPYC3, increased winter mean sea levels were found mainly in the gulfs (Meier 2001; Meier et al. 2004b). These changes follow approximately the wind speed changes averaged over the Baltic Sea surface (see e.g. Räisänen et al. 2004).

However, compared to land uplift and the global average sea level rise, wind induced seasonal sea level changes may be smaller. The downscaling experiments indicate that in the future climate the risk of coastal inundation may be largest in the eastern and southern parts of the Baltic Sea

(Meier et al. 2004b). This agrees with Miętus et al. (2004), who project a Baltic sea level rise of 33–125 cm in the 21st century (75 cm on average) at the Polish coast due to global sea level rise (Church et al. 2001) and changes in atmospheric circulation patterns.

In the scenarios there is no overall agreement whether the intensity or frequency of storm surges will increase in future climate in addition to the mean sea level rise. Using statistical downscaling, Baerens and Hupfer (1999) found that storm surges at the German Baltic Sea coast will not change significantly.

However, regional wind changes could have additional impact on surge heights (Meier 2006). For instance, the 100-year surge in the Gulf of Riga could change from the present 2 m to a future 1.9–3.3 m relative to the mean sea level for the period 1903–1998 (Fig. 3.38). The future range comes from using different scenarios. Although modelled wind speed changes are rather uniform with similar percent changes in mean and extreme wind speeds (Räisänen et al. 2004), extreme sea levels will increase significantly more than the mean sea level (Meier et al. 2004b; Meier 2006).

Miętus (1999) studied an ECHAM3 time-slice experiment using statistical downscaling (under doubled CO_2-concentration as compared to the late 1980s) and found no statistically significant changes in mean wave height but an increase in wave height range (see also Miętus 2000). Increased wind speed variability and increased occurrence of strong and very strong winds were also projected. More frequent north-western and south-western wind may increase the amount of water in the Baltic Sea at the Polish coast and change the gradient of the water surface. This would also lead to a sea level rise of about 0.07–0.09 cm year^{-1} at the Polish coast as well as to increased wave amplitudes and higher levels of storm surges. Increased sea level variability is projected, most notably at the eastern part of the coast. Another approach, using ECHAM1-LSG and transfer functions and a CO_2-trebling scenario for the late 2060s, also shows a clear increase in sea level (Miętus 1999).

Several studies have focused on the assessment of the impact of rising mean sea level and increased storm surge frequency on coastal processes like erosion and sediment transport. For instance, Orviku et al. (2003) concluded that the most marked coastal changes in Estonia result from a combination of strong storms, high sea levels induced by storm surges, ice free seas and unfrozen sediments, all of which enhance erosion and transport of sediments above the mean sea level and inland relative to the mean coastline. Kont et al. (2003) selected seven case study areas characterising all shore types of Estonia for sea level rise vulnerability and adaptation assessment. According to their scenarios the longest coastline section recession (6.4 km) would occur on the western coast of the Estonian mainland. Meier et al. (2006a) combined the results of calculated sea level changes in the Baltic Sea with scenarios of global average sea level rise, land uplift and digital elevation models to estimate flood prone areas in future climate. Regional and local maps of flood prone areas can serve as decision support for spatial planning. The planning of cities located at the eastern and south-eastern coasts of the Baltic proper, the Gulf of Riga and the Gulf of Finland would be especially affected.

Using surface winds from the Rossby Centre scenarios and a simplified wave model, scenario simulations of the wave climate have been performed (Meier et al. 2006a). Meier et al. (2006a) found that the annual mean significant wave height and the 90th percentile may increase by about 0–0.4 and 0–0.5 m, respectively. In all scenarios performed, the largest increases were found in the Gulf of Bothnia and in the eastern Gotland Sea when lateral boundary data from ECHAM4/OPYC3 were applied.

3.8.5 Projected Changes in Salinity and Vertical Overturning Circulation

The long spin-up time and the positive bias of precipitation and runoff in many control simulations of state-of-the-art regional climate models (cf. Sects. 3.5 and 3.7) make it difficult to perform projections for salinity. In several studies, future stratification was spun up in long simulations. Omstedt et al. (2000) and Meier (2002a) carried out 100-year long scenario simulations using the repeated atmospheric forcing of a time slice experiment.

Assuming that the variability of the 20th century will not change, 100-year long delta-change experiments were performed by Meier (2006). Thereby, the negative impact of the positive bias of the freshwater inflow is avoided. However, it was assumed there that the hydrological changes are large compared to the model biases, which is actually not the case.

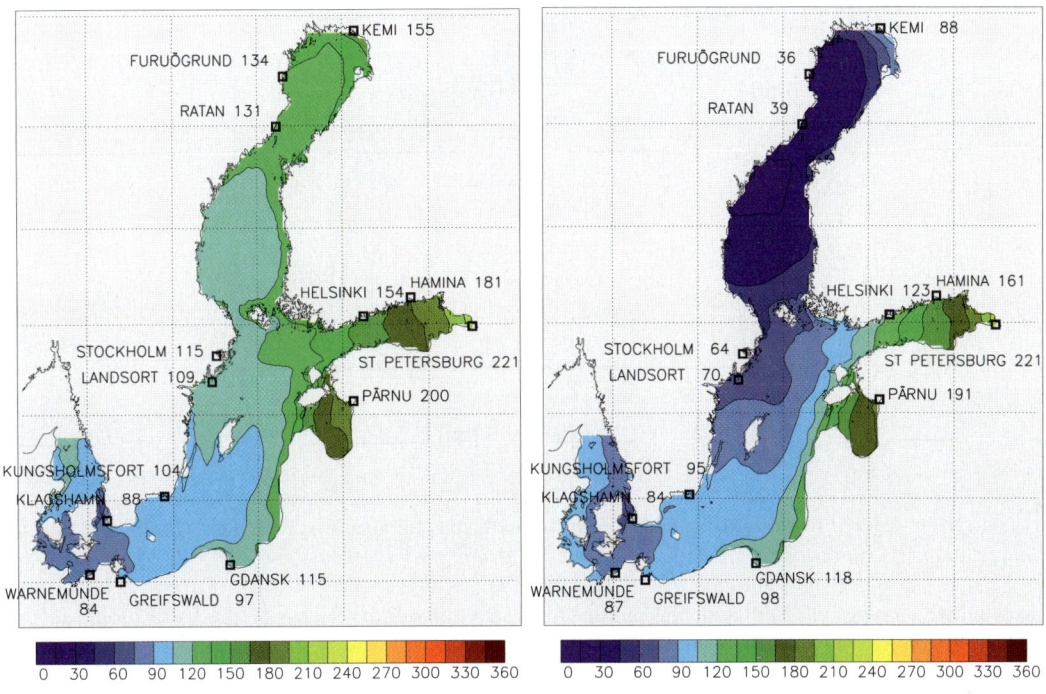

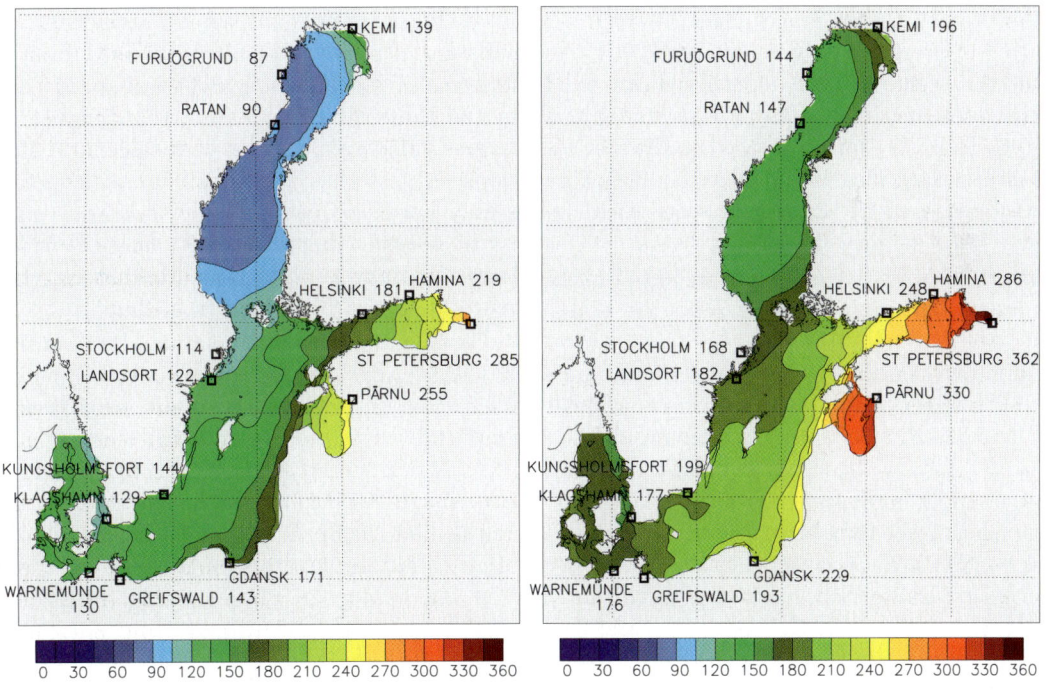

Fig. 3.38. 100-year surge (*in cm*) of the hindcast experiment using RCO (*upper left panel*) relative to the mean sea level for the period 1903–1998 and 3 selected regional scenarios of the 100-year surge (*in cm*): "lower case" scenario (RCO-H/B2) with a global average sea level rise of 9 cm (*upper right panel*), "ensemble average" scenario with a global average sea level rise of 48 cm (*lower left panel*), and "higher case" scenario (RCO-E/A2) with a global average sea level rise of 88 cm (*lower right panel*) (from Meier 2006)

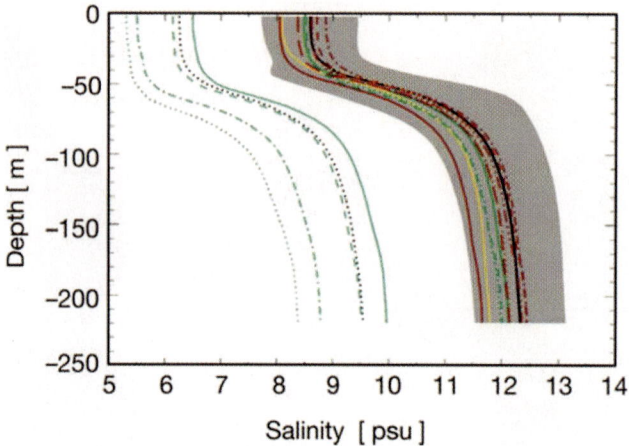

Fig. 3.39. Median profiles of salinity at Gotland Deep for 16 future climate projections from RCO for the period 2071–2100 (*colored lines*) and for the present climate for the period 1961–1990 (*black solid line, shaded areas indicate the ±2 standard deviation band calculated from two-daily values for 1903–1998*). The projections include A2 and B2 emissions scenarios from 7 different RCMs driven by 5 GCMs. Effects of both wind and freshwater inflow changes are included. A detailed discussion of salinity biases of the RCO model is given by Meier (2006) (from Meier et al. 2006b)

In projections performed at the Rossby Centre (Räisänen et al. 2004) the total mean annual river flow to the Baltic Sea changes between −2 and +15% of present-day flow (Graham 2004). In some of the scenarios, the monthly mean wind speed over sea increases, especially in winter and spring, up to 30%. Both increased freshwater inflow and increased mean wind speeds could cause the Baltic Sea to drift into a new state with significantly lower salinity (Fig. 3.39).

However, even with the highest projected freshwater inflow, the Baltic Sea will not be transformed into a freshwater sea, because the relationship between freshwater supply and average salinity of the final steady-state is non-linear (Meier and Kauker 2003b). A pronounced halocline would still be expected to remain and separate the upper and lower layers in the Baltic Proper, limiting the impact of direct wind-induced mixing to the surface layer. Although salinity in the entire Baltic Sea might be significantly lower at the end of the 21st century, stability and deep water ventilation will very likely change only slightly (Fig. 3.40) because only the changing wind-induced mixing alters vertical and horizontal gradients of density on time scales longer than 20 years (Meier 2006).

Sixteen salinity projections were performed by Meier et al. (2006b) using seven RCMs, five driving global models and two emissions scenarios (Fig. 3.39). These results show mean salinity

change by the end of the 21st century (2071–2100) to range between +4 and −45%, although the positive change is not statistically significant. This substantial range in results is mainly due to differences in precipitation and wind speed changes in the Baltic Sea Basin from the different simulations. However, several of the scenario simulations suggest that future salinity will be considerably lower compared to the simulated variability of present climate. Salinity changes will have large impacts on species distributions, growth and reproduction of organisms (see Chap. 5).

Based upon model simulations, the sensitivity of the average steady-state salinity to the external forcing (e.g. freshwater supply, wind speed and amplitude of the sea level in Kattegat) has been estimated in several studies (e.g. Stigebrandt 1983; Gustafsson 1997; 2000b, 2004; Schrum 2001; Meier and Kauker 2003b; Rodhe and Winsor 2002, 2003; Stigebrandt and Gustafsson 2003; Meier 2005).

It was found that the sensitivity of the steady-state salinity to the freshwater supply is non-linear. In different model approaches the results agree rather well. For instance, the sensitivity of the three-dimensional general circulation model RCO (Meier and Kauker 2003b) is only slightly higher than the sensitivity of the process oriented model by Stigebrandt (1983). Even with 100% increased freshwater supply the Baltic Sea cannot be classified as a freshwater sea. Further, the sensitivity in different sub-basins was stud-

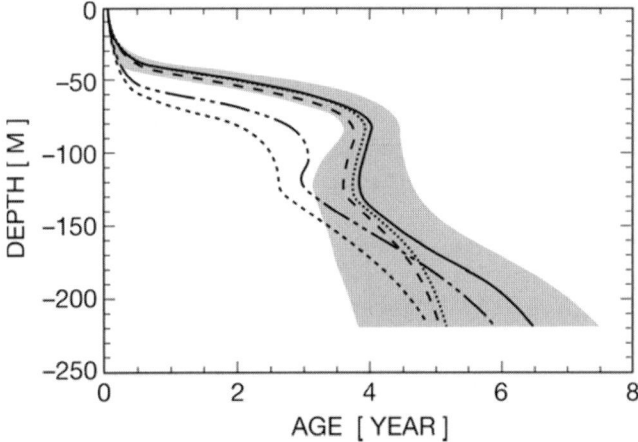

Fig. 3.40. Median profiles of age at Gotland Deep: RCO hindcast simulation for 1961–1990 (*black solid line, the shaded area indicates the range between the first and third quartiles*) and four scenario simulations for 2071–2100 (*dotted line:* RCO-H/B2, *dashed line:* RCO-H/A2, *dash-dotted line:* RCO-E/B2, *long-dashed line:* RCO-E/A2) (adapted from Meier 2006). The age is the time elapsed since a particle left the sea surface and is a measure of deep water renewal

ied when the changing freshwater supply is non-uniform (Stigebrandt and Gustafsson 2003). In contrast to these investigations, Rodhe and Winsor (2002, 2003) found a much larger sensitivity to freshwater inflow. An explanation could be that in the empirical model of Rodhe and Winsor (2002) the freshwater content anomaly depends only on the freshwater supply and not on the wind forcing, as found by Meier and Kauker (2003a).

Meier (2005) investigated not only the sensitivity of modelled salinity but also the sensitivity of modelled age to freshwater supply, wind speed and amplitude of the sea level in Kattegat (Fig. 3.40). In steady-state the average salinity of the Baltic Sea is most sensitive to perturbations of freshwater inflow. Increases in freshwater inflow and wind speed both result in decreased salinity, whereas increases in the Kattegat sea level results in increased salinity. The average age is most sensitive to perturbations of the wind speed. Especially, decreased wind speed causes significantly increased age of the deep water. On the other hand, the impact of changing freshwater or sea level in Kattegat on the average age is comparatively small, suggesting invariance of stability and ventilation in the steady-state. Immediately after the onset of increased freshwater inflow, the saltwater inflow into the Baltic Sea drops significantly due to increased recirculation (Meier 2005). After the typical response time scale, the vertical overturning circulation partially recovers.

3.9 Future Development in Projecting Anthropogenic Climate Changes

Global climate models play a central role in projecting future anthropogenic climate change, both by providing estimates of change on large horizontal scales and by providing the large-scale information needed by various downscaling methods. Further improvement of these models is therefore essential. This concerns especially the parameterisation of sub-grid scale phenomena (such as cloud processes, radiative transfer, convection, small-scale mixing in the atmosphere and the ocean, as well as the exchange of heat, water and momentum between the atmosphere and the other components of the climate system) which are crucially important for the response of climate to external forcing. Success in improving global climate models will require huge efforts from the international community, focusing not only on modelling per se but also on collecting and making widely available detailed observations that will help to guide the development of the models.

Model improvements are also needed for regional climate models. Even if the horizontal resolution is better in those models than in the global climate models, they still have the problem of parameterisation of sub-grid scale processes. It is likely that in the next five to ten years horizontal resolutions of only a few kilometres will become common. Even at such finer scales, many phenomena remain unresolved, including individual clouds. Reanalysis-driven experiments with regional climate models can be compared to high-resolution gridded datasets. Since the horizontal resolution in the models is likely to improve, the observational datasets also need to be represented on finer grids. In this way errors and systematic biases can be identified, leading to further improvements in model formulation. Furthermore, global climate models can benefit from improved parameterisations in regional climate models since the horizontal resolution in the global models will, in a few years time, be close to that of today's regional models.

Climate models should evolve to also include other processes than those of the more "traditional" physical climate system. Complex interaction and feedback processes involving the biogeochemical cycles makes it necessary to introduce other model components to the climate models. Examples of such components are, among others: fully coupled models of atmospheric chemistry, aerosol models, interactive vegetation models, models describing the carbon cycle, and models including marine chemistry.

Despite the efforts put into improving climate models, it is likely that the projections from different models will continue to differ for many years to come, particularly on regional scales. Therefore, it is also important to ask how to best deal with these differences. Is it reasonable to assume that all models give equally likely projections, or should the projections be weighted according to model quality (e.g. in simulating present-day climate)? First steps toward the latter direction have been taken (e.g. Giorgi and Mearns 2002, 2003; Murphy et al. 2004), but as yet there is only limited understanding of which aspects of the present-day climate simulation matter most for simulation of projected anthropogenic climate changes. On the other hand, it is also important to evaluate the possibility that the behaviour of the real climate system might fall outside the range of conventional model results (e.g. Allen and Ingram 2002; Stainforth et al. 2005).

Both statistical and dynamical downscaling have proved to have added value for assessing regional climate in comparison with global models, although they may show different results. Comparison of future scenarios downscaled by the two approaches may form a basis for assessing uncertainties associated with the downscaling. The two approaches have different advantages and disadvantages due the nature of the methods used.

While the dynamical approach may have problems with biases, the statistical approach basically avoids this problem as it is a data-based method. However, being physically based, dynamical methods realistically simulate non-linear effects and other dynamical features, whereas statistical methods often lack the full range of the true variability. Therefore, choice of the approach to be used depends mainly on the questions to be addressed. One important advantage of the statistical approach is that it can deal with local scales, including site scale. In the interim before the resolution of RCMs is considerably improved, it provides an alternative way to create the local anthropogenic climate change scenarios needed for many impact studies.

Hydrological studies have shown that biases in precipitation from climate models make it difficult to directly use meteorological outputs in hydrological and water resources oriented applications. Evapotranspiration is also a source of much uncertainty, both in climate models and hydrological models. The combination of biases in meteorological variables and the need for high resolution has to date precluded using the actual hydrological components from climate models for detailed assessment of hydrological impacts. Continued work to improve the hydrological processes in both global and climate models, including river routing techniques, is needed.

Such development, together with finer resolution, as mentioned above, would serve to enhance the quality and utility of climate model simulations, particularly if it leads to better representation of extreme hydrological events. Recognising that biases may be around for some time to come, hydrological development should also focus on improved methods to filter out biases without misrepresenting the anthropogenic climate change signal coming from climate models.

Regarding ocean processes, the shortcomings of global models are limiting for many variables. For instance, sea level scenarios for the Baltic Sea remain rather uncertain due to the large uncertainty of the global average sea level rise. In addition, overestimation of precipitation by regional climate models considerably affects salinity in the Baltic Sea. Salinity is not only biased by the erroneous freshwater surplus; mixing also has a considerable impact.

As mixing is approximately proportional to the third power of the wind speed, any wind speed biases strongly impact on Baltic Sea climate. Consequently, mixing and other variables, such as sea level extremes, are typically underestimated as RCMs commonly underestimate high wind speeds.

To improve scenarios for the Baltic Sea, the horizontal resolution of ocean models should be increased such that mesoscale processes are included. Description of important processes like saltwater plume dynamics, entrainment and interleaving needs to be improved as well. Thus, further work to improve Baltic Sea models should focus on advanced mixing parameterisations (e.g. including Langmuir circulation and breaking of internal waves) and the coupling between ocean and state–of–the–art wave models.

3.10 Summary of Future Anthropogenic Climate Change Projections

Increasing greenhouse gas (GHG) concentrations are expected to lead to a substantial warming of the global climate during this century. Cubasch et al. (2001) estimated the annual globally averaged warming from 1990 to 2100 to be in the range of 1.4 to 5.8 °C. This range in temperature change takes into account differences between climate models and a range of anthropogenic emissions scenarios, but it excludes other uncertainties (for example, in the carbon cycle) and should not be interpreted as giving the absolute lowest and highest possible changes in the global mean temperature during the period considered.

Projected future warming in the Baltic Sea Basin generally exceeds the global mean warming in GCM (global climate model) simulations. Looking at the annual mean from an ensemble of 20 GCM simulations, regional warming over the Baltic Sea Basin would be some 50% higher than global mean warming. In the northern areas of the basin, the largest warming is generally simulated in winter; further south the seasonal cycle of warming is less clear.

However, the relative uncertainty in the regional warming is larger than that in the global mean warming. Taking the northern areas of the basin as an example, the warming from late 20th century to late 21st century could range from as low as 1 °C in summer (lowest scenario for summer) to as high as 10 °C in winter (highest scenario for winter). The simulated warming would generally be accompanied by an increase in precipitation in the Baltic Sea Basin, except in the southernmost areas in summer. The uncertainty

for precipitation change is, however, larger than that for temperature change, and the coarse resolution of GCMs does not resolve small-scale variations of precipitation change that are induced by the regional topography and land cover.

A more geographically detailed assessment of future anthropogenic climate change in the Baltic Sea Basin requires the use of statistical or dynamical downscaling methods. Yet, as only a limited number of GCM simulations have been downscaled by RCMs (regional climate models) or statistical downscaling methods, the range of results derived from such downscaling experiments does not fully reflect the range of uncertainties in the GCM projections. Accepting this, the range of results from available downscaling studies is presented below as it gives an indication of plausible future changes. All values refer to changes projected for the late 21st century, represented here as differences in climate between the years 1961–1990 and 2071–2100. All references to "northern" and "southern" areas of the Baltic Sea Basin are defined by the subregions shown in Fig. 3.12.

Consistent with GCM studies, all available downscaling studies also indicate increases in temperature during all seasons for every subregion of the Baltic Sea Basin. Combined results show a projected warming of the mean annual temperature by some 3 to 5 °C for the total basin. Seasonally, the largest part of this warming would occur in the northern areas of the Baltic Sea Basin during winter months and in the southern areas of the Baltic Sea Basin during summer months. Corresponding changes in temperatures would be 4 to 6 °C in winter and 3 to 5 °C in summer, as estimated from a matrix of regional climate model experiments.

As noted above, these ranges most probably underestimate the real uncertainty. The diurnal temperature range – the difference between daily maximum and minimum temperature – would decrease, most strongly in autumn and winter months. Such levels of warming would lead to a lengthening of the growing season, defined here as the continuous period when daily mean temperature exceeds 5 °C. Taking an example from one RCM indicates that the growing season length could increase by as much as 20 to 50 days for northern areas and 30 to 90 days for southern areas by the late 21st century. The range depends on the range of different emissions scenarios used.

Projected changes in precipitation from downscaling studies also depend both on differences in

GHG emissions scenarios and differences between climate models. Moreover, precipitation results are more sensitive than temperature results to the statistical uncertainty in determining climatological means from a limited number of simulated years, particularly at regional scales. Seasonally, winters are projected to become wetter in most of the Baltic Sea Basin and summers to become drier in southern areas for many scenarios. Northern areas could generally expect winter precipitation increases of some 25 to 75%, while the projected summer changes lie between −5 and 35%. Southern areas could expect increases ranging from some 20 to 70% during winter, while summer changes would be negative, showing decreases of as much as 45%.

Taken together, these changes lead to a projected increase in annual precipitation for the entire basin. In broad terms, these results are consistent with GCM studies of precipitation change, although the projected summer decrease in the southern areas of the basin tends to be larger and extend further north in the available RCM studies than in most reported GCMs. This difference reflects the fact that the few GCM simulations that have been downscaled by RCMs also show this pattern of summer precipitation change.

Projected changes in wind differ widely between various climate models. Differences in the circulation patterns of the driving GCMs are particularly important for the modelled outcome of this variable. From the RCM results presented here, only those driven by the ECHAM4/OPYC3 GCM show statistically significant changes for projected future climate scenarios. For mean daily wind speed over land areas, this would amount to a mean increase of some 8% on an annual basis and a maximum mean seasonal increase of up to 12% during winter. The corresponding mean seasonal increase over the Baltic Sea in winter, when decrease in ice cover enhances near-surface winds, would be up to

18%. For RCMs driven by the HadAM3H GCM, the changes are small and not statistically significant. Modelled changes in extreme wind generally follow the same pattern as for the mean wind; however, the spatial resolution of both GCMs and RCMs is far too coarse to accurately represent the fine scales of extreme wind.

As the downscaled projections for wind differ widely, there is no robust signal seen in the RCM results. Looking at projected changes in large-scale atmospheric circulation from numerous GCMs, they indicate that an increase in windiness for the Baltic Sea Basin would be somewhat more likely than a decrease. However, the magnitude of such a change is still highly uncertain and it may take a long time before projected GHG-induced changes in windiness, if ever, emerge from background natural variability. It can be noted, moreover, that ECHAM4/OPYC3 is one of the GCMs that gives higher values of change in large-scale wind.

Hydrological studies using climate change projections show that increases in mean annual river flow from the northernmost catchments would occur together with decreases in the southernmost catchments. Seasonally, summer river flows would tend to decrease, while winter flows would tend to increase, by as much as 50%. The southernmost catchments would be affected by the combination of both decreased summer precipitation and increased evapotranspiration. Oceanographic studies show that mean annual sea surface temperatures could increase by some 2 to 4 °C by the end of the 21st century. Ice extent in the sea would then decrease by some 50 to 80%. The average salinity of the Baltic Sea could range between present day values and decreases of as much as 45%. However, it should be noted that these oceanographic findings, with the exception of salinity, are based upon only four regional scenario simulations using two emissions scenarios and two global models.

3.11 References

Achberger C (2004) Recent and future regional climate variations in Sweden in relation to large-scale circulation. Earth Sciences Centre, A92. Göteborg University

Achberger C, Linderson ML, Chen D (2003) Performance of the Rossby Centre regional Atmospheric model in Southern Sweden: Comparison of simulated and observed precipitation. Theor App Climatol 76:219–234

Achberger C, Chen D, Alexandersson H (2006) The surface winds of Sweden during 1999–2000. Int J Climatol 26:159–178

Alexandersson H, Tuomenvirta H, Schmith T, Iden K (2000) Trends of storms in NW Europe derived from an updated pressure data set. Clim Res 14:71–73

Allen MR, Ingram WJ (2002) Constraints on future changes in climate and the hydrologic cycle. Nature 419:224–232

Andersen HE, Kronvang B, Larsen SE, Hoffmann CC, Jensen TS, Rasmussen EK (2006) Climate-change impacts on hydrology and nutrients in a Danish lowland river basin. Sci Total Env 365: 223–237

Andréasson J, Gardelin M, Bergström S (2002) Modelling hydrological impacts of climate change in the Lake Vänern region in Sweden. Vatten 58:25–32

Andréasson J, Bergström S, Carlsson B, Graham LP, Lindström G (2004) Hydrological change – climate change impact simulations for Sweden. Ambio 33:228–234

Arnell NW (1998) Climate change and water resources in Britain. Climatic Change 39:83–110

Arnell NW (1999) The effect of climate change on hydrological regimes in Europe: A continental perspective. Glob Env Change 9:5–23

Arnell NW, Hudson DA, Jones RG (2003) Climate change scenarios from a regional climate model: Estimating change in runoff in southern Africa. J Geophy Res 108:4519, doi: 10.1029/2002JD002782

Arpe K, Bengtsson L, Golitsyn GS (2000) Analysis of changes in hydrological regime on the Lake Ladoga watershed and in the discharge of the Neva River in 20th and 21st centuries using a global climate model. Russ Meteorol Hydrol 12:5–13

Baerens C, Hupfer P (1999) Extremwasserstände an der deutschen Ostseeküste nach Beobachtungen und in einem Treibhausgasszenario (Extreme water levels at the German Baltic Sea coast according to observations and a greenhouse gas scenario). Die Küste 61:47–72 (in German)

Barthelet P, Terray L, Valcke S (1998) Transient CO_2 experiment using the ARPEGE/OPAICE non flux corrected coupled model. Geophys Res Lett 25:2277–2280

Benestad RE (2002a) Empirically downscaled multimodel ensemble temperature and precipitation scenarios for Norway. J Clim 15:3008–3027

Benestad RE (2002b) Empirically downscaled climate scenarios for northern Europe. Clim Res 21: 105–125

Benestad RE (2004) Tentative probabilistic temperature scenarios for northern Europe. Tellus A 56: 89–101

Beniston M, Stephenson DB, Christensen OB, Ferro CAT, Frei C, Goyette S, Halsnæs K, Holt T, Jylhä K, Koffi B, Palutikof J, Schöll R, Semmler T, Woth K (2007) Future extreme events in european climate: An exploration of regional climate model projections. Climatic Change 81:71–95

Bergström S (1976) Development and application of a conceptual runoff model for Scandinavian catchments. Doctoral thesis, Department of Water Resources Engineering, Institute of Technology, Lund University

Bergström S (1995) The HBV Model In: Singh VP (ed) Computer Models of Watershed Hydrology. Water Resources Publications, Highlands Ranch Colorado

Bergström S, Carlsson B (1994) River runoff to the Baltic Sea: 1950–1990. Ambio 23:280–287

Bergström S, Carlsson B, Gardelin M, Lindström G, Pettersson A, Rummukainen M (2001) Climate change impacts on runoff in Sweden – assessments by global climate models dynamical downscaling and hydrological modelling. Clim Res 16:101–112

Bergström S, Andréasson J, Beldring S, Carlsson B, Graham LP, Jónsdóttir JF, Engeland K, Turunen MA, Vehviläinen B, Førland E J (2003) Climate Change Impacts on Hydropower in the

Nordic Countries – State of the art and discussion of principles. CWE Report no 1, CWE Hydrological Models Group, Rejkjavik, Iceland

Bertrand C, Van Ypersele JP, Berger A (2002) Are natural climate forcings able to counteract the projected anthropogenic global warming? Climatic Change 55:413–427

Bjørge D, Haugen JE, Nordeng TE (2000) Future climate in Norway. DNMI Research Report no 103

Blenckner T, Chen D (2003) Comparison of the impact of regional and North Atlantic atmospheric circulation on an aquatic ecosystem. Clim Res 23:131–136

Boer GJ, Yu B (2003) Climate sensitivity and response. Clim Dyn 20:415–429

Boville BA, Gent PR (1998) The NCAR climate system model version One. J Clim 11:1115–1130

Braconnot P, Marti O, Joussaume S (1997) Adjustment and feedbacks in a global coupled ocean–atmosphere model. Clim Dyn 13:507–519

Busuioc A, Chen D, Hellström C (2001a) Temporal and spatial variability of precipitation in Sweden and its link with the large scale atmospheric circulation. Tellus 53A,3:348–367

Busuioc A, Chen D, Hellström C (2001b) Performance of statistical downscaling models in GCM validation and regional climate estimates: Application for Swedish precipitation. Int J Climatol 21: 557–578

Butina M, Melnikova G, Stikute I (1998a) Potential impact of climate change on the hydrological regime in Latvia. In: Raschke E, Isemer HJ (eds) Conference Proceedings of the Second Study Conference on BALTEX, International BALTEX Secretariat Publication No 11

Butina M, Melnikova G, Stikute I (1998b) Potential impact of climate change on the hydrological regime in Latvia. In: Lemmelä R, Helenius N (eds) Proceedings of the Second International Conference on Climate and Water. Espoo, Finland 17–20 August 1998, 3:1610–1617

Carlsson B, Sanner H (1996) Modelling influence of river regulation on runoff to the Gulf of Bothnia. Nord Hydrol 27:337–350

Chen D (2000) A monthly circulation climatology for Sweden and its application to a winter temperature case study. Int J Climatol 20:1067–1076

Chen D, Achberger C (2006) Past and future atmospheric circulation over the Baltic region based on observation reanalysis and GCM simulations. Research Report C74 Earth Sciences Centre Göteborg University, Sweden

Chen D, Hellström C (1999) The influence of the North Atlantic Oscillation on the regional temperature variability in Sweden: Spatial and temporal variations. Tellus 51A,4:505–516

Chen D, Li X (2004) Scale dependent relationship between maximum ice extent in the Baltic Sea and atmospheric circulation. Glob Planet Change 41:275–283

Chen D, Omstedt A (2005) Climate-induced variability of sea level in Stockholm: Influence of air temperature and atmospheric circulation. Adv Atmos Sci 20,5:655–664

Chen D, Achberger C, Räisänen J, Hellström C (2006) Using statistical downscaling to quantify the GCM-related uncertainty in regional climate change scenarios: A case study of Swedish precipitation. Adv Atmos Sci 23,1:54–60

Christensen JH, Christensen OB (2003) Severe summertime flooding in Europe. Nature 421:805–806

Christensen JH, Machenhauer B, Jones RG, Schär C, Ruti PM, Castro M, Visconti G (1997) Validation of present-day regional climate simulations over Europe: LAM simulations with observed boundary conditions. Clim Dyn 13:489–506

Christensen JH, Räisänen J, Iversen T, Bjørge D, Christensen OB, Rummukainen M (2001) A synthesis of regional climate change simulations – a Scandinavian perspective. Geophys Res Lett 28: 1003–1006

Christensen JH, Carter TR, Giorgi F (2002) Prudence employs new methods to assess European climate change. EOS 83:147

Christensen JH, Carter TR, Rummukainen M (2007) Evaluating the performance and utility of regional climate models: The PRUDENCE project. Climatic Change 81:1–6

Christensen OB, Christensen JH (2004) Intensification of extreme European summer precipitation in a warmer climate. Glob Planet Change 44:107–117

Christensen OB, Christensen JH, Machenhauer B, Botzet M (1998) Very-high-resolution regional climate simulations over Scandinavia present climate. J Clim 11:3204–3229

Christensen OB, Christensen JH, Botzet M (2002) Heavy precipitation occurrence in Scandinavia investigated with a regional climate model. In: Beniston M (ed) Climatic Change: Implications for the hydrological cycle and for water management. Kluwer

Church JA, Gregory JM, Huybrechts P, Kuhn M, Lambeck K, Nhuan MT, Qin D, Woodworth PL (2001) Changes in sea level. In: IPCC Climate Change 2001: The scientific basis contribution of working group I to the Third Assessment Report of the Intergovernmental Panel on Climate Change. Cambridge University Press, Cambridge New York

Cubasch U, Meehl GA, Boer GJ, Stouffer RJ, Dix M, Noda A, Senior CA, Raper S, Yap KS (2001) Projections of future climate change. In: IPCC Climate Change 2001: The scientific basis contribution of working group I to the Third Assessment Report of the Intergovernmental Panel on Climate Change. Cambridge University Press, Cambridge New York

De Castro M, Gallardo C, Jylhä K, Tuomenvirta H (2007) The use of a climate-type classification for assessing climate change effects in Europe from an ensemble of nine regional climate models. Clim Change 81:329 341

De Roo A, Schmuck G (2002) Assessment of the effects of engineering, land use and climate scenarios on flood risk in the Oder catchment. European Commission, Joint Research Centre, Institute for Environment and Sustainability

Déqué M, Rowell DP, Lüthi D, Giorgi F, Christensen JH, Rockel B, Jacob D, Kjellström E, De Castro M, Van Den Hurk B (2007) An intercomparison of regional climate simulations for Europe: Assessing uncertainties in model projections. Climatic Change 81:53–70

Déqué M, Marquet P, Jones RG (1998) Simulation of climate change over Europe using a global variable resolution general circulation model. Clim Dyn 14:173–189

Diansky NA, Volodin EM (2002) Simulation of present-day climate with a coupled atmosphere–ocean general circulation model. Izvestia Atmos Ocean Phys 38:732–747

Döscher R, Meier HEM (2004) Simulated sea surface temperature and heat fluxes in different climates of the Baltic Sea. Ambio 33:242–248

Döscher R, Willén U, Jones C, Rutgersson A, Meier HEM, Hansson U, Graham LP (2002) The development of the regional coupled ocean–atmosphere model RCAO. Boreal Env Res 7:183–192

Ekman M (1988) The world's longest continued series of sea level observations. Pure Appl Geophys 127:73–77

Emori S, Nozawa T, Abe-Ouchi A, Numaguti A, Kimoto M, Nakajima T (1999) Coupled ocean–atmosphere model experiments of future climate change with an explicit representation of sulfate aerosol scattering. J Met Soc Jpn 77:1299–1307

Engen-Skaugen T, Roald LA, Beldring S, Førland EJ, Tveito OE, Engeland K, Benestad R (2005) Climate change impacts on water balance in Norway. Research Report No 1/2005 Climate Norwegian Meteorological Institute Oslo

Fenger J, Buch E, Jacobsen PR (2001) Monitoring and impacts of sea level rise at Danish coasts and near shore infrastructures. In: Jørgensen AM, Fenger J, Halsnæs K (eds) Climate change research – Danish Contributions. Danish Climate Centre Copenhagen, pp. 237–254

Ferro CAT (2004) Attributing variation in a regional climate change modelling experiment. PRUDENCE working note available at http://prudence.dmi.dk/public/publications/analysis_of_variance.pdf

Ferro CAT, Hannachi A, Stephenson DB (2005) Simple techniques for describing changes in probability distributions of weather and climate. J Clim 18:4344–4354

Flato GM, Boer GJ (2001) Warming asymmetry in climate change experiments. Geophys Res Lett 28:195–198

Flato GM, Boer GJ, Lee WG, McFarlane NA, Ramsden D, Reader MC, Weaver AJ (2000) The Canadian centre for climate modelling and analysis global coupled model and its climate. Clim Dyn 16:451–467

Forster PM, De F, Blackburn M, Glover R, Shine KP (2000) An examination of climate sensitivity for idealised climate change experiments in an intermediate general circulation model. Clim Dyn 16:833–849

Fossdal ML, Sælthun NR (1993) Energy planning models – climate change. Report from a Nordic expert meeting. NVE-report 08/1993 Norwegian Water Resources and Energy Administration, Oslo

Frei C, Christensen JH, Déqué M, Jacob D, Jones RG, Vidale PL (2003) Daily precipitation statistics in regional climate models: Evaluation and intercomparison for the European Alps. J Geophys Res 108,D3, 4124 doi: 101029/2002JD002287

Gardelin M, Andreasson J, Carlsson B, Lindström G, Bergström S (2002a) Modelling av effekter av klimatförändringar på tillrinningen till vattenkraftsystemet (Modelling of effects of climate change on inflow to the hydropower system). Elforsk Report 02:27 Elforsk Stockholm (in Swedish)

Gardelin M, Bergström S, Carlsson B, Graham LP, Lindström G (2002b) Climate change and water resources in Sweden – analysis of uncertainties. In: Beniston M (ed) Climatic Change: Implications for the Hydrological Cycle and for Water Management. Advances in Global Change Research. Kluwer, Dordrecht

Gellens D, Roulin E (1998) Streamflow response of Belgian catchments to IPCC climate change scenarios. J Hydrol 210:242–258

Gillett NP, Zwiers FW, Weaver AJ, Stott PA (2003) Detection of human influence on sea-level pressure. Nature 422:292–294

Giorgi F, Marinucci M (1991) Validation of a regional atmospheric model over Europe: Sensitivity of wintertime and summertime simulations to selected physics parameterizations and lower boundary conditions. Q J Roy Met Soc 117:1171–1206

Giorgi F, Mearns LO (2002) Calculation of average uncertainty range and reliability of regional climate changes from AOGCM simulations via the "reliability ensemble averaging" (REA) method. J Clim 15:1141–1158

Giorgi F, Mearns LO (2003) Probability of regional climate change based on the Reliability Ensemble Averaging (REA) method. Geophys Res Lett 30 12 1629 (doi:101029/2003GL017130)

Giorgi F, Marinucci M, Visconti G (1990) Use of a limited area model nested in a general circulation model for regional climate simulation over Europe. J Geophys Res 95:18413–18431

Giorgi F, Marinucci M, Visconti G (1992) A 2 × CO$_2$ climate change scenario over Europe generated using a limited area model nested in a general circulation model 2 Climate change scenario. J Geophys Res 97:10011–10028

Giorgi F, Hewitson B, Christensen J, Hulme M, von Storch H, Whetton P, Jones R, Mearns L, Fu C (2001) Regional climate information – evaluation and projections. In: IPCC Climate Change 2001: The Scientific Basis Contribution of Working Group I to the Third Assessment Report of the Intergovernmental Panel on Climate Change. Cambridge University Press, Cambridge New York

Giorgi F, Bi XQ, Pal J (2004a) Mean interannual variability and trends in a regional climate change experiment over Europe I: Present-day climate (1961–1990). Clim Dyn 22:733–756

Giorgi F, Bi XQ, Pal J (2004b) Mean interannual variability and trends in a regional climate change experiment over Europe II: Climate change scenarios (2071–2100). Clim Dyn 23:839–858

Golitsyn GS, Efimova LK, Mokhov II (2002) Izmenenija temperatury i osadkov v bassejne Ladozhskogo ozera po raschetam klimaticheskoj modeli obshchej cirkuljacii v XIX–XXI vekah (Changes of temperature and precipitation on the Lake Ladoga watershed in 19th – 21st centuries simulated by a general circulation model). Izvestia RGO 134,6:80–87 (in Russian)

Gordon C, Cooper C, Senior CA, Banks H, Gregory JM, Johns TC, Mitchell JFB, Wood RA (2000) The simulation of SST, sea ice extents and ocean heat transports in a version of the Hadley Centre coupled model without flux adjustments. Clim Dyn 16:147–166

Grabs WE, Daamen K, Gellens D, Grabs W, Kwadijk JCJ, Lang H, Middlekoop H, Parmet BWAH, Schädler B, Schulla J, Wilke K (1997) Impact of Climate Change on Hydrological Regimes and Water Resources Management in the Rhine Basin. CHR-Report no I–16, International Commision for the Hydrology of the Rhine Basin (CHR) Lelystad

Graham LP (1999a) Modeling runoff to the Baltic Sea. Ambio 28:328–334

Graham, LP (2004) Climate change effects on river flow to the Baltic Sea. Ambio 33:235–241

Graham LP (1999b) Modeling the large-scale hydrologic response to climate change in the Baltic Basin. In: Elíasson J (ed) Proceedings from the Northern Research Basins 12th International Symposium and Workshop, Reykjavík Iceland 23–27 August, pp. 99–110

Graham LP (2002) A simple runoff routing routine for the Rossby Centre Regional Climate Model. In: Killingtveit A (ed) Proceedings from XXII Nordic Hydrological Conference, Røros Norway 4–7 August, Nordic Hydrological Programme Report 47:573–580

Graham LP, Rummukainen M, Gardelin M, Bergström S (2001) Modelling Climate Change Impacts on Water Resources in the Swedish Regional Climate Modelling Programme. In: Brunet M, López D (eds) Detecting and Modelling Regional Climate Change and Associated Impacts. Springer, Berlin Heidelberg New York, pp. 567–580

Graham LP, Andréasson J, Carlsson B (2007a) Assessing climate change impacts on hydrology from an ensemble of regional climate models, model scales and linking methods – a case study on the Lule River Basin. Climatic Change 81:293–307

Graham LP, Hagemann S, Jaun S, Beniston M (2007b) On interpreting hydrological change from regional climate models. Climatic Change 81:97–122

Gregory JM, Mitchell JFB, Brady AJ (1997) Summer drought in northern midlatitudes in a time-dependent CO_2 climate experiment. J Clim 10:662–686

Grigoryev AS, Trapeznikov JA (2002) The water level in the Ladoga Lake in the conditions of potential climate change. Water Resource 29:174–178 (in Russian)

Gustafsson BG (1997) Interaction between Baltic Sea and North Sea. Dt Hydrogr Z 49:163–181

Gustafsson BG (2000) Time-dependent modeling of the Baltic entrance area. 2. Water and salt exchange of the Baltic Sea. Estuaries 23:253–266

Gustafsson BG (2004) Sensitivity of Baltic Sea salinity to large perturbations in climate. Clim Res 27:237–251

Gutry-Korycka M (1999) Ekstremalne stany systemu hydrologicznego w perspektywie ocieplenia klimatu (fakty czy hipotezy). In: Komitet Narodowy PAN IGBP Global Change 1999 Zmiany i zmienność klimatu Polski – ich wpływ na gospodarkę ekosystemy i człowieka (Changes and variability of Poland's climate – their influence on economy, ecosystems and people). Łódź Conference proceedings (in Polish)

Haapala J, Leppäranta M (1997) The Baltic Sea ice season in changing climate. Boreal Env Res 2: 93–108

Haapala J, Meier HEM, Rinne J (2001) Numerical investigations of future ice conditions in the Baltic Sea. Ambio 30:237–244

Hagemann S, Dümenil L (1999) Application of a global discharge model to atmospheric model simulations in the BALTEX region. Nordic Hydrology 30:209–230

Hagemann S, Jacob D (2007) Gradient in the climate change signal of European discharge predicted by a multi-model ensemble. Climatic Change 81:309–327

Hagemann S, Machenhauer B, Christensen OB, Déqué M, Jacob D, Jones R, Vidale PL (2002) Intercomparison of water and energy budgets simulated by regional climate models applied over Europe. Max-Planck-Institute for Meteorology Rep 338. Hamburg, Germany

Hagemann S, Machenhauer B, Jones R, Christensen OB, Déqué M, Jacob D, Vidale PL (2004) Evaluation of water and energy budgets in regional climate models applied over Europe. Clim Dyn 23:547–567

Hall MM, Bryden HL (1982) Direct estimates and mechanisms of ocean heat transport. Deep-Sea Res 29:339–359

Hamlet AF, Lettenmaier D (1999) Effects of climate change on hydrology and water resources in the Columbia River Basin. J Am Water Resour Assoc 35:1597–1623

Hanssen-Bauer I, Tveito OE, Førland EJ (2000) Temperature scenarios for Norway: Empirical Downscaling from the ECHAM4/OPYC3 GSDIO integration. DNMI Report no 24/00 KLIMA Norwegian Meteorological Institure Oslo Norway

Hanssen-Bauer I, Tveito OE, Førland EJ (2001) Precipitation scenarios for Norway Empirical downscaling from ECHAM4/OPYC3 DNMI Report no 10/01 KLIMA Norwegian Meteorological Institute Oslo Norway

Hanssen-Bauer I, Førland E, Haugen JE, Tveito OE (2003) Temperature and precipitation scenarios for Norway: Comparison of results from dynamical and empirical downscaling. Clim Res 25:15–27

Hanssen-Bauer I, Achberger C, Benestad R, Chen D, Førland E (2005) Empirical–statistical downscaling of climate scenarios over Scandinavia: A review. Clim Res 29:255–268

Harvey LDD (2004) Characterizing the annual-mean climatic effect of anthropogenic CO_2 and aerosol emissions in eight coupled atmosphere–ocean GCMs. Clim Dyn 23:569–599

Hay LE, Wilby RL, Leavesley GH (2000) A comparison of delta change and downscaled GCM scenarios for three mountainous basins in the United States. J Am Water Res Ass 36:387–398

Hegerl GC, Zwiers FV, Stott P, Kharin VV (2004) Detectability of anthropogenic changes in annual temperature and precipitation extremes. J Clim 17:3683–3700

Heino R, Kitaev L (2003) INTAS project (2002–2005): Snow cover changes over Northern Eurasia during the last century: Circulation consideration and hydrological consequences (SCCONE). BALTEX Newsletter 5: 8–9

Hellström C, Chen D, Achberger C, Räisänen J (2001) A Comparison of climate change scenarios for Sweden based on statistical and dynamical downscaling of monthly precipitation. Clim Res 19:45–55

Henderson-Sellers A, Hansen AM (1995) Climate Change Atlas. Atmospheric and Oceanographic Sciences Library 17. Kluwer

Hennessy KJ, Gregory JM, Mitchell JFB (1997) Changes in daily precipitation under enhanced greenhouse conditions. Clim Dyn 13:667–680

Heyen H, Zorita E, von Storch H (1996) Statistical downscaling of monthly mean North Atlantic air-pressure to sea level anomalies in the Baltic Sea. Tellus 48A:312–323

Hirst A, O'Farrell SP, Gordon HP (2000) Comparison of a coupled ocean–atmosphere model with and without oceanic eddy-induced advection Part I: Ocean spinup and control integrations. J Clim 13: 139–163

Hoerling MP, Hurrell JW, Xu T, Bates GT, Phillips A (2004) Twentieth century North Atlantic climate change Part II: Understanding the effect of Indian Ocean warming. Clim Dyn 23:391–405

Houghton JT, Callendar BA, Varney SK (eds) (1992) Climate Change 1992 – The Supplementary Report to the IPCC Scientific Assessment. Intergovernmental Panel on Climate Change. Cambridge University Press

Huffman GJ, Adler RF, Arkin A, Chang A, Ferraro R, Gruber A, Janowiak J, Joyce RJ, Mc Nab A, Rudolf B, Schneider U, Xie P (1997) The Global Precipitation Climatology Project (GPCP) combined precipitation data set. Bull Am Met Soc 78:5–20

Hulme M, Barrow EM, Arnell NW, Harrison PA, Johns TC, Downing TE (1999) Relative impacts of human-induced climate change and natural variability. Nature 397:688–691

Huntingford C, Cox PM (2000) An analogue model to derive additional climate change scenarios from existing GCM simulations. Clim Dyn 16:575–586

Hurrell JW, Van Loon H (1997) Decadal variations in climate associated with the North Atlantic Oscillation. Climatic Change 36:301–326

Hurrell JW, Kushnir Y, Ottersen G, Visbeck M (eds) (2003) The North Atlantic Oscillation: Climate Significance and Environmental Impact. Geophys Monogr Series 134

Hurrell JW, Hoerling MP, Phillips A, Xu T (2004) Twentieth century North Atlantic climate change Part I: Assessing determinism. Clim Dyn 23:371–389

Huybrechts P, De Wolde J (1999) The dynamic response of the Greenland and Antarctic ice sheets to multiple-century climatic warming. J Clim 12:2169–2188

ILMAVA Project (2002) Effect of Climate Change on Energy Resources in Finland Final report on the Ilmava Project within the Climtech Programme. Tammelin B, Forsius J, Jylhä J, Järvinen P, Koskela J, Tuomenvirta H, Turunen MA, Vehviläinen B, Venäläinen A, Finnish Meteorological Institute Helsinki (in Finnish)

IPCC (2001a) Climate Change 2001: The scientific basis contribution of Working Group I to the Third Assessment Report of the Intergovernmental Panel on Climate Change. Cambridge University Press, Cambridge New York

IPCC (2001b) Climate Change 2001: Impacts, adaptation and vulnerability. Contribution of Working Group II to the Third Assessment Report of the Intergovernmental Panel on Climate Change. Cambridge University Press, Cambridge New York

Jacob D, Van Den Hurk B, Andrae U, Elgered G, Fortelius C, Graham LP, Jackson S, Karstens U, Köpken C, Lindau R, Podzun R, Rockel B, Rubel F, Sass B, Smith R, Yang X (2001) A comprehensive model inter-comparison study investigating the water budget during the BALTEX-PIDCAP period. Meteorol Atmos Phys 77:19–43

Jansons V, Butina M (1998) Potential impacts of climate change on nutrient loads from small catchments. In: Lemmelä R, Helenius N (eds) Proceedings of the Second International Conference on Climate and Water, Espoo Finland 17–20 August 1998, 2

Järvet A (1998) An assessment of the climate change impact on groundwater regime. In: Kallaste T, Kuldna P (eds) Climate Change Studies in Estonia, Ministry of Environment Republic of Estonia, SEI Tallinn

Johansson MM, Kahma KK, Boman H, Launiainen J (2004) Scenarios for sea level on the Finnish coast. Boreal Env Res 9:153–166

Johns TC, Carnell RE, Crossley JF, Gregory JM, Mitchell JFB, Senior CA, Tett SFB, Wood RA (1997) The second Hadley Centre Coupled ocean–atmosphere GCM: Model description spinup and validation. Clim Dy 13:103–134

Jones CG, Ullerstig A (2002) The representation of precipitation in the RCA2 model (Rossby Centre Atmosphere Model Version 2). SWECLIM Newsletter 12:27–39

Jones CG, Willén U, Ullerstig A, Hansson U (2004) The Rossby Centre Regional Atmospheric Climate Model – Part I: Model climatology and performance for the present climate over Europe. Ambio 33:199–210

Jones RG, Murphy JM, Noguer M (1995) Simulation of climate-change over Europe using a nested regional climate model, 1. Assessment of control climate including sensitivity to location of lateral boundaries. Q J Roy Met Soc 121:1413–1449

Jones RG, Murphy JM, Noguer M, Keen AB (1997) Simulation of climate change over Europe using a nested regional climate model, 2. Comparison of driving and regional model responses to a doubling of carbon dioxide. Q J Roy Met Soc 123:265–292

Joshi M, Shine K, Ponater M, Stuber N, Sausen R, Li L (2003) A comparison of climate response to different radiative forcings in three general circulation models: Towards an improved metric of climate change. Clim Dyn 20:843–854

Jylhä K, Tuomenvirta H, Ruosteenoja K (2004) Climate change projections for Finland during the 21st century. Boreal Env Res 9:127–152

Jylhä K, Fronzek S, Tuomenvirta H, Carter TR, Ruosteenoja K (2007) Changes in frost and snow in Europe and Baltic sea ice by the end of the 21st century. Climatic Change (in press)

Järvet A, Jaagus J, Roosaare J, Tamm T, Vallner L (2000) Impact of climate change on water balance elements in Estonia. Estonia Geographical Studies 8:35–55

Kaas E, Frich P (1995) Diurnal temperature range and cloud cover in the Nordic countries: Observed trends and estimates for the future. Atmos Res 37:211–228

Kaczmarek Z (1993) Water balance model for climate impact analysis. Acta Geophys Pol 41:423–437

Kaczmarek Z (1996) Wpływ klimatu na bilans wody (Impact of climate on water balance). In: Kaczmarek Z (ed) Wpływ niestacjonarności i globalnych procesów geofizycznych na zasoby wodne Polski Oficyna (The impact of nonlinearity and global geophysical processes on water resources in Poland). Wydawnicza Politechniki Warszawskiej Warszawa, 33–53 (in Polish)

Kaczmarek Z (2003) Wpływ klimatu na gospodarkę wodną (Impact of climate on water management). In: Komitet Prognoz "Polska 2000 Plus" (ed) Czy Polsce grożą katastrofy klimatyczne? IGBP PAN Warszawa, 32–52 (in Polish)

Kaczmarek Z (2004) Climate change and European water resources. In: Liszewska M (ed) Potential climate changes and sustainable water management. Publ Inst Geophys Pol Acad Sc 377:33–38

Kaczmarek Z, Jurak D (2003) Assessment and prediction of hydrological droughts. Glob Change 10: 79–95

Kaczmarek Z, Napiórkowski J, Strzepek KM (1996) Climate change impacts on the water supply system in the Warta River Catchment Poland. Int J Water Resour Dev 12:165–180

Kaczmarek Z, Napiórkowski J, Jurak D (1997) Impact of climate change on water resources in Poland. Publ Inst Geophys Pol Acad Sc E-1

Kalinin M (2004) Climate and Water Resources of Belarus. In: Isemer HJ (ed) Conference Proceedings of the Fourth Study Conference on BALTEX, International BALTEX Secretariat Publication No 29

Kallaste T, Kuldna P (1998) Climate Change Studies in Estonia. Ministry of Environment Republic of Estonia SEI Tallinn

Kalnay E, Kanamitsu M, Kistler R, Collins W, Deaven D, Gandin L, Iredell M, Saha S, White G, Woollen J, Chelliah M, Zhu Y, Ebisuzaki W, Higgins W, Janowiak J, Mo KC, Ropelweski C, Wand J, Leetma A, Reynolds R, Jenne R, Joseph D (1996) The NCEP/NCAR 40 reanalysis project. Bull Am Met Soc 77:437–471

Kauker F, Meier HEM (2003) Modeling decadal variability of the Baltic Sea Part 1: Reconstructing atmospheric surface data for the period 1902–1998. J Geophys Res 108(C8) 3267 doi:101029/2003JC001797

Keevallik S (1998) Climate change scenarios for Estonia. In: Tarand A, Kallaste T (eds) Country case study on climate change impacts and adaptation assessments in the Republic of Estonia. Ministry of Environment Republic of Estonia SEI Tallinn

Kilsby CG, O'Connell PE, Fallows CS, Hashemi AM (1999) Generation of precipitation scenarios for assessing climate change impacts on river basin hydrology. In: Balabanis P, Bronstert A, Casale R, Samuels P (eds) Proceedings from Ribamod – River Basin Modelling Management and Flood Mitigation Concerted Action – Final Workshop Wallingford United Kingdom 26–27 February 1998. Office for Official Publications of the European Communities

Kistler R, Kalnay E, Collins W, Saha S, White G, Woollen J, Chelliah M, Ebisuzaki W, Kanamitsu M, Kousky V, Van Den Dool H, Jenne R, Fiorino M (2001) The NCEP/NCAR 50-year reanalysis: Monthly means CD-ROM and documentation. Bull Am Met Soc 82:247–268

Kjellström E (2004) Recent and future signatures of climate change in Europe. Ambio 33:193–198

Kjellström E, Ruosteenoja K (2007) Present-day and future precipitation in the Baltic Sea region as simulated in a suite of regional climate models. Climatic Change 81:281–291

Kjellström E, Döscher R, Meier HEM (2005) Atmospheric response to different sea surface temperatures in the Baltic Sea: Coupled versus uncoupled regional climate model experiments. Nord Hydrol 36:397–409

Kjellström E, Bärring L, Jacob D, Jones R, Lenderink G, Schär C (2007) Modelling daily temperature extremes: Recent climate and future changes over Europe. Climatic Change 81:249–265

Knutson TR, Delworth TL, Dixon KW, Stouffer RJ (1999) Model assessment of regional surface temperature trends (1949–1997). J Geophys Res 104:30981–30996

Kondratyev S (2001) Final Report of the Institute of Limnology RAS to the Project "The Impact of long-term changes in the weather on the dynamics of lakes in the United Kingdom Finland and Russia", No 96–1749

Kondratyev S, Bovykin I (2000) Hydrologic response of small lake and its drainage basin to precipitation and air temperature changes. Proc of Univ of Joensuu Karelian Inst 129:423–427

Kondratyev S, Bovykin V (2003) The effect of possible climatic changes on the hydrological regime of a catchment – lake system (in Russian). Sov Meteorol Hydrol 10:86–96

Kondratyev S, Gronskaya T, Wirkkala RS, Bovykin I, Yefremova L, Ignatieva N, Raspletina G, Chernykh O, Gayenko M, Markova E, Aksenchuk I (1998) Lake Ladoga and its drainage basin: GIS development and application. Univ of Joensuu Karelian Inst Working Papers 5:109–118

Kont A, Jaagus J, Aunap R (2003) Climate change scenarios and the effect of sea-level rise for Estonia. Glob Planet Change 36:1–15

Latif M, Roeckner E, Mikolajewicz U, Voss R (2000) Tropical stabilisation of the thermohaline circulation in a greenhouse warming simulation. J Clim 13:1809–1813

Leckebusch G, Ulbrich U (2004) On the relationship between cyclones and extreme windstorm events over Europe under climate change. Glob Planet Change 44:181–193

Legates DR, Wilmott CJ (1990) Mean seasonal and spatial variability in gauge-corrected global precipitation. Int J Climatol 10:111–127

Lehner B, Henrichs T, Döll P, Alcamo J (2001) EuroWasser – Model-based assessment of European water resources and hydrology in the face of global change. Kassel World Water Series 5. Center for Environmental Systems Research, University of Kassel, Germany

Lenderink G, Buishand A, Van Deursen W (2007) Estimates of future discharges of the river Rhine using two scenario methodologies: Direct versus delta approach. Hydrol Earth Sys Sci 33:1145–1159

Lettenmaier DP, Wood AW, Palmer RN, Wood EF, Stakhiv EZ (1999) Water resources implications of global warming: A US regional perspective. Climatic Change 43:537–579

Linderson ML, Achberger C, Chen D (2004) Statistical downscaling and scenario construction of precipitation in Scania southern Sweden. Nord Hydrol 35:261–278

Lindström G, Gardelin M, Persson M (1994) Conceptual Modelling of Evapotranspiration for Simulations of Climate Change Effects. SMHI Reports Hydrology No 10, Swedish Meteorological and Hydrological Institute, Norrköping, Sweden

Lohmann D, Nolte-Holube R, Raschke E (1996) A large-scale horizontal routing model to be coupled to land surface parameterization schemes. Tellus 48A:708–721

Manabe S, Stouffer RJ, Spelman MJ, Bryan K (1991) Transient responses of a coupled ocean–atmosphere model to gradual changes of atmospheric CO_2 Part I: Annual mean response. J Clim 4: 785–818

McAvaney BJ, Covey C, Joussaume S, Kattsov V, Kitoh A, Ogana W, Pittman AJ, Weaver AJ, Wood RA, Zhao ZC (2001) Model evaluation. In: IPCC: Climate change 2001: The scientific basis. Contribution of Working Group I to the Third Assessment Report of the Intergovernmental Panel on Climate Change. Cambridge University Press, Cambridge New York

Meehl GA, Boer GJ, Covey C, Latif M, Stouffer RJ (2000) The Coupled Model Intercomparison Project (CMIP). Bull Am Met Soc 81:313–318

Meier HEM (2001) The first Rossby Centre regional climate scenario for the Baltic Sea using a 3D coupled ice–ocean model. Reports Meteorology and Climatology No95, SMHI Norrköping Sweden

Meier HEM (2002a) Regional ocean climate simulations with a 3D ice-ocean model for the Baltic Sea Part 1: Model experiments and results for temperature and salinity. Clim Dyn 19:237–253

Meier HEM (2002b) Regional ocean climate simulations with a 3D ice-ocean model for the Baltic Sea Part 2: Results for sea ice. Clim Dyn 19:255–266

Meier HEM (2005) Modeling the age of Baltic Sea water masses: Quantification and steady state sensitivity experiments. J Geophys Res 110 C02006 doi:101029/2004JC002607

Meier HEM (2006) Baltic Sea climate in the late twenty-first century: A dynamical downscaling approach using two global models and two emissions scenarios. Clim Dyn 27:39–68 DOI 101007/s00382–006–0124–x

Meier HEM, Kauker F (2003a) Modeling decadal variability of the Baltic Sea Part 2: Role of freshwater inflow and large-scale atmospheric circulation for salinity. J Geophys Res 108,C11, 3368 doi:101029/2003JC001799

Meier HEM, Kauker F (2003b) Sensitivity of the Baltic Sea salinity to the freshwater supply. Clim Res 24:231–242

Meier HEM, Döscher R, Halkka A (2004a) Simulated distributions of Baltic sea-ice in warming climate and consequences for the winter habitat of the Baltic ringed seal. Ambio 33:249–256

Meier HEM, Broman B, Kjellström E (2004b) Simulated sea level in past and future climates of the Baltic Sea. Clim Res 27:59–75

Meier HEM, Broman B, Kallio H, Kjellström E (2006a) Projections of future surface winds sea levels and wind waves in the late 21st century and their application for impact studies of flood prone areas in the Baltic Sea region. In: Schmidt-Thome P (ed) Special Paper 41, Geological Survey of Finland Helsinki Finland

Meier HEM, Kjellström E, Graham LP (2006b) Estimating uncertainties of projected Baltic Sea salinity in the late 21st century. Geophys Res Lett 33, L15705 doi:101029/2006GL026488

Meleshko VP, Kattsov VM, Govorkova VA, Malevsky-Malevich SP, Nadyozhina ED, Sporyshev PV (2004) Anthropogenic climate change in XXI century in Northern Eurasia. Sov Meteorol Hydrol 7: 5–26 (in Russian)

Middelkoop H, Daamen K, Gellens D, Grabs W, Kwadijk JCJ, Lang H, Parmet BWAH, Schädler B, Schulla J, Wilke K (2001) Impact of climate change on hydrological regimes and water resources management in the Rhine Basin. Clim Change 49:105–128

Miętus M (1999) Rola regionalnej cyrkulacji atmosferycznej w kształtowaniu warunków klimatycznych
i oceanograficznych w polskiej strefie brzegowej Morza Bałtyckiego (The influence of regional atmospheric circulation on climate and oceanographic conditions in the Polish coastal zone). Instytut
Meteorologii i Gospodarki Wodnej Warszawa (in Polish with English Summary)

Miętus M (2000) Climatic and oceanographic conditions in the southern Baltic area under an increasing CO_2 concentration. Geogr Polon 73:89–97

Miętus M, Filipiak J, Owczarek M (2004) Klimat wybrzeża południowego Bałtyku Stan obecny i perspektywy zmian (Climate at the seashore of the southern Baltic Sea. Present conditions and future
changes). In: Cyberski J (ed) Środowisko polskiej strefy południowego Bałtyku – stan obecny i
przewidywane zmiany w przededniu integracji europejskiej. Wydawnictwo Gdańskie Gdańsk, pp. 11–
44 (in Polish)

Ministry of Environmental Protection and Regional Development (2001) The Third National Communication of the Republic of Latvia under the United Nations Framework Convention on Climate
Change

Ministry of the Environment (2003) Lithuania's Second National Communication under the Framework Convention on Climate Change

Mitchell JFB, Manabe S, Meleshko V, Tokioka T (1990) Equilibrium climate change – and its implications for the future In: Houghton JT et al. (eds) Climate Change. The IPCC Scientific Assessment.
Cambridge University Press, Cambridge New York, pp. 131–172

Mitchell JFB, Johns TC, Eagles M, Ingram WJ, Davis RA (1999) Towards the construction of climate change scenarios. Climatic Change 41:547–581

Mitchell JFB, Karoly DJ, Hegerl GC, Zwiers FW, Allen MR, Marengo J (2001) Detection of climate
change and attribution of causes. In: IPCC Climate Change 2001: The Scientific Basis Contribution
of Working Group I to the Third Assessment Report of the Intergovernmental Panel on Climate
Change. Cambridge University Press, Cambridge New York

Mitchell TD (2003) Pattern scaling: An examination of the accuracy of the technique for describing
future climate. Clim Change 60:217–242

Moberg A, Jones P (2004) Regional climate model simulations of daily maximum and minimum near-
surface temperatures across Europe compared with observed station data 1961–1990. Clim Dyn 23:
695–715

Moberg A, Sonechkin DM, Holmgren K, Datsenko NM, Karlén W (2005) Highly variable Northern
Hemisphere temperatures reconstructed from low- and high-resolution proxy data. Nature 433:
613–617

Munich Re Group (1999) Topics 2000 Natural catastrophes – the current position. Münchener Rück-
versicherungs-Gesellschaft, pp. 66, www.munichre.com

Murphy J (1999) An evaluation of statistical and dynamical techniques for downscaling local climate.
J Clim 12:2256–2284

Murphy J (2000) Predictions of climate change over Europe using statistical and dynamical downscaling techniques. Int J Climatol 20:489–501

Murphy JM, Mitchell JFB (1994) Transient response of the Hadley Centre coupled ocean–atmosphere
model to increasing carbon dioxide Part I Control climate and flux correction. J Clim 8:36–47

Murphy JM, Sexton DMH, Barnett DN, Jones GS, Webb MJ, Collins M, Stainforth DA (2004) Quantification of modelling uncertainties in a large ensemble of climate change simulations. Nature
430:768–772

Nakićenović N, Alcamo J, Davis G, De Vries B, Fenhann J, Gaffin S, Gregory K, Grübler A, Jung
TY, Kram T, La Rovere EL, Michaelis L, Mori S, Morita T, Pepper W, Pitcher H, Price L, Riahi
K, Roehrl A, Rogner HH, Sankovski A, Schlesinger M, Shukla P, Smith S, Swart R, Van Rooi-
jen S, Victor N, Dadi Z (2000) Emission scenarios A Special Report of Working Group III of the
Intergovernmental Panel on Climate Change. Cambridge University Press, Cambridge New York

New M, Hulme M, Jones P (1999) Representing twentieth-century space-time climate variability,
Part I: Development of a 1961–90 mean monthly terrestrial climatology. J Clim 12:829–856

Nozawa T, Emori S, Takemura T, Nakajima T, Numaguti A, Abe-Ouchi A, Kimoto M (2000) Coupled ocean–atmosphere model experiments of future climate change based on IPCC SRES scenarios. Preprints 11th Symposium on Global Change Studies. 9–14 January 2000 Long Beach USA

Numaguti A (1999) Origin and recycling processes of precipitating water over the Eurasian continent: Experiments using an atmospheric general circulation model. J Geophys Res 104:1957–1972

Omstedt A, Chen D (2001) Influence of atmospheric circulation on the maximum ice extent in the Baltic Sea. J Geophys Res 106,C3:4493–4500

Omstedt A, Nyberg L (1996) Response of Baltic Sea ice to seasonal interannual forcing and to climate change. Tellus 48A:644–662

Omstedt A, Gustafsson B, Rodhe J, Walin G (2000) Use of Baltic Sea modelling to investigate the water cycle and the heat balance in GCM and regional climate models. Clim Res 15:95–108

Omstedt A, Meuller L, Nyberg L (1997) Interannual seasonal and regional variations of precipitation and evaporation over the Baltic Sea. Ambio 26:484–492

Orviku K, Jaagus J, Kont A, Ratas U, Rivis R (2003) Increasing activity of coastal processes associated with climate change in Estonia. J Coastal Res 19:364–375

Osborn TJ (2004) Simulating the winter North Atlantic Oscillation: The roles of internal variability and greenhouse gas forcing. Clim Dyn 22:605–623

Osborn TJ, Briffa KR, Tett SFB, Jones PD, Trigo RM (1999) Evaluation of the North Atlantic Oscillation as simulated by a coupled climate model. Clim Dyn 15:685–702

Power SB, Colman RA, McAvaney BJ, Dahni RR, Moore AM, Smith NR (1993) The BMRC coupled atmosphere/ocean/sea-ice model. BMRC Research Report No 37, Bureau of Meteorology Research Centre Melbourne Australia

Prentice IC, Farquhar GD, Fasham MJR, Goulden ML, Heimann M, Jaramillo VJ, Kheshgi HS, Le Quéré C, Scholes RJ, Wallace DWR (2001) The carbon cycle and atmospheric carbon dioxide. In: IPCC Climate Change 2001: The Scientific Basis Contribution of Working Group I to the Third Assessment Report of the Intergovernmental Panel on Climate Change: Cambridge University Press, Cambridge New York

Pryor SC, Barthelmie RJ (2003) Long-term trends in near-surface flow over the Baltic. Int J Climatol 23,3:271–289

Pryor SC, Barthelmie RJ (2004) Use of RCM simulations to assess the impact of climate change on wind energy availability. Risø-R-1477(EN) Risø National Laboratory Roskilde Denmark

Pryor SC, Barthelmie RJ and JT Schoof (2005a) The Impact of non-stationarities in the climate system on the definition of a normal wind year: A case study from the Baltic. Int J Climatol 25,6:735–752

Pryor SC, Barthelmie RJ, Kjellström E (2005b) Analyses of the potential climate change impact on wind energy resources in northern Europe using output from a Regional Climate Model. Clim Dyn 25,7–8:815–835

Raab B, Vedin H (1995) Climate Lakes and Rivers. National Atlas of Sweden, vol. 14. SNA Publishing Box 45209 S–10430, Stockholm, Sweden

Rauthe M, Paeth H (2004) Relative importance of Northern Hemisphere circulation modes in predicting regional climate change. J Clim 17:4180–4189

Rimkus E (2001) Prognosis of maximum snow water equivalent changes in Lithuania. In: Meywerk J (ed) Conference Proceedings of the Third Study Conference on BALTEX, International BALTEX Secretariat Publication No 20

Rind D, Healy R, Parkinson C, Martinson D (1995) The role of sea ice in $2 \times CO_2$ climate sensitivity Part I: The total influence of sea ice thickness and extent. J Clim 8:449–463

Roald LA, Beldring S, Væringstad T, Engeset R, Skaugen TE, Førland E (2002) Scenarios of annual and seasonal runoff for Norway based on climate scenarios for 2030–2049. Norwegian Water Resources and Energy Directorate Consultancy Report A 10, Norwegian Meteorological Institute Report no 19/02 KLIMA

Roald LA, Beldring S, Skaugen TE, Førland EJ, Benestad R (2006) Climate change impacts in streamflow in Norway. NVE Consultancy-report A 1-2006 Norwegian Water Resources and Energy Directorate Oslo

Rockel B, Woth K (2007) Extremes of near-surface wind speed over Europe and their future changes as estimated from an ensemble of RCM simulations. Climatic Change 81:267–280

Rodhe J, Winsor P (2002) On the influence of the freshwater supply on the Baltic Sea mean salinity. Tellus 54A:175–186

Rodhe J, Winsor P (2003) Corrigendum: On the influence of the freshwater supply on the Baltic Sea mean salinity. Tellus 55A:455–456

Roeckner E, Bengtsson L, Feichter J, Lelieveld J, Rodhe H (1999) Transient climate change simulations with a coupled atmosphere–ocean GCM including the tropospheric sulfur cycle. J Clim 12: 3004–3032

Roosaare J (1998) Local-scale spatial interpretation of climate change impact on river runoff in Estonia. In: Lemmelä R, Helenius N (eds) Proceedings of the Second International Conference on Climate and Water Espoo Finland 17–20 August 1998, 1

Rubel F, Hantel M (2001) BALTEX 1/6-degree daily precipitation climatology 1996–1998. Meteorol Atmos Phys 77:155–166

Rummukainen M, Räisänen J, Ullerstig A, Bringfelt B, Hansson U, Graham LP, Willén U (1998) RCA – Rossby Centre regional atmospheric climate model: Model description and results from the first multi-year simulation. Reports Meteorology and Climatology 83, Swedish Meteorological and Hydrological Institute, Norrköping Sweden

Rummukainen M, Bergström S, Källén E, Moen L, Rodhe J, Tjernström M (2000) SWECLIM: The first three years. Reports Meteorology and Climatology 94, Swedish Meteorological and Hydrological Institute Norrköping Sweden

Rummukainen M, Räisänen J, Bringfelt B, Ullerstig A, Omstedt A, Willén U, Hansson U, Jones C (2001) A regional climate model for northern Europe: Model description and results from the downscaling of two GCM control simulations. Clim Dyn 17:339–359

Rummukainen M, Räisänen J, Bjørge D, Christensen JH, Christensen OB, Iversen T, Jylhä K, Ólafsson H, Tuomenvirta H (2003) Regional climate scenarios for use in Nordic water resources studies. Nord Hydrol 34:399–412

Rummukainen M, Bergström S, Persson G, Rodhe J, Tjernström M (2004) The Swedish Regional Climate Modelling Programme SWECLIM: A review. Ambio 4–5:176–182

Ruosteenoja K, Tuomenvirta H, Jylhä K (2007) GCM-based regional temperature and precipitation change estimates for Europe under four SRES scenarios applying a super-ensemble pattern-scaling method. Climatic Change 81:193–208

Russell GL, Rind D (1999) Response to CO_2 transient increase in the GISS coupled model Regional coolings in a warmer climate. J Clim 12:531–539

Russell GL, Miller JR, Rind D, Ruedy RA, Schmidt G, Sheth S (2000) Comparison of model and observed regional temperature changes during the past 40 years. J Geophys Res 105:14891–14898

Räisänen J (1994) A comparison of the results of seven GCM experiments in northern Europe. Geophysica 11:3–30

Räisänen J (2000) CO_2-induced climate change in northern Europe: Comparison of 12 CMIP2 experiments. Reports Meteorology and Climatology 87, Swedish Meteorological and Hydrological Institute, Norrköping Sweden

Räisänen J (2001a) CO_2-induced climate change in CMIP2 experiments Quantification of agreement and role of internal variability. J Clim 14:2088–2104

Räisänen J (2001b) Hiilidioksidin lisääntymisen vaikutus Pohjois-Euroopan ilmastoon globaaleissa ilmastomalleissa (The impact of increasing carbon dioxide on the climate of northern Europe in global climate models). Terra 113:139–151 (in Finnish with English abstract, figure and table captions)

Räisänen J (2002) CO_2-induced changes in interannual temperature and precipitation variability in 19 CMIP2 experiments. J Clim 15:2395–2411

Räisänen J, Alexandersson H (2003) A probabilistic view on recent and near future climate change in Sweden. Tellus 55A:113–125

Räisänen J, Joelsson R (2001) Changes in average and extreme precipitation in two regional climate model experiments. Tellus 53A:547–566

Räisänen J, Döscher R (1999) Simulation of present-day climate in Northern Europe in the HadCM2 OAGCM. Reports Meteorology and Climatology 84, Swedish Meteorological and Hydrological Institute, Norrköping Sweden

Räisänen J, Rummukainen M, Ullerstig A, Bringfelt B, Hansson U, Willén U (1999) The First Rossby Centre Regional Climate Scenario: Dynamical Downscaling of CO_2-induced Climate Change in the HadCM2 GCM. Reports Meteorology and Climatology 85, Swedish Meteorological and Hydrological Institute, Norrköping Sweden

Räisänen J, Rummukainen M, Ullerstig A (2001) Downscaling of greenhouse gas induced climate change in two GCMs with the Rossby Centre regional climate model for northern Europe. Tellus 53A:168–191

Räisänen J, Hansson U, Ullerstig A, Döscher R, Graham LP, Jones C, Meier M, Samuelsson P, Willén U (2003) GCM driven simulations of recent and future climate with the Rossby Centre coupled atmosphere – Baltic Sea regional climate model RCAO SMHI Reports Meteorology and Climatology 101, SMHI SE 60176 Norrköping, Sweden

Räisänen J, Hansson U, Ullerstig A, Döscher R, Graham LP, Jones C, Meier M, Samuelsson P, Willén U (2004) European climate in the late 21st century: Regional simulations with two driving global models and two forcing scenarios. Clim Dyn 22:13–31

Sælthun NR (1996) The "Nordic" HBV model – version developed for the project Climate change and Energy Production. NVE publication no 7/1996 Norwegian Water Resources and Energy Administration Oslo

Sælthun NR, Bogen J, Hartman Flood M, Laumann T, Roald LA, Tvede AM, Wold B (1990) Climate change impacts on Norwegian water resources. NVE publication V42 Norwegian Water Resources and Energy Administration Oslo

Sælthun NR, Aittoniemi P, Bergström S, Einarsson K, Jóhannesson T, Lindström G, Ohlsson PE, Thomsen T, Vehviläinen B, Aamodt KO (1998) Climate Change Impacts on Runoff and Hydropower in the Nordic Countries. TemaNord 1998:522 Nordic Council of Ministers Copenhagen

Sælthun NR, Bergström S, Einarsson K, Jóhannesson T, Lindström G, Thomsen T, Vehviläinen B (1999) Potential Impacts of climate change on floods in Nordic hydrological regimes. In: Balabanis P, Bronstert A, Casale R, P Samuels (eds) Proceedings from Ribamod – River Basin Modelling Management and Flood Mitigation. Wallingford United Kingdom 26–27 February 1998. Office for Official Publications of the European Communities

Scaife AM, Knight JR, Vallis GK, Folland CK (2005) A stratospheric influence on the winter NAO and North Atlantic surface climate. Geophys Res Lett 32 L18715

Schaeffer M, Selten FM, Opsteegh JD, Goosse H (2004) The influence of ocean convection patterns on high-latitude climate projections. J Clim 17:4316–4329

Scharling M (2000) Klimagrid – Danmark normaler 1961–90 måneds- og årsværdier Nedbør 10 × 10 20 × 20 & 40 × 40 km temperatur og potentiel fordampning 20 × 20 & 40 × 40 km (Climate grid Denmark. Climate normals 1961–90, monthly and annual values. Precipitation 10 × 10 20 × 20 & 40 × 40 km. Temperature and potential evaporation 20 × 20 & 40 × 40 km). Danish Meteorological Institute Technical Report 00-11 (in Danish)

Schrum C (2001) Regionalization of climate change for the North Sea and Baltic Sea. Clim Res 18: 31–37

Schrum C, Backhaus JO (1999) Sensitivity of atmosphere–ocean heat exchange and heat content in the North Sea and the Baltic Sea. Tellus 51A:526–549

Schär C, Vidale PL, Lüthi D, Frei C, Häberli C, Liniger MA, Appenzeller C (2004) The role of increasing temperature variability for European summer heat waves. Nature 427:332–336

Selten FM, Branstator GW, Dijkstra HA, Kliphuis M (2004) Tropical origins for recent and future Northern Hemisphere climate change. Geophys Res Lett 31 L21205 (doi: 101029/2004GL020739)

Semmler T, Jacob D (2004) Modeling extreme precipitation events – a climate change simulation for Europe. Glob Planet Change 44:119–127

Shindell DT, Schmidt GA, Miller RL, Rind D (2001) Northern Hemispheric climate response to greenhouse gas ozone solar and volcanic forcing. J Geophys Res 106:7193–7210

Shindell DT, Miller RL, Smith GA, Pandolfo L (1999) Simulation of recent northern winter climate trends by greenhouse-gas forcing. Nature 399:452–455

SILMU (1996) The Finnish Research Programme on Climate Change. Final Report (Roos J ed) Edita Helsinki

Skaugen TE, Tveito OE (2002) Heating degree-days – Present conditions and scenario for the period 2021–2050. DNMI Report no 01/02 KLIMA Norwegian Meteorological Institute Oslo Norway

Skaugen TE, Astrup M, Roald LA, Førland EJ (2002) Scenarios of extreme precipitation of duration 1 and 5 days for Norway caused by climate change. Norwegian Water Resources and Energy Directorate Consultancy Report A 7

Stainforth DA, Alna T, Christensen C, Collins M, Fauli N, Frame DJ, Kettleborough JA, Knight S, Martin A, Murphy JM, Pianl C, Sexton D, Smith LA, Spicer RA, Thorpe AJ, Allen MR (2005) Uncertainty in the predictions of the climate response to rising levels of greenhouse gases. Nature 433:403–406

Stephenson DB, Pavan V (2003) The North Atlantic Oscillation in coupled climate models: A CMIP1 evaluation. Clim Dyn 20:381–399

Stephenson DB, Pavan V, Collins M, Junge MM, Quadrelli R and Participating CMIP2 Modelling Groups (2006) North Atlantic Oscillation response to transient greenhouse gas forcing and the impact on European winter climate: A CMIP2 multi-model assessment. Clim Dyn 27:401–420

Stigebrandt A (1983) A model for the exchange of water and salt between the Baltic and the Skagerrak. J Phys Oceanogr 13:411–427

Stigebrandt A, Gustafsson BG (2003) Response of the Baltic Sea to climate change – Theory and observations. J Sea Res 49:243–256

Stocker TF, Schmittner A (1997) Influence of CO_2 emission rates on the stability of the thermohaline circulation. Nature 388:862–865

Stouffer RJ, Manabe S (2003) Equilibrium response of thermohaline circulation to large changes in atmospheric CO_2 concentration. Clim Dyn 20:759–773

Strzepek KM, Yates DN (1997) Climate change impacts on the hydrologic resources of Europe: A simplified continental scale anlaysis. Climatic Change 36:79–92

Suursaar Ü, Jaagus J, Kullas T (2006) Past and future changes in sea level near the Estonian coast in relation to changes in wind climate. Boreal Env Res 11:123–142

Tarand A, Kallaste T (eds) (1998) Country Case Study on Climate Change Impacts and Adaptation Assessments in the Republic of Estonia Ministry of Environment Republic of Estonia. SEI Tallinn

Thodsen H (2007) The influence of climate change on stream flow in Danish rivers. J Hydrol 333:226–238

Thodsen H, Erichensen A, Lumborg U, Edelvang K (2005) Effekt af afstrømnings ændringer som følge af klima ændringer på salinitet og næringsstofforhold i Odense Fjord (The effect of climate change induced changes of river flow on salinity and nutrient conditions in the Odense fjord (in Danish)) Abstracts from Det 13 Danske Havforskermøde Copenhagen

Thompson SL, Pollard D (1995) A global climate model (GENESIS) with a land-surface-transfer scheme (LSX) Part 1: Present-day climate. J Clim 8:732–761

Tinz B (1996) On the relation between annual maximum extent of ice cover in the Baltic Sea level pressure as well as air temperature field. Geophysica 32:319–341

Tinz B (1998) Sea ice winter severity in the German Baltic in a greenhouse gas experiment. Deutsch Hydr Z 50:33–45

Tokioka T, Noda A, Kitoh A, Nikaidou Y, Nakagawa S, Motoi T, Yukimoto S, Takata K (1995) A transient CO_2 experiment with the MRI CGCM Quick Report. J Meteorol Soc Jpn 73:817–826

Tuomenvirta H, Heino R (1996) Climatic changes in Finland – recent findings. Geophysica 32:61–75

Tuomenvirta H, Uusitalo K, Vehviläinen B, Carter T (2000) Ilmastonmuutos mitoitussade ja patoturvallisuus: Arvio sadannan ja sen ääriarvojen sekä lämpötilan muutoksista Suomessa vuoteen 2100 (Climate change, design, precipitation and dam safety: Estimation of the changes of extreme precipitation and temperature in Finland until 2100). Ilmatieteenlaitoksen raportteja 2000:4 Helsinki (in Finnish)

Uppala SM, Kållberg PW, Simmons AJ, Andrae U, Da Costa Bechtold V, Fiorino M, Gibson JK, Haseler J, Hernandez A, Kelly GA, Li X, Onogi K, Saarinen S, Sokka N, Allan RP, Andersson E, Arpe K, Balmaseda MA, Beljaars ACM, Van De Berg L, Bidlot J, Bormann N, Caires S, Chevallier F, Dethof A, Dragosavac M, Fisher M, Fuentes M, Hagemann S, Hólm E, Hoskins BJ, Isaksen L, Janssen PAEM, Jenne R, McNally AP, Mahfouf JF, Morcrette JJ, Rayner NA, Saunders RW, Simon P, Sterl A, Trenberth KE, Untch A, Vasiljevic D, Viterbo P, Woollen J (2005) The ERA-40 Reanalysis. Q J Roy Met Soc 131:2961–3012 doi:101256/qj04176

Varis O, Kajander T, Lemmelä R (2004) Climate and water: From climate models to water resources management and vice versa. Climatic Change 66:321–344

Vehviläinen B (1994) The watershed simulation and forecasting system in the National Board of Waters and the Environment. Publications of Water and Environment Research Institute 17 Helsinki

Vehviläinen B, Huttunen M (1997) Climate change and water resources in Finland. Boreal Env Res 2: 3–18

Vidale PL, Lüthi C, Frei C, Seneviratne S, Schär C (2003) Predictability and uncertainty in a regional climate model. J Geophys Res 108(D18) 4586 doi:101029/2002JD002810

Vidale PL, Lüthi C, Wegmann R, Schär C (2007) European climate variability in a heterogeneous multi-model ensemble. Climatic Change 81:209–232

Volodin EM, Galin VY (1999) Interpretation of winter warming on Northern Hemisphere continents in 1977–94. J Clim 12:2947–2955

Voss R, Sausen R, Cubasch U (1998) Periodically synchronously coupled integrations with the atmosphere–ocean general circulation model ECHAM3/LSG. Clim Dyn 14:249–266

Washington WM, Meehl GA (1989) Climatic sensitivity due to increased CO_2: Experiments with a coupled atmosphere and ocean general circulation model. Clim Dyn 4:1–38

Washington WM, Weatherly JW, Meehl GA, Semtner AJ Jr, Bettge TW, Craig AP, Strand WG Jr, Arblaster J, Wayland VB, James VB R, Zhang Y (2000) Parallel climate model (PCM) control and transient simulations. Clim Dyn 16:755–774

Weaver AJ, Eby M, Fanning AF, Wiebe EC (1998) Simulated influence of carbon dioxide orbital forcing and ice sheets on the climate of the Last Glacial Maximum. Nature 394:847–853

Wetherald RT, Manabe S (1995) The mechanisms of summer dryness induced by greenhouse warming. J Clim 8:3096–3108

Wigley TML, Raper SCB (1987) Thermal expansion of sea water associated with global warming. Nature 330:127–131

Wigley TML, Raper SCB (1992a) Implications for climate and sea level rise of revised IPCC emissions scenarios. Nature 357:293–300

Wigley TML, Raper SCB (1992b) Implications of revised IPCC emissions scenarios. Nature 357:127–131

Wigley TML, Raper SCB (2001) Interpretation of high projections for global-mean warming. Science 293:451–455

World Bank (2002) Republic of Belarus Ministry of Natural Resources and Environmental Protection: Assessment of potential impact of climatic changes in the Republic of Belarus and vulnerability and adaptation of social and economic systems to climate change. Minsk

Xie P, Arkin PA (1997) Global precipitation: A 17-year monthly analysis based on gauge observations, satellite estimates, and numerical model outputs. Bull Am Met Soc 78:2539–2558

Yukimoto S, Endoh M, Kitamura Y, Kitoh A, Motoi T, Noda A (2000) ENSO-like interdecadal variability in the Pacific Ocean as simulated in a coupled GCM. J Geophys Res 105:13945–13963

Zhang X, Shi G, Liu H, Yu Y (eds) (2000) IAP global ocean–atmosphere–land system model. Science Press Beijing China

Zorita E, von Storch H (1997) A survey of statistical downscaling techniques. GKSS 97/E/20

Zwiers FW, Kharin VV (1998) Changes in the extremes of the climate simulated by CCC GCM2 under CO_2 doubling. J Clim 11:2200–2222

4 Climate-related Change in Terrestrial and Freshwater Ecosystems

Benjamin Smith, Anto Aasa, Rein Ahas, Thorsten Blenckner, Terry V. Callaghan, Jacqueline de Chazal, Christoph Humborg, Anna Maria Jönsson, Seppo Kellomäki, Ain Kull, Esa Lehikoinen, Ülo Mander, Peeter Nõges, Tiina Nõges, Mark Rounsevell, Mikhail Sofiev, Piotr Tryjanowski, Annett Wolf

4.1 Introduction

Ecosystems on land, whether in a comparatively natural state or artificially constructed and managed, are a fundamental part of the environment in which most humans live. They also provide or help to control a variety of resources and intangible values vital to the health and economic conditions of human society (e.g. Costanza et al. 1997). These so-called ecosystem services, which are particularly important in a relatively populous region such as that of the Baltic Sea Basin, include the provision of food, fibre and wood products. Ecosystems also contribute to controlling water supplies, air and water quality, and conditions for the maintenance of biodiversity. Through their part in the Earth's carbon, water and energy cycles they may ameliorate – or exacerbate – climate change.

Chapter 2 demonstrates that climatic conditions over much of the Baltic Sea Basin have undergone changes in recent decades. The changes are likely to continue, possibly at an increasing rate, over the coming century. In concert with adjustments in other environmental and socio-economic factors, these changes seem likely to influence the structure and functioning of ecosystems, and to impact the services they provide to society.

In the present chapter we assess the potential impacts of the changing environment on terrestrial and freshwater ecosystems of the Baltic Sea catchment area (Annex 3.2.1). Through a synthesis of available studies from the Baltic region and other climatically-comparable regions of the world, we aim to evaluate two hypotheses:

1. that climate change over recent decades has affected ecosystems within the Baltic Sea Basin, impacting the services they provide; and
2. that ongoing climate change will cause further changes in the regional ecosystems and their services over the remainder of the 21st century.

Where possible we attempt to distinguish impacts of climate from those of other drivers of ecosystem processes, such as atmospheric carbon dioxide (CO_2) concentrations, nutrient deposition rates, as well as changes in human land use and land management; we review the potential role of land use change on ecosystem changes in Sect. 4.2. As terrestrial, freshwater and marine ecosystems do not exist in isolation from each other, we also address one of the major links between them: nutrient fluxes from land ecosystems to the Baltic Sea and the eutrophication of freshwater and marine habitats due to pollution loads from agriculture (Sect. 4.5).

To confine our scope, we do not address all classes of ecosystem impacts, but focus on:

1. processes and indicators of particular diagnostic value for the attribution of ecosystem changes to changing environmental conditions;
2. ecosystems and functions of particular socio-economic relevance; and
3. uncertainty associated with ecological complexity and limitations to process understanding.

Under the first category of impacts, we synthesise the evidence for changes in phenology, focusing particularly on shifts in spring phenological phases in plants, such as budburst and leaf unfolding (Sect. 4.3.1). In Sect. 4.3.2 we evaluate the evidence for shifting species distributions as early indicators of ongoing changes in ecosystem structure. Section 4.3.6 reviews climate-change impacts on arctic and subarctic ecosystems, considered especially sensitive to climate change.

Under the second category of impacts, we synthesise understanding of the effects of changing climate and atmospheric CO_2 concentrations on ecosystem productivity and carbon sequestration capacity (Sect. 4.3.4). Forests account for more than half of the total land area of the Baltic Sea Basin. A separate treatment focuses on potential climate impacts on forest resources (Sect. 4.3.5).

Most research on ecosystem responses to climate to date has focused on the effects of changes in average conditions, yet changes in extremes, such as low temperatures, extended drought periods, and strong winds may be of comparable importance for the overall responses of ecosystems. The state of empirical and mechanistic understanding of climatic extremes and their interactions with the physiological tolerance thresholds of plants is discussed in Sect. 4.3.3.

All three categories of impacts are raised with respect to freshwater ecosystems in Sect. 4.4.

4.2 Non-Climatic Drivers of Ecosystem Changes

While climate and weather exert a controlling influence over many of the component processes underlying ecosystem dynamics (see Annex 3.2.2), ecosystems are also sensitive to variation in a number of non-climatic environmental variables, as well as socio-economic factors leading to changes in land use and land management. Non-climatic environmental variables of significance for ecosystems include: (i) atmospheric CO_2 concentrations, which influence the availability of carbon as a substrate for photosynthesis; (ii) deposition loads of atmospheric pollutants, which influence ecosystem structure and function *inter alia* via changes in soil and water chemistry (acidification, eutrophication, excess of nutrient nitrogen); (iii) near-surface ozone concentrations, which may directly affect vegetation productivity and vigour; and (iv) concentrations and depositions of toxic pollutants, which may have a direct effect on the metabolism of animals and alter the food chain. Many of these factors are changing concurrently with climate, but interrelationships between them may change in the future.

Rising atmospheric CO_2 concentrations are one of the most certain aspects of global change. CO_2 is a plant resource, and increased concentrations are known to stimulate photosynthesis and improve plant water budgets, in both cases tending to augment net carbon uptake (Bazzaz 1990; Ainsworth and Long 2005), even though negative feedbacks via nutrient cycles might reduce "CO_2 fertilisation" on decadal or longer time scales (McGuire et al. 1995; see also Annex 3.2.2). The climate scenarios for 2070–2100 reported in Chap. 3, on which the assessment of potential ecosystem impacts presented in the present chapter is based, have associated assumptions as to the trajectory

of atmospheric CO_2 concentrations over the 21st century. These assumptions are, in turn, based on "storylines" of socio-economic change, with associated assumptions as to fossil fuel emissions and land use (the SRES scenarios; Nakićenović and Swart 2000). Projected CO_2 concentrations by 2100 range among scenarios from ca. 550 to ca. 970 ppmv. The SRES scenarios and their relationship to the modelled scenarios of future climate change are described in Sect. 3.2.1 and further discussed in Annex 6.

4.2.1 Atmospheric Pollutants

This outline provides a short summary of the atmospheric contribution to the pollution load onto the Baltic Sea Basin. More details of the geographic patterns and trends are available in Annex 3.1.2. Issues connected with the availability and quality of modelled and measurement data are outlined in Annex 4.

The atmospheric pathway of pollution delivery to the terrestrial ecosystem of the Baltic Sea Basin is significant for three groups of pollutants: acidifying pollutants (oxidised sulphur and nitrogen); toxic species, such as heavy metals (HMs) and persistent organic pollutants (POPs); and, presently less important, high level of tropospheric ozone.

Acidifying compounds are comparatively well studied by several modelling teams and observational networks. According to them, the deposition of nitrogen oxides to the catchment area of the Baltic Sea amounts to ~ 1.2 Mton N year^{-1}, with the Baltic Proper watershed taking over 50% of the load and no significant trend. The deposition density largely varies from south to north (about an order of magnitude for both sulphur and nitrogen compounds) and it is also subject to noticeable multi-annual variability. In some areas, the difference between the annual concentrations and depositions for two sequential years can be as much as 20–30%.

The ability of the ecosystems to withstand the acidification and nitrogen deposition is generalised via a critical load concept, which quantifies the maximum amount of acidic and nutrient deposition that the ecosystem can sustain without major damage. At present, the critical loads are exceeded in the southern part of the catchment area: in Germany, Denmark and, to a lesser extent, in Poland and southern Sweden.

It is generally agreed that high ozone levels during the vegetation season are not yet a matter of primary concern for Baltic Sea Basin ecosystems, except for the most south-western areas in Germany and Denmark. They can, however, become more significant in a warming climate.

Information about the toxic species is much less precise – both from observational and from modelling points of view. The density of observations is insufficient, while the modelling estimates from different studies can differ greatly. It is, however, plausible that the total deposition onto the Baltic Sea catchment area is at least 1.2–1.5 kt yr^{-1} for lead (Pb) and at least 100 t yr^{-1} for cadmium (Cd).

Persistent toxic species, such as mercury (Hg), polychlorinated biphenyls (PCBs) and hexachlorocyclohexanes (HCHs), are even more difficult to evaluate due to contradictions of different studies and methodologies. For Hg, one can only estimate the order of magnitude of deposition density to be 5–10 μg m^{-2} yr^{-1} in the south of the area and a few times lower in the north. For PCBs, the observed mean deposition flux at the Swedish west coast is about 0.5–1 μg PCB-sum m^{-2} yr^{-1} with a pronounced downward trend. The values for the sum of α- and γ-HCH do not have clear trends, fluctuating between 1 and 4 μg HCH-sum m^{-2} yr^{-1}.

4.2.2 Land Use

Land use and land cover change, including both conversions (complete replacement of one land cover type with another, e.g. forest to cultivated land) and modifications (more subtle changes in cover or management practices, e.g. intensification of agricultural land), in association with climate change, are recognised as primary drivers of global change (Guo 2000; Hansen et al. 2001; Grunzweig et al. 2003; Körner 2003; Lambin et al. 2003; Duraiappah et al. 2005; Lepers et al. 2005). Indeed, some authors have proposed that land use change is likely to be a more important driver than climate change, at least for the near future (Sala et al. 2000; Slaymaker 2001; Sala et al. 2005). Ecosystem changes associated with land use and land cover change are complex, involving a number of feedbacks (Duraiappah et al. 2005; Lepers et al. 2005). For example, conversion of forest to agricultural land works as a driver of climate change, by representing a major contributor to greenhouse gas release to the atmosphere via losses of biomass and soil carbon (Gitz and Ciais 2003; Canadell et al. 2004; Levy et al. 2004)

and, additionally, through a 'land use amplifier effect' (Gitz and Ciais 2003), where the overall global carbon sequestration ('sink') capacity is decreased. In contrast, reforestation, and other land use or land management changes such as modifications to agricultural practices, can work to mitigate climate change through carbon sequestration (Lal 2003; Jones and Donnelly 2004; King et al. 2004; Lal 2004; Wang et al. 2004; de Koning et al. 2005). Land use change also drives other ecosystem changes in combination, or above and beyond climate change, such as biodiversity changes, soil erosion, and land and water pollution (Lambin et al. 2003). Hansen et al. (2001) provide some examples of feedbacks between climate, land use and biodiversity, emphasising the importance of not treating these factors in isolation.

For instance, Hansen et al. (2001) describe situations in which land use change works to modify species response to climatic changes by altering dispersal routes. Alterations can represent both barriers to and facilitation for species dispersal. For example, when native vegetation is cleared for agriculture, a barrier can be created, limiting dispersal for native species, while on the other hand facilitating access for exotic species. Climate and land use can also jointly influence disturbances such as wildfire, flooding and landslides. The two disturbances may amplify each other; for example, logging can further dry fuels and increase the probability of wildfire during periods of drought conditions. Alternatively, the two disturbances may counteract each other, such as is the case where grazing reduces fire frequency and intensity. Land cover change can also directly affect climate: such is the case where deforestation over large areas causes reductions in transpiration, cloud formation, rainfall and increased evaporation. This can lead to vegetation shifts, such as the replacement of forests by shrubs and grassland.

Despite the recognised importance of land use change in driving global change, it is rarely included in current global climate models (Hansen et al. 2001; Holman et al. 2005a; Levy et al. 2004; Zebisch et al. 2004). This limitation is attributed to an insufficient understanding of the underlying causes and prospective consequences of land use and land cover change at the global scale (Hansen et al. 2001; Lambin et al. 2001; Lambin et al. 2003). However, much progress has been made in this area, both in terms of a better understanding of the underlying processes as well as in the development of more sophisticated models (Lambin et al. 2003). Several studies incorporating land use change have been undertaken at local and regional scales, some of which are discussed in further detail in the following section. The explicit inclusion of land use change in analyses of climate change could therefore lead to some unexpected outcomes (Hansen et al. 2001). Consequently, many impact studies of climate change may represent inadequate estimates of projected changes.

4.2.2.1 *Projected Future Changes in European Land Use*

There have been a number of recent developments in future European land use scenarios and their application, often based on the global SRES storylines of the IPCC (Nakićenović and Swart 2000), or alternative scenario frameworks such as the Millennium Ecosystem Assessment scenarios (Carpenter et al. 2005). Other developments have been largely carried out on regional or local scales, either representing down-scaled versions or modifications of the SRES scenarios (Holman et al. 2005a,b; Ewert et al. 2005; Rounsevell et al. 2005; van Meijl et al. 2005; Abildtrup et al. 2006; Verburg et al. 2006).

Worldwide, most regional and global scenarios indicate an expansion of agricultural land over the coming decades due to the trade-off between food supply and demand as moderated by international trade, with the biggest changes occurring in the tropics (Alcamo et al. 2006). Conversely, most European scenarios show agricultural land abandonment (Ewert et al. 2005; Rounsevell et al. 2005; Verburg et al. 2006). Land abandonment scenarios tend to assume that increases in the supply of agricultural goods due to the effect of technological development on productivity will offset changes in food demand (Ewert et al. 2005; Rounsevell et al. 2005; Verburg et al. 2006). Scenarios for which agricultural land abandonment does not occur are often characterised by yield reductions arising from extensification of the agricultural production system. It is likely that both agricultural land abandonment and extensification would create opportunities for ecosystems and conservation (Rounsevell et al. 2005), although the loss of mountain pastures to regrowing forests is regarded as a negative impact in terms of biodiversity (Dirnbock et al. 2003; Giupponi et al. 2006).

Forest scenarios tend to mirror agricultural scenarios in that forested areas are often merely indirectly determined from the assumed expansion or contraction of agricultural land (Alcamo et al. 2006). In Europe, agricultural land abandonment would create the conditions for potential reforestation (Kankaanpää and Carter 2004; Rounsevell et al. 2006a).

4.2.2.2 *Land Use Change in the Baltic Sea Basin*

There have been no climate change impact studies on land use that have focused on the Baltic sea Basin as a geographic region. However, a number of studies that have been undertaken at the European scale allow targeted conclusions to be drawn for the Baltic Sea Basin (for example see Audsley et al. 2006; Ewert et al. 2005; Rounsevell et al. 2005; Rounsevell et al. 2006b). It is important to recognise that the Baltic Sea Basin, like other European regions, is strongly dependent on driving forces that operate at global and European scales, so that examining the response of the Baltic Sea Basin within European-wide studies is an appropriate methodology.

Most studies find that the socio-economic scenario assumptions have a greater effect on future land use than the climate change scenarios. Even for agriculture, which might be expected to be sensitive to crop yield changes, there are winners and losers at the European scale with little overall change in agricultural production potential (Rounsevell et al. 2005). At the regional scale, however, there may be large changes in land use. Most agricultural scenarios present an argument for decreasing agricultural areas, often as a result of crop yield increases due to technological development rather than climate change (Rounsevell et al. 2005; Audsley et al. 2006), and these effects are greatest in scenarios that are characterised by free markets. Reductions in agricultural areas for food production are often compensated somewhat by increases in areas for bio-energy production and forests and small increases in urban and nature conservation areas (Rounsevell et al. 2006b). Such changes are also important for landscapes and ecosystems. Some scenarios, however, suggest that the Baltic Sea Basin will respond differently. Audsley et al. (2006) indicate that Finland's agricultural area may increase significantly in response to the yield gains resulting from climate change. Whether such changes would be realised in practice is debatable when simultaneously the agricultural areas of the rest of Europe are decreasing (Rounsevell et al. 2006b), but it remains a possibility. Policy would probably play a strong role in moderating this change, as increasing agricultural areas have negative environmental consequences not least for biodiversity (Berry et al. 2006).

It would seem logical for the land use of Sweden and Finland to respond in similar ways. However, Audsley et al. (2006) suggest that southern Sweden will be very similar to Denmark (for which most future scenarios suggest constant agricultural land use) and good for agriculture, whereas the north of Sweden remains cold even in the A1FI scenario (Sect. 3.3.4). Although Finland is at a higher latitude than much of Sweden, it has a much warmer summer and a huge area of Finland has higher yields of wheat than the south of Sweden in the A1FI climate. Therefore, whereas the agricultural area in Sweden doubles, the area in Finland increases ten-fold under the A1FI scenario (Audsley et al. 2006).

4.2.2.3 *Projected Combined Impacts of Land Use and Climate Change on Ecosystems*

Relatively few studies have attempted to couple land use and climate change scenarios in the analysis of impacts on ecosystem processes, goods and services. Global scale studies include the Millennium Ecosystem Assessment (Reid et al. 2005) and the studies by Sala et al. (2000) and World Resources Institute (2000). Global scale scenarios are not appropriate, however, for the analysis of ecosystem impacts at local to regional scales. Moreover, there has been increasing interest in regional scale studies to promote development of similarly-scaled policies and to enable stakeholders to engage with climate-related issues at a scale where they interact in terms of daily activities and management (Holman et al. 2005b). Only a few such studies have been undertaken at the regional or local scale, and many of these are very recent, with some of the work still under review or currently in press (Dirnbock et al. 2003; Stefanescu et al. 2004; Holman et al. 2005a,b; Scheller and Mladenoff 2005; Giupponi et al. 2006; Harrison et al. 2006; Lavorel et al. 2006; Metzger et al. 2006; Schröter et al. 2005; Rounsevell et al. 2006b).

The limited available literature suggests that the effects of land use change on species at regional scales work to further fragment the overall climate driven trends (Holman et al. 2005b; del Barrio et al. 2006; Harrison et al. 2006; Rounsevell et al. 2006a). The degree of fragmentation depends on where the land use changes occur, the relative 'hostility' of the new environment to the species, and the species dispersal ability. Habitat fragmentation is exacerbated when species have relatively low dispersal distances (del Barrio et al. 2006). Furthermore, land use change can favour some species over others, depending on the relative hostility of the new environment (Giupponi et al. 2006). For example, projected agricultural abandonment and increases in forest in the Italian Alps resulted in an increase in the distribution of forest species (e.g. *Ostrya carpinifolia*) and a decrease in the distribution of open habitat species (e.g. *Crex crex* and *Chorthippus dorsatus*). Berry et al. (2006) argue that northern latitude countries within Europe will become increasingly important for biodiversity and species conservation as the ranges of species distributions move northward in response to climate change. As northern latitude countries are also the most likely to be sensitive to the effects of climate change on land use (especially agriculture), these regions could play a prominent role in adaptation strategies to mitigate the negative effects of future climate and land use change on Europe's natural heritage.

Two studies examined projected changes at the European level. The ATEAM project (Schröter et al. 2005) focused on a number of goods and services for the region as a whole (e.g. soil organic carbon, carbon sequestration, wood production, carbon storage, water resources, biodiversity changes) while the ACCELERATES project (Rounsevell et al. 2006a) focused on biodiversity changes in agricultural areas only. Three studies examined projected changes at local and/or regional scales (Holman et al. 2005b; Lavorel et al. 2006; Rounsevell et al. 2006a).

Within the ATEAM project, projected changes in forests as a result of climate change, land use and management were modelled using the inventory-based, European Forest Information Scenario Model (EFISCEN) (Karjalainen et al. 2002; see also Annex 3.2.4). Growing stock, increment, age-class distribution and carbon stocks in above- and below-ground biomass were all simulated. Projected changes in carbon balance were estimated using a modified version of the LPJ model (Sitch et al. 2003; Zaehle et al. 2007). The total area of European forests was projected to increase by up to 6% by 2080, particularly in the B2 scenario. All scenarios projected an increase in forest growth, especially in northern Europe and in the economically-oriented A scenarios. The intensity of management had the greatest influence over increment in comparison to the effects of land use or climate change, with increases being the greatest in the most intensively managed areas. The effect of land use change on increment varied substantially across the scenarios, depending on the extent of forest change. In the A1FI scenarios there were little changes in forest area, whereas in the B2 scenario, forest area increased by 32%. Although timber harvesting exceeded forest increment for the A1FI scenario in the second half of the 21st century, all scenarios showed an overall increase in growing stock compared to present levels. Management had the greatest influence on growing stock, whereas, depending on the scenario, management accounted for 60–80% of the difference in growing stock between 2000 and 2100, climate change accounting for 10–30% of the difference, with land use change accounting for only 5–22%.

In terms of carbon balance, the combined land use and climate change scenarios showed an overall increased uptake of carbon up to 2050, with a shift towards carbon release towards the end of the century. In northern Europe, increased warming, mainly in winter, enhanced soil respiration more than net primary production in all scenarios, decreasing the sink, or turning it into a net carbon source. Results from southern Europe were less clear, with a mixture of increases in carbon uptake and carbon neutral projections. Land use change was shown to contribute greatly to the increased carbon uptake in the first half of the century as a result of increased carbon sequestration due to reforestation in former agricultural areas. Socio-economic scenarios also showed a great deal of variation in projected carbon balances. This influence of differing socio-economic scenarios and associated changes in land use suggests that decisions regarding socio-economics and land use are drivers of the European terrestrial carbon balance of comparable importance to climate change.

4.3   Terrestrial Ecosystems

4.3.1   Phenology

Phenology has been in the focus of scientists in the Baltic region since the early studies of Carl Linné. Over the last 300 years, phenologists have observed a wide range of plant, bird, insect, fish and animal species using different observation methods in the Baltic region (Schnelle 1955; Leith 1974; Schults 1981). The long time series of phenological

data that have been recorded have great potential to characterise the impacts of the changing climate on ecosystems. In this section we focus on plant phenology, since phytophenological data are most readily available for different countries and for the best-known indicator species.

Phenological observation programmes most commonly observe phases of spring such as budburst, leaf unfolding and flowering, as well as summer phases such as ripening of fruit and harvesting. There are fewer observations on autumn phenological phases – including leaf colouring and leaf abscission – and the data quality is much lower. It is therefore complicated to use autumn-phase data as climate change indicators (Chuine et al. 2003). Apart from their relevance as climate change indicators, spring and autumn leaf phenological phases are key determinants of growing season length and primary production in deciduous vegetation. The relevance of leaf phenological changes to the productivity and functioning of ecosystems is further discussed in Sect. 4.3.4.

Temperature is the main triggering factor for spring and summer phenological events in temperate zones. Two classes of temperature parameters have a high correlation with both spring and summer phases and have therefore been used in most phenological analyses and models. The first comprises positive ($> 0\,°\mathrm{C}$) temperatures (mean, cumulative etc.) during the growing season, which have a direct triggering influence on plant phenology (Chuine et al. 2003; Schwartz 2003). The second class of parameter is the chilling factor, i.e. the sub-zero temperatures during the previous winter, as spring phenology is very dependent on the temperature regime during winter dormancy. The physiological and biochemical mechanisms controlling onset and release from dormancy have so far not been thoroughly studied (Chuine et al. 2003). Apart from temperature, the phenological development of plants is also influenced by other environmental factors (e.g. radiation, moisture, soil, landscape, ecosystem type) and human impacts such as pollution, land use and urban heat islands (Schnelle 1955; White et al. 2002; Schwartz 2003). In the Baltic Sea Basin, snow cover is an important factor for spring phenology, dampening temperature variation and isolating plants from potentially harmful weather extremes.

In the present climate, the temperature variability is generally determined by atmospheric circulation (Sect. 2.1.1 and Annex 1.2). Analysis of the correlation between phenological phases –

flowering of coltsfoot (*Tussilago farfara*), unfolding of birch leaves (*Betula pendula*), flowering of lilac (*Syringa vulgaris*) – and circulation patterns in Europe (Aasa et al. 2004) has shown that in the Baltic Sea Basin, spring phenology is primarily influenced by the North Atlantic oscillation (NAO) index during winter (December–March) and the Arctic oscillation (AO) index during the first three months of the year (January–March). This was evidenced by a measured correlation (Pearson r) stronger than -0.5 in the Baltic Sea Basin.

4.3.1.1 *Recent and Historical Impacts*

The databases on which the following assessment is based provide relatively good coverage of Germany, Poland, the Baltic States and Russia. Relevant data from the Scandinavian countries, which ceased carrying out plant phenological observation programs in the 1950s and 1960s, are not available for recent decades. The only data publicly available from the Scandinavian countries is from the International Phenological Gardens (IPG) network, which has phenological observation stations in most of Central and Western Europe, and operates some stations in Scandinavia. For Finland, some long-term phenological time series are available from the national Phenology Research Programme (www.fmnh.helsinki.fi/seurannat/fenologia).

Analysis of trends in phenological time series over the last 50 years in the Baltic Sea Basin shows a tendency for spring to arrive earlier and autumn later. This conclusion is based on data from the International Phenological Gardens (IPG) network. Studies have shown that the length of the phenological growing season, as determined by leaf unfolding and leaf fall of various tree species, has lengthened significantly during the period 1951 to 1996 (Menzel and Fabian 1999). On average, the spring arrival trend was negative (arriving earlier) and the autumn trend positive (arriving later). The mean trend of the 616 spring phases studied over all of Europe changed by -2 days per decade; the 178 autumn phases changed by $+1.6$ days per decade during the study period. As a result, the average growing season was extended by 3.6 days per decade in the study area.

Chmielewski and Rötzer (2001) used the IPG dataset to analyse changes in the phenological growing season for 1968-1998 over Europe. They determined the beginning of the growing season by means of a combined leafing index (*Betula pubescens*, *Prunus avium*, *Sorbus aucuparia* and *Ribes alpinum*) and the end of the growing season by the mean leaf fall date of four species (*Betula pubescens*, *Prunus avium*, *Salix smithiana*, *Ribes alpinum*). For Europe as a whole, the analysis identified a significant ($P < 0.05$) lengthening of the growing season, by 2.7 days per decade. For the Baltic Sea Basin (southern Fennoscandia, Denmark and northern Poland) the trends were much stronger in most of the stations: the growing season lengthened by 4.5 days per decade, as the onset of spring shifted significantly earlier.

Phenological trends in central and eastern Europe were studied within the EU project POSITIVE (Menzel 2002), which mapped phenological trends on the southern and eastern coasts of the Baltic Sea using data sets from weather services (DWD, EMHI) and the Russian Geographical Society, covering the period 1951–1998 (Ahas et al. 2002; Scheifinger et al. 2003). Analyses of common plant and tree species in this area showed that the highest rate of change appeared in the early spring phases (e.g. pollination of hazel [*Corylus avellana*] and coltsfoot [*Tussilago farfara*]), which now occur 10–20 days earlier than 50 years ago (an average change of 2–4 days per decade). Spring phases (birch leaf unfolding, lilac and apple trees in full bloom) have undergone smaller changes, now occurring 5–15 days earlier than 50 years ago (Ahas et al. 2002). Linear regression maps of birch leaf unfolding (Fig. 4.1) show the highest statistically-significant slope for a linear regression against time, ranging from -4.2 to -7.3 days per decade in the Baltic region. The slope of lilac (*Syringa vulgaris*) flowering is steepest in the Baltic region (-2.6 days per decade), likewise showing a relatively steep slope over the northern part of the Eastern European Plain (-2.1 days per decade). Lime (*Tilia cordata*) shows a significant trend towards earlier flowering in the summer in the Baltic region. Similar results to the above were mapped within the EU Programme European Phenological Network (van Vliet 2003; van Vliet et al. 2003).

Table 4.1 presents slopes of linear trends (days per year) at selected stations within the Baltic Sea Basin (Ahas et al. 2002). The early spring phases, such as blossoming of hazel or coltsfoot, have greater slopes than the late spring and summer phases. The statistically significant ($P < 0.05$) trends tend to have a negative value.

Trends in Estonian phenological time series have been studied in several projects, taking advantage of the very good available series of obser-

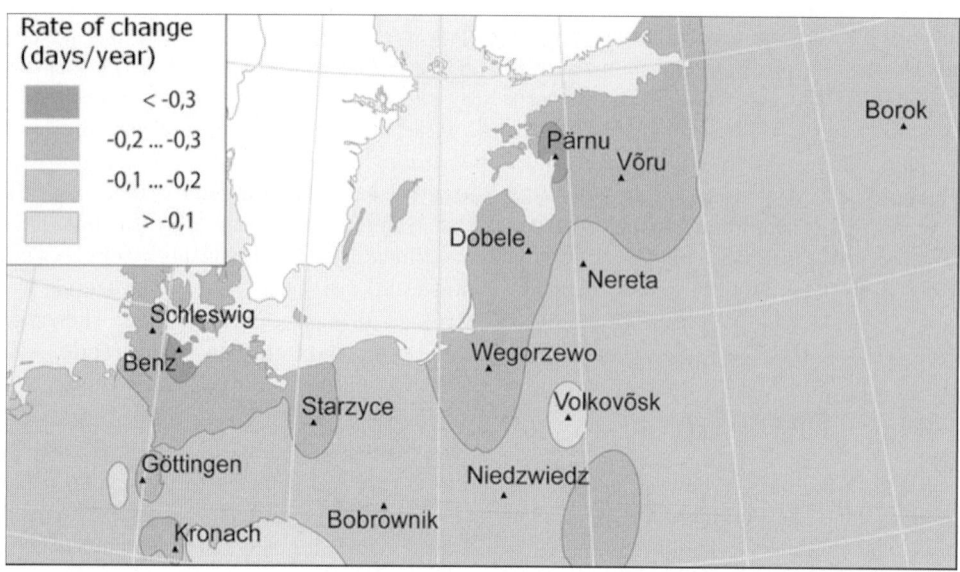

Fig. 4.1. Rate of phenological changes (days per year) in birch (*Betula pendula*) leaf unfolding over southern and eastern parts of the Baltic Sea Basin for 1951–1998 (Ahas et al. 2002)

Table 4.1. Changes in phenological time series (days per year) for 1951–1998 for plant taxa at selected observation sites presented in Fig. 4.1 ($^{*}r$ = significant at $P < 0.05$; F = flowering, L = leaf unfolding, D = defoliation) (Ahas et al. 2002)

Station	Corylus F	Tussilago F	Taraxacum F	Betula L	Malus F	Syringa F	Sorbus F	Tilia F	Betula F
Benz	−0.90	0.01	−0.08	−0.37*	−0.15	−0.14			
Bobrowniki	−0.36	−0.11	−0.27*	−0.14		−0.05			
Borok		−0.45*		−0.17	−0.06	−0.08	−0.13	−0.18	0.31
Dobele	−0.36			−0.21*	−0.25*	−0.16	−0.23*		−0.02
Göttingen	0.06	0.20	−0.30	−0.26*	−0.17	0.01		0.24	
Kronach	−0.32	−0.25	−0.22	−0.34*	−0.11	−0.02	−0.09		
Nereta	−0.91*	0.10	0.05	−0.15	−0.11	−0.12	−0.17		0.15
Niedzwiedz	−0.11	−0.18	−0.20	0.14		−0.13			
Pärnu	−0.46*			−0.33*	−0.18*	−0.26*	−0.27*	−0.20	
Schleswig	−0.40	−0.28	−0.09	−0.06	−0.09	−0.13		0.05	
Starzyce	−0.29	−0.16	−0.45*	−0.23		0.18			
Wegorzewo	−0.40	−0.17	−0.16	−0.26*		−0.15			
Võru	−0.18			−0.21*	−0.24*	−0.23*	−0.23*	−0.09	

vations, collected since 1951 by ENS and EMHI. Trends among 644 phytophenological time series gathered from 1948 to 1999 in Estonia have been described by Ahas (1999), Ahas et al. (2002) and Ahas and Aasa (2006). Results are summarised in Fig. 4.2. Statistically-significant trends show that both spring and summer phases tend to commence earlier than previously: 138 of the studied time series have advanced by 2–3 days per decade; 112 time series by 1–2 days per decade; and 53 time series by 3–4 days per decade over the 49-year study period.

There were 32 time series that advanced by more than 4 days per decade, all of them statis-tically significant ($P < 0.05$). A further 151 time series show a non-significant advance by 0–5 days, while 93 time series show a delay. Twenty of the delayed time series show a statistically-significant trend in the order 10–30 days per decade. These are mainly autumn phases such as the ripening of fruits and the colouring and falling of leaves.

The spatial distribution of phenological trends shows generally stronger advancement in the maritime climate of western Estonia compared with the continental inland areas or northern Estonia. For example, the ear of rye (variety "Sangaste") has advanced by 10–12 days (2.0–2.4 days per decade) in western Estonia, and by 2 to 4

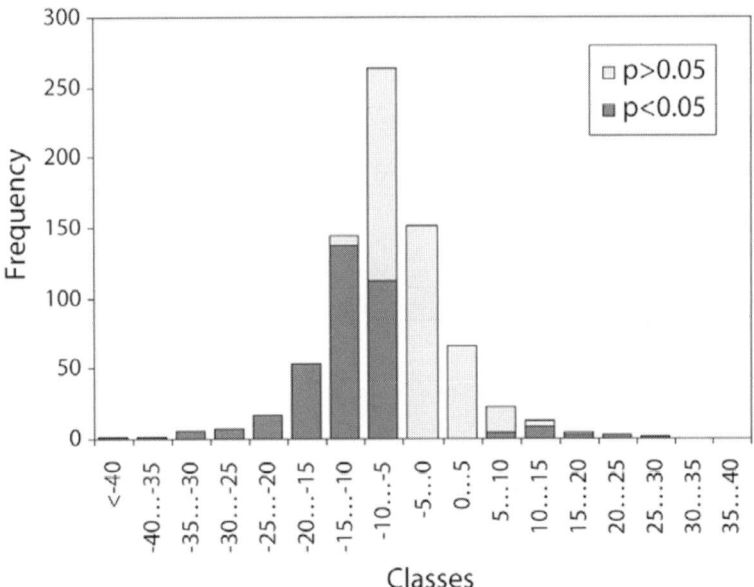

Fig. 4.2. The distribution of trends over the period 1951–1999 in the Estonian phytophenological time series according to the size and significance of changes (Ahas and Aasa 2004). Classes are in days per 49 years

days (0.4–0.8 days per decade) in eastern Estonia (Ahas and Aasa 2006).

In the Estonian phenological dataset, the distribution of phenologically early and late springs was also analysed. These were determined using a threshold of values greater or less than 1 standard deviation in all spring phases, i.e. measurements outside of the $+/-1$ standard deviation threshold were considered either early or late, accordingly.

Results (Fig. 4.3) show that at the beginning of the studied period (1950s) there were more delayed phases, reaching a peak in 1955 with a total of 87 delayed phases. The number of delayed time series decreased between 1955 and 1995. Years that are earlier by more than 1 standard deviation begin to dominate after 1980, reaching a peak in 1989, with 77 advanced phases. In 1990, 1991, 1992 and 1993, a majority of the time series was also advanced (Ahas and Aasa 2004; Ahas and Aasa 2006).

Historical changes in phenological time series in Estonia have also been studied. During the period 1919–1996 the mean change in spring phenophases was an advance of 2–5 days per decade (Ahas 1999; Ahas et al. 1998)

Taken together, phytophenological time series from the Baltic Sea Basin show a marked trend towards an advancement of spring and early spring phases in recent decades, indicative of the effects of a warming climate on plants. The phenological

trends appear to be generally somewhat stronger in the Baltic Sea Basin compared to Europe as a whole, possibly reflecting a stronger warming trend in northern Europe (Sect. 2.1.2). As indicators of ecosystem responses to climate, phytophenological trends for the Baltic Sea Basin clearly support the hypothesis that recent climate change has impacted the functioning of ecosystems there.

4.3.1.2 *Potential Future Impacts*

Extrapolation of the average effect of climate warming suggests that a rise in mean temperature by 1 °C will lengthen the phenological growing season by 5 days in temperate Europe (Chmielewski and Rötzer 2001). Predictions based on GCM-modelled climate scenarios suggested that snow-drop (*Galanthus nivalis*) flowering would occur 2 weeks earlier if atmospheric CO_2 concentrations doubled by 2035, and one month earlier if CO_2 levels tripled by 2085 in northern Germany (Maak and von Storch 1997).

The scenarios of +2 °C and −2 °C temperature rise used in the Estonian analysis showed that in these cases the phenological development of nature remains within the limits of natural variation (Ahas et al. 2000). The case of a 4 °C warmer spring has also been studied, with results suggesting that the common spring phases will still re-

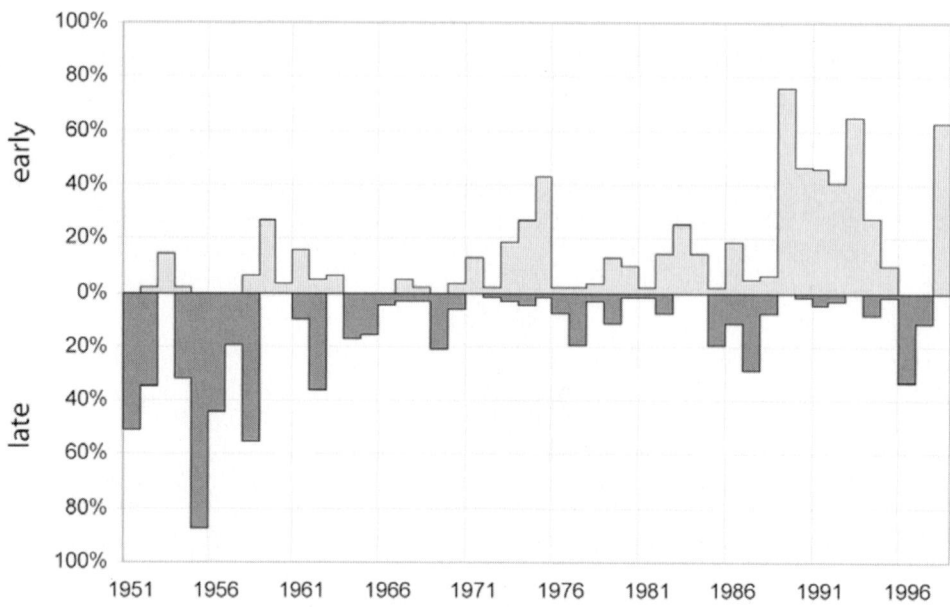

Fig. 4.3. The percentage of values larger or smaller than ±1 standard deviation in the summarised phenological calendar of three selected stations in Estonia (Pärnu, Türi and Võru) (Ahas and Aasa 2004)

main within the limits of extreme (warm) years such as 1975, 1989 and 1990 (Ahas et al. 2000). Several authors have pointed out that the phenological development of nature is relatively flexible since many species and their phenological varieties adapt to the wide natural variations of spring weather in temperate regions. However, many phenological models predict that global warming and earlier springs will result in an increased risk of frost damage for species throughout Europe (see also Sect. 4.3.3).

Linkosalo et al. (2000) studied frost-damage risks with a chill-triggered model in Finland, showing that increased frost risk would occur during earlier (warmer) springs. A light-climate-triggered model did not, however, predict those risks (Linkosalo et al. 2000). Statistical analysis of data from the POSITIVE project showed that the risk of late frost damage for plants in Central Europe has been lower in the last decade compared to previous decades (Scheifinger et al. 2003).

Climate scenarios for the Baltic Sea Basin generally point to a 3–5 °C rise in average temperatures by the end of the 21st century (Chap. 3). For many species, this could lead to an extension of the growing season in the order of weeks, a significant change in conditions for growth, especially in northern areas where the growing season is currently shortest.

4.3.2 Species and Biome Range Boundaries

Species and biome range boundaries are dependent on a wide range of environmental factors, including climate and meteorological events. Therefore, species and biome ranges are potentially very sensitive to climate changes, particularly changes in temperatures, precipitation patterns (especially reduced precipitation leading to drought stress), as well as the incidence of extreme events like storms and floods. Moreover, recent observed species range shifts may be tied to the influence of large-scale weather systems, such as NAO in the case of the Baltic Sea Basin (Annex 7).

In the following assessment, we focus mainly on changes in ecosystems related to temperature and precipitation. However, it should be borne in mind that, even in the case of range shifts correlated with changes in temperature and precipitation, other factors may play a role as driving forces. Examples include UV-B radiation (Cummins 2003), or atmospheric CO_2 concentrations, which may affect the carbon balance of vegetation.

Species responses to environmental gradients typically differ in a community setting compared to the fundamental physiological response due to the secondary effects on individual vigour of competition, impacts of pests and diseases and other biotic factors. Natural disturbances (e.g. wildfires, floods) may likewise modify responses. Finally, the present distribution of many species is more evidently tied to land use, habitat fragmentation (e.g. Hill et al. 1999) or the absence of suitable habitats due to human activities than to species responses to bioclimatic gradients. Anthropogenic factors are likely to play an important role in future range shifts as well as species extinctions (Pounds et al. 1999; Williams et al. 2003).

4.3.2.1 *Methodological Remark*

Evidence for distributional changes at the species, community or population level are often derived by comparing two surveys at discrete points in time. Background changes between the survey events are rarely addressed. Yet without consideration of how much background variation is plausible between years, or even within years, any apparent "trend" cannot be robustly ascribed to climate change or any other specific driving factor (Sparks and Tryjanowski 2005). Species' behaviour at those specific time points can affect conclusions, because changes in the distribution of species are associated not only with climate but also with major changes in habitats (Pounds et al. 1999; Warren et al. 2001; Konvicka et al. 2003). Apparent range margin shifts can be a possible response to the influence of both factors.

4.3.2.2 *Recent and Historical Impacts*

Changes in species distributions

Distributional shifts are the potential result, at geographic scales, of two very different mechanisms operating at the local scale: expansion into new areas; for example, areas that have become favourable as a result of climatic amelioration; and local extinction, which reduces the distribution. We discuss each of these mechanisms in turn, and evaluate their potential role in recent and historical distributional changes of species in the Baltic Sea Basin.

There is evidence of changes in the distribution of terrestrial species throughout the world (Parmesan et al. 1999; Walther et al. 2002) consistent with recent global warming (Grabherr et al. 1994; Pounds et al. 1999), although consensus on a general, causal link has not yet been achieved (Jensen 2003).

Globally, many studies have focused on species abundances and distributions to corroborate predicted systematic shifts related to shifts in climatic regimes, often based on species-specific physiological thresholds of temperature and precipitation tolerance (e.g. Woodward 1987; Sykes et al. 1996; Grace et al. 2002). Some commonly invoked mechanisms are described in Annex 3.2.2. A certain inherent resilience is reported for some community types, such as treeline forests, while the magnitude of elevational shifts of alpine plant species lags behind the isothermal shift. By contrast, some butterflies appear to track decadal warming quickly (Parmesan et al. 1999).

For the Baltic Sea Basin there are a few examples of species showing both latitudinal and altitudinal shifts of the species distributional range. Changing temperatures are known to have produced shifts in species range limits in historical times. For example, the Scandinavian distributions of beech (*Fagus*), lime (*Tilia*), oak (*Quercus*) and spruce (*Picea*) have generally tracked growing-season heat sums over the last 8000 years, based on pollen observations and vegetation model simulations (Bradshaw et al. 2000). The current rate of climate change is rapid compared with changes of similar magnitude in the past. Walther et al. (2005) report new observations of naturally-regenerating holly (*Ilex aquifolium*), a species purported to be a climatic indicator, east and northward of the previously reported natural limit (Iversen 1944). The new observations coincide with an advance of the 0 °C January isoline since the 1940's.

Kullman (2002) reported upslope range shifts of 100–150 m by near-treeline birch (*Betula pubescens* ssp. *czerepanovii*), Scots pine (*Pinus sylvestris*), Norway spruce (*Picea abies*), rowan (*Sorbus aucuparia*) and willow (*Salix* spp.) in the Swedish part of the Scandes mountain range, identifying a 0.8 °C increase in mean temperature since the late 19th century as the most important cause. Increased height growth by existing *krummholz* individuals was a more important contributing factor to apparent treeline advance than establishment of new individuals at higher elevations. After a slight retardation during some cooler decades after 1940, a new active phase of tree-limit advance has occurred with a series of exceptionally mild winters and some warm summers during the 1990s (Kullman 1993, 1997, 2002). The magnitude of total 20th century tree-limit rise varies with topoclimate and is mainly confined to wind-sheltered and snow-rich segments of the landscape. Thickening of birch tree stands in the "advance belt" has profoundly altered the general character of the subalpine/low alpine landscape and provides a positive feedback loop for further progressive change and resilience to short-term cooling episodes. All upslope tree-limit shifts and associated landscape transformations during the 20th century have occurred without appreciable time lags, which constitutes knowledge fundamental to the generation of realistic models concerning vegetation responses to potential future warming. The new and elevated pine tree-limit may be the highest during the past 4000 ^{14}C years. Thus, it is tentatively inferred that the 20th century climate is unusually warm in a late-Holocene perspective.

Animals, within their lifetime, mainly follow changes in habitat structure; for this reason, changes in animal species ranges can be predicted from plants and plant community changes over time. However, the availability of data to compare historical and current distribution are in some cases much better for animal species than for plants, and this is true in the case of the Baltic Sea Basin countries. For instance, changes in the bird fauna of Finland were summarised by Brommer (2004). He analysed changes in range margins of birds breeding in Finland from published atlas data for a 12-year period (1974–1979 to 1986–1989). The change in range margin was statistically corrected for range size change within Finland using linear regression. For species predominantly occurring in southern Finland ($n =$ 116), the corrected range margin shift was 18.8 km northwards, which may be related to climate change. Northerly species ($n = 34$) showed no such significant range margin shift.

In a fresh overview on climate change impacts in Finland, Pöyry and Toivonen (2005) report that over 100 new butterfly and moth species have been recorded in Finland during the last decade (see also Parmesan et al. 1999). Some originally univoltine moths (having a single generation per year) have become multivoltine (more than one generation per year).

Many recent range shifts of animals to the north may be connected with climate warming. For the majority of bird species showing recent range shifts within the Baltic Sea Basin, the shift is to the north, suggesting a climatological explanation (Lehikoinen et al. 2003; Tomiałojć and Stawarczyk 2003; Tryjanowski and Winiecki 2003). However, it should be noted that some southern species – which should be favoured by climate warming – are becoming rarer, while some northern species have extended their distribution southwards (Lehikoinen et al. 2003; Tomiałojć and Stawarczyk 2003; Tryjanowski and Winiecki 2003).

Among mammals, an interesting example is shown by Sachanowicz and Ciechanowski (2006) which links the first winter record of a migratory bat (*Pipistrellus nathusii*) in Poland and records from other countries to climate changes. It is suspected that for migratory bats changes in distribution ranges, relocation of wintering areas, as well as the timing of hibernation may be similar to those well described for birds.

An example of uphill shifts is the distribution of butterflies observed in the Czech Republic (Konvicka et al. 2003) within a study region including areas belonging to the Baltic Sea Basin (northern Moravia, Silesia). Recent butterfly distributions were compared with historical (1950–1980) data, showing that altitude shifts in the distribution of Czech butterflies are already detectable on the coarse scales of standard distribution maps. The species undergoing altitudinal advance show no consistent patterns with regard to habitat affiliations, conservation status or mountain versus non-mountain distribution, which renders climatic change as the most likely explanation for the distributional shifts.

Tryjanowski et al. (2005) demonstrated that the white stork (*Ciconia ciconia*) has increased its elevational range during the last 30 years in the Podhale region, Poland. An answer to the question of why white storks, traditionally wetland species, have increased their altitudinal breeding range by at least 500 m in upland and mountain areas with relatively small water bodies, is likely to be complex: firstly, because the species has the biological potential to colonise mountain areas; secondly, because storks appear to follow habitat and/or climate changes. During the study period, many arable fields in the Podhale region were abandoned and/or converted to pasture following the collapse of communism. This, together with climatological factors (changes in temperatures and runoff), created new habitat and offered new foraging opportunities for storks. Hence, when climatological conditions improved, and after saturation of lower altitude nest sites, the storks expanded to higher ground, and occupied new areas.

Consequences of species range shifts on longer time scales could be extinction (Thomas et al. 2004). Palaeontological data clearly support the hypothesis that major extinction events in the past were related, at least in part, to changes in temperature regimes. For example, pollen data reveal range fluctuations in trees correlated with environmental changes through the Holocene (e.g. Bradshaw et al. 2000). However, individual longevity, high intra-population genetic diversity and the potential for high rates of pollen flow might make tree species comparatively resistant to extinction (Hamrick 2004).

There is some evidence of imminent range retraction via local extinction for tree species. The primeval forest of Białowieża National Park, Poland, is considered to be a unique remnant of a mixed broadleaf-conifer forest type that once dominated lowland regions of central-eastern Europe. Norway spruce occurs as a codominant species in the forest, close to its inferred climatological ("warm") limit (Dahl 1990). The species is currently in decline due to regeneration failure (Walankiewicz 2002).

The historical past of southwest Finnish birds, including the archipelago areas, was studied by Lehikoinen et al. (2003), who found indications of possible past extinctions (cormorant [*Phalacrocorax carbo*], grey heron [*Ardea cinerea*], bittern [*Boturus stellaris*], red kite [*Milvus milvus*]) that co-occur with the coldest period of the "Little Ice Age" and/or intensification of human impacts. Similarly, changes in Polish avifauna during pre- and historical times have been linked to changed climatological regimes in the past (Bocheński 1993; Tomiałojć and Stawarczyk 2003), while Najbar and Szuszkiewicz (2005) linked a strong decline in the European pond turtle (*Emys orbicularis*) with reproduction failure (egg freezing) as a result of declining snow cover in recent years.

Climate-related invasions

The clearest evidence for a climatic trigger for large-scale changes in ecosystem structure occurs where a suite of species with different histories of introduction spread en-masse during periods of climatic amelioration (Walther et al. 2002). With climate change, non-native species from adjacent areas may cross barriers and become new elements of the local biota. In cases where long distances have been covered, such movements have often been mediated by human activity. A permanent establishment in the native ecosystem may not be possible without changes in local conditions.

To our knowledge there are no data on invasions forced by climate change in the Baltic Sea Basin. However, some cases of geographical range expansion in invertebrates could be interpreted as responses, in part, to climate change. An example is the wasplike spider (*Argiope bruennichi*), previously restricted to southeastern Europe, which in recent decades has expanded its geographical range to the northern parts of Europe, colonising many new sites in Germany, Poland, Denmark and Sweden (Scharff and Langemark 1997; Barabasz and Górz 1998; Jonsson and Wilander 1999). The most important factors causing rapid geographical expansion of the wasplike spider are: climate change, especially increase of the number of sunny and dry days in summer; floods of large rivers in Europe; and the establishment of large open habitats due to deforestation and drainage (Scharff and Langemark 1997; Barabasz and Górz 1998). Similarly, weather-regime changes were noted among important factors controlling invasion of parts of Europe – including the Baltic Sea Basin – by the leaf miner moth (*Cameraria ohridella*), an important pest of horse-chestnut trees (Sefrová and Lasuvka 2001), and by moths (e.g. Palmqvist 1999). Expansion of Mediterranean dragonflies into central Europe has likewise been linked to climate warming during recent decades (Ott 2001). It is important to bear in mind that evidence, of the kind presented above, for a climatic driving force for invasions is circumstantial and in many cases not well documented.

Changes in migratory distributional status

Apart from effecting absolute changes in species' geographical ranges, climate changes may influence relationships between wintering and breeding parts of the geographical ranges of migratory animal species, especially birds, but potentially also bats and migratory insects. Examples of the latter include certain butterfly and moth species in Finland (e.g. many Nymphalids), which depend on northward migration from southern populations.

Mild winters during recent years appear to have produced a tendency towards shorter migratory distances to wintering places, and even to wintering in breeding areas. Birds can change migratory behaviour in many ways (Fiedler 2003). Selected examples are increased proportions of sedentary individuals in the local populations, as noted for waterfowl such as mute swan (*Cygnus olor*), great crested grebe (*Podiceps cristatus*), greylag goose (*Anser anser*), and a number of other species (Švažas et al. 2001; Fiedler 2003), and for passerines like stonechat (*Saxicola torquata*), chiffchaff (*Phylloscopus collybita*), starling (*Sturnus vulgaris*) and many others. These findings were obtained by direct visual observations, as well as by strict analysis of ringing data sets (Švažas et al. 2001; Fiedler 2003; Tomiałojć and Stawarczyk 2003). Other effects that have been observed in conjunction with changing climatic patterns include changes in migratory directions and changes in the proportions of individuals with stronger or weaker migratory activity, as well as changes in the timing of movements to breeding and wintering areas (Fiedler 2003).

Changes in biome boundaries

Altitudinal shifts of vegetation are well-documented for many parts of the Earth. Higher temperatures and longer growing seasons, associated with climate change, have released new areas for colonisation by certain plant species (e.g. Parmesan et al. 1999). However, data for the Baltic Sea Basin are scarce and current studies only seem to permit one robust conclusion concerning biome-level range displacements, namely that endemic mountain plant species are threatened by the upward migration of more competitive sub-alpine shrubs and tree species, to some extent as a result of recent climate change (Kullman 2003; see *Changes in species distributions*, above).

4.3.2.3 *Potential Future Impact*

As noted above, the direct impacts of climate changes on species distributional limits (e.g. via physiological tolerance limits) are compounded with indirect effects, changes in non-climatic environmental factors, and – of perhaps overriding importance in the Baltic Sea Basin – the activities of humans. Given this complexity, future changes in species and biome distributions are extremely difficult to predict or to model, although many new technologies are being developed (e.g. Thuiller 2003; Araújo et al. 2005; Sparks and Tryjanowski 2005; Koca et al. 2006). In general terms, however, we may suspect that climate change of the magnitude described in Chap. 3 will accentuate the processes, already underway for some species, of colonising new habitats and localities rendered favourable by warmer temperatures and other forms of climatic amelioration. This might result in increasingly detectable northerly and upslope shifts in species and biome boundaries.

The unprecedented rate of climate modification expected in the coming century has important implications for the likehood of large-scale biogeographical change. Many species, especially trees and large animals, react to environmental variation with lags (Saether et al. 2005). Most trees, for example, have a generation time in the order of decades, and limited dispersal distances, most seeds spreading only a few metres from the parent tree. Palynological records show that many tree species can nevertheless spread by rates of 150–500 m year^{-1} during periods of range expansion (Huntley 1991; Grace et al. 2002), but this still amounts to only a fraction of the average rate of isotherm migration anticipated for the coming century (Chap. 3). Even if the seed dispersal of some species is able to keep up with climate change, local dominance shifts may be delayed by the long generation times, competition with resident species, required changes in soil structure, hydrology, chemistry, litter depth, requirements for mycorrhizae and other factors (Sykes and Prentice 1995; Huntley 1991; Malanson and Cairns 1997). Of course, human intervention, including the planting of "new" species in the context of commercial forestry, may accelerate range displacements for certain species (Lindner 2000).

Patterns of change in distributions are often asymmetrical, with species from warmer climates (lower latitudes or altitudes) invading fastest (Walther et al. 2002). This could produce transient assemblages lacking modern analogues, such as a temporary beech (*Fagus sylvatica*) dominated community within a "window of opportunity" following a warming-driven decline in Norway spruce, and prior to beech chilling failure, simulated by a forest dynamics model for southern Sweden (Sykes and Prentice 1995).

A number of model-based studies have explored potential changes in species distributions under future climate and/or atmospheric CO_2 concentrations. Most such studies of relevance to the Baltic Sea Basin have been concerned with major tree species.

The temperature changes predicted by climate models for the coming 50–100 years represent a general northward and eastward shift of the potential equilibrium ranges of species in climatic space. Using a static bioclimatic model, Sykes et al. (1996) examined the potential effects on tree species distributions of a climate scenario derived from the average of four GCM runs based on doubled CO_2 forcing. The model predicted substantial northward and eastward displacements in the ranges of the boreal forest dominants Norway spruce and Scots pine. The warm range limits of both species were predicted to migrate some 1200 km northwards across Sweden, and to retreat eastwards from southwestern Germany to the Belaussian-Russian border. Broadleaved deciduous tree species, especially birch, hazel, ash, aspen, oak and elm were correspondingly predicted to increase their potential cover in southern Scandinavia and northern parts of central Europe. Beech was predicted to colonise most parts of Scandinavia and the eastern Baltic Sea Basin, areas from which it is currently absent. The climate scenarios used in these studies were of generally stronger warming over the Baltic Sea Basin than recent assessments describe (Chap. 3). More recent results using the same model suggest that range shifts will be qualitatively similar but smaller (Bradshaw et al. 2000).

Koca et al. (2006) used a process-based dynamic vegetation model (Smith et al. 2001) to assess potential vegetation changes for Sweden under four regional climate model-derived scenarios for the late 21st century (Räisänen et al. 2003). The climate scenarios, generated by the RCAO model under SRES A2 and B2 scenarios, may be seen as representative of the range of RCM scenarios described in Sect. 3.5, and for overall projected trends as synthesised in Chap. 3. Lags imposed by tree generation times and competition between resident and colonising species are taken into account by the model. As in earlier studies using static vegetation models (Sykes and Prentice 1995; Bradshaw et al. 2000), expansion of boreal forest into alpine areas, and an increased northern extent of deciduous broadleaved species (beech, lime and oak) are predicted. Spruce and pine exhibit regeneration failure in southern Sweden, but remain present in declining abundance to the end of the 21st Century. The overall pattern throughout the boreo-nemoral and boreal zone is of an increasing regeneration of broadleaved species within ageing conifer-dominated stands. A modern analogy to this type of transient vegetation may be old-growth remnants of Białowieża Forest in Poland, where regeneration failure in Norway spruce has been linked to climate warming, while broadleaved trees (principally lime and ash) are gaining in dominance (Walankiewicz 2002; see *Changes in species distributions*, above).

Using a stand dynamic model, Kozak et al. (2002) predicted increasing dominance of Euro-

pean beech over Norway spruce in conjunction with a 2 °C increase in mean annual temperature in the Bieszczady Mountains in southeastern Poland. However, the simulated dominance shifts were gradual, with a new equilibrium being achieved first after several centuries of simulated vegetation dynamics.

4.3.3 Physiological Tolerance and Stress

Temperature, precipitation and light are the main factors affecting the large-scale distribution of plant species (see Annex 3.2.2). A temperature increase is generally expected to increase the primary production (Saxe et al. 2001; Sect. 4.3.4; Annex 3.2.2). A simultaneously increased frequency of extreme climatic events could, however, lead to an increased number of stressful weather situations, reducing the net primary production via vegetation damage. In this context, it is important to recognise that different species have different physiological tolerance limits. Stressful situations must therefore be defined in relation to species- and provenance-specific adaptations, not from a climatological point of view (Gutschick and BassiriRad 2003).

Plants become stressed when the physiological tolerance is exceeded, and this causes long-term changes in metabolism and resource allocation. The physiological processes can be altered for several months after ambient conditions have returned to normal. Both a sudden and extreme change in weather conditions and a longer period of unfavourable conditions, exceeding the endurance capacity of the organism, can become stressful. The intensity, duration, frequency and combination of stressful weather events determine the long-term impact on plant performance. This makes the impact of climate change on ecosystem structure and functioning inherently difficult to assess. Ecosystem modelling is required in order to develop management strategies, aiming to reduce the risk for damage and achieve long-term sustainability in forestry, agriculture and nature conservation.

4.3.3.1 *Acclimatisation and Adaptability*

One reason that an extreme weather event is not the same as an extreme situation for plants, and vice versa, is that the plants have the ability to acclimatise and adapt to changing weather conditions (Kozlowski and Pallardy 2002). Through acclimatisation, the range of tolerance is adjusted to avoid harmful stress. An example is the hardiness status of plants, a reflection of seasonal changes in temperature and light conditions. Trees from cool temperate climates in Scandinavia can tolerate $-30\,°C$ during midwinter when the trees are fully hardened, but a temperature backlash below $-5\,°C$ can be very harmful in spring when the trees have started to deharden (Larcher 1995). Acclimatisation thus creates temporal thresholds in tolerance to weather conditions.

The impact of climate on the performance of an individual plant will be affected by local conditions in topography, water availability and interaction with neighbouring organisms. Through adaptation, plant individuals can adjust to long-term site conditions. For instance, the root-shoot ratio can change in response to altered water availability. The rate of climate change will affect the adaptation ability, and plants at different life stages may differ in their ability to respond to altered conditions. At any given site, differences in species-specific capacity for acclimatisation and adaptability will affect the risk for weather-induced damage. Events exceeding the acclimatisation capacity of plants act as strong selection forces (Gutschick and BassiriRad 2003). Over time, genetic adaptations of species, as well as changes in species abundance, will result. A prerequisite is, however, that evolutionary adaptations are possible and sufficiently rapid to compensate for the changes in climate (Saxe et al. 2001).

4.3.3.2 Plant Resource Allocation

Damage to plants affects the allocation of carbohydrates and nutrients. Damage causes direct costs for healing and renewal of plant parts. Indirect costs result from a lowered efficiency of resource uptake or utilisation, sometimes mediated via altered symbiotic relationships (McLaughlin and Shriner 1980). In the event of subsequent attacks by pests and pathogens, an increased amount of resources must be allocated to defence. The total cost of chemical resistance depends not only on synthesis, but also on compound turnover rates and costs of transport and storage. These costs and the resultant effects on plant vitality are difficult to estimate. One reason is that the genetically-determined growth rates of plant species covary with stress resistance and resilience to damage by affecting the basic allocation between growth, storage and defence (cf. Chapin et al. 1990; MacGillivray et al. 1995; Heil

2001; Shudo and Iwasa 2001). Another reason is that the costs are regulated by damage severity, determined by the discrepancy between plant tolerance and the weather situation. For instance, a spring frost damage causing disruption of cell membranes can be reversed, but the consumption of energy and water is increased during repair (Linder and Flower-Ellis 1992; Sutinen et al. 2001). Irreversible frost damage, causing losses of foliage, can cause a long-term reduction in photosynthetic capacity. New leaves that are produced in a second flush could substantially reduce the carbon storage (McLaughlin and Shriner 1980). Frost damage can also cause necroses on bark and roots that demand resources for repair, and the necroses will be susceptible to pathogenic fungi until healed (Schoeneweiss 1975).

4.3.3.3 Recent and Historical Impacts

Dieback and decline have been reported during the late 20th century for several nemoboreal tree species, often in relation to the negative impact of air pollution. In general, it has been difficult to trace the cause and effect relationships due to interactions among abiotic stress factors and forest biogeochemistry (Nihlgård 1997). An example related to climatic effects is drought during late summer that can enhance the onset of winter hardening. This can result in an insufficient storage of carbohydrates that increases the risk for frost damage during spring (Sakai and Larcher 1987). Furthermore, a severe frost damage that kills parts of the root system can lower the ability to withstand summer drought, in turn lowering the capacity for winter hardening. This feedback loop has been found to initiate dieback of northern hardwood forests in North America during the 20th century (Auclair et al. 1996). In Sweden, both the resin flow of Norway spruce (Barklund et al. 1995) and dieback of oak (Barklund 1994) have been related to combined effects of mild winters, frost episodes and summer drought. By sorting the stress factors in a chronological order, they can be divided into predisposing, triggering and contributing factors. Predisposing factors increase the susceptibility to damage caused by triggering factors. Climate change can become a very important predisposing factor as the trees may lose their site adaptation, affecting resistance and resilience (Larsen 1995). The acute damage is commonly triggered by an extreme weather event, such as temperature backlashes. The predisposition determines the strength of the triggering factor needed to induce visible damage (Saxe 1993). Opportunistic pests and pathogens can attack trees that are weakened by damage, contributing to declined vitality or even mortality.

Strong winds are a significant agent of disturbance in forests, with a potential link to climate change. For Europe as a whole, disturbance-related damage in forests has generally increased in the last century, and part of this trend might be due to changes in the wind climate, although this conclusion is speculative (Schelhaas et al. 2003). The European trend is reflected, for example, in Swedish (Nilsson et al. 2004) and Danish (Holmsgaard 1986) time series of forest wind damage. In Sweden, changes in forest management practices, forest cover and tree structural properties have been invoked as likely contributing factors (Nilsson et al. 2004). It should be recalled, however, that statistically-significant trends in storm frequency or intensity have not been shown for the last century in the Baltic Sea Basin or its subregions (Sect. 2.1.5.4), nor do current scenarios agree on the direction or magnitude of future changes (Sect. 3.6.2.3).

4.3.3.4 *Potential Future Impacts*

Even though plants have the ability to acclimate and adapt, some of the expected changes described in Chap. 3 will increase the general risk for stressful weather events in comparison with the current climate in the Baltic Sea Basin. These climatic conditions are described below, in a seasonal order, together with their potential impact on vegetation. The impact on forest ecosystems is highlighted, as tree individuals are long-lived: trees that are young today will most probably experience a changing climate during their lifetime, and they will have to endure sequentially stressful and potentially damaging weather situations.

Higher winter temperatures and frost episodes

Due to winter hardening, changes in the risk for frost damage in plants do not equate simply to changes in annual mean temperature, mean and minimum temperatures of the coldest month or number of frost days. Damage occurs during frost episodes when the plants are not adequately hardened. This can happen at any time of the year, but is common after a longer period of warmer temperatures during spring when dehardening has

been initiated. Higher winter temperatures can induce changes in plant phenology, as thermal requirements for dehardening and budburst will be fulfilled earlier (Sect. 4.3.1). The impact will be dependent on species-specific reactions to temperature and light conditions (Heide 1993). Species that are strongly regulated by light will be less affected than species that are strongly regulated by temperature. In a warmer climate, the latitudinal variation in light and temperature can be an important factor for modifying the risk for frost damage. At the end of the 21st century, dehardening and budburst of Norway spruce has been projected to occur already in the beginning of the year in the southern part of Sweden. At that time, the seasonal temperature progression is slow and the frequency of temperature backlashes is high, thus the risk for spring frost damage will increase. The effect will be gradually less pronounced towards the north of Sweden, as the onset of dehardening and budburst will occur closer to the spring equinox, when the seasonal temperature progression is faster (Jönsson et al. 2004).

Heat spells and reduced summer precipitation

An increased frequency of heat spells, reduced summer precipitation, and increased summer evapotranspiration, as anticipated by many scenarios – especially for temperate areas of the Baltic Sea Basin (Sect. 3.5.2) – increase the risk for drought stress. Soil water potential, ambient light intensity, temperature and wind influence the severity. Plants respond immediately to water shortage by osmotic adjustments and closure of stomata (Kozlowsky and Pallardy 2002). This lowers the transpiration, as well as the photosynthesis and uptake of mineral nutrients. Mild drought does not completely inhibit photosynthesis, and this may increase the carbohydrate reserves and levels of defence compounds as the demand for shoot growth is reduced. Long term drought, however, reduces the carbohydrate reserves as maintenance respiration exceeds production (Larcher 1995). This non-linear response, affected by the duration and frequency of drought stress, will have an impact on tree vitality. By affecting leaf area, drought can lower photosynthetic capacity even between seasons (Tesche 1992; Löf and Welander 2000). Furthermore, changes in climatic conditions that increase the severity of heat spells and drought will tend to increase the risk for forest fires (McGuire et al. 2007).

Changes in autumn and winter precipitation

In areas with an increased precipitation during autumn and winter, the risk for waterlogging will increase. This can cause anaerobic root conditions and may kill parts of the root system. As a consequence, the plants will become more susceptible to drought stress and attacks by pathogens. Some woody plant species have the ability to adapt by changes in root structure, promoting oxygen transport to the roots, or by adjusting the metabolism to anaerobic conditions (Kozlowski and Pallardy 2002).

Due to climate change, a reduction in snow cover is expected (Sect. 3.5.2.4). In regions with a thin snow cover and low winter temperatures, the soil could be deeply frozen. In a simulation of soil frost in Finland at the end of the 21st century, the probability of frozen ground during midwinter was projected to increase in the southern part due to a reduced snow cover. Further north, the soil frost depth was projected to be reduced (Venäläinen et al. 2001), an effect of higher winter temperatures in combination with a sufficiently insulating snow cover (Sect. 3.5.2.4). Soil frost increases the risk for winter desiccation, occurring when the trees are exposed to higher temperatures in spring and the transpiration is increased but the frozen water cannot be taken up (Larcher 1995; Sutinen et al. 2001). The transpiration increase is more pronounced in evergreen tree species, making conifers more sensitive than broadleaf trees.

Strong winds

The risk for storm damage is affected by wind properties, in combination with other climatological factors. Several future climate scenarios point to increased average wind speeds and frequencies of strong wind events, especially in southern parts of the Baltic basin. However, uncertainty remains high (Sects. 3.5.2.3 and 3.6.2.3). The tree anchorage capacity is reduced by unfrozen soil conditions and high soil water content (Peltola et al. 1999). Both conditions are expected to increase in most parts of the Baltic Sea Basin due to warmer winter temperatures and increased precipitation during autumn and winter. Wind throw is one of the major threats in forestry, causing large losses in wood production and timber quality. The risk for subsequent damage to living trees by pests, such as the Norway spruce bark beetle (*Ips typographus*), is also increased (Wermelinger 2004). Furthermore,

the remaining trees may incur severe root damage caused by wind-swaying, and thereby a decreased ability to take up water and nutrients. Root damage also reduces future storm stability, and the risk for wind throw by the next storm is markedly increased (Nielsen 2001). The risk for wind throw is related to the forest age structure and management, with older and newly-thinned forests being more sensitive. Conifers are in general more vulnerable than deciduous broadleaved trees, as most storm events occur when the broadleaves have shed their leaves, reducing the wind impact. Differences in root structure make Norway spruce more prone to wind throw than Scots pine.

Geographical and temporal patterns of change

By combining climate modelling with ecosystem modelling, geographical patterns and gradients indicating the risk for weather events exceeding the physiological tolerance of different species can be distinguished. However, as discussed in several review articles, improved understanding of physiological processes is needed in order to reduce model uncertainties. At the individual tree level, the carbon allocation and storage is poorly understood, and current models lack the capacity to account for responses to complex and non-standard situations (Lacointe 2000; LeRoux et al. 2001). Tree nutrient status affects carbon allocation, responses to stressful situations and attacks by pests and pathogens (Nihlgård 1997; Andersson et al. 2000). Thus, the impact of climate change on root uptake ability, mycorrhizal symbiosis, decomposition and weathering of mineral nutrients needs to be considered. Plant performance will also be affected by the direct impact of climate on the performance of pests and pathogens (Ayres and Lombardero 2000; Strand 2000; Volney and Fleming 2000). A major problem is the lack of knowledge regarding dose-response relationships, and the lack of quantitative knowledge about combined effects of biotic and abiotic conditions on ecosystem structure and functioning (Loehle and LeBlanc 1996; Kickert et al. 1999; Hänninen et al. 2001; Karnosky 2003). For instance, the productivity of boreal conifer forests may be overestimated by about 40% if the effects of frozen soil and the energetic costs for repairing reversible frost damage are not considered (Bergh et al. 1998).

In the discussion of adaptation options, such as the selection of species and provenances when replanting forest, it is necessary to consider a range of climate change scenarios (Jönsson 2005). The central question is how the frequency, severity and duration of potentially harmful climatic conditions will change over time. The answer must relate to the species-specific ability for acclimatisation, local adaptation and genetic adaptation by natural selection. It is important to establish whether an increased risk for damage will be permanent and perhaps aggravated over time, or if the effect is temporary. Assessments of impact magnitude and subsequent predisposition to damage by other stress factors are needed. And last, but not least, the impact of climate change on nutrient availability and species-specific interactions with symbionts, pests and pathogens requires attention.

4.3.4 Ecosystem Productivity and Carbon Storage

Climate changes affect biogeochemical cycling within ecosystems by modifying the rates and modes of individual ecosystem processes. The physiological processes of photosynthesis and respiration in plants are of fundamental importance to the overall ecosystem response, controlling the net primary production (NPP), which drives individual growth, vegetation structural development, substrate input to litter, soil organic matter pools, and the entire terrestrial food chain. Non-climatic environmental drivers controlling NPP include atmospheric CO_2 concentrations, tropospheric ozone, soil nutrient status and pH. The latter are in turn subject to modification through deposition of atmospheric pollutants, especially nitrogen oxides and other acidifying compounds (see Annex 3.1.2.1).

Overall responses of ecosystem biogeochemical exchanges to environmental shifts are characterised by the differential temporal signatures of many constituent processes. Short-term physiological responses, such as the direct response of net photosynthesis to a change in temperature, may be modified by longer-term changes in, for example, plant structure, population dynamics, vegetation species composition and soil organic matter stoichiometry (Shaver et al. 2000). Disturbances and stress responses may further modify ecosystem exchanges, both via mortality and adjustments in the vigour of individuals (see Sect. 4.3.3).

The major mechanisms by which climate and other environmental drivers influence terrestrial ecosystem biogeochemistry are outlined in Annex 3.2.2.

4.3.4.1 *Recent and Historical Impacts*

Evidence that vegetation activity has generally increased at high northern latitudes since the early 1980s is apparent in optical measurements of land surface "greenness" (normalised difference vegetation index, NDVI) performed by satellites (Myneni et al. 1997; Tucker et al. 2001). The satellite-based trends have been independently reproduced as changing growing season length and NPP by a climate-driven vegetation model, confirming that the trends are a result of climate forcing and primarily increased average temperatures since the 1980s (Lucht et al. 2002). Changed allocation patterns related to soil resource availability may also play a role in the observed greenness trend, as has been shown for European Russia (Lapenis et al. 2005). Growth trends consistent with NDVI time series have also been seen in tree-ring data from northern forests (Kaufmann et al. 2004). NDVI trends for the Baltic Sea Basin are representative for high northern latitudes in Eurasia, and are strong compared with similar latitudes in North America (Tucker et al. 2001).

Leaf phenological changes were described in Sect. 4.3.1. Long-term observations point to an increased growing season, with spring having advanced by 6 days and autumn delayed by 5 days on average for Europe from the 1960s to the 1990s. For Scandinavia the advance may be smaller but in the same direction. The change is mainly related to temperature (Menzel and Fabian 1999). For forest trees in general, spring phenology has advanced by 3 days in a decade (Root et al. 2003).

Several studies point to recent productivity increases for ecosystems within or near the Baltic Sea Basin. Overall growth trends for European forest ecosystems have been positive during the last 50 years (Spieker et al. 1996). Climate changes (increased temperatures and an extended growing season) are one likely cause, although other factors, including the fertilisation effect of anthropogenic nitrogen deposition, and changed silvicultural management, are likely to play a role (Mund et al. 2002). Alekseev and Soroka (2002) observed up to 78% elevated growth (annual radial increment) for the latter half of the 20th century compared with the previous century based on dendrochronology of Scots pine trees from the northwest Kola Peninsula, the northernmost extent of that species' range. The increases were substantial for young trees (0–20 years) and progressively smaller for older trees, but trees > 100 years still showed 10% growth enhancement. Synergistic effects of increased temperatures and CO_2 concentrations were suggested as a likely explanation for the growth increases. At least part of recent observed treeline advance (see Sect. 4.3.2) is due to height increases and increased vigour on the part of stunted individuals above the main tree front, as opposed to seedling establishment beyond the current species limit (Kullman 2002; Grace et al. 2002).

In some cases, recent climate change may be associated with growth reductions. Based on dendrochronology, Dittmar et al. (2003) showed a declining growth trend for beech at higher altitudes in central Europe during the last 50 years. Correlations with climate parameters suggest that an increased preponderance of wet, cloudy summers coupled with late frosts reduced vigour and growth for this species. The negative growth trend was most pronounced for the period 1975–1995, even though increasing CO_2 levels, N deposition and an extended growing season would all be expected to have positive impacts on production. Interdecadal growth anomalies in Scots pine growing near treeline in northern Sweden have been related to changing growing season length and interference of recent warm, wet winters with the hardening-dehardening cycles (Linderholm 2002). Setbacks in the general pattern of treeline advance in the Swedish Scandes have been attributed to stress associated with short-term climatic extremes in the context of a generally milder winter climate (Kullman 1997; see also Sect. 4.3.3).

4.3.4.2 *Potential Future Impacts*

Changes in ecosystem productivity and carbon storage under future global change scenarios for the Baltic Sea Basin have been explored in a variety of studies. Apart from changing climate, changes in other environmental factors (e.g. CO_2, N deposition) and human management and land use have been considered. Most studies point to generally increasing productivity for the next 50–100 years (Kellomäki and Kollström 1993; Lasch et al. 2002; Nabuurs et al. 2002; Bergh et al. 2007; Koca et al. 2006; Morales et al. 2007; see also Sect. 4.3.5). This general trend for northern Europe is not necessarily reflected in southern Europe, where some studies suggest that increased drought may reduce ecosystem production in the future (Schröter et al. 2005; Morales et al. 2007; Zaehle et al. 2007). Bergh et al. (2007) explored

effects on production of Norway spruce and Scots pine of four regional climate model (RCM) scenarios for 2070–2100 for Sweden and Finland. NPP stimulation, according to a process-based production model, ranged from 10–50% for both species and was especially pronounced for younger stands of Scots pine and under the SRES A2 scenario, which represented higher CO_2 concentrations and stronger greenhouse forcing than the alternative B2 scenarios. An extended growing season was the most important factor underlying the predicted production increases. CO_2-stimulation of production was most important in southern and central Sweden and southern Finland, and coincided with increased summer water deficits there, suggesting a positive influence of increased CO_2 concentrations on plant water economy (Drake et al. 1997). Middle-aged Norway spruce stands showed the lowest increase in NPP, mainly as an effect of water demand caused by larger needle biomass and transpiration.

Koca et al. (2006) used a dynamic vegetation model driven by the same set of RCM scenarios as in the Bergh et al. (2007) study to explore coupled changes in forest species composition, structure, production and carbon storage in Sweden. Overall increases in NPP were predicted to augment biomass stocks, especially under an extended growing season in northern Sweden. In the south, NPP increases were reduced by increased summer soil water deficits. Increased light extinction in a denser canopy and a longer growing season were predicted to shift the competitive balance in mixed coniferous and broadleaved stands in favour of broadleaved trees, especially early successional trees such as birch (*Betula* spp.). This suggests that broadleaved trees may become a more important component of northern European forests under climate change, even if shifts in species range limits are not assumed to occur.

The future productivity increases suggested by most studies may not necessarily be accompanied by net carbon sequestration by ecosystems, since rising temperatures might amplify decomposition and mineralisation processes even more strongly than plant production, depleting soil carbon stocks. Simulations with a biogeochemical model under a range of climate scenarios representative for the regional climate trends discussed in Sect. 3.5 suggest that ecosystems of the southern Baltic Sea Basin may switch from sinks to sources of carbon by the end of the 21st century, while more northerly ecosystems are likely to re-

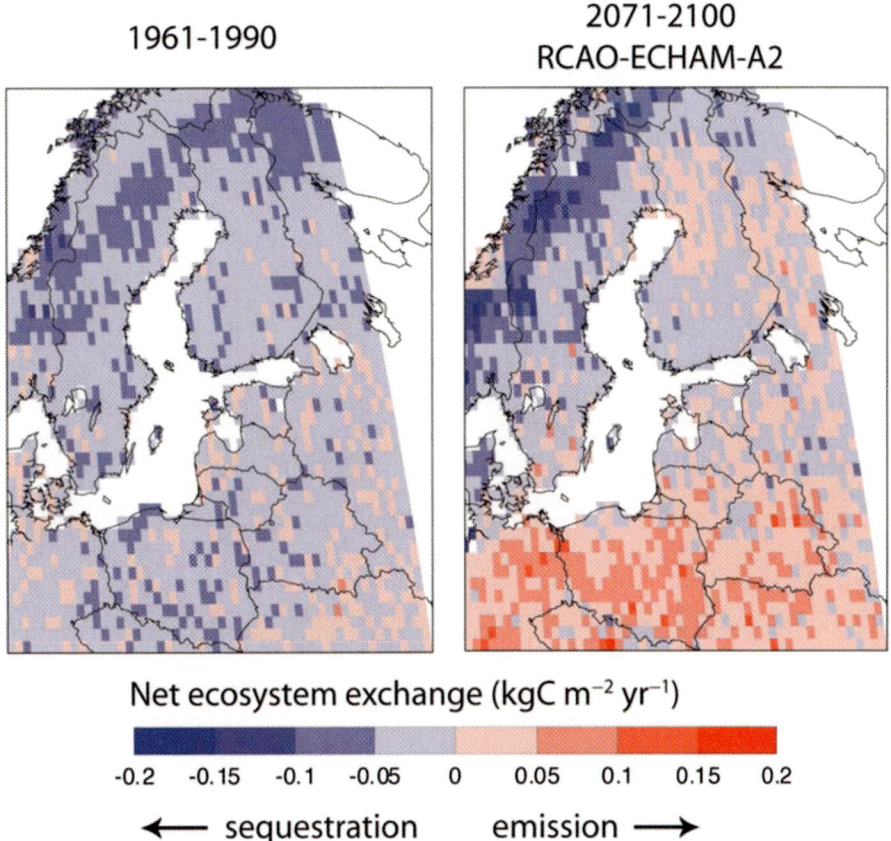

Fig. 4.4. Simulated net carbon exchange (NEE, kgC m^{-2} yr^{-1}) by ecosystems of the Baltic Sea Basin under the modern climate (1961–1990) and a representative regional climate model scenario (A2) for 2071–2100. Negative values represent a net uptake of carbon by ecosystems (sequestration), positive values a net release to the atmosphere (emission) (adapted from Morales et al. 2007)

main sinks (Koca et al. 2006; Morales et al. 2007; Fig. 4.4). Far northerly and mountain areas are projected to increase in sink strength as boreal trees advance onto extant non-vegetated or tundra areas (Fig. 4.4; see also Sect. 4.3.6).

A number of studies have explored the potential for changes in forest management to modify climate effects on forest yield (e.g. Nabuurs et al. 2002; Lasch et al. 2002). Studies incorporating alternative forest management scenarios indicate that adaptive management, including choice of the climatically best-adapted species and changed thinning and felling regimes, can impact productivity as much as climate change alone (Lindner 2000; Lasch et al. 2002).

Potential impacts of ongoing climate change on forest production are addressed in further detail in the following section.

4.3.5 Forest Productivity

4.3.5.1 *Forest Resources in the Baltic Sea Basin*

Following the FAO guidelines (UN-ECE/FAO 2000), forest is defined as a land area covered by natural or planted populations of woody species or trees, which can grow taller than 7 m in stands with a canopy closure of more than 10%. The forests in the Baltic Sea Basin represent mainly the boreal coniferous forests above and the temperate deciduous forests below 60 degrees of latitude (Table 4.2).

These conditions are characterised by a clear south–north gradient in temperature and west–east gradient in humidity. Low temperature and low nitrogen supply are the main limiting factors for forest growth in the boreal forests, whereas water limits forest growth in the temperate parts of the Baltic Sea Basin.

Table 4.2. Country-by-country presentation of the features of the current forests in the Baltic Sea Basin (UN-ECE/FAO 2000, Pisarenko et al. 2001, Federal Forest Inventory 2005, Karvinen et al. 2005)

Country	Forest area, M ha	Total volume, M m³	Total growth, M m³ per year	Total felling, M m³ per year	Felling, % of growth
Boreal coniferous forest zone					
Estonia	2.016	322	10.1	4.0	40
Finland	19.082	1904	75.0	53.0	71
Latvia	2.884	542	17.8	8.2	46
Lithuania	1.978	374	12.8	5.8	45
Russian Federation	16.440	2557	45.2	13.4	30
Sweden	27.264	2994	103.4	67.8	66
Total	69.668	8693	264.3	152.2	58
Temperate deciduous forest zone					
Belarus	2.620	401	12.3	3.2.	26
Denmark	0.223	30	1.9	1.2	65
Germany	0.697	193	6.9	3.2	46
Poland	8.942	1974	58.0	32.2	56
Total	12.483	2597	79.1	39.8	50
Grand total	82.150	11.290	343.4	192.0	56

In the Baltic Sea Basin, most of the forests are managed and forestry is mainly based on native tree species, which invaded this region after the last glacial period. However, many forested parts of the region have been cleared through land use practices. Only in northern Europe (e.g. Sweden, Finland, northwestern Russia) do forests still dominate the landscape (see Annex 3.2.2). Especially in the temperate parts of the Baltic Sea Basin, the current tree species composition is determined by past land use and management activities rather than by natural factors (Ellenberg 1986).

The total forest area in the Baltic Sea Basin is ca. 82 Mha. The total volume of stem wood is ca. 11,290 Mm³, representing mainly Scots pine and Norway spruce (together more than 70%). The proportion of deciduous trees is ca. 30% of the total volume, representing mainly birch. The role of deciduous trees in the species composition is more dominant in the temperate parts of Baltic Sea Basin than in the boreal parts, since the temperate parts coincide with the transition between the temperate deciduous forest zone and the boreal coniferous forests. The role of exotic species is most important in the temperate forest zone, but even there, their role is small compared to the native species. In the central European lowland, Norway spruce is widely used in forestry outside its natural distribution area, but it is currently doing quite well in these conditions. The forests in Finland and Sweden cover ca. 43% (by volume) of the total forest resources in the Baltic Sea Basin.

Throughout the region, growth exceeds cuttings with increasing stocking and maturing of forest resources. In the foreseeable future, the forest resources are expected to increase further due to the afforestation of agricultural land (see Sect. 4.2.2) and enhanced growth. The long time scale of forest production (rotation length of 40–160 years depending on species and region) implies that significant climate change will occur within the life span of the existing tree stands. On the other hand, the forests in the Baltic Sea Basin are also subject to a variety of other anthropogenic influences such as nitrogen deposition, sulphur emission, ozone depletion, and changes caused by unbalanced game populations, which are likely to interact with climatic change to bring about a complex series of responses that will differ from place to place. However, the genetic variability of most common tree species is probably large enough to accommodate the mean changes in temperature and precipitation (Beuker et al. 1996, Persson and Beuker 1997).

4.3.5.2 Objective of the Assessment

The following assessment of the climate change impacts on forests in the Baltic Sea Basin aims at identifying how the climate change may alter the forest growth and forest resources.

Forest growth refers to the rate of stem wood growth excluding other organs of trees (foliage, branches, roots). In this context, forest resources refer to the stocking of stem wood in forests at a specific point of time, and the term indicates the accumulation of stem growth during the period before the specific point of time. When the mass of stem wood plus the mass of other tree organs are multiplied by their fractional carbon content, one obtains the carbon storage in trees. The carbon storage in trees plus the carbon in soil (above- and below-ground litter, humus in the soil profile) gives the total carbon storage in forest ecosystems.

This assessment utilises widely the findings of the SilviStrat project (Kellomäki and Leinonen 2005), which addressed climate change impacts throughout Europe, including the boreal and temperate forests in the Baltic Sea Basin. The assessment builds on simulations by the BIOMASS model used in the SilviStrat project, applying a set of GCM-based climate change scenarios. These climate scenarios correspond closely to the average of the scenarios presented in Chap. 3; e.g. an increase of 3–4 °C in the annual mean temperature, and a 10–30% increase in the annual precipitation in Finland. The climate change impacts on the physiological performance of trees were further translated to the growth of stem wood and the consequent stocking of stem wood by applying the EFISCEN model. The BIOMASS and EFISCEN models and the climate scenarios used in the SilviStrat project, are outlined in Annex 3.2.

4.3.5.3   Sensitivity of Main Tree Species to Climate Change

Coniferous species

The left panel of Fig. 4.5 shows that there is a strong gradient in the growth response to the changes in stem wood growth of Scots pine from north to south. In the northern boreal forests, temperature is the only important driving force for the climate response, with generally positive effects on tree growth (Bergh et al. 2003), whereas in the temperate forests in the southern part of the Baltic Sea Basin, precipitation becomes more important and temperature increase may have both positive and negative effects (Lindner et al. 2005).

However, the regional differences are pronounced. In northern Finland, temperature affected increments very strongly and there was a large positive growth response. In southern Fin-

land, the general growth response was smaller and decreasing precipitation had a clear negative effect on growth (Kellomäki and Väisänen 1997). In northern Germany, the growth decline under high temperature and low precipitation was in relative terms as high as the growth increases in the north (Lasch et al. 2002). However, the response was quite different in the continental sites in Germany compared to the maritime sites in the Netherlands, where the growth of Scots pine was less drought-limited than in northern Germany (Lindner et al. 2005).

Regarding Norway spruce (the right panel of Fig. 4.5), a water limitation of growth and thus sensitivity to changes in precipitation can be expected when moving from colder high latitude to warmer southern boreal forests, as was the case for Scots pine (Bergh et al. 2005). The temperature response of Norway spruce is much more pronounced under a lower level of precipitation (Bradshaw et al. 2000; Lindner et al. 2005). The stronger temperature response of Norway spruce at low precipitation indicates that under water limitation the temperature-induced increase in evapotranspiration demand cannot be met. Water stress was thereby simulated to reduce the productivity of Norway spruce (Lindner et al. 2005). These findings are supported by two large-scale optimal nutrition experiments in northern and southern Sweden. The aim of these experiments was to demonstrate the potential yield of Norway spruce under given climatic conditions and non-limiting soil water by optimising the nutritional status of the stands (Bergh et al. 1999). In the southern study site (Asa), irrigation had a positive effect on stem growth, while no such effect was observed at the northern site Flakaliden. When the temperature response shows an optimum, there is normally also a clear sensitivity to changes in precipitation because of the interactions between water supply, temperature and evapotranspiration (Lindner et al. 2005).

Deciduous species

Regarding the deciduous trees, in the boreal zone the growth of birch was sensitive to the temperature increase, similarly to the response of the coniferous species; i.e. a strong positive effect of increasing temperatures in the north and much smaller temperature sensitivity with a maximum response at the temperature increase of 3–4 °C in the south. There was very little sensitivity to the

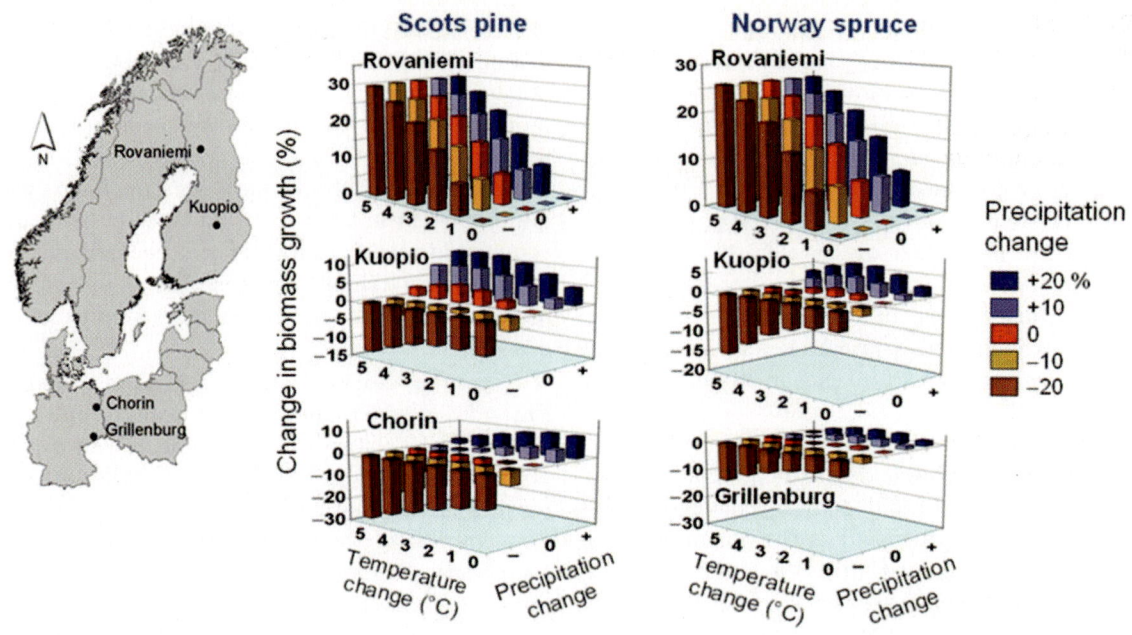

Fig. 4.5. Sensitivity of growth in Scots pine and Norway spruce in different parts of the Baltic Sea Basin compiled from the findings of the SilviStrat project (Lindner et al. 2005). Rovaniemi is representative of the northern boreal and Kuopio of the southern boreal forests in Finland, while Chorin and Grillenburg are representative of temperate forests in northern Germany. In the simulations with the BIOMASS model, the temperature and precipitation representing the current climate (1961-1990) were changed systematically; i.e. the daily mean temperature was increased by 0 °C, +1 °C, +2 °C, +3 °C, +4 °C, +5 °C. The precipitation was increased/decreased by −20%, −10%, 0%, +10%, +20%. A stable CO_2 concentration of 350 ppm was used. Simulations were run over 30 years starting 15 years prior to the culmination of growth and terminating 15 years after the culmination of growth

amount of precipitation. By contrast, in the temperate zone, the growth of oak and beech in northern Germany was clearly increased under higher precipitation and clearly reduced under reduced precipitation, with a larger sensitivity to drought for beech than for oak. Whereas the temperature increase was generally negative for the growth of beech, oak showed a weak temperature response with the optimum at the increase of +1 °C (Lindner et al. 2005).

4.3.5.4 *Impact of Climate Change on Forest Growth and Stocking*

Table 4.3 presents the mean annual increment at the beginning and at the end of the simulation period (2000–2100) used in the SilviStrat project under the current climate and the climate change (Pussinen et al. 2005) based on the EFIS-CEN model. The results are shown for country groups: the boreal group includes, for example, Finland and Sweden; the Baltic group, the Baltic states; the temperate-Atlantic group, maritime temperate conditions such as in Denmark and the northern Netherlands. The temperate continental group is representative, for example, for Poland and eastern Germany.

Under the current climate, the mean annual increment remained almost constant in the boreal region and increased in the Baltic region. The effects of climate change on forest growth were strongest in the Baltic and boreal country groups, where the mean annual increase was 12–13% under the climate change compared to the increment under the current climate. In the temperate-Atlantic country group, the climate change had only little effect on forest growth. The climate change effects on the mean annual increment differed very little between the two climate scenarios applied in the simulations.

Table 4.3. Mean annual increment (m^3 ha^{-1} yr^{-1}) in 2001–2005 (current climate) and 2096–2100 (current climate and two climate scenarios), baseline management (Pussinen et al. 2005; Annex 3.2.5)

Region and country group	2001–2005 current climate	2096–2100 current climate	ECHAM4		HadCM2	
Boreal	3.9	3.9	4.4	(+13%)	4.4	(+13%)
Baltic	5.5	6.2	6.9	(+12%)	7.0	(+13%)
Temperate-Atlantic	8.5	5.8	5.9	(+2%)	5.9	(+3%)
Temperate-continental	7.4	6.2	6.6	(+6%)	6.7	(+7%)

Table 4.4. Average regional growing stock m^3 ha^{-1} in 2000 (current climate) and 2100 (current climate and two climate scenarios), baseline management (Pussinen et al. 2005). The baseline management indicates the cutting rate, which is driven by the demand approximately at the current rate (Annex 3.2.5)

Region and country group	2000 current climate	2100 current climate	ECHAM4		HadCM2	
Boreal	114	151	185	(+22%)	182	(+20%)
Baltic	197	208	239	(+15%)	240	(+15%)
Temperate-Atlantic	209	461	480	(+4%)	485	(+5%)
Temperate-continental	255	389	421	(+8%)	418	(+8%)

Table 4.4 presents the average growing stock for the country groups at the beginning and at the end of the simulation period for current climate conditions and two climate scenarios. Under the current climate, the average growing stock increased throughout the Baltic Sea Basin, e.g. by ca. 20% in the boreal forests. In the temperate-Atlantic country group, the growing stock more than doubled between 2000 and 2100 even under the current climate, while the increase in the Baltic states was smaller (ca. 15%) due to the relatively high fellings there.

Under both of the climate change scenarios used in the simulations, the increase in growing stock was higher than under the current climate, particularly in the boreal and Baltic states. In the temperate-Atlantic region, the climate change had only little effect on growing stock and the main part of the increase was due to cuttings, which were substantially smaller than the growth.

The findings on the climate change impacts of growth and stocking imply that the carbon sequestration in the forests in the Baltic Sea Basin might increase even under the current climate. This sequestration is further enhanced under the climate change in linear relation to the increase in growth and stocking induced by the climate change (Karjalainen et al. 2002). The increasing growth and stocking imply also an increase in the litter input into the soil and thus a potential increase in the amount of carbon in the soil profile. This increase is most probably realised for the northern parts of the Baltic Sea Basin, but in the southern parts the enhancement of decomposition of soil organic matter may exceed the enhancement in litter production, resulting in little change, or a decrease, in soil carbon storage (Karjalainen et al. 2002; Lasch et al. 2005; see also Sect. 4.3.4).

4.3.5.5 Conclusion with the Management Implications

The model simulations show that in the northern boreal zone, temperature is the main limiting factor for tree growth. Changes in precipitation will not affect growth very much (Bergh et al. 1999, 2003). Any systematic increase in temperature will strongly stimulate growth and thus have only positive effects. At the other end of the climatic gradient, in the southern parts of the Baltic Sea Basin, tree growth is strongly water limited (Lasch et al. 2002, 2005).

Under such conditions, any increase in temperature will further enhance the water deficit and will thus lead to decreasing growth. Between these two extremes, it is the relationship between water supply and temperature that determines the overall response to the changes in climate.

Some species are more adapted to dry conditions than others, with a consequence that the simulation results often showed a temperature response optimum; i.e. small increases in temperature were positive for tree growth, but a stronger warming would lead to a water deficit and consequently declining growth (Lindner et al. 2005). Hence, the temperature response optimum is higher and the general impact of rising temperature is more on the positive side, when precipitation is concurrently increased. Under reduced precipitation the temperature response optimum is lower and the main effect of increasing temperatures is negative (Lindner et al. 2005).

In the temperate-continental zone, the forest growth is, in general, more constrained by water than by temperature, but the effect of elevated CO_2 may counteract or offset potential negative effects of changes in the climate (Freeman et al. 2005; Lindner et al. 2005). Problems may also be encountered with changes in the frequency and amplitude of extreme events such as drought events, storms and spring and summer frosts (CCIRG 1996, pp. 84–85; Sect. 4.3.3), with consequent biotic damages in forests.

The question of how climate change will impact major outbreaks of pests and pathogens is still to a large extent open. Nevertheless, many damaging fungi and insects may expand their distribution to the Baltic Sea Basin from central Europe and from further south (Parry 2000; see also Sect. 4.3.2). On the other hand, there is empirical evidence that elevated CO_2 and temperature may increase the resistance of deciduous species to herbivory and thus reduce the risk of forest damages (Mattson et al. 2004).

In the Baltic Sea Basin, there are only a few tree species of economic importance in forestry at present (principally Scots pine, Norway spruce, birch and oak). However, changes in tree species composition may be an appropriate adaptive management strategy. The following changes in tree species composition may be considered in implementing adaptive management strategies.

- Incorporation of other indigenous tree species, currently of minor importance in forestry, but with high potential for timber production or carbon sequestration under climate change.
- An increased share of broadleaved species, because broadleaved species are assumed to perform better under the climate change.
- Substitution of species sensitive to drought and to late spring frosts with more drought-tolerant and frost-resistant tree species or provenances.
- Replacement of low-productivity tree populations with high-productivity ones whenever the current population does not make full use of the potential productivity of a site.

The tree species choice is a basis for an appropriate adaptive management strategy, which further includes the adjustment of thinning (intensity, interval, pattern) to the changing productivity. In this context, the regulation of the rotation period is an effective way to manage the timber production and carbon budget of forests. Over the rotation, the timing and intensity of thinning determine the growth rate and stocking, which control the rate of carbon sequestration and the amount of carbon retained in trees and soils. In most European countries, growing stock is still increasing, because timber harvest (thinning, final felling) is below the increment. This implies that the total carbon storage in the forest ecosystems is increasing. On the other hand, the age-class distribution of the forests in the Baltic Sea Basin is shifting towards the older age classes, and the overall length of rotation is increasing. This implies that the rate of carbon sequestration is declining. However, replacement of over-mature old forests with new fast-growing ones with the objective of carbon sequestration in the biomass of the new stand is not always an appropriate option, as old forests and their specific characteristics are generally thought to have benefits for the conservation of biodiversity.

4.3.6 Arctic Ecosystems

4.3.6.1 *Introduction*

The arctic/subarctic region in northern Fennoscandia, together with comparable environments in high mountain areas of Scandinavia and central Europe, represents a minor proportion of the total Baltic Sea Basin (Annex 3.2.1). However, arctic and alpine areas are of particular interest in an environmental change context because they tend to exhibit a particularly close connectivity between climate, on the one hand, and ecosystem structure and function, on the other, providing sensitive indicators of the effects of climate change on ecosystems (Chapin et al. 2006). One reason for this is the lower direct impact of local human activities, for example through land use, compared with boreal and temperate parts of the region. The rate of climate warming in the arctic region is already faster than elsewhere, and this trend is projected to increase (ACIA 2005). Many aspects of arctic ecosystems are vulnerable to climatic and environmental change (Callaghan et al. 2004b), and arctic ecosystems are expected to undergo large and rapid changes. Some of these changes, for example in biodiversity of migratory species and in climate forcing, will have consequences outside the Arctic.

Arctic terrestrial ecosystems interact with the climate system by exchanging water, energy and trace gases between the biosphere and atmosphere. Historically, these interactions have resulted in cooling. Low, snow-covered vegetation reflects 60–80% of incoming solar radiation in winter, and 8–15% in summer (McGuire et al. 2007). The position of the circumarctic tree line has a substantial influence on global climate, with effects extending to the tropics (Bonan et al. 1992; Thomas and Rowntree 1992; Foley et al. 1994; McGuire et al. 2007). Due to constraints on decomposer activity by low soil temperatures and often anoxic conditions, a sizeable proportion of all carbon sequestered in photosynthesis is stored in tundra soils, which contain about 12% of global soil carbon (Jonasson et al. 2001). Further carbon stores exist in permafrost.

Major concerns about the impacts of climate and other environmental changes on arctic ecosystems focus on changes in biodiversity, which could have implications for the sustainable use of natural resources by northern peoples, and changes in ecosystem function, which could enhance global warming. The tangible services that are provided by the subarctic ecosystems of Finland and Sweden include products from the domesticated reindeer (meat, skins, bone for handicrafts), firewood, berries, tourism, as well as hunting and fishing (Laine et al. 2003). Additional services include the sequestration of CO_2 (that might otherwise enter the atmosphere and contribute to greenhouse forcing), and a catalogue of habitats and biodiversity of great conservation value.

As explained below, "subarctic" is commonly used to denote high-latitude tundra-dominated vegetation in Fennoscandia. For convenience, however, we employ the term "arctic" when referring to arctic and subarctic areas collectively.

4.3.6.2 *Characteristics of Arctic Ecosystems*

Compared with Russia and Canada, the arctic areas of Sweden and Finland are small: together with similar areas of mainland Norway they cover about $300,000 \, \mathrm{km^2}$ (Fig. 1.11, Chap. 1). If a strict definition of arctic vegetation is applied (CAVM Team 2003), then arctic tundra ecosystems do not occur in northern Sweden and Finland despite their high latitudes, because the arctic species found there only occur at higher elevations and are a result of altitudinal effects on climate. However, subarctic vegetation characterised by open forests of low-stature trees and tundra-like vegetation at higher elevations are extensive. The treelines are usually formed by small, polycormic and twisted mountain birch (*Betula pubescens* ssp. *czerepanovii*). The arctic region is characterised by low levels of net primary production, largely because of low nutrient availability and a short growing season, but also because woody plant species are restricted in distribution and stature. Surprisingly, although biodiversity is generally acknowledged to be low in the Arctic, a total of over 7200 terrestrial and freshwater species of animals, plants and fungi occur in an area of $22,705 \, \mathrm{km^2}$, situated between $68°$ and $70°\,\mathrm{N}$ surrounding the Kevo Subarctic Research Institute in northern Finland (Callaghan et al. 2004b). Wetland areas and lakes are extensive: wetlands contain vast stores of carbon and many of the peatlands, particularly in Finland, are exploited.

Recent climate warming has been relatively moderate in subarctic Finland and Sweden: mean annual temperatures have increased by about $1\,°C$ since the beginning of the instrumental record, with the most consistent warming in spring (Kohler and Brandt 2004; Sect. 2.1.2). There have been two discrete periods of warming and, since 2000, some cooling is apparent in some northern areas (Callaghan et al. 2004b), supported by recent observations of phenology in the region (see below).

The warming in northern Sweden is associated with an increase in length of the growing season (defined by temperature thresholds) of up to three weeks (Holmgren and Tjus 1996). Precipitation in the region is extremely variable, particularly in northern Sweden, where there are major gradients from a minimum of about $300\,\mathrm{mm\,year^{-1}}$

in the rain-shadow of the Abisko Alps to about 1000 mm year^{-1} close to the Swedish–Norwegian border (see also Annex 1.2.3). There has been a positive increase in snow depth at Abisko of 20 mm per 10 years over the period 1914–2004 (Kohler et al. 2006).

Contaminant concentrations in the Subarctic are relatively small, but there is some tendency to increase towards the east. Mean annual total deposition of nitrogen varied between 0.01 and 0.06 gN m^{-2} between 1988 and 1998 at Abisko, with no apparent trends. At Kevo, northern Finland, sulphur deposition has clearly decreased in recent decades, while the annual nitrate deposition has remained low throughout the period 1980–1996 (Kulmala et al. 1998).

Infrastructure development over the past 60 years has had a major impact on habitat fragmentation and reindeer pastures, and this trend is expected to accelerate (ACIA 2004), with increasing impacts on wildlife (Nelleman et al. 2001). Exploitation of natural resources has had smaller impacts, and forest expansion during the last century might be more related to reduced dependence on natural resources such as firewood (Emanuelsson 1987; Laine et al. 2003) and abandonment of agricultural land (Eilertsen 2002) than to climate change. Reindeer populations rose in the latter half of the 20th century and caused damage to vegetation in some areas (Bernes 1996) but are generally decreasing at present.

4.3.6.3   Recent Ecosystem Changes

Compared with many other northern areas, monitoring and research into the subarctic ecosystems of northern Finland and Sweden have been extensive, due to the accessibility of the area, a relatively advanced infrastructure for research, and a long history of settlement by indigenous and non-indigenous populations. A number of long time series of data relevant to the ecosystem impacts of climate change exist. Recent changes have been detected in both ecosystem structure and function. Attributing causes of these changes is difficult, but current environmental manipulation experiments are contributing both to understanding of current changes and projection of future changes, as outlined below.

Species performance

Monitoring of plant performance for *Diapensia lapponica* (Molau 1996) and *Carex bigelowii*

(Callaghan et al. 1996) and retrospective analyses of the growth and growing point dynamics of the vascular cryptogam *Lycopodium annotinum* have provided climate-related growth records for at least two decades (Callaghan et al. 1996). These analyses show individualistic responses of the plants to climate variability. *D. lapponica* was retarded by increased precipitation (Molau 1996). *C. bigelowii* showed an apparent cyclical pattern in flowering on a 3–4 year cycle, but no overall trend from 1984 to 2000 (Brooker et al. 2001). *L. annotinum* showed a strong buffering ability against climatic variations, which led to stable clonal growth.

Monitoring records reveal a breakdown in the cyclical nature of lemming and vole populations in Fennoscandia since the 1990s and this has been attributed to changing winter snow conditions that affect feeding, nesting and breeding under the snow during winter (Yoccoz and Ims 1999). Further south, arctic fox and snowy owl have declined severely in the past decades in northern Fennoscandia (Angerbjörn et al. 2001).

While some northern specialist species are declining in the Baltic Sea Area, some species with southern distributions are extending their ranges northwards. Moose and red fox have increased in abundance in the north (Hersteinsson and MacDonald 1992), while North American mink is becoming an invasive species in some areas (Bernes 1996).

There is some evidence that insect pest outbreaks resulting from increased populations and, to some extent, range shifts (Tenow 1996) in response to climate warming (McGuire et al. 2007; Chapin et al. 2005; Juday et al. 2005) are increasing in the boreal forest and forest tundra and are causing major disturbances. Recent large and extensive outbreaks of insect pests such as the autumn moth (*Epirrita autumnata*) in Fennoscandia are associated with warmer winters (Juday et al. 2005; Callaghan et al. 2004a): the winter minimum temperature is critical as eggs are killed by temperatures below −36 °C (Neuvonen et al. 2001).

Ecosystem structure

As noted in Sect. 4.3.2, the tree line has advanced in general in the Swedish part of the Scandes mountain range over the last century (Kullman 2002). In the Abisko area, there has been an increase (1940s to 1980s) in tree line altitude of 20–50 m (Sonesson and Hoogeseger 1983), and a recent study using aerial photography found a ca. 20 m average increase between 1959 and 2000 (Hållmarker 2002), giving rates of increase of about 0.5 m year^{-1}, or 40 m per °C of mean summer air temperature (Callaghan et al. 2004b).

Reindeer herding influences vegetation dynamics, including tree line, locally (Helle 2001; Tømmervik et al. 2004). Large-scale overexploitation by reindeer in the Swedish mountains is not evident (Moen and Danell 2003), and reindeer there have smaller impacts on vegetation than those imposed by small rodents, even in non-peak years for rodent populations (Olofsson et al. 2004), but over-exploitation is evident elsewhere (Bernes 1996).

Ecosystem functioning

Satellite monitoring of the greening of vegetation has recorded a delay in the onset of the growing season in many northern areas of Fennoscandia over the last few decades (Stöckli and Vidale 2004; Høgda et al. 2001; see also Sect. 4.3.1). The onset of autumn has been delayed even more so that there has been an overall increase in the length of growing season in the Scandes, as also noted for high-latitude areas of the northern hemisphere as a whole (Myneni et al. 1997; Høgda et al. 2001). In contrast, some areas in the very north of the Baltic Sea Basin are experiencing a shortening of the growing season. Independent observations on the ground have generally failed to identify any phenological trends in 144 species above the tree-line in the past 10 years in the northern Scandes (Molau et al. 2005). This is also the case for the date of bud burst by mountain birch over the past 70 years in the Scandes (Karlsson et al. 2003) and over the past 20 years in the north of Finland (Høgda et al. 2001).

Analysis of satellite images also reveals areas in the Scandes where there have been increases both in plant production and carbon storage in recent decades (Myneni et al. 2001; Nemani et al. 2003).

In the north of the Baltic Sea Basin, discontinuous/sporadic permafrost, often occurring beneath small, local raised areas (palsas) in mires, is degrading (Christensen et al. 2004; Malmer et al. 2005). A recent analysis of Western Utsjoki (3,370 km^2 study area) in northern Finland suggested that the palsas are disappearing at a high rate (Luoto and Seppälä 2003). Associated with a degradation of discontinuous permafrost in a subarctic Swedish mire, many dry habitats

and their vegetation have been lost, and there has been a shift towards a greater abundance of wet habitats with higher methane (CH_4) emissions over the past 30 years (Christensen et al. 2004; Malmer et al. 2005). This trend contrasts with that found in areas of continuous permafrost in arctic North America and Russia, where warming-induced permafrost thaw leads to drying (Smith et al. 2005) and increased carbon emissions (Oechel et al. 1998).

In Scandinavia, mountain areas have been modelled to be small carbon sinks (Zhuang et al. 2003; Koca et al. 2006; Wolf et al. 2007b). Available measurements of the carbon balance are contradictory: a Finnish mire showed a small release of $32 \pm 10\,\mathrm{gC\,m^{-2}}$ year^{-1} (Saarnio et al. 2003), whereas a subarctic mire in northern Finland was a carbon sink of $7 \pm 5\,\mathrm{gC\,m^{-2}}$ year^{-1} (Aurela et al. 2002). Uncertainties are high, both due to understanding and implementation of carbon cycling processes in models (e.g. respiration), and due to the high spatial variability in carbon sink or source patterns (Christensen et al. 1999). The question of whether arctic areas of the Baltic Sea Basin currently act as a net source or sink of carbon remains unresolved.

4.3.6.4 *Projected Ecosystem Changes*

Within the last 15 years, environmental manipulation experiments, measurements of carbon fluxes in the field, remote sensing and modelling have given important insights into the future responses of ecosystem structure and function to changes in climate and other environmental factors such as atmospheric CO_2 concentrations, UV-B radiation, herbivory and plant nutrient availability.

Species performance

Analyses of current patterns of biodiversity in the Arctic in relation to latitudinal/temperature gradients suggest that species richness will be sensitive to changes in climate and will increase overall as forest displaces tundra and species associated with the forest extend their ranges northwards and upslope (Callaghan et al. 2004a; see also Sect. 4.3.2). At the same time, the dominance and abundance patterns of many species over the landscape will change in accordance with changing mosaics of moisture and snow cover. Some species may be expected to decline near their southern range margins. Specialisations of current arctic plant and animal species to extreme environments are likely to constrain their abilities to compete with species invading from the south. As the latitudinal temperature gradient within the Arctic is steeper than for any other biome, and outlier populations of more southerly species frequently exist in favourable microenvironments far north of their centres of distribution, the migration northwards and local increases in abundance are very likely to occur more rapidly in the Arctic than in other biomes (Callaghan et al. 2004a).

Experiments throughout the Arctic, including the Subarctic of Sweden and Finland, showed that tundra plant species generally responded positively to experimental summer air warming, showing increased growth (Molau 2001; Arft et al. 1999; Press et al. 1998; Parsons et al. 1994; Shaver and Jonasson 1999); changes in phenology (Welker et al. 1997; Alatalo and Totland 1997; Jones et al. 1997; Stenström et al. 1997) and increased reproductive success (Molau 1997). However, some responses of reproductive effort were delayed (Arft et al. 1999; Molau 2001). Increases in the growth of woody plants led to changes in the architecture and height of the vegetation (Arft et al. 1999; Press et al. 1998). Species composition of plant communities changed as species and species groups responded individualistically (van Wijk et al. 2003), as dominance changed (Press et al. 1998), as some species declined and, occasionally, as new species became established (Robinson et al. 1998). Soil warming showed only transitory increases of nitrogen mineralisation and associated increases in plant growth in northern Sweden (Hartley et al. 1999).

Experimental elevation of summer temperature has shown that many invertebrates can increase population growth as long as there is no desiccation. Insects in particular are very likely to quickly expand their ranges (Virtanen and Neuvonen 1999). Some species, such as aphid species on Svalbard, can change their life history to take advantage of summer air warming (Strathdee et al. 1995). Increases in ice-crust formation during warmer winters is likely to reduce winter survival rates of a range of animals, including soil-dwelling springtails (Coulson et al. 2000), small mammals (Aars and Ims 2002), and large herbivores such as reindeer (Aanes et al. 2000; Klein 1999; Syroechovski and Kuprionov 1995).

The effects of increased precipitation on plants are less consistent than those of air warming, and non-significant in a meta-analysis (Dormann and

Woodin 2002). For the Baltic Sea Basin, experiments showed no response for arctic dwarf shrubs and *Calamagrostis lapponica* (Parsons et al. 1994; Press et al. 1998), whereas for evergreen species there was an effect of water addition (Phoenix et al. 2001). Mosses benefit from moderate summer watering (Potter et al. 1995; Phoenix et al. 2001; Sonesson et al. 2002) and nitrogen fixation rates by blue-green algae associated with the moss *Hylocomium splendens* were increased (Solheim et al. 2002).

Shading (simulating competition) for 6 years clearly disadvantaged some plant species in an experiment in Swedish Lapland (Graglia et al. 1997), and flowering of the dwarf heather-like shrub *Cassiope tetragona* ceased when it was shaded (Havström et al. 1995). In a nearby experiment in a birch forest understory community, shading by increased growth of vascular plants probably reduced the cover of the moss *Hylocomium splendens* (Potter et al. 1995). Similar experiments throughout the Arctic suggest that mosses and lichens could become less abundant when vascular plants increase their growth in response to warming (Cornelissen et al. 2001; van Wijk et al. 2003), although warming in summer and winter snow addition increased the biomass of *Sphagnum fuscum* in the Subarctic (Dorrepaal et al. 2004).

Addition of nutrients (simulating increased decomposition as a response to soil warming) increased canopy height of tundra vegetation (Press et al. 1998), but the response was individualistic (Parsons et al. 1994): the grass *Calamagrostis lapponica* was stimulated by a factor of more than 18 (Press et al. 1998); graminoids are generally responsive to nutrient addition (Cornelissen et al. 2001; van Wijk et al. 2003).

CO_2 enrichment experiments have shown that responses of arctic plants are generally early and transient (Gwynn-Jones et al. 1997), although effects on microbial communities (Johnson et al. 2002) and frost hardiness (Beerling et al. 2001) could have longer-term consequences. For example, CO_2 enrichment and increased temperatures influenced mycorrhizal colonisation in a dwarf shrub ecosystem (Olsrud et al. 2004), with potential consequences for plant nutrient uptake, allocation and long-term carbon storage in the soil. The lack of a growth response in mosses (Sonesson et al. 1996) and lichens (Sonesson et al. 1995) may be a result of down-regulation as an aspect of adaptation to the high levels of CO_2 that these organism groups experience, even under present-day ambient CO_2 levels, in their habitual microenvironments close to the ground surface (up to 1000 ppm: Sonesson et al. 1992).

Ecosystem structure

Most information on plant responses to climate warming is limited to the short-term and small-plot scales. Even the longest time scales – up to two decades – of some monitoring studies are relatively short compared to generation times and successional rates in arctic vegetation. Because of great longevity and a prevalence of clonal growth habits among arctic species, it is difficult to extrapolate plant responses from the individual plant to the population. Magnitudes of response may also vary between sites (Henry and Molau 1997; Karlsson et al. 2003; Jonasson et al. 1999). Process-based modelling may offer the greatest potential to explore structural dynamics in the long term and on geographically extensive scales.

Model projections for Finland and Sweden under a scenario of a ca. 5 °C increase in average temperature combined with a 25% increase in precipitation (SRES-B2 scenario from the REMO RCM; Keup-Thiel et al. 2006) suggest that, by 2100, boreal forest could dominate 75% of grid cells on a $0.5 \times 0.5°$ grid now dominated by tundra and/or shrubs (Wolf et al. 2007b). Alpine heath might decrease by as much as 75–84% of its current extent in the Swedish mountain areas in the next 100 years as a consequence of tree line shifts (Moen et al. 2004). However, there is great uncertainty, as the models do not currently have the ability to predict impacts of changing land use, hydrological changes due to thawing permafrost, and insect damage (but see Wolf et al. 2007a, c).

Vegetation models suggest an increase in tree cover and a decrease of tundra vegetation in the next 100 years (Fig. 4.6; Wolf et al. 2007b; Moen et al. 2004). Factors that could enhance or mitigate forest advance into tundra include herbivory (Cairns and Moen 2004), disturbance events such as forest fires (McGuire et al. 2007), and constraints on tree line advance such as inertia (Chapin et al. 2005), human activities (Vlassova 2002), geographical barriers, and inappropriate soil conditions (Callaghan et al. 2004a).

Trophic level structure is simpler in the Arctic than elsewhere, and climate warming is expected to change the abundance of keystone species, which is expected to lead to ecological cascades. The dynamics and assemblages of ver-

tebrate predators in the Arctic are almost entirely based on lemmings and other small rodents. The lack of population peaks since the 1990s in Fennoscandia have already contributed to loss of predators such as the arctic fox and snowy owl (Bernes 1996). Reductions in the abundance of lemmings also affects the populations of geese and waders that are used as alternative prey by foxes (Summers and Underhill 1987).

Ecosystem functioning

Models of net primary production estimate an increase of ca. 44% over the next 100 years in the non-forested area of Sweden and Finland (Wolf et al. 2007b). The projected increases in NPP are likely to be constrained in some areas by increased disturbance as detailed above (McGuire et al. 2007; Callaghan et al. 2004a). In the wider area defined by Greenland in the west and the Russian European Arctic in the east, NPP and carbon storage in vegetation and soils are projected to increase less (46% and 0.04 PgC, respectively) between 1960 and 2080 than in other areas of the Arctic globally (Sitch et al. 2003; Callaghan et al. 2004a).

Expansion of shrubs into tundra, displacement of tundra by the boreal forest, and earlier leaf-out might decrease albedo in both summer and winter (Wolf et al. 2007b) and should cause substantial heating of the atmosphere (a positive feedback to climate) that will also accelerate the subsequent replacement of tundra by boreal forest (McGuire et al. 2007). The advance of shrubs and forest onto the tundra will cause complex interactions between the generally expected negative feedback of longer-term carbon sequestration and the positive feedback of reduced albedo. Models suggest that there will be a clear increase in NPP in the Scandes (Wolf et al. 2007b), but in terms of carbon sequestration this might be offset by increases in soil respiration and increased disturbance (see Sect. 4.3.4). The uncertainties surrounding model predictions of future carbon storage are high, at least in quantitative terms (Callaghan et al. 2004a).

4.3.6.5 Implications

Climate warming is likely to lead to greater production, northwards expansion of agriculture and greater choice of crops in the Arctic as a whole. However, moisture stress and diseases such as

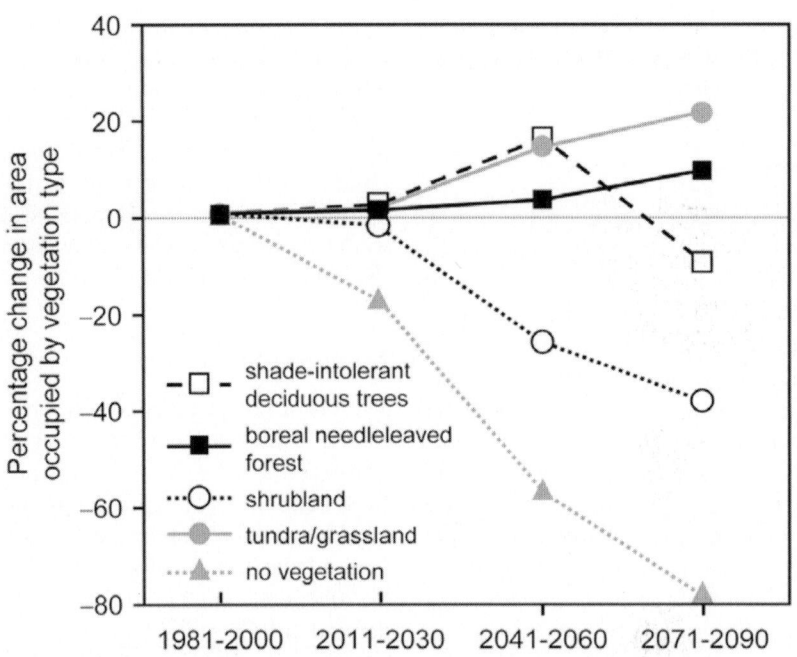

Fig. 4.6. Vegetation changes under a future climate scenario for northern Europe. Projected changes of dominant vegetation within $0.5 \times 0.5°$ grid cells according to a dynamic vegetation model (adapted from Wolf et al. 2007b)

potato blight might reduce yields in some areas (Juday et al. 2005). Increases in the forested area and shrublands, together with increased forest production in the Fennoscandian Subarctic (Wolf et al. 2007b) suggest potential increases in forest industries (although increased pest outbreaks and fire might reduce this potential) while innovation would lead to improved adaptation by local people to ecosystem changes (Chapin et al. 2006).

Reindeer herding is likely to be negatively influenced by future changes in climate conditions. Lichens, which are an important food source for reindeer, will decrease with the predicted increase of biomass of vascular plants (Cornelissen et al. 2001). However, the increase in shrubby vegetation in other areas might provide food for reindeer there. The earlier onset of spring greening will have an impact on the foraging patterns of reindeer. Due to such large changes in vegetation, there will be a need for adaptation by reindeer herders; for example, in the choice of routes to move between winter and summer grazing lands, the timing of migration and the selection of new pastures.

Conservation management will become increasingly important, but decisions need to be made

about the goals and methods of conservation in the future (Usher et al. 2005; Klein et al. 2005). Current concepts of conservation of threatened species need to take into account the likelihood that some currently widespread arctic species are likely to be disadvantaged by climate warming, while some currently rare species might become more common. Current protected areas protect against direct human actions, not against climate change leading to vegetation-zone shifts and changes in biodiversity (Usher et al. 2005; Callaghan et al. 2004a).

Tourism (see also Annex 3.1.3.2), an important activity for the economy of the European Subarctic, might be affected in various ways: while hitherto scenic mountains and ski slopes might be covered by forest – which could deter some tourists – ecotourism might increase as northern landscapes and their wildlife become ever more exotic.

The present uncertainty involved in calculating the magnitude and even sign of future radiative forcing from trace gas emissions for the region, and of balancing the possible opposing feedbacks from trace gas fluxes and albedo (see above), prevents a definitive statement from being made about the future "ecosystem service" of feedbacks to the climate system.

4.3.7 Synthesis – Climate Change Impacts on Terrestrial Ecosystems of the Baltic Sea Basin

A number of observed and potential impacts on terrestrial ecosystems of the Baltic Sea Basin emerge from the above review. These are summarised in Table 4.5. The robustness with which it can be stated that any particular class of impact has occurred, is occurring or is likely to occur in the future naturally depends on the extent of the available data and mechanistic understanding. Some classes of impacts are supported by direct observations, others inferred indirectly from observations, models, or findings in other regions with comparable climates and ecosystems. Some are reported in a single study, others in multiple studies spanning different sites, regions, species or biomes. For some types of impacts, the mechanism of response to climate and other factors is well understood, while for others, the climate connection remains tenuous or speculative.

To help guide policy and adaptation decisions and the setting of research priorities, we characterise robustness in terms of three levels of confidence for recent historical impacts, and two for potential future impacts. Our criteria are as follows:

High confidence – this categorisation is reserved for recent and historical impacts only, in cases where the impact has been described based on *observations or measurements* of *several* species, ecosystems or localities (depending on the nature of the impact) within the Baltic Sea Basin; *and/or* the impact has been observed in other comparable regions *and* the *mechanism* of climate response is well understood.

Medium confidence – in the case of recent and historical impacts, the impact has been described based on *a few/occasional* observations or measurements within the Baltic Sea Basin *and* the mechanism of response to climate is well understood. Projected future impacts are designated this confidence level where the impact has been projected or modelled for the Baltic Sea Basin using several approaches *and* the mechanism of response to climate change is well understood.

Low confidence is assigned to recent or historical impacts for which a mechanism of climate response has been proposed in the absence of irrefutable observations. In the case of projected future impacts, this confidence level is assigned where the impact has been hypothesised for the Baltic Sea Basin, or projected or modelled in a single study.

Major sources of uncertainty emerging from the review are listed in the table. These include technical and methodological aspects of uncertainty (e.g. the quantity or quality of available data), but also uncertainties related to the limited understanding of mechanisms, and to the role of antagonistic processes and non-climatic driving forces with the potential to ameliorate or reverse the predicted impact.

A general conclusion that may be drawn is that climate change during the last 30–50 years has already given rise to measurable changes in communities and ecosystems of the Baltic Sea Basin. Robust indicators of change include the advancement of spring phenological phases in certain plant species, changed migratory patterns and behaviour – for example in migratory birds – upslope displacement of the alpine treeline, and increased land-surface greenness as plants and ecosystems adjust to improved growth conditions and a richer carbon supply.

Many of the current trends may be expected to continue for at least some decades into the future, as atmospheric CO_2 concentrations continue to rise, and the climate continues to warm. However, future projections are intrinsically uncertain, and for a number of impacts, current trends may be broken or reversed due to system-internal feedbacks, shifts in controlling processes, non-climatic drivers and system transformations – perhaps most importantly, in conjunction with the appearance of novel pathogens, pests or invasive species.

Nearly all of the ecosystems of the region are to a greater or lesser extent managed, and the climatic impact may in many cases be ameliorated (or enhanced) through human intervention – for example, via adaptive species choice in forestry. The most important source of uncertainty with regard to many impacts is the future development in non-climatic, anthropogenic drivers of ecosystem processes, including greenhouse gas emissions, deposition rates of atmospheric pollutants, land use changes and their underlying socio-economic controls: human populations, markets, international trade, and technological development.

Table 4.5. Climate change impacts on terrestrial ecosystems of the Baltic Sea Basin. ✓✓✓ = high confidence; ✓✓ = medium confidence; ✓ = low confidence (see main text)

Class of impact	Impact of recent climate change (Hypothesis 1)	Projected impact of future climate change (Hypothesis 2)	Relevant uncertainties
Phenology			
Shift towards earlier spring and summer phases in plants	Advancement by 5–20 days over last 50 years in many species ✓✓✓	Advancement by 1–3 weeks depending on temperature increase ✓✓	Increased risk for temperature backlashes could modify responses
Shift towards later autumn phases in plants	Delay by 5–30 days over last 50 years in some species ✓✓	Delay by 1–3 weeks depending on temperature increase ✓✓	Data quality
Extension of phytophenological growing period	Extension by ca. 20 days over last 50 years in many species ✓✓✓	Extension by 2–6 weeks depending on temperature increase ✓✓	1. As above 2. Effects of pollutants, UV-B, O_3
Biogeography and community structure			
Changed species distributions	Cold limit range boundary shifts tracking isotherm migration for some plant and animal species ✓✓	1. Continued cold limit range boundary shifts ✓✓ 2. Range retraction near warm limit ✓	1. Impacts of changing climate extremes 2. Impacts of non-climate abiotic drivers 3. Population/community lags 4. Trophic interactions, e.g. effects of pests, pathogens, predators, herbivores 5. Changing disturbance regimes 6. Land use, habitat fragmentation, species choice in agriculture and forestry 7. Poorly characterised background variation
Changed migratory patterns and behaviour	Reduced migratory distance, direction, increased proportions of sedentary individuals etc. in some birds, bats and insects ✓✓✓	No studies	
Changed biome distributions	Treeline advance in Fennoscandian Mountains, mainly via increased stature of existing individuals ✓✓✓	1. Treeline advance in mountain areas ✓✓ 2. Lagged northward shift of boreal and nemoral forest zones, with wider transition zone ✓ 3. Increased broadleaved component in boreal forest ✓ 4. Displacement of tundra plants by grasses and trees ✓✓	1. As for species distributions (above) 2. Hydrological transformations in tundra
Exotic species invasions	Invasions of southern European invertebrates into central/northern Europe ✓✓	No studies	
Animal population cycles	Breakdown in population cycles of arctic rodents due to reduced snow cover ✓✓	No studies	

Table 4.5. (cont.) Climate change impacts on terrestrial ecosystems of the Baltic Sea Basin. ✓✓✓ = high confidence; ✓✓ = medium confidence; ✓ = low confidence (see main text)

Class of impact	Impact of recent climate change (Hypothesis 1)	Projected impact of future climate change (Hypothesis 2)	Relevant uncertainties
Physiological tolerance and stress			
Tree stress and forest dieback	1. Climate-related dieback in some temperate species ✓✓ 2. Increased susceptibility to pathogens and insect pests in some species ✓	1. Increased susceptibility to spring temperature backlashes in some regions ✓ 2. Increased susceptibility to drought in temperate parts of Baltic Sea Basin ✓ 3. Increased susceptibility to pathogens and insect pests ✓	1. System complexity and process understanding 2. Impacts of non-climate abiotic drivers 3. Species and provenance choice in forestry
Disturbance	1. Increased vulnerability to and damage by storms in forest ✓✓ 2. Increased frequency of insect pest outbreaks in forest and subalpine woodland ✓✓	1. Increased frequency of windthrow in forest, particularly southern areas ✓ 2. Increased damage by bark beetles (Norway spruce) ✓	1. Role of non-climate factors, e.g. stand structure, species choice, tree structural properties and anchorage, management practices 2. Future trends in storm tracks, strengths and frequency 3. Complexity of secondary trophic responses
Ecosystem functioning			
Net primary productivity and forest growth	1. Increased land surface greenness, NPP and growing season length since 1980's ✓✓✓ 2. Increased forest growth ✓✓✓	1. Further NPP and growth increases in northern Baltic Sea Basin ✓✓ 2. Negative to small positive NPP and growth change in southern Baltic Sea Basin ✓✓	1. Impacts of non-climate abiotic drivers e.g. N-deposition and mineralisation, acidification, CO_2 "fertilisation" 2. Impacts of stress and disturbance (see above) 3. Species choice and management practice in forestry
Carbon sequestration capacity	Net C sink due to increasing NPP and changed forest cover and management ✓✓	Northern areas remain net CO_2 sinks, southern areas generally switch to small-moderate sources ✓✓	1. Future changes in soil respiration and NPP (with uncertainties as above) 2. Carbon exchange of wetlands 3. Land use changes
Greenhouse gas exchange in tundra and boreal soils	Increased soil respiration and CH_4 release associated with permafrost degradation ✓✓	Permafrost degradation/ reduced ground frost lead to liberation of inactive soil carbon ✓✓	Hydrological transformations in tundra

4.4   Freshwater Ecosystems

In temperate aquatic ecosystems, most of the ecologically important factors are affected by climate change. These include changes in temperature; evapotranspiration; the availability of water and limiting nutrients; the frequency and magnitude of precipitation, wind, storm and fire events; the length of the ice-free season; the depth of the mixed layer; the period of nutrient limitation. These factors can be broken down into two categories that in reality are strongly interrelated: changes in the variability of climatic patterns; and changes in the mean trends of climatic patterns (Choi 1998). In general, many water quality problems may be exacerbated by climatic change, particularly those relating to eutrophication and ageing processes, oxygen depletion, salinisation, and acidification. However, extreme uncertainties exist at any case-specific level (Varis and Kuikka 1997). These are often related to special lake or catchment characteristics. In the search for a more general framework to group and structure climatic responses of lake ecosystems, a conceptual model has been developed (Blenckner 2005), which consists of two main components: a *Landscape Filter* comprising the features of geographical position, catchment characteristics and lake morphology, and an *Internal Lake Filter*, comprising the features of lake history and biotic/abiotic interactions. The inclusion of both the landscape and internal interactions help explain different responses of individual lakes to climatic change and variation (for details see Blenckner 2005). We begin by discussing general mechanisms by which climate changes impact freshwater ecosystems, then go on to review recent and potential future changes in lakes of the Baltic Sea Basin.

4.4.1   Mechanisms of Response to Climate Change

4.4.1.1   *Water Temperature*

Water temperatures (Sect. 2.2.1.5) have shown a strong, coherent increase in conjunction with warming air temperatures across European lakes. A faster temperature rise in the spring (due to earlier ice-out dates) precedes the beginning of spring stratification, which may influence nutrient cycling (Schindler et al. 1990; Abgeti and Smol 1995) and phytoplankton dynamics. Consequently, effects of higher water temperatures on the lake mixing regime could lead to dimictic lakes (fully mixed twice a year) turning into warm monomictic lakes (Schindler 1996; Blenckner et al. 2002a).

In general, temperature affects all physiological processes and, therefore, the growth of organisms. The manner in which physiological rates change with respect to temperature is generally in good agreement with an Arrhenius formulation (Sharpe and DeMichele 1977; Lin 1995; Choi 1998). For example, the zooplankton found in lakes can tolerate fairly high summer temperatures, yet small increases in the winter temperature, in some lakes significantly related to the timing of ice break-up (Jassby et al. 1990), strongly affected the overwintering and seasonal dynamics (George and Hewitt 1999) of zooplankton taxa such as *Daphnia* and *Eudiaptomus* through a strong temperature-dependent growth rate (Lampert and Muck 1985).

The survival and growth of fish species depends strongly on temperature (Magnuson et al. 1990; De Stasio et al. 1996; Magnuson et al. 1997). Thermal limits of some fish species, altered by global warming, can induce changes in distribution (Magnuson et al. 1990; Carpenter et al. 1992). Temperature-induced changes in the growth rate of predatory fish may result in cascading effects through the entire food web (Carpenter et al. 1985).

4.4.1.2   *Ice Regime*

The timing of ice freeze and break-up are recorded in some of the longest and most widespread limnological datasets and have been used as indicators for regional climate change (Robertson et al. 1992; Assel and Robertson 1995). Ice dynamics are driven by several meteorological variables, of which air temperature has been shown to be the most important (Palecki and Barry 1986; Robertson et al. 1992; Vavrus et al. 1996; Livingstone 1997). According to Lithuanian data (Bukantis et al. 2001), lakes are the best indicators of thermal conditions in the pre-winter period (November–December) because the early ice-cover may last until spring even if the winter is warm. This is one reason for the scatter in the winter temperature-ice-cover duration relationship.

A long-term trend towards shorter periods of ice-cover due to a later freezing and an earlier ice break-up has been reported for lakes around the Northern Hemisphere (Palecki and Barry 1986; Kuusisto 1987; Assel and Robertson 1995; Livingstone 1997; Magnuson et al. 2000; Assel et al. 2003). The trend towards an earlier ice-out increases the ice-free period and raises lake temperatures in spring in Canadian (Schindler et al. 1990) and European (Gerten and Adrian 2000; Straile 2000; Livingstone and Dokulil 2001) lakes. In northern Europe, the duration of the ice-covered period and the timing of freeze and break-up greatly affect lake systems (Salonen et al. 1984; Weyhenmeyer et al. 1999; Järvinen et al. 2002), in particular the succession events in plankton dynamics (Weyhenmeyer et al. 1999).

4.4.1.3 *Stratification*

Most temperate deep lakes are stratified during summer, with a warm surface water (epilimnion) and cold deep water (hypolimnion). The most crucial triggering factors for the timing of phytoplankton blooms are light and turbulence, both of which are influenced by stratification and therefore directly affected by climatic variation and change. Consequently, strong relationships between the timing of phytoplankton blooms and the winter climate have been found in European lakes (Adrian et al. 1995; Müller-Navarra et al. 1997; Weyhenmeyer et al. 1999; Gerten and Adrian 2000). Differences in the individual lake response depend on the morphometry of the lake (a component of the *landscape filter*). For example, in Lake Constance (Germany), a large and deep lake generally lacking ice-cover, the spring bloom only occurs under stratified conditions, because reduced mixing increases light availability for phytoplankton, extending transport down to only 20 m, as opposed to 100 m or more with no stratification (Gaedke et al. 1998).

Generally, an enlarged mixing depth increases the extinction of incoming light, which is attenuated abiotically within the mixed layer. This leads to a decrease in phytoplankton production, averaged over the mixed layer, with increasing mixing depth (Huisman et al. 1999; Diehl 2002). By contrast, in shallow lakes with a smaller mixing depth, the phytoplankton spring bloom is associated with ice-cover disappearance (determined by winter climate), influencing the light availability. This has been observed for Müggelsee, Germany (Gerten and Adrian 2000) and Lake Erken (Weyhenmeyer et al. 1999). Later in the year, conditions in the lake appear to be more complex and stratification becomes the predominant triggering factor. Longer periods of stratification can promote a dominance of potentially toxic cyanobacteria (George and Harris 1985; George

et al. 1990; Hyenstrand et al. 1998). Higher water temperatures prolong the stratification period, which creates favourable conditions for buoyant phytoplankton such as cyanobacteria. Additionally, large cyanobacterial colonies (besides other phytoplankton groups) are resistant to grazing by zooplankton, especially in eutrophic lakes. These two mechanisms could promote the dominance of cyanobacteria in stratified lakes at warmer temperatures. Longer periods of summer stratification and higher hypolimnetic temperature are also predicted to cause increased hypolimnetic anoxia, or at least lower oxygen concentrations (Magnuson et al. 1997), and this can enhance the nutrient release from the sediment (Pettersson et al. 2004).

4.4.1.4 *Hydrology*

The hydrological water residence time, another main component of the filter concept, determines the persistence of a climatic signal stored in lakes. For example, this factor influences the recycling of phosphorus, the main limiting nutrient in lakes. Stratified lakes with a water residence time longer than one year have a higher potential to undergo a climate-induced eutrophication because additional phosphorus released from the sediment because of hypoxia and/or enhanced microbiological processes is recycled repeatedly. By contrast, in lakes with short residence times (months), phosphorus released from the sediment is flushed out relatively quickly and therefore contributes less to climate-induced eutrophication. Decreases in water residence time, for example through an increase in precipitation with a higher associated runoff, can decrease concentrations of nutrients, as water residence time scales the rate of the ecosystem response to chemical inputs (Carpenter et al. 1992 and literature cited therein) and to climate change and variability. The water level of closed lakes, where the hydraulic load is counterbalanced by evaporation and rather stable seepage, serves as one of the best indicators of long-term climatic variation (Kilkus 1998).

4.4.1.5 *Ecosystem Transformations*

Climatic changes relevant at a species level are likely to vary at the ecosystem scale among lakes (Petchey et al. 1999). Weak links in food web interactions, in particular, dampen oscillations between consumers and resources (McCann et al. 1998). This implies that not all responses at a specific trophic level are propagated to lower trophic levels or have significant impacts on ecosystem processes (Pace et al. 1999). Additionally, a prolongation of the climate signal due to food web interactions is possible as the signal of winter climate can be detected in the clear water phase in early summer (Straile and Adrian 2000) or in summer phytoplankton composition and biomass (Weyhenmeyer 2001; Blenckner et al. 2002b). A system approach is necessary to examine the cascading effects in response to climatic change and variability. The magnitude of a climate-driven response of an autotrophic organism does not necessarily have to be mediated or cascaded to the heterotroph species, or vice versa.

Another example indicates that a consecutive period of five mild winters led to a complete change of the spring phytoplankton bloom, from a dominance of diatoms and cryptophytes to a dominance of cyanobacteria, in a German lake (Adrian et al. 1995). This illustrates, furthermore, that the response of phytoplankton, and probably also other lake biota, to climatic variation might be totally decoupled from a warming trend (climatic change), as many ecosystem processes are non-linear.

Furthermore, the non-linear responses of animals and plants to environmental variables, including climate (May 1986; Mysterud et al. 2001), may cause progressive changes to be interrupted by drastic switches to a contrasting ecosystem state (Scheffer et al. 2001). One such example is the shift between the clear water (macrophyte dominated) and the turbid (plankton dominated) state of shallow lakes. A recent study in a Swedish lake showed that climate variability alone cannot explain such a shift, but is likely to contribute to a multi-causal stress and, therefore, reduces the resilience of the clear-water state in shallow lakes (Hargeby et al. 2004).

4.4.1.6 *Impacts of Non-Climatic Anthropogenic Drivers*

In the past 40–50 years, many lakes, particularly in Europe and North America, have been subjected to anthropogenic increases in the supply of nutrients and other substances (e.g. organic pollutants). This supply decreased somewhat in the late 1980s and early 1990s. While improvements in waste-water treatment over the last few decades have reduced inputs of nutrients from point sources, the increase in nutrients from diffuse sources is still a major problem and, for many

lakes, remains the predominant concern. Additionally, catchment alteration, acidification, dam regulation, agriculture and the water-borne sanitation solution can complicate a separation of the various influences on ecosystems (Andersson and Arheimer 2003; see also the comments of Jeppesen et al. 2003; van Donk et al. 2003). It is therefore often difficult to clearly distinguish between the independent effects of climate and direct human impact on important lake and catchment processes.

Multiple stress factors, such as climate change together with ozone depletion and acidification, may alter ultraviolet light and temperature regimes in freshwater ecosystems. Changes in underwater UV and temperature can significantly influence the composition of the zooplankton community and ultimately food web dynamics (Persaud and Williamson 2005). Phytoplankton communities experience climate change indirectly through changes in lake level, ice-out time, stratification, nutrient inputs and zooplankton grazing (Anneville et al 2004). Among other effects, global warming could accelerate eutrophication processes (Knight and Staneva 2002, Blenckner et al. 2002b, Malmaeus et al. 2006). The key limitation to predicting effects of climate change on ecosystems lies in the understanding of how biotic interactions will respond (Winder and Schindler 2004).

4.4.2 Recent and Historical Impacts

4.4.2.1 *Physical Responses*

Studies on global climate have revealed that the northern hemisphere is becoming warmer, winters are becoming wetter and summers drier (IPCC 2001). The same overall trends are apparent for the Baltic Sea and its catchment, though with some regional variation; e.g. with regard to the summer precipitation trend (Chap. 2.1.3). These changes are partly reflected in the increasing trend of the North Atlantic Oscillation index (NAO; Sect. 2.1.1.1 and Annex 7).

The NAO seems to affect mostly the late-winter and mid-spring temperature period. The winter NAO (NAOw) has a weak, though significant, effect on the regime of spring temperature in the Baltic Sea Basin and explains the most significant fluctuating components embedded in the cryophenological records (Yoo and D'Odorico 2002). From April until November, water temperature in the Estonian lake Võrtsjärv was significantly correlated with air temperature and the water temperature in spring was significantly correlated with the NAOw (Nõges 2004). Comparing the 1960s and the 1980s, the mean surface water temperature in May–October in Lithuanian lakes did not differ, while it increased from the 1980s to 1990s with the greatest difference (0.8 °C) occurring in April (Pernaravičiūte 2003).

4.4.2.2 Ice

A warmer winter air temperature strongly advances the timing of the ice break-up (Yoo and D'Odorico 2002). From Middle Sweden to Germany, the difference in the timing of the ice break-up during the last 30 years has been up to one month (Gerten and Adrian 2001; Weyhenmeyer 2004). In the large Swedish lakes Vänern, Vättern and Mälaren, the number of years without continuous winter ice-cover has increased in the 1990s compared with earlier decades (Weyhenmeyer 2001). In northern Sweden (north of 62° N), the change is weaker or even totally absent (Blenckner et al. 2004). During the last 30 years, there has been a clear trend towards a greater variation in the ice break-up dates in the south of Sweden compared to the north, with a distinct change in the relationship at around 62° N (Blenckner et al. 2004). This tendency was less pronounced for the timing of ice freeze and totally absent for the ice break-up dates of the Finnish lakes (Blenckner et al. 2004).

Since 1880, the number of ice-free days on Lake Onega (Russia) has declined only marginally from 225 to 217 (Filatov et al. 2003). Trends towards a shorter ice-free period have also been observed, for example on the Russian Kola Peninsula over the period 1930-1998 (Kozlov and Berlina 2002) and in the Estonian lake Võrtsjärv over the period 1923–1998 (Järvet 1999). In Võrtsjärv this phenomenon was caused mainly by advancement of the freezing date, while on the Kola peninsula the cause was both a delayed ice break-up in spring and advanced freezing in the autumn/winter period. Further trends are discussed in Sect. 2.2.2. In Võrtsjärv, the duration of ice-cover correlated negatively with the NAOw (Nõges 2004). Additionally, year-to-year variability in ice break-up dates in Europe could be related to climatic (NAOw) variation (Weyhenmeyer et al. 1999; Gerten and Adrian 2000; Yoo and D'odorico 2002), while the NAO signal has been detected even in the ice characteristics of the world's largest (by volume) lake, Lake Baikal (Livingstone 1999).

The tendency towards an earlier ice break-up can have strong ecosystem effects. An example from northern Germany showed that a series of mild winters with earlier ice break-up led to an earlier stratification and a dominance shift from diatoms to cyanobacteria (Adrian et al. 1995). Polar lakes are early detectors of environmental change because snow and ice-cover variation markedly affect all ecological variables (Quayle et al. 2002).

4.4.2.3 Hydrology

As discussed in Chap. 3, global warming in the Baltic Sea Basin may bring about higher amounts of winter and spring precipitation, primarily in the northern part of the catchment. In Estonia and Sweden the amount of spring and winter precipitation correlates with the NAOw, but the relationship was lacking in other seasons (Nõges 2004; Weyhenmeyer 2004). In general, changes in precipitation can influence lake water level dynamics. Long-term changes in water levels with a periodicity of 20–30 years have been reported for many European lakes; e.g. Peipsi and Võrtsjärv in Estonia (Nõges et al. 2005); the majority of Belarussian lakes (UNFCCC 2003b); Saimaa in Finland; Ilmen, Onega and Ladoga in Russia (Masanova and Filatova 1985; Malinina et al. 1985); and Müggelsee in Germany (Behrendt et al. 1987).

Peipsi and Võrtsjärv showed a strong positive relationship between the NAOw and the water level. In the very shallow Võrtsjärv (mean depth 2.8 m), the annual mean amplitude of the water level is equal to half of the mean lake depth, and the absolute range of water level fluctuations (3.2 m) even exceeds mean depth, which clearly has a strong influence on the ecosystem dynamics (Nõges and Nõges 2004). In low-water years, the water column is better illuminated, while both phosphorus release and denitrification increase because of more frequent resuspension of bottom sediments bringing about substantially higher phosphorus concentration and resulting in a lower nitrogen:phosphorus ratio. This favours the development of nitrogen-fixing cyanobacteria (Nõges et al. 2003). In the deeper Peipsi (mean depth 7.1 m), such effects are not as obvious.

Climate-induced changes in hydrology of boreal lakes and rivers may change the growth conditions of aquatic macrophytes as well as decomposition and accumulation of detritus. When a spring flood does not occur, the littoral detritus from previous growing seasons will more likely accumulate at aquatic plant-stands and will therefore be subjected to anaerobic decomposition and probably improve the substrate supply for methanogenic bacteria. In two Finnish lakes, Alinen Rautjärvi and Ekojärvi, methane emission in dense stands of three emergent macrophyte species (*Equisetum fluviatile, Schoenoplectus lacustris* and *Phragmites australis*) was even more limited by temperature than by substrate. Thus, not only changes in the flooding pattern but also higher summer temperatures can accelerate methane emissions from boreal lakes (Kankaala et al. 2005).

4.4.2.4 *Chemical Responses*

Climate change affects the hydrological regime (as described above) and thus also nutrient loadings into lakes. In warm and dry years, the nutrient loading usually decreases while in-lake concentrations may still increase because of increased water evaporation and a more intensive release from sediments at low water stages (Magnuson et al. 1997). In a study of the three largest lakes in Sweden (Vänern, Vättern and Mälaren), synchronous relationships between the winter climate (here NAOw) and water chemistry were restricted to variables closely linked to surface water temperature, i.e. reactive silica and pH in May (Weyhenmeyer 2004).

Additionally, a warm winter was found to increase the winter discharge and the associated load of phosphorus and nitrogen into one basin of Mälaren. In nearby Lake Erken, nitrate concentrations were much lower in late winter during the warm period, the 1990s, compared to the cold period, 1975–1980 (Pettersson et al. 2004). In summer, elevated phosphate and ammonium concentrations in the hypolimnion were observed and resulted in significantly higher phosphate and chlorophyll *a* levels in autumn during the warm period, compared to the cold period (Pettersson et al. 2004).

In the River Vantaanjoki watershed in southern Finland, increased temperature and precipitation in recent decades were responsible for an increased contribution of diffuse nutrient losses to the total nutrient load (Bouraoui et al. 2004). In Lithuanian lakes, the increase in concentrations of total nitrogen and total phosphorus over the period 1993–2002 was insignificant (Anon 2002). Due to the high spatial variability of soil and precipitation in Lithuania, the most vulnerable region is in the west of the country, where the annual precipitation is highest (800–900 mm) and associated with

annual loads of nitrogen up to four times higher compared to Middle Lithuania (Tumas 2000).

Long-term changes in loadings of nutrients and organic matter to the Estonian lake Võrtsjärv resulted both from changes in agricultural practices and from the climate change (Nõges et al. 2007). The analysis of detrended time series revealed significant positive correlations between the amount of precipitation and annual loadings of ammonium, phosphate and dissolved organic carbon (DOC). Both water discharge and solute concentrations increased during wet years.

Peatlands occupy approximately 15% of boreal and subarctic regions, contain approximately one third of the world's soil carbon pool, and supply most of the DOC entering boreal lakes and rivers (see Sect. 4.6). Changes in temperature, water tables and discharge could affect delivery of DOC to downstream ecosystems, where it exerts significant control over productivity, biogeochemical cycles, and attenuation of visible and UV radiation (Pastor et al. 2003; see also Sect. 4.5.2). The annual potential net carbon sequestration rate of forests, peat-producing wetlands and lake sediments in the Baltic Sea Basin has been estimated at 30–60, 8–55 and 10–51 $tC\,km^{-2}$, respectively (Jansson et al. 1999). In Estonian lakes, mean accumulation of CO_2-C is estimated to be only 0.54 $tC\,km^{-2}$ (UNFCCC 1995). It has been shown that lakes in northern Sweden are CO_2-supersaturated and therefore constitute net sources of CO_2 to the atmosphere (Jonnsson et al. 2003).

Per unit of electricity produced, the hydroelectric reservoirs emit much less CO_2 than the Swedish combustion plants, while their emission is also much lower compared to natural lakes (Bergström et al. 2004). In a study of some 80,000 lakes in the boreal zone of Sweden, it was estimated that 30–80% of the total organic carbon entering the lakes was lost, mainly due to mineralisation and subsequent CO_2 emission to the atmosphere, the exact proportion depending on residence time and water temperature (Algesten et al. 2004).

However, despite the controlling effect of water temperature, hydrological changes and catchment characteristics (see also Sobek et al. 2003) can be of greater importance than interannual temperature fluctuations for the rate of carbon cycling (Algesten et al. 2004). The relative emission of CO_2 from boreal lakes (a major CO_2 source) in comparison to boreal forests (currently considered a net CO_2 sink) may be of considerable importance for the CO_2 concentration of the atmosphere and future climatic forcing.

4.4.2.5 Biological Responses

Climate change may clearly affect the timing of the development of ecosystem components in spring. In a German lake (Plusssee), the timing of phyto- and zooplankton maxima (clear-water phase) was negatively related to water temperature: under a temperature increase, both maxima occurred earlier. The intensity of the spring algal maximum was negatively related to its timing, whereas no clear relationship between the zooplankton maximum and its timing could be observed (Müller-Navarra et al. 1997). In Võrtsjärv (Estonia), the phytoplankton biomass in spring was positively related to the NAOw, while during wet summers and autumns, phytoplankton biomass was lower due to an elevated water level (Nõges 2004). In shallow lakes, the effects of winter climate on plankton are short-lived, and are soon overtaken by the prevailing weather and by biotic interactions (Adrian et al. 1999).

In deep lakes, however, the winter climate signal (here the NAOw) can persist until late summer (Gerten and Adrian 2001). Distribution, composition and species diversity of diatoms in subarctic Lapland are markedly regulated by temperature and other climate-related factors (Weckström and Korhola 2001).

The 20th century arctic warming has caused changes in overall species richness of benthic and planktonic diatoms in lakes of Finnish Lapland. A particularly strong relationship was found between spring temperatures and the compositional structure of diatoms, depending on lake type and catchment characteristics and reflected by several other biological indicators, such as chrysophytes and zooplankton, suggesting that entire lake ecosystems have been affected (Sorvari et al. 2002).

The timing of the clear-water phase (CWP) is associated with a warmer winter climate (Müller-Navarra et al. 1997; Gerten and Adrian 2000; Straile and Adrian 2000; Straile 2002). However, this may be difficult to totally separate from management actions (fish removal and nutrient reductions) occurring at the same time (Jeppesen et al. 2003). A strong relationship between the onset of the CWP and NAOw was found in 28 central European lakes and 71 shallow Dutch lakes (Straile and Adrian 2000; Scheffer et al. 2001). This earlier onset was combined with an earlier summer

decline of *Daphnia* (Straile 2000), which illustrates the fact that winter climatic variation affects successional events even in summer.

A higher summer phytoplankton biomass was related to higher winter air temperature in a small lake in northern Germany, probably due to the higher grazing pressure on zooplankton during late spring and the faster development of zooplanktivorous fish (Müller-Navarra et al. 1997). A similar pattern was found in the biomanipulated Bautzen Reservoir (Germany), where the water temperature during early summer probably influenced the mortality of daphnids through predation of young-of-the-year (YOY) fish (Benndorf et al. 2001). In Müggelsee, the annual peak abundance of two thermophile zooplankton species increased significantly in the warm period (Gerten and Adrian 2002). In Lake Ilmen (Russia), low zooplankton biomass has been related to hydrometeorological conditions. Also, zoobenthos have been favoured by high water levels, especially if such conditions last for several years.

Nevertheless, over the period 1952–2001, there have been no major qualitative changes in the zoobenthos, only changes in the quantitative representation of some groups (Andreeva 2003). Ichthyophenological time series from Estonia reveal that the commencement of spawning for pike and bream has advanced by 6 and 8 days, respectively, over the period 1951–1996 (Ahas 1999). Increasing water temperature in spring in Estonian inland waters affected the spawning conditions of roach and bream differently. Over the period 1951–1990, the spawning of bream shifted 10 days earlier with an unchanged spawning temperature.

In contrast, no shift in the spawning time of roach has been observed. The difference between spawning times of roach and bream decreased from 22 to 13 days and the difference in average temperatures at the onset of spawning by about 3 °C. In years with higher water levels during March and April, both fish species started to spawn earlier (Nõges and Järvet 2005).

In Danish lakes, warm early summers most often lead to a higher abundance of planktivorous fish (YOY) and, thereby, increased grazing pressure on large-bodied zooplankton (Jeppesen et al. 2003). Autumn spawners such as vendace in Sweden's largest lakes will have difficulties in adapting to climate warming, whereas spring spawners such as smelt and pike-perch benefit from the higher spring water temperature.

However, the responses depend on the trophic status of the lake: in a eutrophic lake the effects are not as clear, as the predation pressure from other species may become too high (Nyberg et al. 2001). In Peipsi (Estonia/Russia) the stock of smelt may seriously suffer when low water level, prolonged periods of calm, hot weather and strong cyanobacterial blooms coincide, causing night-time hypoxia (Kangur et al. 2003). Additionally, a sharp decline of vendace in Peipsi and other lakes in the region in the 1990s was evidently connected to winter climate (Sarvala et al. 1999; Kangur et al. 2003): mild winters in the late 1980s and early 1990s, when Peipsi had no permanent ice-cover and the bottom was exposed to the action of waves, led to a high mortality of vendace eggs on spawning grounds (Pihu and Kangur 2000). The eggs can be injured mechanically or buried under sediments, which causes mass mortality (Sterligova et al. 1988).

A 200-year sediment core from the alpine Lake Chuna on the Kola Peninsula revealed a decrease of cold-stenothermal chironomid taxa in the uppermost sediment layers, which can be explained by warming during the 20th century (Ilyashuk and Ilyashuk 2001).

4.4.3 Potential Future Impacts

4.4.3.1 Physical Responses

As described in Chap. 3, most climate change scenarios for the eastern part of the Baltic Sea Basin are characterised by increasingly mild winters with a decrease in snow cover (Kondratyev et al. 2002, Filatov et al. 2003, Järvet 2004). The increase of monthly mean air temperature will be in the range of 3–5 °C in the next 100 years, with the winter period showing the largest changes. However, inconsistencies among scenarios are high; for example, the predicted annual temperature change for the Baltic Sea Basin ranges from 0.7 °C to 6.9 °C, and precipitation change from 3% to 22%, among GCMs included in the CMIP2 series of simulations (see Sect. 3.3.3).

A series of combined climate-lake modelling studies of European lakes showed that continuously increasing air temperatures led to similar trends in surface water temperature, resulting in longer stratification periods and shorter ice-cover periods or no ice at all as a result of strong changes in the mixing regime (Blenckner et al. 2004; Persson et al. 2005). Climate change experiments sim-

ulated a 1–2 month shorter lake ice season in the Nordic area for the 2070s. All simulations show the largest changes for southern-central Sweden, southwestern Baltic States and western Norway (Räisänen et al. 2001). These results were in reasonable agreement with the ice season length observed in 37 Swedish lakes (Rummukainen et al. 2001). A shortening of the winter ice-cover period by the year 2050 has also been projected for two Russian lakes, Ladoga and Onega (Filatov et al. 2003) and for Estonian rivers (Jaagus 1997). A more intensive water mixing is projected for Ladoga and Onega, leading to a decrease of the hypolimnion thickness and an associated decrease in dissolved oxygen concentrations by 2050 (Filatov et al. 2003). A deepening of the thermocline as a result of a warmer climate has also been discussed for Scandinavian lakes (Stendera and Johnson 2005).

4.4.3.2 Hydrology

The simulation of hydrological parameters for lakes varies due to a large scatter in the projected simulations of precipitation (see Chap. 3). Simulations for Lakes Ladoga and Onega with two future scenarios (1: doubling of greenhouse gases; 2: doubling of greenhouse gases, taking into account the direct effect of atmospheric aerosols) projected an increase in mean annual precipitation and total evaporation, but a decrease of runoff, over the period 2000–2050 (Filatov et al. 2003). Similar trends were predicted for the Upper Volga catchment area bordering with the Baltic Sea Basin, where the surface runoff in spring was projected to become higher compared to the present day.

In contrast, summer and autumn runoff could be suppressed in the future due to higher transpiration rates associated with a larger proportion of deciduous trees in forests. Decreased summer precipitation in combination with an increased evapotranspiration would result in decreased ground water recharge and a lower water level for the Volga River and Upper Volga lakes (Oltchev et al. 2002). In contrast, projections for the Lake Ilmen-River Volkhov catchment point to runoff increases of 45% and 66% for 2050 and 2100, respectively, compared to the period 1981–1990 (Kondratyev et al. 2002).

In a modelling study of river flows (Lehner and Döll 2001), only a few areas, such as southern Finland and northern Russia, showed consistent decreases in drought frequencies under a range of future climate scenarios (from the ECHAM4 and HadCM3 GCMs). By contrast, Scandinavia, Lithuania, Latvia, Estonia, northern Belarus and Russia, most of Germany and the Alpine region generally tended towards a higher drought risk situation by the 2070s.

A significant redistribution of seasonal runoff has been projected for the eastern part of the Baltic Sea Basin. The most important changes would take place during the colder half of the year. Frequent melting periods decrease snow and ice accumulation during winter that will result in an earlier commencement of snowmelt in spring and a lower maximum runoff, while winter runoff minima will disappear (Jaagus 1997; Kont et al. 2002; Oltchev et al. 2002; UNFCCC 2003a; Ziverts and Apsite 2005). All models simulate drier soils in summer (Oltchev et al. 2002; Kondratyev et al. 2002; Järvet 2004) and an increase in river runoff during autumn. In the western part of Estonia, the runoff maximum in autumn is expected to exceed the spring maximum. On Saaremaa Island in the west-Estonian archipelago, the runoff maximum in spring will disappear, and the annual cycle will consist of a single cold half-year maximum and a warm half-year minimum (Kont et al. 2002; UNFCCC 2003a).

Changes in temperature and precipitation will also affect lake water levels. Simulations of three lakes from northern Europe (Bysjön, Sweden; Karujärv and Viljandi, Estonia) show that the lake water level is more sensitive to decreases than to increases in precipitation. Increased precipitation results in increased runoff, but this is largely compensated by increased outflow and changes in lake level are therefore small (Vassiljev 1998).

Significant changes in lake water levels can be attributed to the displacement of the boundaries of climatic seasons even without visible deviations of the annual averages of temperature and precipitation from their long-term standards (Grigor'ev and Trapeznikov 2002). According to model simulations, reductions in winter (November–April) precipitation have a stronger impact on lake levels than changes in summer (May–October) precipitation (Vassiljev 1998). Despite the prospected decrease in the annual runoff, an increase in lake water level for the coming 50-year period was predicted for Lakes Ladoga and Onega (Filatov et al. 2003).

4.4.3.3 Chemical Responses

A warmer and wetter future climate will affect watershed processes: reduced soil frost and increasing precipitation in winter will increase nutrient leaching from the soil and accelerate the eutrophication of water bodies. Despite abandonment of a substantial proportion of previously cultivated lands and decreased use of fertilisers in Estonia, wintertime nutrient losses are currently almost as high as during former years because of increased water discharge and surface runoff in winter. Increases in water discharge in winter together with the still large nutrient pool in soils may explain the fact that nutrient concentrations in lakes have decreased less than expected in conjunction with reduced fertiliser application (Nõges and Nõges 2004). The time-lag between catchment changes and the responses in lakes is difficult to predict. However, the delay before terrestrial vegetation and soil covers are in balance with the new climatic conditions will presumably amount to at least 100–200 years (Karlsson et al. 2005).

Climate warming simulations for Lake Batorino in Belarus performed with the LakeWeb model (Håkanson et al. 2003) showed a significant increase in total phosphorus concentrations as more phosphorus became bound in organisms with short turnover times (phyto-, bacterio-, and zooplankton). Longer summer stratification will increase hypolimnetic anoxia (Stendera and Johnson 2005) in some lakes, which is a precondition for iron-bound and aluminium-bound phosphorus release from lake sediments (Pettersson 1986). Enhanced remineralisation of nutrients and higher diffusion rates at higher temperatures combined with the longer ice-free period will enhance nutrient availability to algae, especially in lakes with a long water residence time (Blenckner et al. 2002a).

Future projections in Belarus (UNFCCC 2003b) take into account the effect of people spending more time in recreational zones around rivers, lakes, and water reservoirs. This higher anthropogenic load could lead to lower water quality and aggravation of the epidemiological situation in the sense of viral and bacterial contamination. A more intensive eutrophication in the next 50 years as a result of climate change has also been projected for the Lake Ladoga and Lake Onega region (Filatov et al. 2003). In contrast, a study for Lithuania (UNFCCC 2002) showed a decrease of more than 20% in nitrogen washed into surface waters in conjunction with a 2 °C temperature increase, as a result of a reduced meltwater amount, and, consequently, smaller surface and subsurface runoff. Nitrogen washout is not necessarily a simple function of affluence but could also be climate-driven: net nitrification rate and the formation of mineral N increased in experimental drought treatments of N-fertilised forest soils (Smolander et al. 2004).

Malmaeus et al. (2006) used a regional climate model coupled to a physical lake model and a phosphorus model to simulate three lakes in Sweden. Responses of lake phosphorus content to a warmer future climate differed according to water residence time. In lakes with a short residence time, the simulations showed no future changes in the phosphorus concentration, regardless of lake depth. By contrast, for a lake with a longer water residence time (Lake Erken), the concentration of dissolved phosphorus in the epilimnion almost doubled in spring and autumn under the warmest climate scenario, increasing the potential for phytoplankton production accordingly.

The most important flux intensifying the phosphorus cycle is the enhanced diffusion due to a longer stratification period. This leads to a higher mineralisation with elevated bacterial activity, resulting in decreased oxygen concentrations and higher rates of diffusion. As a consequence, additional available phosphate increases the phytoplankton biomass in Lake Erken, which then settles out and contributes to the enhancement of the cycle. The implication would be that in Lake Erken, and in other eutrophic lakes with long water residence times, eutrophication problems may become serious in the future, and that management may need to take action today in order to maintain good water quality in these lakes.

Changes in affluence are likely to affect allochthonous dissolved organic matter inputs (Stendera and Johnson 2005). Experiments with large mesocosms (Pastor et al. 2003) showed that the DOC budget of boreal peatlands was controlled largely by changes in discharge rather than by any effect of warming or position of the water level. At a critical discharge rate, the DOC budget switched from net export to net retention. The same study showed that if the current decreasing trend of the discharge from peatlands continues, DOC retention and concentration in bogs and fens will increase, while export will decrease. The warming trend during the past 20 years has likewise caused an increase in DOC concentrations in United Kingdom peatlands (Freeman et al. 2001),

where, in the absence of changes in discharge, the DOC export increased.

4.4.3.4 Biological Responses

The direct impact of a temperature increase on biota strongly depends on the extent of the increase. In a microcosm experiment, warming by 3 °C had no effect on phytoplankton size structure and very minor effects on chlorophyll *a* and total phytoplankton biovolume (Moss et al. 2003). Biovolumes of some phytoplankton groups decreased and abundance of cyanobacteria did not change, suggesting that fears of dominant cyanobacteria summer blooms in future may not be realistic, at least in shallow, unstratified lakes dominated by macrophytes. In contrast, an experiment with natural plankton communities grown at 10 and 20 °C (Rae and Vincent 1998) showed that a stronger temperature increase could considerably influence the plankton community structure.

A food web simulation of warmer conditions for Lake Batorino in Belarus (the mean annual epilimnetic temperature was raised by 2 °C and the range in weekly epilimnetic temperatures was increased by the exponent 1.1) predicted higher primary and secondary production, increases in algal volumes and bacterioplankton biomass, and a reduced macrophyte cover, suggesting a significant change in the structure of the lake foodweb (Håkanson et al. 2003). Warming is likely to accelerate eutrophication processes in northern lakes, and thereby shift the species composition of phytoplankton assemblages toward species with higher temperature optima (for example, cyanobacteria), posing a substantial risk for deteriorated drinking water quality (UNFCCC 2003b). An increase in cyanobacterial dominance under warmer conditions, increased spring growth and an earlier decline of the summer cyanobacteria bloom was also predicted by a phytoplankton model for Bassenthwaite Lake, UK, while the overall productivity of the lake remained unchanged (Elliott et al. 2005).

It is likely that the impacts of climate change will be more pronounced in littoral zones of lakes than in pelagic zones, because aquatic macrophytes have access to nutrients in the sediment rendered available under higher surface water temperatures. A three-year study carried out in two experimental ponds in Finland, one covered with a greenhouse (Kankaala et al. 2000, 2002; Ojala et al. 2002), showed that the emergent macrophytes emerged earlier and grew better in the warmer conditions (+2.5–3 °C). The released phosphorus from decaying plants supported the growth of filamentous algae in the greenhouse pond, illustrating the uptake of phosphorus by macrophytes as an important accelerator of eutrophication of lakes in the boreal zone. Plant growth efficacy is likely to increase in a warmer climate, as evidenced by high plant coverage in some Belgian and Spanish lakes at total nitrogen concentrations far exceeding the threshold of 1.2–2 mg l^{-1} at which a shift from a clear to a turbid state occurs in northern Danish lakes (Sagrario et al. 2005).

The fauna of sublittoral zones in relatively shallow lakes may be affected by stronger oxygen depletion caused by the deepening of the thermocline and hence a smaller volume of the hypolimnion containing the oxygen pool available during the stagnation period (Stendera and Johnson 2005)

Generally, cold-water species may be extirpated from much of their present range while cool- and warm-water species may expand northward (Chu et al. 2005). In Belarus, an increase in water temperature in the shallow water reservoirs is expected to result in the biomass loss of cold-water fish species and in a decrease of fish species diversity (UNFCCC 2003b). A positive effect is that winter fish-kills due to anoxia documented in Võrtsjärv (Nõges and Nõges 1999) are likely to become less frequent in the future as a reduced ice duration and a higher water level leads to a larger under-ice oxygen pool during winter (Nõges and Nõges 2004).

Besides direct temperature effects, alterations in the humic content of water will affect the biota of lakes. Increased allochthonous dissolved organic matter inputs may negatively affect periphyton communities and, in turn, reduce benthic macroinvertebrate richness (Stendera and Johnson 2005).

4.4.4 Synthesis – Climate Change Impacts on Freshwater Ecosystems of the Baltic Sea Basin

In comparison to the situation for terrestrial ecosystems (Sect. 4.3.7), the evidence for general or consistent patterns of change in freshwater ecosystems of the Baltic Sea Basin attributable to recent climate change is not equally pronounced according to our review of available studies. Climate variables (particularly temperature and insolation, but also precipitation and wind) are clearly important drivers of freshwater ecosystem processes, both via direct biological effects (e.g.

growth rates of fish and plankton) and indirectly via effects on lake hydrology, stratification and nutrient cycling. Responses of lake ecosystems to climate forcings may, however, be highly sensitive to lake and catchment characteristics and may vary depending on system state, resulting in highly individualistic responses in different lakes (Blenckner 2005; Sect. 4.4.1).

A number of studies from the Baltic Sea Basin report evidence of change in freshwater systems attributable to recent climate change or variability. Impacts described from particular lakes or rivers and their ecosystems during recent decades include:

- A longer annual ice-free period and earlier breakup of ice cover leading to community structure changes such as dominance shifts among phytoplankton taxa, changed successions, reduced species diversity, and transformations from a clear-water (macrophyte-dominated) to turbid (phytoplankton-dominated) state.

- Increased water temperatures leading to increased primary production, higher lake biomass and the incidence of phytoplankton blooms in some northern European lakes.

- Changes in fish communities associated with the effect of warmer early summers on zooplankton biomass.

- Increased nutrient loads and/or increased contributions of nutrients from diffuse sources under a warmer and wetter climate, with dominance shifts in phytoplankton assemblages as a possible result.

A description of the potential impacts of future climate changes on freshwater ecosystems emerges from a handful of experimental and modelling studies, extrapolation of recent observed trends, and elucidation of mechanisms. As in the case of terrestrial ecosystems, statements about the likely responses of freshwater ecosystems to future environmental change vary in robustness depending on the available data, process understanding, system complexity, and the degree to which lake- and catchment-specific parameters influence the response variable of interest.

Climate scenarios for the Baltic Sea Basin point to temperature increases in all seasons, while the majority of the scenarios suggest that precipitation will increase over most areas in the future (Chap. 3). These patterns are consistent with an elevated value of the NAO index, a familiar scenario from available impact studies (e.g. Gerten and Adrian 2000, 2001). Potential impacts of such a climate development include the following:

- Warmer water temperatures combined with longer stratified and ice-free periods in lakes could accelerate eutrophication, particularly in the western and southern Baltic Sea Basin. Shallow lakes and littoral zones may be particularly vulnerable.
- Cold-water fish species may be extirpated from much of their current range while cool- and warm-water species expand northwards.
- Altered lake nutrient status: increased remineralisation and higher diffusion rates of nutrients in warmer water would be expected to increase nutrient availability, especially in lakes with longer water residence times. Reduced N:P status combined with higher temperatures could result in phytoplankton community structure shifts favouring N-fixing and warm-temperature species, including cyanobacteria.
- Increased influxes of humic substances to ecosystems downstream of boreal and subarctic peatlands would steepen light attenuation with negative impacts on lake periphyton and benthic communities, while potentially increasing the contribution of northern lakes to regional CO_2 emissions and climate forcing.

Plausible direct and indirect impacts of an increased-NAO scenario on lake and ecosystem properties are summarised in Fig. 4.7, which is based on studies described in the text. The associated uncertainties are characterised on a three point scale, following similar criteria as for terrestrial ecosystems in Sect. 4.3.7, viz:

High confidence – impact has been described based on *observations or measurements* of *several* species, ecosystems or lakes within the Baltic Sea Basin; *and/or* the impact has been observed in other comparable regions *and* the *mechanism* of climate response is well understood; *and* there are no known or suspected antagonistic mechanisms or feedbacks that could mitigate or reverse the impact.

Medium confidence – impact has been projected or modelled for the Baltic Sea Basin using several approaches *and* the mechanism of response to climate change is well understood.

Low confidence – impact is hypothesised for the Baltic Sea Basin; *or* has been projected or modelled in a single study; *or* antagonistic mechanisms

or feedbacks exist which could mitigate or reverse the projected impact.

In general, changes in physical lake properties (higher water temperature, earlier ice-off) and ecosystem functioning (enhanced primary production, earlier spring bloom) may be expected with some confidence, while net effects on lake chemistry, food webs and community structure remain far less certain.

4.5 Nutrient Fluxes from Land Ecosystems to the Baltic Sea

4.5.1 Agriculture and Eutrophication

4.5.1.1 *Pollution Load to the Baltic Sea*

The Baltic Sea is an enclosed sea which is highly sensitive to eutrophication. At the same time, increasing anthropogenic pressure in its catchment area is responsible for a severe nutrient load to the sea (Jansson 1997). Jansson et al. (1999) estimated the spatial appropriation of terrestrial and marine ecosystems – the ecological footprint – of the 85 million inhabitants in the Baltic Sea Basin with regard to consumption of food and timber and waste production of nutrients and carbon dioxide. When the amount of fresh water that the inhabitants depend upon for their appropriation of these ecosystem services was also included, the ecological footprint estimate corresponded to an area as large as 8.5–9.5 times the Baltic Sea and its drainage basin (about $22,500,000\,km^2$) with a per capita ecosystem appropriation of 22 to 25 ha (Jansson et al. 1999).

Nitrogen and phosphorus are considered as main limiting nutrients for primary production in the Baltic Sea; thus agriculture and nutrients runoff from rural areas play an important role in stress on the Baltic Sea ecosystem. Since the 1950s, the surplus nitrogen of agriculture has approximately tripled and only since the mid-1980s has this stopped increasing (Rheinheimer 1998). Likewise, the phosphorus load has increased clearly in the past 20–30 years as a result of the increased fertilisation level and changes in agricultural structure (Valpasvuo-Jaatinen et al. 1997). In the beginning of the 1990s, the annual nitrogen load to the Baltic Sea totalled 1,409,000 t, of which 980,000 t (69%) were waterborne discharges (rivers, point and non-point sources), 134,000 t (10%) were caused by N_2-fixation while the rest, 296,000 t (21%), was at-

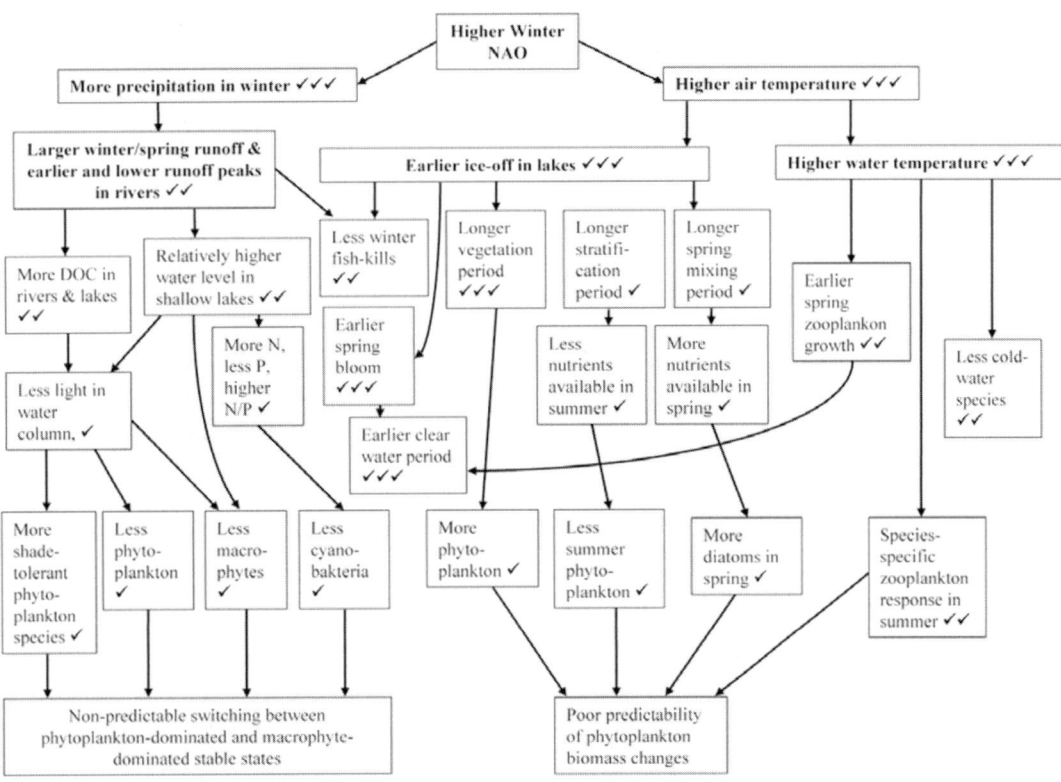

Fig. 4.7. Climate change impacts on freshwater ecosystems of the Baltic Sea Basin. ✓✓✓= high confidence; ✓✓= medium confidence; ✓= low confidence (see text)

mospheric deposition on the sea surface (Enell and Fejes 1995). About two-thirds of the waterborne discharges of nitrogen possibly came from agriculture (HELCOM 1993). In 1980–1993 the mean annual total input of phosphorus to the Baltic Sea has been estimated to be as high as 59,500 t yr^{-1}, of which 41,000 t makes up the riverine load (Stålnacke et al. 1999a). In 2000, 58% of N and 53% of P reaching the sea originated from diffuse pollution, mainly from agriculture (HELCOM 2005; see also Chap. 5).

4.5.1.2 *Change in Land Use and Agricultural Production Intensity*

The present eutrophic situation is caused by an increased external load of nutrients during several decades and the low water exchange capacity of the Baltic Sea. However, in recent decades, the agricultural sector in the Baltic Sea Basin has significantly changed, and agricultural input of nutrients has diminished. In Germany, Denmark, Sweden and Finland, the EU Common Agricul-

tural Policy (CAP) has enforced a restructuring of agriculture and agricultural production. The influence of CAP has been more important in Sweden and Finland, where detailed regulation of the use of land and a system of production quotas have resulted in a tendency to set aside more than 10% of former arable land (Granstedt et al. 2004). For instance, in Finland more than 80% of farmers are participating in the EU programme Agri-Environmental Support Scheme (Rekolainen et al. 1999).

The same tendencies, but in a much more amplified mode, have taken place in Estonia, Latvia and Lithuania. After regaining independence, agriculture in these Baltic countries changed from a state-controlled economy with huge collective farms to a market economy with privately-owned family farms. Due to the financial situation, the use of mineral fertilisers in Estonia, Latvia and Lithuania dropped by more than a factor of 10 (to the 1950s level), achieving the lowest level in 1995–1996 (Löfgren et al. 1999; Mander and Palang 1999, Fig. 4.8). In Latvia, the use of mineral N,

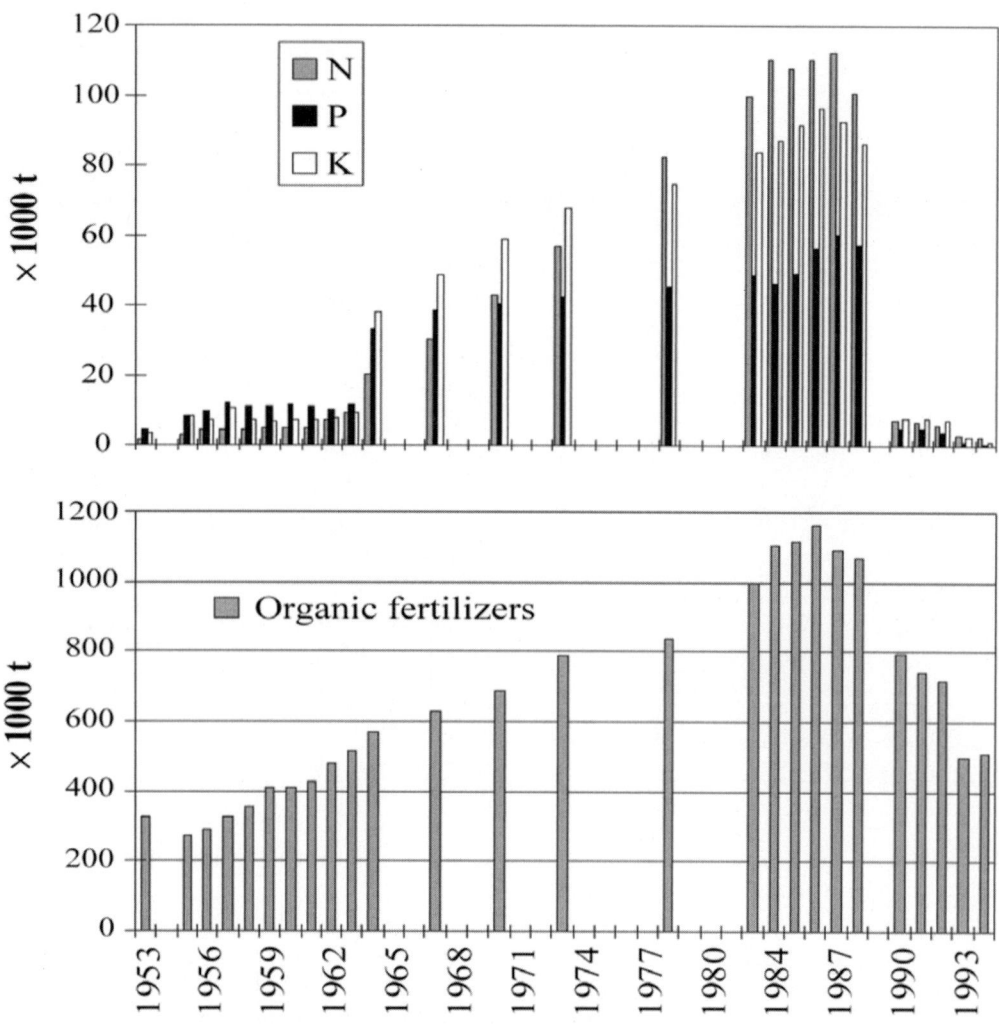

Fig. 4.8. Application of fertilisers in Estonia in 1953–1994 (adapted from Mander and Palang 1999)

P and K fertilisers decreased from 159 to 29, from 108 to 11, and from 172 to 24 kt yr^{-1}, respectively, over the period 1989–1994 (Rivza 1997).

Poland has the largest agricultural area proportion among the Baltic Basin countries (Table 4.6) and contributes the largest amount of nutrients among all the countries in the Baltic Sea Basin (247,000 tN yr^{-1} and 12,200 tP yr^{-1} in 1993, around 60% of total nitrogen and 35% of total phosphorus came from agricultural non-point sources; Dzikiewicz 2000). However, due to the changing economic and socio-political order, consumption of commercial fertilisers in Poland has decreased to less than half since 1988 (Žylicz 1997; Stålnacke et al. 2004).

Although several estimations have been made on changes in land-use structure in the Baltic Sea Basin since the mid-1990s (Table 4.6), we could not find clear data on this situation. At the local level, the increase in the proportion of abandoned agricultural lands has been remarkable (Rivza 1997; Mander et al. 2000a); however, there is strong evidence that during the last 3–5 years many set-aside areas have been taken into agricultural use again. This is especially the case in Lithuania, Latvia and Estonia, where the EU farmer-supporting schemes encourage farmers to intensify their agricultural activities.

Some data suggest that despite changing land use in some countries, no significant decrease in

Table 4.6. Population, total land area and agricultural land use in the Baltic Basin countries in 1993 (after Khalili et al. 1997)

	Population 10^6	Total area 10^3 ha	Arable land 10^3 ha	Pastures 10^3 ha	Total agricultural area 10^3 ha	%
Denmark	5.2	4,300	2,540	197	2,740	64
Sweden	8.6	45,000	2,780	576	3,361	7
Finland	5.1	33,800	2,580	106	2,692	8
Russia*	9.0	31,480	4,699	1,045	5,810	18
Estonia	1.5	4,500	1,114	243	1,400	30
Latvia	2.6	6,500	1,687	819	2,530	39
Lithuania	3.8	6,500	2,300	1,173	3,524	59
Poland	38.5	31,300	14,305	4,047	18,715	60
Germany**	4.5	3,900	1,963	740	2,656	68
Total	78.9	167,280	33,968	8,946	43,428	26

*Data from 2000 after Granstedt et al. 2004
**Data from 1995 about Mecklenburg–Vorpommern and Schleswig-Holstein (Khalili et al. 1997)

Table 4.7. Population, loads of nitrogen, nitrogen load per capita and percent of the total nitrogen load in the Baltic Sea Basin countries in 2000 (after Granstedt et al. 2004)

	Population $\times 10^3$	Total N loads 10^3 kg N yr^{-1}	N load per capita kg N capita^{-1}	Shares of N load % of total
Denmark	5,155	62,240	12	7
Sweden	8,500	175,610	21	21
Finland	4,938	146,560	30	18
Russia*	9,028	53,720	6	7
Estonia	1,595	32,990	21	4
Latvia	2,606	54,070	21	7
Lithuania	3,446	35,560	10	4
Poland	37,764	229,990	6	28
Germany*	3,300	31,510	10	4
Total	76,332	822,250		

total annual nitrogen loads into the Baltic Sea has occurred in the period 1995–2000 (Granstedt et al. 2004). The only exceptions were Latvia, Estonia and Russia, the countries having the probably largest decrease in agricultural and fertilisation intensity. Of the calculated total nitrogen load in 2000, 24% originates from Estonia, Latvia, Lithuania and Russia, 28% from Poland, 21% from Sweden and 18% from Finland (Table 4.7). However, per-capita output levels of nitrogen are almost four times higher for Sweden and Finland than for Poland.

4.5.1.3 Nutrient Losses from Agricultural Catchments

In agricultural catchments, drainage accounts for the major part of both nitrogen and phosphorus loads to rivers. According to Behrendt and Bachor (1998), in Mecklenburg-Vorpommern, Germany, this proportion is up to 90% for nitrogen and about 75% for phosphorus. In some catchments, even storm-water events did not contribute with more phosphorus than the drainage water (up to 0.25 kgP ha^{-1} yr^{-1}; Gelbrecht et al. 2005). Depending mainly on management practice, especially fertilisation and livestock density, land use structure, soil and climatic conditions, the average annual losses of nitrogen in the countries around the Baltic Sea in 1994–1997 ranged from 5 to 75 kgN ha^{-1} yr^{-1} (Vagstad et al. 2004), being higher in the beginning of 1990s (up to 105 kgN ha^{-1} yr^{-1} in Norway; Bechmann et al. 1998; and up to 100 kgN ha^{-1} yr^{-1} in Denmark; Kronvang et al. 1995). In the Baltic states, the value did not exceed 25 kgN ha^{-1} yr^{-1} (Mander et al. 1998; Jansons et al. 2003), except in the case of pig slurry applications (Vagstad et al. 2000).

Phosphorus losses in Lithuania, Latvia and Estonia ranged between 0.1 and 0.32 kgP ha^{-1} yr^{-1} (Mander et al. 1998; Sileika et al. 2002, 2005; Iital et al. 2005). In Denmark, the average annual loss of phosphorus from 270 agricultural watersheds was 0.29 kgP ha^{-1} yr^{-1} (Kronvang et al. 1995). Nitrogen and phosphorus losses similar to those in Denmark and Norway have been found in agricultural watersheds in southern Sweden (Johnsson and Hoffmann 1998), whereas the mean annual total phosphorus losses in agricultural catchments in southern Finland attained 2 kgP ha^{-1} yr^{-1} (Valpasvuo-Jaatinen et al. 1997; Rekolainen et al. 1999).

Comparison of annual median losses of total nitrogen and phosphorus in Danish agricultural catchments and undisturbed catchments revealed an average loss ratio of 14:1 for total N (23.4 and 1.7 kgN ha^{-1}, respectively) and 4:1 for total P (0.29 and 0.070 kgP ha^{-1}). The values of both total N and total P losses from undisturbed catchments in Denmark are at the same level as losses from forested catchments in Finland (see Lepistö et al. 1995; Kortelainen et al. 1997). However, the major part (on average 79%) of the nitrogen transported from the catchments consisted of organic nitrogen (Kortelainen et al. 1997) whereas mineral nitrogen (mainly in nitrate form) is the major part of total N transported from agricultural catchments.

Landscape factors play a significant role in nutrient leaching and transport from catchments. Several studies indicate that in mosaic (in some cases hilly) landscapes, export rates of both total phosphorus and total nitrogen from the diversified small catchments tend to be lower than in non-mosaic areas (Rybak 2002). Uuemaa et al. (2007) found a significant correlation between various landscape metrics and nutrient losses from catchments: landscapes with higher diversity (higher edge and patch density, higher value of

Shannon and Simpson indices, lower value of the contagion index) tended to release less nitrogen and phosphorus than more homogeneous landscapes.

The concept of critical source areas (Weld et al. 2001) helps to identify regions within the catchment area from which the losses of nutrients have the highest probability. Heathwaite et al. (2000) demonstrate that areas most vulnerable to phosphorus loss are limited to small, well-defined areas of watersheds (< 20% area) near the stream channel. In contrast to phosphorus, larger areas contribute to nitrate leaching and generally occur on the upper boundaries of the watershed (60%).

Riparian buffer zones are the most significant parts in watersheds controlling both nitrogen and phosphorus flows (Vought et al. 1994; Mander et al. 1997; Kuusemets and Mander 2002). Weller et al. (1996) demonstrate that a hectare of riparian wetland can be many times more important in reducing phosphorus than an agricultural hectare is in producing phosphorus.

4.5.1.4 *Modelling Approaches*

Various empirical models have been elaborated and used to describe and forecast the nitrogen and phosphorus losses from agricultural catchments. The HBV-model (Graham and Bergström 2001) for water balance modelling and, based on it, the HBV-N model (Arheimer and Wittgren 2002; Arheimer et al. 2004a,b) for nitrogen cycling have been successfully used for large watersheds. A large variety of models have served to describe nutrient flows in small agricultural catchments (Andersen et al. 1999; Rekolainen et al. 1999; Mander et al. 2000a,b; Kronvang et al. 1999; Müller-Wohlfeil et al. 2002; Wendland et al. 2002; Kronvang et al. 2003; Granlund et al. 2004; Andersen et al. 2005).

Mander et al. (2000a) present a simple empirical model which incorporates land-use pattern, fertilisation intensity, soil parameters and water discharge and accurately describes variation of total N and total P runoff in both the whole catchment of the Porijõgi river (258 km^2, southern Estonia) and its agricultural subcatchments (R^2 varies from 0.95–0.99 for N to 0.49–0.93 for P; Fig. 4.9). During the period 1987–1997, abandoned lands of this catchment area increased from 1.7 to 10.5% and arable land decreased from 41.8 to 23.9%. At the same time, the runoff of total inorganic N and total P decreased from 25.9 to 5.1 and 0.32 to 0.13 kg ha^{-1} yr^{-1}, respectively. In small agricultural sub-catchments, the rate of fertilisation was found to be the most important factor for nitrogen runoff, whereas in larger mosaic watersheds, land-use pattern plays the main role (Mander et al. 2000a).

Another approach to the analysis of changes in nutrient flows in agricultural catchments is the concept of potential excess nutrients, the difference between inputs to and outputs from the nutrient pools (see Bechmann et al. 1998; Garten and Ashwood 2003). Kull et al. (2005) assessed the potential excess nitrogen (PEN) under different land-use categories based on a GIS model and compared modelling results with observed data from the Porijõgi river catchment in southern Estonia (see also Fig. 4.9). In this study, negative values of PEN denote land-cover categories as potential nitrogen sinks and positive values as potential nitrogen sources (Fig. 4.10). Due to the significant decrease in agriculture from 1987 to 1997, the average PEN in the Porijõgi river basin has decreased several times (Mander et al. 2000a,b). The source areas embraced 50.1% of the catchment in 1987, while by 2001 this area had dropped to 17.8% of the total area (Kull et al. 2005; Fig. 4.10). In agricultural subcatchments without riparian buffers, the PEN value decreased from 1.5 in 1987 to −14 kg ha^{-1} in 2001, whereas in another agricultural subcatchment with excessive buffer strips, average PEN values decreased from −8.5 to −18.5 kg ha^{-1} (Fig. 4.10). Thus, the potential nutrient excess analysis can provide us with a clear indicator of changes in nutrient flows driven by the land-use changes and/or climatic fluctuations.

4.5.1.5 *Influence of Changes in Land Use and Agricultural Intensity on Nutrient Runoff*

Changes in land use and agricultural intensity influence nutrient runoff in contradictory ways. It is obvious that the effect is different in catchments of different sizes. At the local level, in small agricultural catchments, the decreasing load relatively quickly results in decreasing runoff (Johnsson and Hoffmann 1998; Mander et al. 1998, 2000a,b; Iital et al. 2003, 2005; Kronvang et al. 2005). Decrease in phosphorus losses is often due to decrease in numbers of livestock per catchment area (Sileika et al. 2002).

However, in catchments with relatively low losses and/or well established ecotechnological measures (riparian buffers, constructed wetlands), decreasing load may not result in significant water quality improvement. For instance, lowering inputs into riparian buffer zones may cause outwash of accumulated nutrients (Kuusemets et al. 2001). In larger catchments there is a remarkable lack of response to the significant decrease in agricultural pressure (Grimvall et al. 2000; Tumas 2000; Stålnacke et al. 1999a,b, 2003, 2004; Räike et al. 2003).

4.5.1.6 *Potential Future Impact of Climate Change on Nutrient Runoff*

Changes in climate patterns and related runoff regimes can significantly influence the nutrient losses from catchments. In some cases this influence can play an even more important role than the change in anthropogenic loads in watersheds (see Andersson and Arheimer 2003; Sileika et al. 2005). The most important effects of climate change are expected through changes in the timing of seasonal and annual events (spring runoff, autumn low flow, ice and snow cover formation, etc.), frequency and severity of extreme events (floods, droughts, erosion), thresholds and ranges. Changes to the climate in the Baltic Sea Basin will not only affect the total amount of freshwater flowing into the sea, but also the distribution of the origin and nutrient content of these flows.

Graham (2004) has analysed effects of four climate change scenarios on river runoff to the Baltic Sea. The resulting change in total mean annual river flow to the Baltic Sea ranges from +2 to +15% of present-day flow according to the different climate scenarios for the coming 50-year period. The magnitude of changes within different sub-regions of the basin varies considerably, with the most severe mean annual changes ranging from +30 to +40%. Common to all of the scenarios evaluated is a general trend of reduced river flow from the south of the Baltic Sea Basin together with increased river flow from the north (Sect. 3.7).

Another study on climate change scenarios (HadCM2-IS92a and ECHAM3TR-IS92a as basic, and HadCM2-IS92c and ECHAM3TR-IS92e as extreme versions) show similarly clear trends: snow packs will decrease and snow-melt will occur earlier with climate warming. A reduced influence of snow-melt on stream discharge will increase the synchronisation between precipitation and stream discharge (Kull and Oja 2001). The basic (IS92a) scenarios from the latter study show comparable climate trends to the overall trends among scenarios as synthesised in Chap. 3.

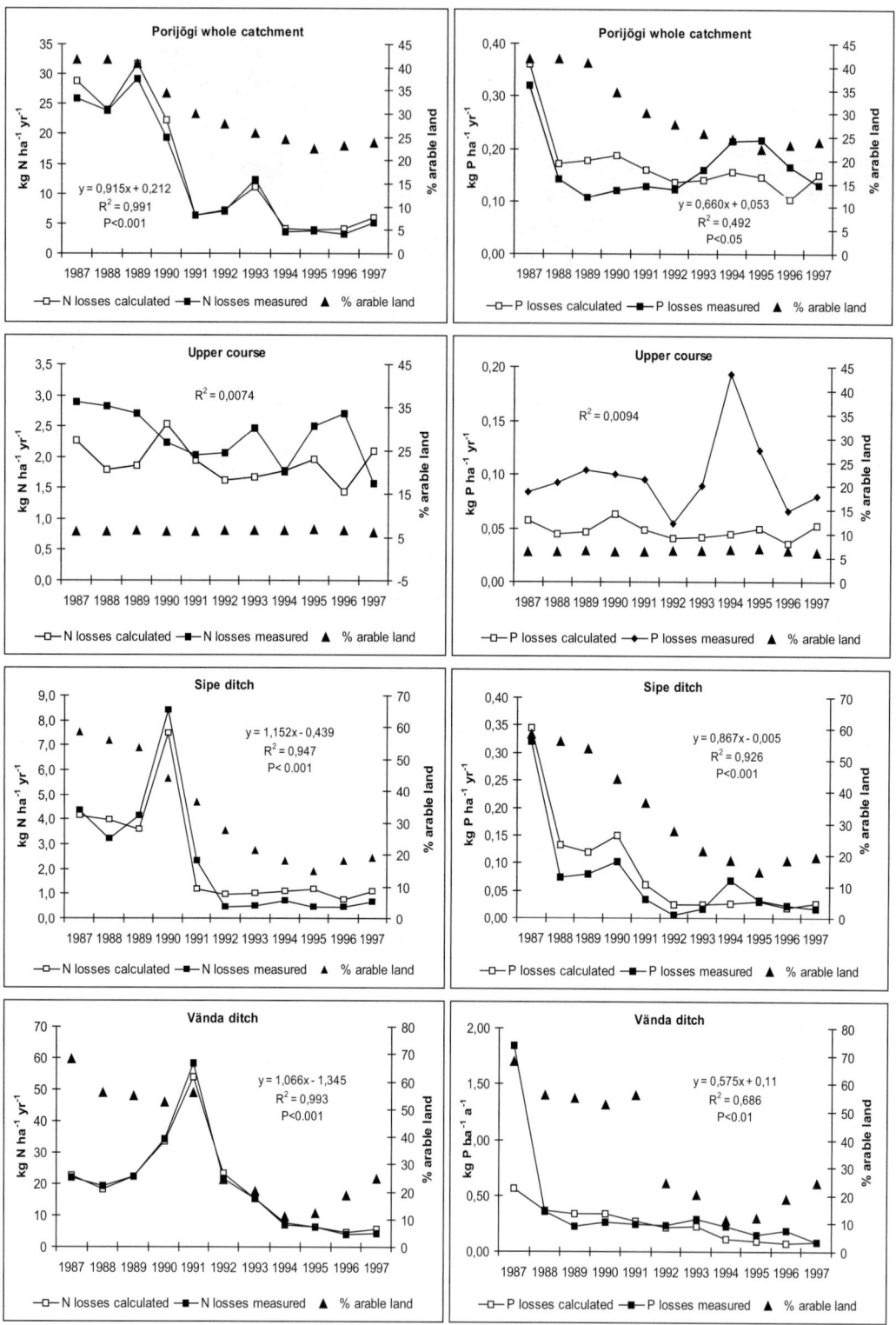

Fig. 4.9. Relationships between the measured (y) and modelled (x) total inorganic nitrogen and total phosphorus flows in subcatchments of a small river in southeastern Estonia over the period 1987–1997 (adapted from Mander et al. 2000a)

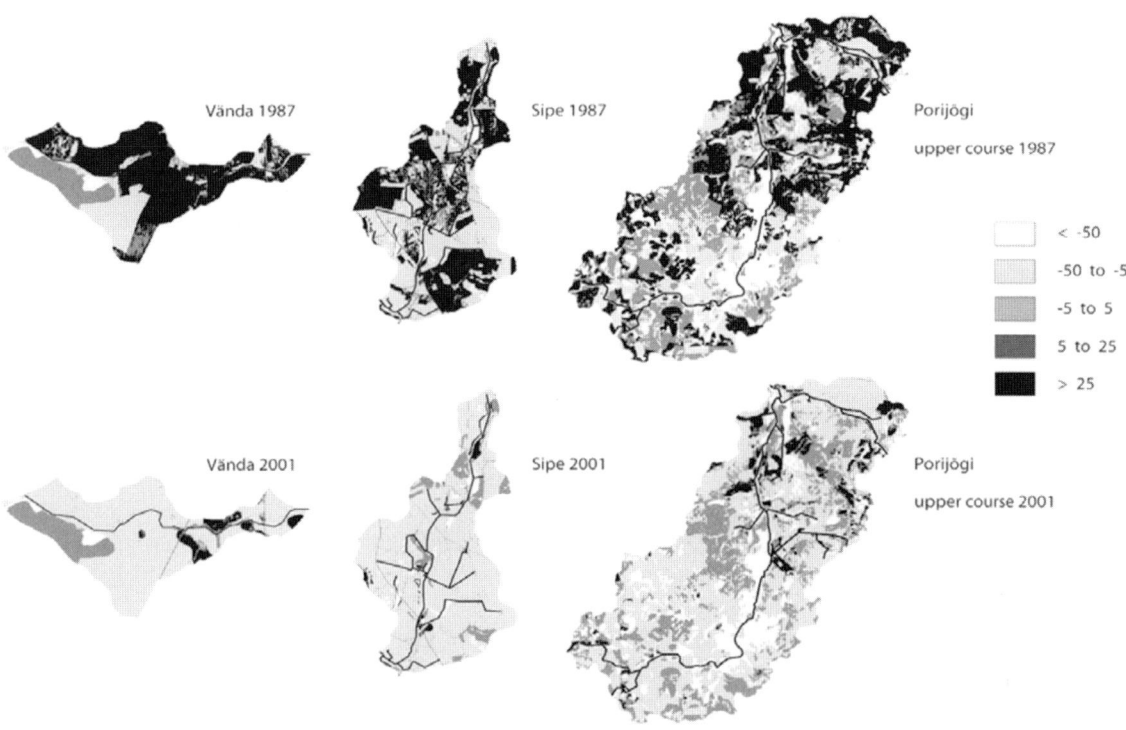

Fig. 4.10. Modelled potential excess nitrogen (PEN) values in three subcatchments of the Porijõgi River in 1987 and 2001. Both Vända (3.8 km^2) and Sipe (8.87 km^2) subcatchments are agricultural, whereas the Sipe subcatchment is characterised by excessive riparian buffer strips. The upper course (22.1 km^2) is mainly covered by forests and semi-natural meadows (adapted from Kull et al. 2005)

Higher cyclonic activity in the northern Atlantic brings more frequent inflow of warm Atlantic air masses in winter to central and northern parts of the Baltic Sea Basin. This will lead to less stable weather with more frequent storms and melting periods in winter (Sect. 3.6). Due to less snow-pack, shorter duration of snow cover and frequent melting periods, the peak flow and floods in early spring will be less intensive in the eastern part of the Baltic Sea Basin. Base flow of the rivers, typical for summer, will be reached earlier and will last until autumn when a significant increase of river runoff will set in because of higher precipitation. For the growing period, soils will become slightly drier for most of the Baltic Sea Basin.

However, the southern part of the Baltic Sea Basin may become more vulnerable to rainstorms and flood events, which would have a severe effect on agricultural activities in former floodplain areas, as well as on nutrient losses. For example, during the flood in 1997 (July 1 to August 28) the total water outflow near the mouth of the Oder River increased by 65% in comparison to 1996. This resulted in increased fluxes of phosphate by 34% and nitrite nitrogen by 88% (Niemirycz 1999). In addition to N and P, some priority substances like total organic carbon (TOC), nitrogen, and the heavy metals Cu, Pb and Zn, the pollutant entries via suspended solids during the flood period are estimated to be approximately one-third of the usual annual load. For the concentrations of the priority pollutants in suspended solids, accumulation factors from 2 to 4 times that of normal conditions were observed (Müller and Wessels 1999).

In the central and northern part of the Baltic Sea Basin, agricultural catchments show a clear annual course of nutrient runoff with moderate losses in winter, peak in early spring, high nutrient losses even in autumn, and the lowest nutrient runoff in summer (low temperatures, a thick snow-pack, frozen surface and low water discharge are the main factors that keep nutrient losses at a low level during the winter period). The highest year-to-year variability of nutrient losses is associated with autumn, while variability is also high in late winter and spring. Under climate change, variability of nutrient losses is expected to increase mainly in winter and spring when the duration of frozen surfaces, amount of snow-pack water, duration and continuity of certain weather types, occurrence of strong night frost events and the number of soil freeze-thaw cycles is changing. Shorter period of frozen surface and increased numbers of soil freeze-thaw cycles will lead to more intensive leaching from arable land.

Responses of nutrient losses to climate change will be faster and more direct in small agricultural catchments, while dynamics in large catchments are more complicated due to internal material cycles (Mander et al. 2000b; Stålnacke et al. 2004).

Milder winters (increase of air temperature in February from −7.9 to −5.5 °C during the period 1950–1997 in Tartu, Estonia) and a change in the precipitation pattern in recent decades have influenced the mean annual water discharge in more continental parts of the Baltic Sea Basin (Mander et al. 2000b). In the Porijõgi River catchment area, southern Estonia (see Figs. 4.9 and 4.10), the change in the annual runoff pattern resulted in more intensive material flow during colder seasons and decreased water runoff in summer. The most significant change in the mean monthly nutrient runoff pattern is that of total inorganic nitrogen (TIN). In all subcatchments except the forested Porijõgi upper course, a significant decrease in inorganic nitrogen runoff has been observed. Due to increased surface flow caused by melting snow during milder winters, the TIN losses are still high, but rapidly decrease in spring. During the growing season, the intensive uptake by plants minimises nitrogen losses. Likewise, nitrogen losses in summer show the lowest variation (Mander et al. 1998; Fig. 4.11).

In the whole Porijõgi River catchment and in its upper course, the annual patterns of monthly runoff of phosphorus show only minor changes (Fig. 4.12). However, winters in the last decade have been milder, resulting in a shorter period with permanent snow cover. Frozen surface and snow melting water with wet precipitation give rise to increased surface flow. As shown in Fig. 4.12, this mostly affects the intensively cultivated Sipe and Vända subcatchments. Due to more frequent periods with thawing soil in winter, the phosphorus particles are exposed to surface flow for a longer period, with total monthly phosphorus runoff in winter months therefore remaining at a high level. By contrast, less intensive

tillage in spring and autumn has reduced phosphorus losses in those periods (Mander et al. 1998).

In addition to clearly water budget-based explanations, long term cyclicity in water discharge, which varies in frequency (from 13 to 43 years) among different rivers (Tsirkunov et al. 1992; Klaviņš et al. 2002), plays a certain role in the nutrient regime of rivers. However, there are no clear explanations for these kinds of cyclicity.

4.5.1.7 Measures for Further Decrease of Nutrient Flows to the Baltic Sea

During the past 15 years or so, EU and national policy actions have been adopted to decrease the nutrient load and threats of eutrophication in the Baltic Sea. On the other hand, there has been a dramatic change in agricultural policy and practices since the beginning of 1990s in the eastern part of the Baltic Sea Basin. Despite all of these measures, the quality of the Baltic Sea is not significantly improving.

In particular, the ecological quality in estuaries and coastal lagoons has not improved, despite reduced nutrient loads in the catchments and decreasing riverine loads. Therefore, additional measures are planned by several authorities. Among them, sustainable agricultural practices (wider practicing of ecological agriculture and implementation of environment-friendly crop rotations; Paulsen et al. 2002; Granstedt et al. 2004) and changes in agricultural practices (high-tech solutions like GPS-supported fertilisation schemes; implementation of shallow injection technique when using liquid manure; Sundell 1997) should be mentioned as measures targeted to decrease loadings in the catchments.

Several ecotechnological measures will help to control nutrient flows from the catchments to the water bodies. Maintenance, sustainable management and restoration of wetlands is the widely-used ecotechnological practice at the catchment level (Weller et al. 1996; Gustafson et al. 1998; Arheimer and Wittgren 2002; Paludan et al. 2002).

In addition, riparian buffers are already common practice in several countries (Kuusemets and Mander 2002). Jansson et al. (1998) have calculated that existing natural wetlands in the Baltic Sea Basin can retain an amount of nitrogen which corresponds to about 5–13% of annual total (natural and anthropogenic) N emissions entering the Baltic Sea.

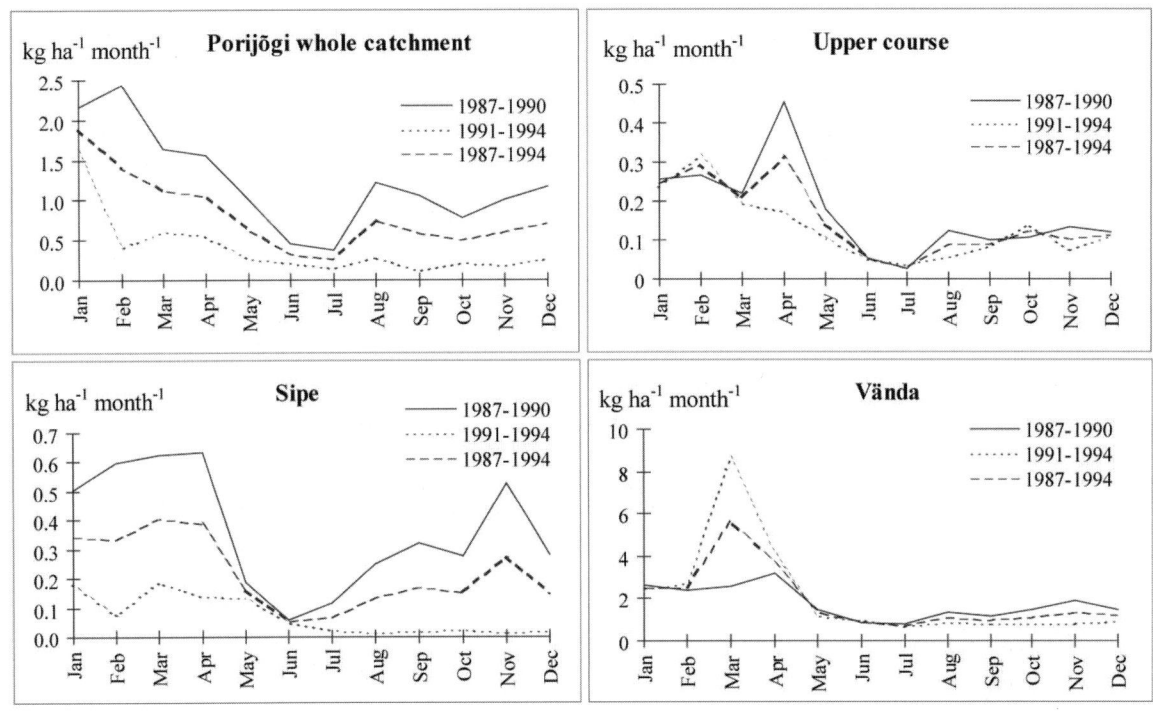

Fig. 4.11. Changes in total inorganic nitrogen (TIN) fluxes in the Porijõgi catchment. For subcatchment characteristics, see Fig. 4.10 (after Mander et al. 1998)

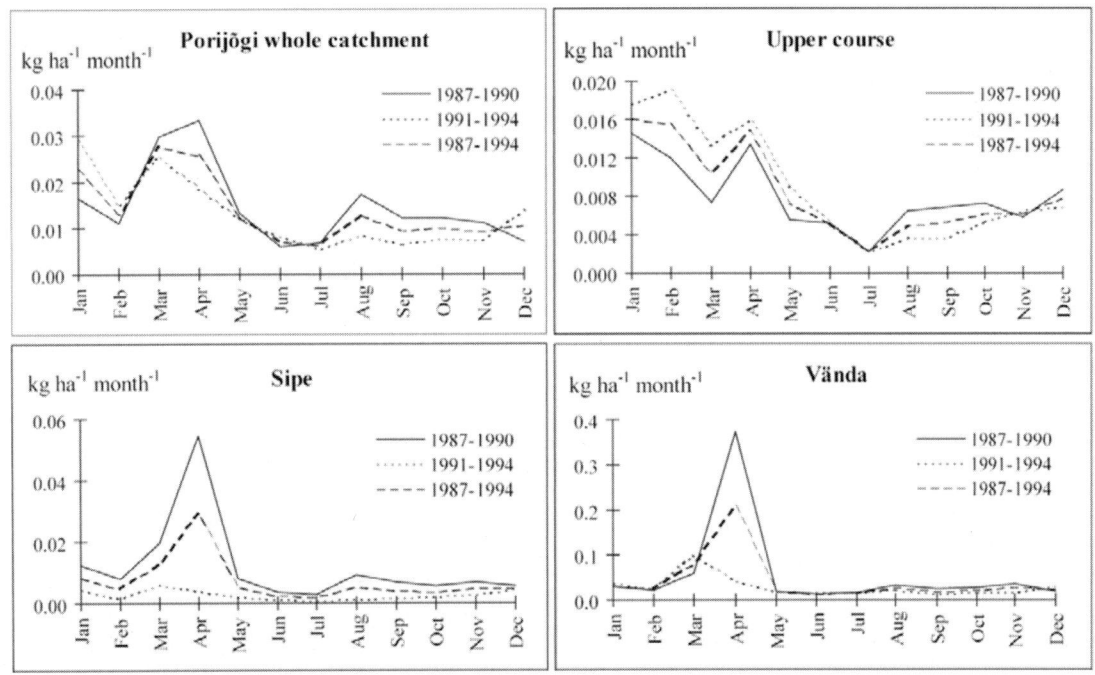

Fig. 4.12. Changes in total phosphorus fluxes in the Porijõgi catchment. For subcatchment characteristics see Fig. 4.10 (after Mander et al. 1998)

Likewise, the existing wetlands can significantly decrease the phosphorus emissions (Weller et al. 1996). Constructed wetlands play an important role while effectively retaining both nitrogen and phosphorus from both diffuse and point sources of pollution (Arheimer et al. 2004b). Constructed wetlands can also help in terms of stabilising the balance between N and P in purified wastewater, while relatively cheap and widely used phosphorus sedimentation in conventional purification systems has increased the proportion of nitrogen relative to phosphorus transported to the Baltic Sea (Gren et al. 1997; Laznik et al. 1999). Nitrogen removal with conventional techniques is, however, a very costly measure, the benefits of which are not always reliable (Turner et al. 1999).

4.5.2 Organic Carbon Export from Boreal and Subarctic Catchments

The catchment area of the Baltic Sea can be roughly divided into a south-eastern part, draining into the Gulf of Riga and the Baltic Proper, that is characterised by a cultivated landscape with a temperate climate, and a northern boreal part, characterised by coniferous forests and peat, draining into the Bothnian Bay, Bothnian Sea and Gulf of Finland. The northern part also includes subarctic watersheds in the very northern catchment of the Bothnian Bay (Fig. 1.11, Chap. 1; Table 4.8). In general, the specific runoff is twice as high in the catchments of the Bothnian Bay and Bothnian Sea as in the south-eastern part of the Baltic Sea catchment, whereas the catchment of the Gulf of Finland, which is dominated completely by the Neva River, has an intermediate specific runoff (Table 4.8).

In this section we describe the potential effects of global warming on the land–sea fluxes of dissolved constituents (C, N, P, Si) in these boreal and subarctic watersheds of the Baltic Sea and discuss their implications for the marine ecosystem of the Baltic Sea, as well as potential positive and negative feedbacks to global warming.

4.5.2.1 *Oligotrophic Character of Boreal Watersheds*

The Baltic Sea is often described as one of the most eutrophicated seas, although it should be emphasised here that > 60% of the river water entering the Baltic Sea comes from the subcatchments of the Bothnian Bay, the Bothnian Sea and the Gulf of Finland (Table 4.8). The water from these

subcatchments has an oligotrophic (Bothnian Bay and Bothnian Sea) to slightly mesotrophic (Gulf of Finland) character, as for example the Neva River. The much lower cover of cultivated landscape in the watersheds of the northern catchments compared to the south-eastern catchment (Table 4.8) is clearly reflected in the trophic status of these rivers, expressed here as dissolved inorganic nitrogen (DIN) and phosphorus (DIP) concentrations, which are in the same range or sometimes even lower than in the respective basins of the Baltic Sea (Humborg et al. 2003). The DIN and DIP concentrations of major rivers entering the Gulf of Bothnia are shown in Fig. 4.13 and compared with those from the major monitoring stations of the Bothnian Bay (BO3) and the Bothnian Sea (USB5).

Especially the phosphorous concentrations in the rivers are extremely low and can explain the P-limitation observed in the receiving waters of the Bothnian Bay (Hagström et al. 2001). The low DIP concentrations in unperturbed rivers can partly be explained by the high iron concentrations of these rivers (Ingri et al. 2005). However, the seasonal variations of the dissolved constituents are much lower in the heavily regulated rivers, as for example the Luleälven, Skellefteälven or Indalsälven, as compared to the unregulated rivers in the area, of which only a few are left – for example the rivers Torneälven, Kalixälven and Råneälven. Damming has reduced the nutrient land–sea fluxes of nearly all dissolved constituents including nutrients and carbon, which can be explained mainly by changes in the weathering regime as a result of changes in hydrology and surface water-groundwater interactions in these heavily dammed rivers (Humborg et al. 2000, 2002).

4.5.2.2 *Potential Future Impacts*

Runoff is the leading environmental variable determining land–sea fluxes of dissolved constituents, as also observed for the more eutrophicated watersheds of the Baltic Sea Basin (Vagstad et al. 2004). Over the next century, runoff is estimated to increase by 15% averaged over the entire Baltic Sea Basin and between 10% and 40% for the catchments of the Bothnian Bay, Bothnian Sea and the Gulf of Finland (Graham 2004; Sect. 3.7). The most prominent effect of global warming with respect to watershed exports of matter is therefore likely to be a significant increase in fluxes of dissolved constituents such as nutrients (N, P, Si) as

Table 4.8. Catchment area, population, land cover and runoff characteristics of the major catchments in the Baltic Sea Basin. Runoff characteristics are based on monthly measurements (1970–2000) at the river mouths of 84 major rivers (see Fig. A.36) (source: MARE-Nest, www.mare.su.se/nest; page visited 20 January 2006)

Basin[1]	BB	BS	BP	GF	GR	DS	KA	All
Catchment area ($\times 10^3 \, \text{km}^2$)	263	227	573	428	134	29	81	1,735
Population ($\times 10^6$)	1.3	2.4	54.8	12.3	4.0	5.2	3.0	83.1
Deciduous forest (%)	3.0	2.0	2.3	3.0	6.0	1.2	2.2	
Coniferous forest (%)	43.4	59.6	17.7	43.3	16.7	2.6	42.2	
Mixed forest (%)	27.3	19.6	10.0	17.6	16.8	2.7	15.6	
Herbaceous vegetation (%)	5.3	3.0	8.4	0.9	6.8	4.7	2.5	
Wetlands (%)	9.0	3.3	0.4	0.4	0.4	0.4	1.0	
Cultivated land (%)	4.0	4.9	57.3	20.0	50.8	78.6	22.9	
Bare ground (%)	2.3	0.7	–	–	–	–	0.1	
Water (%)	5.4	6.4	2.4	14.3	1.9	4.7	12	
Snow and ice (%)	0.2	–	–	–	–	–	–	
Artificial surfaces (%)	0.1	0.5	1.4	0.5	0.6	5.2	1.5	
Runoff ($\times 10^9 \text{m}^3$)	108	93	124	109	33	6	35	507
Specific runoff (mm m^{-2})	409	410	217	252	245	210	442	

[1]BB = Bothnian Bay; BP = Baltic Proper; BS =Bothnian Sea; DS = Danish Straits; GF = Gulf of Finland; GR = Gulf of Riga; KA = Kattega

well as total organic (TOC) and inorganic carbon (mainly alkalinity in the form of bicarbonate; Ludwig et al. 1999).

In Fig. 4.14, the average monthly loads calculated for the period 1969–2002 of nitrogen, phosphorus, dissolved silicate (DSi), alkalinity and total organic carbon (TOC) are shown as a function of the average monthly water discharge calculated for the same period for the three remaining unregulated and unperturbed river systems in northern Sweden, i.e. the Torneälven, Kalixälven and Råneälven. It appears that water runoff explains between 71 and 97% of the variability in nutrient and carbon land sea-fluxes to the Baltic Sea. Thus, a similar range of increases in transports of dissolved constituents can be expected.

In the case of nitrogen and phosphorus, these potentially higher imports will probably not result in dramatic changes in the ecosystem of the northern basins of the Baltic Sea, since the concentrations of DIN and DIP are very close to each other both in the major rivers and in the Gulf of Bothnia, as discussed above (Fig. 4.13). Increased DSi transports will not affect either the ecosystems of the Bothnian Bay or those of the Bothnian Sea, since diatoms are not Si limited in the northern basins; an effect could be an overall increase in DSi concentrations in the southern basins of the

Baltic, where DSi limitations of diatoms have indeed been observed (Yurkovskis 2004; Kristiansen and Hoell 2002).

More important, though more subtle, for global change issues may be the possible effects of changes in the ratio of organic carbon to inorganic carbon (alkalinity) fluxes in these boreal watersheds as a response to global warming: changes in the overall carbon budgets of these watersheds can be expected. It has been argued that a change in climate and hydrology in high latitude regions could liberate large amounts of previously inactive soil carbon, especially from peat (Freeman et al. 2001; Tranvik and Jansson 2002), and new studies from northern Sweden have shown that a great deal of this organic carbon is respired, mainly in lakes, on its way down to the sea (Algesten et al. 2004; Sobek et al. 2003; see Sect. 4.4.2).

However, besides the total amount of runoff – which will probably increase – there will also be a shift in annual water discharge patterns (Graham 2004; Sect. 3.7). At high latitudes the hydrology is characterised by a distinctive peak discharge during May/June driven by rain and, especially, snow melt that contributes in these regions to approximately half of the annual water discharge to the Baltic Sea, at least in unregulated rivers (Bergström and Carlsson 1994).

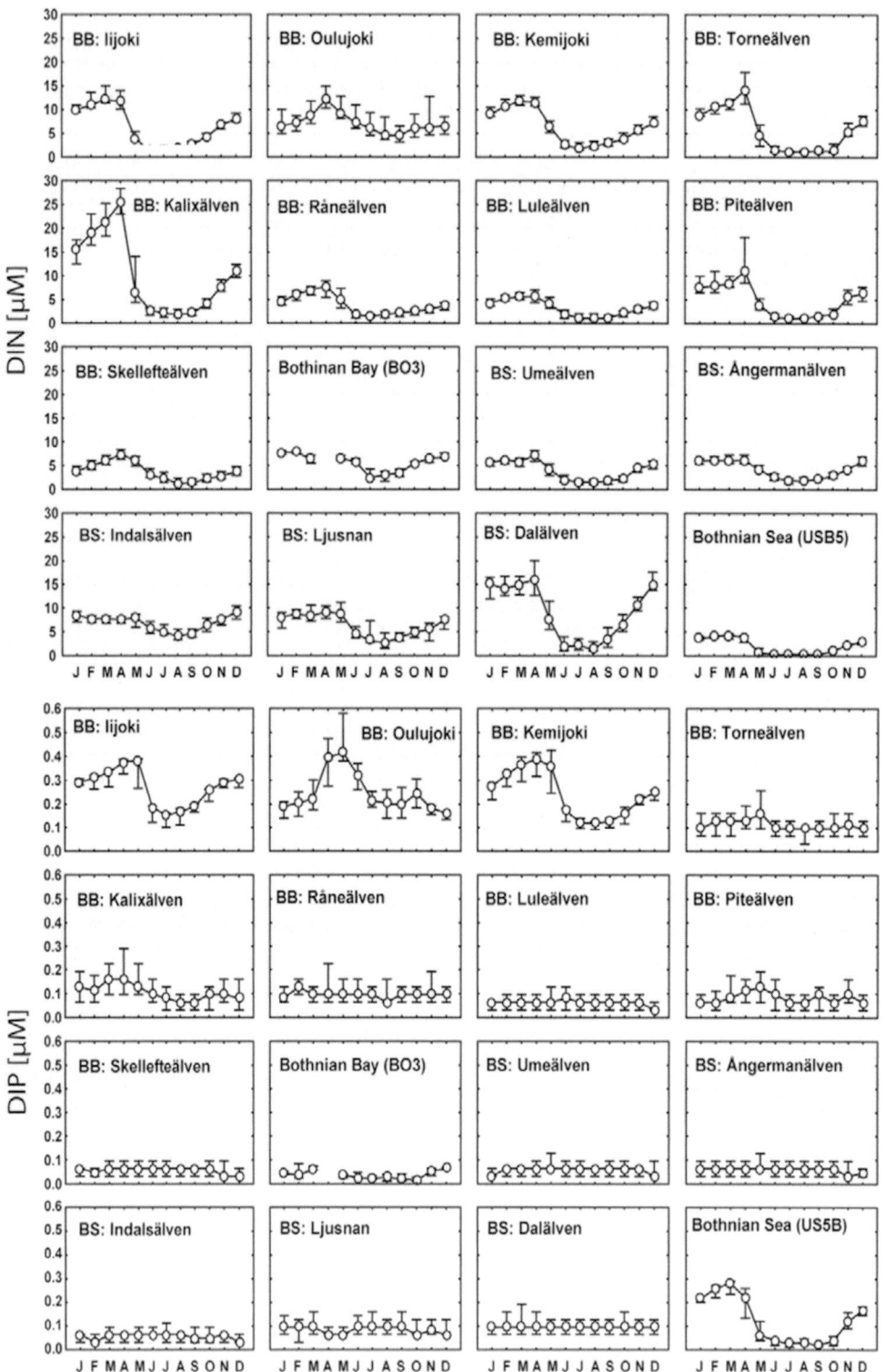

Fig. 4.13. Long-term (1980–2000) annual median river nutrient concentration in major rivers running into the Bothnian Bay and Bothnian Sea: (*upper panels*) dissolved inorganic nitrogen (DIN), (*lower panels*) dissolved inorganic phosphorus (DIP); whiskers indicate 25th and 75th percentile; horizontal axis gives month of year (adapted from Humborg et al. 2003)

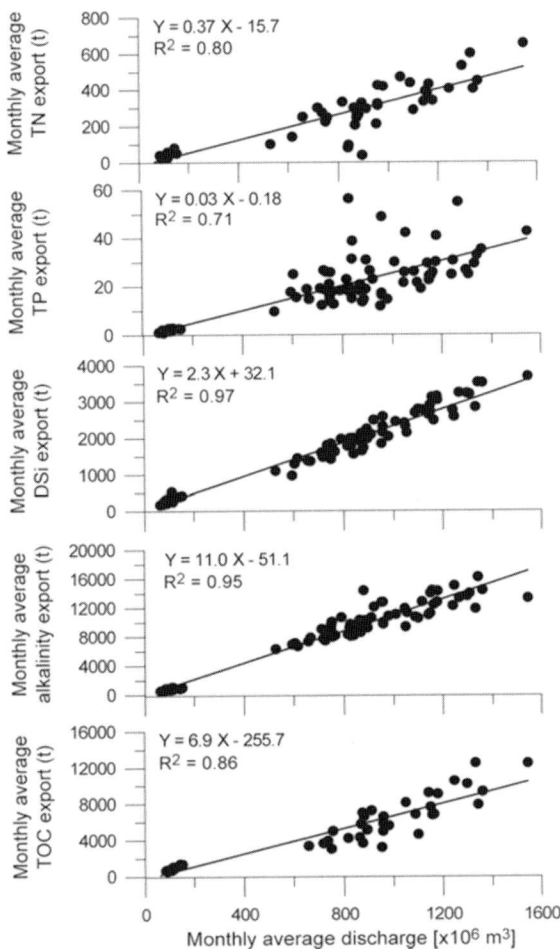

Fig. 4.14. Monthly averaged loads (1969–2002) of dissolved constituents as a function of monthly averaged water discharge for the unperturbed rivers Torneälven, Kalixälven and Råneälven (source: MARE-Nest, www.mare.su.se/nest; page visited 20 January 2006)

Moreover, a climate change in these regions would certainly affect their hydrology, i.e., the peak flow in the northern catchment of the Baltic Sea will be smoothed out (Graham 2004). A warmer and wetter climate would result in more water circulating through the soils, making the exchange of soil water and groundwater more active, and would also increase groundwater discharge over the year (Michel and Vaneverdingen 1994; Yang et al. 2002).

As a result, less TOC might be exported to the Baltic Sea, since less of the top soils that are responsible for most of the TOC transport in alpine and high latitude watersheds will be flushed (Boyer et al. 1997; Carey 2003). At the same time, more weathering-related constituents such as Si, P and alkalinity, that are enriched in groundwater (Drever 1997), might be exported to the Baltic

Sea. Alkalinity can be regarded as a negative feedback to global warming, since atmospheric CO_2 is consumed in weathering reactions (Berner and Berner 1996; Schwartzman and Volk 1989).

For plagioclase, a common mineral in the northern catchments of the Baltic Sea, the weathering reaction has the following equation:

$$2CO_2 + 3H_2O + Ca_2Al_2Si_2O_8 \rightarrow Al_2Si_2O_5(OH)_4 + Ca^{2+} + 2HCO_3^-$$

and for carbonates it can be written as:

$$CaCo_3 + CO_2 + H_2O \rightarrow Ca^{2+} + 2HCO_3^-$$

where HCO_3^- constitutes the major part of alkalinity. Part of the carbon (alkalinity) liberated by weathering is then stored in the soils, and part is transported to the ocean (Ludwig et al. 1999).

Thus, the anticipated increase in TOC export having a potentially positive feedback to global warming (Freeman et al. 2001, 2004) will probably be partly compensated by an increase in alkalinity export. Moreover, increasing temperatures will likely lead to increases in forested areas (e.g. Kullman 2002; Koca et al. 2006; see also Sects. 4.3.2, 4.3.4 and 4.3.5), so that changes in vegetation belts in the northern parts of the Baltic Sea Basin towards wetlands, taiga (boreal coniferous forests) and broadleaved forest can be expected. This anticipated change in land cover would likewise lead to an increase in fluxes of weathering-related dissolved constituents (DSi, P, alkalinity), since it has been found that vegetation cover in watersheds of northern Sweden has a positive effect on weathering-related constituents in river waters (Humborg et al. 2004).

However, an increase in alkalinity export to the oceans in response to changes in rainfall and land use has been observed in the Mississippi watershed (Raymond and Cole 2003) and the possible negative feedback to global warming has been emphasised (Ittekkot 2003). Thus, a major unresolved question appears to be whether a change in alkalinity exports accompanying changes in the landscapes associated with global warming is also an issue in high latitude watersheds, such as those draining into the northern basins of the Baltic Sea.

4.5.3 Synthesis – Climate Change Impacts on Nutrient Fluxes from Land Ecosystems to the Baltic Sea

Nitrogen and phosphorus losses from agriculture in the Baltic Sea Basin have increased markedly over the last 50 years, but have stabilised since the mid-1980s. The main causes for the recent stabilisation are probably reduced fertiliser use and agricultural land abandonment, in both cases associated with more stringent agricultural and environmental policy, and with the transition to a market economy in former eastern bloc countries.

Milder winters and changing precipitation patterns since the 1950s have influenced water discharge to the Baltic Sea. Over continental areas, the net effect seems to be a reduction in nutrient runoff, while a tendency towards longer dry periods interspersed with intensive rain seasons has caused an increase in transport of dissolved organic carbon from forests and wetland areas to the water bodies.

Water runoff explains between 71 and 97% of the variability in nutrient and carbon land-sea fluxes to the Baltic Sea. Future changes in climate patterns and related runoff regimes can significantly influence export rates. Trends will be sensitive to changes in the timing of seasonal and annual events (spring runoff, autumn low flow, ice and snow cover); the frequency and severity of extreme events (floods, droughts, erosion); as well as to climatic thresholds and ranges.

Climate scenarios for the next century point to moderate increases in mean annual river flow to the Baltic Sea (Sect. 3.7). Common to all scenarios is a general trend of reduced river flow from agriculture-dominated southern and continental areas, with increased flows in largely forested northern catchments. In contrast, southern areas may become more vulnerable to rainstorms and flood events, with severe effects on agriculture in former floodplain areas, as well as nutrient losses. The balance between these counteractive forcings in their effects on total and regional nutrient export rates to the Baltic Sea is difficult to predict.

A reduced influence of snow melt will increase the synchronisation between precipitation and stream discharge, reducing and extending summertime base flow while increasing wintertime runoff. Springtime peak flows and floods are likely to become less intense. Shorter periods with frozen surfaces and more frequent freeze-thaw cycles would further promote leaching during winter from arable land, especially in northern areas.

Organic carbon loads in runoff from boreal and subarctic catchments will be sensitive to the counteractive effects of increased total water discharge and changed seasonality, leading to less intensive flushing events of top soils but deeper infiltration, potentially decreasing TOC fluxes and increasing alkalinity fluxes. The net effect of these contrasting mechanisms on net carbon exports to the Baltic Sea remains uncertain.

4.6 Conclusions and Recommendations for Research

This chapter sought to evaluate, through literature review and synthesis, two hypotheses. The first of these asserts that climate change over recent decades has already impacted terrestrial and freshwater ecosystems of the Baltic Sea Basin. This hypothesis receives clear support, even allowing for the many uncertainties. Phenological events, species distributions and migration behaviour are readily-observed indicators of biological changes, with straightforward relationships to large-scale climate.

Numerous species and populations from the Baltic Sea Basin exhibit shifts in these parameters consistent with climate warming and the poleward migration of isotherms. Changes in ecosystem functioning have been registered by satellites, as well as land-based surveys such as forest inventories and treeline studies. It is apparent that trees are growing taller and lusher compared with a few decades ago; net primary productivity has increased, the ecosystems are net carbon sinks. The likely cause is an extended growing season associated with higher average temperatures. With regard to freshwater ecosystems, reductions in winter ice cover and duration, as well as higher water temperatures and secondary effects of these changes on lake nutrient cycles have most likely contributed to biological changes such as dominance shifts of species in some lakes.

The second hypothesis of this chapter is that climate changes simulated by models for the next century will have ongoing ecosystem impacts. This hypothesis likewise receives general support from our analysis. Based on both experimental and modelling studies, we may cautiously conclude that current trends in terrestrial ecosystem functioning will continue for some decades into the future. Based on developments in abiotic variables alone, forest production is likely to increase at least over northern parts of the Baltic Sea Basin. Over the southern continental area, water stress may offset the positive effects of temperature and CO_2 increases on productivity. Northern terrestrial ecosystems will continue to sequester carbon, while southern areas may switch from carbon sinks to sources by the end of the 21st century.

However, uncertainties are relatively high concerning soil carbon turnover, especially for the extensive peatlands of the boreal and subarctic zones. Changes in disturbance regimes, and vulnerability of species and ecosystems to climatic extremes and to pest and pathogen attacks are more difficult to predict but could significantly modify trends, for example in forest productivity. Conclusions concerning future changes in freshwater ecosystems are more speculative. Warmer water temperatures combined with longer stratified and ice-free periods in lakes may be expected to accelerate eutrophication, increasing phytoplankton production and causing potential dominance and successional shifts. Shallow lakes and lake littoral

zones may be particularly vulnerable to the impacts of warming.

For several of the ecosystem functions addressed in this chapter, the trajectory of future changes remains elusive, due either to a limited understanding of the driving mechanisms, the complexity of ecosystem structure and dynamics, or because of uncertainties concerning the future course of non-climatic drivers such as land use and agricultural practices. The complexity of biological and abiotic interactions in lakes prohibits general conclusions about the responses of populations and communities to climate forcing. Land-sea nutrient fluxes will be sensitive to future changes in runoff, but also to regional patterns of agricultural land use, fertiliser loads and ecotechnological practice. Export rates of total organic carbon in runoff from boreal and subarctic peatlands are significant for the carbon balance of the Baltic Sea, but it is unclear whether changing rainfall patterns and temperature will amplify or attenuate TOC loads in the future.

Numerous areas of uncertainty emerge from our synthesis. These fall into three main categories; firstly, limitations to current understanding of the mechanisms by which species and ecosystems respond to climate change; secondly, factors with a potential to modify climate responses that are difficult to characterise because of their episodic or species-specific nature – for example, impacts of pests, pathogens and climatic extremes; and finally, sensitivities to non-climatic covariates of climate change, such as land use, forest management, agricultural practices, pollutant deposition rates and technology development.

The further elucidation and constrainment of these uncertainties constitute fertile ground for research, especially as many of them surround ecosystem functions of sectoral or societal importance: forest growth, net carbon emission, water quality, habitat conservation. We would identify the following key questions for research:

- How will the distribution and abundance of pests and pathogens respond to climate change, and what consequences will this have for the functioning of natural and managed ecosystems?

- Given current greenhouse gas emissions scenarios, their associated socio-economic storylines, and the changes in climate and ecosystem functioning they are expected to lead to, what are the likely trajectories and spatial patterns of land use change within the Baltic Sea Basin? What consequences will these, in turn, have for species distributions, carbon sequestration, water runoff and land-sea nutrient fluxes?

- What are the likely future anthropogenic loads of nitrogen and phosphorus to terrestrial and freshwater ecosystems, and what consequences will they have for ecosystem functioning?

- How will changes in hydrology, soil thermal dynamics, biogeochemistry, microtopography and vegetation affect the contribution of extant peatland areas to greenhouse forcing, not forgetting the carbon exports of these areas to the Baltic Sea?

- What changes in disturbance regimes will result from future climate variability, especially extreme events like storms, spring temperature backlashes, extended drought periods or a succession of extreme summers? What will be the consequences for ecosystems, taking into account feedback mechanisms such as insect pest outbreaks and herbivore pressure in forests? What will be the consequences for water quality in lakes and land runoff?

4.7 Summary

The changing climate and other associated environmental and anthropogenic changes may be expected to affect the structure and functioning of ecosystems and threaten the services they provide to society.

We assess the potential impacts of the changing environment on terrestrial and freshwater ecosystems of the Baltic Sea Basin, aiming to evaluate the hypotheses:

1. that climate change and other associated environmental change over recent decades have affected the ecosystems and their services; and

2. that ongoing climate change will cause (further) changes in the ecosystems and their services over the remainder of the 21st century.

In order to highlight the most compelling and societally-relevant aspects of ecosystem change, the analysis focuses on:

1. processes and indicators of particular diagnostic value for the attribution of ecosystem changes to identifiable forcing factors; for example, changes in phenology, species distributions and the seasonality of physical, chemical and biological phases in lakes;

2. ecosystems and functions of sectorial relevance; for example, productivity and carbon storage in forests; and

3. uncertainty associated with ecological complexity and limitations to process understanding; for example, regarding stress responses to changing climatic extremes.

Significant changes in climate, including increasing temperatures and changing precipitation patterns, have occurred over the Baltic Sea Basin in recent decades (Chap. 2). Other associated changes include the continuously rising atmospheric CO_2 concentrations and increases in deposition loads of atmospheric pollutants, including nitrogen compounds and other acidifying pollutants. A variety of ecosystem impacts of these changes have been identified (hypothesis 1), including the following:

- An advancement of spring phenological phases such as budburst and leaf expansion is apparent for many plant species, likely reflecting increasing mean temperatures. Many species also show delayed autumn phases, but trends are less consistent. Phenological trends are stronger in northern Europe than for Europe as a whole, possibly reflecting stronger climate warming.

- Species distributional shifts tracking isothermal migration are apparent for both plant and animal species. Possibly related changes include weaker migratory behaviour, for example in some bird species. Treeline advance has been observed in the Fennoscandian mountain range. Increased growth and vigour of vegetation at high northern latitudes generally is apparent from satellite observations and can be attributed to increased growing season warmth and an extended growing season. Other observations, such as tree ring data, support the existence of a positive growth trend. The magnitude of the trend within the Baltic Sea Basin is representative for high latitude areas in Eurasia and is strong compared with similar latitudes in North America.

- Physiological stress related to the combined effects of atmospheric pollutants and extreme weather events such as spring frosts and drought are a possible explanation for late 20th-century dieback in boreal and temperate forests.

- Degradation of discontinuous permafrost in the subarctic north may be causing a shift towards a greater representation of wet habitats in tundra. Possible consequences include an increased release of methane through (anaerobic) decomposition, which would accentuate greenhouse forcing.

- Climate-related changes in lakes including higher water temperatures, advancement of ice break-up, lower water levels and increased influxes of dissolved organic matter from land have consequences for lake ecosystems, including dominance shifts in phytoplankton communities, higher summer algal biomass, and shifts in trophic state.

Climate scenarios described in Chap. 3 consistently point to increased temperatures throughout the Baltic Sea Basin by the end of the 21st century, compared with today. Precipitation scenarios are more variable but generally point to increased precipitation in winter, with southern areas experiencing decreased rainfall in summer. Combined with the effect of higher temperatures on evapotranspiration, this suggests that ecosystems of the temperate zone may face increasingly unfavourable growing season water budgets in the future. Potential impacts of these and other associated environmental changes (hypothesis 2) include the following:

- Extrapolation of recent phytophenological trends suggests that extension of the vegetation period by 2–6 weeks, depending on the climate scenario, is likely over much of the Baltic Sea catchment area.
- Further changes in the distributions of some species may be expected, but for many species, lags associated with population and community processes, dispersal limitations etc. are likely. Wholesale biome shifts, such as the northward displacement of the temperate-boreal forest boundary, will be slow compared to the rate of isotherm migration. Natural and semi-natural vegetation of the future may be of a transient character, e.g. ageing conifer stands with an increased representation of broadleaved trees in the younger age classes. Changes may be especially marked in subarctic and alpine areas, with forest invading areas that are currently tundra. Increased local richness is likely as species associated with the forest extend their ranges northward and upslope.
- Modelling studies generally point to increasing ecosystem production and carbon storage

capacity throughout the Baltic Sea Basin in the next 50–100 years, in conjunction with a longer growing season, increased atmospheric CO_2 concentrations and the stimulation of mineralisation processes in warmer soils. However, increased autumn and winter temperatures may be detrimental to hardening processes in trees, increasing susceptibility to spring frost damage. Growing season drought stress may reduce or inhibit production enhancement in temperate parts of the region.

- The potential impacts of climatic change on the incidence of pest and pathogen outbreaks affecting vegetation is still largely open. It seems reasonable to assume that harmful insects and fungi from central and southern Europe may expand into the Baltic Sea Basin in the warming climate.
- Warmer water temperatures combined with longer stratified and ice-free periods in lakes may be expected to accelerate eutrophication, increasing phytoplankton production and shifting the phytoplankton community structure towards species with higher temperature optima, including cyanobacteria. Shallow lakes and lake littoral zones may be particularly sensitive to climate warming. Increasing influxes of humic substances in runoff from boreal catchments would steepen light attenuation, with negative impacts on periphyton and benthic communities in lakes. Cold-water fish species may be extirpated from much of their present range while cool- and warm-water species expand northwards.

Uncertainties associated with the assessment of future ecosystem changes are substantial and include uncertainties due to insufficient understanding of the biological phenomena being modelled or projected including system-internal feedbacks and complexity, as well as variation among climate and greenhouse gas emissions scenarios on which the assessments are based.

The most important source of uncertainty with regard to many impacts are the future development in non-climatic, anthropogenic drivers of ecosystem dynamics, including deposition of atmospheric pollutants, land use changes, changes in forest management and agricultural practices, changes in human populations, markets and international trade, and technological development.

4.8   References

Aanes R, Saether BE, Øritsland NA (2000) Fluctuations of an introduced population of Svalbard reindeer: The effects of density dependence and climatic variation. Ecography 23:437–443

Aars J, Ims RA (2002) Climatic and intrinsic determinants of population demography: The winter dynamics of tundra vole populations. Ecology 83:3449–3456

Aasa A, Jaagus J, Ahas R, Sepp M (2004) The influence of atmospheric circulation on plant phenological phases in central and eastern Europe. Int J Climatol 24:1551–1564

Abgeti MD, Smol JP (1995) Winter limnology: A comparison of physical chemical and biological characteristics in two temperate lakes during ice cover. Hydrobiologia 304:221–234

Abildtrup J, Audsley E, Fekete-Farkas M, Giupponi C, Gylling M, Rosato P, Rounsevell M (2006) Socio-economic scenario development for the assessment of climate change impacts on agricultural land use: A pairwise comparison approach. Env Sci Pol 9:101–115

ACIA (2004) Impacts of a Warming Arctic. Cambridge University Press, Cambridge

ACIA (2005) Arctic Climate Impact Assessment. Cambridge University Press, New York

Adrian R, Deneke R, Mischke U, Stellmacher R, Lederer P (1995) A long-term study of the Heiligensee (1975–1992) Evidence for effects of climatic change on the dynamics of eutrophied lake ecosystems. Arch Hydrobiol 133:315–337

Adrian R, Walz N, Hintze T, Hoeg S, Rusche R (1999) Effects of ice duration on plankton succession during spring in a shallow polymictic lake. Freshwat Biol 41:621–632

Ahas R (1999) Long-term phyto- ornitho- and ichthyophenological time-series analysis in Estonia. Int J Biometeorol 42:119–124

Ahas R, Aasa A (2004) Attribution of Estonian phyto- ornitho- and ichthyophenological trends with parameters of changing climate. 16th Conference on Biometeorology and Aerobiology 23–27 August 2004, Vancouver Canada http://ams.confex.com/ams/pdfpapers/78288.pdf

Ahas R, Aasa A (2006) The effects of climate change on the phenology of selected Estonian plant, bird and fish populations. Int J Biometeorol 51:17–26

Ahas R, Tarand A, Meitern H (1998) Temporal variability of the phenological time series in Estonia. In: Ahas R, Tarand A, Meitern H (eds) Country Case Study on Climate Change Impacts and Adaptation. Assessments in the Republic of Estonia. Report to the UNEP/GEF Project No GF/2200-96-45 Stockholm Environment Institute – Tallinn, pp. 28–30

Ahas R, Jaagus J, Aasa A (2000) The phenological calendar of Estonia and its correlation with mean air temperature. Int J Biometeorol 44:159–166

Ahas R, Aasa A, Menzel A, Fedotova VG, Scheifinger H (2002) Changes in European spring phenology. Int J Climatol 22:1727–1738

Ainsworth EA, Long SP (2005) What have we learned from 15 years of free-air CO_2 enrichment (FACE)? A meta-analytic review of the responses of photosynthesis canopy properties and plant production to rising CO_2. New Phytologist 165:351–372

Alatalo JM, Totland Ø (1997) Response to simulated climatic change in an alpine and subarctic pollen-risk strategist Silene acaulis. Glob Change Biol 3 S1:74–79

Alcamo J, Kok K, Busch G, Priess J, Eickhout B, Rounsevell M, Rothman D, Heistermann M (2006) Searching for the future of land: Scenarios from the local to global scale. In: Lambin EF, Geist H (eds) Land-use and Land-cover Change: Local Processes Global Impacts. The IGBP Series. Springer, Berlin Heidelberg New York

Alekseev A, Soroka AR (2002) Scots pine growth trends in northwestern Kola Peninsula as an indicator of positive changes in the carbon cycle. Climatic Change 55:183–196

Algesten G, Sobek S, Bergstrom AK, Agren A, Tranvik LJ, Jansson M (2004) Role of lakes for organic carbon cycling in the boreal zone. Glob Change Biol 10:141–147

Andersen HE, Kronvang B, Larsen SE (1999) Agricultural practices and diffuse nitrogen pollution in Denmark: Empirical leaching and catchment models. Water Sci Tech 39:257–264

Andersen HE, Kronvang B, Larsen SE (2005) Development validation and application of Danish empirical phosphorus models. J Hydrol 304:355–365

Andersson FO, Ågren GI, Führer E (2000) Sustainable tree biomass production. Forest Ecol Manag 132:51–62

Andersson L, Arheimer B (2003) Modelling of human and climate impact on nitrogen load in a Swedish river 1885–1994. Hydrobiologia 497:63–77

Andreeva EA (2003) The present state of macrozoobenthos in Lake Ilmen. In: Simola H, Terzhevik AY, Viljanen M, Holopainen IK (eds) Proceedings of 4th International Lake Ladoga Symposium 2002, University of Joensuu, Publications of Karelian Institute 138

Angerbjörn A, Tannerfeldt M, Lundberg H (2001) Geographical and temporal patterns of lemming population dynamics in Fennoscandia. Ecography 24:298–308

Anneville O, Souissi S, Gammeter S, Straile D (2004) Seasonal and inter-annual scales of variability in phytoplankton assemblages: Comparison of phytoplankton dynamics in three peri-alpine lakes over a period of 28 years. Freshwat Biol 49:98–115

Anon (2002) State of Environment. The Ministry of Environment of the Republic of Lithuania, Vilnius

Araújo MB, Pearson RG, Thuiller W, Erhard M (2005) Validation of species-climate impact models under climate change. Glob Change Biol 11:1504–1513

Arft AM, Walker MD, Gurevitch J, Alatalo JM, Bret-Harte MS, Dale M, Diemer M, Gugerli F, Henry GHR, Jones MH, Hollister RD, Jónsdóttir IS, Laine K, Lévesque E, Marion GM, Molau U, Mølgaard P, Nordenhäll U, Raszhivin V, Robinson CH, Starr G, Stenström A, Stenström M, Totland Ø, Turner PL, Walker LJ, Webber PJ, Welker JM, Wookey PA (1999) Response patterns of tundra plant species to experimental warming: A meta-analysis of the International Tundra Experiment. Ecol Monogr 69:491–511

Arheimer B, Wittgren HB (2002) Modelling nitrogen removal in potential wetlands at the catchment scale. Ecol Eng 19:63–80

Arheimer B, Andersson L, Larsson M, Lindström G, Olsson J, Pers BC (2004a) Modelling diffuse nutrient flow in eutrophication control scenarios. Water Sci Tech 49:37–45

Arheimer B, Torstensson G, Wittgren HB (2004b) Landscape planning to reduce coastal eutrophication: Agricultural practices and constructed wetlands. Landsc Urban Plann 67:205–215

Assel RA, Robertson DM (1995) Changes in winter air temperature near Lake Michigan 1851–1993 as determined from regional ice records. Limnol Oceanogr 40:165–176

Assel RA, Cronk K, Norton D (2003) Recent trends in Laurentian Great lakes ice cover. Climatic Change 57:185–204

Auclair AND, Lill JT, Revenga C (1996) The role of climate variability and global warming in the dieback of northern hardwoods. Water Air Soil Pollut 91:163–186

Audsley E, Pearn KR, Simota C, Cojocaru G, Koutsidou E, Rounsevell MDA, Trnka M, Alexandrov V (2006) What can scenario modelling tell us about future European scale land use and what not? Agr Ecosyst Env 9:148–162

Aurela M, Laurila T, Tuovinen J (2002) Annual CO_2 balance of a subarctic fen in northern Europe: Importance of the wintertime efflux. J Geophys Res 107:4607

Ayres MP, Lombardero MJ (2000) Assessing the consequences of global change for forest disturbance from herbivores and pathogens. Sci Total Environ 262:263–286

Barabasz B, Górz A (1998) *Argiope bruennichi* (Scopoli 1772) Rare and insufficiently examined spider species in Poland. Fragmenta Faunistica 41:255–267

Barklund P (1994) Svårtolkad ekdöd (Oak dieback). In: Ekfrämjandet 50 år Wallin and Dahlholm boktryckeri. AB Lund, Sweden (in Swedish)

Barklund P, Ericsson A, Gemmel P, Johansson U, Olsson M, Walheim M, Åhman G (1995) Bark och vedskador hos granar med kådflöde – kådflödessjukan hos gran (Bark and wood damage with resin flow in Norway Spruce - Resin flow disease in Norway Spruce). Swedish University of Agricultural Sciences (SLU), Report Info/Skog. 15, Alnarp, Sweden (in Swedish)

Bazzaz FA (1990) The response of natural ecosystems to the rising global CO_2 levels. Ann Rev Ecol Systemat 21:167–196

Bechmann M, Eggestad HO, Vagstad N (1998) Nitrogen balances and leaching in four agricultural catchments in southeastern Norway. Env Poll 102 S1:439–499

Beerling DJ, Terry AC, Mitchell PL, Jones DG, Lee JA, Callaghan TV (2001) Time to chill: Effects of simulated global change on leaf ice nucleation temperatures of sub-Arctic vegetation. Am J Bot 88: 628–633

Behrendt H, Bachor A (1998) Point and diffuse load of nutrients to the Baltic Sea by river basins of northeast Germany (Mecklenburg–Vorpommern). Water Sci Tech 38:147–155

Behrendt H, Stellmacher R, Oldberg M (1987) Long-term changes in water quality parameters of a shallow eutrophic lake and their relations to meteorologic and hydrologic elements. In Soloman SI, Beran M, Hogg W (eds) The Influence of Climate Change and Climatic Variability on the Hydrologic Regime and Water Resour. IAHS Publications 168:535–544

Benndorf J, Kranich J, Mehner T, Wagner A (2001) Temperature impact on the midsummer decline of *Daphnia galeata:* An analysis of long-term data from the biomanipulated Bautzen Reservoir (Germany). Freshwat Biol 46:199–211

Bergh J, McMurtrie RE, Linder S (1998) Climatic factors controlling the productivity of Norway spruce: A model-based analysis. Forest Ecol Manag 110:127–139

Bergh J, Linder S, Lundmark T, Elfving B (1999) The effect of water and nutrient availability on the productivity of Norway spruce in northern and southern Sweden. Forest Ecol Manag 119:51–62

Bergh J, Freeman M, Sigurdsson BD, Kellomäki S, Laitinen K, Niinistö S, Peltola H, Linder S (2003) Modelling the short-term effects of climate change on the productivity of selected tree species in Nordic countries. Forest Ecol Manag 183:327–340

Bergh J, Linder S, Bergström J (2005) Potential production of Norway spruce in Sweden. Forest Ecol Manag 204:1–20

Bergh J, Räisänen J, Freeman M, Linder S (2007) Effects of global change on net primary production in northern Europe – a model-based analysis on regional climate scenarios. Glob Change Biol (in press)

Bergström AK, Algesten G, Sobek S, Tranvik L, Jansson M (2004) Emission of CO_2 from hydroelectric reservoirs in northern Sweden. Arch Hydrobiol 159:25–42

Bergström S, Carlsson B (1994) River runoff to the Baltic Sea: 1950–1990. Ambio 23:280–287

Berner EK, Berner RA (1996) Global Environment: Water Air and Geochemical Cycles. Prentice-Hall, Upper Saddle River, NJ, USA

Bernes C (1996) The Nordic Arctic Environment – Unspoilt Expoited Polluted? The Nordic Council of Ministers, Copenhagen

Berry PM, Rounsevell MDA, Harrison PA, Audsley E (2006) Assessing the vulnerability of agricultural land use and species to climate change and the role of policy in facilitating adaptation. Env Sci Pol 9:189–204

Beuker E, Kolström M, Kellomäki S (1996) Changes in wood production of *Picea abies* and *Pinus sylvestris* under a warmer climate: Comparison of field measurements and results of a mathematical model. Silva Fennica 30:239–246

Blenckner T (2005) A conceptual model of climate related effects on lake ecosystems. Hydrobiologia 533:1–14

Blenckner T, Omstedt A, Rummukainen M (2002a) A Swedish case study of contemporary and possible future consequences of climate change on lake function. Aquat Sci 64:171–184

Blenckner T, Pettersson K, Padisak J (2002b) Lake plankton as tracer to discover climate signals. Verh Proc Trav SIL 28:1324–1327

Blenckner T, Järvinen M, Weyhenmeyer G (2004) Atmospheric circulation and its impact on ice phenology in Scandinavia. Boreal Env Res 9:371–380

Bocheński Z (1993) Catalogue of fossil and subfossil birds of Poland. Acta Biol Cracov 36:329–460

Bonan GB, Pollard D, Thompson SL (1992) Effects of boreal forest vegetation on global climate. Nature 359:716–718

Bouraoui F, Grizzetti B, Granlund K, Rekolainen S, Bidoglio G (2004) Impact of climate change on the water cycle and nutrient losses in a Finnish catchment. Climatic Change 66:109–126

Boyer EW, Hornberger GM, Bencala KE, McKnight DM (1997) Response characteristics of DOC flushing in an alpine catchment. Hydrolog Process 11:1635–1647

Bradshaw RHW, Holmqvist BH, Cowling SA, Sykes MT (2000) The effects of climate change on the distribution and management of *Picea abies* in southern Scandinavia. Can J Forest Res 30:1992–1998

Brommer JE (2004) The range margins of northern birds shift polewards. Annales Zoologici Fennici 41:391–397

Brooker RW, Carlsson BÅ, Callaghan TV (2001) *Carex bigelowii* Torrey ex Schweinitz (*C. rigida* Good., non Schrank; *C. hyperborea* Drejer) Biological Flora of the British Isles. J Ecol 89:1072–1095

Bukantis A, Gulbinas Z, Kazakevicius S, Kilkus K, Mikelinskiene A, Morkunaite R, Rimkus E, Samuila M, Stankunavicius G, Valiuškevicius G, Zaromskis R (2001) Klimato svyravim? poveikis fiziniams geografiniams procesams Lietuvoje (The Influence of Climatic Variations on Physical Geographical Processes in Lithuania) Monograph, Geografijos Inst., Vilnius (in Lithuanian)

Cairns DM, Moen J (2004) Herbivory influences tree lines. J Ecol 92:1019–1024

Callaghan TV, Carlsson BÅ, Svensson BM (1996) Some apparently paradoxical aspects of the life cycles demography and population dynamics of plants from the subarctic Abisko area. Ecol Bull 45: 133–143

Callaghan TV, Björn LO, Chernov Y, Chapin T, Christensen TR, Huntley B, Ims RA, Johansson M, Jolly D, Jonasson S, Matveyeva N, Panikov N, Oechel W, Shaver G, Schaphoff S, Sitch S, Zöckler C (2004a) Climate change and UV-B impacts on Arctic tundra and polar desert ecosystems. Ambio 33: 385–479

Callaghan TV, Johansson M, Heal OW, Saelthun NR, Barkved L, Bayfield N, Brandt O, Brooker R, Christiansen HH, Hoye TT, Humlun O, Järvinen A, Jonasson C, Kohler J, Magnusson B, Meltofte H, Mortensen L, Neuvonen S, Pearce I, Rasch M, Turner L, Hasholt B, Huhta E, Leskinen E, Nielsen N, Siikamäki P (2004b) Environmental changes in the North Atlantic region: SCANNET as a collaborative approach for documenting understanding and predicting changes. Ambio Special Report 13:39–50

Canadell JG, Ciais P, Cox P, Heimann M (2004) Quantifying understanding and managing the carbon cycle in the next decades. Climatic Change 67:147–160

Carey SK (2003) Dissolved organic carbon fluxes in a discontinuous permafrost subarctic alpine catchment. Permafrost & Periglacial Processes 14:161–171

Carpenter SR, Kitchell JF, Hodgson JR (1985) Cascading trophic interactions and lake productivity. BioScience 35:634–639

Carpenter SR, Fisher SG, Grimm NB, Kitchell JF (1992) Global change and freshwater ecosystems. Ann Rev Ecol Systemat 23:119–139

Carpenter SR, Pingali PL, Bennett EM, Zurek MB (2005) Ecosystems and Human Well-being – Scenarios. Island Press, Washington, D.C.

CAVM Team (2003) Circumpolar Arctic Vegetation Map Scale 1:7500000 Conservation of Arctic Flora and Fauna (CAFF). Map No 1, US Fish & Wildlife Service, Anchorage, Alaska

CCIRG (1996) Review of the Potential Effects of Climate Change in the United Kingdom. UK Climate Change Impacts Review Group (CCIRG). HMSO, London

Chapin FS III, Schultze ED, Mooney HA (1990) The ecology and economics of storage in plants. Ann Rev Ecol Systemat 21:423–447

Chapin FS III, Berman M, Callaghan TV, Convey P, Crépin AS, Danell K, Ducklow H, Forbes B, Kofinas G, McGuire AD, Nuttall M, Virginia R, Young O, Zimov SA, Christensen T, Godduhn A, Murphy EJ, Wall D Zöckler C (2005) Polar systems. In: Hassan R, Scholes R, Ash N (eds) Ecosystems and Human Well-Being: Current State and Trends. Volume I: Millennium Ecosystem Assessment Series. Island Press, Washington DC, pp. 717–743

Chapin FS III, Hoel M, Carpenter SR, Lubchenco J, Walker B, Callaghan TV, Folke C, Levin SA, Mäler KG, Nilsson C, Barrett S, Berkes F, Crépin AS, Danell K, Rosswall T, Starrett D, Xepapadeas A, Zimov SA (2006) Building resilience and adaptation to manage Arctic change. Ambio 35:198–202

Chmielewski FM, Rötzer T (2001) Response of tree phenology to climate change across Europe. Agr Forest Meteorol 108:101–112

Choi JS (1998) Lake ecosystem responses to rapid climate change. Env Monit Assess 49:281–290

Christensen TR, Jonasson S, Callaghan TV, Havström M (1999) On the potential CO_2 release from tundra soils in a changing climate. Appl Soil Ecol 11:127–134

Christensen TR, Johansson T, Åkerman JH, Mastepanov M, Malmer N, Friborg T, Crill P, Svensson BH (2004) Thawing sub-arctic permafrost: Effects on vegetation and methane emissions. Geophys Res Lett 31:L04501

Chu C, Mandrak NE, Minns CK (2005) Potential impacts of climate change on the distributions of several common and rare freshwater fishes in Canada. Diversity & Distributions 11:299–310

Chuine I, Kramer K, Hänninen H (2003) Plant development models. In: Schwartz MD (ed) Phenology: An Integrative Environmental Science. Kluwer, pp. 217–235

Cornelissen JHC, Callaghan TV, Alatalo JM, Michelsen A, Graglia E, Hartley AE, Hik DS, Hobbie SE, Press MC, Robinson CH, Henry GHR, Shaver GR, Phoenix GK, Gwynn Jones D, Jonasson S, Chapin FS III ,Molau U, Neill C, Lee JA, Melillo JM, Sveinbjörnsson B, Aerts R (2001) Global change and Arctic ecosystems: Is lichen decline a function of increases in vascular plant biomass? J Ecol 89:984–994

Costanza R, d'Arge R, de Groot R, Farber S, Grasso M, Hannon B, Limburg K, Naeem S, O'Neill RV, Paruelo J, Raskin RG, Sutton P, van den Belt M (1997) The value of the world's ecosystem services and natural capital. Nature 387:253–260

Coulson SJ, Leinass HP, Ims RA, Søvik G (2000) Experimental manipulation of winter surface ice layer: The effects on a high arctic soil microarthropod community. Ecography 23:299–306

Cummins CP (2003) UV-B radiation climate change and frogs – the importance of phenology. Annales Zoologici Fennici 40:61–67

Dahl E (1990) Probable effects of climatic change due to the greenhouse effect on plant productivity and survival in North Europe. In: Holten JI, Paulsen G, Oechel WC (eds) Effects of Climate Change on Terrestrial Ecosystems. Norwegian Institute for Nature Research, Trondheim, pp. 81–83

de Koning F, Olschewski R, Veldkamp E, Benitez P, Lopez-Ulloa M, Schlichter T, de Urquiza M (2005) The ecological and economic potential of carbon sequestration in forests: Examples from South America. Ambio 34:224–229

de Stasio BT, Hill DK, Kleinhans JM, Nibbelink NP, Magnuson JJ (1996) Potential effects of global climate change on small north-temperate lakes: Physics fish and plankton. Limnol Oceanogr 41:1136–1149

del Barrio G, Harrison PA, Berry PM, Butt N (2006) Impacts of climate change on species' distributions in a temperate and a Mediterranean area: Comparison and implications for policy. Env Sci Pol 9:129–147

Diehl S (2002) Phytoplankton light and nutrients in a gradient of mixing depths: Theory. Ecology 83: 386–398

Dirnbock T, Dullinger S, Grabherr G (2003) A regional impact assessment of climate and land-use change on alpine vegetation. J Biogeogr 30:401–417

Dittmar C, Zech W, Elling W (2003) Growth variations of common beech (*Fagus sylvatica* L) under different climatic and environmental conditions in Europe – a dendroecological study. Forest Ecol Manag 173:63–78

Dormann CF, Woodin SJ (2002) Climate change in the Arctic: Using plant functional types in a meta-analysis of field experiments. Funct Ecol 16:4–17

Dorrepaal E, Aerts R, Cornelissen JHC, Callaghan TV, Logtestijn RSP (2004) Summer warming and increased winter snow cover affect *Sphagnum fuscum* growth structure and production in a sub-arctic bog. Glob Change Biol 10:93–104

Drake BG, Gonzalez-Meler MA, Long SP (1997) More efficient plants: A consequence of rising atmospheric CO_2? Ann Rev Plant Physiol Plant Mol Biol 48:609–639

Drever JI (1997) The Geochemistry of Natural Waters. Prentice Hall

Duraiappah A, Naeem S, Agardy T, Ash NJ, Cooper HD, Díaz S, Faith DP, Mace G, McNeely JA, Mooney HA, Oteng-Yeboah AA, Pereira HM, Polasky S, Prip C, Reid WV, Samper C, Schei PJ, Scholes R, Schutyser F, van Jaarsveld A (2005) Ecosystems and Human Well-being: Biodiversity Synthesis. Island Press, Washington DC

Dzikiewicz M (2000) Activities in nonpoint pollution control in rural areas of Poland. Ecol Eng 14: 429–434

Eilertsen SM (2002) Utilization of abandoned coastal meadows in northern Norway by reindeer. PhD thesis, University of Tromsø, Norway

Ellenberg H (1986) Vegetation Mitteleuropas mit den Alpen, 4th ed. Ulmer, Stuttgart, Germany

Elliott JA, Thackeray SJ, Huntingford C, Jones RG (2005) Combining a regional climate model with a phytoplankton community model to predict future changes in phytoplankton in lakes. Freshwat Biol 50:1404–1411

Emanuelsson U (1987) Human influence on vegetation in the Torneträsk area during the last three centuries. Ecol Bull 38:95–111

Enell M, Fejes J (1995) The nitrogen load to the Baltic Sea – Present situation acceptable future load and suggested source reduction. Water Air Soil Pollut 85:877–882

Ewert F, Rounsevell MDA, Reginster I, Metzger MJ, Leemans R (2005) Future scenarios of European agricultural land use: Estimating changes in crop productivity. Agr Ecosyst Environ 107:101–116

Federal Forest Inventory (2005) The Federal Forest Inventory, Bundesministerium für Ernährung Landwirtschaft und Verbraucherschutz, Germany, http://www.bundeswaldinventur.de, site visited in August 2005

Fiedler W (2003) Recent changes in migratory behaviour of birds: A compilation of field observations and ringing data. In: Berhold P, Gwinner E, Sonnenschein E (eds) Avian Migration. Springer, Berlin Heidelberg New York

Filatov NN, Nazarova LY, Salo YA (2003) Climate changes and Water Resour in the region of the largest European lakes. In: Simola H, Terzhevik AY, Viljanen M, Holopainen IK (eds) Proceedings of 4th International Lake Ladoga Symposium, 2002, University of Joensuu, Publications of The Karelian Institute 138:31–36

Foley JA, Kutzbach JE, Coe MT, Levis S (1994) Feedbacks between climate and boreal forests during the Holocene epoch. Nature 371:52–54

Freeman C, Fenner N, Ostle NJ, Kang H, Dowrick DJ, Reynolds B, Lock MA, Sleep D, Hughes S, Hudson J (2004) Export of dissolved organic carbon from peatlands under elevated carbon dioxide levels. Nature 430:195–198

Freeman C, Evans CD, Monteith DT, Reynolds B, Fenner N (2001) Export of organic carbon from peat soils. Nature 412:785–785

Freeman M, Morén AS, Strömmer M, Linder S (2005) Climate change impacts on forests in Europe: Biological impact mechanisms. In: Kellomäki S, Leinonen S (eds) Management of European Forests Under Changing Climatic Conditions. University of Joensuu, Faculty of Forestry Research, Notes 163:46–115

Gaedke U, Ollinger D, Bäuerle E, Straile D (1998) The impact of the interannual variability in hydro-dynamic conditions on the plankton development in Lake Constance in spring and summer. Arch Hydrobiol, Special Issues Advancing Limnology 53:565–585

Garten CT Jr, Ashwood TL (2003) A landscape level analysis of potential excess nitrogen in east-central North Carolina. Water Air Soil Pollut 146:3–21

Gelbrecht J, Lengsfeld H, Pöthig R, Opitz D (2005) Temporal and spatial variation of phosphorus input retention and loss in a small catchment of NE Germany. J Hydrol 304:151–165

George DG, Harris GP (1985) The effect of climate on long-term changes in the crustacean zooplankton biomass of Lake Windermere UK. Nature 316:536–539

George DG, Hewitt DP, Lund JW, Smyly WJP (1990) The relative effects of enrichment and climate change on the long-term dynamics of *daphnia* in Esthwaite Water, Cumbria. Freshwat Biol 23:55–70

George DG, Hewitt DP (1999) The influence of year-to-year variations in the winter weather on the dynamics of *Daphnia* and *Eudiaptomus* in Esthwaite Water, Cumbria. Funct Ecol 13:45–54

Gerten D, Adrian R (2000) Climate-driven changes in spring plankton dynamics and the sensitivity of shallow polymictic lakes to the North Atlantic Oscillation. Limnol Oceanogr 45:1058–1066

Gerten D, Adrian R (2001) Differences in the persistency of the North Atlantic Oscillation signal among lakes. Limnol Oceanogr 46:448–455

Gerten D, Adrian R (2002) Species-specific changes in the phenology and peak abundance of freshwater copepods in response to warm summers. Freshwat Biol 47:2163–2173

Gitz V, Ciais P (2003) Amplifying effects of land-use change on future atmospheric CO_2 levels. Glob Biogeochem Cy 17

Giupponi C, Ramanzin M, Sturaro E, Fuser S (2006) Climate and land use changes biodiversity and agri-environmental measures in the Belluno province, Italy. Env Sci Pol 9:163–173

Grabherr G, Gottfried M, Pauli H (1994) Climate effects on mountain plants. Nature 369:448

Grace J, Berninger F, Nagy L (2002) Impacts of climate change on the tree line. Ann Bot 90:537–544

Graglia E, Jonasson S, Michelsen A, Schmidt IK (1997) Effects of shading nutrient application and warming on leaf growth and shoot densities of dwarf shrubs in two arctic-alpine plant communities. Ecoscience 4:191–198

Graham LP (2004) Climate change effects on river flow to the Baltic Sea. Ambio 33:235–241

Graham LP, Bergström S (2001) Water balance modelling in the Baltic Sea drainage basin – analysis of meteorological and hydrological approaches. Meteorol Atmos Phys 77:45–60

Granlund K, Rankinen K, Lepistö A (2004) Testing the INCA model in a small agricultural catchment in southern Finland. Hydrol Earth Syst Sci 8:717–728

Granstedt A, Seuri P, Thomsson O (2004) Effective recycling agriculture around the Baltic Sea. Ekologiskt Landbruk 41

Gren IM, Söderqvist T, Wulff F (1997) Nutrient reductions to the Baltic Sea: Ecology costs and benefits. J Env Manag 51:123–143

Grigor'ev AS, Trapeznikov YA (2002) Level of Lake Ladoga at possible climate changes. Water Resour 29:155–159

Grimvall A, Stålnacke P, Tonderski A (2000) Time scales of nutrient losses from land to sea – a European perspective. Ecol Eng 14:363–371

Grunzweig JM, Sparrow SD, Chapin FS (2003) Impact of forest conversion to agriculture on carbon and nitrogen mineralization in subarctic Alaska. Biogeochemistry 64:271–296

Guo QF (2000) Climate change and biodiversity conservation in Great Plains agroecosystems. Glob Env Change 10:289–298

Gustafson A, Fleischer S, Joelsson A (1998) Decreased leaching and increased retention potential cooperative measures to reduce diffuse nitrogen on a watershed level. Water Sci Tech 38:181–189

Gutschick VP, BassiriRad H (2003) Extreme events as shaping physiology ecology and evolution of plants: Toward a unified definition and evaluation of their consequences. New Phytologist 160: 21–42

Gwynn-Jones D, Lee JA, Callaghan TV (1997) Effects of enhanced UV-B radiation and elevated carbon dioxide concentrations on a sub-Arctic forest heath ecosystem. Plant Ecol 128:243–249

Hagström Å, Azam F, Kuparinen J, Zweifel UL (2001) Pelagic plankton growth and resource limitations in the Baltic Sea. In: Wulff F, Rahm L, Larsson P (eds) A Systems Analysis of the Baltic Sea. Springer, Berlin Heidelberg New York, pp. 177–210

Håkanson L, Ostapenia A, Parparov A, Hambright D, Boulion VV (2003) Management criteria for lake ecosystems applied to case studies of changes in nutrient loading and climate change. Lakes & Reservoirs: Research & Management 8:141–155

Hållmarker M (2002) Skogsgränsförändringar i Abisko (Forest limit changes in Abisko). Masters thesis in Environmental Science, Gothenburg University, Sweden (in Swedish)

Hamrick JL (2004) Response of forest trees to Global Environ Changes. Forest Ecol Manag 197:323–335

Hänninen H, Beuker E, Johnsen Ø, Leinonen I, Murray M, Sheppard L, Skrøppa T (2001) Impacts of climate change on cold hardiness of conifers. In: Bigras FJ, Colombo SJ (eds) Conifer Cold Hardiness. Kluwer, Dordrecht, pp. 305–327

Hansen AJ, Neilson RP, Dale VH, Flather CH, Iverson LR, Currie DJ, Shafer S, Cook R, Bartlein PJ (2001) Global change in forests: Responses of species communities and biomes. BioScience 51: 765–779

Hargeby A, Blindow I, Hansson LA (2004) Shifts between clear and turbid states in a shallow lake: Multi-causal stress from climate nutrients and biotic interactions. Arch Hydrobiol 161:433–454

Harrison PA, Berry PM, Butt N, New M (2006) Modelling climate change impacts on species' distributions at the European scale: Implications for conservation policy. Env Sci Pol 9:116–128

Hartley AE, Neill C, Melillo JM, Crabtree R, Bowles FP (1999) Plant performance and soil nitrogen mineralization in response to simulated climate change in subarctic dwarf shrub heath. Oikos 86: 331–343

Havström M, Callaghan TV, Jonasson S (1995) Effects of simulated climate change on the sexual reproductive effort of *Cassiope tetragonal*. In: Callaghan TV, Oechel WC, Gilmanov T, Molau U, Maxwell B, Tyson M ,Sveinbjörnsson B, Holtén JI (eds) Global Change and Arctic Terrestrial Ecosystems. Proceedings of Papers Contributed to the International Conference, 21–26 August 1993, Oppdal, Norway. European Commission Ecosystems Research Report 10, pp. 109–114

Heathwaite L, Sharpley A, Gburek W (2000) A conceptual approach for integrating phosphorus and nitrogen management at watershed scales. J Env Qual 29:158–166

Heide OM (1993) Daylength and thermal time responses of budburst during dormancy release in some northern deciduous trees. Physiol Plantarum 88:531–540

Heil M (2001) The ecological concept of cost of induced systemic resistance (ISR). Eur J Plant Pathol 107:137–146

HELCOM (1993) Second Baltic Sea Pollution Load Compilation. Baltic Sea Environment Proceedings 45. HELCOM, Helsinki, Finland

HELCOM (2005) Nutrient Pollution to the Baltic Sea in 2000. Baltic Sea Environment Proceedings 100. HELCOM, Helsinki, Finland

Helle T (2001) Mountain birch forests and reindeer husbandry. In: Wielgolaski FE (ed) Nordic Mountain Birch Ecosystems. Man and Biosphere Series. The Parthenon Publishing Group, New York, pp. 279–291

Henry GHR, Molau U (1997) Tundra plants and climate change: The International Tundra Experiment (ITEX). Glob Change Biol 3, Suppl 1:1–9

Hersteinsson P, MacDonald DW (1992) Interspecific competition and the geographical distribution of red and arctic foxes *Vulpes vulpes* and *Alopex lagopus*. Oikos 64:505–515

Hill JK, Thomas CD, Huntley B (1999) Climate and habitat availability determine 20th century changes in a butterfly's range margin. Proc Roy Soc London, Series B 266:1197–1206

Høgda KA, Karlsen SR, Solheim I (2001) Climatic change impact on growing season in Fennoscandia studied by a time series of NOAA AVHRR NDVI data. Proceedings of IGARSS 9–13 July 2001, Sydney, Australia

Holman IP, Rounsevell MDA, Shackley S, Harrison PA, Nicholls RJ, Berry PM, Audsley E (2005a) A regional multi-sectoral and integrated assessment of the impacts of climate and socio-economic change in the UK: Part I. Methodology. Climatic Change 71:9–41

Holman IP, Rounsevell MDA, Shackley S, Harrison PA, Nicholls RJ, Berry PM, Audsley E (2005b) A regional multi-sectoral and integrated assessment of the impacts of climate and socio-economic change in the UK: Part II. Results. Climatic Change 71:43–73

Holmgren B, Tjus M (1996) Summer air temperatures and tree line dynamics at Abisko. Ecological Bulletins 45:159–169

Holmsgaard E (1986) Historical development of wind damage in conifers in Denmark. In: Communities CotE (ed) Minimizing Wind Damage to Coniferous Stands. Løvenholm Castle, Denmark, pp. 2–4

Huisman J, Jonker RR, Zoneveld C, Weissing FJ (1999) Competition for light between phytoplankton species: Experimental tests of mechanistic theory. Ecology 80:211–222

Humborg C, Conley DJ, Rahm L, Wulff F, Cociasu A, Ittekkot V (2000) Silicon retention in river basins: Far-reaching effects on biogeochemistry and aquatic food webs in coastal marine environments. Ambio 29:45–50

Humborg C, Blomqvist S, Avsan E, Bergensund Y, Smedberg E, Brink J, Mörth CM (2002) Hydrological alterations with river damming in northern Sweden: Implications for weathering and river biogeochemistry. Glob Biogeochem Cy 16:1039

Humborg C, Danielsson Å, Sjöberg B, Green M (2003) Nutrient land–sea fluxes in oligotrophic and pristine estuaries of the Gulf of Bothnia, Baltic Sea. Estuar Coast Shelf Sci 56:783–795

Humborg C, Smedberg E, Blomqvist S, Mörth CM, Brink J, Rahm L, Danielsson Å, Sahlberg J (2004) Nutrient variations in boreal and subarctic Swedish rivers: Landscape control of land–sea fluxes. Limnol Oceanogr 49:1871–1883

Huntley B (1991) How plants respond to climate change: Migration rates individualism and the consequences for plant communities. Ann Bot 67:15–22

Hyenstrand P, Blomqvist P, Pettersson A (1998) Factors determining cyanobacterial success in aquatic systems – a literature review. Arch Hydrobiol. Special Issues Advancing Limnology 51:41–62

Iital A, Stålnacke P, Deelstra J, Loigu E, Pihlak M (2005) Effects of large-scale changes in emissions on nutrient concentrations in Estonian rivers in the Lake Peipsi drainage basin. J Hydrol 304:261–273

Iital A, Loigu E, Vagstad N (2003) Nutrient losses and N &P balances in small agricultural watersheds in Estonia. Nord Hydrol 34:531–542

Ilyashuk BP, Ilyashuk EA (2001) Response of alpine chironomid communities (Lake Chuna Kola Peninsula northwestern Russia) to atmospheric contamination. J Paleolimnol 25:467–475

Ingri J, Widerlund A, Land M (2005) Geochemistry of major elements in a pristine boreal river system; Hydrological compartments and flow paths. Aquat Geochem 11:57–88

IPCC (2001) Climate Change 2001: The Scientific Basis. Cambridge University Press Cambridge

Ittekkot V (2003) A new story from the ol' man river. Science 301:56–58

Iversen J (1944) Viscum, Hedera and Ilex as climatic indicators. A contribution to the study of past-glacial temperature climate. Geologiska Föreningens Förhandlingar 66:463–483

Jaagus J (1997) The impact of climate change on the snow cover pattern in Estonia. Climatic Change 36:65–77

Jansons V, Busmanis P, Dzalbe I, Kirsteina D (2003) Catchment and drainage field nitrogen balances and nitrogen loss in three agriculturally influenced Latvian watersheds. Eur J Agron 20:173–179

Jansson A, Folke C, Langaas S (1998) Quantifying the nitrogen retention capacity of natural wetlands in the large-scale drainage basin of the Baltic Sea. Landsc Ecol 13:249–262

Jansson A, Folke C, Rockström J, Gordon L (1999) Linking freshwater flows and ecosystem services appropriated by people: The case of the Baltic Sea drainage basin. Ecosystems 2:351–366

Jansson BO (1997) The Baltic Sea: Current and future status and impact of agriculture. Ambio 26:424–431

Järvet A (1999) Ice regime of Lake Võrtsjärv and its long-term changes. In: Jaagus J (ed) Uurimusi Eesti Kliimast. Publicationes Instituti Geographici Universitatis Tartuensis 85:171–179

Järvet A (2004) Influence of hydrological factors and human impact on the ecological state of shallow Lake Võrtsjärv in Estonia. Dissertationes Geographicae Universitatis Tartuensis 19:1–119

Järvinen M, Rask M, Ruuhijärvi J, Arvola L (2002) Temporal coherence in water temperature and chemistry under the ice of boreal lakes (Finland). Water Resour 36:3949–3956

Jassby AD, Powell TM, Goldman CR (1990) Interannual fluctuations in primary production: Direct physical effects and the trophic cascade at Castle Lake, California. Limnol Oceanogr 35:1021–1038

Jensen MN (2003) Consensus on ecological impacts remains elusive. Science 299:38

Jeppesen E, Sondergard M, Jensen JP (2003) Climatic warming and regime shifts in lake food webs – some comments. Limnol Oceanogr 48:1346–1349

Johnson D, Campbell CD, Lee JA, Callaghan TV, Gwynn-Jones D (2002) Nitrogen storage (communication arising): UV-B radiation and soil microbial communities. Nature 416:82–83

Johnsson H, Hoffmann M (1998) Nitrogen leaching from agricultural land in Sweden. Ambio 27:481–488

Jonasson S, Michelsen A, Schmidt IK, Nielsen EV (1999) Responses in microbes and plants to changed temperature nutrient and light regimes in the Arctic. Ecology 80:1828–1843

Jonasson S, Chapin FS III, Shaver GR (2001) Biogeochemistry in the Arctic: Patterns processes and controls. In: Schulze ED, Heimann M, Harrison SP, Holland EA, Lloyd JJ, Prentice IC, Schimmel D (eds) Global Biogeochem Cy in the Climate System. Academic Press, San Diego, pp. 139–150

Jones MH, Bay C, Nordenhäll U (1997) Effects of experimental warming on arctic willows (*Salix* spp): A comparison of responses from the Canadian High Arctic, Alaskan Arctic, and Swedish Subarctic. Glob Change Biol 3, Suppl. 1:55–60

Jones MB, Donnelly A (2004) Carbon sequestration in temperate grassland ecosystems and the influence of management climate and elevated CO_2. New Phytologist 164:423–439

Jonnsson A, Karlsson J, Jansson M (2003) Sources of carbon dioxide supersaturation in clearwater and humic lakes in northern Sweden. Ecosystems 6:224–235

Jonsson LJ, Wilander P (1999) Is the wasp like spider, *Argiope bruennichi*, established in Sweden? Entomologisk Tidskrift 120:17–22

Jönsson AM (2005) Tracing the impact of adverse environmental conditions on the frost sensitivity in trees. J Sustain Forest 21:3–73

Jönsson AM, Linderson ML, Stjernquist I, Schlyter P, Bärring L (2004) Climate change and the effect of temperature backlashes causing frost damage in Picea abies. Glob Plan Change 44:195–207

Juday GP, Barber V, Vaganov E, Rupp S, Sparrow S, Yarie J, Linderholm H, Berg E, D'Arrigo R, Duffy P, Eggertsson O, Furyaev VV, Hogg EH, Huttunen S, Jacoby G, Kaplunov VYa, Kellomaki S, Kirdyanov AV, Lewis CE, Linder S, Naurzbaev MM, Pleshikov FI, Savva YuV, Sidorova OV, Stakanov VD, Tchebakova NM, Valendik EN, Vedrova EF, Wilmking M (2005) Forests, land management and agriculture. In: ACIA (ed) Arctic Climate Impact Assessment. Cambridge University Press, New York, pp. 781–862

Kangur K, Möls T, Milius A, Laugaste R (2003) Response of phytoplankton communities to altered nutrient content and water level fluctuations in the large shallow Lake Peipus. In: Simola H, Terzhevik AY, Viljanen M, Holopainen IK (eds) Proceedings of 4th International Lake Ladoga Symposium 2002. University of Joensuu, Publications of Karelian Institute 138:148–153

Kankaala P, Ojala A, Tulonen T, Haapamäki J, Arvola L (2000) Response of littoral vegetation on climate warming in the boreal zone; an experimental simulation. Aquat Ecol 34:433–444

Kankaala P, Ojala A, Tulonen T, Arvola L (2002) Changes in nutrient retention capacity of boreal aquatic ecosystems under climate warming: A simulation study. Hydrobiologia 469:67–76

Kankaala P, Käki T, Mäkelä S, Ojala A, Pajunen H, Arvola L (2005) Methane efflux in relation to plant biomass and sediment characteristics in stands of three common emergent macrophytes in boreal mesoeutrophic lakes. Glob Change Biol 11:145–153

Kankaanpää S, Carter T (2004) Construction of European Forest Land Use Scenarios for the 21st Century. The Finnish Environment Institute, Helsinki

Karjalainen T, Pussinen A, Liski J, Nabuurs GJ, Erhard M, Eggers T, Sonntag M, Mohren GMJ (2002) An approach towards an estimate of the impact of forest management and climate change on European forest sector carbon budget: Germany as a case study. Forest Ecol Manag 162:87–103

Karlsson J, Jonsson A, Jansson M (2005) Productivity of high-latitude lakes: Climate effect inferred from altitude gradient. Glob Change Biol 11:710–715

Karlsson PS, Bylund H, Neuvonen S, Heino S, Tjus M (2003) Climatic response of budburst in the mountain birch at two areas in northern Fennoscandia and possible responses to global change. Ecography 26:617–625

Karnosky D (2003) Impacts of elevated atmospheric CO_2 on forest trees and forest ecosystems: Knowledge gaps. Environ Int 29:161–169

Karvinen S, Välkky E, Torniainen T (2005) Idän Metsätieto Luoteis-Venäjän Metsätalouden Taskutieto Metla, Kopijyvä, Kuopio, Finland (Eastern Forest Information. Handbook on northwestern Russian Forest Economy, in Finnish)

Kaufmann RK, D'Arrigo RD, Laskowski C, Myneni RB, Zhou L, Davi NK (2004) The effect of growing season and summer greenness on northern forests. Geophys Res Lett 31:L09205

Kellomäki S, Kollström M (1993) Computations on the yield of timber by Scots pine when subjected to various levels of thinning under changing climate in southern Finland. Forest Ecol Manag 59:237–255

Kellomäki S, Leinonen S (2005) Management of European Forests Under Changing Climatic Conditions. Final Report of the Project "Silvicultural Response Strategies to Climatic Change in Management of European Forests" funded by the European Union under the Contract EVK2-2000-00723 (SilviStrat). University of Joensuu, Faculty of Forestry Research. Notes 163:1–4–27

Kellomäki S, Väisänen H (1997) Modelling the dynamics of the forest ecosystem for climate change studies in the boreal conditions. Ecol Model 97:121–140

Keup-Thiel E, Göttel H, Jacob D (2006) Regional climate simulations for the Barents Sea region. Boreal Env Res 11:1–12

Khalili M, Ebbersten S, Mosiej J, Kull A, Palang H (1997) The changing face of agriculture – Land use and farm structure. In: Bodin, B, Ebbersten S (eds) Food and Fibres. Sustainable Agriculture, Forestry and Fishery. A Sustainable Baltic Region, Session 4 Baltic University Programme, Uppsala University, pp. 11–14

Kickert RN, Tonella G, Simonov A, Krupa SV (1999) Predictive modeling of effects under global change. Env Poll 100:87–132

Kilkus K (1998) Ezeru vandens lygis klimato svyravimai indikatoriniu savybiu analize. (Lake levels and climatic fluctuations: Analysis of indicatory qualities) Geografijos metrastis 31:53–61 (in Latvian)

King JA, Bradley RI, Harrison R, Carter AD (2004) Carbon sequestration and saving potential associated with changes to the management of agricultural soils in England. Soil Use Manag 20:394–402

Klavinš M, Briede A, Rodionov V, Kokorite I, Frisk T (2002) Long-term changes of the river runoff in Latvia. Boreal Env Res 7:447–456

Klein DR (1999) The roles of climate and insularity in establishment and persistence of Rangifer tarandus populations in the high Arctic. In: Hofgaard A, Ball JP, Danell K, Callaghan TV (eds) Animal responses to global climate change in the North. Ecol Bull 47:96–104

Klein DR, Baskin LM, Bogoslovskaya LS, Danell K, Gunn A, Irons DB, Kofinas GP, Kovacs KM, Magomedova M, Meehan RH, Russell DE, Valkenburg P (2005) Management and conservation of wildlife in a changing arctic environment. In: ACIA (ed) Arctic Climate Impact Assessment, Cambridge University Press, New York, pp. 597–648

Knight CG, Staneva MP (2002) Climate change research in central and eastern Europe. GeoJournal 57:117–137

Koca D, Smith B, Sykes MT (2006) Modelling regional climate change effects on Swedish ecosystems. Climatic Change 78:381–406

Kohler J, Brandt O (2004) Regional Assessment of Climatic Variability for SCANNET Stations. Scandinavian/North European Network of Terrestrial Field Bases Work package 5. Report to the EU, NPI, Tromsø, Norway

Kohler J, Brandt O, Johansson M, Callaghan T (2006) A long-term arctic snow depth record from Abisko, northern Sweden, 1913–2004. Polar Res 25:1–113

Kondratyev SA, Efimova LK, Markova EG (2002) Estimation of hydrological regime of drainage basin changes and inflow in the Lake depend on climatic fluctuations. In: Rumyantsev VA, Drabkova VG (eds) Lake Ladoga Past, Present and Future. Nauka, St. Petersburg, pp. 269–282

Kont A, Jaagus J, Oja T, Järvet A, Rivis R (2002) Biophysical impacts of climate change on some terrestrial ecosystems in Estonia. GeoJournal 57:169–181

Konvicka M, Maradova M, Benes J, Fric Z, Kepka P (2003) Uphill shifts in distribution of butterflies in the Czech Republic: Effects of changing climate detected on a regional scale. Glob Ecol Biogeogr 12:403–410

Körner C (2003) Ecological impacts of atmospheric CO_2 enrichment on terrestrial ecosystems. Phil Trans Roy Soc Lond A 361:2023–2041

Kortelainen P, Saukkonen S, Mattsson T (1997) Leaching of nitrogen from forested catchments in Finland. Glob Biogeochem Cy 11:627–638

Kozak I, Menshutkin V, Jóźwina M, Potaczała G (2002) Computer simulation of fir forest dynamics in the Bieszczady Mountains in response to climate change. J Forest Sci 48:425–431

Kozlov MV, Berlina NG (2002) Decline in length of the summer season in the Kola Peninsula, Russia. Climatic Change 54:387–398

Kozlowski TT, Pallardy SG (2002) Acclimation and adaptive responses of woody plants to environmental stresses. Bot Rev 68:270–334

Kristiansen S, Hoell EE (2002) The importance of silicon for marine production. Hydrobiologia 484:21–31

Kronvang B, Grant R, Larsen SE, Svendsen LM, Kristensen P (1995) Non-point-source nutrient losses to the aquatic environment in Denmark – impact of agriculture. Mar Freshwat Res 46:167–177

Kronvang B, Svendsen LM, Jensen JP, Dørge J (1999) Scenario analysis of nutrient management at the river basin scale. Hydrobiologia 410:207–212

Kronvang B, Bechmann M, Pedersen ML, Flynn N (2003) Phosphorus dynamics and export in streams draining micro-catchments: Development of empirical models. Zeitschrift für Pflanzenernährung and Bodenkunde 166:469–474

Kronvang B, Jeppesen E, Conley DJ, Søndergaard M, Larsen SE, Ovesen NB, Carstensen J (2005) Nutrient pressures and ecological responses to nutrient loading redusctions in Danish streams, lakes and coastal waters. J Hydrol 304:274–288

Kull A, Oja T (2001) Influence of climate change on nutrient flows in boreonemoral floodplain ecosystem. In: Villacampa Y, Brebbia CA, Uso JL (eds) Ecosystems and Sustainable Development III. Advances in Ecological Sciences 10 WIT Press, Southampton, pp. 585–594

Kull A, Kull A, Uuemaa E, Kuusemets V, Mander Ü (2005) Modelling of excess nitrogen in small rural catchments. Agr Ecosyst Environ 108:45–56

Kullman L (1993) Tree limit dynamics of *Betula pubescens* ssp. *tortuosa* in relation to climate variability: evidence from central Sweden. J Veget Sci 4:765–772

Kullman L (1997) Tree-limit stress and disturbance. A 25-year survey of geoecological change in the Scandes mountains of Sweden. Geografiska Annaler 79A:139–165

Kullman L (2002) Rapid recent range-margin rise of tree and shrub species in the Swedish Scandes. J Ecol 90: 68–77

Kullman L (2003) Recent reversal of Neoglacial climatic cooling trend in the Swedish Scandes as evidenced by birch tree-limit rise. Glob Planet Change 36:77–88

Kulmala A, Leinonen L, Ruoho-Airola T, Salmi T, Waldén J (1998) Air quality trends in Finland. In: Air Quality Measurements 1998. Finnish Meteorological Institute, Helsinki, pp. 1–91

Kuusemets V, Mander Ü (2002) Nutrient flows and management of a small watershed. Landsc Ecol 17 Suppl 1: 59–68

Kuusemets V, Mander Ü, Lõhmus K, Ivask M (2001) Nitrogen and phosphorus variation in shallow groundwater and assimilation in plants in complex riparian buffer zones. Water Sci Tech 44:615–622

Kuusisto E (1987) An analysis of the longest ice observation series made on Finnish lakes. Fennica 17:123–132

Lacointe A (2000) Carbon allocation among tree organs: A review of basic processes and representation in functional-structural tree models. Ann Forest Sci 57:521–533

Laine K, Skre O, Wielgolaski FE (2003) Human Interactions with the Mountain Birch Ecosystem: Implications for Sustainable Development (HIBECO) Final report to the EU. Oulu University, Finland

Lal R (2003) Offsetting global CO_2 emissions by restoration of degraded soils and intensification of world agriculture and forestry. Land Degrad Dev 14:309–322

Lal R (2004) Soil carbon sequestration to mitigate climate change. Geotherma 123:1–22

Lambin EF, Turner BL, Geist HJ, Agbola SB, Angelsen A, Bruce JW, Coomes OT, Dirzo R, Fischer G, Folke C, George PS, Homewood K, Imbernon J, Leemans R, Li XB, Moran EF, Mortimore M, Ramakrishnan PS, Richards JF, Skanes H, Steffen W, Stone GD, Svedin U, Veldkamp TA, Vogel C, Xu JC (2001) The causes of land-use and land-cover change: Moving beyond the myths. Global Environ Change – Human and Policy Dimensions 11:261–269

Lambin EF, Geist HJ, Lepers E (2003) Dynamics of land-use and land-cover change in tropical regions. Annu Rev Environ Resour 28:205–241

Lampert W, Muck P (1985) Multiple aspects of food limitation in zooplankton communities: The *Daphnia-Eudiaptomus* example. Arch Hydrobiol Beiheft Ergebnisse Limnologie 21:311–321

Lapenis A, Shvidenko A, Shepaschenko D, Nilsson S, Aiyyer A (2005) Acclimation of Russian forests to recent changes in climate. Glob Change Biol 11:2090–2102

Larcher W (1995) Physiological Plant Ecology 3rd ed. Springer, Berlin Heidelberg New York

Larsen JB (1995) Ecological stability of forests and sustainable silviculture. Forest Ecol Manag 73: 75–84

Lasch P, Lindner M, Erhard M, Suckow F, Wenzel A (2002) Regional impact assessment on forest structure and functions under climate change – the Brandenburg case study. Forest Ecol Manag 162:73–86

Lasch P, Badeck FW, Suckow F, Lindner M, Mohr P (2005) Model-based analysis of management alternatives at stand and regional level in Brandenburg (Germany). Forest Ecol Manag 207:59–74

Lavorel S, Quétier F, Thébault A, Daigney S, Davies ID, De Chazal J, VISTA consortium (2006) Vulnerability to land use change of services provided by alpine landscapes. In: Price MF (ed) Global Change in Mountain Regions. Sapiens Publishing Perth UK, pp. 215–216

Laznik M, Stålnacke P, Grimvall A, Wittgren HB (1999) Riverine input of nutrients to the Gulf of Riga – temporal and spatial variation. J Mar Syst 23:11–25

Lehikoinen E, Gustafsson E, Aalto T, Alho P, Klemola H, Laine J, Normaja J, Numminen T, Rainio K (2003) Varsinais-Suomen Linnut (Birds of Southwest Finland). Turun lintutieteellinen yhdistys ry Turku, Finland (in Finnish)

Lehner B, Döll P (2001) Europe's droughts today and in the future. In: Lehner B, Henrichs T, Döll P, Alcamo J (eds) EuroWasser. Model-based Assessment of European Water. Resour and Hydrology in the Face of Global Change. Center for Environmental Systems Research. University of Kassel, Germany

Leith H (1974) Phenology and Seasonality Modeling. Springer, Berlin Heidelberg New York

Lepers E, Lambin EF, Janetos AC, DeFries R, Achard F, Ramankutty N, Scholes RJ (2005) A synthesis of information on rapid land-cover change for the period 1981–2000. BioScience 55:115–124

Lepistö A, Andersson L, Arheimer B, Sundblad K (1995) Influence of catchment characteristics forestry activities and deposition on nitrogen export from small forested catchments. Water Air Soil Pollut 84:81–102

LeRoux X, Lacointe A, Escobar-Gutiérrez A, LeDizès S (2001) Carbon-based models of individual tree growth: A critical appraisal. Ann Forest Sci 58:469–506

Levy PE, Cannell MGR, Friend AD (2004) Modelling the impact of future changes in climate CO_2 concentration and land use on natural ecosystems and the terrestrial carbon sink. Glob Env Change 14:21–30

Lin P (1995) Adaptations to Temperature in Fish: Salmonids Centrarchids and Percids. PhD thesis. University of Toronto, Canada

Linder S, Flower-Ellis JGK (1992) Environmental and physiological constrains to forest yield. In: Teller A, Mathy P, Jeffers JNR (eds) Responses of Forest Ecosystems to Environmental Changes. Elsevier, London, pp. 149–164

Linderholm HW (2002) Twentieth-century Scots pine growth variations in the central Scandinavian mountains related to climate change. Arctic Antarct Alpine Res 34:440–449

Lindner M (2000) Developing adaptive forest management strategies to cope with climate change. Tree Physiology 20:299–307

Lindner M, Lasch P, Badeck F, Beguiristain P, Junge S, Kellomäki S, Peltola H, Gracia C, Sabate S, Jäger D, Lexer M, Freeman F (2005) SilviStrat model evaluation exercises. In: Kellomäki S, Leinonen S (eds) Management of European Forests under Changing Climatic Conditions. University of Joensuu, Faculty of Forestry. Research Notes 163

Linkosalo T, Carter TR, Häkkinen R, Hari P (2000) Predicting spring phenology and frost damage risk of Betula spp under climatic warming: A comparison of two models. Tree Physiology 20:1175–1182

Livingstone DM (1997) Break-up dates of alpine lakes as proxy data for local and regional air temperatures. Climatic Change 37:407–439

Livingstone DM (1999) Ice-break up on southern Lake Baikal and its relationship to local and regional air temperatures in Siberia and the North Atlantic Oscillation. Limnol Oceanogr 44:1486–1497

Livingstone DM, Dokulil MT (2001) Eighty years of spatially coherent Austrian lake surface temperatures and their relationship to regional air temperature and the North Atlantic Oscillation. Limnol Oceanogr 46:1220–1227

Loehle C, LeBlanc D (1996) Model-based assessments of climate change effects on forests: A critical review. Ecol Model 90:1–31

Löf M, Welander NT (2000) Carry-over effects on growth and transpiration in *Fagus sylvatica* L. seedlings after drought at various stages of development. Can J Forest Res 30:468–475

Löfgren S, Gustafson A, Steineck S, Stålnacke P (1999) Agricultural development and nutrient flows in the Baltic states and Sweden after 1988. Ambio 28:320–327

Lucht W, Prentice IC, Myneni RB, Sitch S, Friedlingstein P, Cramer W, Bousquet P, Buermann W, Smith B (2002) Climatic control of the high-latitude vegetation greening trend and Pinatubo effect. Science 296:1687–1689

Ludwig W, Amiotte-Suchet P, Probst JL (1999) Enhanced chemical weathering of rocks during the last glacial maximum: A sink for atmospheric CO_2? Chem Geol 159:147–161

Luoto M, Seppälä M (2003) Thermokarst ponds as indicators of the former distribution of palsas. in Finnish Lapland. Permafrost and Periglacial Processes 14:19–27

Maak K, von Storch H (1997) Statistical downscaling of monthly mean air temperature to the beginning of flowering of Galanthus nivalis L in Northern Germany. Int J Biometeorol 41:5–12

MacGillivray CW, Grime JP, The ISP Team (1995) Testing predictions of the resistance and resilience of vegetation subjected to extreme events. Funct Ecol 9:640–649

Magnuson JJ, Meissner JD, Hill DK (1990) Potential changes in thermal habitat of Great Lakes fish after global climate warming. Trans Am Fish Soc 119:254–264

Magnuson JJ, Webster KE, Assel RA, Bowser CJ, Dillon PJ, Eaton JD, Evans HE, Fee EJ, Hall RI, Mortsch LR, Schindler DW, Quinn FH (1997) Potential effects of climate changes on aquatic ecosystems: Laurentian Great Lakes and Precambrian Shield Region. In: Cushing CE (ed) Freshwater Ecosystems and Climate Change in North America. A Regional Assessment. Wiley, New York, pp. 7–53

Magnuson JJ, Robertson DM, Benson BJ, Wynne RH, Livingstone DM, Arai T, Assel RA, Barry RG, Card VV Kuusisto E, Granin NG, Prowse TD, Stewart KM, Vuglinski VS (2000) Historical trends in lake and river ice cover in the Northern Hemisphere. Science 289:1743–1746

Malanson GP, Cairns DM (1997) Effects of dispersal population delays and forest fragmentation on tree migration rates. Plant Ecol 131:67–79

Malinina TI, Filatova IV, Filatov NN (1985) Long-term changes in the elements of water budget of Lake Ladoga. In: Problemy Issledovaniya Krupnyh Ozer. Nauka, Leningrad, pp. 79–81

Malmaeus JM, Blenckner T, Markensten H, Persson I (2006) Lake phosphorus dynamics and climate warming: A mechanistic model approach. Ecol Model 190:1–14

Malmer N, Johansson T, Olsrud M, Christensen TR (2005) Vegetation climatic changes and net carbon sequestration in a North-Scandinavian subarctic mire over 30 years. Glob Change Biol 11:1895–1909

Mander Ü, Palang H (1999) Landscape changes in Estonia: Causes, processes and consequences. In: Krönert R, Baudry J, Bowler IR, Reenberg A (eds) Land-Use Changes and Their Environmental Impact in Rural Areas in Europe. MAB Series, Vol 24. The Parthenon Publishing Group Paris, pp. 165–187

Mander Ü, Kuusemets V, Lõhmus K, Mauring T (1997) Efficiency and dimensioning of riparian buffer zones in agricultural catchments. Ecol Eng 8:299–324

Mander Ü, Kull A, Tamm V, Kuusemets V, Karjus R (1998) Impact of climatic fluctuations and land use change on runoff and nutrient losses in rural landscapes. Landsc Urban Plann 41:229–238

Mander Ü, Kull A, Kuusemets V (2000a) Nutrient flows and land use change in a rural catchment: A modelling approach. Landsc Ecol 15:187–199

Mander Ü, Kull A, Kuusemets V, Tamm T (2000b) Nutrient runoff dynamics in a rural watershed: Influence of land use changes climatic fluctuations and ecotechnological measures. Ecol Eng 14: 405–417

Masanova MD, Filatova IV (1985) Probability structure of interannual water level changes in north-western lakes. In: Problemy Issledovaniya Krupnyh Ozer Nauka Leningrad, pp. 81–84 (in Russian)

Mattson WJ, Kuokkanen K, Niemelä P, Julkunen-Tiitto R, Kellomäki S, Tahvanainen J (2004) Elevated CO_2 alters birch resistance to Lagomorpha herbivores. Glob Change Biol 10:1402–1413

May RM (1986) When two and two do not make four: Nonlinear phenomena in ecology. Proceedings of the Royal Society of London B 228:241–266

McCann K, Hastings A, Huxel G (1998) Weak trophic interactions and the balance of nature. Nature 395:794–798

McGuire AD, Chapin FS III, Wirth C, Apps M, Bhatti J, Callaghan T, Christensen TR, Clein JS, Fukuda M, Maximov T, Onuchin A, Shvidenko A, Vaganov E (2007) Responses of high latitude ecosystems to global change: Potential consequences for the climate system. In: Canadell JG, Pataki D, Pitelka LF (eds) Terrestrial Ecosystems in a Changing World. The IGBP Series. Springer, Berlin Heidelberg New York, pp. 297–310

McGuire AD, Melillo JM, Joyce LA (1995) The role of nitrogen in the response of forest net primary production to elevated atmospheric carbon dioxide. Ann Rev Ecol Systemat 26:473–503

McLaughlin SB, Shriner DS (1980) Allocation of resources to defence and repair Plant Disease V: 407–431

Menzel A (2002) Final Report (Feb 2000–June 2002) of the EU project POSITIVE (EVK2-CT-1999-00012). Technische Universiät München, Munich.

Menzel A, Fabian P (1999) Growing season extended in Europe. Nature 397:659

Metzger MJ, Rounsevell MDA, Acosta-Michlik L, Leemans R, Schröter D (2006) The vulnerability of ecosystem services to land use change. Agr Ecosyst Environ 114:64–85

Michel FA, Vaneverdingen RO (1994) Changes in hydrogeologic regimes in permafrost regions due to climatic-change. Permafrost and Periglacial Processes 5:191–195

Moen J, Danell Ö (2003) Reindeer in the Swedish mountains: An assessment of grazing impacts. Ambio 32: 397–402

Moen J, Aune K, Edenius L, Angerbjörn A (2004) Potential effects of climate change on treeline position in the Swedish mountains. Ecol Soc 9,1:16

Molau U (1996) Climatic impacts on flowering growth and vigour in an arctic-alpine cushion plant Diapensia lapponica under different snow cover regimes. Ecol Bull 45:210–219

Molau U (1997) Phenology and reproductive success in arctic plants: Susceptibility to climate change. In: Oechel WC, Callaghan T, Gilmanov T, Holtén JI, Maxwell B, Molau U, Sveinbjörnsson B (eds) Global Change and Arctic Terrestrial Ecosystems. Springer, Berlin Heidelberg New York, pp. 153–170

Molau U (2001) Tundra plant responses to experimental and natural temperature changes. Memoirs of National Institute of Polar Research, Tokyo, Special Issue 54:445–466

Molau U, Nordenhäll U, Eriksen B (2005) Onset of flowering and climate variability in an alpine landscape: A 10-year study from Swedish Lapland. Am J Bot 92:422–431

Morales P, Hickler T, Rowell DP, Smith B, Sykes MT (2007) Changes in European ecosystem productivity and carbon balance driven by Regional Climate Model output. Glob Change Biol 13:108–122

Moss B, McKee D, Atkinson D, Collings SE, Eaton JW, Gill AB, Harwey I, Hatton K, Heyes T, Wilson T (2003) How important is climate? Effects of warming nutrient addition and fish on phytoplankton in shallow lake microcosms. J Appl Ecol 40:782–792

Müller A, Wessels W (1999) The flood in the Odra River 1997 – Impact of suspended solids on water quality. Acta Hydrochimica and Hydrobiologica 27:316–320

Müller-Navarra DC, Güss S, von Storch H (1997) Interannual variability of seasonal succession events in a temperate lake and its relation to temperature variability. Glob Change Biol 3:429–438

Müller-Wohlfeil DI, Jørgensen JO, Kronvang B, Wiggers L (2002) Linked catchment and scenario analysis of nitrogen leaching and loading: A case study from a Danish catchment-fjord system, Mariager Fjord. Phys Chem Earth 27:691–699

Mund M, Kummetz E, Hein M, Bauer GA, Schulze ED (2002) Growth and carbon stocks of a spruce forest chronosequence in central Europe. Forest Ecol Manag 171:275–296

Myneni RB, Keeling CD, Tucker CJ, Asrar G, Nemani RR (1997) Increased plant growth in the northern high latitudes from 1981 to 1991. Nature 386:698–702

Myneni RB, Dong J, Tucker CJ, Kaufmann RK, Kauppi PE, Liski J, Zhou L, Alexeyev V, Hughes MK (2001) A large carbon sink in the woody biomass of northern forests. Proceedings of the National Academy of Sciences, USA 98:14784–14789

Mysterud A, Stenseth NC, Yoccuz NG, Langvatn R, Steinheim G (2001) Nonlinear effects of large-scale climatic variability on wild and domestic herbivores. Nature 410:1096–1099

Nabuurs GJ, Pussinen A, Karjalainen T, Erhard M, Kramer K (2002) Stemwood volume increment changes in European forests due to climate change – a simulation study with the EFISCEN model. Glob Change Biol 8:304–316

Najbar B, Szuszkiewicz E (2005) Reproductive ecology of the European pond turtle *Emys orbicularis* (Linnaeus 1758) (Testudines: Emydidae) in western Poland. Acta Zoologica Cracoviensa 48A:11–19

Nakićenović N, Swart R (2000) Emissions Scenarios. A Special Report of Working Group III of the Intergovernmental Panel on Climate Change. Cambridge University Press, Cambridge

Nellemann C, Kullerud L, Vistnes J, Forges BC, Kofinas GP, Kaltenborn BP, Gron O, Henry D, Magomedova M, Lambrechts C, Bobiwash R, Schei PJ, Larsen TS (2001) GLOBIO – Global Methodology for Mapping Human Impacts on the Biosphere. United Nations Environment Programme

Nemani RR, Keeling CD, Hashimoto H, Jolly WM, Piper SC, Tucker CJ, Myneni RB, Running SW (2003) Climate-driven increases in global terrestrial net primary production from 1982 to 1999. Science 300:1560–1563

Neuvonen S, Ruohomäki K, Bylund H, Kaitaniemi P (2001) Insect herbivores and herbivory effects on mountain birch dynamics. In: Wielgolaski FE (ed) Nordic Mountain Birch Ecosystems. Man and Biosphere Series. The Parthenon Publishing Group, New York, pp. 207–222

Nielsen CN (2001) Vejledning i styrkelse af stormfasthed og sundhed i na letræbevoksninger (Guide to the assessment of storm resilience and health in broadleaved forest stands). Dansk Skovbrugs Tidsskrift 86:216–263 (in Danish)

Niemirycz E (1999) The pollution load from the River Odra in comparison to that in other Polish rivers in 1988–1997. Acta Hydrochimica and Hydrobiologica 27:286–291

Nihlgård B (1997) Forest decline and environmental stress In: Brune D, Chapman DV, Gwynne MD, Pacyna JM (eds) The Global Environment; Science Technology and Management. Scandinavia Science, Oslo, pp. 422–440

Nilsson C, Stjernquist I, Bärring L, Schlyter P, Jönsson AM, Samuelsson H (2004) Recorded storm damage in Swedish forests 1901–2000. Forest Ecol Manag 199:165–173

Nõges P, Järvet A (2005) Climate driven changes in the spawning of roach (*Rutilus rutilus* L) and bream (*Abramis brama* L) in the Estonian part of the Narva River basin. Boreal Env Res 10:45–55

Nõges P, Kägu M, Nõges T (2007) Role of climate and agricultural practice in determining the matter discharge into large shallow Lake Võrtsjärv, Estonia. Hydrobiologia 581:125-134

Nõges T (2004) Reflection of the changes of the North Atlantic Oscillation Index and the Gulf Stream Position Index in the hydrology and phytoplankton of Võrtsjärv, a large, shallow lake in Estonia. Boreal Env Res 9: 401–408

Nõges T, Nõges P (1999) The effect of extreme water level decrease on hydrochemistry and phytoplankton in a shallow eutrophic lake. Hydrobiologia 408/409:277–283

Nõges T, Nõges P (2004) Large shallow temperate lakes Peipsi and Võrtsjärv: Consequences of eutrophication and climate change. In: Wassmann P, Olli K (eds) Drainage Basin Nutrient Inputs and Eutrophication: An Integrated Approach. University of Tromsø, Norway, pp. 290–301

Nõges T, Nõges P, Laugaste R (2003) Water level as the mediator between climate change and phytoplankton composition in a large shallow temperate lake. Hydrobiologia 506:257–263

Nõges T, Järvet A, Laugaste R, Loigu E, Leisk Ü, Tõnno I, Nõges P (2005) Consequences of catchment processes and climate changes on the ecological status of large shallow temperate lakes. In: Ramachandra TV, Ahalya N, Rajasekara Murthy C (eds) Aquatic Ecosystems. Conservation Restoration and Management. Capital Publishing Company, New Delhi, pp. 88–99

Nyberg P, Bergstrand E, Degerman E, Enderlein O (2001) Recruitment of pelagic fish in an unstable climate: Studies in Sweden's four largest lakes. Ambio 30:559–564

Oechel WC, Vourlitis GL, Hastings SJ, Ault RP, Bryant P (1998) The effects of water table manipulation and elevated temperature on the net CO_2 flux of wet sedge tundra ecosystems. Glob Change Biol 4:77–90

Ojala A, Kankaala P, Tulonen T (2002) Growth response of *Equisetum fluviatile* to elevated CO_2 and temperature. Env Exp Bot 47:157–171

Olofsson J, Hulme PE, Oksanen L, Suominen O (2004) Importance of large and small mammalian herbivores for the plant community structure in the forest tundra ecotone. Oikos 106:324–334

Olsrud M, Melillo JM, Christensen TR, Michelsen A, Wallander H, Olsson PA (2004) Response of ericoid mycorrhizal colonization and functioning to global change factors. New Phytologist 162: 459–470

Oltchev A, Cermak J, Gurtz J, Tishenko A, Kiely G, Nadezhdina N, Zappa M, Lebedeva N, Vitvar T, Albertson JD, Tatarinov F, Tishenko D, Nadezhdin V, Kozlov B, Ibrom A, Vygodskaya N, Gravenhorst G (2002) The response of the water fluxes of the boreal forest region at the Volga's source area to climatic and land-use changes. Phys Chem Earth 27:675–690

Ott J (2001) Expansion of Mediterranean Odonata in Germany and Europe – consequences of climatic changes. In: Walther GR, Burga CA, Edwards PJ (eds) Fingerprints of Climate Change – Adapted Behaviour and Shifting Species Ranges. Kluwer Academic/Plenum Publications, New York, pp. 89–111

Pace ML, Cole JJ, Carpenter SR, Kitchell JF (1999) Trophic cascades revealed in diverse ecosystems. Trends Ecol Evol 14:483–488

Palecki MA, Barry RG (1986) Freeze-up and break-up of lakes as an index of temperature changes during the transition season: A case study for Finland. J Appl Meteorol 25:893–902

Palmqvist G (1999) Intressanta fynd av storfjärilar (Macrolepidoptera) i Sverige 1998 (Interesting observations of large butterflies and moths in Sweden). Entomologisk Tidskrift 120:59–74 (in Swedish)

Paludan C, Alexeyev FE, Drews H, Fleischer S, Fuglsang A, Kindt T, Kowalski P, Moos M, Radlowski A, Stromfors G, Westberg V, Wolter K (2002) Wetland management to reduce Baltic Sea eutrophication. Water Sci Tech 45:87–94

Parmesan C, Ryrholm N, Stefanescu C, Hill JK, Thomas CD, Descamon H, Huntley B, Kaila L, Kullberg J, Tammaru T, Tennent WJ, Thomas JA, Warren M (1999) Poleward shifts in ranges of butterfly species associated with regional warming. Nature 399:579–583

Parry M (2000) Assessment of Potential Effects and Adaptation for Climate Change in Europe. The Europe Acacia Project. Report of a Concerted Action of the Environment Programme of the Research Directorate General of the Commission of the European Communities. Jackson Environment Institute, University of East Anglia, UK

Parsons AN, Welker JM, Wookey PA, Press MC, Callaghan TV, Lee JA (1994) Growth responses of four sub-Arctic dwarf shrubs to simulated environmental change. J Ecol 82:307–318

Pastor J, Solin J, Bridgham SD, Updegraff K, Harth C, Weishampel P, Dewey B (2003) Global warming and the export of dissolved organic carbon from boreal peatlands. Oikos 100:380–386

Paulsen HM, Volkgenannt U, Schnug E (2002) Contribution of organic farming to marine environment protection. Landbauforschung Völkenrode 52:211–218

Peltola H, Kellomäki S, Väisänen H (1999) Model computations of the impact of climatic change on the windthrow risk of trees. Climatic Change 41:17–36

Pernaravičiūte B (2003) Peculiarities of the thermal regime of Lithuanian lakes. In: Lake Ecosystems, Biological Processes, Anthropogenic Transformation, Water Quality. Materials of the II International Scientific Conference, September 22–26, 2003. Minsk, Belarus

Persaud AD, Williamson CE (2005) Ultraviolet and temperature effects on planktonic rotifers and crustaceans in northern temperate lakes. Freshwat Biol 50:467–476

Persson B, Beuker E (1997) Distinguishing between the effects of changes in temperature and light climate using provenance trials with *Pinus sylvestris* in Sweden. Can J Forest Res 27:572–579

Persson I, Blenckner T, Dokulil M, Hewitt D, Jones I, Leppäranta M (2005) Modeled thermal response of three European lakes to a probable future climate. Verh Proc Trav SIL 29:667–671

Petchey OL, McPhearson PT, Casey M, Morin PJ (1999) Environmental warming alters food-web structure and ecosystem function. Nature 402:69–72

Pettersson K (1986) The fractional composition of phosphorus in lake sediments of different characteristics. In: Sly PG (ed) Sediments and Water Interactions. Springer, Berlin Heidelberg New York, pp. 149–155

Pettersson K, Grust K, Weyhenmeyer GB, Blenckner T (2004) Seasonality of chlorophyll and nutrients in Lake Erken – effects of weather conditions. Hydrobiologia 506–509:75–81

Phoenix GK, Gwynn-Jones D, Callaghan TV, Sleep D, Lee JA (2001) Effects of global change on a sub-Arctic heath: Effects of enhanced UV-B radiation and increased summer precipitation. J Ecol 89:256–267

Pihu E, Kangur A (2000) Main changes in the ichthyocenosis of Lake Peipus since the 1950s. Proceedings of the Estonian Academy of Sciences. Biology, Ecology 49:81–90

Pisarenko AI, Strakhov V, Päivinen R, Kuusela K, Dyakun FA, Sdobnova VV (2001) Development of Forest Resources in the European Part of the Russian Federation. European Forest Institute Research Report 11. Brill Academic Publishers, Leiden, The Netherlands.

Potter JA, Press MC, Callaghan TV, Lee JA (1995) Growth responses of *Polytrichum commune* and *Hylocomium splendens* to simulated environmental change. New Phytologist 131:533–541

Pounds JA, Fogden MPL, Campbell JH (1999) Biological response to climate change on a tropical mountain. Nature 398:611–615

Pöyry J, Toivonen H (2005) Climate change adaptation and biological diversity. FINADAPT Working Paper 3. Finnish Environmental Institute Mimeographs 333, Helsinki

Press MC, Potter JA, Burke MJW, Callaghan TV, Lee JA (1998) Response of a subarctic dwarf shrub heath community to simulated environmental change. J Ecol 86:315–327

Pussinen A, Meyer J, Zudin S, Lindner M (2005) European mitigation potential. In: Management of European forests under changing climatic conditions. In: Kellomäki S, Leinonen S (eds) Management of European Forests Under Changing Climatic Conditions. University of Joensuu, Faculty of Forestry. Res Notes 163

Quayle WC, Peck LS, Peat H, Ellis-Evans JC, Harrigan PR (2002) Extreme responses to climate change in antarctic lakes. Science 295:645

Rae R, Vincent WF (1998) Effects of temperature and ultraviolet radiation on microbial foodweb structure: Potential responses to global change. Freshwat Biol 40:747–758

Räike A, Pietiläinen OP, Rekolainen S, Kauppila P, Pitkänen H, Niemi J, Raateland A, Vuorenmaa J (2003) Trends of phosphorus nitrogen and chlorophyll? concentrations in Finnish rivers and lakes in 1975–2000. Sci Total Env 310:47–59

Räisänen J, Rummukainen M, Ullerstig A (2001) Downscaling of greenhouse gas induced climate change in two GCMs with the Rossby Centre regional climate model for northern Europe. Tellus 53A:168–191

Räisänen J, Hansson U, Ullerstig A, Döscher R, Graham LP, Jones C, Meier M, Samuelsson P, Willen U (2003) GCM Driven Simulations of Recent and Future Climate With the Rossby Centre Coupled Atmosphere–Baltic Sea Regional Climate Model RCAO. Reports Meteorology and Climatology (RMK) No 101. Swedish Meteorological and Hydrological Institute, Norrköping, Sweden

Raymond PA, Cole JJ (2003) Increase in the export of alkalinity from North America's largest river. Science 301:88–91

Reid WV, Mooney HA, Cropper A, Capistrano D, Carpenter SR, Chopra K, Dagupta P, Dietz T, Duraiappah AK, Hassan R, Kasperson R, Leemans R, May RM, McMichael AJ, Pingali P, Samper C, Scholes R, Watson RT, Zakri AH, Shidong Z, Ash NJ, Bennett E, Kumar P, Lee MJ, Raudsepp-Hearne C, Simons H, Thonell J, Zurek MB (2005) Ecosystems and Human Well-being: Synthesis Island Press, Washington DC

Rekolainen S, Grönroos J, Bärlund I, Nikander A, Laine Y (1999) Modelling the impacts of management practices on agricultural phosphorus losses to surface waters of Finland. Water Sci Tech 39: 265–272

Rheinheimer G (1998) Pollution in the Baltic Sea. Naturwissenschaften 85:318–329

Rivza B (1997) Economic social and environmental conditions in Latvian rural areas. Ambio 26:439–441

Robertson DM, Ragotzkie RA, Magnuson JJ (1992) Lake ice records used to detect historical and future climate changes. Climatic Change 21:407–427

Robinson CH, Wookey PA, Lee JA, Callaghan TV, Press MC (1998) Plant community responses to simulated environmental change at a high Arctic polar semidesert. Ecology 79:856–866

Root TL, Price JT, Hall KR, Schneider SH, Rosenzweig C, Pounds JA (2003) Fingerprints of global warming on wild animals and plants. Nature 421:57–60

Rounsevell MDA, Ewert F, Reginster I, Leemans R, Carter T (2005) Future scenarios of European agricultural land use. II. Projecting changes in cropland and grassland. Agr Ecosyst Env 107:117–135

Rounsevell MDA, Berry PM, Harrison PA (2006a) Future environmental change impacts on rural land use and biodiversity: A synthesis of the ACCELERATES project. Env Sci Pol 9:93–100

Rounsevell MDA, Reginster I, Araújo MB, Carter TR, Dendoncker N, Ewert F, House JI, Kankaanpää S, Leemans R, Metzger MJ, Schmit C, Smith P, Tuck G (2006b) A coherent set of future land use change scenarios for Europe. Agr Ecosyst Environ 114:57–68

Rummukainen M, Räisänen J, Bringfelt B, Ullerstig A, Omstedt A, Willén U, Hansson U, Jones C (2001) A regional climate model for northern Europe: Model description and results from the downscaling of two GCM control simulations. Clim Dyn 17:339–359

Rybak J (2002) Seasonal and long-term export rates of nutrients with surface runoff in the river Jorka catchment basin (Masurian Lakeland Poland). Polish J Ecol 50:439–458

Saarnio S, Järviö S, Saarinen T, Vasander H, Silvola J (2003) Minor changes in vegetation and carbon gas balance in a boreal mire under a raised CO_2 or NH_4/NO_3 supply. Ecosystems 6:46–60

Sachanowicz K, Ciechanowski M (2006) First winter record of migratory bat *Pipistrellus nathusii* (Keyserling and Blasius 1839) (Chiroptera: Vespertilionidae) in Poland: Yet more evidence of global warming? Mammalia 70:168–169

Saether BE, Lande R, Engen S, Weimerskirch H, Lillegaard M, Altwegg R, Becker PH, Bregnballe T, Brommer JE, McCleery R, Merila J, Nyholm E, Rendell W, Robertson RR, Tryjanowski P, Visser ME (2005) Generation time and temporal scaling of bird population dynamics. Nature 436:99–102

Sagrario MG, Jeppesen E, Gomà J, Søndergaard M, Jensen JP, Lauridsen T, Landkildehust F (2005) Does high nitrogen loading prevent clear-water conditions in shallow lakes at moderately high phosphorus concentrations? Freshwat Biol 50:27–41

Sakai A, Larcher W (1987) Frost Survival of Plants Ecological Studies 62. Springer, Berlin Heidelberg New York

Sala OE, Chapin FS III, Armesto JJ, Berlow E, Bloomfield J, Dirzo R, Huber Sanwald E, Huenneke LF, Jackson RB, Kinzig A, Leemans R, Lodge DH, Mooney HA, Oesterheld M, Leroy Poff N, Sykes MT, Walker BH, Walker M, Wall DH (2000) Global biodiversity scenarios for the year 2100. Science 287:1770–1774

Salonen K, Arvola L, Rask M (1984) Autumnal and vernal circulation of small forest lakes in Southern Finland. Verh Proc Trav SIL 22:103–107

Sarvala J, Helminen H, Auvinen H (1999) Portrait of a flourishing freshwater fishery: Pyhäjärvi a lake in SW-Finland. Boreal Env Res 3:329–345

Saxe H (1993) Triggering and predisposing factors in the "Red" decline syndrome of Norway spruce (*Picea abies*). Trees 8:39–48

Saxe H, Cannell MGR, Johnsen Ø, Ryan MG, Vourlitis G (2001) Tree and forest functioning in response to global warming. New Phytologist 149:369–400

Scharff N, Langemark S (1997) *Argiope bruennichi* (Scopoli) in Denmark (Araneae; Araneidae). Entomologiske Meddelelser 65:179–182

Scheffer M, Straile D, van Nes EH, Hosper H (2001) Climatic warming causes regime shifts in lake food webs. Limnol Oceanogr 46:1780–1783

Scheifinger H, Menzel A, Koch E, Peter C (2003) Trends of spring time frost events and phenological dates in Central Europe. Theor Appl Climatol 74:41–51

Schelhaas MJ, Nabuurs GJ, Schuck A (2003) Natural disturbances in the European forests in the 19th and 20th centuries. Glob Change Biol 9:1620–1633

Scheller RM, Mladenoff DJ (2005) A spatially interactive simulation of climate change harvesting wind and tree species migration and projected changes to forest composition and biomass in northern Wisconsin USA. Glob Change Biol 11:307–321

Schindler DW (1996) Widespread effects of climate warming on freshwater ecosystems in North America. Hydrolog Process 11:1044–1069

Schindler DW, Beaty KG, Fee EJ, Cruikshank DR, DeBruyn ER, Findlay DL, Linsey GA, Shearer JA, Stainton MP, Turner MA (1990) Effects of climate warming on lakes of the central boreal forest. Science 250:967–970

Schnelle F (1955) Pflanzen-Phänologie (Plant Phenology). Akademische Verlagsgesellschaft Geest and Portig KG, Leipzig, Germany (in German)

Schoeneweiss DF (1975) Predisposition stress and plant disease. Ann Rev Phytopathol 13:193–211

Schröter D, Cramer W, Leemans R, Prentice C, Araújo MB, Arnell NW, Bondeau A, Bugmann H, Carter TR, Garcia CA, de la Vega-Leinert AC, Erhard M, Ewert F, Glendining M, House JI, Kankaanpää S, Klein RJT, Lavorel S, Lindner M, Metzger MJ, Meyer J, Mitchell TD, Reginster I, Rounsevell M, Sabaté S, Sitch S, Smith B, Smith J, Smith P, Sykes MT, Thonicke K, Thuiller W, Tuck G, Zaehle S, Zierl B (2005) Ecosystem service supply and vulnerability to global change in Europe. Science 310:1333–1337

Schults GE (1981) Obshtshaja Fenologija (Principles of Phenology). "Nauka" Leningradskoje Otdelenije, Leningrad (in Russian)

Schwartz MD (ed) (2003) Phenology: An Integrative Environmental Science Tasks for Vegetation. Science 39 Kluwer Academic Publishers Dordrecht, The Netherlands.

Schwartzman DW, Volk T (1989) Biotic enhancement of weathering and the habitability of Earth. Nature 340:457–460

Sefrová H, Lasuvka Z (2001) Dispersal of the horse-chesnut leaf miner *Cameraria ohridella* Deschka and Dimić 1986, in Europe: Its course, ways and causes (Lepidoptera: Gracillaridae) Entomologische Zeitschrift 111: 194–198

Sharpe PJH, DeMichele DW (1977) Reaction kinetics of poikilotherm development. J Theor Biol 64: 649–670

Shaver GR, Jonasson S (1999) Response of Arctic ecosystems to climate change: Results of long-term field experiments in Sweden and Alaska. Polar Res 18:245–252

Shaver GR, Canadell J, Chapin FS III, Gurevitch J, Harte J, Henry G, Ineson P, Jonasson S, Melillo J, Pitelka L, Rustad L (2000) Global warming and terrestrial ecosystems: A conceptual framework for analysis. BioScience 50:871–882

Shudo E, Iwasa Y (2001) Inducible defence against pathogens and parasites: Optimal choice among multiple options. J Theor Biol 209:233–247

Sileika AS, Kutra G, Berankiene L (2002) Phosphate run-off in the Nevezis River (Lithuania). Env Monit Assess 78:153–167

Sileika AS, Gaigalis K, Kutra G, Smitiene A (2005) Factors affecting N and P losses from small catchments (Lithuania). Env Monit Assess 102:359–374

Sitch S, Smith B, Prentice IC, Arneth A, Bondeau A, Cramer W, Kaplan JO, Levis S, Lucht W, Sykes MT, Thonicke K, Venevsky S (2003) Evaluation of ecosystem dynamics plant geography and terrestrial carbon cycling in the LPJ dynamic global vegetation model. Glob Change Biol 9:161–185

Slaymaker O (2001) Why so much concern about climate change and so little attention to land use change? Canadian Geographer 45:71–78

Smith B, Prentice IC, Sykes MT (2001) Representation of vegetation dynamics in modelling of terrestrial ecosystems: Comparing two contrasting approaches within European climate space. Glob Ecol Biogeogr 10:621–637

Smith LC, Sheng Y, MacDonald GM, Hinzman LD (2005) Disappearing arctic lakes. Science 308:1429

Smolander A, Barnette L, Kitunen V, Lumme I (2004) N and C transformations in long-term N-fertilized forest soils in response to seasonal drought. Appl Soil Ecol 29:225–235

Sobek S, Algesten G, Bergström AK, Jansson M, Tranvik LJ (2003) The catchment and climate regulation of pCO_2 in boreal lakes. Glob Change Biol 9:630–641

Solheim B, Johanson U, Callaghan TV, Lee JA, Gwynn Jones D, Bjorn LO (2002) The nitrogen fixation potential of arctic cryptogam species is influenced by enhanced UV-B radiation. Oecologia 133:90–93

Sonesson M, Hoogesteger J (1983) Recent treeline dynamics (*Betula pubescens* Ehrh ssp *tortuosa* (ledeb) Nyman) in northern Sweden. Nordicana 47:47–54

Sonesson M, Gehrke C, Tjus M (1992) CO_2 environment microclimate and photosynthesis characteristics of the moss *Hylocomium splendens* in a subarctic habitat. Oecologia 92:23–29

Sonesson M, Callaghan TV, Björn LO (1995) Short term effects of enhanced UV-B and CO_2 on lichens at different latitudes. Lichenologist 27:547–557

Sonesson M, Callaghan TV, Carlsson BA (1996) Effects of enhanced ultra-violet radiation and carbon dioxide concentrations on the moss *Hylocomium splendens*. Glob Change Biol 2:67–73

Sonesson M, Carlsson BA, Callaghan TV, Halling S, Björn LO, Bertgren M, Johansson U (2002) Growth of two peat-forming mosses in subarctic mires: Species interactions and effects of simulated climate change. Oikos 99:151–160

Sorvari S, Korhola A, Thompson R (2002) Lake diatom response to recent Arctic warming in Finnish Lapland. Glob Change Biol 8:171–181

Sparks TH, Tryjanowski P (2005) The detection of climate impacts: Some methodological considerations. Int J Climatol 25:271–277

Spiecker H, Mielikäinen K, Köhl M, Skovsgaard JP (1996) Growth Trends in European Forests – Studies From 12 Countries. EFI–Research Report No 5. Springer, Berlin Heidelberg New York

Stålnacke P, Grimvall A, Sundblad K, Tonderski A (1999a) Estimation of riverine loads of nitrogen and phosphorus to the Baltic Sea 1970–1993. Environ Monit Assess 58:173–200

Stålnacke P, Grimvall A, Sundblad K, Wilander A (1999b) Trends in nitrogen transport in Swedish rivers. Environ Monit Assess 598:47–72

Stålnacke P, Grimvall A, Libiseller C, Laznik M, Kokorite I (2003) Trends in nutrient concentrations in Latvian rivers and the response to the dramatic change in agriculture. J Hydrol 283:184–205

Stålnacke P, Vandsemb SM, Vassiljev A, Grimwall A, Jolankai G (2004) Changes in nutrient levels in sone Eastern European rivers in response to large-scale changes in agriculture. Water Sci Tech 49: 29–36

Stefanescu C, Herrando S, Páramo F (2004) Butterfly species richness in the north-west Mediterranean basin: The role of natural and human-induced factors. J Biogeogr 31:905–915

Stendera S, Johnson RK (2005) Does climate change confound lake recovery? Eos Transactions. AGU, 86, Joint Assembly Supplement, Abstract NB32D-04

Stenström M, Gugerli F, Henry GHR (1997) Response of *Saxifraga oppositifolia* L to simulated climate change at three contrasting latitudes. Glob Change Biol 3, Suppl 1:44–54

Sterligova OP, Pavlovskiy SA, Komulainen SF (1988) Reproduction of coregonids in the eutrophicated Lake Sjamozero, Karelian ASSR. Finnish Fisheries Research 9:485–488

Stöckli R, Vidale PL (2004) European plant phenology and climate as seen in a 20-year AVHRR land-surface parameter dataset. Int J Rem Sens 25:3303–3330

Straile D (2000) Meteorological forcing of plankton dynamics in a large and deep continental European lake. Oecologia 122:44–50

Straile D (2002) North Atlantic Oscillation synchronizes food-web interactions in central European lakes. Proceedings of the Royal Society of London B 269:391–395

Straile D, Adrian R (2000) The North Atlantic Oscillation and plankton dynamics in two European lakes – two variations on a general theme. Glob Change Biol 6:663–670

Strand JF (2000) Some agrometeorological aspects of pest and disease management for the 21st century. Agr Forest Meteorol 103:73–82

Strathdee AT, Bales JS, Strathdee FC, Block WC, Coulson SJ, Webb NR, Hodkinson ID (1995) Climatic severity and the response to temperature elevation of arctic aphids. Glob Change Biol 1:23–28

Summers RW, Underhill LG (1987) Factors relating to breeding populations of Brent Geese Branta b bernicla and waders (Charadrii) on the Taimyr Peninsula. Bird Study 34:161–171

Sundell B (1997) The future of agriculture in the Baltic Sea region: Sustainable agriculture and new technological development. Ambio 26:412–414

Sutinen ML, Arora R, Wisniewski M, Ashworth E, Strimbeck R, Palta J (2001) Mechanisms of frost survival and freeze-damage in nature. In: Bigras FJ, Colombo SJ (eds) Conifer Cold Hardiness. Kluwer, Dordrecht, pp. 89–120

Švažas S, Meissner W, Serebryakov V, Kozulin A, Grishanov G (2001) Changes of wintering sites of waterfowl in Central and Eastern Europe. OMPO and Institute of Ecology, Vilnius, Lithuania

Sykes MT, Prentice IC (1995) Boreal forest futures: Modelling the controls on tree species range limits and transient responses to climate change. Water Air Soil Pollut 82:415–428

Sykes MT, Prentice IC, Cramer W (1996) A bioclimatic model for the potential distributions of north European tree species under present and future climates. J Biogeogr 23:203–233

Syroechovski EE, Kuprianov AG (1995) Wild reindeer of the arctic Eurasia: Geographical distribution numbers and population structure. In: Grønlund E, Melander O (eds) Swedish–Russian Tundra Ecology – Expedition 94: A Cruise Report. Swedish Polar Res Secretariat, Stockholm

Tenow O (1996) Hazards to a mountain birch forest – Abisko in perspective. Ecol Bull 45:104–114

Tesche M (1992) Immediate and long-term (memory) responses of *Picea abies* to a single growing season of SO_2-exposure or moderate drought. Forest Ecol Manag 51:179–186

Thomas CD, Cameron A, Green RE, Bakkenes M, Beaumont LJ, Collingham YC, Erasmus BF, De Siqueira MF, Grainger A, Hannah L, Hughes L, Huntley B, van Jaarsveld AS, Midgley GF, Miles L, Ortega-Huerta MG, Townsend Peterson A, Phillips OL, Williams SE (2004) Extinction risk from climate change. Nature 427:145–148

Thomas G, Rowntree PR (1992) The boreal forest and climate. Q J Roy Met Soc 118:469–497

Thuiller W (2003) BIOMOD: Optimising predictions of species distributions and projecting potential future shifts under global change. Glob Change Biol 10:2020–2027

Tomiałojć L, Stawarczyk T (2003) Awifauna Polski Rozmieszczenie, Liczebność i Zmiany (Avifauna in Poland. Distribution, population and variability). Pro Natura, Wrocław, Poland (in Polish)

Tømmervik H, Johansen B, Tombre I, Thannheiser D, Høgda KA, Gaare E, Wielgolaski FE (2004) Vegetation changes in the Nordic mountain birch forest: The influence of grazing and climate change. Arctic Antarct Alpine Res 36:323–332

Tranvik LJ, Jansson M (2002) Climate change – terrestrial export of organic carbon. Nature 415:861–862

Tryjanowski P, Winiecki A (2003) Ptaki jako wskaźnik stepowienia Wielkopolski. In: Banaszak J (ed) Stepowienie Wielkopolski – Pół Wieku Później. Wyd Akademii Bydgoskiej, Bydgoszcz, Poland, pp. 175–184 (in Polish)

Tryjanowski P, Sparks T, Profus P (2005) Uphill shifts in the distribution of the white stork *Ciconia ciconia* in southern Poland: The importance of nest quality. Diversity and Distributions 11:219–223

Tsirkunov VV, Nikanorov AM, Laznik MM, Dongwei Z (1992) Analysis of long-term and seasonal river water-quality changes in Latvia. Water Res 26:1203–1216

Tucker CJ, Slayback DA, Pinzon JE, Los SO, Myneni RB, Taylor MG (2001) Higher northern latitude normalized difference vegetation index and growing season trends from 1982 to 1999. Int J Biometeorol 45:184–190

Tumas R (2000) Evaluation and prediction of nonpoint pollution in Lithuania. Ecol Eng14:443–451

Turner K, Georgiou S, Gren IM, Wulff F, Barrett S, Söderqvist T, Bateman IJ, Folke C, Langaas S, Żylicz Z, Mäler KG, Markowska A (1999) Managing nutrient fluxes and pollution in the Baltic: An interdisciplinary simulation study. Ecological Economics 30:333–352

UN-ECE/FAO (2000) Contribution to the Global Forest Resource Assessment – Forest Resources in Europe, CIS, North America, Australia, Japan and New Zealand. United Nations Publications, Geneva, Timber and Forest Study Papers No 17:1–445

UNFCCC (1995) Estonia's First National Communication Under the UN Framework Convention on Climate Change. UNFCCC, Bonn, Germany, http://unfccc.int

UNFCCC (2002) The First National Communication of the Republic of Lithuania on Climate Change. UNFCCC, Bonn, Germany, http://unfccc.int

UNFCCC (2003a) Estonia's Third National Communication Under the UN Framework Convention on Climate Change. UNFCCC, Bonn, Germany, http://unfccc.int

UNFCCC (2003b) Lithuania's Second National Communication Under the Framework Convention on Climate Change. UNFCCC, Bonn, Germany, http://unfccc.int

Usher MB, Callaghan TV, Gilchrist G, Heal B, Juday JP, Loeng H, Muir MAK, Prestrud P (2005) Principles of conserving the Arctic's biodiversity. In: ACIA. 2005. Arctic Climate Impact Assessment. Cambridge University Press, New York, pp. 539–596

Uuemaa E, Roosaare J, Mander Ü (2007) Landscape metrics as indicators of river water quality at catchment scale. Nordic Hydrology 38:125–138

Vagstad N, Jansons V, Loigu E, Deelstra J (2000) Nutrient losses from agricultural areas in the Gulf of Riga drainage basin. Ecol Eng 14:435–441

Vagstad N, Stålnacke P, Andersen HE, Deelstra J, Jansons V, Kyllmar K, Loigu E, Rekolainen S, Tumas R (2004) Regional variations in diffuse nitrogen losses from agriculture in the Nordic and Baltic regions. Hydrol Earth Syst Sci 8:651–662

Valpasvuo-Jaatinen P, Rekolainen S, Latostenmaa H (1997) Finnish agriculture and its sustainability: Environmental impacts. Ambio 26:448–455

van Donk E, Santamaria L, Mooij WM (2003) Climate warming causes regime shifts in lake food webs: A reassessment. Limnol Oceanogr 48:1350–1353

van Meijl H, van Rheenen T, Tabeau AB, Eickhout B (2005) The impact of different policy environments on land use in Europe. Agr Ecosyst Environ 114:21–38

van Vliet AJH (2003) Toward a multifunctional European phenology network. In: Schwartz MD (ed) Phenology: An Integrative Environmental Science Tasks for vegetation science. Kluwer, Dordrecht, pp. 105–117

van Vliet AJH, de Groot RS, Bellens Y, Braun P, Bruegger R, Bruns E, Clevers J, Estreguil C, Flechsig M, Jeanneret F, Maggi M, Martens P, Menne B, Menzel A, Sparks T (2003) The European Phenological Network. Int J Biometeorol 47:202–212

van Wijk MT, Clemmensen KE, Shaver GR, Williams M, Callaghan TV, Chapin FS III, Cornelissen JHC, Gough L, Hobbie SE, Jonasson S, Lee JA, Michelsen A, Press MC, Richardson SJ, Rueth H (2003) Long term ecosystem level experiments at Toolik Lake, Alaska and at Abisko, Northern Sweden: Generalisations and differences in ecosystem and plant type responses to global change. Glob Change Biol 10:105–123

Varis O, Kuikka S (1997) BENE-EIA: A Bayesian approach to expert judgment elicitation with case studies on climate change impacts on surface waters. Climatic Change 37:539–563

Vassiljev J (1998) The simulated response of lakes to changes in annual and seasonal precipitation: Implication for Holocene lake-level changes in northern Europe. Clim Dyn 14:791–801

Vavrus SJ, Wymne RH, Foley JA (1996) Measuring the sensitivity of southern Wisconsin lake ice to climate variations and lake depth using a numerical model. Limnol Oceanogr 41:822–831

Venäläinen A, Tuomenvirta H, Heikinheimo M, Kellomäki S, Peltola H, Strandman H, Väisänen H (2001) Impact of climate change on soil frost under snow cover in a forested landscape. Climate Res 17:63–72

Verburg PH, Schulp CJE, Witte N, Veldkamp A (2006) Downscaling of land use changes to assess the dynamics of European landscapes. Agr Ecosyst Environ 114:39–56

Virtanen T, Neuvonen S (1999) Climate change and macrolepidopteran biodiversity in Finland. Chemosphere Glob Change Sci 4:439–448

Vlassova TK (2002) Human impacts on the tundra-taiga zone dynamics: The case of the Russian lesotundra. Ambio Special Report 12:30–36

Volney WJA, Fleming RA (2000) Climate change and impacts of boreal forest insects. Agr Ecosyst Env 82:283–294

Vought LBM, Dahl J, Pedersen CL, Lacoursiere JO (1994) Nutrient retention in riparian ecotones. Ambio 23:342–348

Walankiewicz W (2002) The Number and Composition of Snags in the Pine-Spruce Stands of the Bialowieza National Park, Poland. USDA Forest Service General Technical Report PSW-GTR-181:489–500

Walther GR, Post E, Convey P, Menzel A, Parmesan C, Beebee TCJ, Fromentin JM, Hoegh-Guldberg O, Bairlein F (2002) Ecological responses to recent climate change. Nature 416:389–395

Walther GR, Berger S, Sykes MT (2005) An ecological "footprint" of climate change. Proceedings of the Royal Society of London B 272:1427–1432

Wang SQ, Liu JY, Yu GR, Pan YY, Chen QM, Li KR, Li JY (2004) Effects of land use change on the storage of soil organic carbon: A case study of the Qianyanzhou Forest Experimental Station in China. Climatic Change 67:247–255

Warren MS, Hill JK, Thomas JA, Asher J, Fox R, Huntley B, Roy DB, Telfer MG, Jeffcoate S, Harding P, Jeffcoate G, Willis SG, Greatorex-Davies JN, Moss D, Thomas CD (2001) Rapid responses of British butterflies to opposing forces of climate and habitat change. Nature 414:65–69

Weckström J, Korhola A (2001) Patterns in the distribution composition and diversity of diatom assemblages in relation to ecoclimatic factors in Arctic Lapland. J Biogeogr 28:31–45

Weld JL, Sharpley AN, Beegle DB, Gburek VJ (2001) Identifying critical source areas of phosphorus export from agricultural watersheds. Nutrient Cycling in Agroecosystems 59:29–38

Welker JM, Molau U, Parsons AN, Robinson CH, Wookey PA (1997) Responses of Dryas octopetala to ITEX environmental manipulations: A synthesis with circumpolar comparisons. Glob Change Biol 3, Suppl 1:61–73

Weller CM, Watzin MC, Wang D (1996) Role of wetlands in reducing phosphorus loading to surface water in eight watersheds in the Lake Champlain basin. Env Manag 20:731–739

Wendland F, Kunkel R, Grimvall A, Kronvang B, Müller-Wohlfeil DI (2002) The SOIL-N/WEKU model system – a GIS-supported tool for the assessment and management of diffuse nitrogen leaching at the scale of river basins. Water Sci Tech 45:285–292

Wermelinger B (2004) Ecology and management of the spruce bark beetle *Ips typographus* – a review of recent research. Forest Ecol Manag 202:67–82

Weyhenmeyer GA (2001) Warmer winters: Are planctonic algal populations in Sweden's largest lakes affected? Ambio 30:565–571

Weyhenmeyer GA (2004) Synchrony in relationships between the North Atlantic Oscillation and water chemistry among Sweden's largest lakes. Limnol Oceanogr 49:1191–1201

Weyhenmeyer GA, Blenckner T, Pettersson K (1999) Changes of the plankton spring outburst related to the North Atlantic Oscillation. Limnol Oceanogr 44:1788–1792

White MA, Nemani RR, Thornton PE, Running SW (2002) Satellite evidence of phenological differences between urbanized and rural areas of the eastern United States deciduous broadleaf forest. Ecosystems 5:260–277

Williams SE, Bolintho EE, Fox S (2003) Climate change in Australian tropical rainforests: An impending environmental catastrophe. Proceedings of the Royal Society of London Series B 270:1887–1892

Winder M, Schindler DE (2004) Climatic effects on the phenology of lake processes. Glob Change Biol 10:1844–1856

Wolf A, Blyth E, Harding R, Jacob D, Keup E, Goettel H, Callaghan T (2007a) Sensitivity of an ecosystem model to hydrology and temperature. Climatic Change DOI: 10.1007/s10584-007-9339-z

Wolf A, Callaghan TV, Larson K (2007b) Future changes in vegetation and ecosystem function of the Barents Region. Climatic Change (in press)

Wolf A, Kozlov MV, Callaghan TV (2007c) Impact of non-outbreak insect damage on vegetation in northern Europe will be greater than expected during a changing climate. Climatic Change DOI: 10.1007/s10584-007-9340-6

Woodward FI (1987) Climate and Plant Distribution. Cambridge University Press, Cambridge

World Resources Institute (2000) World Resources 2000–2001. People and Ecosystems: The Fraying Web of Life. World Resources Institute, Washington DC

Yang DQ, Kane DL, Hinzman LD, Zhang XB, Zhang TJ, Ye HC (2002) Siberian Lena River hydrologic regime and recent change. J Geophys Res – Atmospheres 107 D23, 4694, doi:10.1029/2002JD002542

Yoccoz NG, Ims RA (1999) Demography of small mammals in cold regions: The importance of environmental variability. Ecological Bulletins 47: 137–144

Yoo J, D'Odorico P (2002) Trends and fluctuations in the dates of ice break-up of lakes and rivers in Northern Europe: The effect of the North Atlantic Oscillation. J Hydrol 268:100–112

Yurkovskis A (2004) Long-term land-based and internal forcing of the nutrient state of the Gulf of Riga (Baltic Sea). J Mar Syst 50:181–197

Zaehle S, Bondeau A, Carter TR, Cramer W, Erhard M, Prentice IC, Reginster I, Rounsevell MDA, Sitch S, Smith B, Smith PC, Sykes M (2007) Projected changes in terrestrial carbon storage in Europe under climate and land use change 1990–2100. Ecosystems 10:380–401

Zebisch M, Wechsung F, Kenneweg H (2004) Landscape response functions for biodiversity – assessing the impact of land-use changes at the county level. Landsc Urban Plann 67:157–172

Zhuang Q, McGuire AD, Melillo JM, Clein JS, Dargaville RJ, Kicklighter DW, Myneni RB, Dong J, Romanovsky VE, Harden J, Hobbie JE (2003) Carbon cycling in extratropical terrestrial ecosystems of the Northern Hemisphere during the 20th century: A modeling analysis of the influences of soil thermal dynamics. Tellus B 55:751–776

Ziverts A, Apsite E (2005) Simulation of daily runoff and water level for the Lake Butrnieks. In: Merkuryev Y, Zobel R, Kerckhoffs E (eds) Proceedings 19th European Conference on Modelling and Simulation. European Council for Modelling and Simulation. Riga, Latvia

Žylicz T (1997) Agriculture and environment in Poland. Ambio 26:445–447

5 Climate-related Marine Ecosystem Change

Joachim W. Dippner, Ilppo Vuorinen, Darius Daunys, Juha Flinkman, Antti Halkka, Friedrich W. Köster, Esa Lehikoinen, Brian R. MacKenzie, Christian Möllmann, Flemming Møhlenberg, Sergej Olenin, Doris Schiedek, Henrik Skov, Norbert Wasmund

5.1 Introduction

This chapter deals with climate-related changes in the marine ecosystem of the Baltic Sea. The Baltic Sea is often described as one of the world's largest brackish water bodies. It has a unique combination of oceanographic, climatic, and geographic features. Most important in this context is: the sea is a nearly enclosed area having a water residence time of 30 years, due to restricted water exchange through the Danish Straits. It is situated in northern Europe and has, therefore, some arctic characteristics and a pronounced seasonality. It is affected alternately by continental and marine climatic effects. It has a catchment area approximately four times larger than the sea itself, while it is as the same time very shallow, with an average depth of only 56 m, having thus a relatively small water body. Seasonal vertical mixing of the water reaches a depth of 30–50 m and contributes to re-suspension of nutrients and pollutants. In deeper parts, a permanent halocline appears, below which anoxia is common and interrupted only by major inflows of North Sea water.

The ecology of the Baltic Sea is often described as highly sensitive, which, again, is due to these interacting environmental factors. For most of the animal and plant species found in the Baltic Sea, the distribution limits are set by horizontal and vertical isohalines, thus the biota is often living close to environmental conditions that are lethal. Further peculiarities are due to the relatively young age of the sea (the recent stage is only about 8,000 years old), which, in turn, contributes to the fact that new species are constantly being reported. The young age of the sea is also seen in the relatively poor adaptation of marine species to the low salinity, causing them often to be smaller than representatives of the same species from the North Sea.

The human-induced change in the marine ecosystem of the Baltic Sea cannot be neglected, since it is caused by ca. 85 million people. A list of the 149 most significant pollution sources ("hot spots") around the Baltic Sea was first drawn up by Baltic Marine Environment Protection Commission (HELCOM) under the Baltic Sea Joint Comprehensive Environmental Action Programme (JPC) in 1992. Today, 95 hot spots or sub-hot spots remain on the list, following the deletion of 54 by the end of 2002. It will be a question of time, investment and remediation projects in environmental policy to continue the reduction of the hot spots (HELCOM 2003b). HELCOM has published four five-year periodic reviews of the state of the Baltic Sea since 1993 (HELCOM 1993b).

Climate induced changes in marine ecosystems include changes in nutrient cycling, contaminant distribution and changes at all trophic levels from bacteria up to sea birds and marine mammals. In preparing this chapter the co-authors were asked to take a look at the whole published monitoring data set and, additionally, to review also other, national, data sets, also those not available in English. One of the most important sources has been the data set of the Baltic Marine Environment Protection Commission (HELCOM). Another large international source of information used by authors of this chapter are fish and fisheries data of the International Council for the Exploration of the Sea (ICES). The authors were asked to consider whether anthropogenic climate change can be seen in their data sets and, secondly, if the predictions presented in Chap. 3 come true, what the ecological consequences will be.

5.2 Sources and Distribution of Nutrients

5.2.1 Current Situation

The sources and distribution of nutrients and other substances are documented in various HELCOM assessment reports and pollution load compilations (HELCOM 1990, 1996, 2002, 2005). Nitrogen and phosphorus enter the Baltic Sea area either as waterborne or as airborne input. The total input in the year 2000 was 1,009,700 tons of nitrogen and 34,500 tons of phosphorus (HELCOM 2005). 75% of the nitrogen entered the Baltic Sea via waterborne inputs and 25% via atmospheric deposition, whereas the estimated airborne input

Table 5.1. Estimates of nitrogen input in tons into surface waters of the Baltic Sea in 2000, partitioned into major source categories (after HELCOM 2005)

Sub-region	Natural background	Diffuse sources	Point sources	Total load
Bothnian Bay	40819	30935	2351	74105
Bothnian Sea	37682	32289	3728	73699
Archipelago Sea	2610	6906	136	9652
Gulf of Finland	67330	44635	14518	126483
Gulf of Riga	16834	40006	1573	58413
Baltic Proper	69699	219957	48534	338190
Belt Sea	7314	40079	2984	50377
Kattegat	17231	69279	4811	91321
Total Baltic Sea	259519	484086	78635	822240

Table 5.2. Estimates of phosphorus input in tons into surface waters of the Baltic Sea in 2000, partitioned into major source categories (after HELCOM 2005)

Sub-region	Natural background	Diffuse sources	Point sources	Total load
Bothnian Bay	2479	1574	67	4120
Bothnian Sea	2078	1445	98	3621
Archipelago Sea	88	723	4	815
Gulf of Finland	1463	2380	1526	5369
Gulf of Riga	380	1089	268	1737
Baltic Proper	3770	12530	5798	22098
Belt Sea	216	822	189	1227
Kattegat	489	1477	266	2232
Total Baltic Sea	10963	22040	8216	41219

of phosphorus was on the order of 1–5%. The sources of waterborne input of nitrogen were 58% from agriculture and managed forestry, 32% from natural background sources and 10% from point sources, whereas the contributors of phosphorus input were 53% from agriculture and managed forestry, 27% from natural backgrounds and 20% from point sources (HELCOM 2005). The sources of nitrogen and phosphorus input for the Baltic Sea sub-regions are displayed in Tables 5.1 and 5.2.

These tables clearly show that the part of the Baltic Sea area with low population density and low agriculture and forestry activities supplies almost no nutrients, with concentrations thought to be near the natural background level. In some areas damming may even have reduced nutrient output by rivers to below the natural level. This has been claimed e.g. for silica (Humborg et al. 2000).

5.2.2 Regional Nutrient Trends

Bothnian Bay and Bothnian Sea

Concerning the hydrochemistry, the two sub-basins of the Baltic Sea, the Bothnian Bay and the Bothnian Sea, are quite different in terms of stoichiometry (Fonselius 1978). In the Bothnian Bay the concentrations of nitrate, nitrite and silicate are rather high, whereas phosphorus concentrations are low. This results in high DSi:DIN:DIP ratios in wintertime, exceeding the Redfield ratio for phytoplankton. In contrast, in the Bothnian Sea the Redfield ratios are "normal" for DIN:DIP but much higher for DSi:DIN, approximately 67:16:1. Long-term changes in concentrations and Redfield ratios are small within natural variations (HELCOM 2002). In the deep water of the Bothnian Bay in the 1980s, the annual mean phosphate concentrations decreased – parallel to the bottom salinity – towards 0.05 μ mol/l and have stabilised around this value since 1993. The nitrate and nitrite concentrations seem to have decreased from 1993 onwards. No trend in silicate concentration is apparent since 1990. In the Bothnian Sea annual mean phosphate concentration in deep water has increased by $0.2\,\mu\,mol/l$, whereas nitrate and nitrite concentrations tended to decrease in the period 1994 to 1998. For the deep water it is typical that phosphorus is out of phase with nitrate and nitrite changes, a pattern seen for the past 20 years (HELCOM 2002).

Archipelago Sea

In the Archipelago Sea the concentration of both N and P increased during the 1970s and 1980s (Pitkänen et al. 1987; Kirkkala et al. 1998) due to intensification of agriculture and fish farming, and eutrophication has proceeded during the 1990s, when winter concentrations of phosphate and total phosphorus slightly increased in the surface layer (Bonsdorff et al. 1997; Kirkkala et al. 1998). In contrast, the P concentration near larger towns has decreased due to the introduction of phosphorus stripping at municipal wastewater treatment plants.

Gulf of Finland

The River Neva is the dominant source of nitrogen. The Neva water spreads along the northern coast of the Gulf of Finland and follows the general circulation pattern as documented in Annex 1.1. This pattern is less distinct for the phosphate distribution. All nutrients show pronounced seasonal variations, with maximum concentrations in winter and minimum concentrations after the spring bloom. The spring bloom is nitrogen-limited and the excess DIP remaining after the spring bloom is available for the development of a summer bloom of e.g. diazotrophic cyanobacteria. The long-term trends in the surface layer show a decrease in DIP concentrations and an increase in DIN concentrations in the Gulf of Finland from the late 1970s to the early 1990s. The opposite trends appear in the mid-to-late 1990s, when an increase in DIP concentrations and a decrease in DIN concentrations have been observed. The same trends have been observed in the bottom layer. It is remarkable that the nutrient reserves in the bottom layers in the Gulf of Finland are much higher than in other basins of the Baltic Sea. Heiskanen et al. (2000) argued that the reason for the high accumulation of nutrients in the bottom layer is the intense production in the euphotic layer and its sedimentation.

Gulf of Riga

The Gulf of Riga is a large estuary which is strongly controlled by the inflow of the River Daugava. This results in the high nutrient concentrations in the southern part near the river mouth and in lower nutrient levels in the northwestern part where the Gulf is influenced by the water exchange with the Baltic Proper. From the mid 1970s to 1991 winter nitrate concentrations increased significantly, followed by a gradual decrease from 1992 onwards. Winter phosphate concentrations increased continuously from 1981–1998. Winter concentrations in silicate show no trend between 1988 and 1998, although an increase can be identified since 1995. The periods of rising and falling nitrate concentrations coincide with corresponding changes in river runoff. This indicates that the river input is the main driving factor of the long-term dynamics of the nitrogen pool in the Gulf of Riga (Yurkovskis 1998) and the corresponding biological response (Dippner and Ikauniece 2001).

Baltic Proper

Observations of nutrient distributions in the Baltic Proper go back to the mid 1930s (Fonselius and Valderrama 2003), for which sporadic observations exist. Systematic observations started in the middle of the 1950s, and the actual situation, trends and changes have been documented in numerous publications (e.g. Nehring 1984, 1989; Nausch et al. 1999; Fonselius and Valderrama 2003; Voss et al. 2005). The Baltic Proper exchanges water in the east with the Bothnian Sea, the Gulf of Finland and the Gulf of Riga and in the west with the Belt Sea. Major rivers influencing the nutrient distributions are the River Oder and the River Vistula.

In the period from 1986–1998 no apparent trend could be observed in the input of the River Oder for the nitrogen concentration, and a decrease in phosphate concentration occurred in the same period. In contrast, the inflow of the River Vistula shows an increase in both nitrate and phosphate concentrations for the period from 1979–1998. The situation in the Baltic Proper is quite different, especially in the central basin, the deep Gotland Sea.

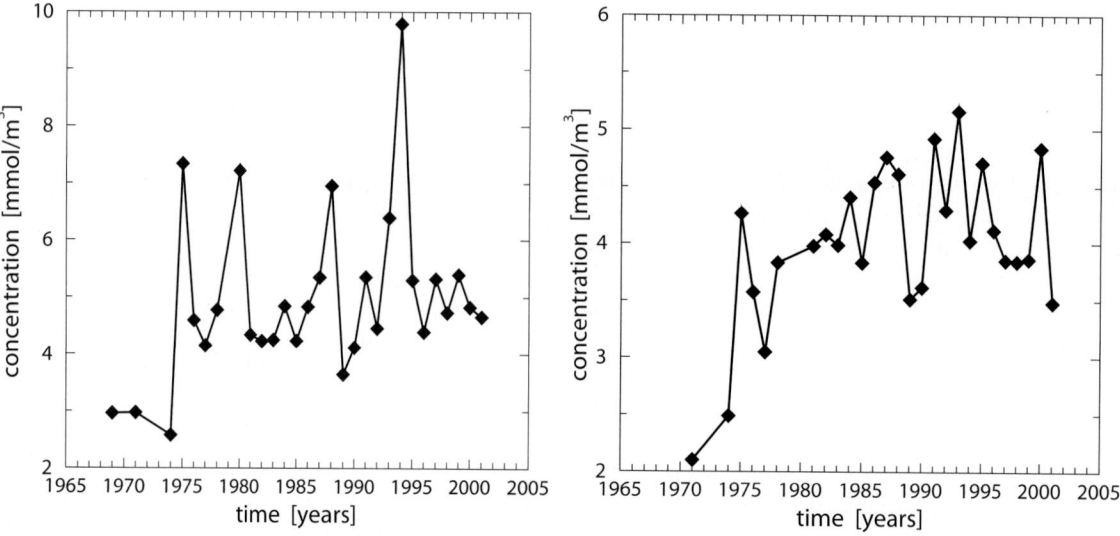

Fig. 5.1. Nitrate concentration in the upper 30 m for winter, January and February average from HELCOM monitoring data for all stations in the Baltic Proper (*left*), and for the stations in the Gotland Sea (*right*)

During the last 40 years in the Baltic Sea, the wintertime phosphorus levels have increased eight-fold and the nitrogen levels four-fold (Larsson et al. 1985; Elmgren 2001). Fonselius and Valderrama (2003) have shown that in the deep basins of the Baltic Proper phosphate, nitrate and ammonium concentrations are increasing at water depths of 80–400 m. Similarly to the Gulf of Finland, an accumulation occurred in the deeper part of the basins. The surface layer is quite different. In a recent paper, Voss et al. (2005) showed for the period 1969–2001 that no trend exists for nitrate concentration during wintertime in the Baltic Proper. Although an increase in nitrate concentration in the deep Baltic Proper and the River Vistula appears, no trend can be detected for the period 1975–2001 in the nitrate concentration in the upper 30 m (Fig. 5.1).

If only the Gotland Sea is considered, a regime shift appears around 1970 with a doubling in nitrate concentration, which can be attributed to a series of salt water intrusions from the North Sea (see Annex 1.1). In contrast, clear trends exist in phosphate concentrations in the Baltic Proper as well as in the Gotland Sea. A significant increase in phosphate concentration occurred in the Baltic Proper until 1983, followed thereafter by a decrease in concentration. The same holds for the Gotland Sea (Fig. 5.2). The open question is: why do the concentration patterns of nitrate and phosphate behave differently in the Baltic Proper and the Gulf of Finland?

Kattegat and Belt Sea

The Kattegat and the Belt Sea are the transition zones between the North Sea and the central basins of the Baltic Sea. Variations in inflow and outflow conditions are the major driving factor which influences inter-annual variability in nutrient concentrations. No long-term trend in nitrogen concentrations can be identified in the period 1980–1998. Extremely high winter values in 1994 an 1995 can be attributed to higher river runoff and the low winter values in 1996 and 1997 in the Kattegat and the Belt Sea can be ascribed to a reduced nutrient load in connection with the lower than normal precipitation during those years. A long-term decrease in phosphate concentration during winter is evident for the period 1980–1998, during which the averaged Kattegat concentration decreased by about ∼ 50% (HELCOM 2002).

Possible consequences of anthropogenic climate change

Anthropogenic climate change scenarios predict increasing temperature and mostly also increasing precipitation during winter for the Baltic Sea Basin (see Chap. 3). Higher than normal air and water temperature during winter have consequences if the water temperature stays above the temperature of maximum water density. In such a case in late winter, the normal deep con-

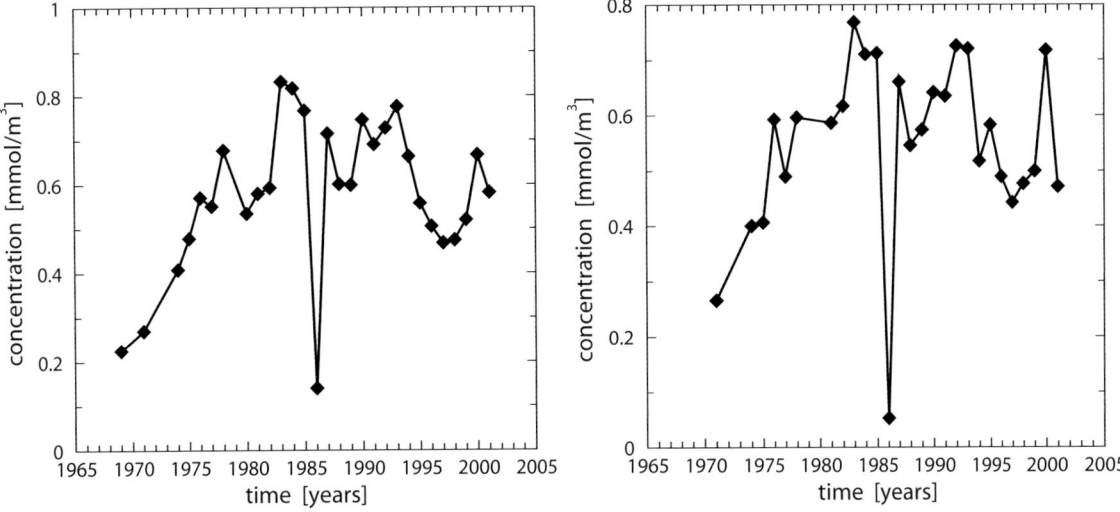

Fig. 5.2. Phosphate concentration in the upper 30 m for winter, January and February, average from HELCOM monitoring data for all stations in the Baltic Proper (*left*), and for the stations in the Gotland Sea (*right*)

vection in brackish water does not occur and fewer nutrients are transported into the euphotic zone. Analyses of a sediment core indicate that this happened during the Medieval Warm Period (Dippner and Voss 2004). An increase in precipitation will result in increasing nutrient input through river runoff and might cause strong eutrophication in the near coastal areas. Hence, an anthropogenic climate change as predicted will change the nutrient concentrations in the euphotic zone and the near coastal area of the Baltic Sea.

5.3 Contaminants (Chemical Pollution)

5.3.1 Chemical Pollution in the Baltic Sea: Past and Present Status

Over the past 50 years there have been substantial inputs of chemical substances into the Baltic Sea via direct discharges from land-based sources (e.g. industrial and municipal wastes), river runoff or draining, atmospheric deposition from local and more distant sources, or due to shipping. As a result, the Baltic Sea ecosystem has become contaminated with numerous substances, including many persistent organochlorines (e.g. DDT, PCB, dioxins) and heavy metals. Meanwhile, the amount of many hazardous substances discharged into the Baltic Sea has been reduced, mainly due to the effective implementation of environmental legislation (e.g. Helsinki Convention), their substitu-

tion by harmless or less hazardous substances, and technological improvements (Selin and VanDeveer 2004). On the other hand, new contaminants (such as pharmaceuticals, agrochemicals and PCB replacements) are being released and can be assumed to have a potential effect on biota. The residence time of chemical pollutants is high because of the persistence of many contaminants, the specific hydrographical conditions (salinity and oxygen gradients) in the Baltic Sea as well as remobilisation processes (HELCOM 2002, 2003b).

During the past 20–30 years, most studies and reports on hazardous substances in the Baltic Sea dealt with the occurrence of chemicals belonging to past industrial practice: the 'dirty dozen' priority pollutants which fall under the global ban of the Stockholm Convention because of their properties as persistent organic pollutants (POPs). Their occurrence has been assessed in several HELCOM reports, and the latest developments are presented in HELCOM (2002, 2003b).

A detailed review concerning "classic contaminants" and occurrence of "novel substances" is also given in Allsopp et al. (2001), UNEP (2002), Bignert et al. (2004) and WWF (2005). When discussing chemical pollutants in the Baltic Sea, their occurrence in the water body, sediment and biota in the different parts of the Baltic has to be considered.

Surface seawater concentrations of halogenated hydrocarbons (PCBs, DDTs, HCHs, HCB) are

Table 5.3. Concentrations of dissolved trace metals (ng/kg) from the North Atlantic and the Baltic Sea. [1]Kremling and Streu (2001), [2]Pohl et al. 1993, [3]Pohl and Hennings 1999, [4]Dalziel 1995 (after HELCOM 2003b)

Element	North Atlantic	Baltic Sea	Factor
Hg	0,15–0,3[4]	5–6[3]	∼ 20
Cd	4 (+ − 2)[1]	12–16[3]	∼ 4
Pb	7 (+ − 2)[1]	12–20[3]	∼ 3
Cu	75 (+ − 10)[2]	500–700[3]	∼ 10
Zn	10–75[1]	600–1,000[3]	∼ 10–50

generally relatively low, thus not allowing for estimation of any temporal trends (HELCOM 2003b); however, regional differences are obvious (Wodarg et al. 2004). Spatial differences are also notable for petroleum and other hydrocarbons (e.g. PAHs), with hot spots mainly in the south-western Baltic Sea. Heavy metal concentrations in Baltic Sea water have decreased in the 1980s (Kremling and Streu 2000), but since the mid 1990s their concentrations have been on the same level and are still clearly higher than in the North Atlantic (Table 5.3). Concentrations of cadmium, lead and zinc are on average higher in the south-western parts of the Baltic Sea, where atmospheric deposition of heavy metals is greater and waste containing high levels of heavy metals has previously been dumped (HELCOM 2003b).

Sediments are a further potential source of organochlorines. PCB measurements in dated sediment cores from the Baltic Proper showed increased or constant concentrations during the last decades (Jonsson 2000). Recently formed sediments from the Baltic Proper can contain 10 times more cadmium and mercury than the deeper layers, 3–5 times more lead and zinc, and about 2 times more copper. Studies of sediments from the Baltic Proper show metal concentrations rising during the 1950s and reaching a peak in the 1960s and 1970s.

Although metal concentrations have decreased since the 1980s, they are still significantly higher than in the 1940s (Borg and Jonsson 1996). It has been suggested that re-suspension of sedimented organic matter and associated pollutants may remain a substantial source in the Baltic Sea for a very long time (Wulff and Rahm 1993; Jonsson et al. 1996; Nilsson et al. 2003). Dioxins accumulate in sediments close to their main sources, such as

old pulp and paper mills, chemical plants including vinyl chloride or biocide manufacturing, and harbours. Information concerning regional distribution of dioxin in surface sediments is limited since data are not available from all parts of the Baltic Sea (HELCOM 2004).

Petroleum-derived hydrocarbons in the Baltic Sea environment have different sources and are composed of numerous single substances. The biologically most harmful group are the polycyclic aromatic hydrocarbons (PAHs). Their spatial distribution patterns suggest that their concentrations in the Baltic Sea are about three times those in the North Sea, and that atmospheric deposition is their main source. Highest PAH concentrations in the sediment have been found in the western and southern Baltic Sea (Lübeck Bight, Mecklenburg Bight, Arkona Basin and Gulf of Gdansk). In sediment cores from Mecklenburg Bight (south-western Baltic Sea), increasing PAH concentrations were detected between 6 and 10 cm (corresponding to the period between 1945 and 1965), reflecting a higher anthropogenic influence during that time (Dannenberger 1996; Kowalewska and Konat 1997; Milukaite and Gulbinskas 1997; Witt and Trost 1999). Another quantifiable PAH source are oil spills caused by accidents causing severe damage in form of mass stranding and mortality of oiled seabirds. They also may result in sublethal effects on various organisms as the Baltic Carrier accident in 2001 has shown (Storstrøms Amt, DK 2002). Oil pollution is expected to increase in the coming 10–20 years due to increasing trade in oil between the Baltic Sea states, increasing fleets, old vessels, inferior control, unqualified crews, and general difficulties in implementing international treaties (HELCOM 2003a, see also Annex 3.1.3.1).

5.3.2 Contaminant Loads and their Effects in Baltic Sea Organisms

In general, it can be stated that since the 1970s, levels of several organochlorines, notably DDTs, have declined in biota as a result of their use being banned or restricted (HELCOM 2002, 2003b). Polychlorinated biphenyls (PCBs) have also declined, though generally at a much slower rate, probably reflecting, in part, their continued release into the environment from old equipment and waste dumps. Levels of dioxins have also decreased but slowly. In recent years the levels of both dioxins and PCBs appear to be the same or have even increased in some species. An overview concerning multiple chemical body burdens in selected fish species, birds, mammals and invertebrates of the Baltic Sea is given in the following reports: Allsopp et al. (2001); HELCOM (2002, 2003b, 2004); WWF (2005).

Owing to the decrease in PCBs and DDTs, the fecundity of the white-tailed sea eagle population has increased since the beginning of the 1980s, and the proportion of successful reproducing pairs has increased to a level close to that prior to 1950 (HELCOM 2002). Shell thickness of guillemot eggs from Stora Karlsö in the central Baltic Proper has increased since the 1970s (Bignert et al. 1995, 2004) and has now returned to the level prior to 1940. In the 1960s the eggshells were much thinner, attributable to severe DDT pollution of the Baltic marine environment.

Baltic marine mammals, on the other hand, continue to exhibit reproductive disorders, indicating that the levels of hazardous substances such as PCBs and dioxin continue to cause ecological harm (Selin and VanDeveer 2004; HELCOM 2003b, 2004; Routti 2005). Ringed and grey seals from the Baltic have been suffering from pathological impairments, including reproductive disturbances, which have resulted in a depressed reproductive capacity (Nyman et al. 2003). Many female seals have been unable to produce pups due to uterine occlusion related to PCBs and dioxins in the environment (Reijnders 2003). Studies on seals fed with PCB-contaminated fish have implied alterations in the immune system induced by contaminants in the fish (Ross et al. 1995; de Swart et al. 1996).

When comparing important fish and shellfish species from the Baltic Sea it is obvious that tissue levels of typical organochlorines have decreased (Table 5.4). For heavy metals, such a clear common trend is not visible: although lead concentrations show significant declining trends, cadmium is still increasing in most of the indicator species and copper concentrations show no changes (Table 5.4). This may partly result from the fact that bioavailability of heavy metals is salinity dependent and that several other factors may have an influence on their distribution (Szefer 2002).

Regarding temporal and spatial trends in pollution loads, differences among species are obvious (Table 5.4). Baltic cod (*Gadus morhua*), for instance, often contains higher amounts compared to other fish species, due to it being predatory, and particularly high fat content, which tends to accumulate contaminants (Schneider et al. 2000; Falandysz 2003; Roots 2003). From a study performed in 2001 it is evident that cod physiology is affected by contaminants and regional differences in the biological responses are apparent probably because of differences in contamination level and patterns (Schnell et al. 2003). An overview of other studies on the effects of pollutants in different fish species and invertebrates is given in Lehtonen and Schiedek (2006) and Lehtonen et al. (2006).

In regard to dioxin, studies have revealed that some of the fish caught in the Baltic Sea, e.g. herring (*Clupea harengus* L.), exceed the new EU limits on concentrations of dioxin in food and livestock feed. The most contaminated herring were found in the Gulf of Bothnia, including herring in the Bothnian Sea (Fig. 5.3) and salmon (*Salmo salar* L.) in the Bothnian Bay (HELCOM 2004). Salmon at the Polish coast also showed elevated levels (Piskorska – Pliszczynska et al. 2004).

In recent years levels of some other substances such as TBT (tributyltin, an antifouling agent), PBDEs (polybrominated diphenyl ethers, primarily used as fire retardants) or synthetic musk compounds have increased (Alaee et al. 2003; HELCOM 2003b; Marsh et al. 2004; Luckenbach et al. 2004). Studies on fish (e.g. herring, flounder or pikeperch), sea birds, harbour porpoises and blue mussels document that TBT is still present in considerable concentrations in the Baltic Sea, despite restrictions in its use (Kannan and Falandysz 1997; Kannan et al. 1999; Strand et al. 2003). Imposex and intersex reproductive and gender disorders induced by TBT are for instance widespread in the Danish Straits and coastal areas (Strand and Jacobsen 2002; HELCOM 2003b). High butyltin concentrations have also been found in sediments and biota from marinas in the Gulf

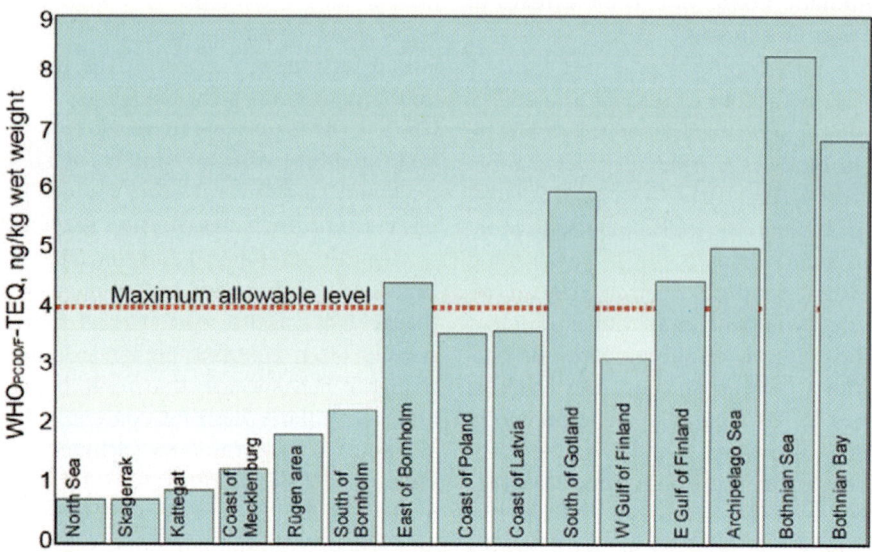

Fig. 5.3. The dioxin content in herring muscle for different fishing grounds. For Latvia the value represents the average of 2002/2003 data from the Gulf of Riga and Latvian coastal waters, the National Food Agency, Finland and the Food and Veterinary Service of Latvia (from HELCOM 2004)

of Gdansk (Falandysz et al. 2002; Albalat et al. 2002).

Residues of hydrocarbons (PAH) have been linked to liver abnormalities and other physiological irregularities in bottom-dwelling fish and other organisms (Bogovski et al. 1998; Ericson et al. 1998). While more information is becoming available about the effects of hydrocarbons and other oil related substances and their metabolites on biota, the full implications for the environment and human health are still not fully understood.

The data and examples presented in this overview clearly indicate that the present contaminant levels in different parts of the Baltic Sea still cause biological effects in various species, resulting in genotoxicity, diseases and reproductive disorders and in some cases in chronic stress, features that may strongly affect populations and communities.

5.3.3 Climate Change Related Implications

A systematic change in the hydrographic conditions due to climate change e.g. towards warmer temperatures and reduced salinity (Räisänen et al. 2004; Gustafsson 2004; Chap. 3) will have a direct impact on the distribution and acclimation capacity of native biota but also on the bioavailability and effects of contaminants. Changes in climate

variables will also alter transport, transfer, and deposition of contaminants.

In a recent report on risks and management strategies of dioxins in Baltic Sea fish (Assmuth and Jalonen 2005) it is stated that climate change may both increase and decrease risks. Climate change may have multi-directional effects on carbon cycling that is important as a carrier of dioxins, particularly in the northern part of the Baltic Sea. However, it was also pointed out that such changes are complex and their influence on dioxins are difficult to predict since they are multifactoral and vary e.g. by season and area. The authors see also some negative impacts concerning the cycling of dioxins in the sea (and sediments), e.g. through increased storms or altered ecosystem structures.

In general, the Baltic Sea appears to be particularly sensitive to persistent toxic substances because of its physical characteristics and the fact that many of the species are not originally adapted to this brackish water environment. Most of them live at the edge of their physiological tolerance range in terms of salinity and many of them are boreal (i.e. cold adapted) organisms (see also Sects. 5.7 and 5.8). Differences among species concerning their thermal tolerance limits and their capacities to adjust to these limits (Somero 2005) may determine how populations are affected by climate change.

Table 5.4: Temporal and spatial trends of contaminant loads (different heavy metals, DDT, PCBs) in different fish species and blue mussels in the Baltic Sea based on data from HELCOM (2002, 2003b)

	Cd	Cu	Pb	Σ DDT	PCB
Herring	**Increase**, 4–6% (annually) but significant decrease in some areas (Gulf of Finland, Bothnian Bay) Spatial distribution: Kattegat < Bothnian Sea	± over past 15 years, highest concentration, Bothnian Sea	**Decrease**, 3–6% (annually) Spatial distribution: Bothnian Bay < Baltic Proper < Kattegat	**Decrease** since 1980s, 9–14% Kattegat and Bothnian Sea; 5–11% Baltic Proper (annually) Spatial distribution: Bothnian Bay < southern Baltic	**Decrease**, 3–10% (annually) in most parts of the Baltic Sea Spatial distribution: Highest concentrations: southern Baltic (German/Polish coast)
Cod	**Decrease**, 10% (annually)	± over past 15 years	**Decrease**, 6% (annually) during past 20 years	**Decrease** since 1980s, 9–14% Kattegat and Bothnian Sea; 5–11% Baltic Proper (annually)	**Decrease**, 3–10% (annually), but over past 10 years decrease has ceased in the Baltic Proper. Influence of local sources
Perch	**Increase** 10% (annually) Baltic Proper	± over past 15 years	**Decrease** during past 20 years	**Decrease** since 1980s, 9–14% Kattegat and Bothnian Sea; 5–11% Baltic Proper (annually)	**Decrease** over past 15 years, Influence of local sources
Blue mussel	**Increase** 10% (annually) Baltic Proper; Spatial distribution: Kattegat < Polish coast, Baltic Proper	± over past 15 years,	Concentrations in the soft tissues are up to one order of magnitude higher than in fish	**Decrease** since 1980s, 9–14% Kattegat and Bothnian Sea; 5–11% Baltic Proper (annually) Spatial distribution: Kattegat < Gdansk Bay	**Decrease** over past 15 years, Spatial distribution: Lowest off Gotland, highest Gdansk Bay, Influence of local sources

The combination of higher temperature and lower salinity will clearly reduce the general fitness of marine species and may favour freshwater species and the invasion of non-native species. The presence of chemical pollutants is likely to add additional stress to biota, as studies from other estuaries have shown (Lanning et al. 2006). Baltic blue mussels clearly respond to changes in temperature or salinity, e.g. in terms of metabolic rates and enzyme activities (Pfeifer et al. 2005). The projected change in abiotic factors may also have an influence on those processes involved in the metabolism of toxic substances: higher temperatures result in increased turnover rates or generally higher metabolic rates. Higher temperatures and/or lower salinity will affect the species' ability to deal with toxic substances and the different physiological regulation processes involved in the detoxification of hazardous substances.

On the other hand, bioavailability of specific contaminants (e.g. metals) is greatly affected by salinity (McLusky et al. 1986; Depledge 1990). Numerous studies have shown an increasing metal uptake by diverse aquatic organisms at reduced salinities (Hall et al. 1995; Wright 1995; Lee et al. 1998). Thus, altered prevailing abiotic conditions will also regulate the extent of exposure of organisms to toxic substances in different areas of the Baltic Sea. The projected changes in the hydrographic conditions will also affect resuspension processes of sediment bound chemical pollutants.

Bearing in mind the complex hydrographic conditions in the Baltic Sea (Annex 1.1), chemical pollution has to be seen as a further factor acting upon the Baltic Sea. In the past years we only have started to understand how multiple stressors (e.g. salinity, temperature and chemical pollution) in combination may affect marine ecosystems and its biota. The different forecasts concerning future changes in key abiotic factors (temperature, salinity) due to climate change will be a new challenge. More experimental studies and modelling efforts are needed to test various scenarios concerning transport, transfer and cycling of chemical pollutants and to assess the counteracting effects on important Baltic species, the impact on their well-being and fitness and, presumably, also on whole populations.

5.4 Bacteria

Heterotrophic bacteria are a natural component of all aquatic ecosystems and play a substantial role in biogeochemical processes both in the pelagic and benthic environment. Indeed, key aspects of the carbon and nitrogen cycles such as anaerobic fermentation of organic carbon and fixation of atmospheric nitrogen are carried out solely by bacteria. A conspicuous feature of the Baltic ecosystem is the regular occurence of summer blooms of nitrogen-fixing cyanobacteria. Their role in the Baltic and their reaction to climate change are dealt with in the following section.

Pelagic bacteria, which occur in densities between 10^5 and 10^7 ml^{-1}, primarily rely on dissolved organic matter released from phytoplankton, but under oligotrophic conditions they are competing with phytoplankton for inorganic nutrients and constitute the basis of a microbial-based food web (Fenchel 1988). A significant part of "pelagic" bacteria may occur attached to particles, and aggregates of *Nodularia spumigena* can support 10^3 times higher concentrations than free-living bacteria (Hoppe 1981), probably due to leaking organic material (Paerl and Gallucci 1985). In the Baltic Sea the intra-annual variation in growth of pelagic bacteria was shown to be due to combined effects of temperature, quality of organic substrate and competition (Wikner and Hagström 1999), and on a smaller time scale diel variation in growth rate in the Bothnian Sea was explained by availability of phosphate (Hagström et al. 2001). Pelagic bacteria are grazed by nanoplanktonic flagellates, and recent studies suggest that viruses represent an additional control agent by infesting bacteria and causing cell lysis (Tuomi and Kuupo 1999). Comparative studies have shown a huge variation in the relative importance of grazing and viral control (Fisher and Velmirov 2002; Bettarell et al. 2003).

The biogeochemical cycle of inorganic nitrogen (the two main species being NO_3^- and NH_4^+) consists of a complex array of processes which are mediated primarily by bacteria. Organic nitrogen is mineralised to ammonia (ammonification), which in turn is oxidised to nitrate (nitrification) which may be denitrified to N_2 (Herbert 1999). The processes take place within sediments, even though denitrification has been demonstrated in the water column (i.e. at the oxic-anoxic interface) in the Baltic Sea (Rönner and Sörensen 1985; Brettar and Rheinheimer 1991). Besides denitrification, also anaerobic ammonium oxidation by bacteria using NO_2^- as oxidised substrate ("anammox") contributes to N_2 production and thus nitrogen excape from sediments and the water column (Dalsgaard et al. 2003; Kuypers et al. 2003; Engström et al. 2005). The anammox process seemingly is ubiquitous in the marine environment, but the importance to nitrogen loss does vary greatly (Dalsgaard et al. 2005). In the Baltic, anammox was measured in sediments at Tväminne in spring and autumn, but the process was not found to contribute significantly to nitrogen reduction (Hietanen and Kuparinen 2005).

Under anoxic conditions denitrification and anammox can be limited by nitrate and nitrite availability, and other bacteria (e.g. *Desulfovibrio*) use $SO_4{}^{2-}$ as electron acceptor to utilise organic compounds (Jørgensen 1977). The product of this respiration is the highly soluble and toxic H_2S. Sulfate reduction occurs widely in anaerobic sediments in the Baltic Sea (Mudryk et al. 2000; Boeschker et al. 2001). Various aerobic or anaerobic autotrophic bacteria can, in turn, utilise H_2S as an energy source, and under oxic conditions H_2S is oxidised to sulphate. An overview of biogeochemical processes including intermediate reactions (reduction of iron and manganese) is given by Jørgensen (1996). The suite of bacteria-mediated processes is to a large extent controlled by availability of oxygen, which, in turn, is determined by the balance between oxygen demand and physical transport processes (e.g. major Baltic saltwater inflows) and vertical mixing that deliver oxygen to the bottom water.

Oxygen demand below the pycnocline (including sediments) is overall regulated by the magnitude of sedimentation of organic particles that originates from primary production. Hence, changes in pelagic primary production, e.g. due to variation in light or nutrient availability, invariably will result in changes in sedimentation rate and thus oxygen demand below the pycnocline.

Effects of temperature increase

Increase in temperature stimulates the metabolic processes of bacteria, provided that other factors such as substrate or oxygen are not limiting (Autio 1998). As bacteria play prominent roles in aquatic ecosystems, small changes in temperature can have disproportionately larger implications in the functioning of marine systems and, ultimately, also global consequences for the carbon cycle (Sarmiento et al. 1998).

Several studies have demonstrated that the activity of pelagic bacteria (and other pelagic heterotrophs) is stimulated more than primary production under temperature increase due to differences in the temperature dependence of photosynthesis and respiration (e.g. Pomeroy and Deibel 1986). This finding is in accordance with observations that the ratio of bacteria biomass to phytoplankton biomass increases with temperature in eutrophic aquatic systems (Laws 2003). In oligotrophic systems (i.e. primary production below $5\,\mathrm{mg\ C\ m^{-3}\ d^{-1}}$) temperature had no influence on the ratio, which may be attributed to substrate limitation (Pomeroy and Wiebe 2001). A higher activity of the heterotrophic communities will increase recycling and mineralisation in surface waters, and, in turn, sedimentation of organic matter may decrease as projected for the open oceans (Laws et al. 2000). Likewise, processes below the pycnocline such as denitrification, and probably anammox in anoxic waters, will be stimulated by temperature increase provided that oxidising agents are not limiting. Other climate-related factors leading to changes in nutrient runoff, vertical mixing, and insolation may confound direct temperature effects (see Chap. 3). Hence, integrated effects cannot be predicted without advanced modelling efforts.

Like their pelagic counterparts, the specific activity (per cell or biomass) of both aerobic and anaerobic benthic bacteria increases with temperature, but due to (seasonal) temperature adaptation effects the change cannot be predicted from short term experiments, e.g. in temperature gradients (Thamdrup et al. 1998; Arnosti et al. 1998). Overall, seasonal changes in oxygen uptake often correlate with temperature (Banta et al. 1995). However, other important controls on benthic oxygen uptake and mineralisation include the input of organic matter from primary production, oxygen concentration in bottom water (which primarily is controlled by physical processes in the Baltic Sea) and irrigating activity of the infauna (Rasmussen and Jørgensen 1992). Hence, direct effects on bacterial activity of a small temperature increase in bottom water probably will be of minor importance compared to changes in sedimentation of organic matter and, especially, physical transport and mixing processes.

5.5 Phytoplankton

Phytoplankton, i.e. the autotrophs suspended in the pelagial, constitute the base of the pelagic food web but also feed benthic areas. Changes in phytoplankton composition and biomass are, therefore, of decisive importance for the whole ecosystem. The detection of these changes is not simple because phytoplankton is subject to high natural variability in time and space (PEX 1989). If the typical spring, summer and autumn blooms are not adequately sampled, no reliable seasonal or annual means can be calculated. Therefore, trend analyses are possible only on the basis of data of sufficient spatial and temporal coverage. The monitoring programme of HELCOM was especially designed for this trend monitoring.

When HELCOM was established in 1974, eutrophication was identified as the main threat for the Baltic Sea (Larsson et al. 1985). Changes in phytoplankton biomass and species composition do, however, not only reflect effects of eutrophication but also of climatic changes. These effects cannot easily be separated. The most important physical and chemical factors influenced by climatic change and affecting phytoplankton growth are discussed.

5.5.1 Physical and Chemical Factors

Nutrient availability is one of the most important factors for phytoplankton growth. An increase in the concentrations of the growth-limiting nutrients (N, P) leads to increased phytoplankton biomass. The heaviest nutrient input into the Baltic Sea occurred in the 1970s, mainly by wastewaters and fertilisers. This caused a stronger phytoplankton growth, which is also reflected in an increase in chlorophyll *a* concentrations (Wasmund and Uhlig 2003) and a reduction of the water transparency (Secchi depth, cf. Sanden and Hakansson 1996). The nutrient conditions are, however, not only dependent on anthropogenic impact, but also on climatic influences, i.e. rainfall and river runoff. Nausch et al. (1999) found a correlation between freshwater runoff and nitrate plus nitrite concentrations in the surface water of the Kattegat and Belt Sea area.

Also, internal processes like re-mineralisation and re-solution, influence the availability of nutrients for phytoplankton growth. Anoxic bottom water in the deep basins is a typical feature of the Baltic Sea. Under these conditions, phosphorus is liberated from the sediment (cf. Nürnberg 1984). Kuparinen and Tuominen (2001) found a negative correlation between mean oxygen and phosphate as well as silicate concentrations within the hypoxic water volume, indicating an anoxic release of phosphorus and silicate. Hypoxic or anoxic condi-

tions allow the reduction of nitrate to ammonium (Kuparinen and Tuominen 2001) and oxidation of ammonium with nitrite (anammox). Sporadic inflows of saline North Sea water (see Annex 1.1) replenish the oxygen in the deep water layers of the Baltic Sea. The re-oxygenation fixes the phosphorus in the sediment by adsorption onto $Fe(OH)_3$ and therefore counteracts eutrophication (Conley et al. 2002).

After a salt water inflow, ammonium is oxidized to nitrate and thus "bursts" of high NO_3 levels appear in deep waters (Kuparinen and Tuominen 2001). Convective and diffusive processes transport the nutrients into the upper water layers where they promote phytoplankton growth. The re-solution of phosphorus during stagnation periods may be more important for the year-to-year variation in the phosphorus pool than the external phosphorus load by eutrophication (Conley et al. 2002).

Carbon dioxide is the main source material for photosynthesis. In the water, it exists as dissolved CO_2, as carbonic acid and as the ions HCO_3^- and CO_3^{2-}. The concentration of free CO_2 decreases with increasing pH. HCO_3^-, which is the dominant total carbon species in moderately alkaline water, can also be used by some algae (Crotty et al. 1994; Tortell et al. 1997). Wasmund (1996) showed in culture experiments that *Nodularia harveyana* grew well at pH 7.5–9.0, whereas the chlorophyceae *Tetrastrum* sp. was retarded at pH 8.5–9.0. Shapiro (1990) discussed the dependency of the cyanobacteria on high pH and/or low CO_2 concentrations comprehensively.

Light is the decisive condition for photosynthesis. Low light intensities limit algal growth, for example during the night, in winter and in greater water depths. If light intensity is very high, e.g. at noontime and near the water surface, it depresses photosynthesis by destruction of the photosynthetic apparatus. Long-term changes in mean light intensities are much lower than diurnal, seasonal or depth-dependent variation and, therefore, of lower influence on biomass trends. Russak (1994) reported on decreasing trends of radiation in the Baltic Sea Basin from 1955 to 1992 (Sect. 2.1.4.2). Also, the decreased water transparency (Sanden and Hakansson 1996) reduced the light availability for the algae. Nevertheless, algal biomass shows an increasing trend, because eutrophication is over-compensating the reduction in light intensity.

Temperature has only little direct effect on algal growth. Phytoplankton grows to high biomasses already at temperatures a little above the water melting point but is inhibited if the temperature exceeds the natural range. Of course, the individual species have their typical temperature preferences (Hegseth and Sakshaug 1983; Wasmund 1994). Therefore, significant global warming may change the species composition.

The indirect effect of temperature via its influence on the water stratification is also of importance. Some algae prefer a stable stratification and others a mixed water column. Deep convective mixing in autumn and spring is of decisive importance for nutrient circulation. It is only possible if the water temperature falls during cold winters below the temperature of maximum density. In mild winters, a complete convective circulation does not occur and the warming of the water in spring leads immediately to a stratification without preceding spring overturn. This is disadvantageous for species which prefer turbulent water (Wasmund et al. 1998).

The Medieval Warm Period (1000–1250 AD) and the modern warm period (since 1850) are correlated with laminated sediments rich in organic carbon in the deep Baltic Sea basins, indicating anoxic conditions caused by a stable halocline and reduced vertical mixing (Leipe et al. 2005). During a cold period, including the "Little Ice Age" (1250–1850), oxic conditions dominated due to lower salinity and vertical mixing, leading to light grey and homogeneous sediment horizons. The oxygenation of the sediments affects the nutrient concentrations as stated above.

Wind causes turbulent mixing that may have a stronger impact on the stratification of the water column than convective mixing. Therefore, since 1989, the reduced convective mixing after "mild" winters may be replaced by increased turbulent mixing. In very cold winters, turbulent mixing is reduced by the ice cover. After the cold period (approximately 1250–1850, according to Leipe et al. 2005), the area covered by ice tends to decrease (FIMR 2003).

5.5.2 Phytoplankton Trends

The above-mentioned factors are the main abiotic factors that influence phytoplankton biomass and composition. They may also have indirect impact via selective stimulation of grazers or competitors. The coincidences of different phytoplankton groups with abiotic factors are compiled in Table 5.5. Of course, the taxonomic groups are not ecological groups and their members may react in different ways. But generalisation on the class level proved to be appropriate and useful. Therefore, the following analyses on trends in phytoplankton composition refer primarily to phytoplankton groups instead of species.

5.5.2.1 *Diatom Spring Bloom*

For historical examinations, the siliceous microfossil stratigraphy in sediments may be used. This may reflect changes in salinity and temperature. The transition from the freshwater Ancylus Lake to the brackish Litorina stage of the Baltic Sea (ca. 7,000–7,500 BP, see also Annex 2) was indicated by a shift from freshwater to brackish diatom species (Bianchi et al. 2000). A change towards a colder climate occurred 700–850 years ago, indicated by a change in abundance from *Pseudosolenia calcar-avis* to *Thalassiosira hyperborea* var. *lacunosa* and *Thalassiosira baltica* (Andrén et al. 2000). This correlates with the onset of the cold period in the 13th century. After a period of warming (1850–1950), diatom composition changed in the colder years 1950–1960: Andrén et al. (2000) found an increase in *Thalassiosira* cf. *levanderi*, *Cyclotella choctawhatcheeana*, *Pauliella taeniata* and *Coscinodiscus granii*, but an extinction of *Pseudosolenia calcar-avis* and *T. hyperborea* var. *lacunosa* in sediment cores of the Gotland Basin. The spring-bloom species *Thalassiosira levanderi* and *Pauliella taeniata* prefer cold water. At the present stage, eutrophication plays an important role especially for phytoplankton growth whereas benthic diatoms decrease due to higher turbidity. Recent monitoring data reveal that spring blooms are dominated by *Skeletonema costatum* and *Thalassiosira baltica* after mild winters or by *Pauliella taeniata* and *Chaetoceros* species after cold winters (Hajdu et al. 1997).

A period of mild winters commenced in 1988, as shown by increasing mean surface water temperatures. Since 1988/89, the typical diatoms disappeared from the spring bloom in the Baltic Proper (Fig. 5.4). Of course, the phytoplankton data show a strong fluctuation due to undersampling (e.g. complete miss of the bloom in 1983). However, the silicate consumption during the spring season (Fig. 5.4) indicates that a diatom bloom regularly occurred until 1988 and failed to appear in the 1990s, except for the cold winter in 1996.

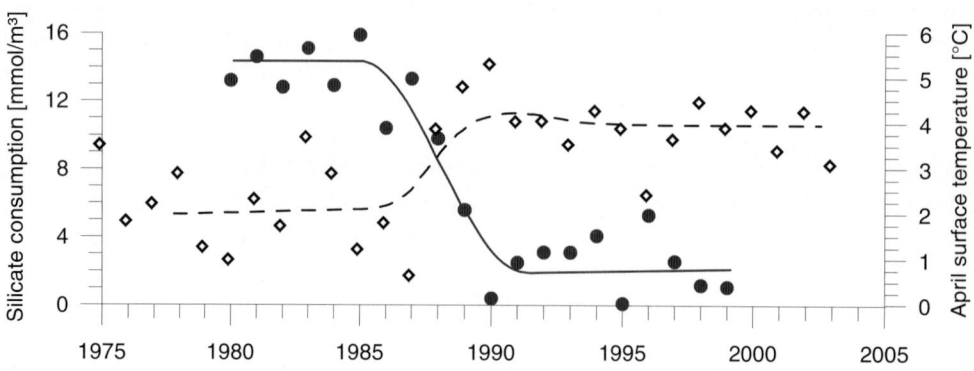

Fig. 5.4. Silicate consumption by the spring bloom and spring surface temperature in the Bornholm Basin (temperature = *diamonds and dashed line*; silicate consumption = *dots and solid line*)

Table 5.5. Generalised preferences of the main taxonomic groups, based on a review by Wasmund (1994).
*High for marine species and low for freshwater species

Factor:	Diazotrophic cyanobacteria	Diatoms	Dinoflagellates	Cryptophyceae	Chlorophyceae
N:P	low	high	?	?	low
Si:P	low	high	low	low	low
pH	high	low	?	low	low
Light intensity	high	low	low	low	high
Temperature	high	low	low	low	high
Stratification	high	low	high	high	low
Salinity	low	low and high*	high	low	low

Wasmund et al. (1998) hypothesised that the lacking convective mixing after mild winters prevents resuspension of diatom spores or floating of the vegetative cells. In contrast to diatoms, dinoflagellates prefer stable water columns (Table 5.5). Indeed, their share in the spring bloom has increased since 1989 (Fig. 5.5). The warming may not only change the species composition but also the timing of the spring bloom, as shown for the North Sea by Wiltshire and Manly (2004).

5.5.2.2 *Cyanobacteria Summer Bloom*

The buoyant surface blooms of diazotrophic cyanobacteria (*Nodularia spumigena*, *Aphanizomenon* sp., *Anabaena* spp.) are the most impressive bloom phenomena in the Baltic Proper. They are of special interest because they are potentially toxic (Wasmund 2002) and enhance eutrophication by their ability to fix nitrogen (Stal et al. 1999; Wasmund et al. 2001). Unlike other

blooms, they are not primarily an indicator of eutrophication, because they grow independently of 'bound' nitrogen (NH_4^+) and (NO_3^-), which is considered the limiting nutrient in the open Baltic Proper (Hübel and Hübel 1980). They are primarily limited by phosphorus or iron (Stal et al. 1999).

Bianchi et al. (2000) and Poutanen and Nikkilä (2001) concluded from pigment analyses in sediment cores that cyanobacteria blooms are nearly as old as the present brackish water phase of the Baltic Sea (7,000–7,500 years BP) and therefore not a modern but a characteristic, natural feature of the brackish Baltic Sea. The inflow of P-rich saltwater into the freshwater Ancylus Lake caused the development of a halocline, limited vertical water circulation and oxygen depletion in bottom water (Bianchi et al. 2000). Increased P release from the anoxic sediments led to N limitation which provided the nitrogen-fixing cyanobacteria with a competitive advantage.

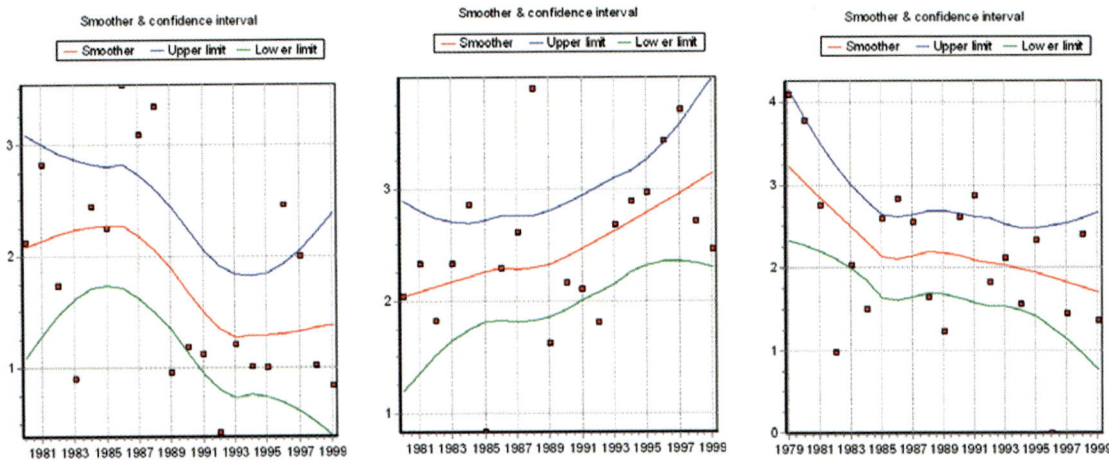

Fig. 5.5. Summary plot of significant trends 1979/1980–1999 in mean spring biomass in the Eastern Gotland Sea (station BMP J1) for diatoms (**a**) and dinoflagellates (**b**) and for summer biomass in the Bornholm Sea (station BMP K2) for cyanobacteria (**c**), exemplified by the LOESS smoother (Cleveland and Devlin 1988). Upper and lower lines represent the approximate pointwise 95% confidence limits for the trend line. Y-axis represents ^{10}log Biomass (in $mg\,m^{-3}$) (from Wasmund and Uhlig 2003)

During the Little Ice Age nitrogen-fixing cyanobacteria might have been depressed. The warming since 1850 has promoted cyanobacteria blooms not only by its direct effect but, again, by stronger stratification, formation of a permanent halocline, oxygen deficit in the bottom water and resulting liberation of phosphorus (Leipe et al. 2005). The same mechanism functioned also in recent stagnation periods.

Cyanobacteria blooms have been reported already in the mid of the 19th century (Lindström 1855; Pouchet and de Guerne 1885), as stated by Finni et al. (2001). In the 1920s, several *Aphanizomenon* blooms were recorded. The first record of an open-sea bloom in which *Nodularia* was identified as dominant is that of Rothe (1941) from July/August 1938 in the southern Baltic Sea. According to Poutanen and Nikkilä (2001), a strong increase occurred in the early 1960s, especially owing to an increase in *Nodularia spumigena* in the northern Baltic Proper.

Hübel and Hübel (1980) noted that *Nodularia* blooms have occurred regularly in coastal waters of the southern Baltic Sea since 1969, when a period of warm summers began. Horstmann (1975) reported on increased mass occurrences of *Aphanizomenon* and *Nodularia* in the Baltic Proper. It was just in this period, from the late 1960s to the early 1980s, that the strongest increase in

phosphorus and nitrate concentrations in the surface water was recorded (see Sect. 5.2). When the Baltic Monitoring Programme of HELCOM started in 1979, the cyanobacteria blooms had already reached a very high intensity (HELCOM 1996).

No further significant increase in cyanobacterial summer biomass could be found in the central Baltic Sea (Wasmund and Uhlig 2003), in contrast to the Gulf of Finland (Kahru et al. 2000; Suikkanen et al. 2007) and the Gulf of Riga (Balode 1999). Struck et al. (2004) found increased nitrogen fixation from 1969–73 to 1989–93, indicating an increase in the importance of diazotrophic cyanobacteria in this eutrophication period. On the other hand, they reported a strong decrease in potential summer nitrogen fixation from the periods 1989–93 to 1994–98.

From interpretations of satellite images, Kahru (1997) found large areas covered by cyanobacteria blooms in 1982–1984 and 1991–1994. Satellite images have the advantage of a synoptic view on large areas but the disadvantage is that they overrate surface accumulations. Depth-integrated samples are less patchy than the eddy-like surface structures suggest. After some wind, these buoyant surface blooms disappear by suspension in the upper mixed layer. Combined satellite and ground truth data suggest that no general increase in cyanobacteria blooms occurred in the open Baltic Proper, but they may have increased in some Gulfs during the last 25 years.

Climatic factors may have a higher impact on cyanobacteria bloom variability than eutrophication. Wasmund (1997) found that strong blooms were formed only if water temperature exceeded 16 °C, provided that global radiation was $> 120\,\mathrm{W\,m^{-2}}$ (daily mean) and wind speed was $< 6\,\mathrm{m\,s^{-1}}$. According to Stal et al. (2003), daily integrals of photon irradiance, rather than temperature, determined the onset of bloom formation. The summers of 1985–1988 had a very low average sunshine duration and therefore no blooms developed, whereas sunny summers promoted cyanobacteria blooms (Kahru et al. 1994). However, Finni et al. (2001) stated that hot summers are not the only reason for high cyanobacteria abundances. Storms may terminate *Nodularia* blooms, whereas Aphanizomenon is affected less (Stal et al. 2003). Therefore, wind conditions may have high relevance for the ratio of *Nodularia* (toxic) to *Aphanizomenon* (non-toxic) in the blooms. Computer simulations of cyanobacteria blooms by Janssen et al. (2004) revealed good agreement with the satellite data of Kahru et al. (1994). The blooms were mainly determined by the excess inorganic phosphorus in the surface layer, which depends on the depth of the mixed layer in winter and on the persistence of coastal upwelling. This, in turn, depends on the winter NAO. High NAO leads to high wind stress, low ice cover, deep mixing, upwelling, high excess inorganic phosphorus and, finally, to a high potential for a cyanobacteria bloom. In contrast, Kuparinen and Tuominen (2001) claimed that stormy winters helped to bind phosphorus to particles and, finally, to the sediments. Salt water inflows give rise to higher overall nitrate levels but lower phosphate levels (Kuparinen and Tuominen 2001), thus inhibiting cyanobacteria growth.

5.5.2.3 *Other Phytoplankton*

Besides diatom and cyanobacteria blooms, there are other components in the phytoplankton which show long-term changes. Nehring (1998) reported 16 non-indigenous phytoplankton species, which have become an integrated part of the pelagic system of the North Sea since the beginning of the 20th century. Among them, 10 thermophilic species were first recorded in the recent period of relatively mild winters. Some of these species

also invaded the Kattegat and the Belt Sea, like *Gyrodinium* cf. *aureolum* (Partensky and Sournia 1986), *Alexandrium minutum* (Nehring 1994) and *Gymnodinium catenatum* (Nehring 1997). The potentially toxic dinoflagellate *Prorocentrum minimum* invaded the Baltic Sea from the Oslofjord in 1979, became established in the southern Baltic Proper in 1982 and formed its first bloom in Kiel Bight in 1983 (Theede 1984). It tolerates a wide salinity and temperature range (Hajdu et al. 2005) and is obviously not directly related to global warming.

5.5.2.4 *Reactions to Future Anthropogenic Climate Change*

Future climatic change will influence several environmental factors which have different impacts on phytoplankton productivity and sometimes may compensate each other. At present, nutrient and light limit phytoplankton growth, but if their availability increases, also CO_2 can become a limiting factor.

The anthropogenic rise in CO_2 may increase primary production, especially in nutrient-rich waters. The acidification of the water due to air pollution and acid rain will enhance the CO_2 availability because of the pH-dependent equilibrium between the different carbon species. The intensified primary production will increase the pH. If the effect of external acidification overcompensates the effect of increased primary production, a decrease in pH in the water will result. As cyanobacteria lose competitiveness in acid water, other algal groups like diatoms may benefit. The influence of acidification on other chemical and biological elements like heavy metal mobilisation or viral activity cannot be assessed here.

Warming will directly inhibit cold-water species (mostly spring-bloom diatoms such as *Thalassiosira levanderi*, *Pauliella taeniata* and *Chaetoceros* spp.) but enhance warm-water species occurrence, like that of the bloom-forming toxic cyanobacteria (*Nodularia spumigena*). New species originating from warmer seas may establish themselves and displace native species. Reduced ice cover and earlier stabilisation of the water column in spring will shift the start of the spring bloom, which will influence the food-supply for zooplankton and the whole food-web.

Both higher temperature and higher freshwater inflow strengthen stratification. If the upper mixed layer becomes shallower, the phytoplankton receives more light and productivity will increase (Doney 2006). On the other hand, stronger vertical stratification of the water column will reduce vertical transport processes (e.g. exchange of nutrients and dissolved gases, upwards transport of cysts/spores, sedimentation, vertical migration of plankton), which may reduce productivity. In any case, cyanobacteria blooms may benefit from this situation.

Although no robust result concerning wind speed exists (see Chap. 3), it can be stated in general that increasing wind speed would counteract these processes. Reduced convective mixing due to warming might be compensated by increased turbulent mixing due to increased wind and reduced ice cover. This may help diatoms but disturb surface accumulations of cyanobacteria. The promotion of the diatoms is supported if the upwelling deep water is rich in CO_2 and low in pH, as explained above.

Stronger freshwater inflow into the northern Baltic Sea and reduced salt water inflow from the North Sea would displace the permanent halocline to greater depths. This would result in a larger area of oxygenated sediments (at 70–100 m depth in the central Baltic Sea Basins), expanding not only the colonisation area for macrofauna (Gerlach 1994) but also the area of the vital seedbeds for phytoplankton. Full circulation of the water column in winter would not only be possible in coastal areas and the Gulf of Riga but also at the slopes of the basins. Phosphorus and silicate would be bound in these oxic sediments, whereas nitrate would be liberated. This might lead to increased nitrogen:silicate and nitrogen:phosphate ratios, which reduce diatoms and nitrogen-fixing cyanobacteria.

Continuing anthropogenic eutrophication will import mainly nitrogen and phosphorus and therefore also raise the nitrogen:silicate ratio, which is disadvantageous for diatoms but beneficial for dinoflagellates. Of course, reduced salinity also has a direct influence on the species composition. Freshwater species will further penetrate into the Baltic Sea. A retreat of polyhalobic species from the southern Baltic Sea might, however, not occur because reduced precipitation would prevent a salinity decrease in this area.

Basically, phytoplankton biomass and species composition are influenced by different mechanisms. Their reaction depends on the overwhelming climatic impact factor. This can be temporally and spatially different. If ongoing eutrophication is enhanced e.g. by stronger weathering and elution of nutrients by precipitation or by increased anoxic conditions in deep water, a higher phytoplankton biomass and frequent blooms may be expected. As some of the blooms are toxic, they have strong impact on fisheries and fish farming. Inherently, blooms decrease the recreational value of the sea and therefore affect tourism (see also Annex 3.1.3.2), which is of high economic value (Wasmund 2002). Changes in the timing of the blooms and in species composition disturb the existing food web, provoking changes in the higher trophic levels.

5.6 Zooplankton

We review studies that combine mesozooplankton, long-term observations, and climatic environmental effects. Relationships between Baltic Sea zooplankton and its physical and chemical environment have been extensively studied (for an early review see, e.g. Ackefors 1981), also using long-term monitoring observations (e.g. Lumberg 1976; Simm 1976; Sidrevits 1980; Kostrichina 1977, 1984; Kostrichina and Sidrevits 1977; Kostrichina and Yurkovskis 1982; Lumberg and Ojaveer 1997), but papers referring to climatic environmental factors only started to appear in 1990s; a single exception by Segerstråle (1969) is discussed later in this section. Periodic assessments by HELCOM (e.g. HELCOM 2002) include chapters on zooplankton temporal change and its environmental causes in various parts of the Baltic Sea, but no effort has been done to combine long-term observations of zooplankton with climatic effects. Pershing et al. (2004) reviewed the studies on climate variability and zooplankton in the North Atlantic. They stated that, following the initial pulse of climatic factors, several interacting mechanisms, directly and indirectly, cause variation of zooplankton populations. In the Baltic Sea, salinity (Ranta and Vuorinen 1987), eutrophication (Vuorinen and Ranta 1988), temperature (Viitasalo et al. 1995), planktivory by pelagic fish (Rudstam et al. 1994) and a non-indigenous alien planktonic invertebrate (Ojaveer et al. 2004), have been considered as possible controllers of long-term changes in abundance. Currently, climatic factors have also been hypothesised to control zooplankton indirectly (Hänninen et al. 2000, 2003; Möllmann et al. 2003b, 2005; Vuorinen et al. 2003, 2004). Surprisingly, there are no long term studies of zooplankton and phytoplankton in relation to each other. We discuss in the following these environmental factors in relation to projected anthropogenic climate changes in the Baltic Sea.

Segerstråle (1969) demonstrated that neritic copepod species (*Acartia bifilosa, Temora longicornis, Centropages hamatus* and *Pseudocalanus acuspes*, see Fig. 5.15) enlarged their distribution area towards the north following a large influx of saline North Sea water into the Baltic Straits in the winter of 1951. Estonian researchers were the first to report a time series study of zooplankton showing a positive response of *Pseudocalanus acuspes* abundance and herring growth to an increase in salinity (and a mirror effect on freshwater runoff) (Lumberg and Ojaveer 1991). They concluded, although without any statistical analysis, that alteration of climatic periods is reflected, via changes in the basic environmental conditions, in the whole pelagic food chain and ecosystem in the Gulf of Finland. Viitasalo et al. (1995) confirmed that conclusion for the northern side of the Gulf of Finland using a proper statistical approach. Furthermore, Viitasalo et al. (1995) divided the plankton community into two arbitrary groups, which are used in this section in order to make projections of predicted climate change.

Cladocerans, smaller copepods, and rotifers (characterised by Viitasalo et al. 1995 as "surface community") are representatives of the "microbial loop", which is often presented as a functional alternative to the "grazing chain", whose representatives in the Baltic Sea are the large neritic copepods (see Annex 3.1.1), which in the Baltic Sea are often submergent. Their relationships have been illustrated e.g. by stating that the microbial loop is rapidly regenerating nutrients in the stratified surface layer with cyanobacteria important as primary producers, and the food chains end up in jellyfish, while in the grazing chain the copepods are moving nutrients from the primary producers, diatoms, in a non-stratified environment towards pelagic fish and other top predators (for a recent study, and more references, see Molinero et al. 2005).

Hänninen et al. (2000, 2003) and Vuorinen et al. (2003, 2004) used Transfer Functions (TF) to model a chain of events from the North Atlantic Oscillation (NAO) to changes in freshwater runoff and salinity and, eventually, to several species of mesozooplankton in the whole Baltic Proper. With that exercise they demonstrated a clear cause and effect relationship between climatic factors, changes in the Baltic Sea hydrography and, finally, a biological outcome in the mesozooplankton. The effect of climate through salinity and temperature is straightforward, but the relative importance of these and contributing factors is complicated by different adaptations of zooplankton species. Complications are further enhanced by the capacity of different species and developmental stages for daily, seasonal and ontogenetical migration between the surface and deeper water. Some rough ideas can, however, be presented on the basis of our knowledge about relationships between zooplankton and its environment.

Due to a decrease in salinity, distribution of mesozooplankton will change, not only horizontally, but also vertically. Marine species in the Baltic Sea are more common near to the Danish straits (see Chap. 1, Fig. 1.12), but they also exhibit brackish water submergence (the increase of the salt concentration with depth causes a corresponding increase of the marine element of the fauna and flora downwards, Segerstråle 1957). Vertical differences between species and developmental stages will modify the effects of salinity and temperature (Lassig and Niemi 1978; Burris 1980), therefore it is important to consider their effect on the species level (as discussed by e.g. Hansen et al. 2006; Renz and Hirche 2006 for *Pseudocalanus*). Interestingly, the species in question has recently been shown (Renz and Hirche 2006), to be a glacial relict, *Pseudocalanus acuspes*, and not *P. elongatus* as named before. Thus, even the late quartenary development of the Baltic Sea may modify the expected outcome of climate change in the Baltic Sea (see Annex 2, Chap. 1).

Among projections for climate change is an increase of temperature (see Chap. 3). Temperature changes are expected to immediately affect growth and reproduction of mesozooplankton (e.g. Pershing et al. 2004). Thus, the expected effects of increased temperature are likely to manifest themselves both in wintertime survival and summertime growth and reproduction of zooplankton (Viitasalo et al. 1995; Möllmann et al. 2000, 2005; Dippner et al. 2001). Temperature is of more importance for the "surface community" of smaller crustaceans, rotifers and cladocerans (Viitasalo et al. 1995; Möllmann et al. 2000, 2003, 2005), as compared to neritic and submergent *Pseudocalanus acuspes* (and possibly *Temora*) which are affected more by salinity (Hansen et al. 2006).

Viitasalo et al. (1995) were able to statistically combine long-term changes of mesozooplankton

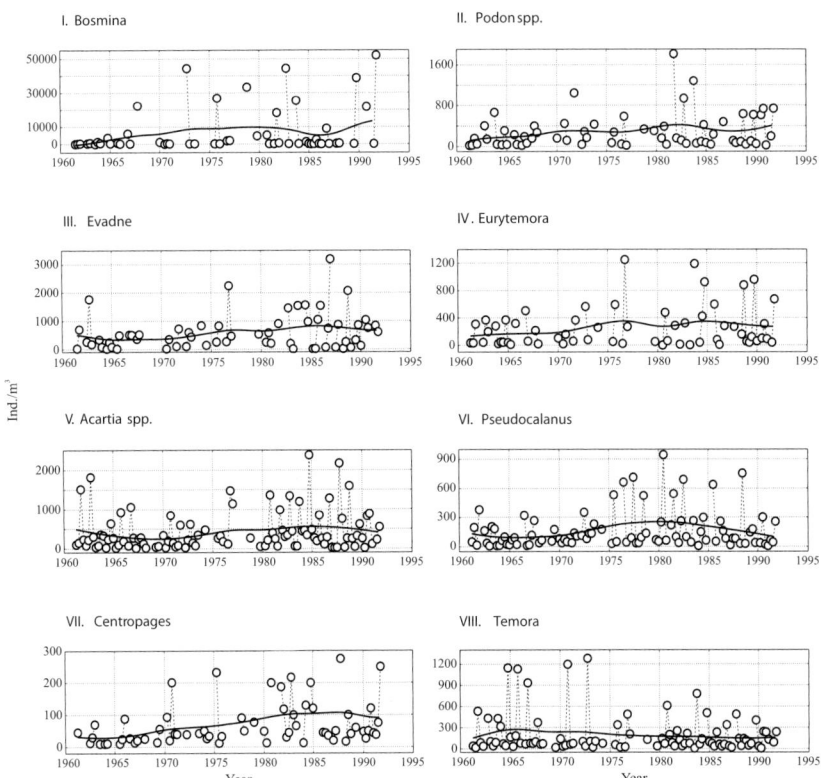

Fig. 5.6. Observed changes in crustacean mesozooplankton in the Baltic Proper from the 1960s to the 1990s. Smooth lines are distance weighted least squares (modified from Vuorinen et al. 2003, 2004)

and temperature in the Gulf of Finland. Multivariate regression modelling by Dippner et al. (2001) suggested somewhat later that small copepods are controlled by wintertime climate variability. Recent studies by Möllmann et al. (2000, 2002, 2003a) point out the importance of temperature in the control of small copepods in the surface zooplankton community (as opposed to the submergent neritic copepod community), such as *Acartia* spp. and *Temora* (see also Sect. 5.8 for a discussion on zooplankton and planktivore growth). What can generally be projected is that in summer, increasing warm periods with high surface water stability and low salinity, smaller mesozooplankton such as cladocerans, rotifers and *Acartia* spp. copepods (Viitasalo et al. 1995; Möllmann et al. 2000) will have an increasing importance in the pelagic food web. Möllmann et al. (2002) have shown, using correlation analyses, an affinity to higher temperature in summer for the predominant Baltic Sea cladoceran species *Bosmina*, and in spring for *Evadne nordmanni* and *Podon* spp. In winter, the temperature effect will come through effects on overwintering copepod survival,

since cladocerans and rotifers overwinter as resting stages in the sediment.

It has been suggested that biological control plays a part in the long-term variation of the Baltic Sea plankton-based food chain (Rudstam et al. 1994; Möllmann and Köster 1999). Suggested controllers could be found in the trophic cascades, either bottom-up from factors affecting productivity (possibly modified by eutrophication), or top-down through selective predation (possibly modified by fisheries). During the last 40 years in the Baltic Sea, the wintertime phosphorus levels have increased eight-fold and the nitrogen levels four-fold (Larsson et al. 1985; Elmgren 2001). Thus, increased primary production could hypothetically be expected to enhance the production of zooplankton. Indeed, in seven species out of eight Vuorinen et al. (2004) found an increase of abundance in the analysis of long-term change of Baltic Proper mesozooplankton (Fig. 5.6).

Line and Sidrevics (1995) report a general increase of crustacean zooplankton biomass in the Gulf of Riga between the early 1950s and late 1985. All mesozooplankton species or sites studied did

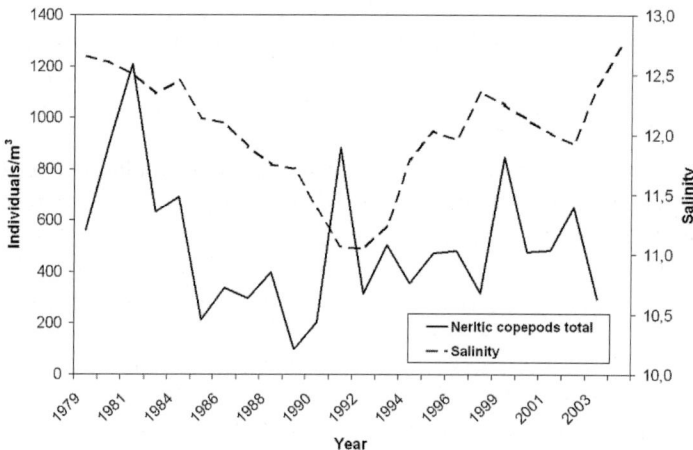

Fig. 5.7. Abundance of two neritic copepod species (Pseudocalanus and Temora, total) and seawater salinity (psu) (Gotland deep, 200 m) in the central Baltic Proper from 1979 to 2003 (Finnish Institute of Marine Research)

not, however, show a simultaneous increase, leaving some room for further speculation. There are possible negative effects also by e.g. cyanobacteria (for an extensive list of examples from the Baltic See, see Karjalainen 2005), therefore it would be simplistic to assume a linear increase of zooplankton due to enhanced primary production.

The importance of fluctuations in crustacean mesozooplankton species composition and, especially, in the abundance of the neritic copepods, *Pseudocalanus* and *Temora*, lies in their impact as fish food. We want to stress the difference between zooplankton species in the interaction between zooplankton and fish growth; for a full discussion see Sect. 5.8. Extensive changes in the numbers of pelagic predatory fish, *Clupea harengus membras* and *Sprattus sprattus*, the Baltic herring and sprat, have been hypothesised to affect zooplankton, top-down (Rudstam et al. 1994; Möllmann et al. 2000; Möllmann and Köster 2002), because they are selective predators of *Pseudocalanus* and *Temora*, respectively (Flinkman et al. 1992; Kornilovs et al. 2001). This could eventually lead to density dependent growth limitation of clupeids as hypothesised by Cardinale and Arrhenius (2000) and Cardinale et al. (2002). A third form of regulation of planktivores, bottom up, could be established by low salinity driven scarcity of neritic copepods (as hypothesised by e.g. Flinkman et al. 1998), this also leading to growth limitation of planktivores. It has been shown (e.g. Sjöblom 1961; Aneer 1975; Flinkman et al. 1992, 1998) that neritic copepods are not only the most pre-

ferred food of Baltic herring and sprat but also contain the most energy. Therefore, the shifts in abundances of these species, driven by salinity changes, are the most important factors affecting growth and condition of the most important fish stocks in the Baltic. Locally, the match between neritic copepods and salinity may be incomplete (Fig. 5.7), however, it has been firmly established by independent researchers over large areas. Recent work by Rönkkönen et al. (2004) and Möllmann et al. (2005) further establishes the fact that growth and condition of Baltic forage fish are in fact mostly bottom-up controlled, as suggested by Flinkman (1999).

It may well be that all these regulating effects (top-down control of zooplankton, density-dependent food limitation of clupeids, and low salinity driven scarcity of large neritic copepods) are acting simultaneously, but so far nobody has been able to measure their relative importance. In addition to controlling available food, the salinity changes of course determine the osmoregulation stress to clupeids. It is well known that suboptimal salinities increase energy expenditure in clupeids (Hettler 1976), which in turn may result in diminished body size (Hoar 1952; Canagaratnam 1959). Salinity also dictates the cod reproduction window in the Baltic Sea (Wieland et al. 1994). Cod is the most important predator of Baltic clupeids; hence a top-down regulatory pathway has been suggested as a major contributor to clupeid growth (Bagge 1989; Bagge et al. 1994; Aro 2000). For a long time, quality of available food, i.e. crustacean mesozooplankton, could be statistically and consistently shown to be the most important factor in clupeid growth and condition in the Baltic Sea. However, recent work by Casini et al. (2006) reveals the effects of density dependent factors as well. This means that the clupeids are indeed capable of reducing their favourable food by predation to an extent that it affects their own growth.

Conclusion

Salinity generally controls Baltic Sea biodiversity (Kändler 1949; Remane and Schlieper 1958) and is projected to decrease with expected climate change due to increased rainfall and runoff. Segerstråle (1969) demonstrated neritic copepod species (*Acartia bifilosa, Temora longicornis, Centropages hamatus* and *Pseudocalanus acuspes*, see Fig. 5.15) to enlarge their distribution area towards the north following a large amount of saline North Sea water penetrating the Baltic Straits in the winter of 1951. A reversal of the neritic species distribution is expected as a consequence of the predicted decrease in Baltic Sea salinity. The physiological mechanism is most likely the osmotic advantage provided for the neritic species by the higher salinity of the incoming seawater. Zajaczkowski and Legezynska (2001) demonstrated the lethality of low salinity to *Calanus* spp. in a nice combination of field sampling (in the freshwater runoff of a melting glacier) and laboratory experiments. The importance of salinity as a controlling factor has been restated by developing statistical methods using time series data from various national monitoring programmes. Finnish data were used by Vuorinen and Ranta (1987, 1988); Ranta and Vuorinen (1990); Viitasalo (1992); Viitasalo et al. (1994); Viitasalo et al. (1995); Vuorinen et al. (1998). Data from the Gulf of Finland, the Gulf of Riga, and the Baltic Proper were analysed somewhat later using Estonian (Lumberg and Ojaveer 1991; Ojaveer et al. 1998) and Latvian time series (Dippner et al. 2000; Möllmann et al. 2000; Ikauniece 2001; Kornilovs et al. 2001; Möllmann and Köster 2002; Möllmann et al. 2002; Hänninen et al. 2003; Möllmann et al. 2003b).

Another projected large scale environmental change is an increase in temperature, especially in winter. The plankton community close to the surface consists generally of smaller species and groups than the above-mentioned submergent neritic community. Therefore, we may expect sub-

stantial changes, probably increased production, also in other than neritic zooplankton, such as small estuarine copepods (*Acartia, Eurytemora*), cladocerans (*Podon, Pleopsis, Evadne* but also the newly established *Cercopagis*) and rotifers (*Synchaeta* and *Keratella*). Concluding, both salinity decrease and temperature increase will have a profound effect on zooplankton in the way described in this section. Several complicating factors can, however, change the pelagic ecosystem in unexpected ways. Radiation changes may, in addition to the complicating factors discussed above, cause a new mismatch between phyto- and zooplankton production, because temperature and light have different effects on timing of phytoplankton and zooplankton reproduction. There might also be changes in predator-prey relationships, caused by stock changes in planktivores, herring and sprat, and even new introduced predator species, such as the comb jelly *Mnemiopsis leidyi*.

5.7 Benthos

5.7.1 Large Scale Benthic Zonation of the Baltic Sea

In the Baltic Sea, salinity has a strong impact on species distribution and therefore on structure and composition of benthic communities. Many marine species reach their limit of distribution along the salinity gradient on the way from the entrance towards the inner, oligohaline parts of the Baltic Sea. Not only species richness, but also diversity of benthic functional guilds (defined by feeding behaviour, motility and sediment modifying ability, *sensu* Pearson 2001) decline along the Baltic Sea salinity gradient (Bonsdorff and Pearson 1999).

On the other hand, there is a pronounced vertical benthic zonation in the Baltic Sea due to strong stratification of the water column, relevant bathymetric changes in dissolved oxygen concentrations and other environmental factors (Demel and Mulicki 1954; Shurin 1968; Theede et al. 1969; Elmgren 1978; Zmudzinski 1978; Cederwall 1979; Järvekülg 1979; Brey 1986; Rumohr 1987; Leppäkoski and Bonsdorff 1989; Olenin 1997; Laine et al. 1997; Laine 2003). The shallow part of the Baltic above the halocline with well-oxygenated water is comparatively rich in bottom macrofauna, while the deeper anoxic part below the halocline represents a "benthic desert" (*sensu* Zmudzinski 1978) without macrobenthic organisms. The shallow part and the benthic desert are separated

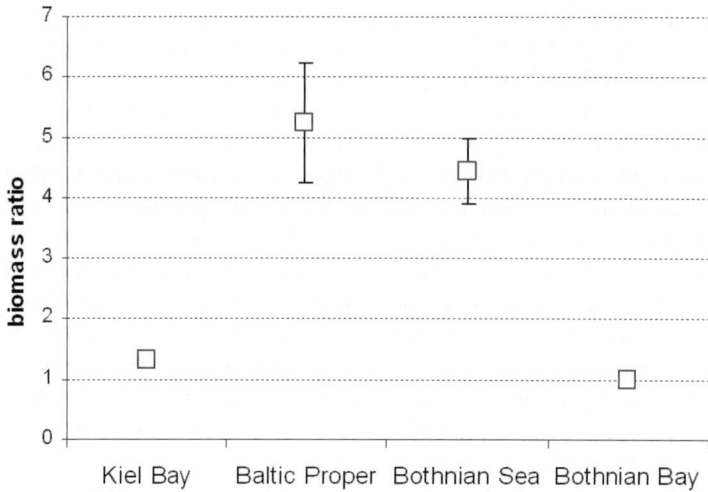

Fig. 5.8. Average ratio of contemporary (1990s) to pristine (1920s) biomass of the bottom macrofauna above the halocline in different areas of the Baltic Sea

by a transitional zone (Olenin 1997) with impoverished and specific bottom fauna, consisting mainly of bentho-pelagic species such as the polychaete *Antinoella sarsi* and the crustaceans *Saduria entomon*, *Diastylis rathkei* and *Pontoporeia femorata*. All these species are able to settle in the transitional zone during their larval stage or as adults, migrating actively from the upper, better oxygenated sections of the sea bottom. Such pronounced benthic zonation is characteristic for the south-western areas of the sea and for the major part of the Baltic Proper. In other areas, where the permanent halocline is absent (Bothnian Bay and Gulf of Riga) or weakened due to absence of major inflows (northern Baltic Proper), and mixing and ventilation of the water column is comparatively good, the bottom macrofauna is present along the entire depth gradient (Järvekülg 1979; Laine et al 1997; Laine 2003; Bonsdorff 2006).

The functional role of bottom macrofauna changes within the vertical benthic zones (Olenin 1997 and references therein). Both diversity and quantitative abundance of benthic functional guilds are highest in shallow waters and decline gradually along the depth gradient. In the benthic desert, burrowing macrofauna which would modify bottom sediments is lacking. Moreover, the vertical benthic zones differ in terms of sedimentation and resuspension rates, current velocity and bottom sediments, temperature fluctuations, salinity and oxygen conditions (Kalejs and Tamsalu 1984; Järvekülg 1979; Nehring 1990; Jonsson et al 1990). Due to these fundamental dissimilari-

ties in biological characteristics and abiotic properties of the benthic zones, their "response" to natural and human-induced ecosystem changes, such as climate fluctuation or eutrophication is also different (see below).

5.7.2 Long-term Trends in the Shallow Benthic Zones

Based on historical records and contemporary data presented in comparative benthic studies, changes in macrofauna biomass have been reviewed for the upper, well oxygenated, benthic zone. Most of these studies are based on resampling at historical sampling sites, taking into consideration the differences in sampling methodology (Arntz and Brunswig 1975; Cederwall and Elmgren 1980; Brey, 1981; Zmudzinski and Osowecki 1981; Elmgren et al 1984; Persson 1987; Aschan 1988; Bonsdorff et al 1991; Kotta and Kotta 1995; Andersin et al 1996; Leonardsson et al 1996; Kube et al 1997; Bonsdorff 2004). The average ratio of the benthic macrofauna biomass in the 1990s to that of the 1920–30s was 4.5 ± 0.9, when including all coastal areas of the Baltic Sea. When calculating this ratio for different Baltic Sea regions (Fig. 5.8), the highest average ratio (5.3 ± 1.0) was obtained for the Baltic Proper, followed by the Bothnian Sea (4.5), Kiel Bay (1.3) and Bothnian Bay (∼ 1). The dominance of two major bivalve species, *Mytilus edulis/trossilus* and *Macoma balthica*, which benefit most from eutrophication, is likely to be mainly responsible for the biomass increase found in many shallow areas of the Baltic Sea. Their absence in the Bothian Bay may explain the almost unchanged situation in this part of the Baltic.

In the southern part of the Baltic Sea, reduced predation of benthos-feeding fish on *M. balthica* and *Mytilus edulis* has been suggested as an explanation for increasing trends in zoobenthos biomass (Persson 1981, 1987; Brey 1986). This hypothesis was not tested quantitatively, however a decline in benthic-feeding fish stocks could blur the eutrophication effect.

5.7.3 Long-term Trends in the Sub-halocline Areas

The deep basins of the Baltic Sea are frequently exposed to hypoxia and anoxia, causing periodic extinction of the bottom macrofauna. In the Bornholm Basin, the previously bivalve-dominated benthic community progressively become polychaete-dominated (Leppäkoski 1975). In the early 1950s, the Arctic element clearly dominated below 85 m, this depth forming a boundary layer between the Arctic relict community and the zoogeographically more heterogeneous *M. balthica* community, typical for most of the Baltic Sea. In the early 1970s, cosmopolitan and Atlantic-boreal species comprised more than 90% of the total density (Leppäkoski 1975). Thus, the deeper parts of the southern Baltic have lost their most fascinating biogeographical peculiarity, namely that of offering hospitable conditions for the marine postglacial relicts, and thereby also their close zoogeographical connection with Arctic shallow-water biomes. At present, the Bornholm Basin (if not denfaunated due to oxygen deficiency) is linked with boreal areas by the most euryhaline representatives of the Atlantic and North Sea soft-bottom fauna (Leppäkoski and Olenin 2001 and references therein).

Drastic changes in species and functional guild composition took place also in the central and northern sub-halocline areas of the Baltic Sea. Figure 5.9 shows a gradual disappearance of the infaunal polychaete worm *Scoloplos armiger* from the inner sub-halocline parts of the Baltic Proper during the past five decades. In 1954–1956, *S. armiger* was the most abundant polychaete in the Northern Basin in areas below 100 m depth, being found even at 450 m in the Landsort Deep (Andersin et al. 1978). In the 1960s, the species was biomass dominant in the eastern Gotland Basin

within the depth range of 100–140 m (Järvekülg 1979); however, it had already vanished from the more northern areas. By the end of the 1980s to the beginning of the 1990s, the distribution of *S. armiger* was restricted to depths of 99–124 m on the south-eastern slope of the Gotland Basin (an area around station BY 9, Olenin 1997).

Long-term observations showed that *S. armiger* can not survive in the central parts of the Baltic Sea if salinity is less than 10.5 psu and oxygen is less than 0.5 ml/l (Järvekülg 1979; Olenin 1997). Recolonisation of defaunated bottoms depends on larvae supply from the "healthier" areas with established populations. Thus, the most distant penetration limits of *S. armiger* ever recorded in the sub-halocline areas of the Baltic Proper, in the mid-1950s, (Northern Basin and Landsort Deep) most probably were caused by the exceptionally intensive inflow of saline water in November–December of 1951 (Matthäus and Frank 1992). Although very little is known about the reproduction strategy of *S. armiger* in the Baltic Sea, it may be suggested that the 1951 inflow could have brought pelagic larvae of *S. armiger* from southern sub-halocline areas. Since then, a gradual decline in salinity, oxygen depletion and enlargement of hydrogen sulphate zones eliminated *S. armiger* from the northern and central sub-halocline areas of the Baltic Proper.

5.7.4 Conceptual Model of Natural and Human-induced Changes in the Sub-halocline Areas of the Baltic Sea

In contrast to fully marine ecosystems, in the inner parts of the Baltic Sea the benthic functional guilds are often represented by a single species (Olenin 1997; Bonsdorff and Pearson 1999). Disappearance of such a key species would result in the loss of a functional group which, in turn, may essentially change the biogeochemical cycling of the system, affecting microbial life, bioturbation activity, release of nutrients and formation of laminated sediments. Analysis of data on distribution of bottom macrofauna species (Järvekülg 1979; Olenin 1997; Laine et al. 1997; Laine 2003) and their functional guilds (Olenin 1997) shows that during recent decades the polychaete worm *S. armiger* was the only infaunal species in the large hypoxic areas below the halocline in the central and northern parts of the Baltic Proper which was able to perform bioturbation of bottom sediments. This polychaete feeds in "conveyor belt" mode,

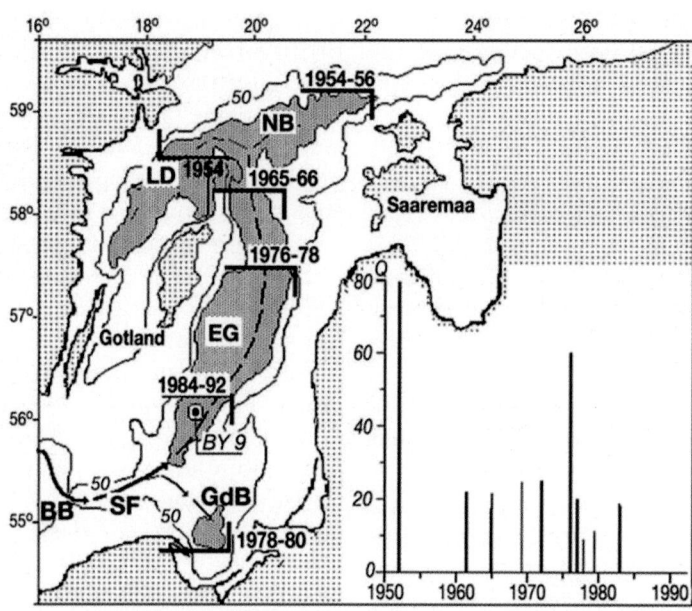

Fig. 5.9. Distribution limits of *Scoloplos armiger* in the sub-halocline areas of the Baltic Proper from the 1950s to the early 1990s: BB – Bornholm Basin, SF – Slupsk Furrow, GdB – Gdansk Basin, EG – Eastern Gotland Basin, LD – Landsort Deep, NB - Northern Basin; BY 9 – location of a sampling station in the south-eastern part of the eastern Gotland Basin; isobaths of 50 m (*line*) and 100 m (*dark areas*) are indicated. The saline water inflow paths are shown with black arrows and the major inflow intensities, Q are given in the diagram on the right (from Franck and Matthäus 1992)

translocating particles from some cm depth within the sediment up onto the surface (Rice 1986).

Disappearance of that species from the northern and central sub-halocline areas of the Baltic Proper coincided with the increased development of the laminated sediments (Jonsson et al. 1990) in the same areas with all ensuing consequences, like trapping of PCBs (Jonsson 2000; Skei at al. 2000). This allows the assumption that the evolution of laminated sediments largely depends on the absence/presence of that key-species. A very approximate estimation based on the particle reworking rate of *Scoloplos* spp. (0.51 g per individual day^{-1}; Lee and Swartz 1980), mean density (200 ind. m^{-2}, Järvekülg 1979; Olenin 1997 and references therein) and area occupied by the *S. armiger* population in 1950s (ca. 50,000 km^2, based on Andersin et al. 1978; Järvekülg 1979; Olenin 1997) results in 1.9×10^9 tons of reworked sediments annually. This number gives a rough estimate on the extent of the bioturbation process performed by a single species and its importance in biogeochemical cycling of elements.

The causal chain analysis of events in the subhalocline areas of the Baltic Sea is presented in

Fig. 5.10. This scheme shows that the North Atlantic Oscillation governs the frequency and intensity of the saline water inflows into the Baltic Sea (e.g. Hänninen et al 2000 and references therein), which, through the chain of related events, may effect the success of the recolonisation of the sub-halocline bottoms by the polychaete *S. armiger*. An "inoculating" inflow should occur during the spawning season of *S. armiger* (usually September–December), when its pelagic larvae appear in the plankton (Järvekülg 1979 and references therein). The inflow should be strong enough to bring the larvae pool up to the Gotland Basin and even further, to the inner sub-halocline areas of the Baltic Proper. Recolonisation of the previously defaunated sub-halocline bottoms by the *S. armiger* population would result in sediment reworking activity, which, in turn, could cause release of PCBs into the water column (Skei at al. 2000).

In contrast, the decrease in number and intensity of inflows reduces the opportunities to recolonise the benthic desert; the absence of macrofaunal organisms able to perform bioturbation leads to formation of laminated sediments, which

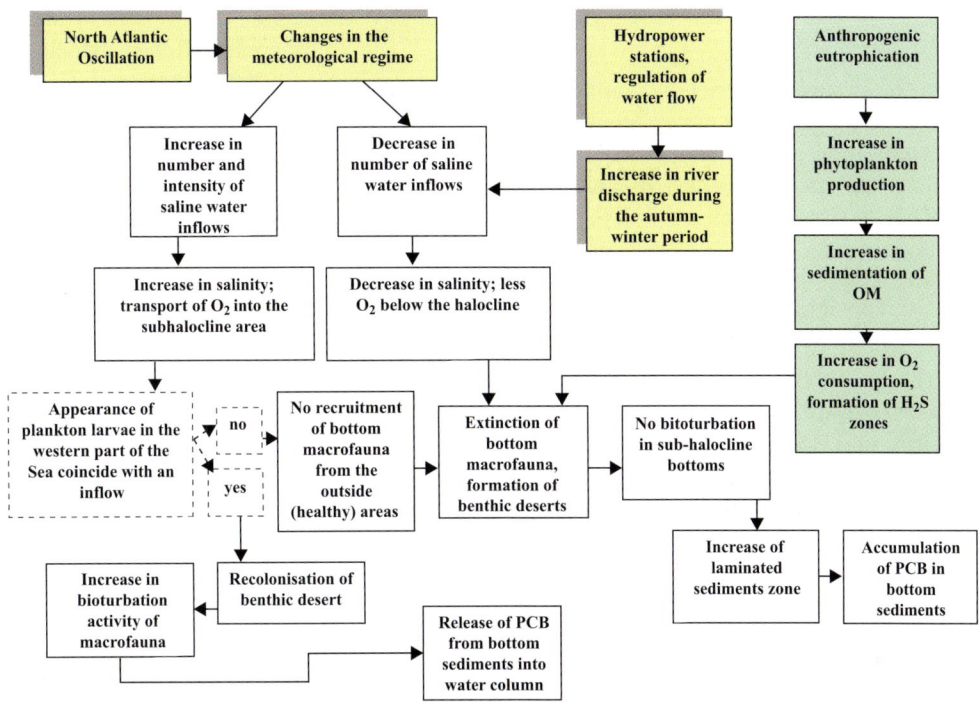

Fig. 5.10. The causal chain analysis of events in the sub-halocline areas of the Baltic Sea

accumulate PCBs (Jonsson 2000; Skei at al. 2000). As was shown recently, there are signs that anthropogenic changes in runoff due to river regulation may be causing changes in the frequency of major inflows (Schinke and Matthäus 1998). Eutrophication is another anthropogenic factor largely contributing to impoverishment of living conditions for bottom macrofauna in the sub-halocline areas (e.g. Karlson et al. 2002).

5.7.5   Climate Change Related Implications

As pointed out in the previous paragraphs and shown in Fig. 5.10, the distribution of macrobenthic species and their functional role within the ecosystem depend on many environmental factors and their fluctuations. Presence or absence of key species are directly linked with ecosystem functions such as bioturbation activity, filtration activities and oxygen release to the water column, nutrients and other biogeochemical matter cycling or formation of laminated sediments.

A systematic change in the hydrographic conditions due to climate change, e.g. towards warmer temperatures and reduced salinity (Räisänen et al. 2004; Gustafsson 2004), will have a direct impact on the distribution patterns of many native species and thus on their functional role. The deeper parts of the southern Baltic Sea have already lost their most fascinating biogeographical peculiarity, in offering hospitable conditions for the marine postglacial relicts.

On the other hand, the projected changes in hydrographic features may facilitate settling of non-native species in various parts of the Baltic Sea. Invasions of species have already occurred during the past centuries (Leppäkoski et al. 2002). The importance of non-native species will probably further increase in the future, when the anticipated changes in e.g. temperature or salinity will be beyond the acclimation capacity of native biota. As a result, boreal species might be replaced by more thermotolerant taxa able to settle successfully in meso- to oligohaline habitats.

Presently, we have only started to understand whether and how such changes in species inventory affect ecosystem functioning. Long term data on benthic species' distribution and composition are available, covering more than a century. These data have been used to explain past trends in the different benthic zones. A future task will be to further monitor and assess changes in the benthic fauna and to evaluate the possible effects on the ecosystem, taking into account its complexity.

5.8   Fish

Fishing is the most direct impact by humans on the Baltic Sea ecosystem. The annual removal of hundreds of thousands of tonnes of fish from the Baltic has direct effects on populations and their dynamics (ICES 2004b), as outlined below.

As described above, oceanographic conditions in the brackish Baltic Sea are strongly linked to atmospheric forcing. The unusual period of a persistently high North Atlantic Oscillation (NAO) index since the late 1980s, with a dominance of westerly weather, resulted in an increase in water temperatures and decreasing salinity and oxygen conditions due to increased runoff and a low frequency of water intrusions from the North Sea (Fonselius and Valderrama 2003; Matthäus and Nausch 2003). Climate has been shown to affect recruitment of Baltic fish stocks, e.g. Brander and Mohn (2004) demonstrated the negative association of the high NAO-period during the 1990s and cod recruitment, and MacKenzie and Köster (2004) showed the positive association with sprat recruitment. Similarly, herring recruitment is related to the NAO (Axenrot and Hansson 2003). Furthermore, climate has been shown to cause fluctuations in growth of Baltic herring and sprat (Möllmann et al. 2003a; 2005; Rönkkönen et al. 2004).

This section will review the primary impacts of fishing and climate variability and change on the Baltic Sea ecosystem. Much more detailed summaries of how fishing and climate impact the Baltic Sea and other marine ecosystems can be found in the reports by HELCOM, ICES advisory and working groups (Advisory Committee on Ecosystems, Working Group on Ecosystem Effects of Fishing Activities; www.ices.dk) and in the scientific literature (Jennings and Kaiser 1998; MacKenzie et al. 2007).

5.8.1   Baltic Fisheries, their Management and their Effects on Exploited Populations and the Ecosystem

The fish community in the Baltic Sea has much fewer species than other marine areas of comparable size. The main reason for the low number of species is low salinity, which imposes a physiological stress on marine organisms and on freshwater species inhabiting coastal areas (Voipio 1981; HELCOM 2002).

The few species in the ecosystem are reflected also in the commercial catches, which are dominated by sprat, herring, and cod (ICES 2004). Several other species are exploited such as flounder, plaice, salmon, eel, and near shore coastal species such as whitefish, pikeperch and smelt, but their catches are much lower (ICES 2004b).

Baltic fisheries were managed by the International Baltic Sea Fisheries Commission (IBFSC) until 2005, after which bilateral negotiations between the EU Commission and Russia have replaced IBFSC. The most important commercial species have quotas and other regulations (e.g. mesh sizes, closed seasons/areas) intended to promote long-term sustainability of the populations and the ecosystem. The scientific advice for making management decisions about international fisheries primarily comes from the International Council for the Exploration of the Sea (ICES 2004a; Daw and Gray 2005). Management of the fisheries is partly disaggregated spatially to reflect local differences in both the fisheries and biology of the target species (e.g. growth rates, maturity, migration patterns, and genetic stock differentiation). As a result, separate quotas are assigned for different sub-divisions of the Baltic (Fig. 5.11).

Effects of fishing on populations and the ecosystem

The most direct impact of fishing on fish populations is the mortality of large numbers and biomass of fish. Fish landings increased substantially during the 20th century (Fig. 5.12), primarily due to increases in fishing effort and technology (Bagge et al. 1994). Combined catches for the three most important species now exceed several 100,000 tonnes annually. Adult mortality rates due to fishing for some populations (e.g. cod in subdivisions 22–24; cod and sprat in subdivisions 25–28) are typically at least 2–4 fold higher than estimates of natural mortality rates (ICES 2003). Fishing is therefore the dominant mortality source for adults of the most abundant species (ICES 2003).

High fishing mortalities reduce and limit the growth of populations, which will decrease if fishing mortality rates remain too high for long. In extreme cases, biomass levels can fall to levels which may not be sustainable. Such a situation presently exists for many Baltic cod and salmon populations, yet they continue to be exploited (ICES 2004b).

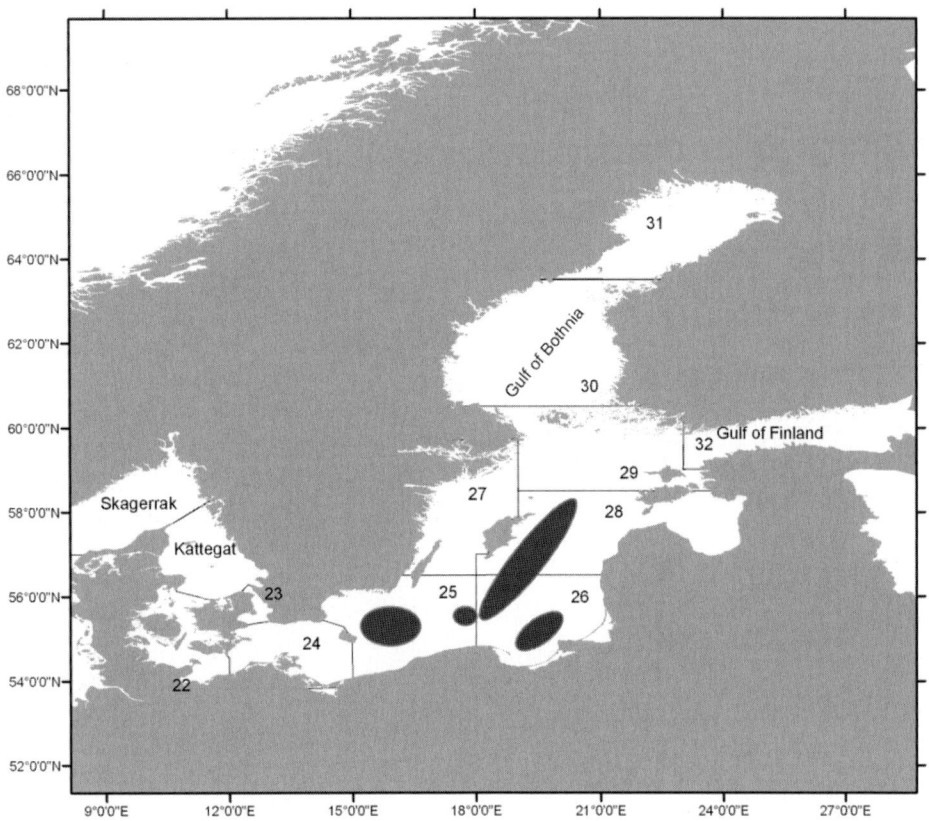

Fig. 5.11. Map of the Baltic Sea showing the ICES subdivisions. The HELCOM definition for the Baltic Sea includes the Kattegat and subdivisions 22-32. Black areas show deep basins of the eastern Baltic Sea where cod spawning occurs (from Bagge et al. 1994)

Fishing also affects many second-order attributes of populations (HELCOM 2002). These include size and age structure, genetic composition and fish life-history (Olsen et al. 2004). Fishing typically removes the largest and oldest individuals within populations (Pauly et al. 1998). In addition, since fishing is a size-selective activity, fishing tends to preferentially kill the fastest growing individuals, which are the first within a population to attain a size which is retained by a given mesh size. After several years of fishing, the older, larger fish become rare and populations are typically dominated by small fish.

These changes in population structure have consequences for stock productivity because many biological attributes such as spawning time and location, migratory behaviour, and individual fecundity are size and age dependent (Marshall et al. 2003). For example, spawning time and location largely determine the environmental conditions (e.g. temperature, salinity, food supply) subsequently experienced by eggs, larvae and juveniles. As a result the removal of specific size and age components from a population can reduce the probability that eggs and larvae encounter favourable environmental conditions for survival (Heath and Gallego 1998; Begg and Marteinsdottir 2002). Moreover, the characteristics of eggs and larvae (e.g. size, buoyancy, yolk sac size) produced by females of different sizes and ages are not equal (Nissling et al. 1994; Nissling et al. 1999), and these variations can affect survival when environmental conditions vary. Interactions between female characteristics, egg characteristics and environmental conditions appear to exist for eastern Baltic cod, with higher recruitment in years with a higher proportion of older fish (Jarre-Teichmann et al. 2000; Köster et al. 2005), particularly in years when oxygen conditions in the deeper layers of cod spawning areas are low (Vallin and Nissling 2000). Similar interactions are evident in other fish stocks (Marshall et al. 2003).

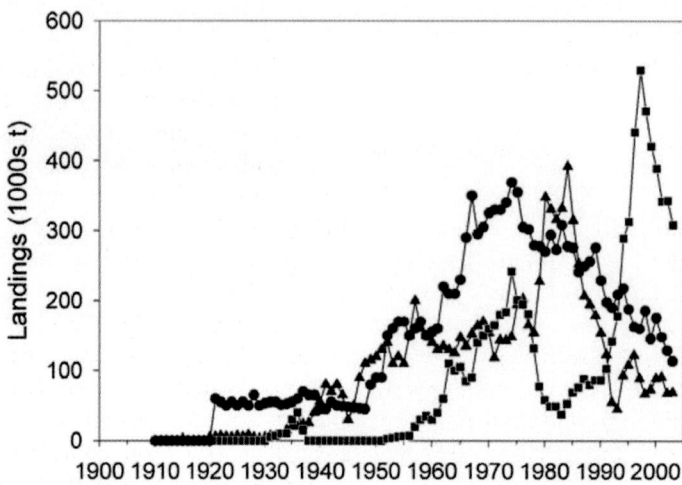

Fig. 5.12. Total international landings of the three commerically most important fish species (cod, herring, sprat) in the Baltic Sea during the 20th century. *Triangles:* cod; *squares:* sprat; *circles:* herring (data source: Sparholt 1994; ICES 2004b)

Fishing can also reduce the genetic variability within a fish species. For example, the eastern Baltic cod population is genetically distinct from other cod populations in the Atlantic Ocean (Nielsen et al. 2001) and physiologically adapted for reproduction in low salinity (Nissling and Westin 1997). However in some years even these adaptations are insufficient to ensure reproductive success (Köster et al. 2001; MacKenzie et al. 2002a). Continued high fishing pressure in combination with frequent periods of environmental conditions detrimental for egg survival suggest that this population will collapse (Jonzen et al. 2002; ICES 2004b). If the eastern Baltic cod population were to collapse, then recovery would be very slow or perhaps impossible even under low or no fishing mortality. Long recovery time is expected because the local genotypes (Nielsen et al. 2001) would be eliminated and because cod immigrating from more saline areas (e.g. Belt Sea, Kattegat) are not adapted to reproduce in the brackish conditions of the more easterly parts of the Baltic Sea (Nissling and Westin 1997; Vallin et al. 1999).

Another species whose genetic diversity in the Baltic Sea is threatened is salmon. Many Baltic salmon populations which spawn in specific rivers are greatly reduced and may not recover under present levels of exploitation (ICES 2004b). Salmon populations in the Baltic Sea are stressed not only by fishing mortality, which has been reduced in recent years (ICES 2004c), but also by

ecological changes associated with habitat damage in rivers and outbreaks of the disease M74 (ICES 2004c). This disease caused many fatalities of juvenile salmon and is believed to be partly related to the nutritional value of prey organisms.

In addition to these direct effects of fishing on target populations, fishing has impacts on non-target species and the ecosystem. These effects include the bycatch of non-target species (e.g. fish species, marine mammals and seabirds; (HEL-COM 2002) and changes in food web structure and species interactions (Pauly et al. 1998; Jackson et al. 2001). Food web changes in the Baltic Sea resemble a trophic cascade as observed in other ecosystems (e.g. Frank et al. 2005). The removal of cod biomass has affected the production and survival of cod prey such as herring and sprat (Sparholt 1994; Gislason 1999; ICES 2003a). The abundance of these two species increased during the late 1980s and 1990s partly because the cod population and predation rate fell (Köster et al. 2003). Further, increased planktivore populations, especially of sprat, have negatively affected the dynamics of major Baltic Sea zooplankton populations, i.e. the copepods *Pseudocalanus acuspes* and *Temora longicornis* (Rudstam et al. 1994; Möll-mann and Köster 2002). Despite the importance of these fishery-induced interactions, a large part of the changes in Baltic Sea food web structure was related to the opposing effects of climate variability on decreasing cod and increasing sprat recruitment (see below). Additionally, zooplankton development is strongly affected by climate-induced changes in hydrography (see Sect. 5.6).

Multi-decadal and century scale variability: a view from the past

Fish populations in Baltic Sea coastal areas and rivers have been exploited by local fishermen since at least the medieval period (Enghoff 1999; MacKenzie et al. 2002a). For example, major herring fisheries near Rügen and in the Øresund supported significant economies and trade among Baltic towns in the medieval period (Sahrhage and Lundbeck 1992; Holm and Bager 2001; Alheit and Hagen 2001).

These fisheries were based primarily on the fall-spawning Rügen herring. This subpopulation is now essentially extinct although the reasons for this disappearance are not clear (Parmanne et al. 1994). Large marine and estuarine fauna such as sturgeon, many salmon populations, seals and harbour porpoises used to be common but over-exploitation and habitat loss have reduced abundances, in many cases (e.g. sturgeon) to near extinction (Elmgren 1989; HELCOM 2002; MacKenzie et al. 2002a). As a result, the present fish and marine mammal community differs considerably from that which existed only 100–150 years ago.

Perceptions of biodiversity and biomass, and consequently the past effects of exploitation, that are based on information from only the last few decades of the 20th century can therefore be highly biased when considered in a longer historical perspective (Jackson et al. 2001). Fisheries management strategies which aim to sustain and recover populations should be aware of historical levels and variations in species abundances to avoid the "shifting baseline syndrome" (Pauly 1995; Jackson et al. 2001).

A comparison of the Baltic Sea ecosystem of today with that in the early 1900s shows that it experienced two major changes during the 20th century. As noted above, the abundance of marine mammals has been drastically reduced (Elmgren 1989) and is only slowly recovering (HELCOM 2002). These animals consumed fish as prey, and as a result of the decline in their biomass, predation by marine mammals must now be lower. Reduced predation by these species may have led to an increase in the biomass of the Baltic fish community (Thurow 1997), particularly if the reduction in mammal predation rates has not been compensated by increased predation among the remaining species.

Second, the Baltic Sea underwent eutrophication during the 20th century, and this process has stimulated primary production (HELCOM 2002). Presumably some of this production has been transferred to higher trophic levels (Nixon 1995), including the fish community. For example, production of benthic fish species in the Kattegat increased between the 1950s and 1990s as a result of eutrophication (Nielsen and Richardson 1996). Similar changes may have occurred in the central and northern Baltic Sea (ICES subdivisions 25-32; Fig. 5.11) although the links have not been quantified (Hansson 1985; HELCOM 2002). Nevertheless, these two impacts (eutrophication, reduction of marine mammal populations) on the Baltic ecosystem suggest that part of the increase in fish landings (Fig. 5.12) may be due to changes in the structure and productivity of the ecosystem (Elmgren 1989; Thurow 1997; MacKenzie et al. 2002a).

Quantifying the consequences of these changes and of long-term variability in climate on Baltic fish populations requires longer time series of fish biomass and landings data than those presently available. New inter-disciplinary (marine ecological-maritime historical) approaches are starting to address this issue (Holm et al. 2001; MacKenzie et al. 2002a; MacKenzie et al. 2002b). One approach is to extend analytical biomass estimates back from the 1960s and 1970s to earlier decades of the 1900s (Eero et al. 2005, 2007). This technique has been applied successfully to other fish stocks (Pope and Macer 1996; Toresen and Østvedt 2000). The second approach is to recover and interpret historical fishery data (landings, effort, fishery taxes, and regulations) from various archival and archaeological sources. These methods can potentially provide valuable information about the Baltic fisheries since at least the 1500s (Otterlind 1984; Enghoff 1999; MacKenzie et al. 2002b; MacKenzie et al. 2002a).

The two approaches will enable investigation of how abundances and distributions varied when exploitation was lower and during combinations of environmental conditions (e.g. Little Ice Age, Medieval Warm Period) which differ from those observed in the late 20th century and which may occur again as a consequence of anthropogenic climate change. For example, archaeological evidence shows that fishermen on Bornholm captured cod during the Atlantic Warm Period (ca. 8,000–

5,000 BP) when both temperature and salinity were considerably higher in the Baltic Sea than at present (Enghoff et al. 2007). Knowledge of reasons for past fluctuations could help understand and forecast how fish populations might react to future combinations of exploitation and anthropogenic climate change.

Conclusions

Fishing and hunting have been impacting targeted populations for centuries and have influenced Baltic Sea biodiversity and food webs. Some populations have been severely reduced in size. The combination of high fishing and environmental stresses makes populations of some species (cod, salmon) vulnerable to collapse and has reduced others to near extinction (sturgeon). In contrast, abundances of some other fish populations are high. These include herring and sprat stocks, whose main predator (cod) is now relatively rare, and whose production has increased due to favourable environmental conditions in the last 5–10 years.

5.8.2 Effects of Climate Variability and Anthropogenic Climate Change

This review concentrates on the effect of climate variability on recruitment of Eastern Baltic cod and sprat (both being the target of extensive multidisciplinary studies during the last decade; STORE 2003) and growth of herring and sprat. As climate variability can affect animals both directly (through physiology) and indirectly through changes in their biological environment (Ottersen et al. 2004), the review is grouped according to direct and indirect responses of recruitment and growth to climate-induced changes in the physical environment. Effects of climate change on some other Baltic fish populations, including flatfish, migratory and coastal fish species and glacial relict species are summarized elsewhere (MacKenzie et al. 2007).

Direct effects on cod and sprat recruitment – influences of hydrographic conditions on egg and larval survival

The survival of fish early life stages in the Baltic Sea is known to be sensitive to hydrographic conditions in the spawning areas (Bagge et al. 1994; Parmanne et al. 1994; Wieland et al. 1994). Eggs

of Eastern Baltic cod successfully develop only in deep water layers with oxygen concentrations > 2 ml/l and a salinity of > 11 psu. These thresholds are the basis for the so-called reproductive volume (RV), i.e. the water volume sustaining cod egg development (Plikshs et al. 1993; MacKenzie et al. 2000). The climate-induced decrease in RV since the 1980s has caused high cod egg mortality, especially in the eastern basins, i.e. Gdansk Deep and Gotland Basin (MacKenzie et al. 2000; Köster et al. 2003).

Due to a different specific gravity, sprat eggs float shallower than cod eggs (Nissling et al. 2003), and consequently their survival is less affected by poor oxygen conditions. However, sprat eggs occur at depths where the water temperature is affected by winter cooling (Wieland and Zuzarte 1991), and egg and larval development is influenced by extremely low water temperatures. Consequently, weak year classes of Baltic sprat have been associated with severe winters (Nissling 2004), resulting in temperatures of below 4 °C in the intermediate water layer during spawning. The absence of severe winters since 1986/1987 and related favourable thermal conditions for sprat egg survival thus contributed to the generally high reproductive success of Baltic sprat during the 1990s (Köster et al. 2003).

Behaviour studies show that cod larvae exposed to oxygen concentrations below 2 ml/l are mostly inactive or moribund (Nissling 1994). Although an impact of the environment on larval survival can thus be expected, no direct effect of hydrography on observed larval abundance could be detected (Köster et al. 2001). Consequently, other factors such as food availability might be critical for larval survival (see below). The same may be true for sprat, where the relationship between temperature and larval survival during the 1990s is most likely a result of enhanced plankton production at higher temperatures (Köster et al. 2003; Baumann et al. 2006; Voss et al. 2006, see below).

Indirect effects on cod and sprat recruitment – influences of mesozooplankton abundance on larval survival

The effect of food availability on growth and survival of cod larvae has been investigated using a coupled hydro/trophodynamic individual-based model (Hinrichsen et al. 2002). Model results suggest that the co-occurrence of peak prey and larval abundances is critical for high survival rates.

The decline of the *P. acuspes* stock during the 1980s/90s, a result of low salinity and oxygen conditions (Möllmann et al. 2000, 2003b), probably caused a food-limitation for early cod larvae. Model simulations including *P. acuspes* nauplii as prey resulted in high survival rates, whereas omitting *P. acuspes* resulted in low survival (Hinrichsen et al. 2002). Thus, low *P. acuspes* availability has contributed to the low recruitment of cod since the late 1980s, and to the lack of recovery despite improved egg survival after the major inflow in 1993.

In contrast to cod, sprat larvae prey mainly on the copepod *Acartia* spp. (Voss et al. 2003). Baumann et al. (2006) observed August temperatures in surface waters to explain most of the variability in sprat recruitment. These correlations may be due to food availability and match-mismatch with the food production. Voss et al. (2006) computed an index of larval mortality which suggests a higher survival of sprat larvae born in spring than in summer. This survival pattern could be linked to the seasonal variability in prey abundance. The advantage for summer-born larvae was due to the progressed state of the *Acartia* spp. population. While in spring mostly nauplii are abundant and only very low concentrations of larger copepodites and adult *Acartia* spp. can be found, the situation is the reverse in summer. These results suggest that the increased *Acartia* spp. availability during the critical stage of late larvae has contributed to the high reproductive success and, eventually, to the unusually high sprat stock during the 1990s (Voss et al. 2006).

Indirect effects on cod and sprat recruitment – influences of predation by clupeids on egg survival

A substantial predation on cod eggs by clupeids has been observed in the Bornholm Basin. Egg predation is most intense by sprat at the beginning of the cod spawning season (Köster and Möllmann 2000a). After cessation of their spawning in spring, the largest part of the sprat population leaves the basin, resulting in a reduced predation pressure on cod eggs. In parallel, herring return to the Bornholm Basin from their coastal spawning areas to feed, including on cod eggs (Köster and Möllmann 2000a).

The drastic increase in the sprat stock during the 1990s has increased the potential of cod egg predation mortality. However, the shift in cod peak spawning time from spring to summer (Wieland et al. 2000) resulted in a decreasing predation pressure by sprat. Additionally, a decline in individual sprat predation on cod eggs was observed from 1993–1996, despite the relatively high cod egg abundance in the plankton. This is explainable by a reduced vertical overlap between predator and prey. Due to the increased salinity after the 1993 major Baltic inflow (Matthäus and Lass 1995), cod eggs were floating in shallower water layers, while clupeids dwelled deeper, due to enhanced oxygen concentration in the bottom water (Köster and Möllmann 2000a). Thus, predation pressure on cod eggs is higher in stagnation periods and contributed to the low reproductive success since the 1980s. Similarly, egg cannibalism was found to be an important source of sprat egg mortality in the Bornholm Basin, thus representing a self-regulating process for the sprat stock (Köster and Möllmann 2000b).

Direct and indirect effects on herring and sprat growth – influences of salinity/oxygen and meso-zooplankton abundance

Historically, three different hypotheses have been tested as explanation for the observed decrease in age-specific weights of Baltic herring, namely (i) selective predation of cod on herring (Sparholt and Jensen 1992; Beyer and Lassen 1994), (ii) mixing of sub-stocks with different growth rates (ICES, 1997), and (iii) a real decrease in growth rates due to changes in the biotic environment (Cardinale and Arrhenius 2000). Recently, increased evidence was found for the last hypothesis (Möllmann et al. 2003b, 2005; Rönkkönen et al. 2004).

The central Baltic Sea feeding areas of herring and sprat overlap in winter as well as spring and early summer, when both species feed in the halocline of the deep basins during daytime (Köster and Schnack 1994). Here the clupeids compete for the calanoid copepod *P. acuspes* reproducing in the high salinity layer (Möllmann et al. 2004). The reduced availability of *P. acuspes* has resulted in a lowered food intake of herring and can be related to the decrease in herring condition (Möllmann et al. 2003b, 2005). Recently, Rönkkönen et al. (2004) supported this by showing the growth rates of herring in the northern Baltic Sea to be explicitly dependent on the abundance of the copepod *P. acuspes*.

Food availability, especially the low abundance of *P. acuspes*, has also been hypothesised to have caused the decrease in sprat growth during the

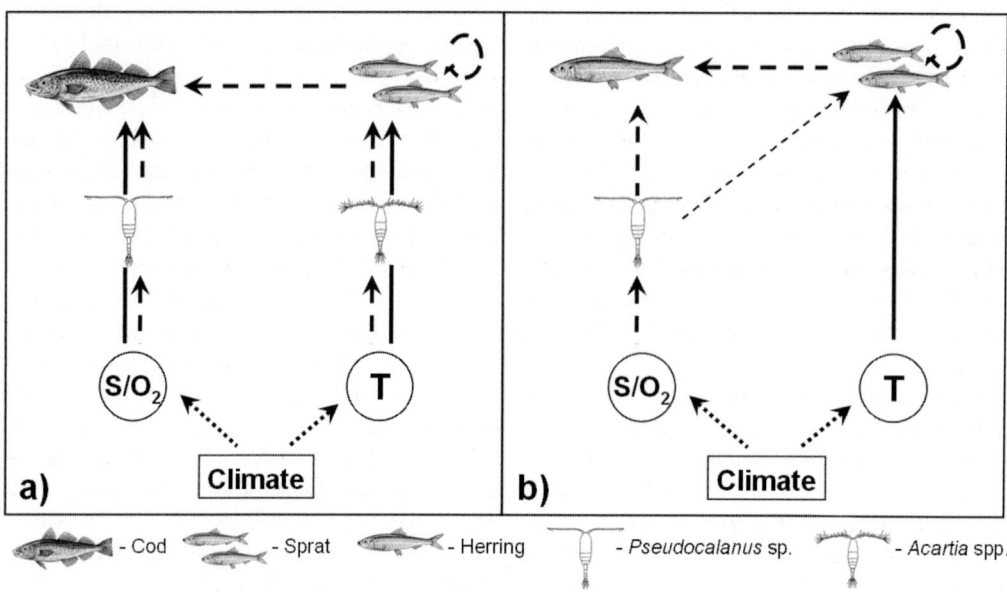

Fig. 5.13. Conceptual model of climate effects on recruitment (**a**) and growth (**b**) of Baltic fish stocks. *Dotted arrows* – effect of climate on hydrography, *dashed arrows* – indirect effects, and *solid arrows* direct effects of climate; S – salinity, O_2 – oxygen, T – temperature (*explanations, see text*) (modified from MacKenzie et al. 2007)

1990s (Cardinale et al. 2002; Möllmann et al. 2004). However, correlations between the abundance of the copepod and sprat condition in the central Baltic Sea are weak (Möllmann et al. 2005). It is more likely that strong intraspecific competition at high stock size caused the lowered growth in sprat, explaining a significant relationship between sprat condition and sprat stock size. Similarly, intraspecific competition within the large sprat stock has contributed to the reduced herring growth (Möllmann et al. 2005; Casini et al. 2006).

A direct relationship between salinity and herring and sprat growth was shown by some authors (Cardinale and Arrhenius 2000; Cardinale et al. 2002; Rönkkönen et al. 2004). However, this relationship, most probably reflects the change in mesozooplankton community structure rather than a direct physiological effect on these euyhaline fish species.

Conceptual model of climate effects on recruitment and growth of Baltic fish stocks – implications for the future

Figure 5.13 gives a summary of our present understanding of direct and indirect effects of variability in climate on cod and sprat recruitment as well as herring and sprat growth. Climate affects salinity and oxygen (S/O_2) through runoff and inflows of North Sea water, and temperature (T) through direct air-sea interaction. Changes in S/O_2 affect directly cod recruitment via egg survival, and indirectly via *P. acuspes* abundance, which influences larval survival. High temperatures are directly supportive for sprat recruitment via egg survival, and indirectly via *Acartia* spp. availability on larval survival (Fig. 5.13). Further, egg production mediated by hydrography regulates cod and sprat recruitment.

Herring growth appears to be affected by the indirect effect of S/O_2 on *P. acuspes* and the increased competition with the enlarged sprat stock. The latter, a result of high reproductive success during the 1990s (see above) and reduced cod predation pressure, has caused a density-dependent decrease in sprat growth. This intra- and interspecific competition may have been amplified by low *P. acuspes* availability.

Conclusions

Anthropogenic climate change projections for the Baltic basin predict higher temperature and, with a lower certainty, decreasing salinity. Low salinity in the deep water, often but not always accompanied by low oxygen concentrations, could mean an even lower reproductive volume and carrying capacity of the ecosystem for the eastern Baltic cod stock which is adapted physiologically and genetically for reproduction in this environment. High exploitation and even lower salinity due to expected climate change will further challenge the sustainability of this population. In contrast, increased temperatures would increase the carrying capacity for the sprat stock. Consequently, the present clupeid-dominated regime in the central Baltic fish community would be stabilized.

However, changes in the exploitation level have a strong potential to alter food web structure and thus to modify the outcome of anthropogenic climate-induced changes. For example, a reduced exploitation of cod would increase the chance of high reproductive success despite a generally low carrying capacity. Because of higher cod predation pressure, the sprat stock would be reduced and, consequently, the predation on cod eggs and *P. acuspes* would be reduced. As a feedback this could lead to higher reproductive success of cod and enhanced feeding conditions for herring and sprat, increasing their growth rate.

5.9 Marine Mammals

5.9.1 Introduction

The Baltic marine mammal fauna is a subset of the north Atlantic temperate/subarctic and arctic marine mammal fauna. A distinctive feature is that the permanent cetacean fauna is restricted to a single species, the harbour porpoise (*Phocoena phocoena*). The contemporary Baltic marine mammal fauna is dominated by three species of phocid seals: the common or harbour seal (*Phoca vitulina*), the Baltic ringed seal (*Phoca hispida botnica*), and the grey seal (*Halichoerus grypus*). A fourth species, the harp seal (*Phoca groenlandica*) was present and relatively abundant for several millennia in the Litorina period (Storå and Ericsson 2004; Storå and Lougas 2005). Of the four or five fairly well defined subspecies of the ringed seal (Reeves 1998), the Baltic Sea encompasses also the two isolated lake subspecies, the Ladoga ringed seal (*P. h. ladogensis*) and the Saimaa ringed seal (*P. h. saimensis*).

As the North Atlantic seal species currently missing from the Baltic Sea are all Arctic species,

species introductions caused by climate warming are highly improbable. Of the Baltic seal species, only the harbour seal breeds exclusively on land. The now regionally extinct harp seal breeds on drift ice (Lavigne 2002), which has also been the preferred breeding substrate for Baltic grey seals (Hook and Johnels 1972). Ringed seals mostly use landfast ice for breeding (Smith and Stirling 1975).

The scientific literature related to marine mammals and climate and, especially, climate change is surprisingly scarce. A recent comprehensive review of marine ecosystems and climate variation in the North Atlantic (Stenseth et al. 2004) mentions marine mammals only briefly when discussing El Niño/Southern Oscillation (ENSO)-related studies (Lima and Jaksic 2004). Studying climatic effects on marine mammal populations can be complicated, as the response of populations is in part determined by the food web structure (Berryman 1999). The effects can be conveyed e.g. via food shortages as in the case of the effects of ENSO on sea lions and elephant seals (Lima and Jaksic 2004).

The existing studies of marine mammals and climate change have largely focused on Arctic and Antarctic species which depend on sea ice (Tynan and DeMaster 1997; Kelly 2001; Ferguson et al. 2005). This regional bias may reflect the vulnerability of the region. Clarke and Harris (2003) state in their review of major threats to polar marine ecosystems that climate change is likely to present the "single greatest long-term threat to the two polar marine ecosystems".

The recent Arctic climate impact assessment (ACIA 2005) considers a general poleward shift in species distribution to be "very likely" for marine mammals. More specifically, an increased distribution of harbour and grey seals is projected, and ringed seal is among the species for which a decline in distribution is "very likely". The assessment infers (p. 504) "dramatic declines in production by ice associated marine mammals and increases by more temperate species".

5.9.2 History of Baltic Seals

The first species to migrate into the Baltic Sea after the retreat of the ice was the ringed seal. The species was present at the Finnish coast at 9,500 BP at the latest (Ukkonen 2002). The grey seal may have entered the basin already at this stage (Lindqvist and Possnert 1997), but it seems to have been present at least from the beginning of the Litorina period (Lõugas 1997). Even harbour seals and porpoises probably belonged to the Baltic fauna quite soon after the beginning of the Litorina period (Lõugas 1997).

The most enigmatic feature of the history of Baltic seals has been the appearance and disappearance of the harp seal from the fauna. First finds of the species from the Baltic Proper are from about 5,000 BP (Lindqvist and Possnert 1997; Lõugas 1997; Ukkonen 2002), and it is assumed that it entered the basin at the end of the Atlantic period, "sometime between 5000 and 4000 cal B.C." (Storå and Ericson 2004). Towards the end of the Subboreal the frequency of harp seal finds starts to diminish in the northern Baltic Sea. They disappear almost completely at the beginning of the Subatlantic (Ukkonen 2002), but the species was still commonly hunted in the Aland islands and on Saaremaa Island (Estonia) in the Iron age (Storå and Lõugas 2005).

One of the proposed explanations for the presence of this pelagic seal species has been the possible enrichment of Baltic fish fauna caused by the elevated salinity in the Litorina period (e.g. Forstén and Alhonen 1975; Lõugas 1997). As the harp seal is a highly migratory species (Lavigne 2002), many researchers (see Lepiksaar 1986; Lõugas 1997 and references therein) have hypothesised that the species was not breeding in the area but was performing feeding migrations originating from populations breeding in the north Atlantic.

New archaeo-osteological evidence has yielded support for a breeding population. Ukkonen (2002) showed that the harp seal was common in the Subboreal, even in the northernmost part of the Baltic Sea. According to Ukkonen (2002), the high abundance and long presence of harp seals in the northern Baltic Sea suggest a stable residence in the area. Storå and Ericsson (2004) show, based on osteometric data, that seals of current newborn size have been present in the middle parts of the Baltic. They conclude that a local harp seal breeding area existed in the Baltic Sea during the Subboreal period (Storå and Ericsson 2004).

The extinction of the harp seal from the Baltic Sea has been repeatedly attributed to climatic factors. Following Lepiksaar (1986), Storå (2002) connects it to a "general warming of winters" and "resulting differences in ice formation" during the Subboreal. This hypothesis of gradually warmer Subboreal winters does not, however, get support from modern winter temperature reconstructions

with modelling (Renssen et al. 2005) or pollen (Davis et al. 2003) data.

Less dramatic, but substantial changes in historical distribution concern the ringed seal, grey seal, the harbour seal and the harbour porpoise. The southern limit of the ringed seal breeding distribution has moved northwards: according to Storå (2002), ringed seals bred regularly in the area of Gotland in the middle Neolithic period, as most of the archaeological bones originate from very young ringed seals.

The grey seal is now the most abundant seal in northern Baltic Sea, but very rare in the Finnish subfossil and prehistoric refuse faunas as well as in the prehistoric northern Baltic Sea in general (Ukkonen 2002; Storå 2002). Ukkonen (2002) postulates that this apparent prehistoric scarcity in the northern Baltic Sea might be related to the possibly slow adaptation of the grey seal to the ice breeding habit; it has been repeatedly discussed whether the grey seal was originally an ice or land breeder (Davies 1957; Bonner 1972).

It is well documented that the prehistoric and historic breeding distribution of the grey seal covered the southern Baltic Sea, the Danish straits and the Kattegat (Møhl 1970). The grey seal is (during periods together with the harp seal) the most abundant seal in Danish refuse faunas; the dominance of the harbour seal in the Danish waters is a phenomenon of the last few hundred years (Møhl 1970; Aaris-Sørensen 1998; Härkönen et al. 2005).

In the early the 20th century, the harbour porpoise was rather common even in the middle Baltic Sea. Kinze (1995) suggested 25,000 as a possible original porpoise population size for the Baltic Proper. Its distribution is now concentrated or almost limited to the southeast Baltic Sea, the Danish Straits and the Kattegat (Hammond et al. 2002). The reasons for the decline of the harbour porpoise are not clear. The harbour seal was distributed more to the north in the Baltic Sea during the Bronze and Iron Age than today (Storå and Lõugas 2005).

Also, Baltic seal species were much more numerous in the early 1900s than they are today. The most abundant species was almost certainly the ringed seal, but also grey seal numbers have been estimated to be at least 88,000–100,000 (Harding and Härkönen 1999), and the eastern Baltic population of harbour seals to be 5000 individuals (Harding and Härkönen 2005). The most important factor behind the decline of seals was hunting, but environmental contaminants contributed strongly to the negative trend, especially in the 1960s and 1970s. Bycatch has been and still is an important mortality factor for all marine mammal species in the Baltic. Harbour porpoise and young seals are especially vulnerable to bycatch.

5.9.3 Climate Change Consequences for the Baltic Ice Breeding Seals

The ringed seal is emerging as a favourite subject of marine mammal connected climate change studies (Tynan and de Master 1997; Kelly 2001; Stirling and Smith 2004; Barber and Iacozza 2004; Ferguson et al. 2005; Meier et al. 2004). The ACIA assessment states that of the high arctic pinnipeds the ringed seal is likely to be most affected by climatic change "because many aspects of their life history and distribution are linked to sea ice".

The Baltic ringed seal has been treated as a subspecies *Phoca hispida botnica*, distinct from the nominate *P. h. hispida* in the Arctic and the Ladoga and Saimaa seals. Amano et al. (2002) studied the skull morphology of ringed seals and state that the subspecies status should be retained, but Palo et al. (2001) found only weak allele frequency differences between Arctic and Baltic ringed seals and no difference between the different breeding populations of the Baltic.

The Baltic ringed seal population exceeded 180,000 in the early 20th century but declined to about 5000 by the 1980s due to hunting and environmental pollutants (Härkönen et al. 2006). The latest estimate for the Baltic ringed seal population is 5500 in 1996 (Härkönen et al. 1998, 2006).

Ringed seals are stationary compared to harp and grey seals. They are able to live in fast ice areas, where they maintain a special breathing hole system and excavate subnivean lairs (nests under the snow cover) on the ice. The typical ice habitat is fast ice with pressure ridges or consolidated pack ice, as accumulated snow is needed for lair construction (Smith and Stirling 1975).

Baltic ringed seals are found mainly in the northern and eastern parts of the Baltic Sea. The present population forms three distinct breeding subpopulations: one in the Gulf of Finland, one in the Gulf of Riga and one centered in the Gulf of Bothnia (Härkönen et al. 1998). Miettinen et al. (2005) present evidence for a fourth small breeding population in the Archipelago Sea, a traditional breeding area of the ringed seal (Bergman 1956; Hook and Johnels 1972).

Meier et al. (2004) have published the only study directly dealing with the possible consequences of climate change for the ringed seal breeding habitat in the Baltic. The most important ice parameter for ringed seals is the length of the ice season, as the seals construct a subnivean lair on top of ice (Smith et al. 1991). The lair hides adults and pups from predators, and offers the pup thermal protection. The pups are suckled for 5–7 weeks. If there is not enough snow for lair formation, pups are born openly on the ice. In the Baltic Sea area pups are born in February–March.

Meier et al. (2004) modelled the duration of the ice cover in the four breeding areas. They used future (2071–2100) ice cover simulated with the GCM-driven Rossby Centre regional Atmosphere–Ocean model RCAO. Ice parameters were calculated for a control climate (1961–1990) and two scenario runs (2071–2100). Two global models (HadAM3H of Hadley Centre/UK and ECHAM4/OPYC3 from the Max Planck Institute for Meteorology/Germany) and two scenarios (IPCC SRES A2 and B2) were used. The IPCC A2 scenario assumes a relatively large increase in greenhouse gas emissions. The increase in B2 is more modest.

Ice cover is drastically reduced in the scenario runs. The future number of ice days is reduced to 18–48 (2071–2100) for the southern breeding areas. In the northernmost part of the Bay of Bothnia the mean number of scenario ice days is 123 or about four months (Meier et al. 2004) and so still exceeds the present ice cover duration of the southern breeding areas. Meier et al. (2004) hypothesise that the southern subpopulations of Baltic ringed seals are in danger of extinction, but more studies are needed to specify the minimum ice-winter requirements of ringed seals.

The grey seal is currently the most abundant marine mammal of the Baltic Sea, with an estimated total population size of about 21,000 (Hiby et al. 2007). According to genetic analysis by Boscovic et al. (1996), Baltic grey seals may form an isolated population. The recent distribution of Baltic grey seals is concentrated in the middle and northern parts of the sea, with most of the population north of 58 degrees latitude (Halkka et al. 2005; Harding et al. 2007).

Baltic grey seals have been treated as typical ice breeding pinnipeds (Hook and Johnels 1972), but extensive breeding on land was revealed in the 1990s. Land breeding has been documented on islets and skerries in the northern part of the

Gulf of Riga, Stockholm Archipelago, and south-western Finland. Land breeding is not a new phenomenon; it has been documented earlier in the 20th century on the Swedish coastline and even mentioned in ethnographic sources (e.g. Ahlbäck 1955). The preferred ice breeding habitat of the grey seal is drift ice in the eastern, central and northern Baltic (Hook and Johnels 1972).

Meier et al. (2004) hypothesised that, because of the land breeding option, the consequences of climate change might not be as severe for the grey seal as for the ringed seal. Land seems, however, to be a suboptimal breeding habitat for Baltic grey seals. According to Jüssi and Jüssi (2001), pup mortality due to high pup densities is common and can reach up to 20–30% in dense breeding colonies on land in Estonia. The causes of mortality and morbidity are not well understood, but the pups often die of starvation as a result of breaking of the mother-pup bond. There are even indications of a lowered weaning weight for pups born on land (Harding et al. 2007). Lowered weaning mass is associated with lowered first-year survival (Hall et al. 2001).

Hansen and Lavigne (1997) found that the mean lower critical temperature of grey seal pups corresponds closely with the −7.5 °C breeding season isotherm that delineates the northern breeding limit of the western Atlantic. As even the northernmost parts of the Baltic Sea have mean breeding period (March) temperatures above this limit (Hansen and Lavigne 1997), and as the Bay of Bothnia is one of the traditional breeding places of grey seal, possible northward extension of the breeding distribution for physiological reasons is not relevant for the Baltic. However, drift ice suitable for breeding may in some years be found mostly/only in the northernmost parts of the Baltic Sea because of the projected changes in ice climate (see Chap. 3). The projected increase in the length of the ice-free period may affect grey seal winter distribution, as it will be able to forage in presently ice-covered parts of the Baltic Sea. The projected lengthened open-water period may also have an effect on seal-fisheries interactions in the northern Baltic Sea, as seals can move freely and fishing with open-water gear is increasingly possible in areas covered by ice in present-day climate. An elevated bycatch mortality of ringed seal and grey seal pups was noted in the Gulf of Finland in the extremely mild winter of 1988–1989 (Stenman 1990).

5.9.4 Harbour Porpoise and Harbour Seal

According to subfossil and archaeo-osteological evidence, the harbour seal has probably never belonged to the northern Baltic seal fauna (Lõugas 1997; Ukkonen 2002). The northernmost population of the Baltic harbour seal is currently situated at Kalmar sound between the island of Öland and mainland Sweden. This small "eastern Baltic" population of harbour seals is genetically distinct from the much larger Kattegat-Skagerrak population (Goodman 1998).

Grey and ringed seals have their pups in winter, whereas the young of harbour porpoise and harbour seal are born during the summer half of the year. There may be physiological reasons for the northern limit of harbour seal distribution. Sufficiently high water temperatures are important for harbour seal pups and may be one of the most important factors affecting pupping time, pup growth and survival (Harding et al. 2005b), as small harbour seal pups can be cold stressed even in their natural range of water temperatures (Harding et al. 2005b).

The lower critical temperatures for harbour seal pups are as high as 3 °C for pups (Miller and Irving 1975) and −2.3 °C for juveniles (Hansen et al. 1995). Lower critical temperature is the limit under which metabolic rate starts to increase: for harbour seal juveniles the increase seems to be linear at least to −10 °C (Hansen et al. 1995). These physiological results are interesting but as such do not indicate a possible future range shift of the harbour seal, as species distributions are affected by many factors other than physiology (Davis et al. 1998).

The harbour porpoise is relatively abundant in the Kattegat and the Danish Straits, but the population of the inner Baltic Sea might be as low as 600 (Hammond et al. 2002). It has been much discussed whether this population of the inner Baltic Sea is separate from the porpoises of the Belt Sea and the Kattegat. According to the recent review of Koschinski (2002), the existence of a distinct Baltic subpopulation of harbour porpoises "appears to be a valid concept". It has been proposed that severe ice winters might be one of the reasons for the decline of the Baltic harbour porpoises (e.g. Teilmann and Lowry 1996). If this is the case, the projected changes in annual ice extent due to climate warming (see Meier et al. 2004; Chap. 3) should be in this respect favourable for this species.

5.9.5 Future Projections

Potential long term trends, linked to climate change in the Baltic Sea Basin, are most likely largely associated with the projected decline of sea ice extent and reduced length of the ice season. The projected increase in surface water temperature might have relevance for the harbour porpoise and the harbour seal. The projected increases in sea level and wind waves might have an impact on the haul-out and breeding distribution of the grey seal and harbour seal in the southern and middle part the Baltic Sea where sea level rise is not compensated by isostatic rebound. Even quite small changes in sea level might render many of the haul-out and breeding sites at least temporally unsuitable as these are typically very low skerries and reefs.

The extent of occurrence and area of occupancy of the **ringed seal** is likely to decline and to shift northwards, with the possible eradication of breeding populations of the Gulfs of Finland and Riga and the Archipelago Sea. Changes in distribution may be so large that the Baltic subspecies of ringed seal *Phoca hispida botnica*) may meet the IUCN criteria of threatened species.

The Baltic population of **grey seal** has a clear potential for a shift to land breeding, but it is not yet possible to say how this may affect the abundance and distribution of the species. There are indications of increased pup mortality for land breeding grey seals, and the possibly limited amount of suitable breeding skerries might induce density dependent mortality. There is a potential for a substantial increase in the winter foraging distribution of the grey seal, as it may be able to forage in coastal and other areas covered with fast ice in the current climate.

The inner Baltic populations of **harbour seal** and **harbour porpoise** are very small, and their abundance and range have declined over the course of the last century. Projected reduction of ice cover and elevated water temperatures may be potentially favourable for these species, but it is not possible to say if they are likely or not likely to extend their range, as other factors than climate are of a governing importance for these threatened populations. The possibility of projecting impacts is limited by inadequate knowledge of seal and porpoise population structure and minimum ice requirements of ringed seals in the Baltic Sea. Single species studies should be complemented by a more ecosystem-based research, as, for example, possi-

ble predation on seal pups and prey availability might be important factors.

5.10 Sea Birds

5.10.1 Introduction

Since the work of Berthold (1990) and Burton (1995), several authors have predicted that distributions and population sizes of birds will change as a response to anthropogenic climate change. In addition, and probably prior to these profound effects, the timing of the events of the annual cycle of birds will be modified (e.g. Fiedler et al. 2004). This section on climate-related ecosystem changes in birds associated with the Baltic Sea is composed of three parts; one section on impacts of historical climate variability, one section on recently documented changes in population sizes and a section on recent changes in the phenology of birds. Changes in bird populations and phenological patterns are interpreted in the light of possible indirect and direct influences of recent climate variability. Although most studies referred to are drawn from the Baltic Sea Basin, experiences from studies outside the Baltic Sea Basin are used to improve the interpretation.

5.10.2 Prehistorical and Historical Bird Communities of the Baltic Sea

Archaeological data on bird remains have revealed some of the early history of bird species occurring in the central Baltic Sea (Mannermaa 2002). The bird fauna 5,400–4,800 BP was quite similar to the present one. Eider and velvet scoter were the most common species. Other species that occurred at the time were the cormorant, white-tailed sea eagle, turnstone, common gull, lesser black-backed gull and herring gull, black guillemot and carrion/hooded crow. The most interesting missing species at this location were the goldeneye, tufted duck and the long-tailed duck. The period covered by Jettböle data corresponds to the late Littorina Sea stage, when salinity was approximately double the present and the sea was probably rather eutrophic (Ericson and Tyrberg 2004).

Ericson and Tyrberg (2004) summarised Swedish paleoecological observations on bird remains and, combining these observations with deductions from climate and vegetation variation, constructed a probable timetable of arrival of different bird species in Scandinavia. According to them the Scandinavian avifauna has been surprisingly

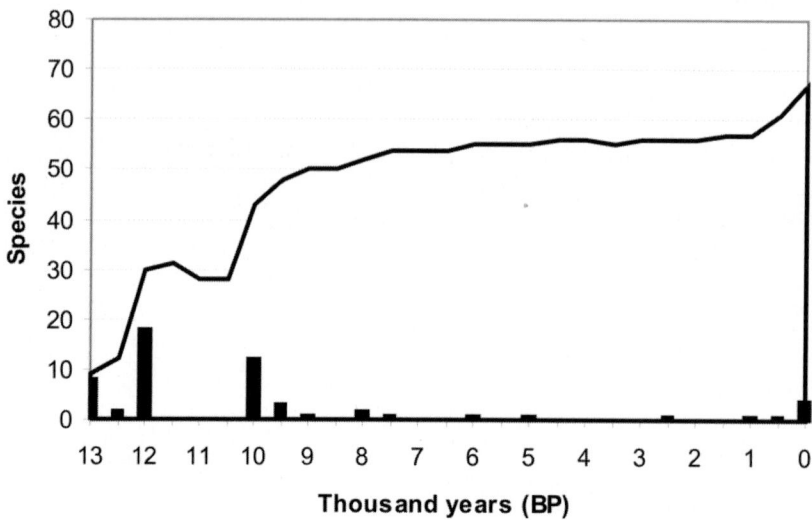

Fig. 5.14. Accumulation of Baltic marine and coastal species since the ice age (*curve*), and the estimated or documented arrival times of species for which there is fairly reliable evidence (*columns*, number of new species/500 y) (modified from Ericson and Tyrberg 2004)

stable during most of the postglacial period. The basic bird community of the Baltic Sea was established very early – of the current 67 species commonly occurring in the Baltic Sea and its coastal areas, at least 56 were in the area already 5,000 years ago (Fig. 5.14, compiled from Ericson and Tyrberg 2004, Fig. 4). Only four species seem to have invaded the Baltic Sea naturally in the last two hundred years or so – *Xenus cinereus, Larus minutus, Sterna sandvicensis, Sterna albifrons*, and, taking into account the scanty paleo-ecological data, even some of these may only reflect lack of observations. In the light of old historical and prehistorical data, most of the "newcomers" of recent times are actually reinvasions, indicating shifts of distribution ranges with climate variation. The same appears to hold for some terrestrial species as well (e.g. Lehikoinen et al. 2003a).

5.10.3   Impacts of Earlier Climate Variability

Most earlier large-scale climate oscillations took place so early in history that we have no quantitative data to rely on. During the "revival" of climate after the Little Ice Age from the mid 19th century until 1940, spring temperatures (March–May) increased (Stockholm) from 1900–1940 by 0.026 °C/year, with an average temperature of 3.8 °C, while during the most recent warming period they have increased even faster, 0.036 °C/year in 1970–1999, with an average of 4.5 °C. During the early warming period the ornithological "knowledge base" started to increase and range shifts, large population changes and timing shifts of phenological events could be studied. For Finland some preliminary reinterpretations of former range shifts are included in Lehikoinen et al. (2003b).

Fennoscandian distribution ranges of bird and mammal populations were studied intensively also immediately after the period. Examples worth remembering are the articles by Kalela (e.g. 1946, 1949). He reported that of the 25 species whose northern border was in southern Finland 11 had increased in number and expanded northwards between the late 19th century and the 1940s and six had decreased or reduced their distribution. He also reported that in Germany the situation was the opposite, with six "winners" and 14 "losers". He attributed the changes mainly to anthropogenic climate change and tended to view other human influence (e.g. habitat changes) as a reason for decreases only. Quite interestingly, the species benefiting from climate amelioration were those that spend their winters in Europe, especially in areas of westerly winds. A warning note when using old time series is indicated: in both old and recent warming periods raw data is affected by factors other than climate – a relevant source of bias in ornithological data is increasing observer activity and efficiency, which has been unaccounted for, so far. In future studies it should be taken into account.

5.10.4 Anthropogenic Climate Change Impacts on Birds – Predictions

Anthropogenic climate changes may influence Baltic Sea bird populations by changes in:

- Distribution ranges during the breeding and non-breeding seasons;
- Abundances during the breeding and non-breeding seasons;
- Traits;
- Migratory routes and stopover sites;
- Timing of spring and autumn migration;
- Migratory tendency within species/ populations.

In the following a short general description of the current knowledge on each of these impact types is given. This summary is based on recent reviews, e.g. those by Møller et al. (2004), Crick (2004) and Rehfisch et al. (2004):

1. Distribution changes

Range changes may be related to changes of habitat, but they need not be so, since factors other than suitable habitat may have constrained the distribution of bird species in the past. Extension of range is a different thing than a range margin shift, but both are expected to require a surplus of recruits in species where breeders show high site-tenacity. In the earlier warming period (in Fennoscandia from 1870 to 1940) many range shifts in birds were observed, both at the northern and the southern borders.

2. Changing abundances and community structures

Change of abundance is a direct consequence of the difference between death and birth rates. The understanding of the reasons and linkages between long term changes in abundance of birds and anthropogenic climate change is still very limited.

3. Evolutionary trait changes

Individual birds first react behaviourally to changes in climate. The responses are phenotypic, i.e. plasticity of individuals is sufficient to fit the behaviour to environmental changes. If the rate of change increases, selection comes into play. Rapid environmental change may cause strong selection effects, and, provided there is enough genetic variability, combinations may exist that adapt the birds evolutionarily to the changed condition. It is only at this stage that population changes and range shifts are likely to occur. For natural selection to be efficient in adapting the population to changed conditions there may be constraints that arise from other counteracting selection pressures, even in a situation where anthropogenic climate change is the driving force. For example, if the initiation of spring migration is constrained either by conditions or by the internal clock (which it need not be) selection effects can not be realised in optimising timing of breeding.

4. Migratory pathways

Routes that the birds use between breeding and wintering areas are often habitual and are likely to change if conditions at favoured stopover sites deteriorate. For species which require specialised habitat patches for refuelling, e.g. arctic waders, water level changes and other factors which change the type and quality of coastal habitats may be critical. The intensively studied knot (*Calidris canutus*) is one model species which may currently suffer locally from climate induced habitat changes (Baker et al. 2004). Within the Baltic Sea especially arctic geese and seaducks may have to respond to anthropogenic climate change impacts.

5. Timing of migration

Timing of migration should be analysed in its entirety from initiation to end in both directions. Impacts at different phases may be different, because anthropogenic climate change trends and stop-over conditions change both spatially and temporally (e.g. Ahola et al. 2004). These highly detailed analyses are only possible for species studied by remote sensing techniques, and for such species there are no long time series available yet. However, timing has been in the focus of research on climate effects simply because it is rather easy to observe and long term data exist starting from the mid 18th century. Timing of critical annual events is an important parameter, as time windows for optimal performance, e.g. in breeding, may change in length. It has been generally expected that the length of suitable breeding period will increase, but this need not necessarily be the case. Interactions between temperature and precipitation may affect food availability, which may be more important than the expected longer breeding periods.

6. Migratory tendency

Short distance, intra-European migrants are a special case which has not been the object of intensive studies. However, analyses of Fennoscandian winter census data and ringing recovery analysis will provide more knowledge on these effects in the years to come (for the potential and problems of ringing recovery analyses, see Fiedler et al. 2004).

The documented recent changes in the populations of birds in the Baltic Sea are discussed in relation to anthropogenic climate change and evaluated in relation to their possible impacts on the population dynamics of the bird population in question. The impacts may be on adult survival, breeding effort, nestling survival/breeding success and dispersal.

5.10.5   Available Studies

Monitoring of population sizes covers the period of recent climate variability rather well in Finland, Sweden and many other countries. European monitoring programs have recently united in a pan-European monitoring program (see http://birdlife.cz/index.php?a=cat.101). Monitoring programs usually only report the trends of population sizes and do not interpret or analyse the causes. Therefore, although we have a lot of information on population variation, rather little can be said about the causes, because simultaneous changes of many environmental factors may have parallel impacts.

Distributions are not monitored annually. In some countries (e.g. Great Britain and Finland) bird atlas work (mapping of distributions) has been done twice with similar methods in 1960–2000. In a recent paper Brommer (2004) suggested that in Finland range border shifts took place between the end of the 1970s and the end of the 1980s, and these might have been due to climate impact. Southern species shifted their northern range borders northwards by 18 km in ca. ten years, approximately double the speed shown by a similar British analysis (Thomas and Lennon 1999). Rigid statistical analysis has yet to be done to separate anthropogenic climate change impacts

from other causes in order to understand better the direct and indirect influences.

With respect to water birds, monitoring of breeding colonies has been carried out in most countries around the Baltic Sea during the period of recent climate variability. However, few studies have attempted to link trends to anthropogenic climate change. Extensive monitoring of water birds breeding in the northern archipelagoes is taking place in selected areas. In the Kaliningrad and St. Petersburg regions of Russia only selected water bird species are monitored, and the same holds for Estonia. Monitoring of breeding water birds is carried out annually in Latvia, Lithuania and Poland, and in all three countries counts are organised in relation to specific areas. In Germany, monitoring of breeding water birds is organised by the state administrations of Schleswig–Holstein and Mecklenburg–Vorpommern, with a focus on protected areas. Yearly monitoring data on breeding water birds in Denmark is currently organised by one state only. In addition, game reserves and the breeding colony of auks at Christiansø are counted regularly. The wetlands along the eastern part of the Gulf of Finland are not currently covered by breeding monitoring programs.

Most recent counts of wintering water birds in the Baltic Sea have been connected to the midwinter counts of Wetlands International. These counts generally cover birds of the coastal zone and lagoons, while offshore areas are surveyed only infrequently. In 1992 and 1993, the first surveys covering all major offshore areas were carried out (Durink et al. 1994). Since 1994, line transect counts in offshore areas have not been undertaken as part of an internationally co-ordinated census.

5.10.6   Fluctuations in Population Levels

Direct influences of climate variability on Baltic bird populations vary between terrestrial birds, water birds and seabirds, as well as between breeding and wintering bird faunas. Direct influences on breeding bird populations are generally of limited significance as compared to indirect influences. The direct influences on breeding bird populations include reduced food availability due to rising sea levels, increased mortality of chicks in low-lying colonies due to rising sea levels, reduced development of the embryo due to low temperatures, abnormal development or death of the embryo due to excessive exposure to high temperatures and increased chick mortality due to overheating and dehydration during warmer weather. The first of these influences impacts on the survival rate, while the other influences impact on the birds' breeding success. Rising sea levels in coastal areas and the impact of decreasing salinity on bivalves may reduce the available food supplies for benthos-feeding water birds such as swans and geese by decreasing the area of shallow water where they are able to reach the bottom vegetation from the surface.

Global warming is likely to directly affect migrating and wintering birds in the Baltic Sea Basin. Extreme winter temperatures have long been documented to influence water bird mortality in the Baltic Sea, and winter conditions in the Baltic Sea Basin are known to determine the range of land- as well as water birds (Nilsson 1980; Oswald et al. 2004). Although the migratory and wintering bird fauna of the Baltic Sea represents a wide range of groups and ecotypes, the large (> 10 million) populations of wintering water birds are probably the component of the Baltic bird fauna most susceptible to changes in winter conditions. Documented trends in the winter distribution of wintering water birds suggest a close relationship between water bird winter populations and winter climate (ICES 2003b; HELCOM Habitat 2004). Trends in the number of wintering birds among coastal species like mute swan (*Cygnus olor*), tufted duck (*Aythya fuligula*), goldeneye (*Bucephala clangula*) and goosander (*Mergus merganser*) between 1987 and 2002 show a large-scale shift in the distribution of the core population of these water bird species from south to north.

The translocation of the core of the winter distribution of the approximately 10 million water birds in the Baltic Sea currently experienced will affect the stocks of the prime food resources. The major part of the wintering fauna is composed of benthic herbivores and carnivores. As a result, the shift to a more northerly distribution may have altered the stocks of benthic vegetation in the coastal zone as well as the bivalve stocks in coastal and offshore areas of the Northern Baltic Sea, because bivalve stocks may shift south due to lower salinity.

Although alterations in the populations of seabirds as a direct result of winter climate variability have been suggested, seabirds are less likely to be affected by variations in the temperature regime of the Baltic Sea, due to their wide thermoneutral zone. The wide thermoneutral zone enables seabirds like auks (*Alcidae*) to make deep dives into cold waters to exploit food resources there. Further, by remaining in contact with seawater, seabirds may avoid effects of extreme bouts of very hot or very cold air temperatures.

Forecasts for areas susceptible to global warming have suggested that the Eurasian tundra environment is likely to experience significant environmental alterations, which may induce large-scale declines in populations of several species of water birds (ACIA 2004). This potential, however, has not been confirmed by modelling work or links to trends in numbers of migrating and wintering water birds. Still, the major water bird species wintering in the Baltic Sea are recruited from the breeding grounds of the Siberian tundra. Thus, changes in numbers of wintering water birds in the Baltic Sea as a direct consequence of anthropogenic climate change on the tundra are a possible scenario.

Indirect influences on bird population sizes may be of significance both for the breeding and non-breeding components of the Baltic terrestrial and marine bird fauna. These effects work through the food chain, where even subtle changes in food supply or available habitat may cause food limitation for birds (Arnott and Ruxton 2002). Accordingly, even if current anthropogenic climate changes may not affect breeding bird population sizes directly at the level of biogeographic populations, they may still be able to alter both the breeding success and survival rates significantly via effects on the birds' prey.

Effects on breeding success are mainly related to the same climatic factors as the timing of breeding (see below, Aebischer et al. 1990; Ramos et al. 2002). The breeding success of auks in the North Atlantic has been documented to be related to sea surface temperature (Gjerdrum et al. 2003; Diamond and Devlin 2003), and many species of petrels and shearwaters (*Procellariiformes*) have shown similar trends. Trends, however, seem to differ between groups and species, and so far no studies on the influence of sea surface temperature on seabirds have been carried out for the Baltic Sea. Prey alterations as a result of climatic variability constitute a well-known factor controlling the breeding success of seabirds. This link has been documented outside the Baltic Sea in relation to the supply of capelin in Newfoundland and the Barents Sea (Nakashima 1996; Regehr and Rodway 1999), supply of herring in northern Norway (Anker-Nielssen 1992; Anker-Nielssen and Aarvak

2002) and supply of sandeel in the western North Sea (Rindorf 2000). Similar controls via food resources have not been reported for water birds or terrestrial birds. The same prey alterations that cause breeding success also may affect the survival rate of adult birds. Mass mortality as a result of potentially climate-induced large-scale fluctuations of prey has been reported in piscivorous seabirds feeding on schooling fish in the Barents Sea (Barrett and Krasnov 1996), the Bering Sea (Baduini et al. 2001) and the North Sea (Blake 1984). Climatic effects were usually coupled to anomalous oceanographic conditions that change the distribution and abundance of prey (Harris and Wanless 1996; Piatt and Pelt 1997). Changes in the availability of benthic prey for water birds as a result of anomalous oceanographic conditions have not been documented in the Baltic Sea or neighbouring waters.

Many pelagic seabirds show a strong affinity to areas of strong stratification and stable frontal processes (Hunt and Harrison 1990; Skov and Durinck 2000). Changes in the stability of foraging areas for seabirds in the Baltic as a result of increased precipitation and runoff may alter the possibilities for diving seabirds to find prey. The impact of changes in stratification and water column structure in the Baltic Sea, which may be induced by global warming, on feeding conditions for seabirds could be both positive (enhanced stability) and negative (deepening of the pycnocline).

5.10.7   Phenological Changes

Birds use many cues to time the phases of annual cycle appropriately with the seasonally predictable environmental variation. In addition to this use of cues, birds also possess a circannual clock, which is photoperiodically regulated. In most species use of cues other than photoperiod allows birds to plastically change their response to the environment. Temperature, precipitation, phenology of vegetation and prey items can be such cues. Since 1990, a large number of studies have shown that spring migration of birds has become earlier in areas where winter/spring climates have become warmer, and no change has happened where climate has not ameliorated. Several reviews are now available that serve as a background to the following general observations (Root et al. 2003; Lehikoinen et al. 2004; Crick 2004; Rehfisch et al. 2004).

Spring migration has generally become earlier, but there is a high degree of variability between and within species. The phenology of migration changes relative to anthropogenic climate change is also connected with (indirect) effects of NAO (e.g. Forchammer et al. 2002; Hüppop and Hüppop 2003; Vähätalo et al. 2004). Advancement of spring migration to earlier dates does not mean that fitness consequences exist. A frequent but inaccurate generalisation is that earlier migration leads to earlier breeding that leads to higher breeding success. On average Southwest-Finnish breeding birds have advanced their spring migration by 10 days in the period from 1965 to 2000 (when measured by FAD, first arrival date). Largest advances relate to short range migrants and partial migrants; the smallest but still significant ones relate to late arriving long distance migrants. Baltic marine and coastal species do not differ from terrestrial ones in rate of advancement of FAD (Lehikoinen and Rainio, unpublished). Since FAD is a rather problematic measure of spring arrival, other measures have been used (e.g. Lehikoinen et al. 2004; Sparks et al. 2004). Bird observatories and intensively studied breeding populations offer data with which arrival of the whole population can be studied. These data suggest that not only are the earliest arriving individuals of a population now arriving earlier, but also the "median bird" in many species arrives earlier, although the rate of change is significantly lower (Table 5.6)

Response of the earliest arriving individuals is generally stronger than the response of whole populations and that of the end tail of populations. FAD-data may be biased towards too high rates of change, because it is likely that in more recent years early individuals are observed earlier because of more intense observations and a denser network of observers. MMT data (Table 5.6) are less biased and provide a better basis for determining changes in spring migration patterns.

There is less information on autumn migration. SW-Finnish birds depart on average five days later now than in the 1970s. There is, however, large variability and quite a few species now leave earlier than before. In Rybachi (Kola Peninsula), timing of autumn migration of long-distance migrants was found to be best explained by spring temperatures (Sokolov et al. 1998). This suggests that in these species arrival in spring determines the whole summer schedule including autumn departure – the whole summer season is shifted earlier in these species after warm springs. This is, however, not a general rule. Short distance migrants, which in small passerine bird species can have two clutches

Table 5.6. Trend and response to target area temperature in different bird species. FAD = first arrival date, MMT = mean migration time of the monitored population. Samples include 1–12 time series per species (source: Lehikoinen et al. 2004)

Variable	upper 95% CI	Average	lower 95% CI	n
Trend, FAD (days/year)	−0.342	−0.373	−0.403	590
Local temperature (days/ °C)	−2.472	−2.901	−3.331	203
Trend, MMT	−0.137	−0.180	−0.223	225
Local temperature	−1.433	−1.761	−2.089	153

per summer, tend to extend their summer stay in breeding areas (Jenni and Kery 2003). Their results are in agreement with the Russian ones for long-distance migrants, since also the Swiss study showed that they tend to start autumn migration earlier. This, at first sight unexpected, result may have a biological reason, which is connected with the best time to cross the difficult ecological barriers of the Mediterranean and Sahara. Breeding of long-distance migrants in high latitudes may be time-constrained even in future climate conditions. If short-distance species can extend their breeding period, but the long-distance cannot, the competitive balance between these groups may favour the former group when climate becomes warmer. This hypothesis is further considered below. The aforementioned studies concerned mainly small terrestrial passerines. The situation may be different in coastal and marine species, which usually have a longer breeding period. If the potential breeding period is sufficiently lengthened, a change from single-brooded to double-brooded breeding strategy is possible, but only for small sedentary and short distance migrant passerines. Larger species may have some benefit by having better opportunities to renest after failure. However, time is not the only constraint.

5.10.8 Breeding and Population Sizes

Breeding times of birds have shifted to earlier dates in many, but not all cases (Crick and Sparks 1999; Dunn 2004). This is probably largely what plasticity of timing allows and temperature as a timing cue helps birds to do. The strong dependency of annual breeding time on prevailing temperatures is well known. In a summary by Dunn (2004), which concerns mainly terrestrial birds, laying dates of most species were significantly earlier in warm springs (50 out of 63 cases). A good candidate for studying the anthropogenic

climate change impacts in marine species is the eider *Somateria mollissima*. An early study by Hario and Selin (1986) showed that breeding of the eider became earlier from the 1950s to 1980s although the disappearance of ice cover did not. A recent study by A. Lehikoinen et al. (2006) showed that females are in better condition after mild winters, and the breeding success is therefore also better. The authors expect that amelioration of winter climate may improve breeding success in the eider. Also the mute swan benefits clearly from mild winters (Koskinen et al. 2003). In an analysis covering most of the important marine bird species breeding in the SW-Archipelago of Finland, several species were sensitive to winter climate but also to eutrophication (Rönkä et al. 2005), the effect being strongest in those species which have a tendency to stay as near the breeding area as possible.

5.10.9 What Might Happen in the Future?

We can try to predict the changes of the avifauna in the Baltic from several starting points: (1) historical comparison with earlier periods of similar climate, (2) prediction from population trends, (3) prediction from changes of community structure, (4) prediction from predicted habitat changes and (5) prediction from predicted food availability changes etc. Due to space limitations and scantiness of finished analyses we can only cover part of these. As far as expected life cycle and phenology changes are concerned, we refer to the articles in Møller et al. (2004).

What appeals to us most is to look at data on past warm periods (9,000–5,000 years BP, Boreal and Atlantic chronospheres (Ericson and Tyrberg 2004) and the period 800–1,300), as it provides an opportunity to identify major changes in the avifauna of the Baltic Sea in the absence of major human impact on habitats, which started to take place around 5,000 BP. The unavoidably incomplete species list indicates a surprisingly stable Baltic (Swedish) avifauna (Ericson and Tyrberg 2004, p. 35). Practically all species currently breeding in the Baltic Sea were present already during the Atlantic period. Some, if not most or even all, recent changes are reinvasions and reflect the climate-dependent variability of distribution ranges. On this basis there seems to be no major species turnover to be expected, but the population sizes, regional distribution patterns and community structures are likely to change. However these characteristics are affected by many other factors, not the least anthropogenic, and a thorough evaluation of likely relative changes as a result of anthropogenic climate change is not currently possible, due to the lack of comparative analyses of weather and habitat dependence in most Baltic bird populations.

Since the early 'predictions' or rather educated guesses by Berthold (1990), who classified a number of European bird species in "winners" and "losers" separately into short- and long-distance migratory categories, we have made little progress in predicting future development of bird populations by species. Berthold's general prediction (Berthold 1990) was that sedentary and short-distance species should win and long-distance migrants lose. Only a few Baltic marine, coastal and estuarine species were included in his study, but cormorant, heron, coot and black-headed gull were listed as short distance migrants with high chances of population increase, while crane and lapwing had less chance of population increase according to Berthold (1990). Among long-distance migrants he predicted strong declines for the little bittern, white stork and garganey.

Although recent changes in the distribution and community structure of wintering water birds seem to be driven by decadal climate changes, the future long-term changes of the Baltic water bird populations are very hard to predict. The relationship between the warming of the winter climate and the tendency for more water birds to winter in the northern parts is a good example for the uncertainties regarding future water bird scenarios (ICES 2003b; HELCOM Habitat 2004). It seems a plausible prediction that more water birds will concentrate in the north with amelioration of winter climate. However, this prediction depends on the availability of a surplus of benthic food, especially bivalves in the northern Baltic Sea. This surplus may, in fact, be unlikely to be available,

as bivalve conditions generally will tend to worsen, also in the northern parts, as a function of increasing winter temperatures. Further, it is still uncertain whether the carrying capacity of benthic food stocks for water birds has already reached its limit due to changes in salinity, and adding to that the future growth of bivalve stocks is uncertain due to the lack of reliable predictions about the level of eutrophication in the northern Baltic Sea, especially in the coastal waters of the Baltic States and Russia. At the same time, warming of the tundra breeding areas may cause declines in the populations of water birds visiting the Baltic Sea, enabling a higher proportion of birds to winter in the north. Accordingly, as for the terrestrial bird fauna only monitoring and comparative habitat-weather dependence studies will provide us with the basis for making sound predictions on the development of the water bird fauna in the future.

5.11 Summary

The Baltic Sea is not a steady state system, and, since its formation, it never has been. External drivers acting on different time scales force major changes in the marine ecosystem structure and function. Postglacial isostatic and eustatic processes have shaped the Baltic Sea's coastline, topography, basic chemistry and sedimentary environment on millennial scales (see Annex 2). Climate variability acts on centennial and decadal scales, and, at least over the last 150 years (see Chap. 2), overlaps with human activities in the drainage basin and the coastal zone, leading to considerable changes in the biogeochemistry of this semi-enclosed sea.

Changes in processes and functions in lagoons, estuaries and coastal waters observed over the last 100 years have been generally assumed to be solely driven by anthropogenic forcing. However, to achieve a balanced classification of all drivers acting on the Baltic Sea and to understand the internal mechanisms and ecosystem response to climate variability and anthropogenic climate change on different trophic levels, the reactions to possible anthropogenic climate change scenarios through all trophic levels of the marine ecosystem are described in Chap. 5. Anthropogenic climate change scenarios for the Baltic Sea Basin (see Chap. 3) predict an increase in temperature especially during wintertime and an increase in rainfall in the northern part of the runoff area. The consequence of increasing precipitation is twofold. Increasing precipitation results in a decrease in salinity and in an increase of nutrient leakage and associated eutrophication. The results of Chap. 5 are summarised in the following.

5.11.1   Increase of Nutrients

This is the most difficult topic to summarise, since so much has been written and is constantly being published on it. The topic has been central to HELCOM activities and has interested Baltic Sea researchers and managers for decades. Many hypothetical ideas have been presented about limiting nutrients in different areas and circumstances. The topic is also of large economic interest, e.g. sewage water purification is certainly one of the most expensive outlays for cities and industry. Recently, also blooms of cyanobacteria have been presented as possible threat to recreation and the tourism industry.

Theoretically eutrophication is expected to enhance the production and biodiversity in the ecosystem up to a certain point, after which a collapse will occur due to several mechanisms such as chemical (anoxia) and biotic interactions (competition, predation, exploitation). After this, a new ecological balance will develop, which is characterised by low biodiversity and high variability due to episodic outbursts of dominant species. Some effects of eutrophication are clear and predictable, such as a general increase of primary production (HELCOM 2002), but others, such as interactions between species and between individuals are extremely hard to predict. There is evidence, however, that some of this increase is transferred to higher trophic levels.

Increasing eutrophication is an expected consequence of the anthropogenic climate change in the region due to freshwater runoff determining most of the external nutrient load entering the Baltic Sea, especially in the near coastal areas (Stålnacke et al. 1999; HELCOM 2002). A further effect of eutrophication is the release of phosphorus from anoxic sediments due to internal loading (see Sect. 5.5). The intensity of buoyant surface blooms of cyanobacteria might be enhanced due to such internal eutrophication. They themselves promote it by nitrogen fixation (see Sect. 5.5). This hypothesised situation means that, in order to return to a previous state, nutrient removal from the Baltic Sea would need to be larger than expected on the basis of nutrient concentrations

at the time when symptoms of eutrophication first became apparent.

The hypothesis that buoyant surface blooms of cyanobacteria are a consequence of eutrophication is under debate. Recently, Dippner and Voss (2004) have shown with stable isotopes of ^{15}N that strong blooms of cyanobacteria appeared in a preindustrial period during the Medieval Warm Period, and Bianchi et al. (2000) demonstrated that cyanobacterial blooms are nearly as old as the present brackish water phase of the Baltic Sea, starting as far back as $\sim 7,000$ BP. In addition, the Baltic Proper, which of course receives nutrients from direct precipitation, has a more or less closed basinwide anticyclonic circulation cell (see also Annex 1.1) which reduces the transport of riverborne nutrients into the central Baltic Proper (Voss et al. 2005). The consequence is that increasing inputs of nutrients with higher river runoff will primarily remain in the near coastal areas rather than feed the cyanobacteria blooms in the central Baltic Proper.

Available time-series studies suggest an overall increase in mesozooplankton biomass, both in the open sea and in the gulfs (see Sect. 5.6); however at species level some deviations of this general trend are found due to the influence of hydrographic changes and selective planktivory.

Changes in the level of eutrophication due to anthropogenic climate change will affect the fisheries indirectly through e.g. planktonic food supply to developing larvae of practically all fish species (see Sect. 5.8). Furthermore, adult benthic fish (e.g. flounders and sculpins) will be indirectly affected by anoxia changing their environment into a "benthic desert" (see Sect. 5.7). Generally concerning benthos, biomasses are expected to increase further in areas that are situated above the permanent pycnocline and to decrease in deep water due to anoxia. Anoxia will also be partly responsible for some changes appearing in pelagic fish. Examples are the buoyancy characteristics of the Baltic cod eggs rendering them subject to lethal deep-water concentrations of hydrogen sulphide and low oxygen (see Sect. 5.8).

Predator-prey interactions are likely to change fish species relations e.g. between cod, herring and sprat (see Sect. 5.8). Decreased cod predation pressure on sprat during the last ten years has led to substantially larger sprat stocks and catches. This development was apparently supported by better feeding conditions for sprat larvae feeding on small brackish water copepods (see Sect. 5.8).

Indirect effects of eutrophication are also expected due to changes in phytoplankton composition and biomass as well as decreased transparency of water, which are believed to subsequently cause lack of vitamins in the pelagic food chain, leading to the reproduction failure called M74 in juvenile salmonid fish (see Sect. 5.8).

There already are signs that filamentous algae in the littoral zone have replaced the bladder wrack, that anoxic bottoms with leaking of nutrients exist even in shallow water and that cyanobacteria will be responsible for changes in pelagic nutrient budgets. Generally speaking, nutrient budgets will change with increasing eutrophication and we are likely to see annual plants replace the perennials (HELCOM 1993b).

5.11.2 Increase of Temperature

Predicted increased temperatures, especially during winter, will change growth and reproduction of fauna and flora. It is important to note that a significant part of the Baltic Sea biota, e.g. birds, is mostly of boreal origin, thus adapted to low temperatures in general. In the following we will shortly summarise main conclusions arrived at for different taxa by the co-authors of this chapter.

Inter-annual variation of pelagic bacteria growth is partly dependent on temperature (along with the quality of available organic substrate and competitive relationships, Wikner and Hagström 1999). Furthermore, the activity of pelagic bacteria is stimulated even more than primary production (Pomeroy and Deibel 1986), thus the ratio between bacteria biomass to phytoplankton is expected to increase with temperature in eutrophic waters (Laws 2003).

Recent monitoring data shows that diatom spring blooms are subject to species change when winters become milder (Hajdu et al. 1997). Furthermore, it has been suggested that the diatom bloom itself will be reduced or perhaps even disappear after milder winters, to be replaced by dinoflagellates (see Sect. 5.5, Wasmund et al. 1998). There are signs that increasing summertime temperatures may enhance cyanobacterial blooms. In addition, higher than normal winter temperatures may prevent convection in late winter and early spring with the result that nutrients are not mixed into the upper euphotic zone. In the Baltic Proper, with a salinity of 7 psu, the maximum density of water occurs at $\sim 2.5\,°\mathrm{C}$. If the winter temperature is below $2.5\,°\mathrm{C}$, seasonal surface warming in early spring will result in an unstable water column with convective overturning. If the water temperature is higher than $2.5\,°\mathrm{C}$, warming will result in the development of a thermocline and no redistribution of nutrients due to convection will occur. This process might also result in a shift in species composition of phytoplankton in spring and could counteract cyanobacteria blooms.

The natural annual range of variation in the Baltic Sea surface temperature is from below zero up to over $20\,°\mathrm{C}$. Therefore, it is understandable that so far only a few long-term studies have provided examples of fauna reacting to temperature changes. In an 18-year time series from the northern Baltic Sea, annual peaks of the most abundant cladoceran species co-varied with the seasonal fluctuation of the surface water temperature (Viitasalo et al. 1995, see Sect. 5.6). A shift in dominance was observed within the open sea copepod community from *Pseudocalanus acuspes* to *Acartia* spp., coinciding with higher temperatures during the 1990s (Möllmann et al. 2000, 2003a). Increased production and survival rates of sprat and herring populations during the last 5–10 years covaried with high temperatures and high NAO indices (Kornilovs 1995; Axenrot and Hansson 2003; MacKenzie and Köster 2004). Essentially the same conclusions are given in Sect. 5.8.

In the earlier warming period in Fennoscandia between 1870–1940, many range shifts in birds were observed, both at the northern and southern borders and for spring as well as autumn migration (see Sect. 5.10). Furthermore, extreme winter temperatures have long been documented to influence water bird mortality in the Baltic Sea, and winter conditions in the Baltic Sea Basin are known to determine the ranges of land- as well as water birds (Nilsson 1980). Spring migration now generally occurs earlier, although there is a high variation between and within species, and the phenology of migration changes are also connected with effects of the NAO (e.g. Forchammer et al. 2002; Hüppop and Hüppop 2003; Vähätalo et al. 2004). A recent report places the ringed seal among the species for which a global decline in distribution is "very likely" (ACIA 2005). Modelling studies project the probable extinction of southern subpopulations of the Baltic ringed seal if ice suitable for breeding declines drastically during this century. The grey seal, which has been shown to have the capability to breed extensively on land even in the Baltic (see Sect. 5.9), should be less sensitive.

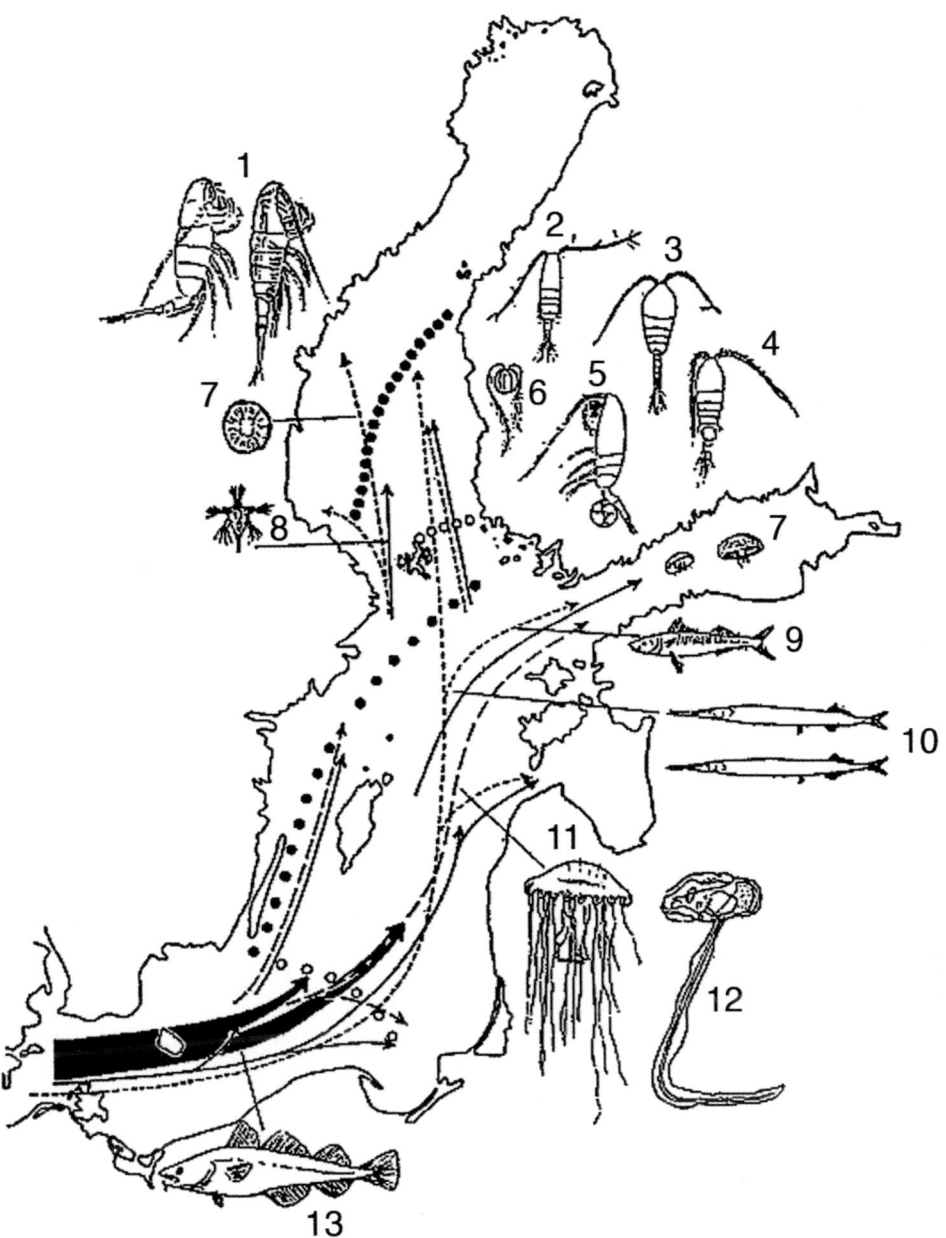

Fig. 5.15. Map of the Baltic Sea, with examples of the most important marine animals which extended/narrowed their range as a consequence of the salinity rise during the first half of the century (from Leppäkoski et al. 1999, modified from Segerstråle 1969). Arrows indicate the direction of spread. Continuous lines: widening of range with reproduction. Broken lines with long strokes: widening of range without reproduction. Broken lines with short strokes: occasional invasion, "guests". **1**: As a brackish water copepod (with low salinity tolerance) *Limnocalanus macrurus* suffered from the rise of the salinity (open circles and black dots indicate roughly limit of range prior to and after the salinity increase, respectively) and moved northward. **2–5**: Large neritic copepods (*Acartia bifilosa*, *Temora longicornis*, *Centropages hamatus* and *Pseudocalanus sp.*, respectively) became more common in the northern part of the Baltic Sea. **6–8**: The ctenophore *Pleurobranchia pileus*, the jellyfish *Aurelia aurita*, and the barnacle *Balanus improvisus* (nauplius larva depicted): same feature. **9–12**: The mackerel *Scomber scombrus*, the garfish *Belone belone*, the jellyfish *Cyanea capillata*, and the tunicate *Fritillaria borealis* became incidental in the northern Baltic Sea. **13**: The cod *Gadus morhua* increased in abundance far and wide in the Baltic Sea

5.11.3 Decrease of Salinity

A decrease of salinity of the Baltic Sea due to increasing precipitation will modify the ecosystem
in several ways. The most important changes will
probably be seen in the distribution (both horizontal and vertical), although growth and reproduction are also likely to be affected. The physiological effect of salinity variation was among the
first problems to be studied in the early years of
Baltic Sea marine biology. Remane (1940) gives
an overview of early osmoregulation studies and
also reports lower salinity limits for several species
of the Baltic Sea fauna. Examples from various
phyla illustrate clearly the expected overall decrease of Baltic Sea marine fauna. The decrease
will be more pronounced and will first be seen
in the northern Baltic Sea surface area, the river
mouths and the Gulfs of Finland and Riga, if the
increase in rainfall in the northern part of the watershed predicted in Chap. 3 of this book takes
place.

According to Remane (1940) the approximate
lower limit of salinity tolerance is 2 psu for *Praunus
flexuosus*, *Neomysis vulgaris*, and *Gammarus locusta*, 3 psu for *Corophium volutator*, 5.5 psu
for *Palaemon adspersus* and *Idotea baltica*, 6 psu
for *Pontoporeia femorata* and *Harmothoe sarsi*,
7 psu for *Pygospio elegans* and *Laomedea lovéni*,
and 7.5 psu for *Terebellides strömii* and *Fabricia sabella*. In the western Baltic Sea the common starfish (*Asterias rubens*) and common shore
crab (*Carcinus maenas*) are among the species
expected to decrease if salinity decreases to less
than 25–15 psu (Remane 1940, Segerstråhle 1957,
Järvekülg 1979). Thus we can expect changes in
species distribution along the complete range of
Baltic Sea surface salinity (for marine species a
decrease) (see also Fig. A.1).

An increase in salinity during the first half of
the 20th century resulted in a spread of several
marine species (e.g. mesozooplankton, barnacle,
jellyfish, larvaceans) towards the north and the
east in the Baltic Sea (Fig. 5.15, Segerstråle 1969).
Correspondingly, the decrease in salinity after the
late 1970s in the northern Baltic Sea (Fig. 5.16)
was reflected in biomass decline of the large neritic copepod species and increase of the freshwater
cladoceran species (Vuorinen et al. 1998). In the
deep basins of the open Baltic Sea, the decrease
in salinity resulted in reduced standing stocks of
Pseudocalanus sp., an important component in
the pelagic food web. In contrast, populations of
species favored by high temperature (e.g. *Acartia*
spp.) increased.

We are likely to see a reversal of the changes of
the 1950s. Some of this expected trend has already
been documented as one of the key species in the
Baltic Proper, the cod, which needs a certain level
of salinity during a certain life stage, now displays
low reproductive success in the Baltic Sea Basin.
Cod eggs need a minimum salinity of 11.5 psu for
buoyancy, which they usually find in the halocline
regions of the deep Baltic Sea basins. Due to the
reduced marine inflow in the last decades, which
resulted in low salinity but also low oxygen concentrations in the deep water, cod eggs are frequently exposed to lethal oxygen conditions in the
layer in which they are neutrally buoyant (Köster
et al. 2005). It has further been indicated that
the decrease in herring and sprat growth is due to
a salinity-mediated change in the copepod community (Rönkkönen et al. 2004; Möllmann et al.
2003a, 2005). A parallel retreat towards the south
has been found in some benthic fauna, e.g. *Scoloplos armiger* (see Sect. 5.7). In addition, the combination of decreasing salinity and increasing temperature will clearly reduce the general fitness of
native benthic species and their adaptability to
cope with other stressors, e.g. low oxygen or chemical pollution (see Sect. 5.3). Finally, decreasing
salinity enables all freshwater species to enlarge
their area of distribution in the Baltic Sea.

One of the key species in the Baltic Proper,
the cod, has already been affected by decreasing
salinities. Fishing is the most important direct
anthropogenic effect on the Baltic Sea, and over-
fishing of cod has contributed to the decrease of
stocks. Again, we are not able to evaluate the
relative importance of decreasing salinity vs. over-
fishing, when discussing the ultimate cause of the
decline in cod stocks. As a top predator in the
pelagic food chain it used to control the sprat and
herring stocks. Due to its decrease and climatically induced enhanced sprat reproduction, the
ecosystem has switched from cod-dominated to
sprat-dominated (recent situation) states. We can
point out, on the basis of zooplankton time series
studies, that there also is a deterioration of zooplankton quality in terms of larval food for cod,
which further contributes to decreased survival
and growth of cod. Finally, increased sprat stocks
may compete for plankton food with cod larvae,
which again would limit cod in the Baltic Sea.

The interaction of all these contributing factors
is, actually, result of peculiarities of the Baltic Sea

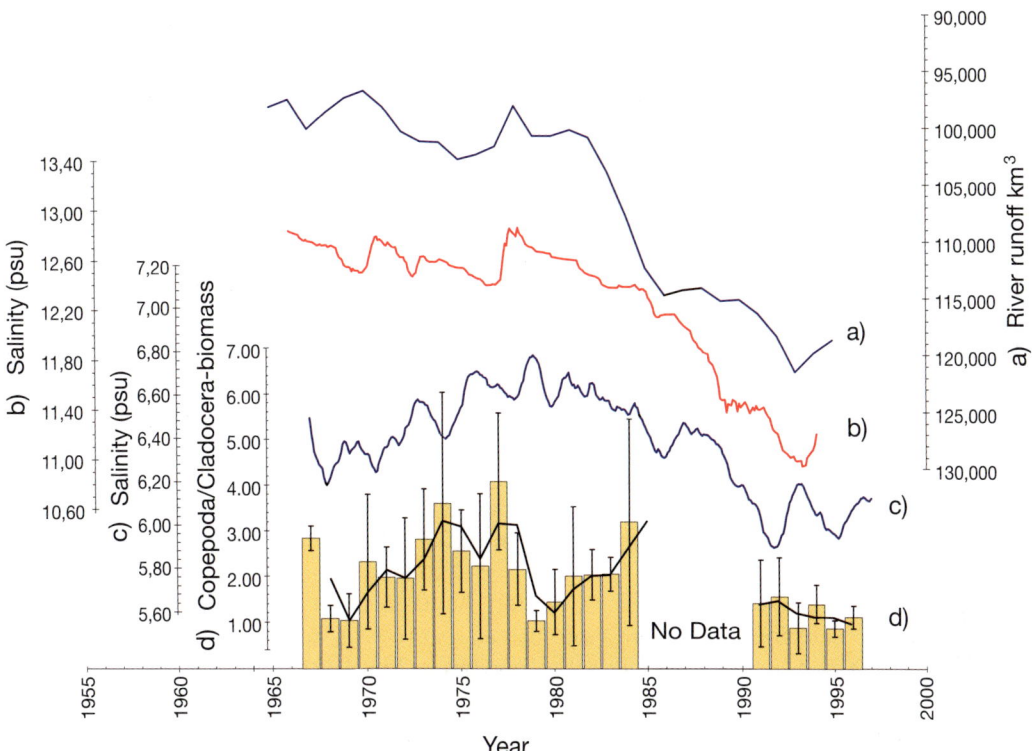

Fig. 5.16. (**a**) Total freshwater runoff to the Baltic Sea (*inverted scale*), (**b**) variation of salinity at 200 meters depth off Gotland, and (**c**) off the Island of Päiväluoto in Northern Baltic, and relation of copepoda/cladocera biomass off Päiväluoto at the depth of 20 meters (modified from Vuorinen et al. 1998)

itself: low salinity, species living near lethally low concentrations of salt, a small water body rendering cod subject to overfishing and the high impact of climatic factors.

Because of its ecological and evolutionary history, the Baltic Sea predominantly is a receiver area for introduced species, donor areas of which are to be found both in the adjacent inland waters and oceanic coasts but also in remote seas. Most of the recent invaders in the Baltic Sea originate from warmer climates (Leppäkoski and Olenin 2001). In conditions of increasing water temperature, not only spontaneously spreading European invaders but also more exotics from warmer regions of the world can be expected to establish themselves in the Baltic Sea.

Two target species known to cause severe changes in invaded ecosystems most likely will spread with climatic warming. The zebra mussel *Dreissena polymorpha* may penetrate to the Gulf of Bothnia, into the areas presently devoid of large biofiltrators. The North American jelly comb *Mnemiopsis leidyi*, which recently invaded

the Black and Caspian Seas, has also invaded the Baltic Sea and possible changes in the pelagic system are not clear at the moment.

Finally, we would like to point out that most of the animal and plant species in the Baltic Sea are incompletely, or not at all, monitored. The examples of time-series presented above are often based on commercial (fish) or recreational (birds) observations, and are in this respect anecdotal and only indicative of biotic consequences of the anthropogenic climate change. It is noteworthy that all climate indications: increase in winter temperatures, precipitation and eutrophication have been changing in the predicted direction during the last decades.

We also want to point out that even with this limited set of relatively short time series we see clear indications of climate change, most likely at least partly of anthropogenic origin in the Baltic Sea animals and plants. Natural and anthropogenic climate change tend to affect biota in the same way, and future effects may turn out to be surprisingly rapid and extensive.

5.12   References

Aaris-Sørensen K (1998) Danmarks forhistoriske dyreverden (Prehistoric animal world of Denmark). Gyldendal, Copenhagen (in Danish)

ACIA (2004) Arctic Climate Impact Assessment. Scientific Report, www.acia.uaf.edu/pages/scientific.html

Ackefors H (1981) Zooplankton. In: Voipio A (ed) The Baltic Sea. Elsevier Oceanography Series 30. Elsevier, Amsterdam, pp. 238–254

Aebischer N, Coulson J, Colebrook J. (1990) Parallel long-term trends across four marine trophic levels and weather. Nature 347:753–755

Ahlbäck R (1955) Kökar. Näringslivet och dess organisation i en utskärssocken (Island of Kökar: Economic life and it's organisation in an outer archipelago parish) Skrifter utgivna av Svenska Litteratursällskapet i Finland 351 (in Swedish)

Ahola M, Laaksonen T, Sippola K, Eeva T, Rainio K, Lehikoinen E (2004) Variation in climate warming along the migration route uncouples arrival and breeding dates. Glob Change Biol 10:1610-1617

Alaee M, Arias P, Sjödin A, Bergman Å (2003) An overview of commercially used brominated flame retardants, their applications, their use patterns in different countries/regions and possible modes of release. Env Int 29:683–689

Albalat A, Potrykus J, Pempkowiak J, Porte C (2002) Assessment of organotin pollution along the Polish coast (Baltic Sea) by using mussels and fish as sentinel organisms. Chemosphere 47:165–171

Alheit J, Hagen E (2001) Climate variability and northwest European fisheries. In: Proceedings of the Hanse Conference on "Past climate and its significance for human history in northwest Europe, the last 10,000 years", Delmenhorst Germany, October 1999, pp. 9–14

Allsopp M, Erry B, Santillo D, Johnston P (2001) POPs in the Baltic – a review of persistent organic pollutants (POPs) in the Baltic Sea. Greenpeace International London, pp. 92

Amano M, Hayano A, Miayzaki N (2002) Geographic variation in the skull of the ringed seal *Pusa hispida.* J Mammal 83,2:370–380

Andersin A-B, Sandler H (1991) Macrobenthic fauna and oxygen deficiency in the Gulf of Finland. Memoranda Societatis Pro Fauna et Flora Fennica 67:3–10

Andersin A-B, Lassig J, Parkkonen L, Sandler H (1978) The decline of macrofauna in the deeper parts of the Baltic Proper and Gulf of Finland. Kieler Meeresforschungen 4:23–52

Andrén E, Andrén T, Kunzendorff H (2000) Holocene history of the Baltic Sea as a background for assessing records of human impact in the sediments of the Gotland Basin. Holocene 10,6:687–702

Aneer G (1975) Composition of food of the Baltic herring (*Clupea harengus* v. *membras* L), fourhorn sculpin (*Myoxocephalus quadricornis* L), and eel-pout (*Zoarces vivparus* L) from deep soft bottom trawling in the Askö-landsort area in two consecutive years. Merentutkimuslait. Julk./ Havsforskningsinst. Skrifter 239:146–154

Anker-Nielssen T (1992) Food-supply as a determinant of reproduction and population development in Norwegian Puffins Fratercula arctica. PhD Thesis, University of Trondheim, Norway

Anker-Nielssen T, Aarvak T (2002) The population ecology of Puffins at Røst. Status after the breeding season 2001. Norsk Institutt for Naturforskning: Oppdragsmelding 736:1–40

Arnosti C, Sagemann J, Jørgensen B, Thamdrup B (1998) Temperature dependence of microbial degradation of organic matter in marine sediments: Polysaccharide hydrolysis, oxygen consumption, and sulfate reduction. Mar Ecol Prog Ser 165:59–70

Arnott SA, Ruxton GD (2002) Sandeel recruitment in the North Sea: Demographic, climatic and trophic effects. Mar Ecol Prog Ser 238:199–210

Arntz WE, Brunswig D (1975) Studies on structure and dynamics of macrobenthos in the western Baltic carried out by the joint research programme "Interaction sea–sea bottom" (SFB 95-Kiel). Proceedings of 10th European Symposium on Marine Biology. Ostend, Belgium, 17–23 Sept 1975, vol. 2

Aro E (2000) The spatial and temporal distribution patterns of cod (*Gadus morhua callaris* L.) in the Baltic Sea, and their dependence on environmental variability – Implications for fishery management. Finnish Game and Fisheries Research Institute, Helsinki.

Aschan M (1988) Soft bottom macrobenthos in a Baltic archipelago: Spatial variation and optimal sampling strategy. Ann Zool Fennici 25:153–164

Assmuth T, Jalonen P (2005) Risks and Management of Dioxin-like Compounds in Baltic Sea Fish: An Integrated Assessment. TemaNord 2005:568, Nordic Council of Ministers, Copenhagen 2005

Autio R (1998) Response of seasonally cold-water bacterioplankton to temperature and substrate treatments. Estuar Coast Shelf Sci 46:465–474

Axenrot T, Hansson S (2003) Predicting herring recruitment from young-of-the-year densities, spawning stock biomass, and climate. Limnol Oceanogr 48:1716–1720

Baduini C, Hyrenbach K, Coyle K, Pinchuk A, Menenhall V, Hunt G (2001) Mass-mortality of short-tailed shearwaters in the south-eastern Bering Sea during summer 1997. Fish Oceanogr 10:117–130

Bagge O (1989) A review of investigations of the predation of cod in the Baltic. Rapp P-V Réun Cons Int Explor Mer 190:51–56

Bagge O, Thurow F, Steffensen E, Bay J (1994) The Baltic cod. Dana 10:1–28

Baker AJ, González PM, Piersma T, Niles LJ, de Lima Serrano do Nascimento I, Atkinson PW, Clark NA, Minton CDT, Peck MK, Aarts G (2004) Rapid population decline in red knots: Fitness consequences of decreased refuelling rates and late arrival in Delaware Bay. Proc Roy Soc London, B 271:875–882

Banta GT, Giblin AE, Hobbie JE, Tucker J (1995) Benthic respiration and nitrogen release in Buzzards Bay, Massachusetts. J Mar Res 53:107–135

Barber DG, Iacozza J (2004) Historical analysis of sea ice conditions in M'Clintoc Channel and the Gulf of Boothia, Nunavut: Implications for ringed seal and polar bear habitat. Arctic 57,1:1–14

Barrett R, Krasnov Y (1996) Recent responses to changes in stocks of prey species by seabirds breeding in the southern Barents Sea. ICES J Mar Sci 53:713–722

Baumann H, Hinrichsen HH, Möllmann C, Köster FW, Malzahn AM, Temming A (2006) Recruitment variability in Baltic Sea sprat (*Sprattus sprattus*) is tightly coupled to temperature and transport patterns affecting the larval and early juvenile stages. Can J Fish Aquat Sci 63:2191–2201

Begg GA, Marteinsdottir G (2002) Environmental and stock effects on spawning origins and recruitment of cod *Gadus morhua*. Mar Ecol Prog Ser 229:263–277

Bergman G (1956) Sälbeståndet vid våra kuster (The seal population at our coast). Nordenskiöld-samfundets tidskrift XVI:49–65 (in Swedish)

Berryman AA (1999) Principles of population dynamics and their application. Stanley Thornes Publishers, Cheltenham, UK

Berthold P (1990) Patterns of avian migration in light of current global "greenhouse" effects: a Central European perspective. Acta XX Congressus Internationalis Ornithologici, vol. II:780–786

Bettarell Y, Amblard C, Sime-Ngando T, Carrias J-F, Sargos D, Garabetian F and Lavandier P (2003) Viral lysis, flagellate grazing potential, and bacterial production in lake Pavin. Microb Ecol 45:119–127

Beyer J, Lassen H (1994) The effect of size-selective mortality on size-at-age Baltic herring. Dana 10:203–234

Bianchi TS, Engelhaupt E, Westman P, Andrén T, Rolff C, Elmgren R (2000) Cyanobacterial blooms in the Baltic Sea: Natural or human-induced? Limnol Oceanogr 45:716–726

Bignert A, Litzen K, Odsjö T, Olsson M, Persson W, Reutergårdh L (1995) Time-related factors influence the concentration of sDDT, PCBs and shell parameters in eggs of Baltic Guillemot (*Uria aalge*). Env Pollut 89:1861–1989

Bignert A, Asplund L, Willander A (2004) Comments Concerning the National Swedish Contaminant Monitoring Programme in Marine Biota. Rapport till Naturvårdsverket, 2004-04-30

Boeschker HTS, de Graf W, Köster M, Meyer-Reil L-A, Cappenberg TE (2001) Bacterial populations and processes involved in acetate and propionate consumption in anoxic brackish sediment. FEMS Microbiol Ecol 35:97–103

Bogovski S, Sergeyev B, Muzyka V, Karlova S (1998) Cytochrome P450 system and heme synthase enzymes activity in flounder liver as biomarkers of marine environments pollution. Mar Env Res 46:13–16

360   5. Climate-related Marine Ecosystem Change

Bonner WN (1972) Grey seal and common seal in European waters. Oceanogr Mar Biol Ann Rev 10: 461–507

Bonsdorff E (2006) Zoobenthic diversity-gradients in the Baltic Sea: Continuous post-glacial succession in a stressed ecosystem. J Exp Mar Biol Ecol 330:383–391

Bonsdorff E, Pearson TH (1999) Variation in the sublittoral macrozoobenthos of the Baltic Sea along environmental gradients: A functional-group approach. Austral Ecol 24:312–326

Bonsdorff E, Aarnio K, Sandberg E (1991) Temporal and spatial variability of zoobenthic communities in archipelago waters of the northern Baltic Sea – consequences of eutrophication? Int Rev Ges Hydrobiol 76,3:433–449

Bonsdorff E, Blomquist EM, Mattila J, Norkko A (1997) Coastal Eutrophication: Cause, consequences and perspectives in the archipelago areas of the northern Baltic Sea. Estuar Coast Shelf Sci 44:63–72

Borg H, Jonsson P (1996) Large-scale metal distribution in Baltic Sea sediments. Mar Pollut Bull 32: 8–21

Brander K, Mohn B (2004) Effect of the North Atlantic Oscillation on recruitment of Atlantic cod (Gadus morhua). Can J Fish Aquat Sci 61:1558–1564

Brettar I, Rheinheimer G (1991) Dinitrification in the Central Baltic: Evidence for H2S-oxidation as motor of denitrification in the oxic-anoxic interface. Mar Ecol Prog Ser 77:157–169

Brey T (1986) Increase in macrozoobenthos above the halocline in Kiel Bay comparing the 1960s with the 1980s. Mar Ecol Prog Ser 28:299–302

Brommer J (2004) The range margins of northern birds shift polewards. Ann Zool Fennici 41:391–397

Burris JE (1980) Vertical Migration of Zooplankton in the Gulf of Finland. Am Midland Nat 103: 316–321

Burton JK (1995) Birds and Climate Change. Helm, London

Canagaratnam P (1959) Growth of fishes in different salinities. J Fish Res Board Can 16:121–129

Cardinale M, Arrhenius F (2000) Decreasing weight-at-age of Atlantic herring (Clupea harengus) from the Baltic Sea between 1986 and 1996: A statistical analysis. ICES J Mar Sci 57:882–893

Cardinale M, Casini M, Arrhenius F (2002) The influence of biotic and abiotic factors on the growth of sprat (Sprattus sprattus) in the Baltic Sea. Aquat Living Resour 15:273–281

Casini M, Cardinale M, Hjelm J (2006) Inter-annual variation in herring Clupea harengus, and sprat, Sprattus sprattus, condition in the central Baltic Sea: What gives the tune? Oikos 112:638–650

Cederwall H (1979) Energy flow and fluctuations of deeper soft bottom communities in the Baltic Sea. Askö Laboratory and Department of Zoology, University of Stockholm, Sweden, pp. 1–23

Cederwall H, Elmgren R (1980) Biomass increase of benthic macrofauna demonstrates eutrophication of the Baltic Sea. Ophelia,Suppl 1:287–304

Clarke A, Harris CM (2003) Polar marine ecosystems: Major threats and future change. Env Conservat 30,1:1–25

Cleveland WS, Devlin SJ (1988) Locally weighted regression: An approach to regression analysisi by local fitting. J American Statistical Assoc 85:596–610

Conley D, Humborg C, Rahm L, Savchuk OP, Wulff F (2002) Hypoxia in the Baltic Sea and basin-scale changes in phosphorus biogeochemistry. Env Sci Technol 36:5315–5320

Crick HQP (2004) The impact of climate change on birds. Ibis 146 (Suppl):48–56

Crick HQP, Sparks TH (1999) Climate change related to egg-laying trends. Nature 399:423–424

Crotty CM, Tyrrell PN, Espie GS (1994) Quenching of Chlorophyll a Fluorescence in Response to Na+ – Dependent HCO3− Transport-Mediated Accumulation of Inorganic Carbon in the Cyanobacterium Synechococcus UTEX 625. Plant Physiol 104:785–791

Dalsgaard T, Canfield DE, Petersen J, Thamdrup B, Acuna-Gonzales L (2003) N-2 production by the anammox reaction in the anoxic water column of Golfo Dulce, Costa Rica. Nature 442:606–608

Dalsgaard T, Thamdrup B, Canfield DE (2005) Anaerobic ammonium oxidation (anammox) in the marine environment. Res Microbiol 145:457–464

Dalziel JA (1995) Reactive mercury in the eastern North Atlantic and southeast Atlantic. Mar Chem 49:307–314

Dannenberger D (1996) Chlorinated microcontaminants in surface sediments of the Baltic Sea: investigations in the Belt Sea, the Arkona Sea and the Pomeranian Bight. Mar Pollut Bull 32:772–781

Davies JH (1957) The geography of the grey seal. J Mammal 38:297–310

Davis AJ, Jenkinson LS, Lawton JH, Shorrocks B, Wood S (1998) Making mistakes when predicting shifts in species range in response to global warming. Nature 391:783–786

Davis BAS, Brewer S Stevenson AC, Guiot J, Data Contributors (2003) The temperature of Europe during Holocene reconstructed from pollen data. Quat Sci Rev 22,15-17:1701–1716

Daw T, Gray T (2005) Fisheries science and sustainability in international policy: a study of failure in the European Union's Common Fisheries Policy. Mar Pol 29:189–197

Demel K, Mulicki Z (1954) Studia ilosciowe nad wydajnoscia biologiczna dna poludniowego Baltyku (Quantitative studies on biological productivity of the southern Baltic bottom). Prace Morsk Inst Ryback w Gdyni 7:75–126 (in Polish)

Depledge M (1990) Interactions between heavy metals and physiological processes in estuarine invertebrates. In: Chambers PL, Chambers CM (eds) Estuarine Ecotoxicology. Japaga, pp. 89–100

de Swart RL, Ross PS, Vos JG, Osterhaus ADME (1996) Impaired immunity in harbour seals (*Phoca vitulina*) exposed to bioaccumulated environmental contaminants: Review of a long-term feeding study. Env Health Perspect 104:823–828

Diamond AW, Devlin CM (2003) Seabirds as indicators of changes in marine ecosystems: Ecological monitoring of Machias Seal Island. Env Monit Assess 88:153–175

Dippner JW, Ikauniece A (2001) Long-term zoobenthos variability in the Gulf of Riga in relation to climate variability. J Mar Sys 30:155–164

Dippner JW, Voss M (2004) Climate reconstruction of the MWP in the Baltic Sea area based on biogeochemical proxies from a sediment record. Baltica 17:5–16

Dippner JW, Kornilovs G, Sidrevics L (2000) Long-term variability of mesozooplankton in the Central Baltic Sea. J Mar Sys 25:23–31

Dippner JW, Hänninen J, Kuosa H, Vuorinen I (2001) The Influence on Climate Variability on zooplankton abundance in the Northern Baltic Archipelago Sea (SW Finland). ICES J Mar Sci 58: 569–578

Doney SC (2006) Plankton in a warmer world. Nature 444:695–696

Dunn P (2004) Breeding dates and reproductive performance. In: Møller A, Fiedler W, Berthold P (2004) Birds and Climate Change. Adv Ecol Res 35:69–85

Durink J, Skov H, Jensen FP, Pihl S (1994) Important Marine Areas for wintering Birds in the Baltic Sea. Report to European Commission DG XI Research Contract 2242/90-09-01. Ornis Consult, Copenhagen, Denmark

Eero M, MacKenzie BR, Köster FW (2005) Developing biomass estimates for the eastern Baltic cod population for the entire 20th century. Havforskermødet 3–5 Feb 2005, Copenhagen, Denmark

Eero M, MacKenzie BR, Karlsdottir HM, Gaumiga R (2007) Development of international fisheries for the eastern Baltic cod (*Gadus morhua*) from the late 1880s until 1938. Fish Res (in press)

Elmgren R (1978) Structure and dynamics of Baltic benthos communities, with particular reference to the relationship between macro- and meiofauna. Kieler Meeresforsch 4:1–22

Elmgren R (1989) Man's impact on the ecosystem of the Baltic Sea: Energy flows today and at the turn of the century. Ambio 18:326–332

Elmgren R (2001) Understanding Human Impact on the Baltic Ecosystem, Changing views in Recent Decades. Ambio 30:222–231

Elmgren R, Rosenberg R, Andersin A-B, Evans S, Kangas P, Lassig J, Leppäkoski E, Varmo B (1984) Benthic macro- and meiofauna in the Gulf of Bothnia (Northern Baltic) Fin Mar Res 250:3–18

Engström P, Dalsgaard T, Huluth S, Aller RC (2005) Anaerobic ammonium oxidation by nitrite (anammox): Implications for N-2 production in coastal marine sediments. Geochim Cosmochim Acta 69:2057–2065

Enghoff IB (1999) Fishing in the Baltic region from the 5th century BC to the 16th century AD: Evidence from fish bones. Archaeofauna 8:41–85

Enghoff IB, MacKenzie BR, Nielsen EE (2007) The Danish fish fauna during the warm Atlantic period (ca. 7,000–3,900 BC): forerunner of future changes? Fish Res (in press)

Ericson G, Lindesjöö E, Balk L (1998) DNA adducts and histopathological lesions in perch (*Perca fluviatilis*) and northern pike (*Esox lucius*) along a polycyclic aromatic hydrocarbon gradient on the Swedish coastline of the Baltic Sea. Can J Fish Aquat Sci 55:815–824

Ericson PGP, Tyrberg T (2004) The early history of the Swedish avifauna. A review of the subfossil record and early written sources. Kungl Vitterhets Historie Och Antikvitets Akademien, Stockholm, Sweden

Falandysz J (2003) The Baltic Sea – Poland. Background document for the 1st Technical Workshop of UNEP/GEF Project Regional Based Assessment of Persistent Toxic Substances- Region III- Europe. TOCOEN REPORT No 241, Brno

Falandysz J, Brzostowski A, Szpunar J, Rodriguez-Pereiro I (2002) Butyltins in sediments and three-spined stickleback (*Gasterosteus aculleatus*) from the marinas of the Gulf of Gdansk, Baltic Sea. J Env Sci Health A 37:353–363

Fenchel T (1988) Marine plankton food chains. Ann Rev Ecol Syst 19:19–38

Ferguson SH, Stirling I, McLoughlin P (2005) Climate change and ringed seal (*Phoca hispida*) recruitment in western Hudson bay. Mar Mamm Sci 21,1:121–135

Fiedler W, Bairlein F, Köppen U (2004) Using large-scale data from ringed bird for the investigation of effects of climate change on migrating birds: pitfalls and prospects. In: Møller A, Fiedler W, Berthold P (eds) Birds and Climate Change. Adv Ecol Res 35: 49–68

FIMR (2003) Ice winter 2002/2003, www.fimr.fi/en/palvelut/jaapalvelu/jaatalvi2002-2003.html

Finni T, Kononen K, Olsonen R, Wallström K (2001) The History of Cyanobacterial Blooms in the Baltic Sea. Ambio 30:172–178

Fisher UR, Velmirov B (2002) High control of bacterial production by viruses in an eutrophic oxbow lake. Aquatic Microb Ecol 27:1–12

Flinkman J (1999) Interactions between plankton and planktivores of the northern Baltic Sea: Selective predation and predation avoidance. Walter and Andrée de Nottbeck Foundation Scientific Reports 18

Flinkman J, Vuorinen I, Aro E (1992) Planktivorous Baltic Herring (*Clupea harengus*) prey selectively on reproducing copepods and cladocerans. Can J Fish Aquat Sci 49:75–77

Flinkman J, Aro E, Vuorinen I, Viitasalo M (1998) Changes in northern Baltic zooplankton and herring nutrition from 1980s to 1990s: Top-down and bottom-up processes at work. Mar Ecol Prog Ser 165:127-136

Fonselius SH (1978) On nutrients and their role as production limiting factors in the Baltic. Acta Hydrochim Hydrobiol 6:329–339

Fonselius SH, Valderrama J (2003) One hundred years of hydrographic measurements in the Baltic Sea. J Sea Res 49:229–241

Forchammer MC, Post E, Stenseth NC (2002) North Atlantic Oscillation timing of long- and short-distance migration. J Anim Ecol 71:1002–1014

Forstén A, Alhonen P (1975) The subfossil seals of Finland and their relations to the history of the Baltic Sea. Boreas 4:143–155

Franck H, Matthäus W (1992) The absence of effective major inflows and the present changes in the hydrographic conditions of the central Baltic deep water. In: Bjornestad, E, L Hagerman, K Jensen (eds) 12th Baltic Mar Biol Symp, Olsen and Olsen, Fredensborg

Frank KT, Petrie B, Choi JS, Leggett WC (2005) Trophic cascade in a formerly cod-dominated ecosystem. Science 308:1621–1623

Gerlach SA (1994) Oxygen conditions improve when the salinity in the Baltic Sea decreases. Mar Poll Bull 28:413–416

Gislason H (1999) Single and multispecies reference points for Baltic fish stocks. ICES J Mar Sci 56: 571–583

Gjerdrum CG, Vallée AMJ, St Clair CC, Bertram DF, Ryder JL, Blackburn GS (2003) Tuften puffin reproduction reveals ocean climate variability. Proceedings of the National Academy of Science of the USA 100:9377–9388

Goodman SJ (1998) Patterns of extensive genetic differentation and variation among Europaean harbour seals (*Phoca vitulina vitulina*) revealed using mitochondrial DNA polymorphisms. Mol Biol Evol 15:104–118

Gustafsson BG (2004) Sensitivity of Baltic Sea salinity to large perturbations in climate. Clim Res 27: 237–251

Hänninen J, Vuorinen I, Hjelt P (2000) Climatic factors in the Atlantic control the oceanographic and ecological changes in the Baltic Sea. Limnol Oceanogr 45:703–710

Hänninen J, Vuorinen I, Kornilovs G (2003) Atlantic climatic factors control decadal dynamics of a Baltic Sea Copepod, Temora longicornis. Ecography 26:672–678

Härkönen TO, Stenman O, Jüssi M, Jüssi I, Sagitov R, Verevkin M (1998) Population size and distribution of the Baltic ringed seal (*Phoca hispida botnica*). In: Heide-Jørgensen MP, Lydersen C (eds) Ringed seals in the North Atlantic NAMMCO Scientific publications, vol. I, Tromsø, Norway

Härkönen T, Harding KC, Goodman SJ, Johannesson K (2005) Colonization history of the Baltic harbor seals: Integrating archaeological, behavioural, and genetic data. Mar Mamm Sci 21: 695–716

Hagström Å, Pinhassi J, Zweifel UL (2001) Marine bacterioplankton show bursts of rapid growth induced by substrate shifts. Aquat Microb Ecol 24:109–115

Hajdu S, Larsson U, Skärlund K (1997) Växtplankton. In: Elmgren R, Larsson U (Eds) Himmerfjärden. Förändringar i ett näringsbelastat kustekosystem i Östersjön. (Phytoplankton. In: Himmerfjärden. Changes in a nutrient-enriched coastal ecosystem of the Baltic Sea). Naturvårdsverket (SEPA) 4565:63-79 (in Swedish)

Hajdu S, Pertola S, Kuosa H (2005) *Prorocentrum minimum* (Dinophyceae) in the Baltic Sea: morphology, occurrence – a review. Harmful Algae 4:471–480

Halkka A, Helle E, Helander B, Jüssi I, Karlsson O, Soikkeli M, Stenman O, Verevkin MF (2005) Number of grey seals counted in censuses in the Baltic Sea 2000–2004. Symposium on Biology and Management of Seals in the Baltic Area. Riista- ja kalaraportteja 346

Hall AJ, McConnell BJ, Barker RJ (2001) Factors affecting first-year survival in grey seals and their implications for life history strategy. J Anim Ecol 70,1:138–149

Hall LW Jr, Anderson RD (1995) The influence of salinity on the toxicity of various classes of chemicals to aquatic biota. Crit Rev Toxicol 25:281–346

Hammond PS, Berggren P, Benke H, Borchers DL, Collet A, Heide-Jørgensen MP, Heimlich S, Hiby AR, Leopold MF, Øien N (2002) Distribution and abundance of the harbour porpoise and other cetaceans in the North Sea and adjacent waters. J Appl Ecol 39:361–376

Hansen FC, Möllmann C, Schütz U, Neumann T (2006) Spatio-temporal distribution and production of calanoid copepods in the Central Baltic Sea. J Plankton Res 28:39–54

Hansen S, Lavigne DM (1997) Temperature effects on the breeding distribution of grey seals (*Halichoerus grypus*). Physiol Zool 70,4:436–443

Hansen S, Lavigne DM, Innes S (1995) Energy metabolism and thermoregulation in juvanile harbor seals (*Phoca vitulina*) in air. Physiol Zool 68,2:290–315

Hansson S (1985) Effects of eutrophication on fish communities with special reference to the Baltic Sea – a literature review. Rep Inst Freshw Res Drottningholm 62:36–56

Harding KC, Härkönen TJ (1999) Development in the Baltic grey seal (*Halichoerus grypus* and ringed seal (*Phoca hispida*) populations during the 20th century. Ambio 28,7:619–625

Harding KC, Fujiwara M, Axberg Y, Härkönen T (2005) Mass-dependent energetics and survival in harbour seal pups. Funct Ecol 19:129–135

Harding KC, Härkönen T, Helander B, Karlsson O (2007) Population assessment and risk analysis of Baltic grey seals. In: Haug T, Hammill M. Olafsdottir D, Pike DG (eds) Grey Seals in the North Atlantic and the Baltic. NAMMCO Scientific Publications 6, in press

Hario M, Selin K (1986) Mitä pesinnän ajoittuminen kertoo haahkan menestymisestä Suomenlahdella (A 30 year change in the breeding time of the common eider in the Gulf of Finland). Suomen Riista 33:19–25 (in Finnish)

Harris M, Wanless S (1996) Differential responses of Guillemots (*Uria aalge*) and Shag (*Phalacrocorax aristotelis*) to a late winter wreck. Bird Study 43:220–230

Heath MR, Gallego A (1998) Bio-physical modelling of the early life stages of haddock in the North Sea. Fish Oceanogr 7:110–125

Hegseth EN, E Sakshaug (1983) Seasonal variation in light- and temperature-dependent growth of marine planktonic diatoms in in-situ dialysis cultures in the Trondheimsfjord, Norway (63° N). J Exp Mar Biol Ecol 67:199–220

Heiskanen A-S, Gran V, Lehtoranta J, Pitkänen H (2000) Fate of nutrients (N,P) along the estuarine gradient of the River Neva in the eastern Gulf of Finland, Baltic Sea. Short communication. ICES J Mar Sci (Suppl) 56:161–164

HELCOM (1990) Second Periodic Assessment of the State of the Marine Environment of the Baltic Sea, 1984-1988. Background document – BSEP No 35B

HELCOM (1993) First assessment of the State of the Coastal Waters in the Baltic Sea. Baltic Sea Environment Proceedings No 54

HELCOM (1996) Third periodic assessment of the state of the marine environment of the Baltic Sea, 1989–93. Background document. Baltic Sea Environment Proceedings 64B

HELCOM (2002) Fourth periodic assessment of the state of the marine environment of the Baltic Sea, 1994–98. Background document. Baltic Sea Environment Proceedings 82B

HELCOM (2003a) Proceedings of the joint IMO/HELCOM/EU Workshop "Environmental impacts due to the increased density of shipping in the Baltic Sea area – Copenhagen plus 1" Baltic Sea Environment Proceedings No 86

HELCOM (2003b) The Baltic Marine Environment 1999-2002. Baltic Sea Environment Proceedings No 87

HELCOM (2004) Dioxin in the Baltic Sea Helsinki Commission. Baltic Marine Environment Protection Commission

HELCOM (2005) Nutrient Pollution to the Baltic Sea in 2000. Baltic Sea Environment Proceedings No 100

HELCOM HABITAT (2004) Development of Baltic waterbird monitoring strategy – Pilot Phase: evaluation of available data and conclusion on necessary follow-up activities Document for HELCOM HABITAT 6, 2004

Herbert R (1999) Nitrogen cycling in coastal marine ecosystems. FEMS Microbiol Rev 23:563–590

Hettler WF (1976) Influence of temperature and salinity on routine metabolic rate and growth of young Atlantic menhaden. J Fish Biol 8:55–65

Hiby L, Lundberg T, Karlsson O, Watkins J, Jüssi M, Jüssi I, Helander B (2007) Estimates of the size of the Baltic grey seal population based on photo-identification data. In: Haug T, Hammill M. Olafsdottir D, Pike DG (eds) Grey Seals in the North Atlantic and the Baltic. NAMMCO Scientific Publications 6 (in press)

Hietanen S, Kuparinen J (2005) Nitrogen fluxes at the sediment-water interface (SEGUE-N). Tvärminne Studies 10:33

Hinrichsen HH, Möllmann C, Voss R, Köster FW, Kornilovs G (2002) Bio-physical modelling of larval Baltic cod (*Gadus morhua*) survival and growth. Can J Fish Aquat Sci 59:1958–1873

Hoar WS (1952) Thyroid function in some anadromous and landlocked teleosts. Transactions of the Royal Society of Canada. VOL XLVI Ser III:39–53

Holm P, Bager M (2001) The Danish fisheries c. 1450–1800. Medieval and early modern sources and their potential for marine environmental history. In: Holm P, Smith TD (eds) Exploited Seas: Directions for Marine Environmental History. St John's, Newfoundland

Holm P, Smith TD, Starkey DJ (2001) The exploited seas: New directions for marine environmental history. Int Maritime Econ Hist Assoc, Maritime Studies Research Unit, Memorial University of Newfoundland, St John's, Newfoundland, Canada

Hook O, Johnels AG (1972) Breeding and distribution of the grey seal (*Halichoerus grypus Fab*) in the Baltic Sea with observations of other seals of the area. Proc R Soc Lond B 182:37–58

Hoppe HG (1981) Blue-green algae agglomeration in surface water: A biotope of high algal activity. Kieler Meeresforsch 5:291–303

Horstmann U (1975) Eutrophication and mass production of blue-green algae in the Baltic. Merentutkimuslaitos Julk/Havsforskningsinst Skr 239:83–90

Hübel H, Hübel M (1980) Nitrogen fixation during blooms of *Nodularia* in coastal waters and back-waters of the Arkona Sea (Baltic Sea) in 1974. Int Rev Ges Hydrobiol 65:793–808

Hüppop O, Hüppop K (2003) North Atlantic Oscillation and timing of spring migration in birds. Proc R Soc Lond, Ser B: Biol Sci 270:233–240

Humborg C, Conley DJ, Rahm L, Wulff F, Cociasu A, Ittekkot V (2000) Silicon retention in river basins: Far-reaching effects on biogeochemistry and aquatic food webs in coastal marine environments. Ambio 29:45–50

Hunt GL Jr, Harrison NM (1990) Foraging habitat and prey taken by least auklets at King Island, Alaska. Mar Ecolol Prog Ser 65:141–150

ICES (1997) Report of the Working Group on the Assessment of Pelagic Stocks in the Baltic ICES CM 1997/Assess: 12

ICES (2003a) Report of the Study Group on Multispecies Assessment in the Baltic ICES CM 2003/H:03

ICES (2003b) A review of the status and trends of seabirds in the Baltic Sea. ICES CM 2003/C: 03 Copenhagen

ICES (2004a) Report of the Advisory Committee on Fisheries Management and Advisory Committee on Ecosystems. ICES Advice 1:1–1544

ICES (2004b) Report of the Baltic Fisheries Assessment Working Group ICES CM 2004/ACFM:22

ICES (2004c) Report of the Baltic Salmon and Trout Assessment Working Group. ICES CM 2004/ACFM: 23

Ikauniece A (2001) Long-term abundance dynamics of coastal zooplankton in the Gulf of Riga. Env Internat 26:175–181

Jackson JBC, Kirby MX, Berger WH, Bjorndal KA, Botsford LW, Bourque BJ, Bradbury RH, Cooke R, Erlandson J, Estes JA, Hughes TP, Kidwell S, Lange CB, Lenihan HS, Pandolfi JM, Peterson CH, Steneck RS, Tegner MJ, Warner RW (2001) Historical overfishing and the recent collapse of coastal ecosystems. Science 293:629–638

Järvekülg A (1979) Donnaja fauna vosochnoj chasti Baltijskogo morja (Bottom fauna of the Eastern part of the Baltic Sea). Valgus, Tallinn (in Russian)

Janssen F, Neumann T, Schmidt M (2004) Inter-annual variability in cyanobacteria blooms in the Baltic Sea controlled by wintertime hydrographic conditions. Mar Ecol Prog Ser 275:59–68

Jarre-Teichmann A, Wieland K, MacKenzie BR, Hinrichsen HH, Plikshs M, Aro, E (2000) Stock-recruitment relationships for cod (*Gadus morhua callarias* L) in the central Baltic Sea incorporating environmental variability. Arch Fish Mar Res 48:97–123

Jennings S, Kaiser MJ (1998) The effects of fishing on marine ecosystems. Adv Mar Biol 34:201–352

Jenni L, Kéry M (2003) Timing of autumn bird migration under climate change: Advances in long-distance migrants, delays in short-distance migrants. Proc R Soc Lond B 270:1467–1471

Jørgensen BB (1977) Bacterial sulfate reduction within reduced microniches of oxic marine sediments. Mar Biol 41:7–17

Jørgensen BB (1996) Material flux in the sediment. In: Jørgensen BB, Richardson K (eds) Eutrophication in coastal marine ecosystems. American Geophysical Union Washington DC

Jonsson P (2000) Sediment Burial of PCBs in the offshore Baltic Sea. Ambio 29:260–267

Jonsson P, Carman R, Wulff F (1990) Laminated sediments in the Baltic – A tool for evaluating nutrient mass balances. Ambio 19: 152–158

Jonsson P, Grimvall A, Cederlof A, Hilden M (1996) Pollution threats to the Gulf of Bothnia. Ambio Special Report 8:21–26

Jonzen N, Cardinale M, Gardmark A, Arrhenius F, Lundberg P (2002) Risk of collapse in the eastern Baltic cod fishery. Mar Ecol Prog Ser 240:225–233

Jüssi I, Jüssi M (2001) Action plan for Grey Seals in Estonia 2001–2005. Eesti Ulikud (Estonian Game) 7:64

Kändler R (1949) Die Häufigkeit pelagischer Fischeier in der Ostsee als Maßstab für die Zu- und Abnahme der Fischbestände (The abundance of pelagic fish eggs in the Baltic Sea as a measure for the rise and decline of fish stocks). Kieler Meeresforsch VI:73–98 (in German)

Kahru M (1997) Using satellites to monitor large-scale environmental change: a case study of cyano-bacteria blooms in the Baltic. In: Kahru M, Brown CW (eds) Monitoring algal blooms. Springer, Berlin Heidelberg New York

Kahru M, Horstmann U, Rud O (1994) Satellite detection of increased cyanobacteria blooms in the Baltic Sea: Natural fluctuations or ecosystem change? Ambio 23:469–472

Kahru M, Leppänen JM, Rud O, Savchuk OP (2000) Cyanobacteria blooms in the Gulf of Finland triggered by saltwater inflow into the Baltic Sea. Mar Ecol Prog Ser 207:13–18

Kalejs M, Tamsalu R (1984) Ocherki po biologicheskoj produktivnosti Baltijskogo morja (The salinity regime of the Baltic Sea. The temperature regime of the Baltic Sea, T.1). In: Essays on the Baltic Sea biological productivity. Moscow, pp. 33–67 (in Russian)

Kalela O (1946) Zur Ausbreitungsgeschichte der Vögel vegetationsreicher Seen (On the migration his-tory of birds from lakes with strong vegetation). Annales Academiae Scientiarum Fennicae. A.IV. Biologica 12:1–81 (in German)

Kalela O (1949) Changes in geographic ranges in the avifauna of northern and central Europe in relation to recent changes in climate. Bird Banding 20:77-103

Kannan K, Falandysz J (1997) Butyltin residues in sediment, fish, fish-eating birds, harbour porpoise and human tissues from the Polish coast of the Baltic Sea. Mar Pollut Bull 34:203–207

Kannan K, Senthilkumar K, Duda CA, Villeneuve DL, Falandysz J, Giesy J (1999) Butyltin com-pounds in sediment and fish from the Polish coast of the Baltic Sea. Env Sci Pollut Res 6:200–206

Karjalainen M (2005) Fate and effects of *Nodularia spumigena* and it's toxin, Nodularin, in Baltic Sea planktonic food webs. Contributions Finnish Institute of Marine Research 10:2005

Karlson K, Rosenberg R, Bonsdorff E (2002) Temporal and spatial large-scale effects of eutrophica-tion and oxygen deficiency on benthic fauna in Scandinavian and Baltic waters: A review. Oceanogr Mar Biol Ann Rev 40:427–489

Kelly BP (2001) Climate change and ice breeding pinnipeds. In: Walther GR, Burga CA, Edwards PJ (eds) "Fingerprints" of climate change: Adapted behaviour and shifting species' ranges. Kluwer, New York London, pp. 43–55

Kinze CC (1995) Exploitation of harbour porpoises (*Phocoena phocoena*) in Danish waters: a historical review. Rap Int Whal Comm Spec Issue 16:141–153

Kirkkala T, Helminen H, Erkkilä A (1998) Variability of nutrient limitation in the Archipelago Sea, SW Finland. Hydrobiologica 363:117–126

Köster FW, Möllmann C (2000a) Trophodynamic control by clupeid predators on recruitment success in Baltic cod? ICES J Mar Sci 57:310–323

Köster FW, Möllmann C (2000b) Egg cannibalism in Baltic sprat *Sprattus sprattus*. Mar Ecol Prog Ser 196:269–277

Köster FW, Schnack D (1994) The role of predation on early life stages of cod in the Baltic. Dana 10: 179–201

Köster FW, Hinrichsen HH, St John MA, Schnack D, MacKenzie BR, Tomkiewicz J, and Plikshs M (2001) Developing Baltic cod recruitment models. II. Incorporation of environmental variability and species interaction. Can J Fish Aquat Sci 58:1534–1556

Köster FW, Möllmann C, Neuenfeldt S, Vinther M, St John MA, Tomkiewicz J, Voss R, Hinrichsen HH, Kraus G, Schnack D (2003) Fish stock development in the Central Baltic Sea (1976-2000) in relation to variability in the physical environment. ICES Marine Science Symposia 219:294–306

Köster FW, Möllmann C, Hinrichsen HH, Tomkiewicz J, Wieland K, Kraus G, Voss R, MacKenzie BR, Schnack D, Makarchouk A, Plikshs M, Beyer JE (2005) Baltic cod recruitment – the impact of climate variability on key processes. ICES J Mar Sci 62:1408–1425

Kornilovs G (1995) Analysis of Baltic herring year-class strength in the Gulf of Riga. ICES CM 1995/J:10

Kornilovs G, Sidrevics L, Dippner JW (2001) Fish and zooplankton interaction in the Central Baltic Sea. ICES J Mar Sci 58: 579–588

Koschinski S (2002) Current knowledge on harbour porpoises (*Phocoena phocoena*) in the Baltic Sea. Ophelia 55,3:167–197

Koskinen P, Saari L, Nummi P, Pellikka J (2003) Kannan tiheys ja sääolot vaikuttavat lisääntymismenestykseen kyhmyjoutsenella (Stock density and weather conditions affect the reproduction success of the mute swan). Suomen Riista 49:17–24 (in Finnish)

Kostrichina EM (1977) Long-term dynamics of zooplankton in the Baltic sea in connection with the changes in water regime. In: Rybokhozyajstvennye issledovanija (BaltNIIRKH) Issue 13 Riga, "Zwaigzne":70–77 (in Russian with English summary)

Kostrichina E (1984) The dynamics of zooplankton abundance and biomass in the south-eastern, eastern and north-eastern Baltic Articles on biological productivity of the Baltic Sea. Moscow 2:204–241 (in Russian)

Kostrichina EM, Sidrevits LL (1977) Some regularities in zooplankton distribution in the Baltic Sea by regions. In: Rybokhozyajstvennye issledovanija (BaltNIIRKH) Issue 13 Riga, "Zwaigzne":178–87 (in Russian with English summary)

Kostrichina EM, Yurkovskis AK (1982) On the relation of the dynamics in zooplankton abundance in the Baltic with phosphorus and Kattegat waters advection. In: Rybokhozyajstvennye issledovanija (BaltNIIRKH). Riga, Avots 17:3–10 (in Russian with English summary)

Kotta J, Kotta I (1995) The state of macrozoobenthos of Pärnu Bay in 1991 as compared to 1959–1960. Proc Estonian Acad Sci Ecol 5:26–37

Kousa H, Kuparinen J, Wikner J (1996) Pelagic Biology of the Gulf of Bothnia. In HELCOM 1996, Third Periodic Assessment of the State of the Baltic Sea, 1989–1993; Background document – Balt Sea Env Proc No 64 B, p. 40

Kowalewska G, Konat J (1997) Distribution of polynuclear aromatic hydrocarbons (PAHs) in sediments of the southern Baltic Sea. Oceanologia 39:83–104

Kremling K, Streu P (2000) Further evidence for a drastic decline of potentially hazardous trace metals in Baltic Sea surface waters. Mar Poll Bull 40:674–679

Kremling K, Streu P (2001) The behavior of dissolved Cd, Co, Zn and Pb in North Atlantic near-surface waters (30° N/60° W to 60° N/2° W). Deep-Sea Res I, 48:2541–2567

Kube J, Gosselck F, Powilleit M, Warzocha J (1997) Long-term changes in the benthic communities of the Pomeranian Bay (southern Baltic Sea). Helgoländer Meeresuntersuchungen 51:399–416

Kuparinen J, Tuominen L (2001) Eutrophication and self-purification: Counteractions forced by large-scale cycles and hydrodynamic processes. Ambio 30:190–194

Kuypers MMM, Sliekers AO, Lavik G, Schmid M, Jørgensen BB, Kuenen JG, Damste JSS, Strous M, Jetten MSM (2003) Anaerobic ammonium oxidation by anammox bacteria in the Black Sea. Nature 422:608–611

Laine AO (2003) Disrtibution of soft-bottom macrofauna in the deep open Baltic Sea in relation to environmental variability. Estuar Coast Shelf Sci 57:87–97

Laine AO, Sandler H, Andersin AB, Stigzelius J (1997) Long-term changes of macrozoobenthos in the Eastern Gotland Basin and the Gulf of Finland (Baltic Sea) in relation to the hydrographical regime. J Sea Res 38:135–159

Lanning G, AS Cherkasiv, Sokolova IM (2006) Temperature-dependent effects of cadmium on mitochondrial and whole-organism bioenergetics in Oysters. Mar Environ Res 62:S79–82

Larsson U, Elmgren R, Wulff F (1985) Eutrophication and the Baltic Sea: Causes and consequences. Ambio 14:9–14

Lassig J, Niemi A (1978) Vertical distribution and diurnal fluctuations of zooplankton in the Gotland Deep, June 1969. A Baltic year study. Kieler Meeresforsch 4:188–193

Lavigne DM (2002) Harp seal. In: Perrins WF, Würsig B, Thewissen JGM (eds) Encyclopedia of Marine Mammals. Academic Press, New York

Laws EA (2003) Partitioning of microbial biomass in pelagic aquatic communities: Maximum resiliency as a food web organizing construct. Aquat Microb Ecol 32:1–10

Laws EA, Falkowski PG, Smith WO, Ducklow H, McCarthy JJ (2000) Temperature effects on export production in the open ocean. Glob Biogeochem Cy 14,4:1231–1246

Lee BG, Wallace WG, Luoma SN (1998) Uptake and loss kinetics of Cd, Cr and Zn in the bivalves *Poamocorbula amurenis* and *Macoma balthica:* Effects of size and salinity. Mar Ecol Prog Ser 175: 177–189

Lee H, Swartz RC (1980) Biological processes affecting the distribution of pollutants in marine sediments. Part II. Biodeposition and Bioturbation. In: Baker RA (ed) Contaminants and Sediments, vol 2. Ann Arbor Science Publ, Ann Arbor, Michigan

Lehikoinen E, Gustafsson E, Aalto T, Alho P, Laine J, Klemola H, Normaja J, Numminen T and Rainio K (2003a) Varsinais-Suomen Linnut Turun (Birds of SW-Finland). Lintutieteellinen Yhdistys ry, Turku

Lehikoinen E, Lemmetyinen R, Vuorisalo T (2003b) Linnuston ja lintututkimuksen historia Varsinais-Suomessa (History of birdlife and bird research in SW-Finland) In: Lehikoinen E, Gustafsson E, Aalto T, Alho P, Laine J, Klemola H, Normaja J, Numminen T, Rainio K (Eds) Varsinais-Suomen Linnut Turun (Birds of SW-Finland). Lintutieteellinen Yhdistys ry, Turku (in Finnish)

Lehikoinen E, Sparks T, Zalakevicius M (2004) Arrival and departure dates. In: Møller A, Fiedler W, Berthold P (2004) Birds and Climate Change. Adv Ecol Res 35:1–31

Lehikoinen A, Kilpi M, Öst M (2006) Winter climate affects subsequent breeding success of common eiders. Glob Change Biol 12:1355–1365

Lehtonen KK, Schiedek D (2006) Monitoring biological effects of pollution in the Baltic Sea: Neglected – but still wanted? Mar Poll Bull 53:377–386

Lehtonen KK, Schiedek D, Koehler A, Lang T, Vuorinen PJ, Förlin L, Baršien? J, Pempkowiak J, Gercken J (2006) The BEEP project in the Baltic Sea: Overview of results and outlines for a regional biological effects monitoring strategy. Mar Pollut Bull 53:523–537

Leipe T, Hille S, Voss M, Bartholdy J, Christiansen C (2005) Sedimentary records of environmental changes of the central Baltic Sea during the past 1000 years. Sopot (Poland), 5th Baltic Sea Science Congress

Leonardsson K, Andersin AB, Mäkinen A, Rönnberg O (1997) Benthic biology. In: HELCOM (ed) Third periodic assessment of the State of the Baltic Sea, 1989–1993; Background document. Balt Sea Env Proc No 64 B, pp. 42–46

Lepiksaar J (1986) The Holocene history of theriofauna in Fennoscandia and Baltic countries. Striae 24:51–70

Leppäkoski E (1975) Assessment of degree of pollution on the basis of macrozoobenthos in marine and brackishwater environments. Acta Acad Aboensis B 35 2:1–90

Leppäkoski E, Bonsdorff E (1989) Ecosystem variability and gradients: Examples from the Baltic Sea as a background for hazard assessment. In: Landner L (Ed) Chemicals in the aquatic environment: Advanced Hazard Assessment. Springer, Berlin Heidelberg New York, pp. 6–58

Leppäkoski E, Olenin S (2001) The meltdown of biogeographical peculiarities of the Baltic Sea: The interaction of natural and man made processes. Ambio 30,4–5:202–209

Leppäkoski E, Helminen H, Hänninen J, Tallqvist M (1999) Aquatic biodiversity under anthropogenic stress: An insight from the Archipelago Sea (SW Finland). Biodiversity and Conservation 8: 55–70

Leppäkoski E, Gollasch S, Gruszka P, Ojaveer H, Olenin S, Panov V (2002) The Baltic – A sea of invaders. Can J Fis Aquat Sci 59:1175–1188

Lima M, Jaksic FM (2004) The impacts of ENSO on terrestrial ecosystems: A comparision with NAO. In: Stenseth NC, Ottersen G, Hurrell JW, Belgrano A (eds) Marine Ecosystems and Climate variation. Oxford University Press

Lindqvist C, Possnert G (1997) The subsistence economy and diet at Jakob/Ajvide, Eksta parish and other prehistoric dwelling and burial sites on Gotland in long-term perspective. In: Burenhult G (ed) Remote Sensing: Applied techniques for the study of cultural resources and the localization, identification and documentation of sub-surface prehistoric remains in Swedisch Arcaealogly. I Thesis and Papers in North-European Archaeology 19:a Hässleholm

Lindström G (1855) Bidrag till kännedomom Östersjöns invertebratfauna (Contribution to knowledge of invertebrate fauna of the Baltic Sea). Öfversigt KVA:s Förh Arg 12 Stockholm:49–73 (in Swedish)

Line R, Sidrevics L (1995) Zooplankton in the Gulf of Riga. In: Ojaveer E (ed) Ecosystem of the Gulf of Riga between 1920 and 1990. Estonian Academy Publishers, Tallinn, pp. 175–186

Lõugas L (1997) Post-glacial development of vertebrate fauna in Estonian water bodies. A paleozoological study. Dissertationes Biologicae Universitatis Tartuensis 32. Tartu University Press

Luckenbach T, Corsi I, Epel D (2004) Fatal attraction: Synthetic musk fragrances comprise multi-xenobiotic defence systems in mussels. Mar Env Res 58:215–219

Lumberg AJ (1976) On Zooplankton in the Guld of Finland. In: Rybokhozyajstvennye issledovanija (BaltNIIRKH) 12. Riga, "Zwaigzne" (in Russian with English summary)

Lumberg A, Ojaveer E (1991) On the Environment and Zooplankton Dynamics in the Gulf of Finland in 1961–1990. Eesti Teaduste Akadeemia Toimetised. Ecology 40:131–140

Lumberg A, Ojaveer H (1997) Zooplankton dynamics in Muuga and Kolga Bays in 1975–1992, with particular emphasis to the summer aspect. In: Ojaveer E (Ed) Proceedings of the 14th Baltic Marine Biologists Symposium, Pärnu, Estonia, 5–8 August 1995. Estonian Academy Publishers, Tallinn, pp. 139–148

MacKenzie BRM, FW Köster (2004) Fish production and climate: Sprat in the Baltic Sea. Ecology 85,3:784–794

MacKenzie BR, Hinrichsen HH, Plikshs M, Wieland K, Zezera AS (2000) Quantifying environmental heterogeneity: Estimating the size of habitat for successful cod egg development in the Baltic Sea. Mar Ecol Prog Ser 193:143–156

MacKenzie, BR, Alheit J, Conley DJ, Holm P, and Kinze CC (2002a) Ecological hypotheses for a historical reconstruction of upper trophic level biomass in the Baltic Sea and Skagerrak. Can J Fish Aquat Sci 59:173–190

MacKenzie, BR, Awebro K, Bager M, Holm P, Lajus J, Must A, Ojaveer H, Poulsen B, and Uzars D (2002b) Baltic Sea fisheries in previous centuries: Development of catch data series and preliminary interpretations of causes of fluctuations. ICES CM 2002 (2002/L: 02)

MacKenzie BR, Gislason H, Möllmann C, Köster FW (2007) Impact of fishing on fish biodiversity and the ecosystem of the Baltic Sea during climate change. Glob Change Biol 13:1–20 doi:10.1111/j.1365-2486.2007.01369.x

Mannermaa K (2002) Bird bones from Jettböle I, a site in the Neolithic Åland – Archipelago in the northern Baltic. Acta Zoologica Cracoviensia 45:85–98

Marsh G, Athanasiadou M, Bergman A, Asplund L (2004) Identification of hydroxylated and methoxylated polybrominated diphenyl ethers in Baltic Sea salmon (*Salmo salar*) blood. Env Sci Technol 38:10–18

Marshall CT, O'Brien L, Tomkiewicz J, Köster FW, Kraus G, Marteinsdottir G, Morgan MJ, Saborido-Rey F, Blanchard JL, Secor DH, Wright PJ, Mukhina NV, Björnsson H (2003) Developing Alternative Indices of Reproductive Potential for Use in Fisheries Management: Case Studies for Stocks Spanning an Information Gradient. J Northw Atl Fish Sci 33:161–190

Matthäus W, Franck H (1992) Characteristics of major Baltic inflows – a statistical analysis, Cont Shelf Res 12:1375–1400

Matthäus W, Lass HU (1995) The recent salt inflow into the Baltic Sea. J Phys Oceanogr 25:280–286

Matthäus W, Nausch G (2003) Hydrographic-hydrochemical variability in the Baltic Sea during the 1990s in relation to changes during the 20th century. ICES Marine Science Symposia 219:132–143

McLusky DD, Bryant V, Campell R (1986). The effect of temperature and salinity on the toxicity of heavy metals to marine and estuarine invertebrates. Oceanogr Mar Biol Ann Rev 24:481–520

Meier HEM, Döscher R, Halkka A (2004) Simulated distribututions of Baltic Sea-ice in warming climate and consequences for the Winter habitat of the Baltic Ringed Seal. Ambio 33,4–5:249–256

Miettinen M, Halkka A, Högmander J, Keränen S, Mäkinen A, Nordström M, Nummelin J, Soikkeli M (2005) The ringed seal in the Archipelago sea, SW-Finland, population size and survey techniques. Symposium on Biology and Management of Seals in the Baltic area, Riista- ja kalaraporteja 346

Miller K, Irving I (1975) Metabolism and temperature regulation in young harbor seals (*Phoca vitulina richardi*) in water. Am J Physiol 229:509–511

Milukaite A, Gulbinskas S (1997) Application of investigations of polycyclic aromatic hydrocarbons for the evaluation of sediments pollution in the Klaipeda Strait and the Baltic Sea coastal zone. Ekologija 2:44–49

Møhl U (1970) Fangstdyrene ved de danske strande (Seal and whale hunting on the danish coasts). Den zoologiske baggrund for harpunerne. KUML. Årbok for Jysk arkæologisk selskab 1970. København 1971 (in Danish with English summary)

Møller A, Fiedler W, Berthold P (eds) (2004) Birds and Climate Change. Advances in Ecological Research 35. Elsevier

Molinero JC, Ibanez F, Nival P (2005) North Atlantic climate and northwestern Mediterranean plankton variability. Limnol Oceanogr 50:164–171

Möllmann C, Köster FW (1999) Food consumption by clupeids in the Central Baltic: Evidence for top-down control? ICES J Mar Sci 56:100–113

Möllmann C, Köster FW (2002) Population dynamics of calanoid copepods and the implications of their predation by clupeid fish in the Central Baltic Sea. J Plankton Res 24:959–978

Möllmann C, Kornilovs G, Sidrevics L (2000) Long-term dynamics of main mesozooplankton species in the central Baltic Sea. J Plankton Res 22:2015-2038

Möllmann C, Köster FW, Kornilovs G, Sidrevics L (2002) Long-term trends in abundance of cladocerans in the Central Baltic Sea. Mar Biol 141:434–452

Möllmann C, Köster FW, Kornilovs G, Sidrevics L (2003a) Interannual variability in population dynamics of calanoid copepods in the Central Baltic Sea. ICES Marine Science Symposia 219:220–230

Möllmann C, Kornilovs G, Fetter M, Köster FW, Hinrichsen HH (2003b) The marine copepod, *Pseudocalanus elongatus*, as a mediator between climate variability and fisheries in the Central Baltic Sea. Fish Oceanogr 12,4–5:360–368

Möllmann C, Kornilovs G, Fetter M, Köster FW (2004) Feeding ecology of central Baltic Sea herring and sprat. J Fish Biol 65:1563–1581

Möllmann C, Kornilovs G, Fetter M, Köster FW (2005) Climate, zooplankton and pelagic fish growth in the Central Baltic Sea. ICES J Mar Sci 62:1270–1280

Mudryk JM, Podgórska B, Bolalek J (2000) The occurrence and activity of sulphate-reducing bacteria in the bottom sediments of the Gulf of Gdansk. Oceanologia 42:150–117

Nakashima B (1996) The relationship between oceanographic conditions in the 1990s and changes in spawning behaviour, growth and early life history of capelin (*Mallotus villosus*). NAFO Scientific Council Research Document 94/74:18

Nausch G, Nehring D, Aertebjerg G (1999) Anthropogenic nutrient load of the Baltic Sea. Limnologica 29:233–241

Nehring D (1984) The further development of the nutrient situation in the Baltic Proper. Ophelia Suppl 3:167–179

Nehring D (1989) Phosphate and nitrate trends and the ratio oxygen consumption to phosphate accumulation in central Baltic deep waters with alternating oxic and anoxic conditions. Beiträge zur Meereskunde Berlin 59:47–58

Nehring D (1990) Die hydrographisch-chemischen Bedingungen in der westlichen und zentralen Ostsee von 1979 bis 1988 – ein Vergleich (The hydrographic-chemical conditions in the western and central Baltic Sea from 1979 to 1988 – a comparison). Meereswiss Berichte des Instituts für Meereskunde, Warnemünde 2:3–45 (in German)

Nehring S (1994) First living *Alexandrium minutum* resting cysts in Western Baltic. Harmful Algae News 9:1–2

Nehring S (1997) Giftalgen. Der Dinoflagellat *Gymnodinium catenatum* (Toxic Algae. The dinoflagellate *Gymnodinium catenatum*). Mikrokosmos 86:151–156 (in German)

Nehring S (1998) Establishment of thermophilic phytoplankton species in the North Sea: Biological indicators of climatic changes? J Mar Sci 55:818–823

Nielsen E, Richardson K (1996) Can changes in fisheries yield in the Kattegat (1950–1992) be linked to changes in primary production? ICES J Mar Sci 53:988–994

Nielsen E, Møller Hansen M, Schmidt C, Meldrup D, and Grønkjær P (2001) Genetic differences among cod populations. Nature 413:272

Nilsson L (1980) Wintering diving duck populations and available food resources in the Baltic Wildfowl 31:131–143

Nilsson P, Jansson M, Brydsten L (2003) Retention and long term accumulation of EOCl from pulp mill effluents in a Baltic Sea recipient. Water Air Soil Pollut 143:225–243

Nissling A (1994) Survival of eggs and yolk-sac larvae of Baltic cod (*Gadus morhua* L) at low oxygen levels in different salinities. ICES Marine Science Symposium 198:626–631

Nissling A (2004) Effects of temperature on egg and larval survival of cod (*Gadus morhua*) and sprat (*Sprattus sprattus*) in the Baltic Sea – implications for stock development. Hydrobiologia 514:115–123

Nissling A, Westin L (1997) Salinity requirements for successful spawning of Baltic and Belt Sea cod and the potential for cod stock interactions in the Baltic Sea. Mar Ecol Prog Ser 152:261–271

Nissling A, Kryvi H, Vallin L (1994) Variation in egg buoyancy of Baltic cod *Gadus morhua* and its implications for egg survival in prevailing conditions in the Baltic Sea. Mar Ecol Prog Ser 110:67–74

Nissling A, Larsson R, Vallin L, Frohlund K (1999) Assessment of egg and larval viability in cod, *Gadus morhua* – methods and results from an experimental study. Fish Res 38:169–186

Nissling A, Müller A, Hinrichsen HH (2003) Specific gravity and vertical distribution of sprat (*Sprattus sprattus*) eggs in the Baltic Sea. J Fish Biol 63:280–299

Nixon SW (1995) Coastal marine eutrophication: A definition, social causes, and future concerns. Ophelia 41:199–219

Nürnberg GK (1984) The prediction of internal phosphorus load in lakes with anoxic hypolimnia. Limnol Oceanogr 29:111–124

Nyman M, Bergknut M, Fant ML, Raunio H, Jestoi M, Bengs C, Murk A, Koistinen J, Backman C, Pelkonen O, Tysklind M, Hirvi T, Helle E (2003) Contaminant exposure and effects in Baltic ringed and grey seals as assessed by biomarkers. Mar Environ Res 55:73–99

Ojaveer E, Lumberg A, Ojaveer H (1998) Highlights of zooplankton dynamics in Estonian waters (Baltic Sea). ICES J Mar Sci 55:748–755

Ojaveer H, Simm H, Lankov A (2004) Population dynamics and ecological impacts of the non-indigenous *Cercopagis pengoi* in the Gulf of Riga (Baltic Sea). Hydrobiologia 522:261–269

Olenin S (1997) Benthic zonation of the Eastern Gotland Basin. Neth J Aquat Ecol 30,4:265–282

Olsen EM, Heino M, Lilly GR, Morgan MJ, Brattey J, Ernande B, Dieckmann U (2004) Maturation trends indicative of rapid evolution preceded the collapse of northern cod. Nature 428:932–935

Oswald S, Huntley B, Hamer KC (2004) Exploring the impact of climate on the distribution of great skuas breeding in the UK. Abstract, 8th International Seabird Group Conference, Aberdeen, 2–4 April 2004

Otterlind G (1984) On fluctuations of the Baltic cod stock. ICES CM 1984/J:14

Ottersen G, Stenseth NC, Hurrell JW (2004) Climatic fluctuations and marine systems: A general introduction to the ecological effects. In: Stenseth NC, Ottersen G, Hurrell JW, Belgrano, A (eds) Marine Ecosystems and Climate Variation. Oxford University Press

Paerl H, Gallucci K (1985) Role of chemotaxis in establishing a specific cyanobacterial-bacterial association. Science 227:647–649

Palo JU, Mäkinen HS, Helle E, Stenman O, Väinölä R (2001) Microsatellite variation in ringed seals (Phoca hispida): Genetic structure and history of the Baltic population. Heredity 86:609–617

Parmanne R, Rechlin O, Sjöstrand B (1994) Status and future of herring and sprat stocks in the Baltic Sea. Dana 10:29–59

Partensky F, Sournia A (1986) Le dinoflagellé *Gyrodinium* cf *aureolum* dans le plancton de l'Atlantique nord: identification, écologie, toxicité (The dinoflagellate *Gyrodinium* cf *aureolum* in the plankton of the North Atlantic: Identification, ecology, toxicity). Cryptogamie Algologie 7:251–275 (in French)

Pauly D (1995) Anecdotes and shifting baseline syndrome of fisheries. TREE 10:430

Pauly D, Christensen V, Dalsgaard J, Froese R, Torres FJr (1998) Fishing down marine food webs. Science 279:860–863

Pearson TH (2001) Functional group ecology in soft-sediment marine benthos: The role of bioturbation. Oceanogr Mar Biol Ann Rev 39:233–268

Persson LE (1981) Were macrobenthic changes induced by thinning out of flatfish stocks in the Baltic Proper? Ophelia 20,2:137–152

Persson LE (1987) Baltic eutrophication: A contribution to the discussion. Ophelia 27,1:31–42

Pershing AJ, Greene CH, Planque B, Fromentin JM (2004) The influences of climate variability on North Atlantic zooplankton populations. In: Stenseth NC, Ottersen G, Hurrell JW, Belgrano A (eds) Marine Ecosystems and Climate Variation. The North Atlantic: A Comparative Perspective. University Press, Oxford

Perus J, Bonsdorff E (2003) Long-term changes in macrozoobenthos in the Aland archipelago, northern Baltic Sea. J Sea Res 52:45–56

PEX (1989) Baltic Sea Patchiness Experiment – PEX 86 Part 1 (1,2) General report. ICES Cooperative Research Report 163

Pfeifer S, Schiedek D, Dippner JW (2005) Effect of temperature and salinity on acetylcholinesterase activity, a common pollution biomarker, in Mytilus sp from the south-western Baltic Sea. J Exp Mar Biol Ecol 320:93–103

Piatt J, Pelt T (1997) Mass-mortality of Guillemots (Uria aalge) in the Gulf of Alaska in 1993. Mar Pollut Bull 34:656–662

Piskorska-Pliszczynska J, Grochowalski A, Wijaszka T, Kowalski B (2004) Levels of PCDD and PCDF in fish edible tissues from Polish coastal waters. Organohalogen Compounds 66:1947–1951

Pitkänen H, Kangas P, Miettinen V, and Ekholm P (1987) The state of the Finnish coastal waters in 1979–1983. Vesi- ja ympäristöhallinnon julkaisuja 8

Pohl C, Hennings U (1999) The effect of redox processes on the partitioning of Cd, Pb, Cu, and Mn between dissolved and particulate phases in the Baltic Sea. Mar Chem 65:41–53

Pohl C, Kattner G, Schulz-Baldes M (1993) Cadmium, copper, lead and zinc on transects through Arctic and Eastern Atlantic surface and deep waters. J Mar Syst 4:17–29

Pomeroy LR, Deibel D (1986) Temperature regulation of bacterial activity during spring bloom in Newfoundland coastal waters. Science 233:359–361

Pomeroy LR, WJ Wiebe (2001) Temperature and substrates as interactive limiting factors for marine heterotrophic bacteria. Aquatic Microb Ecol 23:187–204

Pope JG, Macer CT (1996) An evaluation of the stock structure of North Sea cod, haddock, and whiting since 1920, together with a consideration of the impacts of fisheries and predation effects onthe biomass and recruitment. ICES J Mar Sci 53:1157–1169

Pouchet G, de Guerne J (1885) Sur la faune pelagique de la mer Baltique et du Golfe de Finlande (On the pelagic fauna of the Baltic Sea and the Gulf of Finland). Compt Rend Seances Acad Sciences 100:919–921 (in French)

Poutanen EL, Nikkilä K (2001) Carotenoid pigments of tracers of cyanobacterial blooms in recent and post-glacial sediments of the Baltic Sea. Ambio 30:179–183

Räisänen J, Hansson U, Ullerstig A, Döscher R, Graham LP, Jones C, Meier HEM, Samuelsson P, Willén U (2004) European climate in the late twenty-first century: Regional simulations with two driving global models and two forcing scenarios. Clim Dyn 22:13–31

Ramos JA, Maul AM, Ayrton V, Bullock I, Hunter J, Bowler J, Castle G, Mileto R, Pacheco C (2002) Influence of local and large-scale weather events and timing of breeding on tropical roseate tern reproductive parameters Mar Ecol Prog Ser 243: 271–279

Ranta E, Vuorinen I (1990) Changes in species abundance relations in marine meso- zooplankton at Seili, Northern Baltic Sea, in 1967–1975. Aqua Fennica 20:171–180

Rasmussen H, Jørgensen BB (1992) Microelectrode studies of seasonal oxygen uptake in a coastal sediment: Role of molecular diffusion. Mar Ecol Progr Ser 81:289–303

Reeves RR (1998) Distribution, abundance and biology of ringed seals (Phoce hispida): An overview. NAMMCO Sci Publ 1:9–45

Regehr H, Rodway M (1999) Seabird breeding performance during two years of delayed Capelin arrival in the Northwest Atlantic: A multi-species comparison. Waterbirds 22:60–67

Rehfisch MM, Feare CJ, Jones NV, Spray C (eds) (2004) Climate Change and Coastal Birds. Ibis 146 (Suppl 1)

Reijnders PJH (2003) Reproductive and developmental effects of environmental organochlorines on marine mammals. In: Vos J, Bossart G, Fournier M, O'Shea T (eds) Toxicology of Marine Mammals. Taylor & Francis LTD, London, pp. 55–56

Remane A (1940) Einführung in die zoologische Ökologie der Nord- und Ostsee. Die Tierwelt der Nord- und Ostsee (Introduction to Zoological Ecology of the North and Baltic Sea), vol. I. Becker & Ehler, Leipzig, pp. 1–238 (in German)

Remane A, Schlieper K (1958) Die Biologie des Brackwassers. Die Binnengewässer (The Biology of the Brackish Water. The Inland Waters), vol. 22. E Schweizerbart'sche Verlagsbuchhandlung, Nägele u Obermiller, Stuttgart (in German)

Renssen H, Goosse H, Fichefet T, Brovkin V, Driesschaert E, Wolk F (2005) Simulating the Holocene climate evolution at northern high latitudes using a coupled atmosphere–sea ice–ocean-vegetation model. Clim Dyn 24:23–43

Renz J, Hirche HJ (2006) Life-cycle of *Pseudocalanus acuspes* Giesbrecht (Copepoda, Calanoida) in the central Baltic Sea: I. Seasonal and spatial distribution. Mar Biol 148:567–580

Rice DL (1986) Early Diagenesis in Bioadvective Sediments: Relationships between the Diagenesis of Beryllium-7, Sediment Reworking Rates, and the Abundance of Conveyor-Belt Deposit-Feeders. J Mar Res 44,1:149–184

Rindorf A, Wanless S, Harris MP (2000) Effects of changes in sandeel availability on the reproductive output of seabirds. Mar Ecol Prog Ser 202:241-252

Rönkä M, Saari L, Lehikoinen E, Suomela J, Häkkilä K (2005) Environmental changes and population trends of waterfowl. Ann Zool Fenn 42:587–602

Rönkkönen S, Ojaveer E, Raid T, Viitasalo M (2004) Long-term changes in Baltic herring (*Clupea harengus membras*) growth in the Gulf of Finland. Can J Fish Aquat Sci 61:219–229

Rönner U, Sörensen F (1985) Denitrification rates in the low-oxygen waters of the stratified Baltic proper. Appl Environ Microbiol 50: 801–806

Root TL, Price JT, Hall KR, Schneider SH, Rosenzweig C, Pounds JA (2003) Fingerprints of global warming on plants and animals. Nature 421:57–60

Roots O (2003) Environmental levels of PTS in Estonia. Background document for the 2nd Technical Workshop of UNEP/GEF Project Regional Based Assessment of Persistent Toxic Substances – Region III – Europe TOCOEN REPORT No 243, Brno

Ross PS, De Swart RL, Reijnders PJH, van Loveren H, Vos JG, Osterhaus ADME (1995) Contaminant-related suppression of delayed type hypersenisitivity and antibody responses in harbour seals fed herring from the Baltic Sea. Environ Health Perspectives 103:162–167

Rothe F (1941) Quantitative Untersuchungen über die Planktonverteilung in der östlichen Ostsee (Quantitative investigations on the plankton distribution in the eastern Baltic Sea). Berichte der deutschen wissenschaftlichen Kommission für Meeresforschung, Neue Folge 103:291–368 (in German)

Routti H, Nyman M, Bäckman C, Koistinen J, Helle E (2005) Accumulation of dietary organochlorines and vitamins in Baltic seals. Mar Env Res 60:267–287

Rudstam LG, Aneer G, Hildén M (1994) Top down control in the pelagic Baltic ecosystem. Dana 10: 105–129

Rumohr H (1987) A. Hameier's contribution to our knowledge of the benthos of the Baltic. Mitteilungen aus dem Zoologischen Museum der Universität Kiel, II, 5:1–32

Russak V (1994) Is the Radiation Climate in the Baltic Sea Region Changing. Ambio 23:160–163

Sahrhage D, Lundbeck J (1992) A history of fishing. Springer, Berlin Heidelberg New York

Sandén P, Håkansson B (1996) Long-term trends in Secchi depth in the Baltic Sea. Limnol Oceanogr 4:346–351

Sarmiento JL, Hughes TMC, Stouffer RJ, Manabe S (1998) Simulated response of the ocean carbon cycle to anthropogenic climate warming. Nature 393:245–249

Schinke H, Matthäus W (1998) On the causes of major Baltic inflows – an analysis of long time series – trends, errors and discontinuities. Cont Shelf Res 18,1:67–97

Schneider R, Schiedek D, Petersen GI (2000) Baltic cod reproductive impairment: Ovarian organochlorine levels, hepatic EROD activity, development success of eggs and larvae, challenge tests. ICES CM 2000/S:09

Schnell S, Schiedek D, Schneider R, Balk L, Vuorinen PJ, Vuontisjärvi H, Lang T (2003) Some indications of contaminant effects on Baltic cod (*Gadus morhua* L). ICES CM2003/M:09,1–12

Segerstråle SG (1957) Baltic Sea. In: Hedgepeth JW (Ed) Treatise on Marine Ecology and Paleoecology. Geol Soc America Memoir 67:751–800

Segerstråle SG (1969) Biological fluctuations in the Baltic Sea. Prog Oceanogr 5:169–184

Selin H, VanDeveer SD (2004) Baltic Sea hazardous substances management: Results and challenges. Ambio 33:153–160

Shapiro J (1990) Current beliefs regarding dominance by blue-greens: The case for the importance of CO_2 and pH. Verh Int Verein Limnol 24:38–54

Shurin AT (1968) Status of bottom fauna during period from 1900 to 1962 in the changing conditions of the Baltic Sea. In: Fishery Research (Rybokhozyaistvennye Issledovaniya, Baltniirkh) Riga, Zvaigzne 4:61–88 (in Russian)

Sidrevits LL (1980) Investigation on ecological characteristics of main zooplankton species in the central Baltic. In: Rybokhozyajstvennye issledovanija (BaltNIIRKH), Riga, Avots 65–70 (in Russian with English summary)

Simm MA (1976) On Zooplankton. Ecology in the Bay of Pyarny. In: Rybokhozyajstvennye issledovanija (BaltNIIRKH). Issue 12. Riga, "Zwaigzne", pp. 29–43 (in Russian with English summary)

Sjöblom V (1961) Wanderungen des Strömlings (*Clupea harengus* L) in einigen Schären- und Hochseegebieten der Nördlichen Ostsee (Migration of the herring (*Clupea harengus* L) in archipelagos and open waters of the northern Baltic Sea). Merentutkimuslaitoksen Julk./Havsforskningsinstitutets Skr 199 (in German)

Skei J, Larsson P, Rosenberg R, Jonsson P, Olsson P and Broman D (2000) Eutrophication and Contaminants in Aquatic Ecosystems. Ambio 29,4:184–194

Skov H, Durinck J (2000) Seabird distribution in relation to hydrography in the Skagerrak. Cont Shelf Res 20,2:169–187

Smith TG, Stirling I (1975) The breeding habitat of the ringed seal (*Phoca hispida*). The birth lair and associated structures. Can J Zool 53:1297–1305

Smith TG, Hammill MO, Taugbøl G (1991) A review of the developmental, behavioural and physiological adaptations of the ringed seal, *Phoca hispida*, to life in Arctic winter. Arctic 44,2:124–131

Sokolov LV, Markovets MY, Shapoval AP, Morozov YG (1998) Long-term trends in the timing of spring migration of passerines on the Courish Spit of the Baltic Sea. Avian Ecol Behav 1:1–21

Somero G (2005) Linking biogeography to physiology: Evolutionary and acclimatory adjustments of thermal limits. Front Zool 2, doi:10.1186/1742-9994-2-1

Sparholt H (1994) Fish species interactions in the Baltic Sea. Dana 10:131–162

Sparholt H, Jensen IB (1992) The effect of cod predation on the weight-at-age of herring in the Baltic. ICES Marine Science Symposia 195:448–491

Sparks TH, Bairlein F, Bojarinova JG, Hüppop O, Lehikoinen EA, Rainio K, Sokolov LV, Walker D (2004) Examining the total arrival distribution of migratory birds. Glob Change Biol 11:22–30

Stal LJ, Albertano P, Bergman B, von Bröckel K, Gallon JR, Hayes PK, Sivonen K, and Walsby AE (2003) BASIC: Baltic Sea Cyanobacteria. An investigation of the structure and dynamics of water blooms of cyanobacteria in the Baltic Sea – responses to a changing environment. Cont Shelf Res 23: 1695–1714

Stal LJ, Staal M, Villbrandt M (1999) Nutrient control of cyanobacterial blooms in the Baltic Sea. Aquat Microb Ecol 18:165–173

Stålnacke P, Grimvall A, Sundblad K, Tonderski A, (1999) Estimation of Riverine Loads of Nitrogen and Phosphorus to the Baltic Sea, 1970–1993. Env Monit Assess 58:173–200

Stenman O (1990) Hyljekuolemat itäisellä Suomenlahdella vähäjäisen talven 1988–1989 seurauksena (Seal deaths in the eastern Gulf of Finland as a consequence to scarcity of of ice in the winter 1988–1989). Kalamies 4:12–13 (in Finnish)

Stenseth NC, Ottersen G, Hurrell JW, Belgrano A (2004) Marine Ecosystems and Climate variation. Oxford University Press

Stirling I, Smith TG (2004) Implications of warm temperatures, and an unusual rain event for the survival of ringed seals on the coast of southeastern Baffin Island. Arctic 57,1:59–67

Storå J (2002) Seal hunnting on Ajvide: A taphonomic study of seal remains from a pitted ware culture site on Gotland. In: Burenhult G (ed) Remote sensing, vol. II. Theses and papers in North-European Archaeology 13:b

Storå J, Ericson PGP (2004) A prehistoric breeding population of harp seals (*Phoca groenlandica*) in the Baltic sea. Mar Mamm Sci 20,1:115–133

Storå J, Lõugas L (2005) Human exploitation and history of seals in the Baltic during the late Holocene. In: Monks G (ed) The exploitation and cultural importance of sea mammals. Oxbow Books, Oxford, pp. 95–106

STORE (2003) Environmental fisheries influences on fish stock recruitment in the Baltic Sea. EU-Project FAIR CT98 3959, Final Report

Storstrøms Amt, Teknik- og Miljøforvaltningen, Vandmiljøkontoret (2002) The "Baltic Carrier" Oil Spill Monitoring and Assesment of Environmental Effects in Grønsund (DK)

Strand J, Jacobsen JA (2002) Imposex in two sublittoral neogastropods from the Kattegat and Skagerrak: The common whelk *Buccinum undatum* and the red whelk *Neptunea antiqua*. Mar Ecol Prog Ser 244:171–177

Strand J, Jacobsen JA, Pedersen B, Granmo A (2003) Butyltin compounds in sediment and molluscs from the shipping strait between Denmark and Sweden. Environ Pollut 124:7–15

Struck U, Pollehne F, Bauerfeind E, von Bodungen B (2004) Sources of nitrogen for the vertical particle flux in the Gotland Sea (Baltic Proper) – results from sediment trap studies. J Mar Syst 45:91–101

Suikkanen S, Laamanen M, Huttunen M (2007) Long-term changes in summer phytoplankton communities of the open northern Baltic Sea. Estuar Coast Shelf Sci 71,2-4:580–592

Szefer P (2002) Metal pollutants and radionuclides in the Baltic Sea – an overview. Oceanologica 44: 129–178

Teilmann J, Lowry N (1996) Status of the harbour porpoise (*Phocoena phocoena*) in Danish waters. Rep Int Whal Commn 46:619–625

Thamdrup B, Hansen JW, BB Jørgensen (1998) Temperature dependence of aerobic respiration in a coastal sediment. FEMS Microbiol Ecol 25:189–200

Theede H (1984) Ökosystem Ostsee verändert sich ständig. (The Baltic Sea ecosystem is constantly changing) Naturwiss Rundschau 37:225–227 (in German)

Theede H, Ponat A, Hiroki K, Schlieper C (1969) Studies on the resistance of marine bottom invertebrates to oxygen-defiency and hydrogen sulphide. Mar Biol 2:325–337

Thomas LD, Lennon JJ (1999) Birds extend their ranges northwards. Nature 399:213

Thurow F (1997) Estimation of the total fish biomass in the Baltic Sea during the 20th century. ICES J Mar Sci 54:444–461

Toresen R, Østvedt OJ (2000) Variation in abundance of Norwegian spring spawning herring (*Clupea harengus, Clupeidae*) throughout the 20th century and the influence of climatic fluctuations. Fish Fish 1:231–256

Tortell PD, Reinfelder JR, Morel FMM (1997) Active uptake of bicarbonate by diatoms. Nature 390:243–244

Tuomi T, Kuupo P (1999) Viral lysis and grazing loss of bacteria in nutrient- and carbon-manipulated brackish water enclosures. J Plankton Res 21:923–937

Tynan CT, DeMaster DP (1997) Observations and predictions of Arctic climate change: Potential effects on marine mammals. Arctic 50,4:308–322

Ukkonen P (2002) The early history of seals in the northern Baltic. Ann Zool Fenn 39:187–207

UNEP (2002) Europe Regional Report: Regionally based assessment of persistent substances. GE03-00167-January 2003-500, UNEP/CHEMICALS/2003/3, pp. 142

Vähätalo A, Rainio K, Lehikoinen A, Lehikoinen E (2004) Spring arrival of birds depends on North Atlantic Oscillation. J Avian Biol 35:210-216

Vallin L, Nissling A (2000) Maternal effects on egg size and egg buoyancy of Baltic cod, *Gadus morhua:* Implications for stock structure effects on recruitment. Fish Res 49:21–37

Vallin L, Nissling A, Westin L (1999) Potential factors influencing reproductive success of Baltic cod, *Gadus morhua:* A review. Ambio 28:92–99

Viitasalo M (1992) Mesozooplankton in the Gulf of Finland and Northern Baltic Proper – a review of monitoring data. Ophelia 35:146–168

Viitasalo M, Katajisto T, Vuorinen I (1994) Seasonal dynamics of *Acartia bifilosa* and *Eurytemora affinis* (Copepoda, Calanoida) in relation to abiotic factors in the northern Baltic Sea. Hydrobiol 292/293:415–422

Viitasalo M, Vuorinen I, Saesmaa S (1995) Mesozooplankton dynamics in the northern Baltic Sea: Implications of variations in hydrography and climate. J Plankton Res 17:1857–1878

Voipio A (1981) The Baltic Sea. Elsevier, Amsterdam

Voss M, Emeis KC, Hille S, Neumann T, Dippner JW (2005) The nitrogen cycle of the Baltic Sea from an isotopic perspective. Glob Biogeochem Cy 19, GB3001, doi:10.1029/2004GB002338

Voss R, Köster FW, Dickmann M (2003) Comparing the feeding habits of co-occuring sprat (*Sprattus sprattus*) and cod (*Gadus morhua*) larvae in the Bornholm Basin, Baltic Sea. Fish Res 63:97–111

Voss R, Clemmesen C, Baumann H, Hinrichsen HH (2006) Baltic sprat larvae: Coupling food availability, larval condition and survival. Mar Ecol Prog Ser 308:243–254

Vuorinen I, Ranta E (1987) Dynamics of marine mesozooplankton at Seili, Northern Baltic Sea, in 1967–1975. Ophelia 28,1:31–48

Vuorinen I, Ranta E (1988) Can signs of eutrophication be found in the mesozooplankton of Seili, Archipelago Sea? Kieler Meeresforsch Sonderh 6:126–139

Vuorinen I, Hänninen J, Viitasalo M, Helminen U, Kuosa H (1998) Proportion of copepod biomass declines together with decreasing salinities in the Baltic Sea. ICES J Mar Sci 55:767–774

Vuorinen I, Hänninen J, Kornilovs G (2003) Transfer-function modelling between environmental variation and mesozooplankton in the Baltic Sea. Prog Oceanogr 59:339–356

Vuorinen I, Hänninen J, Kornilovs G (2004) Erratum to: Transfer-function modelling between environmental variation and mesozooplankton in the Baltic Sea. Prog Oceanogr 59:339-356

Wasmund N (1994) Phytoplankton Periodicity in a Eutrophic Coastal Water of the Baltic Sea. Int Rev Ges Hydrobiol 79:259–285

Wasmund N (1996) Periodicity and trends in the phytoplankton of a shallow coastal water. Proceedings of the 13th Symposium of the Baltic Marine Biologists: 63–66

Wasmund N (1997) Occurrence of cyanobacterial blooms in the Baltic Sea in relation to environmental conditions. Int Rev Ges Hydrobiol 82:169–184

Wasmund N (2002) Harmful algal blooms in coastal waters of the south-eastern Baltic Sea. In: Schernewski G, Schiewer U (eds) Baltic coastal ecosystems. Springer, Berlin Heidelberg New York, pp. 93–116

Wasmund N, Uhlig S (2003) Phytoplankton trends in the Baltic Sea. ICES J Mar Sci 60:177–186

Wasmund N, Nausch G, Matthäus W (1998) Phytoplankton spring blooms in the southern Baltic Sea – spatio-temporal development and long-term trends. J Plankton Res 20:1099–1117

Wasmund N, Voss M, Lochte K (2001) Evidence of nitrogen fixation by non-heterocystous cyanobacteria in the Baltic Sea and re-calculation of a budget of nitrogen fixation. Mar Ecol Prog Ser 214: 1–14

Wieland K, Zuzarte F (1991) Vertical distribution of cod and sprat eggs and larvae in the Bornholm Basin (Baltic Sea) 1987–1990. ICES CM 1991/J:37

Wieland K, Waller U, Schnack D (1994) Development of Baltic cod eggs at different levels of temperature and oxygen content. Dana 10:163–177

Wieland K, Jarre-Teichmann A, Horbowa K (2000) Changes in the timing of spawning of Baltic cod: Possible causes and implications for recruitment. ICES J Mar Sci 57:452–464

Wikner J (2006) Bacterioplankton growth rate. HELCOM Indicator Fact Sheets 2006, www.helcom.fi/environment2/ifs/ifs2006/en_GB/cover/

Wikner J, Hagström Å (1999) Bacterioplankton intra-annual variability: Importance of hydrology and competition. Aquat Microb Ecol 20:245–260

Wiltshire KH, Manly BFJ (2004) The warming trend at Helgoland Roads, North Sea: Phytoplankton response. Helgol Mar Res 58:269–273

Witt G, Trost E (1999) Polycyclic aromatic hydrocarbons (PAHs) in sediments of the Baltic Sea and of the German coastal waters. Chemosphere 38:1603–1614

Wodarg D, Komp P, McLachlan MS (2004) A baseline study of polychlorinated biphenyl and hexachlorobenzene concentrations in the western Baltic Sea and Baltic Proper. Mar Chem 87:23–36

Wright DA (1995) Trace metals and major ion interactions in aquatic animals. Mar Pollut Bull 31: 8–18

Wulff F, Rahm L (1993) Accumulation of chlorinated organic matter in the Baltic Sea from 50 years of use – A threat to the environment. Mar Pollut Bull 26:272–275

WWF (2005) Clean Baltic within reach? Baltic Ecoregion Action Programme

Yurkovskis A (1998) Course and environmental consequences of eutrophication in the Gulf of Riga. Proc Latvian Acad Sci, Section B (suppl) 52:56–61

Zajaczkowski MJ, Legezynska J (2001) Estimation of zooplankton mortality caused by an Arctic glacier outflow. Oceanologia 43:341–351

Zmudzinski L (1978) The Evolution of Macrobenthic Deserts in the Baltic Sea. XI Conference of Baltic Oceanographers, vol. 2. Rostock, pp. 780–794

Zmudzinski L, Osowiecki A (1991) Long-term changes in macrozoobenthos of the Gdansk Deep. Int Rev Ges Hydrobiol 76,3:465–471

A Annexes

A.1 Physical System Description

A.1.1 Baltic Sea Oceanography

Jüri Elken, Wolfgang Matthäus

A.1.1.1 *General Features*

The Baltic Sea is an intracontinental dilution basin with a total area of 415,000 km^2 (including Kattegat). Water exchange with the North Sea is restricted by the narrow straits (Little Belt, Great Belt, Sound, with channel width 0.8, 16 and 4 km, respectively) and the shallow sills (Darss and Drogden Sills, with maximum depths of 18 and 8 m, respectively). While saline water enters the Baltic Sea in the southwestern strait area, the freshwater surplus is concentrated in the large gulfs located in the opposite northeastern part of the sea. This leads to the general estuarine gradients, both in salinity (Fig. A.1) and ecosystem variables. In the Baltic Proper, deep water exchange is restricted by submarine sills and channels connecting deep basins. Since the baroclinic Rossby deformation radius is small (1.3–7 km, Fennel et al. 1991), advection-diffusion dynamics of basins and connecting channels is rather complex.

Despite the relatively small depths (mean depth is about 55 m), the water column of the central Baltic Proper is permanently stratified (Fig. A.2). During winter, the permanent halocline (C, Fig. A.2) separates the less saline cold winter water (B) from the more saline and warmer deep water (D). In the shallow western area, there is a change between stratification and well-mixed conditions. The depth of the halocline increases from about 40 m in the Arkona Basin to 60–80 m in the eastern Gotland Basin. During summer, a seasonal thermocline develops at 25–30 m depth (A_2) separating the warm upper layer (A_1) from the cold intermediate water (A_3). During mild and normal winters, ice cover occupies 15–50% of the sea area in its northeastern part, but may extend to the whole sea during the infrequently occurring severe winters (Omstedt and Chen 2001).

The permanent halocline isolates the Baltic Proper deep layers to a great extent from the surface waters and their ventilation occurs mainly by lateral advection of transformed North Sea water. Frequent but weak inflows (10–20 km^3) interleave just beneath the permanent halocline and prevent stagnation there but they have little impact on the deep and bottom waters. Episodic inflows of larger volumes (100–250 km^3) of highly saline (17–25) and oxygenated water – termed major Baltic inflows (MBIs) – represent the only mechanism by which the Baltic Sea deep water is displaced and renewed to a significant degree (Matthäus and Franck 1992; Schinke and Matthäus 1998).

A total of 113 major inflows have been identified since 1880 excluding the periods of the two World Wars (Fig. A.3). All inflows have occurred between the end of August and the end of April. The seasonal frequency distribution of major inflows (Fig. A.3, top right corner) shows that such events are most frequent between October and February. They occur in clusters of several years, but some have been isolated events.

During the first three-quarters of the past century, MBIs were observed more or less regularly (Fig. A.3). Since the mid-1970s, their frequency and intensity has changed, and only a few major events have occurred since then. Oceanographic conditions in the central Baltic deep water changed drastically during this period which culminated between 1977 and 1992 in the most significant and serious stagnation period (Nehring and Matthäus 1991; Matthäus and Franck 1992; cf. also Fig. A.4). Moreover, the major inflows in January 1993 (Håkansson et al. 1993; Jakobsen 1995; Matthäus and Lass 1995) and January 2003 (Feistel et al. 2003) were only isolated events, and conditions in the central Baltic Sea deep water have soon started to stagnate again although in 1997 a small inflow (by the MBI index) led to an unusual increase of deep water temperature (Fig. A.4) after an exceptionally hot summer.

However, a similar stagnation period occurred earlier during 1920–1932 (cf. Fig. A.4). This natural variation is well described by the mean Baltic Sea salinity (Winsor et al. 2001, 2003) which is strongly related to the large-scale atmospheric variability and the accumulated freshwater inflow (Stigebrandt and Gustafsson 2003; Meier and Kauker 2003).

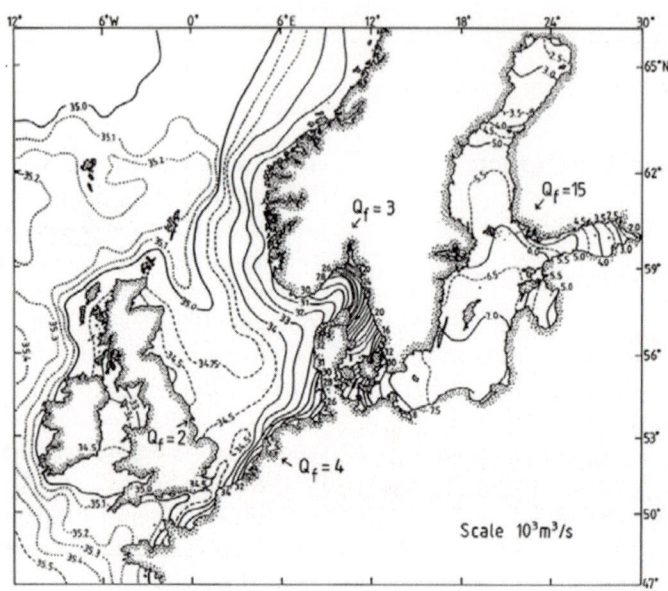

Fig. A.1. Surface salinity distribution. Also indicated is runoff, Qf, to the Baltic Sea $(15,000\,\mathrm{m}^3\,\mathrm{s}^{-1})$, to Kattegat and Skagerrak $(3,000\,\mathrm{m}^3\,\mathrm{s}^{-1})$, to the Danish, German, Dutch, and Belgian coasts $(4,000\,\mathrm{m}^3\,\mathrm{s}^{-1})$, and to the east coast of England $(2,000\,\mathrm{m}^3\,\mathrm{s}^{-1})$ (adapted from Rodhe 1998)

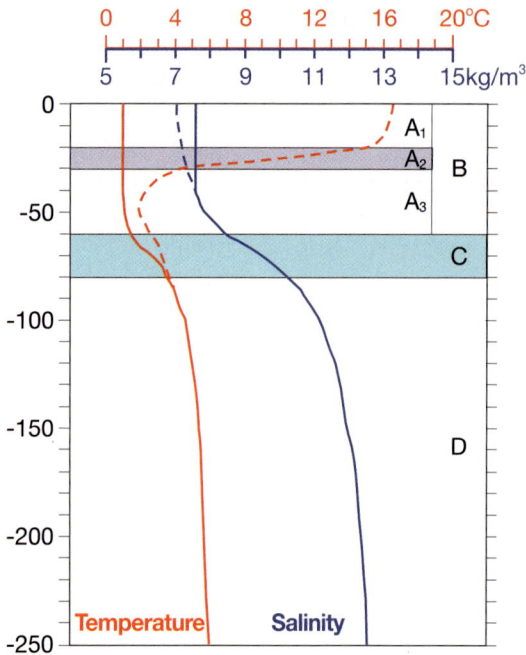

Fig. A.2. Typical thermohaline stratification in the central Baltic Sea during winter (*full line*) and summer (*partly hatched*)

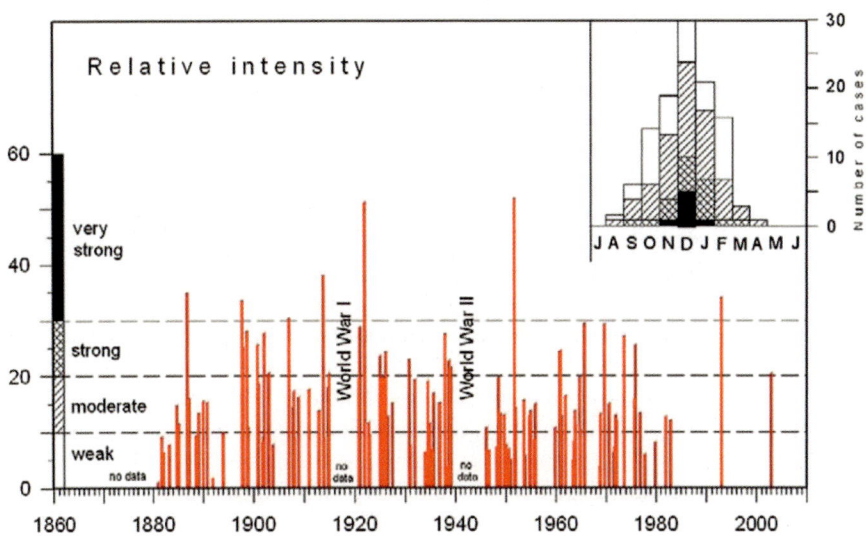

Fig. A.3. Major Baltic inflows (MBIs) between 1880 and 2005 and their seasonal distribution (*upper right*) shown in terms of their relative intensity (Matthäus and Franck 1992; Fischer and Matthäus 1996; supplemented and updated)

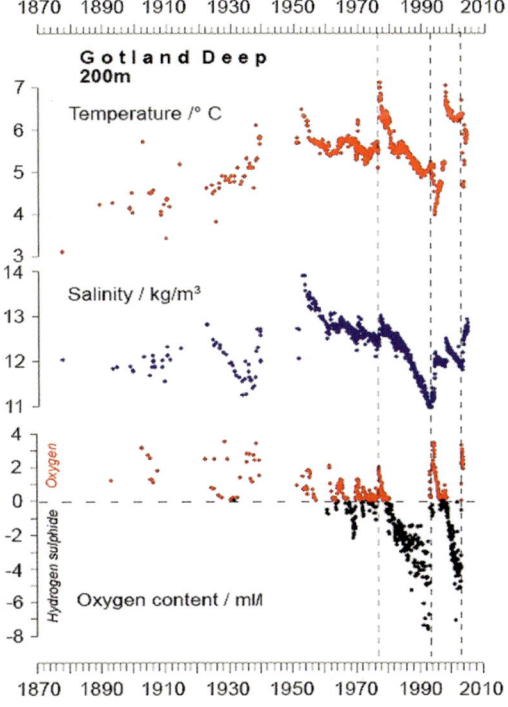

Fig. A.4. Long-term variation of temperature, salinity, oxygen and hydrogen sulphide (expressed in terms of negative oxygen equivalents) concentrations in the deep water of the central Baltic Sea (Gotland Deep)

A.1.1.2   *External Water Budget and Residence Time*

The average volume of the Baltic Sea – about $21,000\,\text{km}^3$ (excluding Kattegat and Belt Sea, e.g. HELCOM 2002) – is maintained by the external water budget, where dominating terms are water import by river discharge, inflowing North Sea water and net precipitation (precipitation minus evaporation), and export by outflowing Baltic Sea water. Volume change due to thermal expansion contributes some 10% (order of magnitude) of river discharge in the seasonal heating and cooling cycle (Stigebrandt 2001) but is small on the annual scales. Minor terms in the long-term budget are volume change by groundwater inflow (Peltonen 2001), salt contraction (Omstedt and Nohr 2004), land uplift and ice export (Omstedt and Rutgersson 2000). The water budget is closed by water storage variation due to the change of mean sea level, which is important on time scales of weeks and months (Lehmann and Hinrichsen 2001).

The major water budget components have recently been reviewed by Omstedt et al. (2004). Magnitudes of river discharge, net precipitation and resulting net outflow to the North Sea are well established (Table A.1) and various estimates differ mainly by the period involved in the study and the amount of data available. Regionally about 80% of the river runoff and 85% of the net precipitation enter the large gulfs (Gulfs of Bothnia, Finland and Riga), which thus represent the major source of freshwater input into the Baltic Sea and control the low salinity in the Baltic Sea surface water (Omstedt and Axell 2003).

Water exchange with the North Sea through the Sound and the Belt Sea (average flow ratio 3:8 according to Jakobsen and Trebuchet 2000) is highly variable in direction and magnitude ($\pm 100,000\,\text{m}^3\,\text{s}^{-1}$) even over short time periods (e.g. Mattson 1996; Jakobsen and Trebuchet 2000). During MBIs the accumulated volume may exceed $200\,\text{km}^3$ (50% of yearly river discharge) during a few weeks, e.g. in January 1993 (Håkansson et al. 1993; Matthäus and Lass 1995; Jakobsen 1995) and in January 2003 (Feistel et al. 2003; Piechura and Beszczynska-Möller 2004; Meier et al. 2004; Lehmann et al. 2004). In general, the water entering the Baltic Sea may flow out again within a short time, thus not affecting the conditions in the larger sea area.

According to modelled Lagrangian trajectories, only 6% of the Great Belt water and 32% of the Sound water remains after one year in the Baltic Sea (Döös et al. 2004). Therefore, summing up the individual inflow and outflow events, calculated from sea level difference along the straits without considering salinity, gives little information about water renewal in the sea. Flow in the straits is often separated according to its salinity and the term "inflow" usually covers the waters which form the deep layers below the primary halocline, i.e. waters entering the Arkona Basin with $S > 8$–9. This treatment also includes Baltic Sea water that is entrained to the inflow. The estimates of salinity-weighted inflows vary from 19,000 to $43,000\,\text{m}^3\,\text{s}^{-1}$ (Stigebrandt 1987; Kõuts and Omstedt 1993; Omstedt and Rutgersson 2000; Gustafsson 2001; Lehmann and Hinrichsen 2002; Meier and Kauker 2003; Omstedt and Nohr 2004) mainly depending on the methods/models used and how the salt fluxes are adjusted.

The above inflow estimates yield the Baltic Sea water residence times of 11 to 22 years. Stigebrandt and Gustafsson (2003) have argued that deep water entering from the Kattegat and Belt Sea is composed from the "true" Kattegat deep water and recirculated Baltic Proper surface water. Only Kattegat deep water, with mean inflow rate $5,000\,\text{m}^3\,\text{s}^{-1}$ contributes to the Baltic Sea water renewal, together with the freshwater supply, and the resulting residence time is 33 years. This latter estimate is consistent with the results from climatic scale runs with 3D models (Meier and Kauker 2003; Döös et al. 2004; Meier 2005).

A.1.1.3   *Processes and Patterns of Internal Water Cycle*

The water effectively recirculates in the Baltic Sea, even with the relatively impermeable halocline. This overturning circulation may be called Baltic haline conveyor belt (Döös et al. 2004), analogous to the Wold Ocean climatic water cycle. Understanding of the water cycle details is rapidly advancing in the present period due to the developments in high resolution measurements and 3D climatic scale modelling.

Calculation of water mass age, as refinement to the bulk residence time, has started only recently based on the method by Deleersnijder et al. (2001). By that, additional Eulerian tracer is embedded in the model to handle the age of seawater. It is defined as time elapsed since a water particle has left the source region that is kept constant in time. Meier (2005) investigated the spreading of

Table A.1. Water budget components of the Baltic Sea (Kattegat and Belt Sea excluded) according to Omstedt et al. (2004)

Water budget component	Long-term mean $(\mathrm{m^3\,s^{-1}})$	Interannual variability $(\mathrm{m^3\,s^{-1}})$
River discharge	14,000	$\pm 4,000$
Net precipitation	1,500	$\pm 1,000$
Volume change	0	$\pm 2,000$
Net outflow to the North Sea	15,500	$\pm 5,000$

surface water into the layers below by putting the constant source on the surface. Median ages of the bottom water between one year in the Bornholm Basin and 7 years in the northwestern Gotland Basin were found.

During 1903–1998 the oldest bottom water of about 11 years appeared at Landsort Deep. A secondary age maximum was calculated in the halocline of the deeper basins. In the eastern Gotland Basin, three stagnation periods (in the 1920/1930s, 1950/1960s, and 1980/1990s, cf. Fig. A.4) with residence times exceeding 8 years were found. Andrejev et al. (2004b) studied spreading of the Neva River water in the Gulf of Finland. The highest water ages (2 years) were found in the southeastern part of the Gulf. It takes around 5 years to renew 98% of the water masses of the Gulf of Finland.

Inflowing saline water (most frequently with salinity 12–16, but during MBIs up to 22–25), driven barotropically by along-strait sea level difference (e.g. Gustafsson and Andersson, 2001) is spread and transformed in the Baltic Proper in the cascade of deep sub-basins. This process is controlled by the flow regime in the connection areas (sills, deep channels) and depends on the "old" stratification of downstream basins relative to the variable density of new incoming water. Sinking water masses (levels determined by the buoyancy of the downstream basin) entrain ambient surface waters, reducing their salinity and increasing the flowrate (e.g. Stigebrandt 1987; Kõuts and Omstedt 1993). Saline water flowing to the Arkona Basin, the first in the basin sequence, forms a thin near-bottom dense water pool (Stigebrandt 1987) that leaks along the northern flanks as a baroclinic geostrophic boundary current to the Bornholm Strait and further on into the Bornholm Basin (Liljebladh and Stigebrandt 1996; Lass and Mohrholz 2003).

Starting from the Bornholm Basin, the basins work as buffers where incoming water may be trapped by the sill depth. Classical flow description in the buffering Bornholm Basin distinguishes three different modes of salt water intrusion (e.g. Grasshoff 1975): (1) regular inflow just below the primary halocline interleaving on the level of neutral buoyancy; (2) occasional inflow of saline water, sinking to the bottom and exchanging the Bornholm Basin deep water; (3) rather infrequent occasional (major) inflow of large amounts of saline water, filling the whole Bornholm Basin above Stolpe Sill level (60 m) and exchanging the Gotland Deep water. New observation techniques which have become available during the recent decade demonstrate the complex dynamics of the inflow process, which contains internal fronts with fine-scale intrusions, surface and subsurface eddies etc. The flow of higher-salinity water over the Stolpe Sill frequently has a splash-like nature (Piechura et al. 1997). Behind the sill, the deep layer often gets contracted, reflecting the internal hydraulic jump (hydraulically controlled transport over the sill). An example of a regular intruding water mass (identifiable by higher temperatures at 50–70 m depths) is given in Fig. A.6 (Zhurbas et al. 2004).

The response of the Southern Baltic Sea to the major inflow in January 2003 has been well documented (e.g. Piechura and Beszczynska-Möller 2004). The inflow water passed the western and southern slopes of the Bornholm Deep (as after the 1993 inflow, Jakobsen 1996) and detached in the cyclonic eddy into the central part. In the central and western parts of the Bornholm Deep, the usual layering of temperature became totally disrupted. Instead, chaotic distribution of patches of old-warm ($> 10\,°\mathrm{C}$) and new-cold ($< 2\,°\mathrm{C}$) water was clearly visible. By the end of April 2003, the new-cold water mass reached the Gdansk Deep,

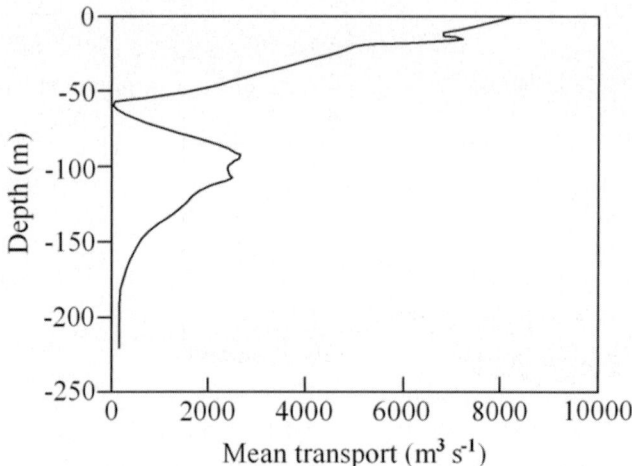

Fig. A.5. Total mean horizontally integrated transport ($m^3 s^{-1}$) across a basin-wide section in the Eastern Gotland Basin between Gotland and Latvia for the period 1902–1998. Northward transports are counted positive. Results from RCO model, redrawn from Meier and Kauker (2003)

generating patchy small-scale intrusions there. After the 1993 inflow, a well-defined front of the intrusive region was found to be propagating north from the Stolpe Channel to the Gotland Deep with a speed of $2\,cm\,s^{-1}$ or more (Zhurbas and Paka 1997). A substantial horizontal intermittence of intrusion intensity, related to mesoscale eddies, was observed behind the front. Such intrusions survive several weeks and months before smearing out by diffusive processes, as recorded after the 2003 major inflow (Zhurbas and Paka 1997, 1999) but also during the stagnation period in the Gotland Deep (Elken et al. 1988; Kõuts et al. 1990; Elken 1996). As shown recently by Zhurbas et al. (2003, 2004), energetic eddies that accompany intruding larger deep water masses are generated by change of potential vorticity in the receiving basin.

Saline water passing the Stolpe Channel (although wind-dependent reversals may take place; e.g. Jakobsen 1996; Elken 1996; Golenko et al. 1999; Lehmann and Hinrichsen 2002) flows on average towards northeast along the eastern slope of the Hoburg Channel, making an occasional cyclonic loop along the slopes of Gdansk Basin. In the Eastern Gotland Basin, the flow forms a semi-enclosed cyclonic circulation cell, with a leakage towards the Northern Basin. This overall flow pattern is confirmed by observations (Elken 1996; Hagen and Feistel 2004; Zhurbas et al. 2004) as well as results from 3D models (Lehmann and Hinrich-

sen 2000; Lehmann et al. 2002; Döös et al. 2004). Gotland Deep receives saline water interleaving preferably at depths of 80–130 m (Elken 1996; Meier and Kauker 2003), with maximum northward flow across the basin reaching $2{,}500\,m^3\,s^{-1}$ around 100 m (Fig. A.5).

This ventilation depth range explains why hydrogen sulfide does not frequently appear at depths above 140–150 m, even during the long stagnation periods (e.g. HELCOM 1996, 2002). While short-term transport of intruding waters in the interior of the basins is mainly isopycnal, then diapycnal mixing (Stigebrandt et al. 2002) is an important stratification control mechanism on the longer time scales.

The Northern Basin is a region which splits into the two terminal areas of the saline water route – the entrance to the Gulf of Finland and the Western Gotland Basin. Due to the continuity requirements, the deep flow has to be converted into upward vertical advection. Since bidirectional diffusive mixing does not restore the high vertical gradients in the halocline, unidirectional upward entrainment has to be more effective in the terminal region than in other areas. Besides the ordinary "mixers" like wintertime convection and wind-driven turbulence (halocline erosion occurs at wind speeds above 14 m/s, Lass et al. 2003), wind waves reach highest significant heights in November (5 m) and December (9 m) due to the long fetch for dominating southwesterly winds

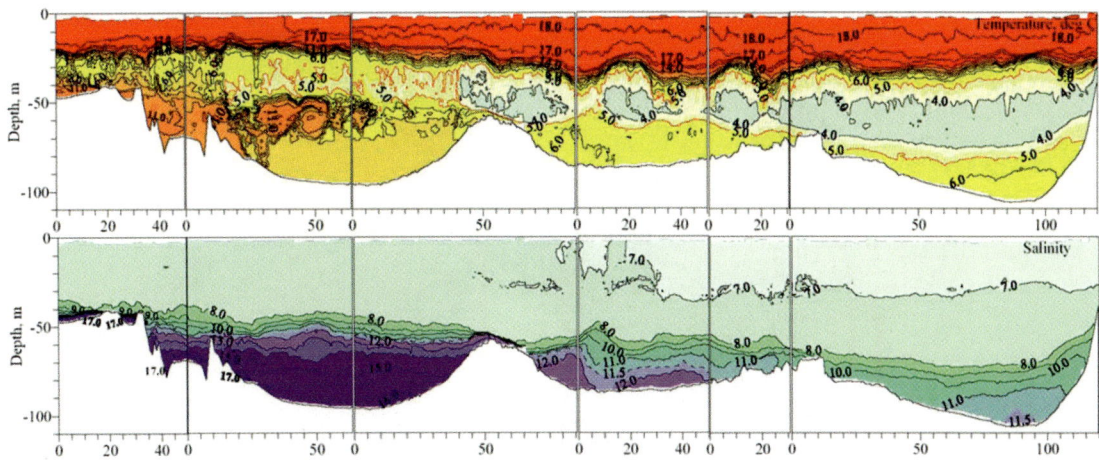

Fig. A.6. Temperature (*top*) and salinity (*bottom*) transect Bornholm Gate – Bornholm Deep – Stolpe Channel – Gdansk Deep in September 1999 (Zhurbas et al. 2004). Note the following features: (1) intruding warm water interleaving the Bornholm Deep at 50–70 m depths; (2) deep layer contraction (internal hydraulic jump) at the Stolpe Still followed by an eddy (halocline displacement) in the Stolpe Channel; (3) northward (cyclonic) deep water jet along the slope of Gdansk Basin

(Jönsson et al. 2003). Significant vertical advection (partly due to the seasonal deep flow reversal in the Gulf of Finland; Elken et al. 2003) coupled with stronger halocline erosion (because the vertical stability is less than in the upstream basins) leads to the highest seasonal amplitudes in the saline water (Matthäus 1984).

The Gulfs of Bothnia and Riga are topographically isolated from the saline water below the halocline and they receive only the Baltic Proper surface water. It is denser than the surface water of the gulfs and sinks behind the sills. During the summer, deep water spreading is similar to that of the Baltic Proper – cyclonic flow along the slopes (e.g. Marmefelt and Omstedt 1993; Omstedt et al. 1993; Håkansson et al. 1996; Lips et al. 1995; Raudsepp 2001; Lehmann et al. 2002) either on the bottom towards the greater depths ("major" inflows for the gulfs) or interleaving on the level of neutral buoyancy. However, during the late autumn before icing, the whole water column is usually mixed due to low haline vertical stability, leaving the horizontal gradients characteristic to the well-mixed estuary. The latter is also true for the eastern half of the Gulf of Finland (Alenius et al. 1998).

Motions of surface waters are strongly affected by variable wind forcing. Drift currents in the off-shore areas are converted into up- and downwelling features in the coastal areas (Lehmann et al. 2002; Myrberg and Andrejev 2003) that are affected by Kelvin waves (Fennel and Strum 1992; Lass and Talpsepp 1993; Fennel and Seifert 1995) and topographic waves of different origin (Raudsepp 1998; Pizzaro and Shaffer 1998; Raudsepp et al. 2003). The water is laterally mixed by mesoscale eddies (Elken et al. 1994; Stigebrandt et al. 2002; Zhurbas et al. 2003) and inertial motions (e.g. Nerheim 2004).

Regular, basin-guided cyclonic flow cells are evident from observed surface salinity distributions (Rodhe 1998) and spreading patterns of juvenile freshwater originating from the spring maximum of discharge (Eilola and Stigebrandt 1998; Stipa et al. 1999), despite the sporadic nature of instantaneous currents. These flow patterns can be also seen from the model results (Lehmann et al. 2002; Stipa 2003; Andrejev et al. 2004a). In the converging flow areas, bi-directional currents feed quasipermanent but migrating and self-restoring salinity fronts (Pavelson 1988; Elken 1994; entrance to the Gulf of Finland: Pavelson et al. 1997; Gulf of Riga: Lilover et al. 1998) that are similar to the Kattegat–Skagerrak front controlling the major inflows to the Baltic Sea (e.g. Stigebrandt and Gustafsson 2003).

A schematic of the internal water cycle in the Baltic Sea is presented in Fig. A.7.

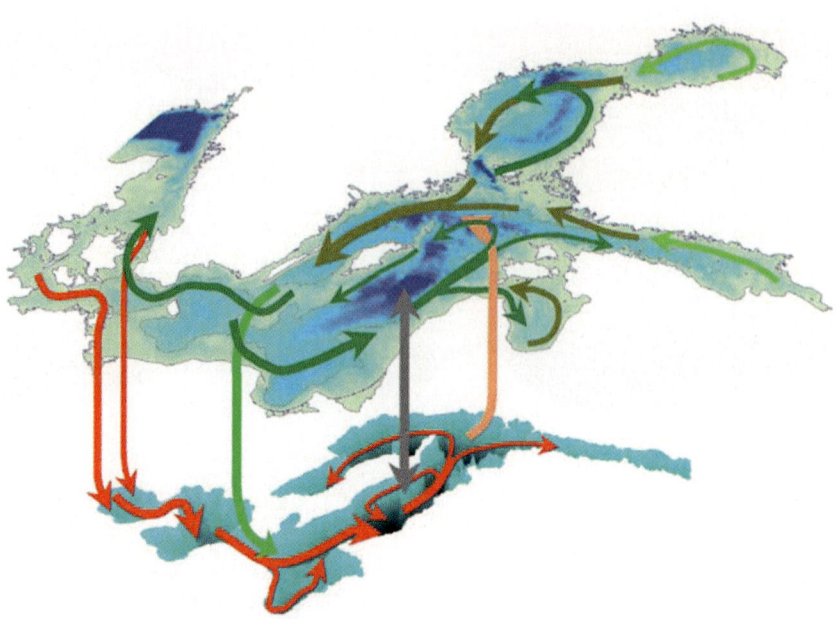

Fig. A.7. A schematic of the large-scale internal water cycle in the Baltic Sea. The deep layer below the halocline is given in the lower part of the figure. Green and red arrows denote the surface and bottom layer circulation, respectively. The light green and beige arrows show entrainment, the gray arrow denotes diffusion

A.1.2 Atmosphere

Hans-Jörg Isemer, Viivi Russak, Heikki Tuomen-virta

A.1.2.1 *Atmospheric Circulation*

The climate of the Baltic Sea Basin, located between the 50th and 70th northern parallels in the Eurasian continent's coastal zone, is embedded in the general atmospheric circulation system of the northern hemisphere, with its cyclonic circumpolar vortex providing for mean troposheric westerly air flow with annually varying intensity (e.g. Defant 1972). Strong westerly air flow provides for maritime, humid air mass transport in particular into the southwestern and southern parts of the basin, while in the east and north the maritime westerly air flow is weakend due to friction and drying processes providing for increasing continental climate conditions.

The following two climatic types according to Köppen's climate classification scheme dominate much of the Baltic Sea Basin: 1) Most of the middle and northern parts of the basin are dominated by the temperate coniferous-mixed forest zone, with cold, wet winters, where the mean temperature of the warmest month is not lower than 10 °C and that of the coldest month not higher than −3 °C, and where the rainfall is, on average, moderate in all seasons. 2) Much of the southwestern and southern region belongs to the marine west coast climate, where prevailing west winds constantly bring in moisture from the oceans, and the presence of a warm ocean current (the North Atlantic current system) provides for, in particular, moist and mild winters. Due to the influence of the warm ocean currents on parts of mid- and northern Europe, the mean temperature of the Baltic Sea Basin is, on average, several degrees higher than that of other areas located in the same latitudes. In addition to the two climate types mentioned above, the northeastern and eastern regions of the basin are influenced by the moderate sub-arctic continental climate.

Major air pressure systems known to affect the weather and circulation in the Baltic Sea Basin are the low-pressure system usually found near Iceland (Icelandic Low) and the high-pressure system in the Azores Island region (Azores High). Also, the continental anticyclone over Russia may influence climate and circulation in the basin. The position and strength of these systems vary on synoptic time scales, and any one of them can dominate

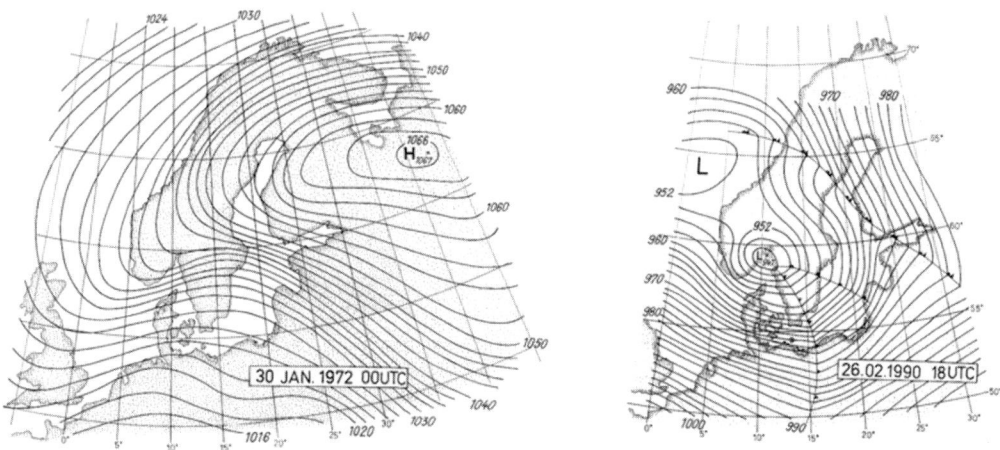

Fig. A.8. Examples of both a major continental anti-cyclone (*left*, core surface air pressure in excess of 1060 hPa) and a violent storm cyclone (*right*, less than 950 hPa at the centre) over the Baltic Sea Basin in winter (from Miętus 1998, with permission by WMO)

the weather for a period of days to weeks. These systems also dominate the long-term mean surface air pressure and related mean circulation patterns over northern Europe, showing a distinct annual cycle, see Chap. 1, Fig. 1.6. The following short description is largely based on Miętus (1998) and Uppala et. al (2005). The latter describe results of the ERA-40 re-analysis project for the period 1979 to 2001, which are used in Fig. 1.6.

In the cool season of the year, beginning in September, southwesterly air flow prevails, intensifying in October and becoming more cyclonic in November and December. The mean flow is especially intensive in January and in February, when the core pressure of the Icelandic Low is deepest and the anticyclone over Russia as well as the Azores High are well developed. The strongest mean horizontal air pressure gradient forms over the Baltic Sea Basin in this season. Note the pressure range in January (Fig. 1.6, Chap. 1), with mean surface air pressures of 1004 hPa and 1020 hPa in the far north and south of the basin, respectively.

A particular feature in winter (see again the January map) is the mean surface pressure trough forming leeward of the Scandinavian mountains along the main north–south axis of the Baltic Sea. In March the intensity of mean air flow over the Baltic Sea Basin decreases, becoming even weaker in April. The Azores High starts to stretch into parts of mid-Europe, and the mean flow over the southern Baltic Sea Basin becomes weakly anti-cyclonic. The mean pressure fields in April and

May represent the weakest mean pressure gradients in the course of the year. During June and July, the direction of the mean air flow is northwesterly to westerly and is rather anticyclonic in character in the south, while in the north of the basin it is weak and hardly specified. Here, an extended but weak low pressure system extends over much of the area between Iceland and the White Sea, covering the northern Baltic Sea Basin as well. The summer months are therefore dominated by meridional weather patterns, in contrast to the winter months, with dominating zonal circulation patterns (e.g. Keevallik et al. 1999).

In September, the Icelandic Low deepens again and the mean pressure gradient starts to increase with the related mean air-flow becoming cyclonic again over much of the basin. Thus, the pressure gradient is rather strong, and the related air flow mostly cyclonic over much of the basin during October to March, with varying magnitude of the pressure gradient and both direction and strength of the flow. It is the Icelandic Low which dominates the basin during this period of the year, while during particularly May to August, especially the southern part of the basin is influenced by an extension of the Azores High with related anti-cyclonic mean flow patterns. The strength of the surface air pressure gradient between the Icelandic Low and the Azores High, the North Atlantic Oscillation (NAO) Index (see Annex 6), has frequently been used to characterise the circulation pattern and strength over northern Europe, and in particular the winter-time NAO has been

388

A. Annexes

shown to correlate with weather and climate in the basin (e.g. Busuicoc et al. 2001; Cheng and Hellström 1999; Jacobeit et al. 2001).

In summary, when westerly winds prevail, the weather may be warm and clear in much of, particularly, the northern part of the basin due to the 'föhn' phenomenon caused by the Scandinavian mountains. Despite the moderating effect of the ocean, the Asian continental climate also extends to the area at times, manifesting itself as severe cold in winter and extreme heat in summer. Since the area is located in the zone of prevailing westerlies where sub-tropical and polar air masses meet, weather types can change quite rapidly, particularly in winter. At synoptic time scales the individual air pressure and flow systems are drastically variable and both storm cyclones and major continental, sometimes long-lasting anticyclones may dominate the weather patterns in the region, see Fig. A.8.

A.1.2.2 Surface Air Temperature

The distribution of surface air temperature, T_a, is closely linked to the general climate and circulation regimes mentioned above. The general north–south gradient is modulated by the southwest/northeast contrast of maritime versus continental climate influences. In Fig. 1.9 (see Chap. 1), we show the annual cycle of T_a at four selected stations which form a transect from north to south through the basin. Mean annual T_a differs by more than 10 °C in the Baltic Sea Basin. The coldest regions are northeast Finland and the upper regions in the Scandinavian mountains, with mean annual surface air temperatures well below 0 °C (e.g. −1 °C in Sodankylä, and regions in northern Sweden and Finland with mean annual T_a even below −2 °C). These are also the regions with the largest amplitudes of the annual cycle (note the mean July minus January difference of about 29 °C at Sodankylä, Fig. 1.9). The "most maritime" region in the basin is the southwestern part (Northern Germany and Denmark) of the basin, which is less sheltered from the North Atlantic Ocean by the Scandinavian mountains, where mean monthly values of T_a exceed 0 °C throughout the year (Miętus 1998, exemplified by the station Schleswig in Fig. 1.9), but July mean temperatures are lower than in continental regions such as eastern Poland.

A.1.2.3 Precipitation

Precipitation in the Baltic Sea Basin shows both a distinct mean annual cycle and considerable regional variations. The latter are caused by the regionally varying circulation systems and the orographic influence of the land surface. As for clouds, precipitation patterns over the Baltic Sea may differ considerably compared to land areas of the basin.

Recent estimates of the annual mean precipitation for the entire Baltic Sea Basin (both land and sea) vary between 620 mm/y (based on the Climate Prediction Centre Merged Analysis of Precipitation (CMAP) climatology, see e.g. Xie and Arkin 1997) and 790 mm/y (based on the Global Precipitation Climatology Project (GPCP) climatology, see e.g. Huffman et al. 1997). Both estimates given for the Baltic Sea Basin are referenced and discussed by Arpe et al. (2005), who concluded that the CMAP data are significantly underestimated while the GPCP estimates are slightly overestimated.

Both climatologies use a blend of gridded rain gauge and satellite data. Earlier estimates based exclusively on rain gauge data are mostly within the above-given range, e.g. the 728 mm/y estimate of Kuusisto (1995), which is based on corrected direct observations. Re-analysis products such as NCEP (e.g. Ruprecht and Kahl 2003) and NCEP-RII (e.g. Roads et al. 2002) yield 730 mm/y and 640 mm/y, respectively; however, re-analysis products are known for several deficiencies which may cause significant biases in precipitation estimates (Ruprecht and Kahl 2003). Therefore, a reasonable current (i.e. for the recent 30 years) mean annual precipitation estimate is 750 mm/y for the entire Baltic Sea Basin, including land and sea.

Rutgersson et al. (2001) give various estimates of precipitation over the Baltic Sea. The SMHI (Swedish Meteorological and Hydrological Institute) 1 degree gridded data set on the one hand and estimates based on COADS (Comprehensive Ocean Atmosphere Data Set, see also Lindau 2002) on the other hand, which are based on totally different data sources, agree astonishingly well (600 and 606 mm/y, respectively) and exhibit a very similar annual cycle. Rutgersson et al. (2001) however present evidence that the SMHI data may underestimate precipitation, because of the neglect of the rain gauges' flow distortion and evaporation error correction. Based on the findings of Rubel and Hantel (1999, 2001), an annual

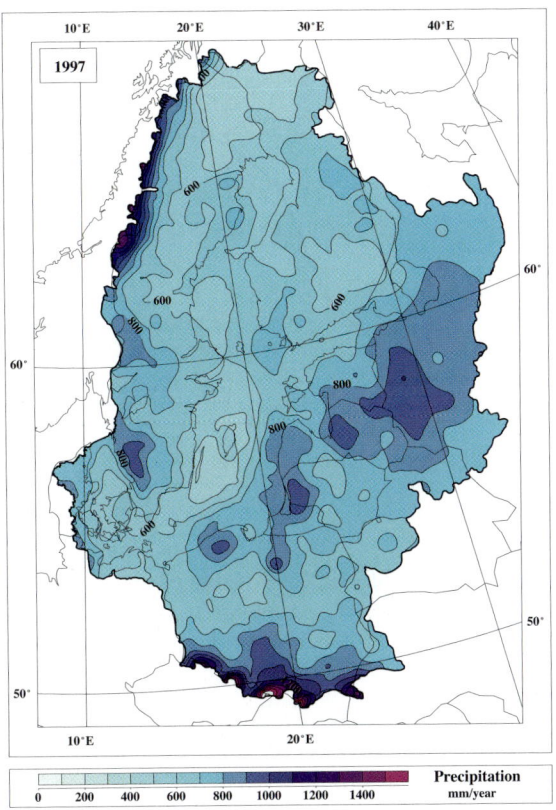

Fig. A.9. Annual precipitation field for the year 1997 in the Baltic Sea Basin (from Rubel and Hantel 2001)

mean correction factor may be in the range of +10 to +20% for stations in the Baltic Sea Basin, with, however, significant annual and monthly variation.

Applying the correction model proposed by Rubel and Hantel (2001) to the CRU (Climate Research Unite University, East Anglia) precipitation product, Jones and Ullerstig (2002) and Räisänen et al. (2003) quantified the correction effect to be about 19 % in the annual and up to 40% in the winter mean, respectively, for the entire land area of the Baltic Sea Basin (see also Sect. 3.3.2 and Fig. 3.6b). Combining the above findings would lead to a mean precipitation estimate over the Baltic Sea of between 600 and 660 mm/year. This estimate is distinctly lower than the combined land and sea estimates given above for the entire basin.

Regional variations of annual mean precipitation are large in the Baltic Sea Basin. The report by Miętus (1998), which is confined to the Baltic Sea and surrounding coastal regions, noted averages for 1961–1990 varying between 927 mm/y in Schleswig (Germany) and 433 mm/y in Oulu (Finland). Annual values in the mountain regions in

Scandinavia and southern Poland may even exceed 1,500 mm/y. Rubel and Hantel (2001) have conducted the first objective analysis of a unique gauge data set for the Baltic Sea Basin defined on a 1/6 degree grid scale (roughly 18 km), which is, however, limited so far to a three-years period only.

The regional distribution for the year 1997 (Fig. A.9) is fairly typical, with maxima in the Scandinavian and Sudeten (South Poland) Mountains exceeding 1,500 mm/y, while minima with less than 600 mm/y occur in the northern and northeastern part of the basin as well as over the central Baltic Sea. Mean annual precipitation may vary even stronger, in particular in orographically structured regions, such as mountains. Tveito et al. (2001) found large spatial gradients in the average amount of precipitation in the Scandinavian mountains, ranging from more than 3,000 mm/y at windward sites to less than 500 mm/y in sheltered mountain valleys.

Mean monthly precipitation is highest during July and August, with up to 80 mm in August, and lowest during February to April, with less than

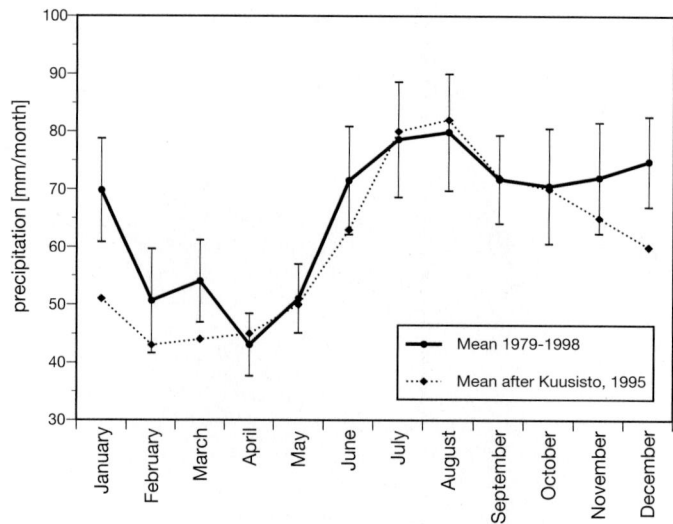

Fig. A.10. Annual cycle of precipitation calculated for the entire Baltic Sea Basin based on GPCP data (Huffman et al. 1997, *thick line*) and after Kuusisto (1995, *dotted line*), taken from Rubel and Hantel (2001). Vertical bars at the GPCP data indicate standard deviations of individual monthly values against the long-term monthly means

45 mm on average, see Fig. A.10. Figure A.10 also indicates inherent uncertainties even in our knowledge on long term means: While both data sets agree well during April to October, differences are noticeable during winter months, with difference peaks in December and January, which may be attributed at least partly to the different correction algorithms used for snow (Rubel and Hantel 2001). The interannual variability is large (Fig. A.10); the detailed analyses of Rubel and Hantel (2001), for example, yielded 646, 721 and 847 mm/y in the years 1996, 1997 and 1998, respectively.

A.1.2.4 *Clouds*

Changes and variability of clouds, as discussed in Sect. 2.2 of this book, relate almost entirely to total cloud cover, as "measured" by eye-observations at synoptic stations. As with many other parameters, information on mean annual and monthly cloud conditions is available either for individual stations (e.g. Keevallik and Russak 2001; Matuszko 2003) or only parts of the Baltic Sea Basin (e.g. Karlsson 2001; Raab and Vedin 1995).

Figure A.11 shows the annual cycle of total cloud cover at different stations across the basin and for the Baltic Proper. For much of the basin, in particular the eastern continental part and also for much of the Baltic Sea, a prominent annual cycle with highest cloud amounts during winter and lowest amounts during summer is clearly evident. Parts of the western and northern regions (mid- and northern Sweden and northern Finland in particular) exhibit a reduced or almost no annual cycle (e.g. at Haparanda and Östersund, Fig. A.11). Karlsson (1999, 2001), using NOAA AVHRR data over Scandinavia for 1991 to 2000, concluded that "with increasing distance from the central part of the Baltic Sea, the amplitude of the annual cycle of cloudiness decreases for inland stations in Scandinavia". This is mostly due to the fact that during summer months no or little convective clouds form over the Baltic Sea, in contrast to the surrounding land areas. Areas in the Swedish mountains even show slightly higher values in summer compared to winter. Karlsson (2001) showed distinct diurnal and inter-annual (Fig. 1.10, Chap. 1) variations of total cloudiness.

A.1.2.5 *Surface Global Radiation*

Global radiation, the solar radiation received by a unit horizontal surface, is usually measured at 1.5 to 2 m above the Earth's surface. Its variation in time and space depends largely on solar elevation and length of day, both linked to the geographical latitude of the site, and is considerably large in the Baltic Sea Basin. For example, in June, at 50° N, the sun is over the horizon for almost 16 hours per day, while in the north at 70° N the polar day

Total Cloud Cover

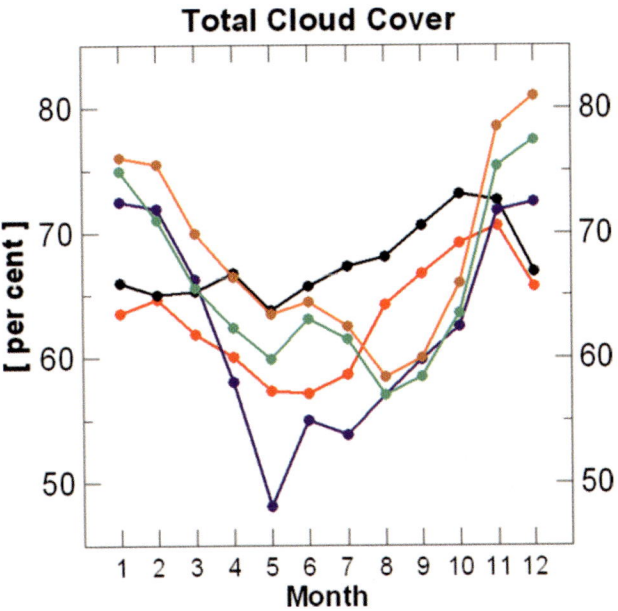

Fig. A.11. Mean annual cycle of total cloud cover (per cent) at various stations or regions in the Baltic Sea Basin. *Red:* Haparanda, northern Sweden; *black:* Östersund, mid-Sweden, both for 1951–2000 obtained from SMHI (H. Alexandersson, pers. comm.); *blue:* for the Baltic Proper based on ship data for 1980 to 1992 (data taken from Isemer and Rozwadowska 1999); *green:* Lindenberg, eastern Germany (52.2° N/14.1° E) for 1951–2003 obtained from German Weather Service DWD (F. Beyrich, pers. comm.); and *brown:* Cracow, southern Poland for 1906–2000, redrawn from Matuszko (2003)

exists in this time. In winter, when in the southern part of the basin the daylight lengths is about eight hours, the solar disk remains below the horizon in the northern Baltic Sea Basin for weeks.

Besides these regular diurnal and annual cycles, the amount and genera of clouds are major factors determining the variations of global radiation across the basin. Totals of global radiation during the warm season are higher in the coastal areas than in the hinterland. This is caused by less intensive formation and development of convective clouds over the sea and coastal region. Another factor influencing the distribution of clouds and, hence, global radiation is topography of the ground. Favourable conditions for cloud formation on slopes of mountain chains may result in a decrease of global radiation. On the other hand, on the elevated areas in mountains located higher than the height of low clouds, global radiation usually exceeds its value at lowland areas. The latter is more evident during winter months. In the cold half year, global radiation may increase, resulting from multiple reflection of solar radiation between the surface and the atmosphere (base of the clouds) in regions with snow cover. Due to the high albedo of snow, monthly totals of global radi-

ation may increase by a factor of 1.4 to 1.8 (Tooming 2002). Global radiation may also be affected by atmospheric transparency. Direct and diffuse radiations are highly sensitive to changes in atmospheric turbidity. But due to their opposite dependencies on transparency, the increase in global radiation with increasing transparency is smaller. However, in highly polluted regions, totals of radiation may be noticeably smaller.

Unfortunately, long-term climatological solar radiation data for the Baltic Sea Basin are available only for a small number of locations and particularly few measurements are regularly made in the northern regions. In December, monthly mean global radiation flux density varies between less than $15\,MJ/m^2$ north of about 65° N to more than $90\,MJ/m^2$ in southern Poland with a rather strong orientation of isolines along latitudes. In June, the mean regional variation is between 550 and $750\,MJ/m^2$, with a much stronger variation according to the difference between Baltic Sea versus land surfaces, and also according to the mountain effect described above. Figure A.12 depicts 4 examples of annual cycles of global radiation flux densities at different locations in the Baltic Sea Basin.

Global Radiation

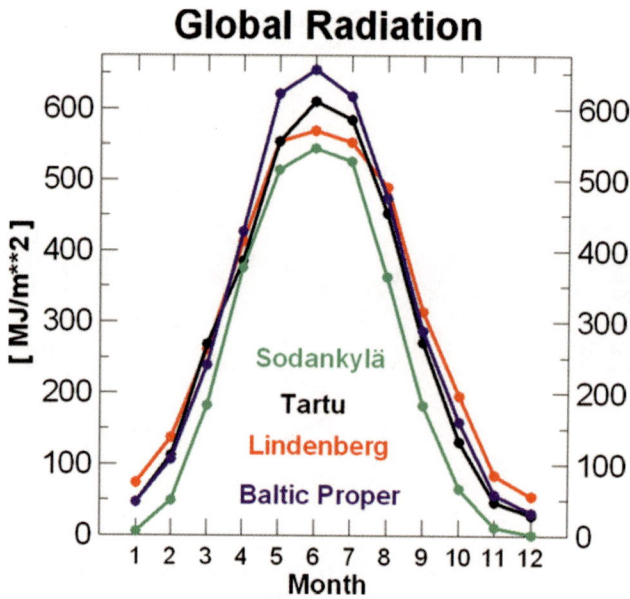

Fig. A.12. Average global radiation (MJ/m^2) at Sodankylä (northern Finland, 1971 to 2000), Tartu (Estonia, 1981 to 2000), Lindenberg (eastern Germany, 1981 to 2000) and at the surface of the Baltic Proper (1980 to 1992). The station data at Lindenberg, Tartu and Sodankylä are based on radiation measurements, while the Baltic Proper data set (taken from Rozwadowska and Isemer 1998) is based on parameterisations applied to cloud cover observations and humidity and air temperature measurements made aboard of voluntarily observing ships

The contribution of diffuse radiation depends on cloudiness, surface albedo and turbidity of the atmosphere. The mean annual percentage of diffuse radiation to global radiation at the surface increases from about 50% in the southern part of the basin to about 60–70% in the higher latitudes. This percentage of diffuse radiation varies seasonally: It is about 40–50% at 50–60° N in June, but may reach about 80–90% in December (European Solar Radiation Atlas, Palz and Greif 1996).

A.1.3 Hydrology and Land Surfaces

Esko Kuusisto, Valery Vuglinsky, Raino Heino, Lev Kitaev

A.1.3.1 *General Characteristics of the Baltic Sea Basin*

The drainage area of the Baltic Sea (in this section also referred to as 'Baltic Drainage' or simply 'Drainage'), which is the land surface region of the Baltic Sea Basin, covers 1.74 million km^2. It includes territories from altogether 14 countries, the largest areas belonging to Sweden (25.3%), Russia (19.0%), Poland (17.8%) and Finland (17.4%). Three countries – Latvia, Lithuania and Estonia –

are completely within the Baltic Sea Basin, while only minor parts of Czech Republic, Germany, Norway, Slovakia and Ukraine drain towards the Baltic Sea.

The Baltic Sea Basin has about 80 lakes with a surface area larger than 100 km^2. The number of lakes larger than 1 km^2 totals almost 10,000; of them 4,300 are located in Sweden and 2,300 in Finland. The total area of all lakes in the Baltic Sea Basin is around 123,000 km^2 – one third of the area of the Baltic Sea – and their volume 2,100 km^3.

Forests cover about 54% of the Baltic Drainage. Agricultural land amounts to 26%, buildup land to 4% (ECE 1993). Wetlands are a hydrologically important feature of the Drainage; they still account for 20% of the total land area, although a considerable proportion of them have been drained and are today classified as forests or agricultural lands.

The climate varies considerably in the Baltic Drainage (see also Annex 1.2.2). Long, cold winters dominate in the north, mean annual temperatures being −2...0 °C. In the southern part of the Drainage, thawing periods are frequent even in midwinter, annual mean temperatures reaching up to 9 °C. Precipitation is highest in the Scandes mountains, locally up to 2,000 mm/y, and exceed-

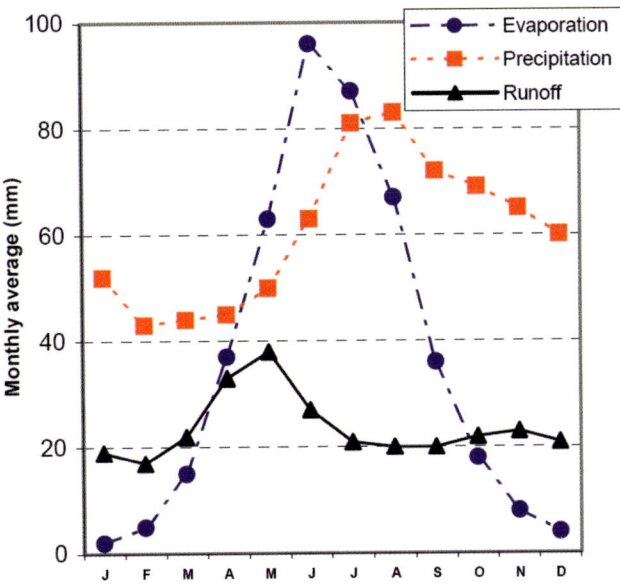

Fig. A.13. The average monthly water balance of the Baltic Drainage, as compiled by Kuusisto (1995) from various sources

ing 1,000 mm also in the Tatra mountains. Over most of the Drainage precipitation ranges between 500 and 750 mm/y. In the north, over 60% of precipitation arrives as snow, in the lowlands south of the Baltic Sea only 10–20%.

Figure A.13 shows the mean monthly values of the water balance components of the whole Baltic Drainage. The annual estimate of corrected precipitation is 728 mm, actual evaporation being 449 mm and runoff 279 mm. The rainiest month is August (82 mm), while evaporation is greatest in June (96 mm) and runoff in May (38 mm).

It is interesting to compare the evaporation from the Baltic Drainage with that of the Baltic Sea itself. In May, 74 mm evaporates from the drainage area, but the evaporation from the sea lingers near the annual minimum of 10 mm. In January, the cold land surface has a minimal vapor flux, while over 40 mm evaporates from the sea more than in July. The energy required to maintain the total evaporation is about $2.5 \times 10^{21} \, \mathrm{J \, y^{-1}}$. Of this energy, 21% is consumed by evaporation from the Baltic Sea, and 5% by lake evaporation (Kuusisto 1995).

A.1.3.2 *River Basins*

The Baltic Sea can be divided into six subbasins (including the Danish Belts and Sound and the Kattegat). Accordingly, it is natural to divide

the land part of the Baltic Sea Basin as shown in Fig. A.14. The Baltic Proper has the largest share of the total drainage area, one third, followed by the Gulf of Finland (24%). Even the smallest drainage subbasins have an area in excess of 100,000 km².

The ten largest river basins draining into the Baltic Sea are given in Table A.2, with the characteristics of their mean runoff. These ten rivers account for 59% of the total Baltic Drainage. The next 10 basins have a total area of 251,000 km², 14% of the total. Eight of these basins are in Sweden, two in Finland. The hundred largest basins cover about 86% of the Baltic Drainage. The remaining 14% or a quarter of a million square kilometers are divided into numerous small catchments along the coastal regions and on the Baltic Sea islands. The total area of the Baltic islands is almost 40,000 km² and their number of the order of 200,000.

As to the mean annual flow, the ten largest river basins are not the top ten. The specific runoff is largest in the northwestern parts of the Baltic Drainage; therefore three rivers from that region, Ångermanälven, Luleälven and Indalsälven cover the positions 8–10, displacing Narva, Torne and Kymi rivers. The drainage area of Lule River is only 25,200 km², but the specific runoff, 19.0 l s⁻¹ km⁻² leads to a mean annual flow of 486 m³ s⁻¹.

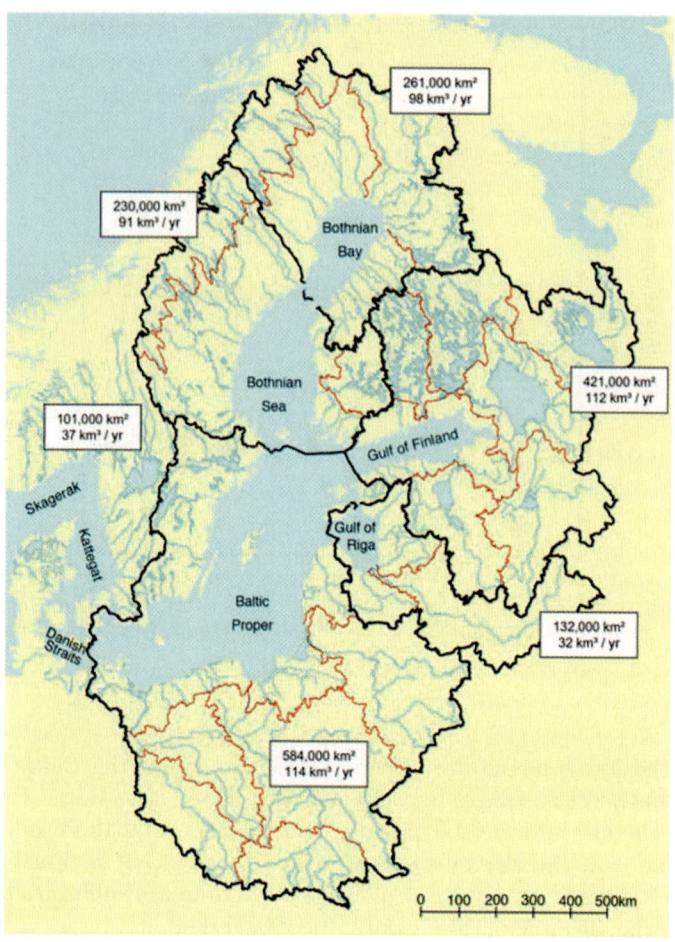

Fig. A.14. The subdivision of the Baltic Sea Basin, with areas and mean annual flows in 1950–1990 (from Bergström and Carlsson 1994)

Table A.2. The ten largest river basins of the Baltic Drainage. The runoff values refer to the period 1950–1990

River	Drainage area (km^2)	Mean annual flow (m^3 s^{-1})	Specific runoff (l s^{-1} km^{-2})
Neva	281,000	2,460	8.8
Vistula	194,400	1,065	5.5
Odra	118,900	573	4.8
Neman	98,200	632	6.4
Daugava	87,900	659	7.5
Narva	56,200	403	7.2
Kemi	51,400	562	11.0
Göta	50,100	574	11.5
Torne	40,100	392	9.8
Kymi	37,200	338	9.1

A.1.3.3 Lakes and Wetlands

The total number of lakes in the Baltic Sea Basin might be almost 400,000, most of them in Sweden, Finland and Russia. Poland has some 9,300 lakes with surface areas over 1 ha; their total area is over 8,000 km^2. Estonia has about 1,200 lakes, the largest completely within Estonian territory is Võrtsjärv, 270 km^2. On the border of Estonia and Russia is the Lake Peipsi (3,555 km^2), the largest international lake in Europe. Lithuania has 2,850 lakes larger than 0.5 ha, covering 914 km^2.

Looking into individual river basins, the River Neva has by far the largest lake area, almost 50,000 km^2, including two largest lakes in Europe, Ladoga (18,130 km^2) and Onega (9890 km^2). As to the lake percentage, the basin of Motala Ström (Sweden) is number one (22.3%), followed by Kymijoki (18.9%) and Göta älv (18.6%).

In addition to natural lakes, there are thousands of man-made ponds and reservoirs in the Baltic Sea Basin. Most of them are small ponds in Poland and the Baltic States, but the largest reservoirs are in Sweden, with the exception of Narva Reservoir (200 km^2) on the Estonian-Russian border.

For centuries ago, all the countries around the Baltic Sea had large natural wetland areas. In Denmark, Germany, Latvia, Lithuania and Poland most of the wetlands have been drained for agricultural purposes. Estonia and particularly Finland still have rather large natural wetlands. In northern Sweden exploitation has had a relatively small impact on wetlands, while utilisation has been much more comprehensive farther south.

Even today the wetlands comprise one fifth of the Baltic Drainage. During the last few decades, the focus has changed from exploitation to conservation and preservation. Also the use of wetlands in water quality issues, such as the retention of nutrient leaching and cleaning of wastewater, has been increasingly recognised.

A.1.3.4 Ice Regimes on Lakes and Rivers

Ice regimes in the water bodies (rivers and lakes) of the Baltic Sea Basin are formed predominantly by the impact of Atlantic air masses producing a warming effect on the study area during the cold season. The Baltic Sea itself stores much heat in wintertime and also warms the adjacent areas. Therefore, the closer the water body to the sea coast, the later ice cover is formed and the earlier the ice break-up occurs.

In the north, rivers are typically frozen in the middle of October; rivers discharging to Lakes Ladoga and Onega are frozen in mid-November, and in southern and southeastern regions only late in December. In the rivers flowing towards the south-western coast of the Baltic Sea, permanent ice cover may not be formed during some warm winters. The duration of complete ice coverage in the rivers flowing in the northern extremity of the Baltic Sea Basin may last 180 to 200 days. Ice break-up in the rivers in the north usually occurs early in May; in the southeast of the area it happens late in March. Mean long-term maximum ice cover thickness on the rivers within the Baltic Sea Basin also differs greatly, depending on location. In the rivers discharging to the Gulf of Bothnia in the north it may be 90 to 100 cm thick; in other regions such as the southeast of the Baltic Sea Basin it is no thicker than 30–40 cm.

Changes in ice regimes in lakes within the study area are similar to those in the rivers. But ice on lakes is usually formed later than that on the rivers; ice break-up in lakes also occurs later.

On lakes in Poland (see e.g. Fig. 2.31), the ice cover is formed in mid-November at the earliest and in mid-February at the latest. A considerable diversity was observed in mean dates of the ice cover freeze-up depending on the lake depth. The ice cover formed earliest in the shallowest lakes (Lake Jeziorak – 12 December, Lake Lebsko – 19 December) and at the latest in the deeper lakes (Lake Hancza – 2 January, Lake Charzykowskie – 4 January). It is worth noting that the ice cover on Lake Studzieniczne froze up on the average 17 days earlier than on Lake Hancza, which is only 65 km away.

Ice characteristics of the Russian lakes differ greatly. This difference is mainly connected with the lakes' morphometry. The considerable influence of the lakes' morphometry is evident from the comparison of the ice characteristics of the lakes Ladoga and Onega. Mean ice cover duration, for example, of the shallower Lake Onega is 20 days longer, and ice cover is 5 cm thicker.

A.1.3.5 Snow Cover

Snowfalls occur every winter in the Baltic Sea Basin and seasonal snow cover is formed except in the southwestern regions. Typical durations of snow cover over most areas are between four and

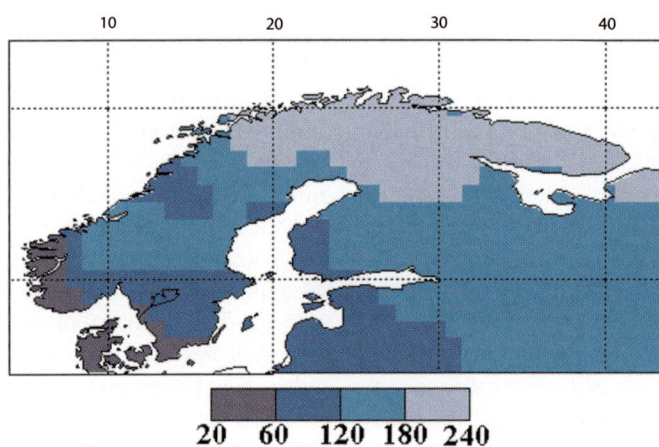

Fig. A.15. Mean regional variability of duration of snow cover (days) for the period 1936–2000 (from Kitaev et al. 2006)

six months (except for regions on the southern coast of the Baltic Sea). Snow cover is a regularly varying feature of the land areas. It affects the winter and spring climate in several ways, the two most important being (Kuusisto 2005):

- because of its high albedo, snow absorbs much less solar radiation than bare soil or vegetated surface;
- melting snow acts as a heat sink, keeping the ground temperature near 0 °C despite high daytime radiative fluxes.
- In the Baltic Sea Basin, 10–60% of annual precipitation occurs in form of snow. Snow is also the origin of a considerable proportion of runoff, its share of average annual runoff being typically higher than its share of annual precipitation. As to the floods, snowmelt is also a major agent almost all over the Baltic Sea Basin.

Although winter precipitation is quite evenly distributed, there are several factors that lead to significant regional variability of snow cover. The orographic gradient for solid precipitation tends to be larger than that for liquid precipitation, because the altitude of the cloud base is low in wintertime. A more important factor is the vertical temperature gradient; rain may fall at lower altitudes, while it snows on higher slopes. Wind redistributes snow particularly in open terrain, leading to extra accumulation in terrain depressions or on leeward sides of ridges or different obstacles. Finally, snow may melt at lower altitudes and on sunny slopes, while no melting occurs at higher sites or in shady places.

Snow Cover Season

The length of the snow cover season varies in the Baltic Sea Basin Basin within wide limits – from several days on average in the western part of the Scandinavian Peninsula to 7–8 months in the territories north of 65° N. Thus, the smooth increase in duration of snow cover from southwest to northeast goes along with the smooth decrease of mean air temperatures during the cold period – from about 0 °C in the west of the Scandinavian peninsula down to −10 °C in the north-eastern part of the eastern European plain (Fig. A.15). However, the correlation between the variation of the duration of the snow period and the air temperature, both at seasonal and at long-term levels, is not significant. In forests, the duration of snow cover is 10–30 days longer than on open ground. The difference depends mainly on the density of the forest canopy, which effectively reduces the rate of snowmelt, thus delaying the disappearance of snow cover (Kuusisto 1984; Kitaev et al. 2005a).

In Finland and north of the eastern European plain, the minimum average snow duration is around 100 days. In southwestern Sweden the duration increases from less than 50 days to more than 100 days within a distance of 100 kilometers; in Finland the gradient hardly reaches 30 d/100 km anywhere in the country. In forests, the duration of snow cover is 10–30 days longer than on open ground.

There are significant spatial differences in duration of snow cover in the territory of Estonia. Its lowest mean values, less than 80 days, are observed on the western coast of the West Estonian Archipelago, i.e. on the open coast of the Baltic Proper (Jaagus 1996, 1997; Tooming and Kadaja 1999, 2000a,b). The highest mean duration of snow cover, more than 130 days, is typical for north-eastern Estonia and for the uplands of southern Estonia. Generally, snow cover duration increases from west to east, similar to decreasing winter air temperature.

Over the territory of Latvia there are remarkable spatial differences in snow cover duration: in the western part of Latvia along the coast of the Baltic Sea and in the proximity of the Gulf of Riga the duration is 70–90 days; it increases to around 110 days moving away from the coast, and the longest snow duration, 114–134 days, is observed in the uplands of eastern Latvia (Draveniece 1998). In general, the duration of snow cover increases from west to east-southeast, following the descending winter air temperature isotherms.

Snow Accumulation

In southern Sweden, maximum snow depths are below 20 cm, and it is below 40 cm in southwestern Finland. Values exceeding 80 cm are reached everywhere north of the latitude 64° except along the coasts of the Gulf of Bothnia. In Finland and Russia the largest region with snow depth exceeding this value extends from Northern Karelia to south-eastern Lapland. The upper slopes of the Scandinavian mountains in Sweden typically have mean maximum snow depths of 100–130 cm, while in Finland one meter is generally exceeded only in the Kilpisjärvi area.

The smallest average maximum depths of snow cover in Poland are recorded in the western part of the country, at no more than 15 cm. The values grow towards the northeast to more than 30 cm; in the mountains they generally increase with altitude (depending strongly on the local topography). In the Tatra Mountains above the tree line, the average maximum seasonal snow depth exceeds 150 cm and 200 cm at summits (Falarz 2004).

The water equivalent of snow is a hydrologically much more important variable than the snow depth. Although weather conditions during the melting period vary considerably from year to year, the correlation coefficients between the maximum areal water equivalent of snow and the volume and peak of the spring flood are significant in almost all major river basins in Fennoscandia.

In Finland in the period 1961–75, the mean maximum water equivalent ranged from 80 to 140 mm in the southern part and from 140 to 200 mm in the northern part (Kuusisto 1984). Up to the latitude of 66° N, the values were 40–60 mm larger in the eastern part of Finland than on the western coast. The year-to-year variation of maximum water equivalent is highest in southwestern Finland, where the ratio between the high and low maximum water equivalent with a return period of 20 years was 4–5, according to data used by Kuusisto (1984). In northern Finland the corresponding value was 1.6–2.0.

In Sweden, about one quarter of the country has mean maximum water equivalent in excess of 200 mm, while the corresponding fraction in Finland is only around 5 per cent. On the other hand, roughly one quarter of Sweden has less than 80 mm of water bound in snow cover during the maximum accumulation; in Finland this fraction is also around 5 per cent. In Norway, the countrywide snow accumulation is still considerably more uneven than in Sweden. The variation of snow conditions from year to year is also quite large throughout Norway.

In western Estonia the mean maximum water equivalent on fields remains below 50 mm, exceeding 70 mm in upland areas, and 90 mm in the Haanja region (Tooming and Kadaja 2000b). The mean maximum water equivalent in Poland extends from below 30 mm in the west to above 75 mm in the north-eastern part of the country; it is above 100/200 mm in the highest part of the Sudeten/Carpathians, respectively (Sadowski 1980).

Extreme Snow Conditions

In Finland the official record snow depth, 190 cm, was measured at the Kilpisjärvi climatological station on April 30th, 1997. It is clear that this value is much lower than the true maximum in Finnish nature; accumulations of three to four meters are possible in narrow gorges in the fjells of Lapland above the tree line. Going further back in history, the snow cover of the winter of 1898–1899 was very probably much thicker than in any winter of the 20th century. Snow depths exceeding 150 cm were reported from several observation sites in northern Savolax.

In Karelia (Russia), the maximum measured snow depth was recorded in the winter of 1983/84, when the average snow depth in March reached 72 cm. Individual extreme measured values of snow depth in Karelia may reach 150 cm, which is more than twice the station standard deviation in this region. Both an increase in snow storage and in the number of winters with extreme snow depth conditions were observed during the last two decades (Kitaev et al. 2005b).

In Swedish lowlands, the highest official snow depth is the same, 190 cm, as the record for Finland as a whole. This value was measured at Degersjö in Ångermanland, only 20 km from the coast of the Gulf of Bothnia. Another coastal region in Sweden with severe snowstorms is northeast Uppland (Dahlström 1995). In the Swedish fjells, snow depths in excess of two metres have been measured at all climatological stations between the latitudes 62–68° N. The highest value, 327 cm, was observed at Kopparåsen, 15 km east of Riksgränsen, on the 28th of February in 1926. However, the variation of maximum snow depth is very high: There are winters when the maximum remains clearly below one meter even at the snowiest observations sites (Pershagen 1981).

Although the maximum snow depth of 77 cm as mean of route observation during the period 1962–2001 was recorded in Estonia on 20 December 1988 in the Haanja upland, the winter periods 1981/82 and 1965/66 can be considered as the snowiest, when the mean spatial (on the basis of land points of a 5×5 km grid) maximum water equivalent in February was 135.3 and 130.3 mm, respectively (Tooming and Kadaja 2006). In Latvia, with a background of average snow depth of 6–29 cm over the territory, a high value, 126 cm, was observed in Gureli, Vidzeme Upland in the third decade of March 1931. The variation of maximum snow depth is very high: almost half of all winters are warm and the others are colder than average or even severe. There are years in Latvia with a very thin snow cover: going further back in history such winters were 1948–1949, 1960–1961 and 1971–1972. In the winter of 1972–1973 the depth of transitional snow cover was only 3–8 cm.

The absolute maximum of snow cover depth in Poland was observed in Dolina Pięciu Stawów Polskich (Tatras, 1670 m a.s.l.; Falarz 2001). It was 503 cm on March 26th 1967. It is highly probable that the true maximum in highly located shielded valleys in the Tatras was much higher. Outside the mountains, the absolute maximum of snow cover depth was 85 (84) cm in Krakow (Suwałki) in February 1963 (1979). During the lowest-snow winter seasons, the maximum snow depth did not exceed 10 cm in the lowlands, with some western stations recording only 1–2 cm (Falarz 2004). In

the winter season 1991/92 in Słubice no snow cover of at least 1 cm depth was observed.

The maximum snow cover duration in southwestern Poland (except for the mountains) reached 100 days, more than 140 days in the northeastern region and 230–260 days in the Tatra Mountains above the tree line. In southern and eastern Poland, these maxima were recorded in the winter of 1995/96, while in the rest of the country, these maxima occured in the season 1969/70. There are permanent snow patches in the Tatra Mountains. The shortest snow cover duration in western Poland was just a few days per winter season. Extremely short snow cover duration was observed in the winters of 1924/25, 1974/75 and 1988/89.

A.2 The Late Quaternary Development of the Baltic Sea

Svante Björck

A.2.1 Introduction

Since the last deglaciation of the Baltic Sea Basin, which began 15,000–17,000 cal yr BP (calibrated years Before Present) and ended 11,000–10,000 cal yr BP, the Baltic Sea Basin has undergone many very different phases. The nature of these phases was determined by a set of forcing factors: a gradually melting Scandinavian Ice Sheet ending up into an interglacial environment, the highly differential glacio-isostatic uplift within the basin (from 9 mm/yr to −1 mm/yr; Ekman 1996), changing the geographic position of the controlling sills (Fig. A.16), varying depths and widths of the thresholds between the sea and the Baltic basin, and climate change. These factors have caused large variations in salinity and water exchange with the outer ocean, rapid to gradual paleographic alterations with considerable changes of the north-south depth profile with time. For example, the area north of southern Finland-Stockholm has never experienced transgressions, or land submergence, while the development south of that latitude has been very complex. The different controlling factors are also responsible for highly variable sedimentation rates, both in time and space, and variations of the aquatic productivity as well as faunal and floral changes. The basic ideas in this article follow the lengthy, but less up-dated version of the Baltic Sea history (Björck 1995), a more complete reference list and, e.g., the

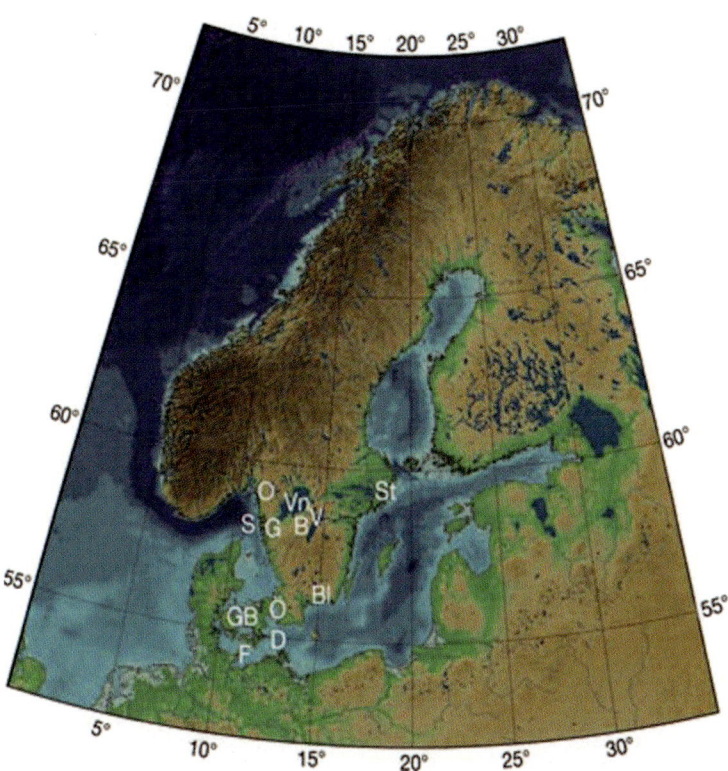

Fig. A.16. The Baltic Sea Basin, showing both land and submarine topography. The letters denote geographical names used in the text. B = Billingen, Bl = Blekinge, D = Darss sill, F = Fehmarn Belt, G = Göta Älv river valley, GB = Great Belt, O = Otteid/Steinselva strait, S = Skagerrak, St = Stockholm, V = Lake Vättern and Vn = Lake Vänern (by courtesy of Martin Jakobsson)

calendar year chronology of the different Baltic Sea phases can be found on the Internet[1]. Although I will focus on the postglacial history of the Baltic Sea in this restricted review, I think it is important to inform the reader about the preceding stages to the more modern Baltic Sea setting.

A.2.2   The Glacial to Late-Glacial Baltic Sea

Due to repetitive, more or less erosive, glaciations during the last glacial cycle, little detailed evidence exists about the glacial conditions in the Baltic before 15,000–14,000 cal yr BP. Based on lithostratigraphic correlations and a large set of OSL (optically stimulated luminescence) and ^{14}C dates, Houmark-Nielsen and Kjær (2003) have, however, indicated several 'embryonic' glacial stages of the Baltic Sea during MIS3 (Marine Isotope Stage 3 dated to about 25,000–60,000 cal yr BP). According to their model the dynamic behavior of the southwestern part of the

Scandinavian Ice Sheet between about 40,000–17,000 cal yr BP, produced several proglacial Baltic Ice lakes before the last Baltic Ice Lake proper between about 15,000–11,600 cal yr BP (Björck 1995; Björck et al. 1996; Andrén et al. 1999). In-between glacial advances these proglacial stages have been dated to about 40,000–35,000 and 33,000–27,000 cal yr BP, but with changing configurations during these stages (Houmark-Nielsen and Kjær 2003). It is also postulated that the deep northwest–southeast trending Esrum-Alnarp valley, through Sjælland and Skåne, often functioned as the main connection between the Baltic Sea basin and the (glacio)marine waters of the Kattegatt-Skagerrak. After the final advance some time between 17,000–16,000 cal yr BP – the Öresund lobe (Kjær et al. 2003) – a rapid deglaciation of the southern Baltic Sea basin seems to have taken place (Björck 1995; Lundqvist and Wohlfarth 2001; Houmark-Nielsen and Kjær 2003).

A proglacial lake – the Baltic Ice Lake (BIL) – was developed in front of the receding ice sheet. Due to glacial in-filling of the Esrum-Alnarp val-

[1]www.geol.lu.se/personal/seb/Maps%20of%20the-%20Baltic.htm

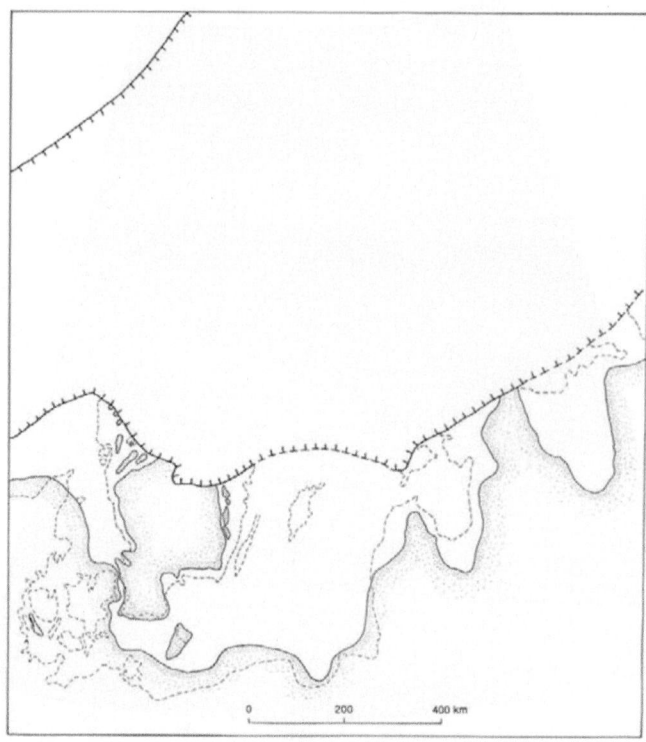

Fig. A.17. The configuration of the Baltic Ice Lake at about 14,000 cal yr BP. Note that the drainage through Öresund and that today's coast line is marked with a stiple line (from Björck 1995)

ley, the lowland in the Öresund region developed into the connecting channel between the Baltic Sea and the sea in the northwest. Glaciolacustrine sediments, e.g., varved clays, were laid down in the Baltic as the ice sheet retreated northwards.

At this early stage of the BIL global sea level was situated at −100 m (Lambeck and Chappell 2001); more than 2/3 of the last glacial maximum ice sheets still remained to be melted. Since the remaining rebound of the loading effect from the ice sheet was fairly small in the southernmost Baltic, the coast line was situated below today's sea level in southern Denmark, Germany and Poland. However, further north both the total and remaining unloading effect – glacial isostatic uplift – was larger than 100 m, and therefore the coast lines of Sweden and the Baltic republics were above today's sea level; the further north the higher.

As a consequence of the uplift the Öresund area, which was now the threshold of the BIL, emerged faster than the rising sea level. This gradual shallowing of the outlet increased the velocity of the out-flowing water and thus also the erosion of the sill area. As long as loose Quaternary deposits could be eroded, the erosion continued, and the present Öresund Strait was possibly shaped, with the island of Ven being an erosional remnant of a previous till-covered landscape. However, when the bedrock sill of flint-rich bedrock between Malmö and Copenhagen was exposed, erosion ceased. The consequence of this was that the continuing uplift made the threshold gradually shallower, until a critical water velocity was reached. At this stage, about 14,000 cal yr BP (Fig. A.17), the water level inside the threshold, i.e. the BIL level, had to rise to compensate for the decreased water depth of the sill. This caused the BIL to rise above sea level; a gradually higher water fall was created between the BIL level south of the threshold and sea level north of it. It also meant that coastal areas situated north of the Öresund isobase (isobases connect areas with the same uplift/shoreline) continued to emerge, while areas south of it submerged, the latter causing a transgression.

The melting Scandinavian Ice Sheet had a strong impact on the aquatic and sedimentary conditions in the Baltic; freshwater with a strong glacial influence produced clayey-silty sediments often of varved (annual layering) type and with-

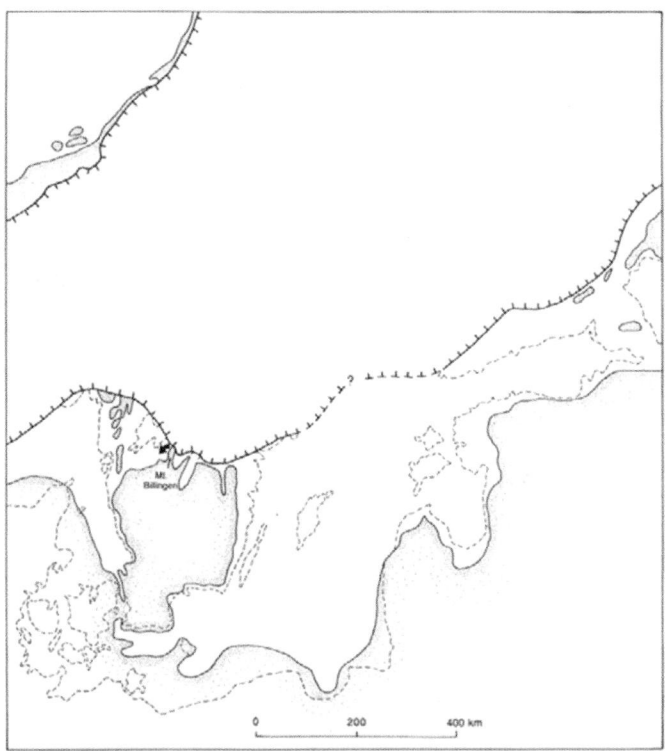

Fig. A.18. The configuration of the Baltic Ice Lake at about 13,000 cal yr BP. Note that it was drained north of Mt. Billingen and that Öresund was dry land. The arrow marks a possible subglacial drainage before Mt. Billingen was completely deglaciated (from Björck 1995)

out organic material. Diatoms are rarely found in these sediments and it is doubtful if any fauna at all existed in this glacial lake. As the melting continued northwards, an important watershed melted out of the retreating ice sheet: the Billingen bedrock ridge in south central Sweden between the two large lakes Vättern and Vänern. The Billingen area almost formed a 'wall' between the sea in the west and the up-dammed BIL in the east.

However, when the ice retreated to the northern tip of Billingen around 13,000 cal yr BP (Fig. A.18), the BIL was drained west, initially beneath the ice. We think the BIL was lowered some 10 m at this event, but any morphologic evidence about this drainage would have been destroyed by the ice. The effect of the sudden lowering was that Öresund was abandoned as the outflow and the BIL water flowed through the, at the time glaciomarine, Vänern basin and out into the Skagerrak through several different valleys/fiords. However, there is no evidence from sediments that saline water managed to penetrate into the Baltic, east of Billingen.

At about 12,800 cal yr BP the North Atlantic region experienced a fairly abrupt climatic change, the so-called Younger Dryas cooling. One effect of the lowered temperatures, especially in winter, was that the previously receding ice sheets of the region began to expand again, and the Scandinavian Ice Sheet advanced southwards to block Billingen again. This would have dramatic effects on the BIL: the water level would once again rise above sea level until Öresund began to function as the outlet (Fig. A.19). This quick transgression would have continued slowly in areas south of the sill, while the remaining Baltic coasts experienced regression, or emergence. The outlet/sill area was still rising quicker than the rising sea, which meant that the BIL rose more and more above sea level; the waterfall in Öresund became gradually higher. During this time the sediments in the Baltic were still very influenced by the glacial input, even in the southern Baltic.

At the end of the Younger Dryas cool period the ice sheet started to retreat again, and sometime between 11,700 and 11,600 cal yr BP a second, and very dramatic, drainage occurred at Billin-

Fig. A.19. The configuration of the Baltic Ice Lake just prior to the final drainage, 11,700–11,600 cal yr BP, which was to lower the Baltic by 25 m (from Andrén 2003a, by courtesy of Stockholm Marine Research Centre)

gen when the ice sheet receded north of the barrier. Since the Öresund threshold at this time had risen by about 25 m above sea level the water level within the Baltic Sea Basin fell by the same amount. It has been calculated that this drainage took 1–2 years and the main traces of it, huge sediment complexes of pebbles and boulders, can be found 5–7 km west of Billingen. As a consequence of the drainage, the coast around the Baltic emerged out of the water and 'fresh' coasts were suddenly exposed. Especially in the southern Baltic, large areas emerged and became land areas, which was of course also the case with the Öresund sill. A large land bridge between Skåne and Själland was established, which favoured a rapid northward plant and animal colonisation during the imminent Holocene interglacial period.

A.2.3 The Post Glacial Baltic Sea

The Yoldia Sea stage

Obviously the final drainage of the BIL at 11,600 cal yr BP was a turning point in the late geologic development of the Baltic Sea: a sudden pa-leogeographic change, a warmer climate, a rapidly retreating ice sheet and direct contact with the saline sea in the west, incl. Vänern. This is also the starting point for the next Baltic Sea stage, the *Yoldia Sea* stage, which would last for about 900 years.

The straits between Vänern and the Baltic were initially narrow, and saline water could not enter into the Baltic mainly due to the large amount of outflowing water. It would take 250 years (Andrén et al. 1999) until the straits had opened up enough to allow eastward penetration of salt water (Fig. A.20). This slightly brackish phase had its highest salinities in the low-lying areas between Vänern and Stockholm. However, brackish bottom-water also managed to penetrate down to the southern Baltic, creating periodically anoxic bottom conditions. Brackish conditions are shown by occurrences of foraminifera and the bivalve mollusk *Portlandia (Yoldia) arctica* as well as by the diatom flora of the sediments. This slightly saline phase only lasted for some 150 years until the straits between the marine water in Vänern and the Baltic became too shallow to allow saline inflow. Although the brackish conditions turned

Fig. A.20. The configuration of the Yoldia Sea stage at 11,400–11,300 cal yr BP, when a short saline phase is about to start. Note the large paleogeographic changes between Figs. A.19 and A.20 with the huge land-bridge in the south and the Närke Strait in the north (from Andrén 2003b, by courtesy of Stockholm Marine Research Centre)

into freshwater the Baltic was still at level with the sea, and the sediments during the complete *Yoldia Sea* stage were characterized by low organic content. As a contrast the western part of Vänern was a fairly fauna-rich marine embayment (Fredén 1986).

Owing to the on-going and still rapid uplift in south central Sweden, the straits between Vänern and Skagerrak became gradually shallower, and even some of them even emerged above sea level. The outflowing water from the Baltic had to pass through Vänern and these straits, and in the end only two straits functioned: the Göta Älv strait, which today is the Göta Älv river valley between Vänern and Göteborg, and the Otteid/Steinselva strait at the Swedish–Norwegian border east of Idefjorden.

The Ancylus Lake stage

The gradual shallowing and narrowing of these straits resulted in increased water velocity in these outlets until a maximum was reached when they could not 'swallow' the amount of water entering the Baltic Sea Basin (including meltwater from the melting ice sheet); the water level inside the narrow straits had to rise to compensate for the decreasing outflow area in the straits. Similar to the up-damming of the Baltic Ice Lake, the water level had to rise in pace with the uplift of the sills/straits. South of the isobases for the outlet region this would result in a transgression, since the uplift here was smaller than the forced water level rise, while to the north the situation would be the opposite; a northwards increasing regression. This tilting effect is the onset of the *Ancylus Lake* transgression, which started around 10,700 cal yr BP.

The possibly already submerging coasts in the southernmost Baltic experienced an increased flooding, while the previously emerging coasts of southern Sweden and the northern Baltic republics now changed into submergence. This sudden submergence, or transgression, is witnessed by, e.g., drowned pine forests east of Skåne and tree-ring analyses show rapidly deteriorated living condi-

Fig. A.21. The configuration of the Ancylus Lake stage at about 10,300 cal yr BP at the culmination of the Ancylus transgression. Note the outlets west and southwest of Lake Vänern (from Andrén 2003c, by courtesy of Stockholm Marine Research Centre)

tions. The transgression is also clearly displayed by the *Ancylus* beach; a raised beach found in many places in, e.g., southeast Sweden, on the island of Gotland, and in Latvia/Estonia showing transgressive features.

The freshwater conditions with low primary productivity at the end of the *Yoldia Sea* and during the *Ancylus Lake*, named after the freshwater limpet *Ancylus fluviatilis*, resulted in good mixing of water without permanent stratification. The sediments of the *Ancylus Lake* are also poor in organic material, and the further north the more glacially influenced they are. The amount of the *Ancylus* transgression varies between areas/regions depending on the local uplift. The maximum transgression probably occurred outside the Polish coast and amounted to about 20 m, while a transgression of 10–12 m characterised the *Ancylus* coast in the southwest, Denmark–Sweden–Germany. The latter amount was probably also how much the *Ancylus Lake* was finally dammed-up above sea level. The transgressive phase of the *Ancylus Lake* lasted ca. 500 years, and was obviously governed by the possibility for

the Baltic water to find an alternative outflow area. This meant that the transgression in the south continued as long as the constrained sills west of Vänern functioned as outlets (Fig. A.21).

From independent studies we know that the *Ancylus* transgression ended abruptly with a fairly sudden lowering of the Baltic water level at about 10,200 cal yr BP. The rate of the lowering, or regression, strongly implies that it is not only a gradual isostatic effect that shows up in the shore displacement curves, but rather a forced regression; the absolute water level fell. The most likely explanation for such a regression would be that the *Ancylus Lake* level fell due to a lowering of the base level. Since the base level was determined by the sills/outlets, it would mean that the sill(s) was eroded.

We do, however, also know that the sills west of Vänern consist of crystalline bedrock, and it is hardly possible for water to suddenly erode 10 m of hard bedrock. It has therefore been assumed that the water found a new outlet. Because of the transgression in the south, the most obvious threshold/outlet candidate should be situated in

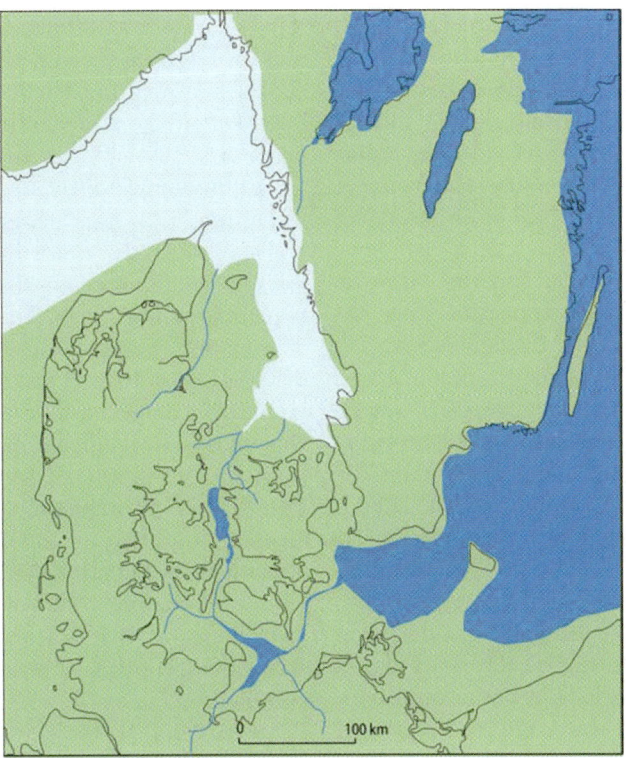

Fig. A.22. The configuration of the Ancylus Lake just prior to the first minor saline ingression at about 10,000 cal yr BP. In comparison with Fig. A.21, note the regressive shore line as a consequence of both isostatic uplift and the lowering caused by the Ancylus drainage. Also note that Lake Vänern was no longer a part of the Ancylus Lake (from Jensen et al. 2002)

low-lying areas of the Danish–German area, which is also characterized by lose Quaternary deposits. For a long time it has therefore been postulated that the water found its way over Darss Sill, into Mecklenburg Bay, over Fehmarn Belt and finally through the deep Great Belt, between the islands of Själland and Fyn, the so-called Dana River. Parts of the submarine morphology along this path have been interpreted to be a remnant of such an erosive event.

However, recent German and Danish studies (e.g. Bennike et al. 1998; Jensen et al. 1999; Lemke et al. 1999) partly contradict such a scenario, although Bennike et al. (2004) recently dated river deposits in the Great Belt channel to about 10,200 cal yr BP, surrounded by levée and lake sediments. In fact, a compromise between the rather dramatic picture of the Ancylus drainage presented by Björck (1995) and Novak and Björck (1998) and the calm Danish–German solution may be possible: an initial regression was caused by a few meters of erosion of the Darss Sill and possibly also in the Great Belt channel lowering the

Baltic by up to 5 m. This was followed by a fairly calm fluvial phase; the gradient between the Kattegatt sea level and the *Ancylus Lake* level was only perhaps 5 m. As the region was now characterized by a rising sea level/base level this gradient decreased and conditions gradually became even calmer. We also know that the area outside the mouth of the Great Belt channel in Kattegatt at this time was not characterised by marine conditions, but was very influenced by freshwater (Bennike et al. 2004). Since the uplift of the area had more or less ceased, conditions were now controlled by the rapidly rising sea level (2–2.5 cm/yr). It would therefore last only 200–300 years before sea level was at level with the *Ancylus Lake*, i.e. at about 10,000 cal yr BP (Fig. A.22).

The transitional stage between the freshwater of the *Ancylus Lake* and the following brackish-marine *Littorina Sea* (named after the marine gastropod *Littorina littorea*), named the Early Littorina Sea by Andrén et al. (2000), is also partly an enigma. The first signs of marine influence have usually been seen in sediments with an age of about 9,000 cal yr BP, usually in the southern Baltic area. This is also in accordance with the time when we think that the Öresund Strait was flooded by the global marine transgression; at that time the sea level rise had exceeded the uplift rate in South Sweden. Therefore the younger limit of the *Ancylus Lake*, has often been set at about 8,500 cal yr BP, although geologists have been aware that the transition into the *Littorina Sea* is complex; the end of this transition stage has occasionally been named the Mastogloia Sea.

Lately, however, data from the Bornholm Basin and the archipelago of Blekinge in southeastern Sweden imply that the first saline influence may have occurred already at about 9,800 cal yr BP (Andrén et al. 2000; Berglund et al. 2005). This indicates that saline water from Kattegatt, through the Great Belt–Fehmarn–Mecklenburg–Darss channel, could occasionally penetrate into the Baltic fairly soon after the point in time when the rising sea level began to rise the Baltic level. However, this narrow and long outlet never did allow large amounts of salt water into the Baltic. It would take another 1,500 years before a real marine influence was felt inside the Baltic; then sea level had finally reached up to the Öresund threshold between Limhamn and Dragør and a wide outlet/inlet area was created.

The Littorina Sea

If we disregard some of the uncertainties about the initial outlet/inlet area during the Ancylus-Littorina transition, we can at least clearly document a rapid spread of saline influence throughout the Baltic basin around 8,500 cal yr BP. When the first clear signs of marine water appear, usually defining the onset of the *Littorina Sea* in the related sediments, this is also usually reflected in the sediment composition as an increased organic content. This implies that with the increased saline influence the aquatic primary productivity clearly increased in the Baltic. In the beginning of this phase salinities were very low in the north, but between 8,500–7,500 cal yr BP the first and possibly most significant *Littorina* transgression set in.

The reason that the southern Baltic, up to approximately the Stockholm–south Finland area, would experience transgressions during the forthcoming 2,500–3,000 years was that isostasy in this whole region was less rapid than the ongoing sea level rise. While the causes for most of the separate *Littorina* transgressions can possibly be re-

lated to sudden collapses of the Antarctic Ice Sheet and its huge ice shelves, the more or less steadily rising sea level until 6,000 cal yr BP was mainly an effect of the still melting North American ice sheets. At this point in time the last remnants of the Labrador Ice Sheet melted.

During the first three successive *Littorina* transgressions the water depths increased considerably (Berglund et al. 2005) in the, at the time, two functioning inlets, Öresund and Great Belt. The extent of these transgressions was in the order of at least 10 m in the inlet areas, with a large increase in water depth at any critical sill. In turn, this allowed a significant increase in the amount of inflowing saline water into the Baltic, with higher salinities as an important consequence. The increasing salinity, in combination with the warmer climate of the mid-Holocene, generated a fairly different aquatic environment, compared to before.

In terms of richness and diversity of life, and therefore also primary productivity, the biological culmination of the Baltic Sea was possibly reached between 7,500–6,000 cal yr BP (Fig. A.23). The high productivity, in combination with increased stratification due to high salinities in the bottom water, caused anoxic conditions in the deeper (> 100 m) parts of the Baltic Sea (Sohlenius et al. 2001), especially in the central and northern parts with its larger distance to 'fresh' oxygenated Atlantic water.

A turning point in the Baltic development can be seen after about 6,000 cal yr BP: the transgressive trend is broken around a large part of the Baltic coastline, although minor transgressions may have occurred until about 5,000 cal yr BP. While sea level had, generally speaking, ceased to rise, the uplift pattern in the Baltic area was complex. In the southernmost Baltic area the submergence continued, which meant that the transgression rate became less extensive than during peak Littorina time. In Sweden and along the coast from northern Lithuania and northwards uplift was still going on, which meant that the end of sea level rise resulted in a renewed regression. The regression caused shallower sills, which meant that a gradually less amount of marine water could enter the basin. Since the main sill areas, Öresund and Great Belt, are situated today around the isobase -0.4 mm/y (Ekman 1996), it implies that this shallowing ended perhaps a few hundred years ago. Since then the inlet areas have become slightly deeper, causing more inflow of Atlantic waters. Another consequence of the differential

Fig. A.23. The Littorina Sea stage at about 7,000 cal yr BP. Note the wide straits in the south and the still much remaining uplift in the north, especially conspicuous in lowland areas (from Andrén 2004, by courtesy of Stockholm Marine Research Centre)

uplift is that a large part of the Baltic is rising, causing a shallowing effect, and a small part is sinking. If we are worried about a future reduced circulation/ventilation in the Baltic, it is fortunate, at least in the long-time perspective, that the deepest parts of the basin are situated in areas with fairly high uplift rates.

Generally speaking, the Baltic sediments tell us that since peak *Littorina* time, some 6,000 years ago, salinities in the Baltic Sea have gone down. Three possible reasons for this decline can be postulated: less inflow of Atlantic waters from Skagerrak/Kattegatt, and decreased summer temperatures and increased precipitation in most of the Baltic Sea Basin. All these three factors would by themselves have triggered reduced salinities, and it is very likely that they all are responsible for the long term trend.

According to several independent studies the last millennium of the Baltic Sea seems to have been characterised by at least two phases of high productivity and anoxic bottom conditions, the Medieval Warm Period and the last century, and one period with decreased productivity and oxygenated sediments, the Little Ice Age. Thus,

the natural climatic variability seems to be a key player for the Baltic Sea and its often profound changes. Owing to the sudden appearance of the North American soft celled clam *Mya arenaria*, the youngest part of the Baltic Sea history have often been named the Mya Sea; it was thought to have been introduced in the bilges of European trade ships. However, Petersen et al. (1992) dated such clams in Danish coastal deposits and found that they predate Columbus' discovery of America by several hundreds of years. This would be additional evidence that Vikings discovered America before Columbus and that their ships brought this 'stranger' and newcomer into the Baltic Sea environment. This is a good example of how humans have been part of, and often were an important player of the natural environment. Today this is truer than ever.

Acknowledgment

With this article I would like to honor the friend and colleague Wolfram Lemke, who very sadly suddenly passed away when this article was being written.

A.3   Ecosystem Description

A.3.1   Marine Ecosystem

A.3.1.1   *The Seasonal Cycle of the Marine Ecosystems*

Maiju Lehtiniemi

The specific characteristics of the Baltic Sea, its shallowness, brackish water and partial ice cover during winter, form a rather unique environment for the biota living in the sea. In addition, changes in salinity from south (almost seawater salinity) to north and east (near freshwater salinity) regulate effectively the distribution of fauna and flora of low diversity. Permanent stratification, i.e. a halocline based on salinity differences between surface and deep water, prevails in most of the Baltic Sea affecting the distribution of fauna. Water exchange is restricted under the halocline, and thus bottom oxygen is often depleted, affecting animals living in benthic habitats (Matthäus 1995). The Baltic biota is a mixture of marine and freshwater species with some genuine brackish water species (Remane 1934). In addition to native species, the Baltic Sea hosts about 100 nonindigenous species, more than half of which have established reproducing populations (Leppäkoski et al. 2002). The ecosystem changes along latitudes and becomes simpler towards the north.

The location of the Baltic Sea at high latitudes affects the whole ecosystem through seasonality. Temperature and light conditions fluctuate along the seasons, which regulate primary production and the length of the growing period, which becomes much shorter from the southern to the northern Baltic Sea. Annual ice cover in the northern parts, which shows large year-to-year variations, requires special adaptations from all organisms.

Most of the invertebrates are specialised to live in either littoral, pelagic or benthic habitats, each of which forms a distinct environment. However, food chains often overlap and these interrelated food chains are part of the broader food web. Certain invertebrates, like mysid shrimps, utilise effectively different systems, enhancing the energy transfer between them. In addition, fish, birds and mammals migrate/move regularly between different subsystems utilising them as feeding, wintering or spawning habitats. Millions of Arctic birds (water birds, geese, divers) migrate through the Baltic

Sea Basin twice a year on their migration between Siberia and Europe. Hundreds of thousands of them also overwinter in the Baltic Sea Basin, while most of the birds breeding in the Baltic Sea Basin, overwinter outside the area. Some of them, like Arctic Tern (*Sterna paradisaea* Pontoppidan), overwinter as far as on the Antarctic Ocean.

Spring Period – Intensive Primary Production and Reproduction

Primary production and growth increase rapidly during spring when enough light reaches the surface waters, in the northern parts after the ice break-up. Nutrients stored in the water column during winter are effectively utilised by the spring bloom, which is mainly dominated by chain forming diatoms and dinoflagellates (Edler 1979; Larsson et al. 1986). The highest biomasses are commonly reached after a weak thermocline has developed, which gives rise to a warmer well-lit surface layer. The development of thermal stratification prevents mixing with deeper water layers and thus suppresses cell losses from the euphotic layer and gives the phytoplankton better survival opportunities.

The spring bloom declines after most of the nitrate and phosphate are consumed from the water layer above the thermocline. After the bloom, most of the organic matter produced sinks out of the euphotic layer to the bottom because grazer numbers are still very low (Lignell et al. 1993). The most abundant species in the zooplankton community, which forms the next trophic level (Fig. A.24), are microprotozoans that have their maximum densities soon after the peak of the phytoplankton bloom (Kivi 1986; Johansson et al. 2004) and a few species of copepods and rotifers (Viitasalo et al. 1995; Möllmann et al. 2000).

The sedimenting organic material forms the main energy source for benthic secondary production (Kuparinen et al. 1984). The heterotrophic benthic community is coupled to the pelagic and littoral systems through the seasonal pulse of sedimenting organic material. Suspension feeders like barnacles (*Balanus improvisus* Darwin) and blue mussels (*Mytilus edulis* L.) on hard substrates or algae, and amphipods (*Monoporeia affinis* Lindström, *Pontoporeia femorata* Kröyer), isopods (*Saduria entomon* L.), bivalves (*Macoma balthica* L.) and polychaetes (e.g. *Harmothoe sarsi* Kinberg, *Marenzelleria viridis* Verrill, *Scoloplos armiger* O.F. Müller) in/on the sed-

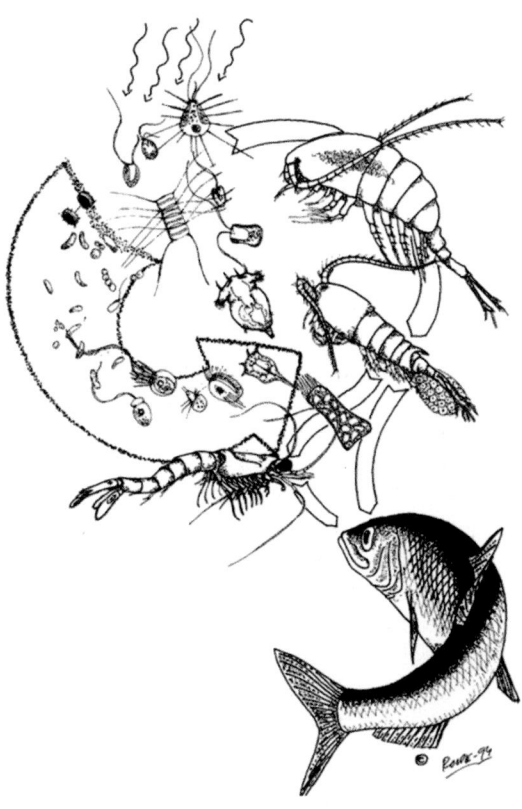

Fig. A.24. Baltic Sea pelagic food web, grazing chain on the right and microbial loop on the left. The grazing chain prevails mainly under turbulent nutrient-rich circumstances e.g. during bloom periods (Kiørboe 1993) while the microbial loop dominates during summer in stratified conditions (Uitto et al. 1997) (from Vuorinen 1994, figure by Juha Flinkman)

iment utilise the sinking organic material effectively. Mysid shrimps (*Mysis mixta* Lilljeborg, *M. relicta* Lovén), which are nektobenthic crustaceans, feed omnivorously on abundant phytoplankton and detritus and start to release their young after the spring bloom (Salemaa et al. 1986; Rudstam and Hansson 1990; Viherluoto et al. 2000). Also, many other invertebrates in the littoral (e.g. *Jaera* sp. and prawns) and benthic habitats (e.g. *M. affinis*) release their young after the spring bloom (Segerstråle 1950). Zooplanktivorous fish like Baltic herring (*Clupea harengus membras* L.) and sprat (*Sprattus sprattus* L.) increase their feeding, which concentrates mainly on large copepods (Fig. A.24; Möllmann et al. 2004). Salmon smolts (*Salmo salar* L.) migrate from spawning rivers to the coastal sea areas to feed on insects and invertebrates (Jutila and Toivonen 1985).

Spring is the most active spawning time for Baltic fish. Sprat and cod (*Gadus morhua* L.) live their whole life in the pelagial. They do not need coastal waters, even at spawning, because their eggs and larvae are buoyant and thus survive in the pelagial (Nissling and Westin 1991). Shallow coastal areas are, however, favoured spawning grounds of many littoral and pelagic fish species. For example herring, perch (*Perca fluviatilis* L.), roach (*Rutilus rutilus* Berg), pike (*Esox lucius* L.) and gobies (e.g. *Pomatochistus* spp.) lay eggs on gravel or among macroalgae and macrophytes, which start their active growing period after winter. The littoral zone provides fish larvae with shelter, a diverse food supply and, thus, a good start for future growth (Urho 2002). Littoral areas are characterised by clear zonation, which gradually changes from land to outer archipelagos and the pelagial.

In the northern Baltic, shallow bays with stands of *Phragmites australis* (Cav.) and Charophytes are dominated by limnic fauna, while outer areas with rocky shores covered with green and brown algal belts are inhabited by brackish and marine species of invertebrates and fish (Kautsky 1995; Munsterhjelm 1997). The southern and southeastern shores greatly differ, with sandy beaches turning into pelagial without the zone of archipelagos.

Early Summer Period – Secondary Production Increases

After the spring bloom phytoplankton production, which is at a low production stage, is maintained by recycling of ammonia and phosphate. The efficiency of the nutrient recycling is regulated by temperature, excretion of zooplankton and bacterial metabolism. The time of low primary production occurs in June–July but the timing varies spatially and between years. The dominant species are pico- and nanoplankton, which have a high turnover rate (Niemi 1975). At this time the main pathway for energy transfer from primary producers to top predators starts with the microbial loop (Fig. A.24; Azam et al. 1983; Uitto et al. 1997). It is a micro-food chain that works within (or alongside) the classical food chain. In the microbial loop the smallest organisms, heterotrophic bacteria and picoplankton, use dissolved organic material (DOM) excreted by phytoplankton as carbon and energy sources. When these bacteria are later eaten by micrograzers such as flagellates and ciliates, the formerly "lost" carbon and energy are recycled back into the food web. After this the process continues up the classical food chain, energy is finally transferred up to fish, birds and seals.

Although each stage in the microbial food web involves consumption of organic matter, it also releases ammonium and phosphate into the water. These recycled nutrients can be taken up by algae, stimulating additional primary production (Azam et al. 1983). At this time zooplankton reproduces and biomass increases to its maximum regulating effectively primary production (Uitto et al. 1997). The zooplankton community is diverse, consisting of surface dwelling rotifers, cladocerans and copepod nauplii, and migrating copepods and larger cladocerans (Ackefors 1969; Viitasalo et al. 1995; Möllmann et al. 2000). Rotifers and cladocerans reproduce very rapidly, building up high densities due to parthenogenetic reproduction. Sinking organic matter is minimal, which increases intra- and interspecific competition among species of zoobenthos feeding on organic detritus (Lehtonen 1997). Mysids feed mainly on abundant zooplankton during their vertical migrations from the

bottom to upper water column (Fig. A.24; Rudstam and Hansson 1990; Viherluoto et al. 2000).

Late Summer Period

During late summer, primary production again increases due to dinoflagellates (Kononen et al. 2003) and nitrogen-fixing cyanobacteria, which form large surface blooms all around the Baltic (Kuparinen et al. 1984; Kahru et al. 1994). Massoccurrences are formed in warm water and calm weather conditions (Kononen et al. 1996). The optimum temperatures for cyanobacteria growth are higher than those of diatoms and green algae partly explaining their dominance during late summer (Robarts and Zohary 1987).

Organic matter produced by a cyanobacteria bloom is largely mineralised and consumed by zooplankton during the bloom's growth period, thus the sedimentation rate stays low (Kuparinen et al. 1984). The decaying bloom favours the growth of various bacteria, which may be consumed by ciliates, which further can be consumed by mesozooplankton (Engström-Öst et al. 2002). Thus, a decaying bloom may offer a good food source for the zooplankton community, although actively growing cyanobacteria are considered low quality food (Sellner et al. 1996).

The zooplankton community is still diverse, copepods and cladocerans being the dominant members. The most abundant species include surface dwelling cladocerans *Bosmina longispina* Leydig, *Podon spp.* and *Evadne nordmanni* Lovén and copepods *Acartia spp.* (mainly *A. bifilosa* Giesbrecht), *Eurytemora affinis* Poppe, neritic copepods *Pseudocalanus elongatus* Boeck, *Temora longicornis* Müller and *Centropages hamatus* Lilljeborg (Ackefors 1969; Viitasalo et al. 1995; Möllmann et al. 2000). Pelagic larvae of bivalves, gastropods and polychaetes also appear in the zooplankton community. The zooplanktivorous medusa *Aurelia aurita* L. is very abundant, although year-to-year variations are high. During the warm water period, a recent invasive predatory cladoceran *Cercopagis pengoi* Ostroumov, builds up dense populations rapidly by parthenogenetic reproduction. Herring, sprat and three-spined sticklebacks have included this cladoceran as a part of their diet (Gorokhova et al. 2004; Peltonen et al. 2004). Fishes have their most active growing period. Young–of–the–year fish have to attain a certain length to be able to survive over their first winter.

Autumn – Primary and Secondary Production Decline

After summer temperature decreases and the thermocline breaks down during autumnal turnover, mixing of the water column extends below the critical depth for net production. Primary as well as secondary production by zooplankton decrease due to water mixing, decreasing temperature and light intensity. Zooplankton biomass is partly reduced through intensive predation by Baltic herring, sprat and mysid shrimps (Hansson et al. 1990). Mysids and amphipods increase in the diet of herring, while sprat stays strictly zooplanktivorous (e.g. Möllmann et al. 2004).

Winter Period – Low Production and Activity

During winter, primary production is negligible in the water at the open sea areas, especially in the northern parts where ice cover prevents light penetration to the water. However, many species of phytoplankton, of which diatoms are dominant, occur in the ice (Ikävalko and Thomsen 1997). The biomass and species diversity of zooplankton are at their lowest. Rotifers and cladocerans winter mostly as resting eggs, which sink to the bottom, and copepods as different life stages or resting eggs (Viitasalo and Katajisto 1994; Werner and Auel 2004). However, rotifers and copepods are also found living in the ice as a part of the wintering community (Werner and Auel 2004). The zooplankton community is thus mainly formed of copepods with spatially varying dominant species and the appendicularian *Fritillaria borealis* Lohmann (Schneider 1990). During winter, copepods allocate energy mainly for maintenance, while reproduction starts later in spring. Mysids and amphipods overwinter as females carrying their young in the brood pouch (Segerstråle 1950, Salemaa et al. 1986). Fishes migrate to deeper areas also from the littoral habitats, which are characterised by perennial macrophytes (*Fucus vesiculosus* L., *Ascophyllum nodosum* (L.) Le Jolis, *Zostera marina* L.) that maintain visible canopies through winter. The activity level of fish during winter is usually low but species specific. Exceptionally active of the fish fauna is burbot (*Lota lota* Oken), which feeds actively and spawns during winter (Scott and Crossman 1973).

A.3.1.2 External Input

A.3.1.2.1 Atmospheric Load

Mikhail Sofiev

This section provides an overview of atmospheric input of anthropogenic pollutants into the Baltic Sea ecosystem. The atmospheric pathway is significant for two large groups of pollutants: nutrients (nitrogen oxide, ammonia and, to a much less extent, phosphorus), and toxic species, such as heavy metals (HMs) and persistent organic pollutants (POPs). Additionally, sulphur oxides contribute to acidification of the Baltic Sea Basin and tropospheric ozone adds to the load on vegetation.

The issues connected with availability and quality of modelled and measurement data are outlined in Annex 4.5.

Acidifying Compounds (Oxidised and Reduced Nitrogen, Sulphur Oxides)

According to the EMEP (European Monitoring and Evaluation Program) computations with the Lagrangian model (Bartnicki et al. 2002), the total deposition of the oxidised nitrogen onto the sea surface is fairly stable in time and amounts to 146 kton year^{-1} (see Table A.3), (Bartnicki et al. 2002, 2003, 2004; EMEP 2000, 2003, 2004, 2006). The corresponding value reported by the Eulerian model (Bartnicki et al. 2003, 2004, 2006) is 115–130 kton year^{-1}. The values for ammonium differ more strongly: 153 and 98 kton year^{-1}, respectively. Standard deviations of the values are about 10% for oxidised and 15% for reduced nitrogen for both models. A slight downward trend as visible but its absolute value is small.

The deposition of nitrogen oxides to the catchment area in 2002 (Bartnicki et al. 2004) was $\sim$ 1.2 Mton N year^{-1}, with uneven distribution between the sub-basins (Table A.4).

To evaluate the load onto the ecosystems of the catchment area, more appropriate characteristics are the deposition density (Fig. A.25) and its exceedances over the critical load values for acidification (Fig. A.26) and nutrient nitrogen (Fig. A.27). The deposition density largely varies from south to north (about an order of magnitude for both sulphur and nitrogen compounds), also being subject to noticeable multi-annual variability. In some areas, the difference between the annual concentrations and depositions for two sequential years can be as much as 20–30% according to Bartnicki et al

Table A.3. Multi-annual deposition of pollutants onto the Baltic Sea surface (kton N year^{-1}). L denotes results from the EMEP Lagrangian, and E for the Eulerian model

	1996L	1997L	1998L	1999L	2000L	2000E	2001E	2002E	Aver.L	Aver.E
NO$_x$ Kton N	140	138	158	150	145	147	128	113	146	129
NH$_x$ Kton N	148	129	168	150	171	115	96	83	153	98
N total. Kton N	288	267	326	300	316	262	224	196	299	227

Table A.4. Total nitrogen deposition onto the Baltic Sea catchment in 2002

N total, Kton N/year	
Gulf of Bothnia Catchment	137
Gulf of Finland Catchment	164
Gulf of Riga Catchment	97
Baltic Proper Catchment	698
Belt Sea Catchment	54
Kattegat Catchment	83
Baltic Sea Catchment	1233

(2004, 2006). With regard to deposition variability, the precipitation amount and its distribution between the seasons plays one a major role.

For exceedances of acidification (Fig. A.26), the downward trend is stronger – mainly due to reduction of sulphur emission and, consequently, deposition during the last 20 years.

Comparison with HELCOM observations reveals some spatial tendencies in the models' quality, which can be used to adjust the above estimates or at least evaluate their uncertainty.

For the Lagrangian model, comparison with measurements shows a spatially stochastic pattern but with an indication that the model overstates the deposition values for all years. The computed wet deposition is higher than the measured one for 9 to 10 out of 14 HELCOM stations for both NO$_x$ and NH$_x$. The extent of over-estimation varies but can be as much as a factor of 2 to 3 for some specific stations and years. Therefore, the overall load is probably also over-estimated.

For the Eulerian model, comparison for 2002 shows that for most of southern Baltic Sae stations (DK5, DK8, DE9, EE9, PL4, SE11) the concentrations of both oxidised and reduced nitrogen in precipitation are overstated by the model by 20–50%. Such over-estimation is not visible for northern stations, which are located in less polluted areas. There is no such tendency for concentrations in air, which generally demonstrate a good agreement with observed levels.

There can be two reasons for such a trend: spatially inhomogeneous quality of precipitation information and some nuances in parameterisation of scavenging. According to the EMEP status report 1/2004 (EMEP 2004), accumulated precipitation amount does not have a pronounced bias (although the difference for some stations can exceed a factor of 2). It is therefore possible to suggest that the Eulerian model over-estimates by ∼ 20% the deposition onto the most-polluted sub-basins of the Baltic Sea (Kattegat, Belt Sea, Gulf of Riga, southern-most part of the Baltic Proper) and is on average unbiased for the other sub-basins. Due to comparatively small areas with such overestimation (Kattegat, Belt Sea and Gulf of Riga combined take ∼ 20% of the total deposition onto the sea surface), the overall over-estimation can hardly be more than 5%. Comparison for 2001 leads to an even smaller number.

Within the scope of the EU project Baltic Sea System Study (BASYS) and a follow-up research, similar-type modelling assessments were made using a semi-independent Eulerian model HILATAR (Hongisto et al. 2003; Hongisto and Joffre 2005). This model, accepting the EMEP chemistry module, emission totals and, for part of the simulations, boundary conditions from the EMEP Lagrangian model, used a different advection method, meteorological data and own time variation of emission. The results appeared to be quite similar (Table A.5): mean total nitrogen de-

Table A.5. Results of a semi-independent model study within the BASYS project

	1993	1994	1995	1996	1997	1998	2000	2001	2002	Average
NO$_x$ Kton N	175	169	150	144	139	179				159
NH$_x$ Kton N	115	103	105	114	99	117				109
N total. Kton N	290	272	255	258	238	296	274	263	224	268

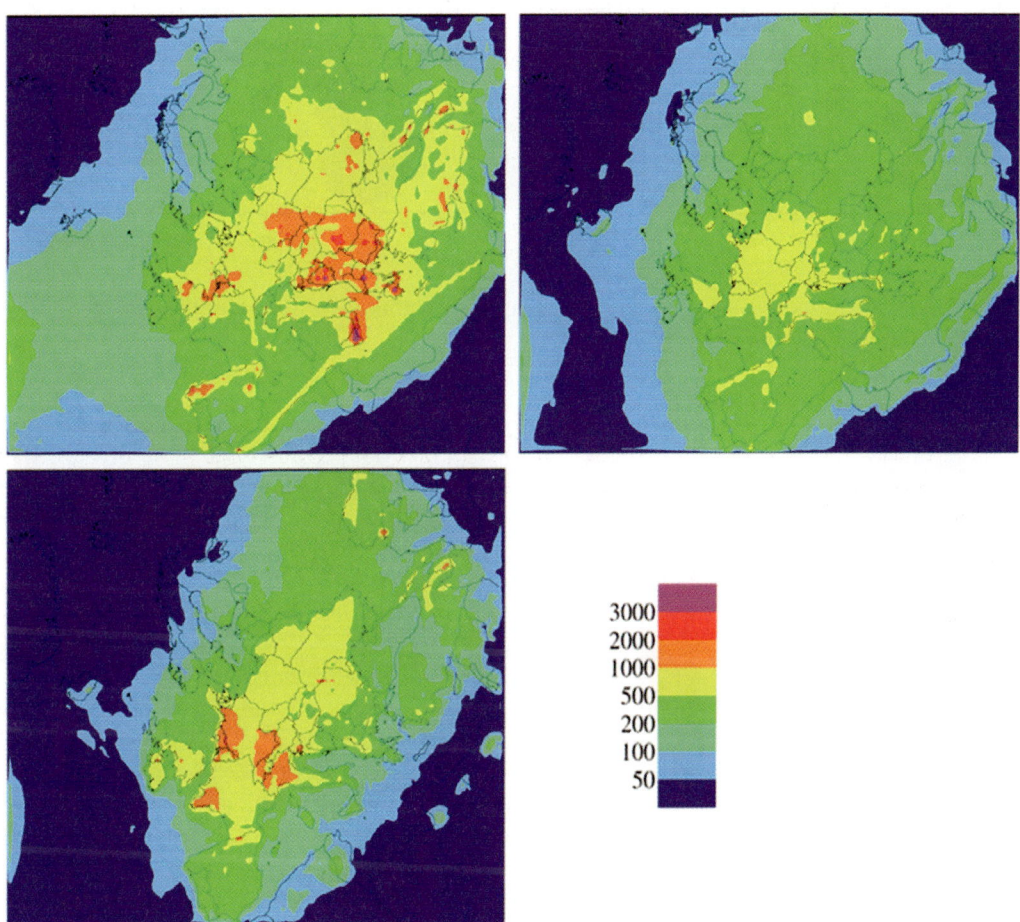

Fig. A.25. Map of total sulphur (*upper left*), oxidised nitrogen (*upper right*) and reduced nitrogen (*lower*) deposition in 2004. Unit is (mg S/N m^{-2} year^{-1}) (from EMEP 2006)

position of 263 ktons year^{-1} for the period 1993–2002. The values are in between the results of the EMEP Eulerian and Lagrangian models, with NO$_x$ load being almost exactly on the half-way and NH$_x$ staying closer to the Eulerian model results.

Regarding the deposition density, the estimates for 1993–1998 showed practically the same mean value for sulphur oxides (for the Polish coastline, the HILATAR model showed about $1\,\mathrm{g\,S\,m^{-2}}$ year^{-1}) but somewhat higher nitro-

gen values (total nitrogen amounted to 1.2–1.4 g N m^{-2} year^{-1}). The standard deviations of the monthly deposition values were 30–40% with a tendency to grow towards the north. The comparison with measurements showed a limited overestimation of the NO$_x$ wet deposition by HILATAR, while the ammonia fluxes are practically unbiased.

Another model-based estimate of Hertel et al. (2003) for 1999 resulted in slightly over 300 ktons N year^{-1} of total nitrogen deposited onto

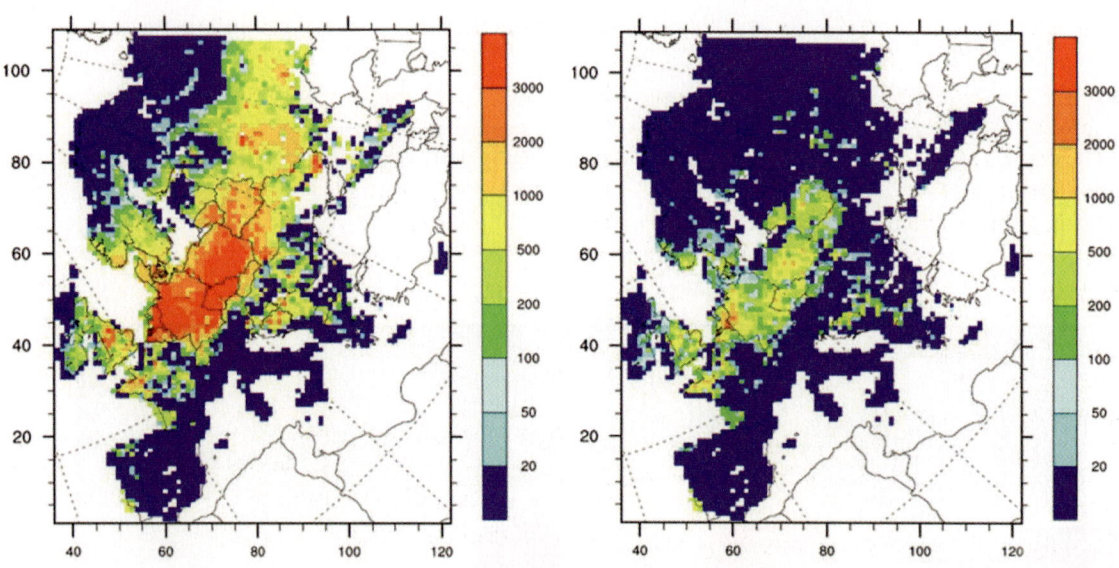

Fig. A.26. Maps of critical loads exceedances for acidification computed for 1990 (*left*) and 2004 (*right*). Unit is (equivalent ha^{-1} year^{-1}) (from EMEP 2006)

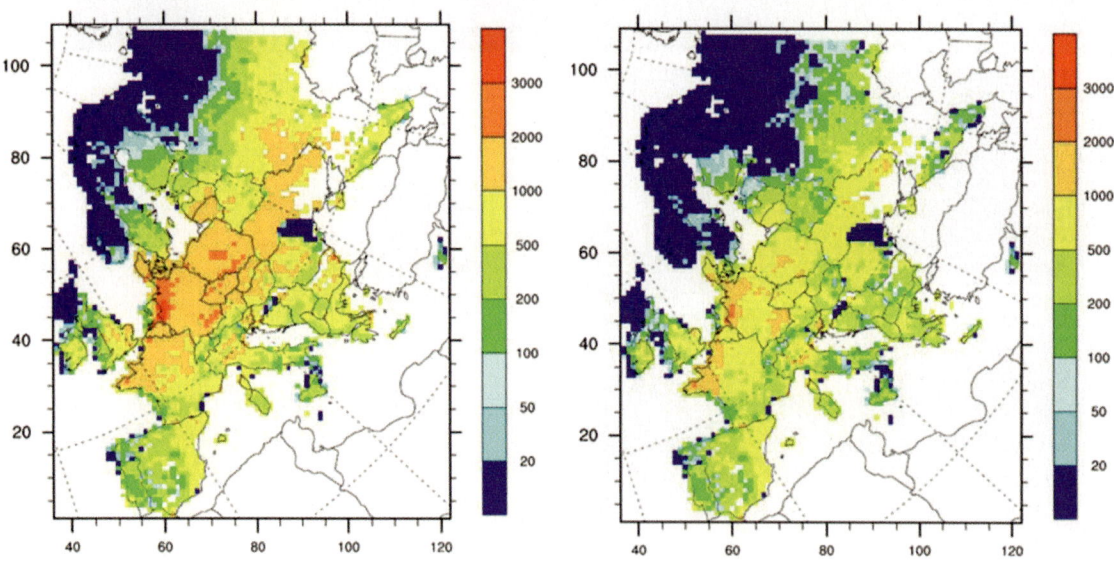

Fig. A.27. Maps of critical loads exceedance for nutrient nitrogen computed for 1990 (*left*) and 2004 (*right*). Unit is (equivalent ha^{-1} year^{-1}) (from EMEP 2006)

the sea surface. The results were obtained with the backward Lagrangian model, which is similar to the old EMEP model.

None of the long-term studies found a significant trend in nitrogen depositions, which allows consideration of some older works (with certain care due to still possible long-term trends and lower reliability of historical results). Thus, Ded-kova et al. (1993) gave the preliminary values of total nitrogen deposition onto the sea surface as varying from 298 up to 343 ktons N year^{-1}. The load was computed for the period 1987–1991.

Another work of Lindfors et al. (1993) combined the extrapolated observations of wet deposition and model computations of dry deposition, resulting in over 400 ktons N year^{-1} and

1.4–$1.9\,\mathrm{g\,N\,m^{-2}}$ year^{-1} for total nitrogen deposition at the southern Baltic coast (average over 7 years 1980–1986). The total deposition onto the sea surface, however, can be challenged due to unclear representativeness of coastal deposition measurements offshore.

The deposition of $0.5\,\mathrm{g\,N\,m^{-2}}$ year^{-1} of oxidised nitrogen at the southern Baltic coast was also reported by MATCH model computations (Laurila et al. 2004). That work has also considered potential developments of the situation in the future climate and predicted small and irregular changes in the pattern – the difference of the deposition values computed for various emission and climate scenarios from the reference estimates was mainly within 10%.

Studies of Sofiev and Grigoryan (1996) and Sofiev et al. (1996) gave the values at the Polish coast of 0.6–$0.7\,\mathrm{mg\,N\,m^{-2}}$ year^{-1} for oxidised and $0.5\,\mathrm{mg\,N\,m^{-2}}$ year^{-1} for reduced nitrogen. The estimates were computed by three different versions of an independent model for the period 1991–94 and appeared quite coherent.

It is seen that the results of the EMEP Lagrangian model and decade-old independent assessments are quite close to each other, while the Eulerian-model is closer to more modern applications.

The following can therefore be concluded:

(i) Annual total deposition of oxidised nitrogen onto the Baltic Sea surface is $125 \pm 20\,\mathrm{kton\,N}$ year^{-1} (a standard deviation range); for ammonia this estimate is $100 \pm 16\,\mathrm{kton\,N}$ year^{-1};

(ii) The deposition density varies from 0.5–$0.6\,\mathrm{g\,N\,m^{-2}}$ year^{-1} at the southern coast down to about $0.05\,\mathrm{g\,N\,m^{-2}}$ year^{-1} in the north – for both oxidised and reduced nitrogen (though ammonia trend is more pronounced). For oxidised sulphur compounds the present-time values are $\sim 1\,\mathrm{g\,S\,m^{-2}}$ year^{-1} and $0.1\,\mathrm{g\,S\,m^{-2}}$ year^{-1}, respectively;

(iii) Inter-annual variability is comparatively small and has a standard deviation of 10–15%, thus accounting for the most of above uncertainty range;

(iv) No statistically significant trend for nitrogen oxides was observed by the above studies, while sulphur emission and deposition have strongly decreased;

Table A.6. A share of deposition obtained by each sub-basin of the Baltic Sea

	NO$_x$	NH$_x$
Belt Sea	8	14
Kattegat	8	10
Baltic Proper	58	54
Gulf of Riga	4	4
Gulf of Finland	6	4
Gulf of Botnia	16	13

(v) A large spatial gradient of depositions is noticed by all computations. The typical distribution of deposition to the sea sub-basins is presented in Table A.6

Tropospheric Ozone

There are several European models regularly computing the ozone level and its exceedance of the thresholds. The model formulations and their results were extensively compared with each other, which allowed to derive consensus-based assessments of the present ozone level and its importance for the Baltic Sea Basin (e.g. Roemer et al. 2003; Hass et al. 2003; Laurila et al. 2004; Zlatev et al. 2002).

It is generally agreed that a high ozone level during the vegetation season is not yet a matter of primary concern for the Baltic Sea ecosystems, except for the most southwestern areas (Germany and Denmark). It can, however, become more significant in future climate (Laurila et al. 2004). At present, the maximum value for the AOT-40 ozone index (Accumulated Ozone above the Threshold of 40 ppb) reaches 10 ppm-hours only in central Poland and in Germany, which is still several times lower than in the other parts of Europe. However, there are warnings of a possible change of the situation due to warmer climate and increasing emissions. In particular, this problem can become severe if strongly increasing amounts of ozone in other parts of the Northern Hemisphere will be advected towards the region (Derwent et al. 2002).

Toxic Species

Information about the toxic species is much less precise than that for nitrogen – both from observational and from modelling points of view. All data can be criticised for problems with quality and reli-

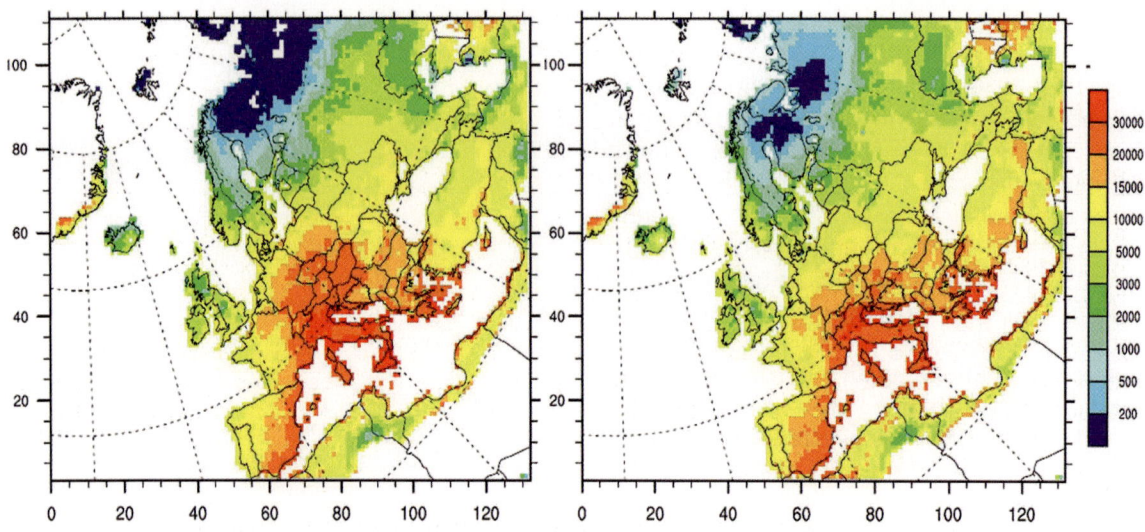

Fig. A.28. AOT-40 as computed by the EMEP model at the canopy level for the years 2000 (*left*) and 2004 (*right*). Note the new threshold of 5000 ppb-hours. Unit = ppb-hour (from EMEP 2006)

ability. and their mutual agreement usually leaves a large area for improvement. The emission of some toxic species is believed to have a strong reduction trend, so historical data cannot be taken into account without the trend correction. There is also an uncertainty in the absolute emission levels, especially for organic pollutants. As a result, historical and recent assessments based on models and measurements show large differences.

One of the ways to improve the load estimates by reducing at least the mean bias is to scale the modelled patterns with the relative bias of the wet deposition or concentration in precipitation. The resulting pattern would keep the simulated spatial distribution and simultaneously be non-biased (in average) from the available measurements. This method is crude and does not eliminate errors in observations and in spatial distributions of the load. However, the measurement-scaled patterns contain fewer problems than the original ones, which sometimes manifest a systematic deviation from observations by a factor of 3 to 10.

Trace Metals: Lead (Pb) and Cadmium (Cd)

Computations of EMEP for Lead (Pb) and Cadmium (Cd) (Bartnicki et al. 2002, 2003, 2004, 2005, 2006), as well as other comparatively recent results are summarised in Table A.7. Depositions of Pb and Cd onto the catchment area of the Baltic Sea are summarised in Table A.8

The deposition flux in the southern Baltic Sea coast is $0.5–1 \, mg \, Pb \, m^{-2} \, year^{-1}$ in 2002 according to Bartnicki et al. (2004, Fig. A.29). The estimates for earlier years, such as Rodhe et al. (1980), HELCOM (1989, 1991, 1997), Duke et al. (1989), Schneider (1993) and Petersen and Krueger (1993) show that the total Pb deposition onto the sea surface ranges from 640 tons $year^{-1}$ (HELCOM 1997, a mean for 1991–1995 computed with emission data for 1990) up to 2400 tons for 1980. The respective deposition flux over the southern coast was about $2.5 \, mg \, Pb \, m^{-2} \, year^{-1}$. Due to an uncertain but probably significant trend in Pb emission in the late 1980s and 1990s, these values are of little interest for modern load estimations, but still useful as indicators of uncertainties and past levels. More recent estimates of Sofiev et al. (2001) for 1998 were about $1 \, mg \, Pb \, m^{-2} \, year^{-1}$. All results show a factor of 4 to 5 drop of fluxes towards the northern end of the Bothnian Bay.

As seen in Table A.7, there are large differences between the EMEP estimates and independent computations, with the model-measurement comparison favouring the latter ones. The EMEP heavy metal model demonstrated strong and systematic under-estimation of the concentration in precipitation and wet deposition. The recent simulations of Bartnicki et al. (2006), however,

Table A.7. Total deposition of Pb / Cd onto the Baltic Sea surface, tons year^{-1}

Pb/Cd			Measurements / model ratio	
Reference	Year	Total deposition	Range	comments
EMEP 2006	2004	235 / 5.7	?	only comp. charts provided
EMEP 2005	2003	134 / 7.0	1.5–5.7 / 1.6–5.8	comparison in precip.
EMEP 2004	2002	149 / 7.4	1.7–5.5 / 1.4–4.8	
EMEP 2003	2001	143 / 8.3	1.1–4.3 / 0.5–4.1	
Sofiev et al. 2001	1997–98	596, 680 / 9.0, 9.4	0.6–1.2 / 1.2–4.7	2 models; emission 1990
Schneider et al. 2000	1997–98	550 / 33, 18		Observ. extrap., 1/2 methods
HELCOM 1997	1990	640 / 27	0.5–4 / ?	mean 1991–95, compr. 1990

Table A.8. Total deposition of Pb / Cd onto the Baltic Sea Basin, in tons year^{-1}

	EMEP 2003	EMEP 2004
Gulf of Bothnia catchment	144 / 6.9	68 / 2.9
Gulf of Finland catchment	230 / 9.2	184 / 6.5
Gulf of Riga catchment	83 / 4.0	58 / 3.0
Baltic Proper catchment	757 / 52.4	573 / 38.9
Belt Sea catchment	12 / 0.7	18 / 0.8
Kattegat catchment	32 / 1.6	31 / 1.4
Baltic Sea total catchment	1,262 / 74.8	932 / 53.6

demonstrated a sharp increase of Pb (but not Cd) total deposition onto the sea surface in comparison with previous estimates (the input emission values have not changed much). The simulations were also performed retrospectively with a similar outcome: the new model version showed a significantly larger deposition of Pb but kept nearly the same Cd load.

The computations done within the of the EU BASYS project (Sofiev et al. 2001) agreed with the observations quite well and provided a load estimate close to the complementary observation-based assessment of Schneider et al. (2000). The BASYS values are also in good agreement with 640 tons Pb year^{-1} (HELCOM 1997), computed for 1991–95 with the same constant emission of 1990.

However, the simulations covered only 4 months during the years 1997 and 1998. These months included both winter and summer seasons, thus being representative for the whole period, but the extrapolation still increased uncertainty. Secondly, these simulations used the Pb emission data for

1990, thus missing a reduction during the 1990–1997 period. The significance of the latter limitation is unclear because the uncertainty in the absolute level of emission compares well with the reported trend. The validation of that model for 1990 by Sofiev et al. (1996) demonstrated $\sim 40\%$ under-estimation of deposition values in comparison with available, albeit limited, observations.

To explain the above contradictions, one can consider the following:

- A systematic underestimation of the Pb emission, which was moderate in 1990 (a factor of ~ 1.4), but practically doubled by 2002, due to an overestimated downward trend;
- Difficulties in the some EMEP heavy metal model versions, resulting in underestimations;
- Processes missing in the most of current models, such as re-suspension of particles after their deposition. This is very difficult to quantify but, according to Bartnicki et al. (2005), its contribution can mount up to 50% of total deposition to some sub-basins.

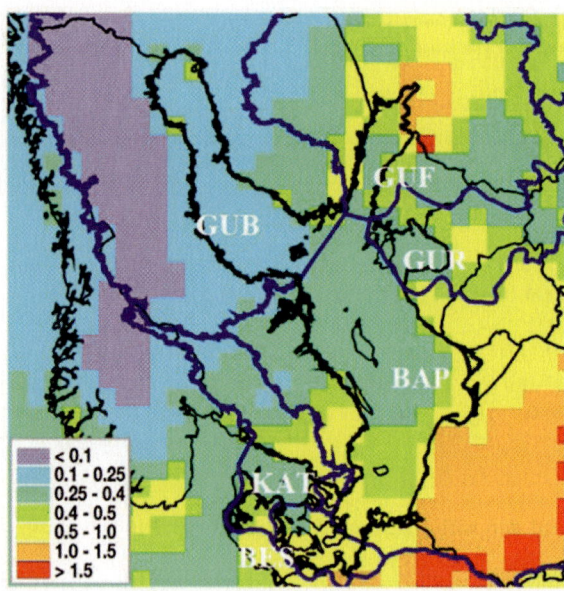

Fig. A.29. Spatial distribution of total lead deposition flux ($\mathrm{kg\,km^{-2}\,year^{-1}}$) in the Baltic Sea Basin in 2002 with resolution $50 \times 50\,\mathrm{km^2}$ (adapted from Bartnicki et al. 2004)

Evaluation of these considerations requires further investigations.

Based on the above considerations, one can assume that current deposition of Pb onto the Baltic Sea surface accounts for at least 350–400 tons $\mathrm{year^{-1}}$ but the uncertainty is large and its origin is not fully clear. The Pb deposition flux at the southern coast does not exceed $1\,\mathrm{mg\,Pb\,m^{-2}\,year^{-1}}$.

The case of Cd is similar to that of Pb: the under-estimation of the current models is a factor of a few times and the differences between the independent estimates are large (Tables A.7 and A.8). There is an agreement between almost all models, except for the old data of HELCOM (1997), for which the model-measurement comparison was not presented. All models show the total load onto the sea surface to be somewhat below 10 tons Cd $\mathrm{year^{-1}}$ and a factor of 2 to 3 of underestimation of the wet deposition or concentration in precipitation in comparison with observations. Scaling with this factor leads to the lower of the two observation-based estimates of Schneider et al. (2000).

It is therefore possible to suggest an uncertain emission as the main cause for the underestimation and conclude that the most probable deposition of Cd onto the Baltic Sea surface is about 20 tons $\mathrm{year^{-1}}$. There was no strong trend in the reported Cd emission during last decade.

Mercury (Hg)

Persistent toxic pollutants constitute the most difficult and the most uncertain set of species. Their emission data are typically not very accurate and complete, nor is the knowledge on physical and chemical transformations in the environment. Therefore, the estimates given below only show an order of magnitude of the phenomena.

According to Bartnicki et al. (2004), Petersen et al. (2001) and Munthe et al. (2003), it is possible to suggest that the order of magnitude of Hg deposition onto the Baltic Sea surface is about 3 tons $\mathrm{year^{-1}}$, but the input from the so-called "natural background", usually arbitrarily introduced into the models in order to meet the observations, can constitute over 50% of this value.

The wet deposition flux of all mercury species strongly varies over the region reaching its maximum in Germany – over $5\,\mathrm{mg\,Hg\,m^{-2}\,year^{-1}}$ – and falling below $0.1\,\mathrm{\mu g\,Hg\,m^{-2}\,year^{-1}}$ in the northern part of the region (Lee et al. 2001). The dry deposition flux could be a few times larger, mainly due to high dry deposition flux of the reactive Hg species. Munthe et al. (2003), to the contrary, suggested that dry deposition is about 3 to $6\,\mathrm{\mu g\,Hg\,m^{-2}\,year^{-1}}$ all over the region with practically no spatial trends. The estimates of Bartnicki et al. (2004) for the total deposition fluxes vary from over 15 to below $5\,\mathrm{\mu g\,Hg\,m^{-2}\,year^{-1}}$,

Table A.9. Example of the budget of HCH isomers fort the Baltic Sea, in tons year^{-1}

| | POPCYCLING model | | EMEP |
	α–HCH	γ–HCH	γ–HCH
Bothnian Bay	1.7	0.4	
Bothnian Sea	4.3	1.9	−0.2
Gulf of Finland	1.1	0.5	−0.1
Gulf of Riga	−0.5	0.0	−0.08
Baltic Proper	7.5	5.3	−0.7
Danish Straits	0.5	0.4	−0.1
Kattegat	1.3	0.7	−0.08
Baltic Sea	16.0	9.2	−1.26

with a strong decrease from south to north. A possible explanation for the differences can again refer to the model formulations and, in particular, to treatment of the "background level". All three models agree well with the mean air concentrations of elemental mercury (the main observed Hg-related parameter), as well as the few available observations of reactive Hg components.

Persistent Organic Pollutants (POP)

For the persistent organic pollutants, a verified quantitative information for the Baltic Sea Basin is available only for α- and γ-isomers of hexachlorocyclohexane (HCH) and some polychlorinated biphenyls (PCB). Recent simulations, such as those of (Bartnicki et al. 2006), also provide lump estimates for dioxins/furans. The main sources of the information are: regional observations (e.g. HELCOM 1997) for HCH and PCB, the EU POPCYCLING project results (Breivik and Wania 2002) for α- and γ-HCH and the EMEP computations of Bartnicki et al. (2004, 2006) (Table A.9). The POPCYCLING project considered an actual emission development over 30 years (1970–2000) and computed the mean flows (Table A.9). The EMEP POP model was run over 12 years (1990–2002) to reach an equilibrium stage and then the last-year flux was taken as a result.

The HELCOM (1997) measurements of 7 PCB congeners suggest that the mean deposition flux at the Swedish west coast decreased from 0.9 to 0.6 μg PCB-sum m^{-2} year^{-1} during the period from 1980 to 1996. Corresponding values for sum of α– and γ–HCH do not show a clear trend, fluctuating between 1 and 4 μg HCH-sum m^{-2} year^{-1}.

Two of the above models disagree on the directions of the net air-surface flux. The EMEP POP model claimed that the region is a strong source of γ-HCH (Bartnicki et al. 2004): it estimated that between 3.2 to 4.5 tons of γ-HCH go into the air from the catchment area of the Baltic Sea. The POPCYCLING model showed that net volatilization occurs only from agricultural soils in the source areas: the Gulf of Riga for both isomers and the Danish Straits for Lindane. Forest soils are always a strong sink for HCHs and canopy and coastal water compartments are close to equilibrium with overlying air, still being net receptors of HCH. According to Breivik and Wania (2002b) these 30-years mean fluxes are also representative for the latest years, such as 1998–2000, with some correction towards equilibrium between deposition and evaporation.

Validation of both models against measurements was done in a different manner as well. The POPCYLING model was verified against observations in air, water, pine needles and sediments, with a general agreement to be within a factor of 2 to 5 (Breivik and Wania 2002a). The EMEP model output was compared only with concentrations in air (very good agreement – within a factor of 1.5 to 2) and in precipitation (strong underestimation by a factor of 3 to 12). Therefore, the results of the POPCYCLING model are probably more reliable, and they show that, on average, the Baltic Sea is a receptor of the HCH compounds. Over the catchment area of the Baltic Sea, only some agricultural fields serve as net emitters of HCH. Absolute levels are uncertain but do not exceed the values shown in Table A.9.

A.3.1.2.2 Aquaculture and Eutrophication

Marianne Holmer, Lars Håkanson

Aquaculture in the Baltic Sea is a relatively recent activity, which was initiated at larger scales in the late 1970s. In the Baltic Sea, the most important cultured species is rainbow trout, which is grown in net cages anchored to the sea floor in sheltered as well as open areas. Aquaculture is considered an important industry in the coastal zones around the world, supporting the growing demand for marine food products, but in the Baltic Sea marine aquaculture is from a global perspective of rather low importance because of unfavourable biological and economical conditions. Due to low salinities only a few species, such as rainbow trout and some brackish water species, such as pike and perch, are commercially attractive, and shellfish cultures are limited to the Swedish west coast and Danish waters. Whereas aquaculture production increased rapidly in the Baltic Sea during the 1980s, the production levelled out and even declined in most Baltic Sea countries, such as Denmark, Sweden and Finland, in the 1990s due to falling economical benefits caused by competitive imports of salmon, primarily from Norway and by environmental restrictions in some countries, such as Denmark (Håkanson et al. 1988; de Pauw et al. 1989; Mäkinen 1991). Total annual production was somewhat less than 57,000 tons in 2002 (Fig. A.30; FAO 2005).

The industry is now facing new challenges as new species have been successfully cultured under experimental conditions and market prices on fish and shellfish for consumption have increased so much that still more species have become attractive for cultivation. Aquaculture production in the Baltic Sea is expected to increase during the next decade, as several "new" countries in the Baltic Sea Basin are investing in aquaculture (e.g. Estonia), and established countries are expanding through culturing new species (e.g. cod, plaice and oyster) and increasing production of rainbow trout and blue mussels encouraged by increasing market prices. Nevertheless, the overall growth rate in aquaculture is expected to be lower than predicted for other parts of the EU, such as the Mediterranean, where fish farming is increasing very rapidly (GESAMP 2001).

Whereas shellfish culturing is considered to be neutral or even to reduce the eutrophication of the coastal zones (Petersen and Loo 2004), the intensive culturing of finfish on artificial feeds leads to a variety of environmental impacts, including the release of dissolved nutrients which may increase eutrophication of the Baltic Sea (Hall et al. 1990, 1992). Most of the nitrogen-containing waste products are released in the dissolved form (primarily through fish excretion) and may lead to eutrophication events in the water column, whereas the phosphorus compounds are primarily released as particles and settle in the vicinity of the fish farms, contributing to the "footprint" of the farms (Brooks and Mahnken 2003). A Danish survey shows that the release of waste nutrients per kilo produced fish has been reduced by 54% for N and 64% for P, and although the production of fish has been stable since the early 1990s, the total release of N and P has been reduced by 26% and 52% since 1987 (Havbrugsudvalget 2003). A similar trend can be expected for the Baltic Sea due to improvements of feed and feeding practices stimulated by increasing expenses and environmental regulations.

Due to high organic contents of the settling waste products the oxygen demand in the sediments increases and may impact the benthic fauna below the net cages (Håkanson and Boulion 2002). The organic loading of the sediments increases the bacterial activity and regeneration of nutrients back to the water column, and both direct release and regenerated nutrients may thus contribute to eutrophication (Hall et al. 1990, 1992). There have been quite a few environmental studies of single farms, which have been able to show benthic effects, and there are also several studies of water column parameters in the Baltic Sea (Wallin et al. 1992; Nordvarg and Håkanson 2002; Gyllenhammar 2004; Håkanson et al. 2004).

Most fish farms are placed at locations with rapid water renewal, and the nutrients released are diluted to low concentrations within hours of release. As the doubling rates of phytoplankton are in the order of days, blooms of phytoplankton caused by nutrient release from single farms have not been observed. The benthic effects are quite similar to observations from salmon farms, as the feeding practices and general management of the rainbow trout farms are similar to salmon farms (Hall et al. 1990, 1992; Holmer and Kristensen 1992; Brooks and Mahnken 2003), and estimates have shown that the overall nutrient release from fish farms will only decrease by 2% by shifting from rainbow trout to salmon culture, whereas the economical consequences probably will be much larger due to the low prices

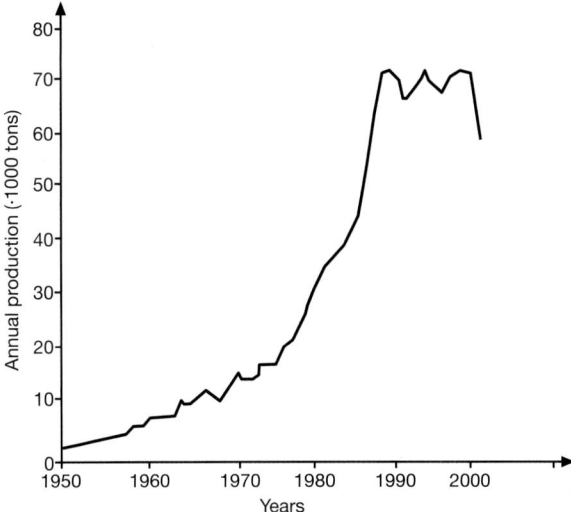

Fig. A.30. Production of rainbow trout (in MT × 1000) in the Baltic States during 1950–2002 (incl. freshwater production) (modified from FAO 2005)

on salmon (Aquaflow 2002). Most farms in the Baltic Sea are relatively small (100–1,000 tons annual production) compared to Atlantic farming (> 500 tons annually), and, as the local impacts among other factors are directly related to loading (Gyllenhammer and Håkanson 2005), local effects in the Baltic Sea are modest compared to the large farms in the Atlantic (Holmer and Kristensen 1992; Christensen et al. 2000).

A review to evaluate how emissions from fish cage farms cause eutrophication effects in coastal areas has just been published (Gyllenhammar and Håkanson 2005). The review focuses on four different scales, (i) the conditions at the site of the farm, (ii) the local scale related to the coastal area where the farm is situated, (iii) the regional scale encompassing many coastal areas, and (iv) the international scale including regional coastal areas (e.g. the Baltic Sea).

The aim was to evaluate the role of the fish farm emissions in a general way, but all selected examples were derived from the Baltic Sea. An important part of this evaluation concerned the method for defining the boundaries of a given coastal area. If this is done arbitrarily, one would obtain arbitrary results in the environmental impact analysis. In this work, the boundary lines between the coast and the sea are drawn using GIS methods (Geographical Information Systems) according to the topographical bottleneck method, which opens a way of determining many fundamental characteristics in the context of mass-balance calcula-

tions. In mass-balance modelling, the fluxes from the fish farm should be compared with other fluxes to, within and from coastal areas.

The study concluded that:

1. At the smallest scale (< 1 ha), one can conclude that the "footprint", expressing the impact areas of fish cage farms often corresponds to the size of a "football field" (50–100 m), if the annual fish production is low (about 50 tons).

2. At the local scale (1 ha to 100 km^2), there exists a simple load diagram (effect-load-sensitivity) to relate the environmental response and effects of a specific load from a fish cage farm. This makes it possible to obtain a first estimate of the maximum allowable fish production in a specific coastal area.

3. At the regional scale (100–10,000 km^2), it has been shown that it is possible to create negative nutrient fluxes, i.e. to use fish farming as a method to reduce the nutrient loading to the sea. The breaking point is to use more than about 1.3 g wild fish from the area per gram feed for the cultivated fish.

4. At the international scale (> 10,000 km^2), related to the Baltic proper, the contribution from fish farms to the overall nutrient fluxes is very small. It has been estimated that, in 1989, fish farming accounted for 1.5% of the total phosphorus and 0.4% of total nitrogen loads to the Baltic Sea and Skagerrak (HELCOM 1993). A Danish study has estimated that marine

aquaculture in Denmark contributes with 0.4% and 1.3% of total Danish N and total P loading (including riverine load), respectively (Ærtebjerg et al. 2003). Of the direct point sources, marine aquaculture contributed with 9.2% and 8.1% N and P, respectively, but when the Danish nutrient release from marine aquaculture was compared with the total inputs to the Danish coastal waters from the Baltic Sea, Sweden and Germany it only accounted for 4‰ and 9‰ of the inputs of N and P, respectively (Ærtebjerg et al. 2003). Although the release of nutrients from marine aquaculture only has minor effects on the total budgets, it is important to note that it, together with other sources, contributes to the high loading of the Baltic Sea.

Due to the complexity of marine ecosystems, including spatial and temporal variabilities, complete studies using an ecosystem based approach to examine eutrophication effects of fish farming in the Baltic Sea are difficult and expensive to carry out. Håkanson (2005) used a modelling approach to demonstrate ecosystem effects of fish farm emission. The main questions in this study were: How would emissions of feed spill and faeces from a fish farm influence the production and biomass of key functional organisms and how long would such changes remain if the fish farm is closed down?

The work is based on a comprehensive lake ecosystem model, LakeWeb, which accounts for production, biomass, predation, abiotic/biotic interactions of nine key functional groups of organisms, phytoplankton, bacterioplankton, two types of zooplankton (herbivorous and predatory), two types of fish (prey and predatory), as well as zoobenthos, macrophytes and benthic algae. The LakeWeb model gives seasonal variations (the calculation time is one week), and it has been calibrated and critically tested using empirical data and regressions based on data from many lakes. These tests have demonstrated that the model can capture typical functional and structural patterns in lakes very well, lending credibility to the results, and the model suggests that fish farm emissions cause significant increases in the biomass of wild fish, without corresponding increases in algal volume.

Thus, it is concluded that the fish farm emissions influence the secondary production more markedly than primary production. Although this finding might seem to be a paradox, it is related to the fact that wild fish directly consume food

spill and faeces from the fish farm, thereby creating a specific foodweb pathway. Similar findings have been done in the oligotrophic Mediterranean (Dempster et al. 2002), but remains to be validated for fish farms in the Baltic Sea, although many fish farmers support this hypothesis through observations of wild fish around their farms (K.M. Kjeldsen, pers. comm.).

The primary production in the Baltic Sea is limited by nutrients during summer, and phosphate is the typical limiting nutrient in the north, whereas nitrogen is argued to be limiting in the south (Elmgren 2001). Fish farming in the Baltic Sea may thus, due to the release of dissolved nitrogen directly through excretion of ammonium from the fish during summer and through the settling and regeneration of phosphorus, lead to a stimulation of the nutrient-limited primary producers. One important aspect when discussing mass-balances is that the nutrients released from fish farms are immediately bioavailable, whereas up to 50% of the nitrogen in the Baltic Sea is considered to be nonbioavailable, as it is bound in humus compounds (Elmgren 2001).

Increased water column chlorophyll-a concentrations and periphyton growth has been linked to fish farming activities in the Archipelago Sea in SW Finland (Honkanen and Helminen 2000). The study of the Archipelago Sea in Finland is one of the first to show an ecosystem effect on primary production under eutrophic conditions, and suggests that attached algae may be more sensitive indicators of nutrient release from fish farms. Although the nutrient release from fish farms in eutrophic regions is low compared to the other sources of nutrients as discussed above, the nutrients are released during the period, when the primary producers often are limited by the availability of nutrients, which may explain the eutrophication effects in the Finnish archipelago during summer. Most of the run-off from land occurs in the late autumn, winter and early spring (Håkanson and Boulion 2002).

In conclusion, fish farming in the Baltic Sea contributes to the general eutrophication, and although the overall contributions are low ($< 2\%$ of total load), calculations from Denmark show that fish farms account for up to 8–9% of the point sources. Whereas large investments have been done to reduce loads from point sources, only limited research has been undertaken to develop technology to reduce the environmental impacts of fish farms. The loss of nutrients can be minimized by capture of wild fish for feed production or by introduction of extractive cultures such as shellfish and macroalgae.

At the local scale, nutrient release should be compared with imports and exports from the area and the water residence time. If the water residence time is short, most nutrients will be exported to the coastal area, whereas most of the nutrients may be turned over within the area if the residence time is long. This may lead to increased primary and secondary production, in particular in periods with natural nutrient limitation in summer and autumn. Such increased production may result in enhanced settling of organic matter with risks of oxygen depletion events, in particular in stratified water bodies. The location of future aquaculture units has to be evaluated based on local conditions, such as nutrient limitation, water residence time and stratification.

A.3.1.3 *External Pressures*

A.3.1.3.1 Sea Traffic

Gerhard Dahlmann

Traffic and Navigational Situation

Shipping is steadily increasing in the Baltic Sea, reflecting intensifying international co-operation and economic prosperity. According to a more detailed statistical evaluation of the Baltic Sea shipping traffic, which was commissioned by Finland (Rytkönen et al. 2002), the sea-borne volume of transported goods is expected to roughly double by the year 2017.

Concerning ferry traffic and passenger transports the picture was a bit vague in the end of the 1990s because of the impact from the Belt-crossings and the abolition of duty free sales. Given data on the development during the last few years figures of ferry transport are still rising.

Around 2,000 sizeable ships are normally at sea at any time in the Baltic Sea. The Baltic Sea has some of the busiest shipping routes in the world (HELCOM 2005)

Generally, it is difficult to obtain a reliable up-to-date overview of ship traffic in the Baltic Sea. The situation is expected to change when a monitoring system for ships, based on AIS (Automatic Identification System) signals, which started in 2005, will be routinely explored. This will allow all the relevant statistics on shipping in the Baltic Sea

to be stored at the HELCOM server in Denmark and displayed on the HELCOM website.

The AIS network is one of a package of 16 measures adopted by the extra-ordinary HELCOM meeting at ministerial level in Copenhagen in September 2001, known as the "Copenhagen Declaration". It includes incentives for IMO (International Maritime Organisation), such as enhancement of pilots, routing measures and introduction of traffic separation schemes, as well as regional and national activities, such as improved hydrographic services and use of Electronic Navigational Charts. The Copenhagen Declaration has to be seen as one of the most important milestones of the HELCOM work in the field of improving maritime safety and thus preventing pollution from ships.

Maritime shipping can clearly be regarded as one of the most environmentally friendly means of transportation. Nevertheless, steps to further enhance safety of navigation have to be considered, among them enforced use of AIS onboard ships and for monitoring systems in order to control traffic, improved availability of electronic data to be used for sophisticated ships' navigation systems in order to support ships' management, further needs of routing systems and traffic separation schemes in order to support ships' management and avoid as far as possible conflict situations.

Impacts from Maritime Shipping

In order to minimise detrimental impacts of maritime shipping to the environment of the Baltic Sea, energetic measures have been taken by HELCOM as well as by the IMO. The most important instruments for preventing marine pollution from ships are the world-wide applicable MARPOL Convention (International Convention for the Prevention of Pollution from Ships), and the Helsinki Convention, which reflects, with respect to shipping, MARPOL Regulations. These rules lead to a significant reduction of *inter alia* oil discharges form ships during operation and from oil tankers, discharge of sewage and waste and emissions from ships by reducing sulphur content of bunker fuel oil. A further international convention which has to be mentioned in this context is the Ballast Water Convention, aiming at the prevention of transfer of alien species through ships' ballast water (not yet in force).

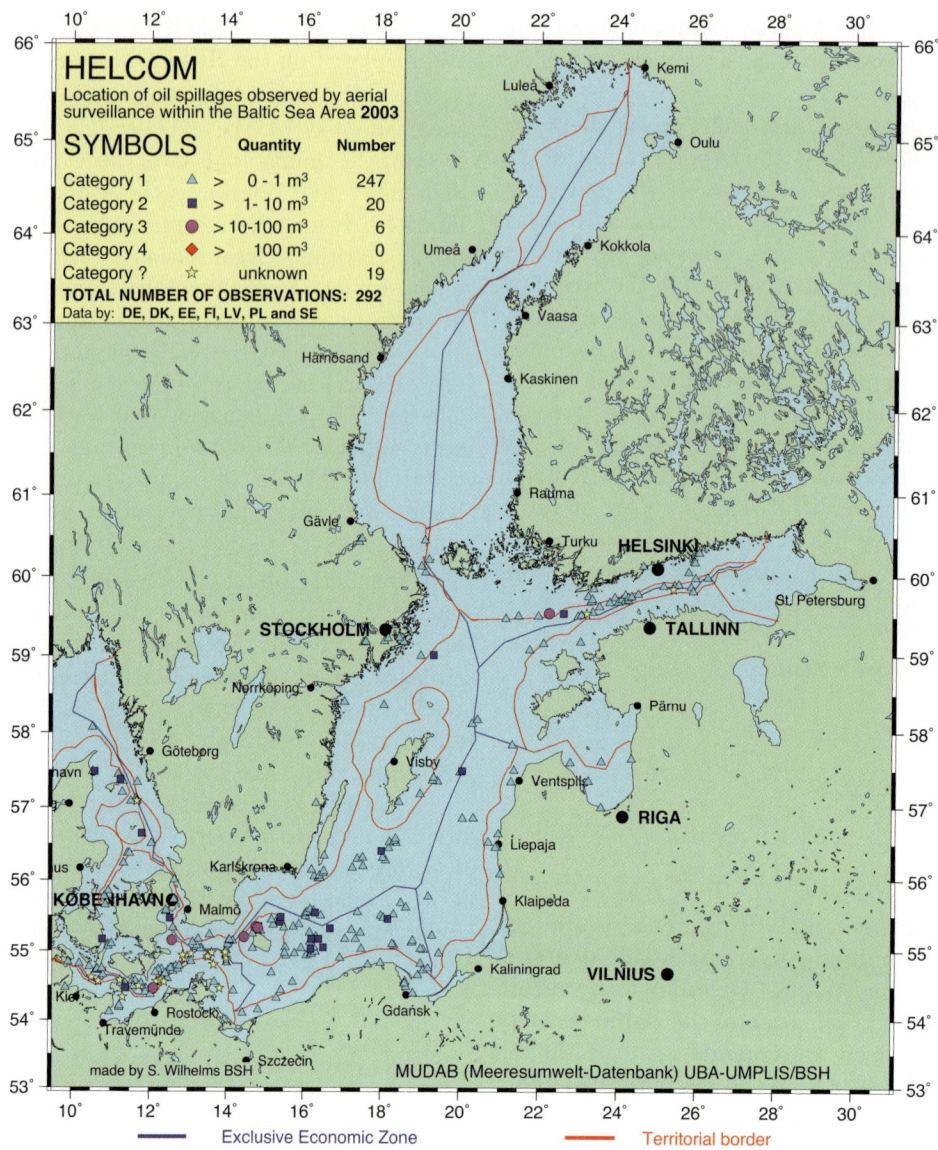

Fig. A.31. Location of oil spillages observed by aerial surveillance in 2003 (from HELCOM 2003b)

Oil Pollution

Oil is a serious threat to Baltic Sea ecosystems and wildlife. The most visible effect of oil spills is the oil contamination on the surface which causes smothing and death of birds and seals. As many of the chemicals in the oil spills are toxic, oil pollution causes serious accumulative effects, for example on plankton, fish and animals living on the seafloor, as it builds up in the ecosystems. Fauna and flora close to the shipping lanes are also negatively affected. In the Baltic Sea, where the average water temperature is only about 10 degrees, oil decomposes slowly due to the cold waters.

Illegal Discharges

About 10% of all the oil hydrocarbons in the Baltic Sea originate from intentional illegal discharges from the machinery spaces or cargo tanks of vessels sailing in the Baltic Sea.

During 1988 to 2001, surveillance aircraft detected an average of 400 illegal oil discharges per year in the Baltic Sea (HELCOM 2003a). In most cases, the amount of oil discharged is less than 1 cubic meter. Deliberate illegal oil discharges from ships are regularly observed within the Baltic Sea since 1988. As from 1999 the number of observed illegal oil discharges has decreased from ap-

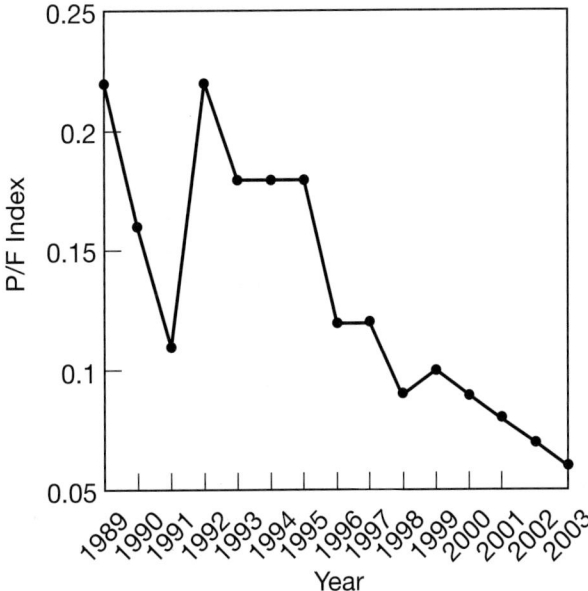

Fig. A.32. Annual number of detected oil slicks per flight hour (adapted from HELCOM 2003b)

Table A.10. Baltic Sea maritime traffic in 1995 (*second column from left*) and prognosis for 2017 (*3ʳᵈ column from left*) (from Rytkönen et al. 2002)

Comodity	Volume in Baltic Sea (million tons)	Estimated future Volume in Baltic Sea (million tons)	Growth from 1995 to 2017
Break Bulk	29	82	186%
Dry Bulk	61	113	84%
General Cargo	22	64	186%
Liquid Bulk	1	2	84%
Oil	81	112	39%
Total	194	372	92%

proximately 500 to less than 300 in 2003. This trend is reflected also in a decrease in the number of observed oil discharges per flight hour.

Accidents

Increasing shipping traffic also means that ship accidents causing marine pollution become more probable. To reduce these risks, the Copenhagen Declaration will lead, besides the enhancement of maritime safety, *inter alia*, to improved emergency and response capacities and an improved network of places of refuge.

A.3.1.3.2 Tourism

Ralf Scheibe, Wilhelm Steingrube

Tourism is a comparatively recent social phenomenon. It developed very slowly at first, after the Industrial Revolution had radically changed all Western, Central and Middle European countries. Only after World War II did tourism start to become a mass phenomenon in the industrialised nations. Today it is one of the world's largest economic sectors (ranking fourth, according to www.world-tourism.org/facts/trends/economy.htm).

General development of tourism in the Baltic Sea Basin

Tourism developed properly in the 19th century. Its sudden rise was caused by the coincidence of several factors, e.g. the developments in the European transportation (steam boats, railways, road systems) and communication, and – decisively – the economic consequences of industrialisation which brought a certain economic wealth to new social groups and thus enabled them to travel. This new civic upper class immediately embraced the developing spas, for new scientific and medical insights confirmed their health benefits. In the course of the 19th century, the health interests of the bathers were soon joined by the economic interests of the local spa operators. However, this early form of tourism was restricted to relatively few spas and sea baths, and until the 20th century, the seaside repose remained a strict privilege of the upper social classes.

In Germany, this quickly changed when the national socialist system established the "Kraft durch Freude" movement and state-controlled recreation and holidays. Only the outbreak of the war in 1939 prevented the plans for mass tourism from being realised.

In the German Democratic Republic (GDR) of the 1970s and 80s, the state newly created a rising pressure on the southern Baltic Sea coast. As a result, that region had about 400,000 beds and 37 million overnight stays in 1989 – about 40% of the GDR's total volume (cf. Benthien 1996, p. 12). 1990 saw a huge slump in this sector caused by the reunification of Germany. The infrastructure of the East German socialistic tourism was not competitive, and new organisational and economic structures had to be built from scratch. Although the new infrastructure was quickly set up, tourism has failed to reach the pre-Unification volume. Today, the number of overnight stays in the German "Land" Mecklenburg-Vorpommern is about 22 million per year.

The development at the coasts of Poland, Lithuania, Latvia and Estonia took a similar course. But there, even more than in the GDR, the state installed large tourism infrastructures which were alien to the landscape and, today, create considerable utilisation problems.

The development of tourism in Scandinavia took a somewhat retarded and structurally different course. Although there, too, its early stage was shaped by the behaviour of the nobility, Scandinavian noblemen began already in the early 19th century to use their manors as summer residences, exhibiting a particular preference for waterside locations. As elsewhere, the assurgent civic circles imitated upper-class behaviour and looked for cheap summer lodgings in the countryside (Jäderholm and Steingrube 1996, p. 140 ff.). Thus, at an early time they already induced the beginning of the cabins and related facilities which are so typical of Scandinavia today.

Present and future situations in different tourism sectors

The present situation in the Baltic Sea Basin (cf. the up-to-date overview in Breitzmann 2004) is characterised by relative stability and can be summarised in the following points:

- Domestic tourism is dominant in all countries boardering the Baltic Sea, i.e. most of the tourists come from the respective country.
- The cabin holiday has become an integral part of life in the Scandinavian countries and, therefore, dominates domestic tourism. Because of the keen demand, domestic tourism is extending into the "dry" countryside. Otherwise, there is a clear local and regional trend: "to the water", i.e. especially to the seaside, but also to the lakes.
- In line with this trend, the southern Baltic Sea coast, with its typical sand beaches receives many bathing tourist.
- The intensively used areas are expanding continuously. On the whole, however, there is less of an addition in tourist numbers than a spatial redistribution:
  - The pattern of locally centralised facilities in old, long-standing sea baths is replaced by a scatter of resorts along the entire coastline.
  - Poland and the three Baltic states are slowly joining this market.
- The predominance of marine and other water-based activities creates a pronounced seasonality.
- Business tourism provides some counter-balance, but concentrates on the tourist centres and cities.
- The Baltic Sea is of increasing interest for cruise boat operators.

Medium-term trends:

- Domestic visitors will continue to dominate Baltic Sea Basin tourism;
- The pronounced seasonality will hardly abate, because the many efforts to extend the season will hardly have any quantitative effect.
- There are many current projects to diversify tourism. In rural regions, this will lead to the construction of a number of new facilities and to the provision of new visitor activities.
- In business tourism, the EU enlargement will cause an increase in travel volume and a shift into Poland and the Baltic States.
- The focus of the new developments will be on maritime tourism which may cause visible effects: creation of new marinas, more boat traffic, and an increasing number of "mini-cruises".
- Ferry traffic will further increase in spite of newly built or planned bridges.
- In the Baltic Sea Basin in general, there will be a slightly increasing travel volume caused by the recent opening of the borders to the new EU member states.
- There will be a net redistribution at the expense of the established destinations like Denmark, Schleswig-Holstein and Mecklenburg-Vorpommern, leading to more tourism in Poland and the Baltic States.

Environment, climate, and tourism – general observations

Of course there is an intimate interrelationship between tourism and the environment. On the one hand, the natural assets of a region are generally recognised as the basis of tourism; and on the other hand, (mass) tourism contributes to the destruction of its own basis.

Nature, however, is a wide term, and climate is usually understood to be a part of it. However, it makes sense to treat climate on its own. This is due to, in the global dimension of climatic effects, the atmosphere's high susceptibility to change, as compared to other environmental factors, and the special impact of climate on the different prerequisites and aspects of tourism. As a schematic, we thus yield the simple relation shown in Fig. A.33.

The largest, i.e. most intensively discussed intersection (A) is doubtlessly covered by those issues that deal with global climate change. It is not caused by specifically tourism-related activities, but rather by other human economic activities. Intersections B and C include all kinds of in-

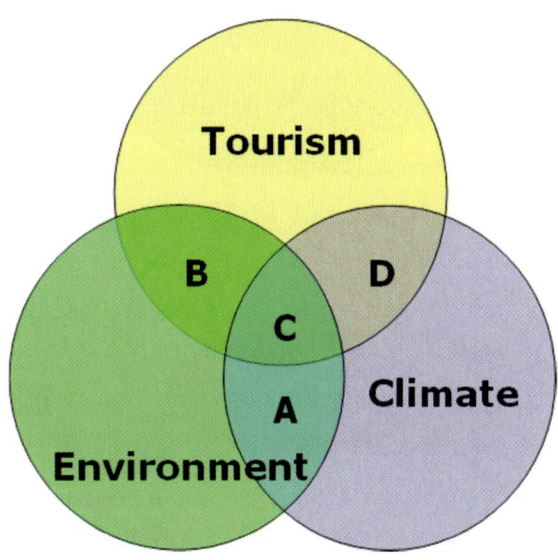

Fig. A.33. Tourism, Environment and Climate – Areas of Overlap

teractions between tourism and the environment, e.g. the waste problem in general, pollution caused by car traffic, as well as water supply and sewage treatment, or direct impacts on animals and plants (not only in reserves) that are caused directly by tourists. It must be emphasised that these interactions occur invariably in both directions.

That tourism depends on the general natural and environmental conditions is self-evident. Tourism requires nature both in the form of the local natural assets, as a space of action for activities, and simply as a backdrop. "Beautiful scenery", "exotic nature" and similar attributes describe the predominantly desired framework for pleasant, successful holidays.

Trend sports, in particular, which play an increasingly important role in tourism, make use of natural areas that until recently were completely uninteresting for the leisure industry.

The opposite relationship – the impact of tourism on the environment – is easy to understand on a general level (Fig. A.34). The necessary physical infrastructure is the most obvious connection. It is also self-evident that leisure activities, the traffic they induce, and finally, the behaviour of the tourists themselves are always accompanied by environmental impacts. Equally easy to understand is the observation that tourism consumes resources like space, water, or energy.

The following, simple example gives an impression of the complexity that characterises this re-

Fig. A.34. Environmental implications of leisure and tourism

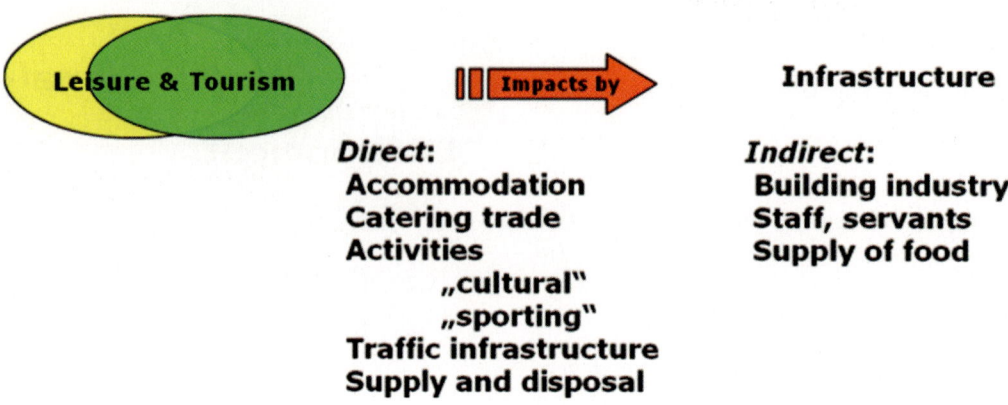

Fig. A.35. Infrastructure as a source of environmental impacts

lationship: To avoid tall hotel buildings that spoil the scenery, new projects are subject to recommendations or regulations that demand an architectural style in harmony with and typical of the scenery. But these buildings require, for the same number of beds, a floor space several times larger than do multi-storey hotels.

Admittedly, this rather general list of the most important environmentally relevant "pressure zones" needs to be spelled out in greater detail. Figure A.35 gives some examples for the environmental impacts created by tourism infrastructure, like lodging, catering trade, traffic infrastructure, as well as supply and waste facilities. But also the activities themselves – ranging from sports (swimming pools, bicycle lanes, golf courses, etc.) to culture (museums, theatres, etc.) – require a specific infrastructure.

These are the direct, immediately obvious environmental impacts. But we should not forget the indirect impacts:

• Tourism infrastructure requires a construction industry which, in turn, entails its own facilities, ranging from the gravel pit over the car pool to the business building.

• Many seasonal workers in tourism require additional lodgings and supply facilities.
• The gastronomic supply of the tourists implies a row of special production plants, ranging from fields or glasshouses to entire cattle herds and the meat-processing businesses.

Of course, many of the implications mentioned are desiderates of the economic policy of the respective country, and of course not every environmental impact is in itself destructive. The point here is rather to demonstrate the intimacy and complexity of these relationships.

Interactions between climate change and tourism

The central issue of this paper is part of intersection D (Fig. A.33), i.e. of the direct intersection between climate and tourism. It is not intended to separate it strictly from aspects which would be placed in its vicinity but still within intersection C.

The "First International Conference on Climate Change and Tourism" of the World Tourism Organization (WTO) in Djerba, Tunisia (1993) identified this subject area meanwhile as a sphere of problems too and focused on climate change related impacts on water resources, at coastal and island destinations, as well as mountain areas. The *Djerba Declaration on Climate Change and Tourism* recognises that climate change impacts are already occurring at some tourism destinations and the effects are expected to spread in the future[2]. Besides this, two other international meetings were held: at Milan (Italy) in 2003 on "Climate Change, the Environment and Tourism: The Interactions" and at Genoa (Italy) in 2004 on "Climate Change, the Environment and Tourism in Europe's Coastal Zones"[3].

In the context of climate and tourism we meet both directions of interdependencies: Consequences of climate change for tourism on the one hand, and the impact of tourism on climate change on the other hand.

Consequences of climate change for tourism

The climate, as well as the weather, crucially influences the choice of the holiday destination and duration. The decision pattern is shaped both by the objective opportunity to conduct certain activities (skiing, sailing, etc.) and by very subjective whims. Astonishingly, people often attach more importance to weather forecasts than to the actual weather (cf. Lohmann 2003, who also provides a comprehensive review of the literature). On the other hand, many tourists fail to catch up on the weather conditions in their holiday destination, or they do so only shortly before they set off (Hamilton and Lau 2004). Referring to selected European countries, Agnew and Palutikof (2001) conducted a study of climate impacts on the demand for tourism. In general, temperature is more strongly perceived and considered in the decision-making process than sunshine. Precipitation ranks only third. A certain "pleasant minimum temperature" should prevail, but the weather should also not be too warm. In general, an average temperature of 21 °C can be regarded as attractive for tourists. In all countries under investigation, a one-degree rise in temperature resulted in an increase of the number of domestic tourists by one to five percent.

Because the many existing studies are, for the greater part, not comparable, it became necessary

to establish an independent index which is free especially of the interviewees' subjective influence. The Tourism Climate Index (TCI) fulfills this requirement. It combines seven weighted monthly averages as mean and maximum temperatures, humidity, sum of precipitation, sun hours and wind speed. The TCI is also an appropriate tool for estimating the attractiveness of a destination after climate change.

Scott and McBoyle (2001), who also provide a more detailed description of the TCI, attempted a forecast for selected North American cities, which may be quantified using the TCI in combination with models of climate change. They concluded that approximately from 2070, many Canadian and northern United States cities will be affected positively. As a contrast, cities located further to the south, which even today stands out with unpleasant summer TCIs, will become even less attractive. It is an obvious step to expect a similar development for Northern Europe and the Mediterranean.

The studies by Bigano et al. (2005) are also based on simulations, but this time using the Hamburg Tourism Model (HTM[4]), which predicts the development of demand patterns in tourism. These authors' predictions resemble those of Scott and McBoyle (2001): Tourism demand will double in colder regions and decrease by 20% in warmer countries. In some regions, visitor numbers will even drop to half their current level. There will be an increase, especially of international tourism, in the currently colder regions.

Besides the direct consequences of climate change, which can be immediately attributed to the respective climate elements, there are numerous indirect effects. These may be quite diverse, effects on e.g. terrestrial and marine ecosystems are covered in Chaps. 4 and 5 of this volume.

The most important effect for tourism, especially for infrastructure, is the expected rise of the sea level. Its particular consequences, though, appear to be less crucial than those of the general climate change (Bigano et al. 2005). At first glance, it seems to be a great advantage that the sea level will not rise unexpectedly and within a very brief time, but that this process will occur very slowly in the course of many years. This leaves ample opportunity to meet the effects with technical constructions (dykes, barrages, etc.).

---

[2]cf. www.world-tourism.org/sustainable/climate/final-report.pdf

[3]cf. www.e-clat.org; more scientific information is available via the very actual bibliography of "Climate, Tourism and Recreation" on www.fes.uwaterloo.ca/u/dj2scott/Documents/CTREC%20Bibliography_FINAL.pdf

[4]www.uni-hamburg.de/Wiss/FB/15/Sustainability/htm.htm

Tourism-relevant general impacts of a sea level rise are (after Sterr et al. 1999):

- *More available water surface.* Not all extremely low areas will be "defended" by dykes against the sea. This new water surface will be very attractive destinations for many people, i.e. these areas create additional tourism.
- *Increasing likelihood of (storm) flood events.* Its impact concentrates on coastal facilities like marinas, near-beach buildings, and the water vehicles themselves, even though the events appear (at present) to occur more often in the less tourism-relevant winter half year.
- *Loss of land in floods and enforced erosion.* This impact may extend to tourism-relevant areas, e.g. because buildings were erected in known flood-prone and erosion-endangered areas. Admittedly, most facilities are affected by sediment set free by erosion, the costs for the unsilting of harbour accesses would rise.
- *Penetration of seawater into surface- and groundwater bodies.* This may especially concern isolated water catchments on islands, where water consumption increases considerably during the main tourist season.
- *Biological changes in the coastal ecosystems.* For example, they might affect the number and distribution of fish species relevant for fishing, or, in combination with changes in water temperature and salinity, promote neozoa like the piddock which causes considerable damage to harbour infrastructure and coastal protection facilities.

Against the background of these results, we may expect the following trends in the Baltic Sea Basin: The summer tourism will profit from the expected climate change, because the climate will generally become more tourism-friendly. The TCI (Tourism Climate Index) will reach attractive levels even in the northern part of the Baltic Sea Basin, as well as the presently rather short tourist summer season will be extended. There will be negative impacts on the winter tourism. Especially in the more continental regions of Northern Europe, winter temperatures will rise while snow certainty decreases.

However, we should not forget that these changes will be much less effective than the socio-economic factors which define the general framework. The decision for the time and destination of a holiday depends crucially on income, education, and age. Especially the habits of younger holiday-makers (and, thus, their decision patterns) change very quickly. These factors entail a much less predictable risk than climate change (cf. Lise and Tol 2005).

Impact of tourism on climate

Tourism is not only a "victim" of climate change, but itself influences climate development. However, there is only one sector besides indirect effects translated through general pressure on the environment (resource consumption, waste, soil sealing, etc.) that directly affects the climate: tourism-based traffic. It produces considerable amounts of exhaust gases and thus greatly contributes to climate change.

For many years, the trend in tourism has been towards long-distance travel, which in turn causes an increase in flight traffic. Flight traffic is one of the strongest and most immediate sources of climate change. In the Baltic, however, long-distance traffic will increase "only" proportional to the general increase in long-distance travel. Because, on a global scale, the population size in this region is comparatively low, this cause of environmental pressure will be rather low-ranking. The increase in temperature could, under certain circumstances, even have a contrary effect. A certain – though probably rather small – fraction of the population in the Baltic Sea Basin will not travel to long-distance destinations but rather spend their holiday in the region itself. Thus, their contribution to the further increase of CO_2 emissions will be less than average. The present phase of cheap flight travel causes very sizeable new tourist streams. This form of tourism also occurs in the Baltic Sea Basin, which is both source area and destination. However, we can safely presume that the market will soon put an end to this phase and thus limit the present extreme emissions.

The primary form of tourism-related traffic in the Baltic Sea Basin is individual transport, i.e. car traffic. Even in the medium-term, it will not lose its predominant position.

In general, the tourism-based traffic in the Baltic Sea Basin will grow – as a continuing effect of the political opening of the East. Even in the medium-term, we will continue to sense that new source areas of travel traffic with an increased action range have entered the market. The population of Poland and the Baltic States still feel they have to catch up in terms of travel, and they will fulfil their desire as their economic wealth increases. On the other hand, these countries present new destinations which – lacking alternatives – aggressively promote tourism in their turn.

On a regional and local level, tourism- and leisure-based transport plays an especially important role. In Germany, 54% of the total traffic volume is caused by leisure traffic, which, in turn, includes a share of problematic individual traffic of more than 80% (UBA 2003, p. 19). Especially the rise of the trend sports is regularly linked to an increase in traffic volume. Among these particularly traffic-intensive leisure activities are diving, wind surfing, skiing, and climbing (Stettler 1998). Thus, in the medium-term tourism-based individual traffic will considerably contribute to the increase of CO_2 emissions.

Another source of increasing CO_2 emissions is the marine tourism. This form of tourism will gain even more importance in future. Not only will the number of private motor boats rise, but the already booming cruise market will continue to grow in the Baltic Sea as it does elsewhere. In addition, the river cruise market has been developing with a higher than average rate for several years. Thus, in the medium-term marine tourism will also create considerable emissions.

Conclusions

The traffic volume in the Baltic Sea Basin will continue to rise in the medium-term. This development is primarily still an effect of the political opening of the East and the consequent EU enlargement. All of the new EU member states place great hopes in the tourism sector.

As a result, we will witness shifts of the tourist destinations within the Baltic Sea Basin - at the southern Baltic Sea coast the shift will be from west to east, i.e. from Denmark and Germany to Poland and to some select seaside resorts in the Baltic States. Large cities such as Gdansk, Riga, and St. Petersburg are also attracting more tourists.

This development will lead to more traffic – predominantly individual traffic, but also ferry traffic – which can contribute considerably to climate-relevant emissions.

The trend towards marine tourism (more motor boats, more cruises) will also strongly increase emissions.

The remaining consequences are of a rather more general, ecological nature (consumption of space and resources, especially water, as well as waste disposal problems). However, they will occur predominantly as problems on the local and regional level – the total sum will hardly rise and could perhaps be balanced by infrastructural and technical counter-measures.

The opposite relationship – the impact of climate change on tourism in the Baltic Sea Basin – raises positive expectations. The probable change of the sea level will hardly have any negative effect. On the contrary, it might enhance the marine potential, and the rise in temperature will improve the natural basis for tourism. Consequently, the total volume of tourism will grow. But on the whole, the changes will lead to more domestic and less long-distance tourism, thus contributing to a reduction of the growth of climate-relevant emissions caused by tourism.

A.3.2 Terrestrial and Freshwater Ecosystems

Benjamin Smith, Thorsten Blenckner, Christoph Humborg, Seppo Kellomäki, Tiina Nõges, Peeter Nõges

A.3.2.1 Catchment Area of the Baltic Sea

The catchment area of the Baltic Sea (1,735,000 km^2), which constitutes the land surface region of the Baltic Sea Basin, comprises watersheds draining the Fennoscandian Alps in the west and north, the Erzgebirge, Sudetes and western Carpathians in the south, uplands along the Finnish-Russian border and the central Russian Highlands in the east. Politically, this includes most of Sweden, Finland, the Baltic States, Poland and part of Russia, Belarus, Ukraine and northern Germany (Fig. A.36). The Baltic Sea Basin spans some 20 degrees of latitude and climate types ranging from nemoral to alpine and subarctic. It can be roughly divided into a south-eastern part, draining into the Gulf of Riga and Baltic Proper, that is characterized by a cultivated landscape with a temperate climate and a northern boreal part characterized by coniferous forests and peat. The natural vegetation is mainly broadleaved deciduous (nemoral) forest in lowland areas of the temperate southeast, and conifer-dominated boreal forest (taiga) in northern Scandinavia. Cold-climate shrublands and tundra occur in mountainous areas and in the subarctic far north of the catchment region. Wetlands are a significant feature of the boreal and subarctic zones.

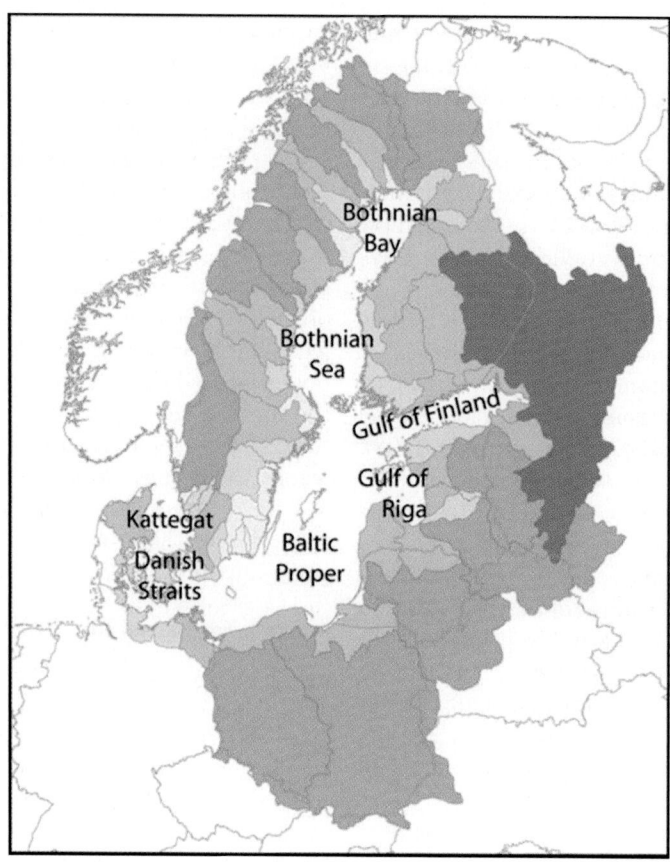

Fig. A.36. The Baltic Sea Basin showing sub-watershed drainage basins, national boundaries and the major basins of the Baltic Sea (source: UNEP/ GRID-Arendal, www.grida.no/baltic, page visited 26 February 2007)

Much of the nemoral forest zone has long been converted to agriculture (Ledwith 2002); only in northern parts of the Baltic Sea Basin (e.g. Sweden, Finland, north-western Russia) do forests still dominate the landscape. Approximately half of the total catchment area (about 8,200,000 km²) consists of forest, most of the remainder being agricultural land (Chap. 1, Fig. 1.11). Most of the forests are managed, and forestry is mainly based on native tree species; particularly Norway Spruce (*Picea abies*) and Scots Pine (*Pinus sylvestris*). Especially in the temperate parts of the Baltic Sea Basin, the current tree species composition is determined by past land use and management activities rather than by natural factors. Areas of vegetation largely unaffected by direct human management are presumably of very limited extent, but some such areas occur in upland parts of northern Sweden and Finland.

A.3.2.2 Climate and Terrestrial Ecosystems

Climate control of large-scale vegetation patterns

Poleward or upper altitudinal range boundaries of many plant species and biomes correlate with isolines of absolute minimum temperature, and are assumed to be the result of ice formation in plant tissues leading to tissue and plant death, either directly or via secondary mechanisms such as dessication (Woodward 1987). Some range boundaries, including the alpine treeline in Scandinavia, are more closely correlated with growing season heat sums, rather than absolute minimum temperatures, which may suggest growth limitations associated with low temperatures and a short available period for carbon assimilation as possible limits to survival above treeline (Körner 1998; Grace et al. 2002). The mechanisms determining southern (warm) range limits are more varied and less well understood. For Norway spruce (*Picea abies*), limited snow cover and repeated freeze-thaw cycles have been suggested to interfere with the development and maintenance of winter-hardiness in seedlings, inhibiting natural regeneration of this species in oceanic climates (Dahl 1990; Sykes et al. 1996; Bradshaw et al. 2000). In many temperate tree species, the timing of budburst appears to be coupled to the duration and intensity of low temperatures in the winter. Budburst may thus be delayed by several weeks following a warm winter, protecting the frost-sensitive new shoots from damage by late frosts (Murray et al. 1989; Sykes et al. 1996). This chilling mechanism has been proposed to, for example, restrict the distribution of European beech (*Fagus sylvatica*) into the most oceanic climates (Sykes and Prentice 1995).

In reality, temperature is only one of several factors controlling large-scale distributions of tree species. Even if the seed dispersal of some species is able to keep up with climate change, local dominance shifts may be delayed by the long generation times of trees, competition with resident species, required changes in soil structure, hydrology, chemistry, litter depth, requirements for mycorrhizae and other factors (Sykes and Prentice 1996; Huntley 1991; Malanson and Cairns 1997).

Environmental control of ecosystem biogeochemistry

Climate affects biogeochemical cycling within ecosystems by modifying the rates and modes of individual ecosystem processes. The most important processes in terms of overall control of ecosystem functioning are the physiological processes underlying net primary production (NPP): photosynthesis, autotrophic (plant) respiration, stomatal regulation and carbon allocation in plants. Carbon assimilated in photosynthesis is eventually either respired or otherwise emitted (e.g. as biogenic volatile organic compounds) directly to the atmosphere, or fed into the soil organic matter pool as litter, root exudates, or residues from disturbance by fire. Productivity changes thus tend to propogate to litter and soil pools and impact decomposition processes. Productivity changes may also modify vegetation structure (e.g. leaf area index, LAI), changing the competitive balance among individuals and species, and leading to further structural changes and feedbacks on production. Changes in disturbance regimes, for example, damage due to wind storms and wildfires, can be a further consequence of changed vegetation structure.

The direct physiological effect of a temperature increase on plant carbon balance may be either positive or negative, depending on the relative kinetic stimulation of photosynthesis and plant respiration. However, NPP tends to be a constant proportion of GPP across biomes and climate types, which suggests that respiration rates are in the long term dependent on the supply of assimilates (i.e. net photosynthesis), and that acclimation to a temperature change may occur (Waring et al. 1998; Dewar et al. 1999).

Temperatures also affect annual productivity via growing season length. Phenological events such as budburst and leaf abscission in temperate and boreal plants are closely correlated with climate indices, particularly temperature sums (Kramer 1995; Badeck et al. 2004). Both temperature cues and photoperiod may determine growth cessation and the onset of hardening in the autumn. A uniform increase in average temperatures throughout the year implies that threshold temperatures for phenophase transitions in spring may be achieved a number of days earlier, while later autumn frosts may delay leaf abscission and winter hardening. However, many tree species require an extended period of cold temperatures (chilling) to initiate budburst and, in these species, warmer winter temperatures may delay the onset of growth even once normal spring temperatures have been achieved (Cannell 1989; Sykes et al. 1996). The effects of temperature changes on plant phenology may be expected to impact NPP particularly at high latitudes, where the growing season is shortest (Walker et al. 1995).

Soil nutrient status, particularly nitrogen availability, is considered to limit productivity in many temperate and boreal ecosystems (McGuire et al. 1995; Bergh et al. 1999; Hungate et al. 2003). Increased soil temperatures will tend to stimulate microbial activity and N-mineralisation rates, so long as soils remain moist, potentially releasing ecosystems from N-limitation (Melillo et al. 2002). This effect may be particularly important in the coldest boreal, arctic and alpine environments (Jonasson et al. 1999; Strömgren and Linder 2002).

Soil water deficits occasionally limit production, even in many mesic ecosystem types. Although water availability is not currently a major limiting factor for forest or agricultural production in the Baltic Sea Basin, it does occur, e.g. on well-drained soils in southern Sweden (Bergh et al. 1999). Increased temperatures lead to increased

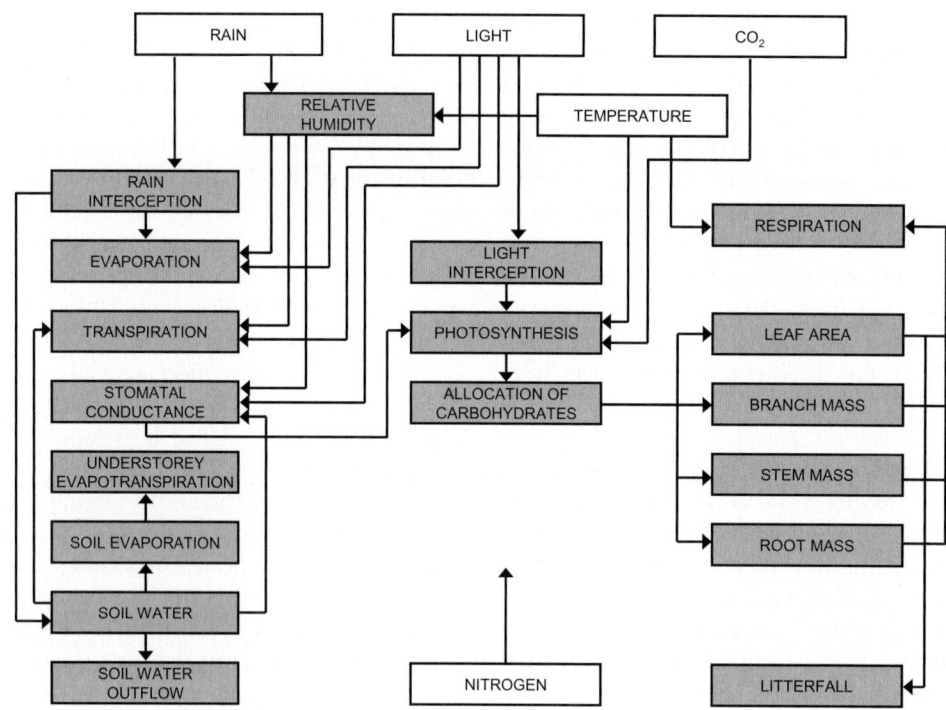

Fig. A.37. Conceptual diagram for the BIOMASS model (adapted from Freeman et al. 2005)

evaporative demand and depletion of soil water via increased evapotranspiration, while changes in precipitation patterns may exacerbate or ameliorate moisture deficits and effects on production.

Rising atmospheric carbon dioxide (CO_2) concentrations are one of the most certain aspects of global change. CO_2 is a plant resource, and it has long been assumed that increased ambient CO_2 levels would stimulate photosynthesis via reduced photorespiration and improve plant water budgets through reduced stomatal conductance, in both cases tending to augment net carbon uptake (Bazzaz 1990). Laboratory and field experiments exposing plants or whole ecosystems to elevated CO_2 levels confirm that production usually increases by 15–50% for a doubling of CO_2 above pre-industrial levels (Poorter and Navas 2003); for temperate forest ecosystems, the response across broad productivity gradients is rather conservative at a median of $23 \pm 2\%$ (Norby et al. 2005). However, it has been suggested that, in many ecosystems, negative biogeochemical feedbacks may inhibit plants from fully utilising the additional assimilates resulting from CO_2-fertilisation on multidecadal time scales (McGuire et al. 1995); reduced quality (lower N:C ratio) of the litter produced by CO_2-fertilised plants could lead to immobilisation

of N by soil microbes, reducing plant N uptake and NPP (McGuire et al. 1995). To a certain extent, plants may compensate for nutrient limitations by an increased relative investment in below ground structures (roots) and functions (e.g. root exudates, investment in mycorrhizae, Lloyd and Farquhar 1996).

Overall responses of ecosystem biogeochemistry to environmental changes are characterised by the differential temporal signatures of many constituent processes. Short-term physiological responses, such as the direct response of net photosynthesis to a change in temperature, may be modified by longer-term changes in, for example, plant structure, population dynamics, vegetation species composition and soil organic matter stoichiometry (Shaver et al. 2000).

A.3.2.3 *Outline of the BIOMASS Forest Growth Model*

The BIOMASS model is based on sub-models describing radiation interception, canopy photosynthesis, phenology, allocation of photosynthates among plant organs, growth, litter fall and water balance, including nitrogen effects on processes (Fig. A.37; Freeman et al. 2005). BIOMASS simulates tree growth on a daily time step, which requires daily meteorological inputs of short-wave radiation, maximum and minimum air temperature, precipitation, and humidity of the air. Gross primary production is calculated from a radiation interception model requiring information on canopy architecture and a biochemically-based model of leaf photosynthesis by C_3 plants. Net primary production is obtained by subtracting the autotrophic respiration.

A.3.2.4 *Outline of the EFISCEN Forest Resource Model*

EFISCEN is a large-scale matrix model, which uses forest inventory data as input (Sallnäs 1990; Pussinen et al. 2001). EFISCEN can be used to compile information on forest resources in Europe and to produce projections of the possible future development of forests. The state of a forest is depicted in the model as an area distribution over age and volume classes in matrices. Growth is described as area changes to higher volume classes and ageing of forest is incorporated as a function of time up to the point of regeneration. The user defines the level of fellings and the model implements cuts according to predefined management regimes.

The basic input data include forest area, growing stock and increment by age classes, i.e. the data available from national forest inventories. European-wide data are compiled in the EFISCEN European Forest Resource Database (EEFR) at the European Forest Institute (Schelhaas et al. 1999). Country-level data consist of forest types, which are distinguished by region, owner class, site class and tree species, depending of the aggregation level of the provided data.

The current version of EFISCEN can be used to study the carbon balance of the whole forest sector. Stem-wood volumes are converted to carbon in the compartments stems, branches, leaves, and coarse and fine roots using dry wood density, carbon content and an age-dependent biomass distribution. Litter production is estimated using age-dependent turnover rates of each compartment and is used as input to a dynamic soil carbon module. A dynamic wood products model enables the flow of carbon to be followed further through processing and wood products up to the time point at which carbon is released back to the atmosphere. The EFISCEN version used in the calculations presented in Sect. 4.3.5 takes into account the impact of changes in temperature and precipitation on forest growth, and thus forest structure, carbon budget and fellings, as detailed by Pussinen et al. (2005).

A.3.2.5 *Climate Scenarios Used in SilviStrat Calculations*

Three climate scenarios were used in the SilviStrat project (Kellomäki and Leinonen 2005) to yield the predictions of change in forest growth presented in Sect. 4.3.5.

- *Current climate.* Temperature and precipitation values are based on monthly time series 1901–1995 and the CRU monthly climatology 1961–1990, with a spatial resolution of 0.5 degrees of latitude and longitude (Michell and Jones 2005). The warming trend in the interpolated monthly time series 1961–1990 was removed to generate 30 years of climate data with the typical interannual variability between years. From this sample, individual years were randomly drawn to generate the baseline climate ('current climate') for the simulation period of 110 years.
- *ECHAM4 climate.* Values for the period between 1990 and 2100 were used, taking the CRU monthly climatology 1961–1990 as a baseline. It was based on output from the ECHAM4-OPYC3 GCM (cf. Chap. 3) which is available as monthly means at a spatial resolution of $2.81 \times 2.76°$. The GCM simulation does not include the cooling effects of sulphate aerosols on climate.
- *HadCM2 climate.* The values were estimated for the period between 1990 and 2100 taking the CRU monthly climatology 1961–1990 as a baseline. It was based on output from the HadCM2 GCM (Chap. 3) at a spatial resolution of $2.5 \times 3.75°$. The GCM simulation includes the cooling effects of sulphate aerosols on climate.

Climate scenarios from both GCMs were based on the IS92a 'business as usual' emission scenario, which assumes a doubling of atmospheric CO_2 concentrations in the 21st century. To generate monthly surface climate data, GCM results were downscaled to the sites by calculating the difference for each parameter between the time period 1990–2100 and the average values of the reference period 1931–1960 using monthly time steps. Spatial interpolation of GCM data on the sites was performed with the Delaunay triangulation. The time series of these anomalies were then added to the average values of baseline (CRU) data for the study sites.

A.4 Observational Data Used

A variety of data has been used for the description and analyses of the atmosphere, the ocean, the runoff from rivers, and the ecology of the Baltic Sea and its surrounding area. Below follows a short survey.

A.4.1 Atmosphere

Øyvind Nordli

Regular measurement with mercury in glass thermometers and barometers started at a few sites in the Baltic Sea Basin already in the 18th century. As standard instruments at the meteorological services they were much improved as time passed. Now, the mercury instruments have been replaced in large numbers by sensors for automatic registration. For long-term temperature trend analyses, the homogeneity of the series is a challenge for several reasons. Examples are improvements of the thermometers themselves, and also, a better sheltering from short wave radiation (Nordli et al. 1997).

As site measurements grew denser, reliable grid box mean values could be calculated. Advantage is taken of a work done by the Climate Research Unit (CRU) at the University of East Anglia (IPCC 2001, Chap. 2). For this study of the Baltic Sea Basin (see Chap. 2), a subset of the CRU data set with solely land stations is chosen. The dataset starts in 1851, but only data after 1870 are used in this book. The CRU data set includes many well homogenised series from the Nordic countries (Jones and Moberg 2003).

Glass barometers with mercury are of about the same age as the mercury thermometers. A challenge has been to keep the calibration of the instruments constant during transportation and also to keep them in good shape over long time spans. The readings of the barometers have to be adjusted to standard gravity and temperature. As standards, 45° latitude and 0 °C have been chosen, respectively. The pressure at station level is also adjusted to sea level. This operation may be challenging for mountain stations and stations situated in valleys where temperature inversions (temperature increase with height) tend to develop. Thanks

to a network of radiosond stations also the height of distinct pressure surfaces, as for example the 500 hPa surface is mapped. During World War II, the sonds were much improved in an increasing number, and a global data network from the upper atmosphere was established.

The pressure data sets have undergone two very important procedures, namely reanalysis and gridding. This enables the study of circulation and circulation changes. Data sets often used for daily data are the ERA-40 (1957–2002) for Europe and a global data set from NCEP (1948–present).

The analyses of precipitation changes are based upon measurements by rain gauges. Precipitation has larger spatial variability than temperature and, in particular, pressure. This is often sought to be compensated by a denser network. During wind a main problem is that the gauges fail to catch all precipitation. The under-catch, which is largest under solid precipitation, is sometimes adjusted for by empirical formulae (see more in Annex 1.2.3).

Cloud cover is manually observed at the meteorological stations. The coverage of the sky is assessed in oktas, which has been standard since the 1st of January 1949. Solar radiation is also measured at some stations (see Annex 1.2.5); the length of reliable time series is now up to 40–50 years. For Europe as a whole they amount to about 300 (Kallis 1995).

A.4.2 Ocean

Philip Axe

A.4.2.1 *Hydrographic Characteristics*

Systematic temperature observations in the Baltic Sea extend back to measurements from lightships during the 1880's. Early observations of sea temperature were made using mercury thermometers and buckets. The invention of the reversing thermometer in 1874 made accurate (potentially better than 0.01 °C) in-situ measurements possible. These thermometers continued in use until very recently, though are now largely replaced by electronic ones offering better accuracy and ease of use. More detailed temperature profiles were obtained using bathythermographs – first mechanical, and later electronic. Claimed accuracies were better than 0.06 °C though Emery and Thomson (1997) suggest an accuracy of $+/-0.1$ °C is optimistic. Profiles (accurate to 0.002 °C) are now obtained by a combination of platinum resistance thermometers and thermistors mounted in a CTD (standing for 'Conductivity, Temperature and Depth'). The CTD is conventionally lowered through the water column, though can be towed behind vessels. Sea surface temperature is measured both by ships and from satellite borne radiometers.

Baltic Sea lightships also measured seawater density using an aerometer – a form of hydrometer. Knowing the sample temperature, it is possible to estimate the salinity. Salinity was also estimated from the chlorinity of seawater, determined by titration. Now salinity is determined by comparing the conductivity of a sample compared to the conductivity of a reference solution. This is done both for water samples, and *in situ* using the CTD. A modern CTD is accurate to about 0.005 psu (practical salinity unit). For more information on the salinity scale, see for example, UNESCO 1985.

Together with changes in salinity, oxygen content is used for identifying inflow events and for estimating seawater ventilation. The principle method of determination (Winkler titration) has not changed since 1889, and is capable of an accuracy of 1%.

A.4.2.2 *Sea Level Variability*

Harbour authorities have collected sea level data since at least the beginning of the eighteenth century. First measurements consisted of scratching the sea level on rocks, some of which are still apparent. Sea level observations in harbours were made by visual observation of a graduated staff, before being superseded by float gauges. Float gauges remain widely used. When well maintained and used in conjunction with a properly designed stilling well they are accurate. More recently, systems based on pressure gauges, acoustic time-of-flight and radar have been introduced. Pressure gauge systems require knowledge of the seawater density to calculate the height of the sea surface above the sensor. Acoustic and radar based systems measure the distance between sensor and sea surface from the time taken by a signal to return after reflection from the sea surface. Early acoustic systems had problems with changes in the speed of sound in air due to temperature changes. Acoustic and pressure gauge systems can suffer from sensor drift, requiring good datum control. When regularly checked, these systems are also

capable of delivering high quality sea level data. Since 1991, the coastal observations have been augmented by satellite altimetry. These space-borne sensors provide sea level measurements over the offshore Baltic Sea.

The world's longest continuous sea level record is from Stockholm, starting in 1774 (Ekman 1988), while data from Kronstadt extend back to 1777 (Bogdanov et al. 2000). The Intergovernmental Oceanographic Commission's GLOSS programme (IOC 1997) have identified a set of 175 tide gauges with sea level records of at least 60 years suitable for climate research. Of these, 45 are located in the Baltic.

Sea level recorded at Landsort, Sweden (established 1886) is considered a good proxy for the volume of the Baltic Sea as a whole. The volume of barotropic exchanges through the Danish Straits can be estimated using sea level gradients in the Sound. These methods are now supplemented by current (and hydrographic) measurements from automatic stations in the straits.

Though sea state observations have been made from ships, long-term wave measurements were uncommon before the 1970's. Swedish measurements started in 1978 (B. Broman, pers. comm). Methods used have included upward looking echo sounders deployed near lighthouses (e.g. Almagrundet). Systems based on pressure transducers, as well as accelerometer buoys, are currently in use.

A.4.2.3 *Optical Properties*

A dataset consisting of more than 40,000 Secchi depth measurements, collected in the North and Baltic Seas between 1902 and 1999, was assembled by Aarup (2002). This data set is available from ICES (International Council for the Exploration of the Sea).

A.4.3 Runoff

Göran Lindström

Systematic runoff observations in the Baltic Sea Basin have been carried out since the early 1800s. The longest continuous data series is from the outlet of Lake Vänern, where observations started in 1807. Since around the mid-1800s, data is available from several other rivers, including Neva, Vistula, Neman, Daugava, Dalälven, Emajogi and Vuoksi.

Several systematic runoff datasets covering mainly the last fifty years have been collected. The BALTEX Hydrological Data Centre at SMHI in Norrköping has collected runoff data from the different national institutes. The database extends to the year 1950 and consists of data from over 200 river flow stations, most of them near the mouths of the rivers. The pan-Nordic dataset created in the CWE (Climate, Water and Energy, www.os.is/cwe) and CE (Climate and Energy, www.os.is/ce) Programmes consists of over 150 streamflow records with an average length of 84 years of daily data.

In addition, separate observation series from different countries have been utilised. These include runoff, water level and water temperature data. Some historical data concerning major floods before systematic observations started have also been available.

A.4.4   Marine Ecosystem Data

Ilppo Vuorinen, Joachim Dippner, Darius Daunys, Juha Flinkman, Antti Halkka, Friedrich W. Köster, Esa Lehikoinen, Brian R. MacKenzie, Christian Möllmann, Flemming Møhlenberg, Sergej Olenin, Doris Schiedek, Henrik Skov, Norbert Wasmund

Observational evidence used in Chap. 5 (*Climate-related marine ecosystem change*) is mainly, but not completely, based on long term monitoring data from various sources. Information on nutrients, contaminants and phytoplankton is exclusively based on data from the "Convention on the Protection of the Marine Environment of the Baltic Sea Area" – more usually known as the Helsinki Convention, or HELCOM[5]. HELCOM is also the source for most of the data used on zooplankton. Information on fish, fisheries and mammals rely on data from the International Council for the Exploration of the Sea, ICES[6] Bird data is based on a Pan-European monitoring programme established by the Royal Society for the Protection of Birds (RSPB[7]).

Below is a short summary of both HELCOM and ICES monitoring programmes based on these organisations' websites. In most cases also other, national data or data sets from different research projects have been used in Chap. 5. Their use is explained in Annex 4.4.3 below for sections of Chap. 5.

A.4.4.1   *HELCOM*

HELCOM works to protect the marine environment of the Baltic Sea from all sources of pollution through intergovernmental co-operation between Denmark, Estonia, the European Community, Finland, Germany, Latvia, Lithuania, Poland, Russia and Sweden. The Monitoring and Assessment Group (MONAS) looks after one of HELCOM's key tasks by assessing trends in threats to the marine environment[8], their impacts, the resulting state of the marine environment, and the effectiveness of adopted measures. MONAS aims to ensure that HELCOM's monitoring programmes[9] are efficiently used through horizontal co-ordination between HELCOM's five permanent working groups.

HELCOM's oldest monitoring program for physical, chemical and biological variables of the open sea started in 1979, followed in 1984 by the monitoring of radioactive substances in the Baltic Sea. Monitoring of inputs of nutrients and hazardous substances was initiated in 1998.

The HELCOM monitoring system consists of several complementary programmes:

- The Pollution Load Compilation[10] (PLC) programmes (PLC-Air and PLC-Water) quantify emissions of nutrients and hazardous substances to the air, discharges and losses to inland surface waters, and the resulting air and water-borne inputs to the sea.
- The COMBINE programme[11] quantifies the impacts of nutrients and hazardous substances in the marine environment, also examining trends in the various compartments of the marine environment (water, biota, sediment). The programme also assesses physical forcing.
- Monitoring of radioactive substances (MORS) quantifies the sources and inputs of artificial radionuclides, as well as the resulting trends in the various compartments of the marine environment (water, biota, sediment).

---

[5]www.helcom.fi
[6]www.ices.dk/ocean/
[7]www.rspb.org.uk/science/survey/2004/Europe.asp

[8]www.helcom.fi/environment2/en_GB/cover/
[9]www.helcom.fi/groups/monas/en_GB/monas_monitoring/
[10]www.helcom.fi/groups/monas/en_GB/plcwaterguide/
[11]www.helcom.fi/groups/monas/CombineManual/en_GB/main/

- HELCOM also coordinates the surveillance of deliberate illegal oil spills around the Baltic Sea, and assesses the numbers and distribution of such spills on an annual basis.

A.4.4.2 *ICES*

ICES provides advice on the status of fish and shellfish stocks in the North Atlantic Ocean. Last year's advice can be found in the ICES Advice report series[12]. The information forming the basis of this advice is collected by marine scientists in the ICES 20 member countries. They collect data through sampling landings of fish at fish markets, sampling the amount of fish discarded from fishing boats and by targeted surveys with research vessels. The data is used by ICES Working Groups[13] to assess the status of fish and shellfish stocks. There also is a Working Group of Marine Mammal Ecology (WGMME), which has reported annual changes in marine mammal species since 2001.

This information is then collated into advice by the Advisory Committee on Fishery Management (ACFM). This Committee has representatives from each of the member countries and meets every year in summer and autumn. ICES advice covers over 135 separate fish and shellfish stocks. The advice for each stock usually includes:

- An estimate of historical trends in landings, spawning stock biomass, recruitment and fishing mortality rate;
- A description of the 'state of the stock' in relation to historical levels;
- The likely medium term development of the stock using different rates of fishing mortality;
- A short term forecast of spawning stock biomass and catch.

A.4.4.3 *Other Observational Data used in Chap. 5*

Nutrients, Contaminants (chemical pollution) and Phytoplankton

Information on sources and distribution of nutrients, contaminants (chemical pollution), and phytoplankton is exclusively based on HELCOM data, see Annex 4.4.1 above.

---

[12]www.ices.dk/products/icesadvice.asp
[13]www.ices.dk/iceswork/workinggroups.asp

Bacteria

Data and information on bacteria is based on research projects on both the Baltic Sea and other marine areas. Regular monitoring of bacteria biomass and production is carried out in the Bothnian Bay and the Bothnian Sea by the Umeå Marine Sciences Centre, Sweden.

Zooplankton

Zooplankton has been monitored in a variety of ways in the Baltic Sea. There is no comprehensive assessment of this data prior to 1979. However, after the commencement of HELCOM monitoring programs, the zooplankton has been a standard issue. A recently established working group under the leadership of Dr. Lutz Postel at the Baltic Sea Research Institute Warnemünde (IOW), has just made an assessment on the current status of zooplankton monitoring in the Baltic Sea.

Altogether, there are 122 regularly visited stations where vertical net samples are taken. The monitoring has been carried out once to several times per annum. Sampling gear consists of vertical Juday nets (Baltic States) and WP-2 nets, which is the HELCOM standard today. The mesh size is $100\,\mu$m. Hauls are usually taken separately from close to the bottom to halocline, halocline to thermocline, and thermocline to surface. Ship-of-opportunity techniques have been applied to zooplankton monitoring in the Baltic Sea since 1998, when the Finnish Institute for Marine Research (FIMR), in co-operation with the Sir Alister Hardy Foundation for Ocean Science (SAHFOS) Plymouth and supported by the shipping company Transfennica Ltd, undertook Continuous Plankton Recorder (CPR) experiments.

After experimental and gear comparison studies conducted during 1998 to 2004, the Baltic Sea CPR survey is now operational on a route between Lübeck, Germany and Hamina, Finland on a monthly basis. Another one is just being set up by the Baltic Sea Regional Programme (BSRP), running across the Baltic Sea between Sweden and Poland.

Benthos

Quantitative methods to study bottom macrofauna in the Baltic Sea have been used since the early 1910s. Data used are referenced in Chap. 5. No ICES or HELCOM materials on bottom macrofauna in the Baltic Sea were used because they are not available in a form accessible for international use.

Fish

Only ICES data (see Annex 4.4.2 above) were used for fish and fisheries.

Marine Mammals

Until recently, marine mammal monitoring in the Baltic Sea was not well coordinated. An informal working group on grey seals has coordinated annual surveys since 1999. Ringed seal monitoring in the Bothnian Bay started in the 1970s and is now regularly performed by the Swedish Museum of Natural History, which also has the responsibility for Swedish harbor seal surveys. Lack of ice has caused problems for ringed seal surveys in the Gulf of Finland and Riga as aerial counts should preferably take place in peak melting season (end of April to beginning of May). Surveys in these areas have not been possible on an annual basis, and, in the Gulf of Finland, the border zone permission system has been additionally demanding.

Methodically sound data on ringed seal abundances in the Gulfs of Finland and Riga exist since the mid-1990s, and more limited data from the Archipelago Sea since 2001. A reliable trend estimate for the ringed seal exists currently only for the Bothnian Bay. Projected changes in ice climate are very challenging for ringed seal surveys in especially the southern breeding areas, as well timed aerial surveys of seals hauled on ice have been the standard method. Harbor porpoise abundances have been only sporadically estimated in the Baltic Sea. The extensive SCANS (1994) and SCANS II (2005) surveys (biology.st-andrews.ac.uk/scans2/) have only partial coverage, and the number of observed animals in transects is necessarily very small because of limited numbers.

Marine mammals data are based on information from the Working Group on Marine Mammal Ecolgy under the auspices of ICES[14], which has reviewed information on marine mammal species since 2001, and on advice on seal and harbor porpoise populations in the Baltic marine area (ICES 2005). There are also national monitoring data included in the analysis. An expert group on seals

has also recently been established in HELCOM, and analysis initiated by this group will be available in the coming years.

Sea Birds

The Pan-European Common Bird Monitoring Project was launched in January 2002 by the European Bird Census Council[15]. Its main project goal is to use common birds as indicators of the general state of nature using scientific data on changes in breeding populations across Europe. It is a collection of national monitoring programs. In 2006, 20 countries reported monitoring results of 244 species. In this review published results of Swedish[16] and Finnish[17] monitoring programs and those directed to Baltic marine birds by Danish authorities were used. Phenological changes were described based on unpublished databases obtained from the Jurmo and Hanko bird stations (run by Turku Ornithological Society and Helsinki Ornithological Society Tringa, respectively) and phenology programs of ornithological societies of Turku and Kemi-Tornio (Xenus).

A.4.5 Observational and Model Data for Anthropogenic Input

Mikhail Sofiev

Two most important sources of information about the atmospheric pollution of the Baltic Sea Basin are the EMEP (European Monitoring and Evaluation Program) and HELCOM programmes, which provide regular assessments of concentration and deposition of several species over Europe and over the Baltic Sea Basin, respectively. Databases of these programmes contain both observations and model data (the latter ones are based on EMEP models). An advantage of these datasets is their internal consistency and long period of time covered.

However, the data quality strongly depends on considered species. Some of the species are comparatively well studied and there are both observational and model assessments of their input to the Baltic Sea, with uncertainties to be within a factor of 2 (first of all, oxidised nitrogen and ammonia). Some others are less known and available

---

[14]www.ices.dk/iceswork/wgdetailace.asp?wg=WGMME

[15]www.ebcc.info/

[16]www.biol.lu.se/zooekologi/birgmonitoring/Eng/index.htm

[17]www.fmnh.helsinki.fi/english/zoology/vertebrates/info/birds/index.htm

estimates have an uncertainty as large as a factor of 3 to 10 (most of toxic metals). Finally, some species are known poorly and the corresponding estimates are based on crude considerations showing an order of magnitude as the best or even serving only as indicators of presence of the effect (most of persistent multi-media pollutants). Therefore, the estimates from one source have to be cross-verified with (semi-) independent sources of information, such as dedicated scientific studies. Such studies are usually more concentrated on specific problems and processes and do not provide that universal and long-term datasets as the EMEP/HELCOM ones: they cover limited period of time, possibly only part of the sea, etc. Generalization of their output might lead to large uncertainties in the final estimates and thus has to be done with care.

Due to permanent development of all models, the estimates given by formally the same model in different years can vary significantly. For example, the nitrogen deposition onto the Baltic Sea surface estimated by Bartnicki et al. (2001) and Bartnicki et al. (2002) for the same year 1998 using the same EMEP Lagrangian model differ by as much as 36%. Similar concerns are valid for other modelling sources of information, although the corresponding long-term data are rarely available. Similar problems with observational data are discussed in Annex 5. Therefore, a special attention has to be paid to a synchronous consideration of model, observations and their mutual fitting. Should several estimates are available for the same period made by the same tool the chronologically last ones should be taken.

A.5 Data Homogeneity Issues

Raino Heino

In practice it is difficult to obtain long homogeneous data records. Various factors, such as changes in (i) instruments and their exposure, (ii) observation times and averaging methods, and (iii) observation sites and their environments, introduce inhomogeneities into the data.

The inhomogeneities of climatological time series may be in the form of (i) impulsive (or step-like) change of central tendency, (ii) progressive change (or trend) or (iii) some kind of oscillation. Most of the inhomogeneities fall typically in the first category of impulsive change (including changes in instrumentation, observers or averag-

ing methods and station relocations) and typically alter the average value only, usually leaving the higher statistical moments unchanged. An inhomogeneity, however, may also contain changes in variability or in other distribution parameters. In practice, the inhomogeneity of a longer-term time series is usually a combination of many factors.

Climatic records, of course, contain variations that are due to several causative factors, such as variations in incoming solar radiation and changes in atmospheric transparency, which may take any of the above forms. Climatic records, at least those that are in their original form, are normally complex mixtures of both apparent and real variations. It is obvious that the apparent variations should be detected and eliminated before proceeding too far in the detection of real variations and their causes.

Several statistical methods to study the homogeneity can show whether any bias is included in the data records. However, these do not provide any indication of its location or cause. Information on the history of the measurements and stations is thus essential for a successful study of the data inhomogeneity. The importance of this "metadata" should also be emphasised.

A straightforward way to identify possible points of inhomogeneity in records is a careful study of the "methodological history" of the country in question (e.g. country-wide changes in instrumentation or times of observations and averaging methods). The background of each observing station should also be checked from station inspection reports or other relevant documents. Any changes in instrumentation or observing methods occurring at a particular station should be checked as a possible source of inhomogeneity.

Even if the observations were free of instrumental or observational inhomogeneities, the records may still show local step-like or progressive changes. Major, as well as some minor relocations of the stations, typically introduce severe inhomogeneities. In addition to horizontal moves, a station relocation often includes a change in elevation and environment. Information on station histories is of primary importance to the homogenisation process and can only be assessed station by station.

Progressive changes in the surroundings of the observation station also represent a frequent source of inhomogeneity. Many of them are connected with urbanisation and/or industrialisation and they include (i) increase in artificial heat (thermal pollution), (ii) increase in gases, smoke and dust (atmospheric pollution) and (iii) decrease in natural surfaces, evaporation and snow.

In addition to urban and industrial effects, there are many other obvious consequences of man's activities for local climates (e.g. de/reforestation, artificial lakes, etc.). Their detection and correction, however, are more difficult, because these, like real variations in climate, are generally trend-like. Apparent cyclic changes are also generally gradual and hidden amongst the real changes.

A.5.1   Homogeneity of Temperature Records

The "true" temperature of the air (i.e. the thermometer bulb in thermodynamic equilibrium with the surrounding air) cannot be measured realistically in free air because of the effect of direct, diffuse and reflected solar radiation as well as long-wave radiation from the ground and surrounding objects. Thus, thermometers need to be sheltered from radiation disturbances as well as from rain and snow. However, any shelter introduces its own disturbing factors.

Although the first international meteorological congress (held in Vienna, 1873) took a somewhat pessimistic view concerning the achievement of uniform, reliable arrangements for temperature measurements, the majority of temperature observations have been made in sufficiently similar and uniform ways to guarantee reasonably comparable results in time and space. It is also noteworthy that the present regulations for temperature measurements do not differ much from the situation in the latter half of the 1800s.

Information on station relocations is of primary importance in estimating the homogeneity of temperature records at individual stations. A site change normally causes systematic changes in temperature. Parallel observations would help to estimate the corrections needed. Unfortunately such measurements have been more an exception than the rule. Therefore, comparisons between the new and old sites have to be made with other stations, which are usually quite distant. Thus the results of comparisons, at least in the case of individual years, cannot be wholly reliable.

A site change generally involves more than a geographical shift in location, however. In many cases, a change in height or a modification in the screen/shelter may also occur. In addition, although the station might have remained at the same location (according to the coordinates and height information), minor relocations of the thermometer or screen sometimes take place, and may have an even larger effect than a major relocation.

The effects of site changes on a record are usually found alongside urbanisation effects; the two sometimes counterbalancing each other, especially if the airport is situated near the sea or a large lake.

Considering that some of the longest-term (> 100 years) Baltic Sea Basin temperature records are combinations of town and airport records, correcting for station relocations to airports in the 1940 and 1950s was one of the major steps in the homogenisation process. In all cases the correction applied was negative on an annual basis, but for monthly means the corrections were more complicated. In addition, all these records contain one or two corrections due to relocations of earlier town sites.

Despite relocations to airports, some of the longer continuous observational records still come from towns. These records are thus expected to contain a local apparent trend attributable to the development rate of the town.

A.5.2   Homogeneity of Precipitation Records

Measurements of precipitation are highly dependent on the structure of the precipitation gauge and its exposure, and consequently introduce more complications than temperature measurements. The records of precipitation amounts are always underestimates of the real amounts and they can usually be expected to contain great inhomogeneities in both space and time. In addition, most of the random-type inhomogeneities are also negative, resulting from reduction in the catch (e.g. due to a leaking gauge, spilling of water).

Of particular concern is the measurement of snowfall from conventional gauges, where large errors are known to occur. When precipitation errors are expressed as a percentage of the true precipitation, they tend to be largest in windy high latitude climates.

The principal errors in measuring precipitation are due to the following three causes: (i) wetting, i.e. the loss when transferring precipitation from the gauge (ii) evaporation from the gauge and (iii) wind (aerodynamic) effects.

The first loss is an instrument error and can be estimated quite accurately for different amounts of precipitation. Until now, very few of such corrections (see e.g. Hantel and Rubel 2001 and Annex 1.2.3) have been applied to the Baltic Sea Basin precipitation data and thus changes in the wetting error are due only to gauge changes. Evaporation from the gauge depends on inter alia instrument design and material, location of the station and other meteorological parameters (e.g. air temperature, saturation deficit, wind). This error is not accounted for in the Baltic Sea Basin routine precipitation records, either. The largest errors in precipitation measurements are due to aerodynamic causes.

The effect of instrumental errors on the homogeneity of long-term precipitation records can be considerable for winter precipitation, although the magnitude of the error depends strongly on the amount of precipitation occurring in solid forms and on the openness of the observing site. Since this information was taken into account only very crudely, the corrections should be regarded as preliminary until more exact information is available (especially regarding the openness of each station).

Precipitation measurements are extremely dependent on the local exposure of the gauge, so any changes in measuring sites or their surroundings may introduce severe inhomogeneities into the data records. This section describes relocations to airports, other types of relocations and urbanisation.

A.5.3   Homogeneity of Other Climatic Records

Generally speaking, **air pressure** records are likely to have been made with sufficiently good instruments and exposures. Relocations of stations as well as environmental changes have no significant effect on pressure records, providing that the exact heights of the barometers (plus temperature data) are available. Specific points to note in the evaluation of the homogeneity of pressure records include (i) the date of introduction of the correction to standard gravity, and (ii) reduction to mean sea-level and information on the barometer height of each station.

Wind observations have been based on various wind vanes and anemometers. However, the most serious sources of inhomogeneities in long-term records arise from the relocation of observation sites or from environmental changes in the vicinity of stations. Routine wind observations measured about 10 m above the surface are very sensitive to local topography and obstacles. Any changes in these conditions should be checked as a possible break in data homogeneity.

Observations of total **cloudiness** are expected to contain serious inhomogeneities. A change in the estimation scales should also be noted when using the data. The considerable inhomogeneities that occur in individual records are due to the subjectivity of cloudiness observations as well as to the openness of the station.

It is also possible to study changes in cloudiness with the help of measurements of **sunshine** duration. Indeed, sunshine information provides a better indication of the incoming radiation to the surface than cloudiness, because the cloudiness observations do not discriminate between different types and thicknesses of cloud, which can have a large effect on the transmission of radiation. Compared with the other climatic elements, the inhomogeneities of sunshine duration data, however, are relatively small and the data since the 1960s are quite reliable.

The examples selected above are from atmospheric elements, but terrestrial and oceanographic elements contain similar types of inconsistencies.

A.6   Climate Models and Scenarios

Burkhardt Rockel

Climate models are based on mathematical equations that describe the physical behaviour and evolution of the atmosphere and ocean, including more or less complex parameterisations of physical processes. With these models, future climate can be predicted and past climate can be hindcasted to a certain precision. Besides the effect on the accuracy that lies in the concept of the model itself (e.g. numerical schemes, parameterisation of processes), there are factors which influence the results of the model simulations but cannot be predicted by the model on its own. These factors are called "external forcings" and can be of natural (e.g. changes in incident solar radiation due to variations in the activity of the sun, volcanic eruptions) or anthropogenic type.

The natural external forcings have been, at least to some extent, known for the past centuries and can be used in model simulations of the past climate. However, they are unknown for the future and can hardly be predicted. Anthropogenic external forcings are also unknown for future times, but they can at least be assessed by assumptions of different kinds of future behaviour of man. The Intergovernmental Panel on Climate Change (IPCC)

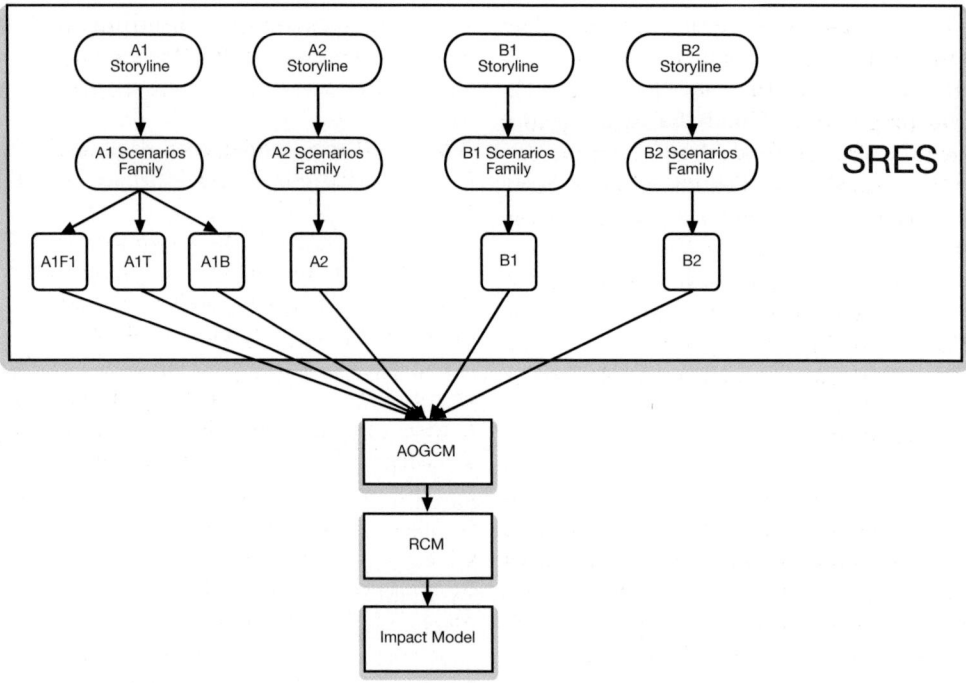

Fig. A.38. SRES scenarios and climate models

developed different kinds of qualitative assumptions, called story lines, for the future and deduced several quantitative future scenarios published in the Special Report on Emission Scenarios (SRES). Figure A.38 shows a schematic sketch for the relationship between story lines, scenarios, and climate models. In the following sections a concise overview is given on scenarios and climate models.

A.6.1 The SRES Emissions Scenarios

Emissions scenarios are plausible representations of the future development of emissions of greenhouse gases and aerosol precursors, based on coherent and internally consistent sets of assumptions about demographic, socio-economic, and technological changes in the future.

The SRES scenarios (Nakićenović and Swart 2000) were built around four narrative storylines, A1, A2, B1 and B2, each based on different assumptions about the factors that drive the development of human society in the 21st century. Several more detailed scenarios were formulated within each storyline. Six of these, commonly referred to as A1FI, A1T, A1B, A2, B1 and B2, were chosen by the IPCC as illustrative marker scenarios. In general, in the world described by the A storylines people strive after personal wealth

rather than environmental quality. In the B storylines, by contrast, sustainable development is pursued.

A.6.1.1 *A1FI, A1T and A1B Scenarios*

The A1 storyline describes a world of very rapid economic growth and efficient international cooperation. Technological development is rapid and new innovations are distributed to developing countries faster than today. Increasing economical well-being leads to decreasing fertility in the developing world, and the global population declines to 7.1 billion in the year 2050 after peaking at about 8.7 billion after the year 2050. The A1FI, A1T and A1B illustrative scenarios describe alternative directions of technological change in the energy system, and are therefore quite different in terms of the greenhouse gas emissions. In A1FI, energy production remains highly dependent on fossil fuels throughout the century, whereas A1T represents a rapid migration toward non-fossil energy sources. A1B is intermediate between these extreme cases.

A.6.1.2 *A2 Scenario*

In the A2 storyline scenarios, the world is characterised by economical blocks that are more inclined to defend their own special interests than to co-operate with each other. As a result, economical growth is slower than in A1, particularly in the developing world. The distribution of new environmentally efficient technologies to the developing world is also slower. The global population increases continuously, reaching 15 billion in the year 2100.

Although the *per capita* economic growth is relatively slow, the increasing population and slow introduction of non-fossil energy sources lead to a large increase in global greenhouse gas emissions.

A.6.1.3 *B1 Scenario*

The B1 storyline is characterized by efficient international co-operation and rapid distribution of new technologies and by the same evolution of global population as A1. However, technological development is driven more strongly by environmental values than in the A1 and A2 storylines. Economical growth is slightly slower, and the gap between the developing and the industrialised world decreases more slowly than in A1, but the introduction of clean and resource-efficient technologies is faster. Furthermore, there is a rapid change in economic structures toward a service and information society. As a result of these changes, greenhouse gas emissions are reduced below the present-day level by the end of the 21st century.

A.6.1.4 *B2 Scenario*

The B2 storyline scenarios share features from both A2 and B1. International co-operation is less efficient and the distribution of new technologies is slower than in A1 and B1. The global population increases continuously but less rapidly than in A2, reaching 10.4 billion in 2100. Like B1, the B2 scenario is also oriented towards environmental protection and social equity, but the development of environmentally friendly technologies proceeds more slowly than in B1. As a result, greenhouse gas emissions continue to grow throughout the 21st century, although at a substantially slower rate than in the A2 and A1FI scenarios.

A.6.2 Climate Models

The behaviour of the climate system can be studied and simulated by using various types of climate models. The results represented in this assessment report are mainly based on simulations made with coupled atmosphere–ocean general circulations models (AOGCMs or GCMs, see also Chap. 3) and regional climate models (RCMs).

A.6.2.1 *Atmosphere–Ocean General Circulation Models (GCMs)*

GCMs are the most advanced tool developed for studying climate change on global and large regional scales. These models simulate the three-dimensional time evolution of atmospheric and oceanic conditions based upon physical laws expressed by mathematical equations. The submodels for the atmosphere and the ocean interact with each other and with separate model components simulating the sea ice and land surface conditions. The atmospheric components of the GCMs used for this assessment report typically have a horizontal resolution of about 300 km with some 10 to 30 levels in the vertical. The resolution of the ocean models is similar or somewhat better. Some but not all GCMs use so-called flux adjustments to artificially add or remove energy, freshwater and momentum at the atmosphere-ocean interface. Flux adjustments improve the simulation of present-day climate, but many modellers find them undesirable because of their unphysical nature. Although the adjustments are kept constant with time, they may also indirectly modify the simulated climate changes. However, there is little evidence of any systematic differences in climate change between flux-adjusted and non-flux-adjusted models. A list of global models referred to in this book is provided in Table 3.1.

A.6.2.2 *Regional Climate Models (RCMs)*

RCMs are used to simulate the climate in some area with a higher horizontal resolution (typically 20–50 km) than is computationally feasible in global GCMs. This allows a more detailed representation of the local physical geography, such as mountain ranges and the land-sea distribution, as well as a more detailed representation of weather systems. An RCM only covers a limited part of the world and is therefore dependent on boundary conditions provided by a global climate model. For this atmospheric quantities (typically temperature, wind, moisture and cloud water) and surface quantities (typically pressure, temperature, moisture, snow amount, sea ice and others) of the GCM are at first interpolated onto the RCM grid. These boundary data are then used in the RCM in mainly two ways. The most common one is

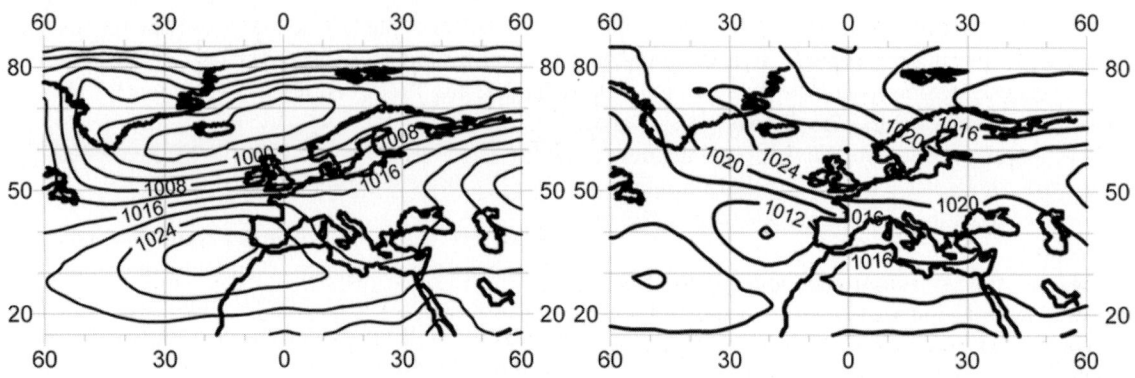

Fig. A.39. Mean sea level pressure SLP (hPa) in the North Atlantic – European sector (20° N to 80° N, 60° W to 60° E) during a positive (*left*, January 1995) and a negative (*right*, January 1963) phase of the NAO. SLP data from CRU (Jones 1997)

the lateral boundary formulation by Davis (1976), where a weighted mean of GCM and RCM data replaces the actual RCM data in a zone of typically 8–10 grid boxes at the lateral boundaries of the RCM area. This zone is called boundary or sponge zone. The weight for the GCM data decreases from 1 at the outermost grid boxes to 0 at the innermost of the boundary zone. The weight for the RCM is vice versa. Additional information of the GCM can be transferred to the RCM by the spectral nudging method (e.g. Waldron et al. 1996; von Storch et al. 2000; Miguez-Macho et al. 2004). In a first step GCM and RCM grid data are transferred into spectral data. The low wave numbers of the RCM data fields are then replaced by those of the GCM. In the last step the combined spectral data are transferred back into grid data. The surface height and land/sea mask of a RCM grid box are determined from high resolution observational data sets. A list of regional models referred to in this report is provided in Table 3.2 (Chap. 3)

GCMs and RCMs provide future assessments of quantities like temperature, wind, precipitation and so forth. However, they cannot describe the influence of these changes e.g. on the environment or the consequences for agriculture. This can be performed by impact models (e.g. crop models, hydrology models) which generally run on a local scale and take the quantities provides by the climate models as input.

A.7 North Atlantic Oscillation and Arctic Oscillation

Joanna Wibig

The North Atlantic Oscillation (NAO) is a leading mode of circulation variability over the North Atlantic mid-latitudes. At sea level it manifests itself as a large scale mass alternation between the Subtropical High and the Polar Low (Walker 1924; van Loon and Rogers 1978). Because in the Northern Hemisphere (NH) air flows counterclockwise around cyclones and clockwise around anticyclones, the high pressure gradient between the Icelandic Low and the Azores High results in strong westerly air flow over the eastern North Atlantic and Europe (Fig. A.39; January 1995). In the negative phase of the NAO, both pressure systems are weak and so are the westerlies. Complete reversal of the pressure pattern, with pressure near Iceland higher than in the vicinity of the Azores, sometimes occurs but is very rare. Such a situation is connected with easterly winds in the midlatitudes of the North Atlantic, blocking episodes and extremely severe winters in Europe (Fig. A.39, January 1963).

There is a great variety of concepts on how to measure the strength of the NAO. The two point normalized pressure difference is the one most often used. Rogers (1984) used the SLP series from Ponta Delgada at the Azores and Stykkisholmur or Akureyri at Iceland, Hurrell (1995) made use of Lisbon and Stykkisholmur (Fig. A.40). Jones et al. (1997) used series from Gibraltar and compiled records from the vicinity of Reykjavik and extended the NAOI record back to 1821.

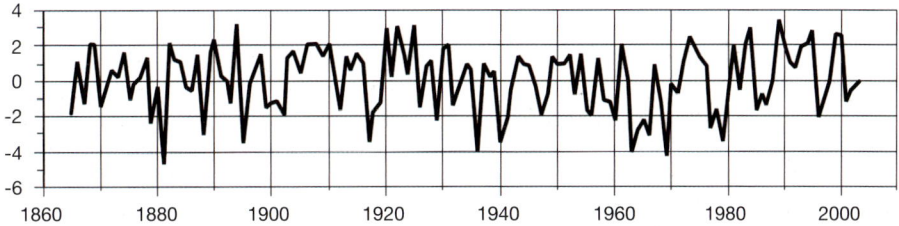

Fig. A.40. Winter (DJFM) index of the NAO based on the difference of normalized sea level pressure (SLP) between Lisbon, Portugal and Stykkisholmur/Reykjavik, Iceland since 1864. The SLP anomalies at each station were normalized relative to the 120-year period 1864–1983. NAO Index Data provided by the Climate Analysis Section, NCAR, Boulder, USA, see Hurrell (1995)

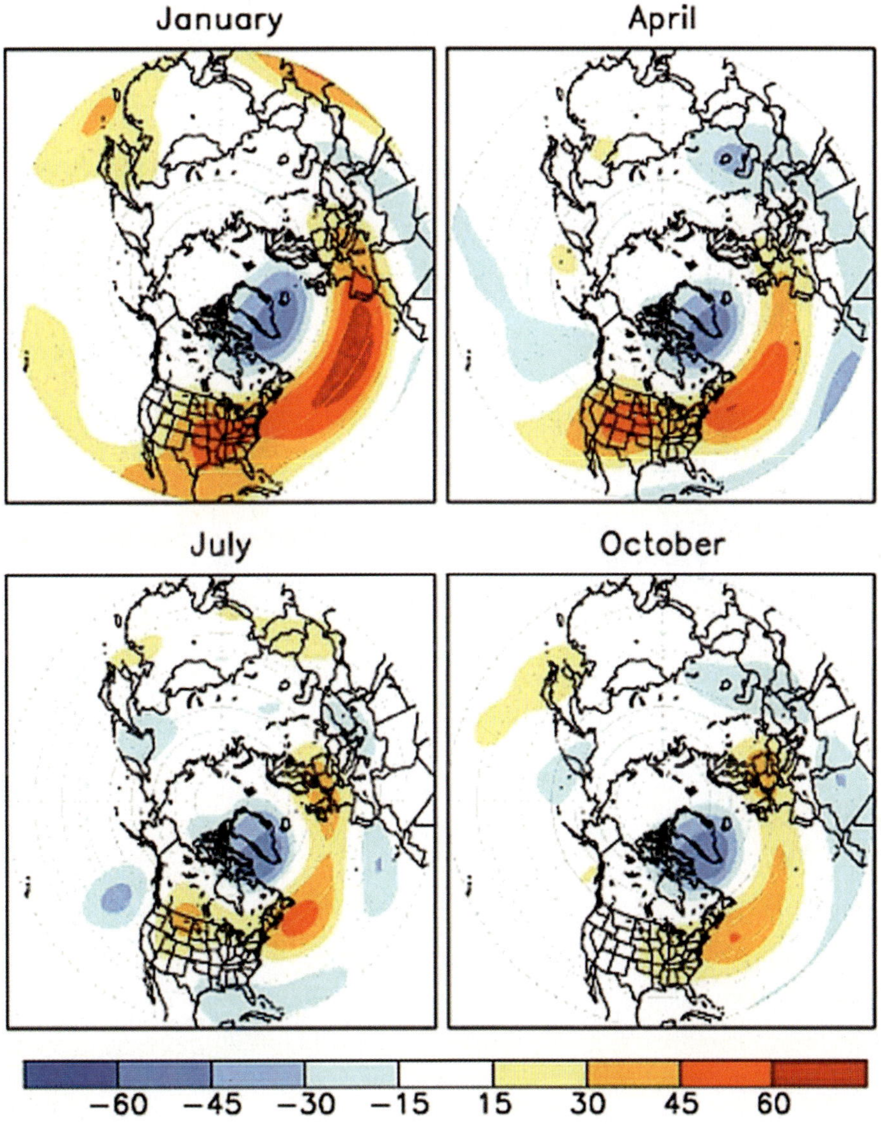

Fig. A.41. The monthly patterns of NAO presented as maps of correlation coefficients (×100) between principal component related to NAO and geopotential heights at the 500 hPa level (from www.cpc.ncep.noaa.gov/data/teledoc/; see also Bell 2007)

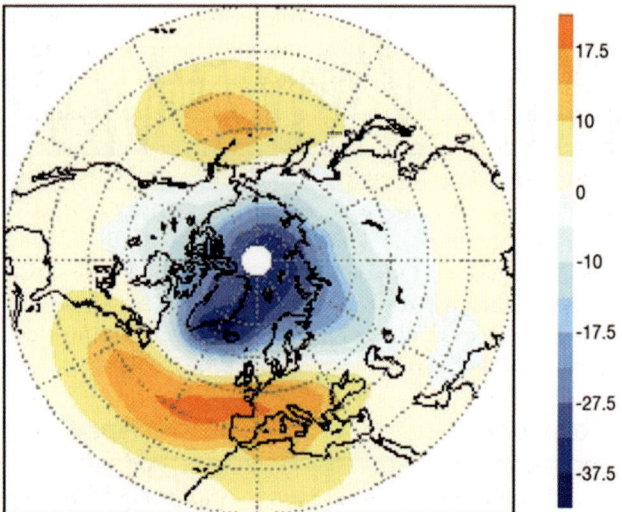

Fig. A.42. The surface signature of the Northern Hemisphere annular node (NAM). The NAM is defined here as the leading EOF of the Northern Hemisphere monthly mean 1000 hPa height anomalies. Units are m/std of the principal component time series (adapted from Thompson and Wallace 2000)

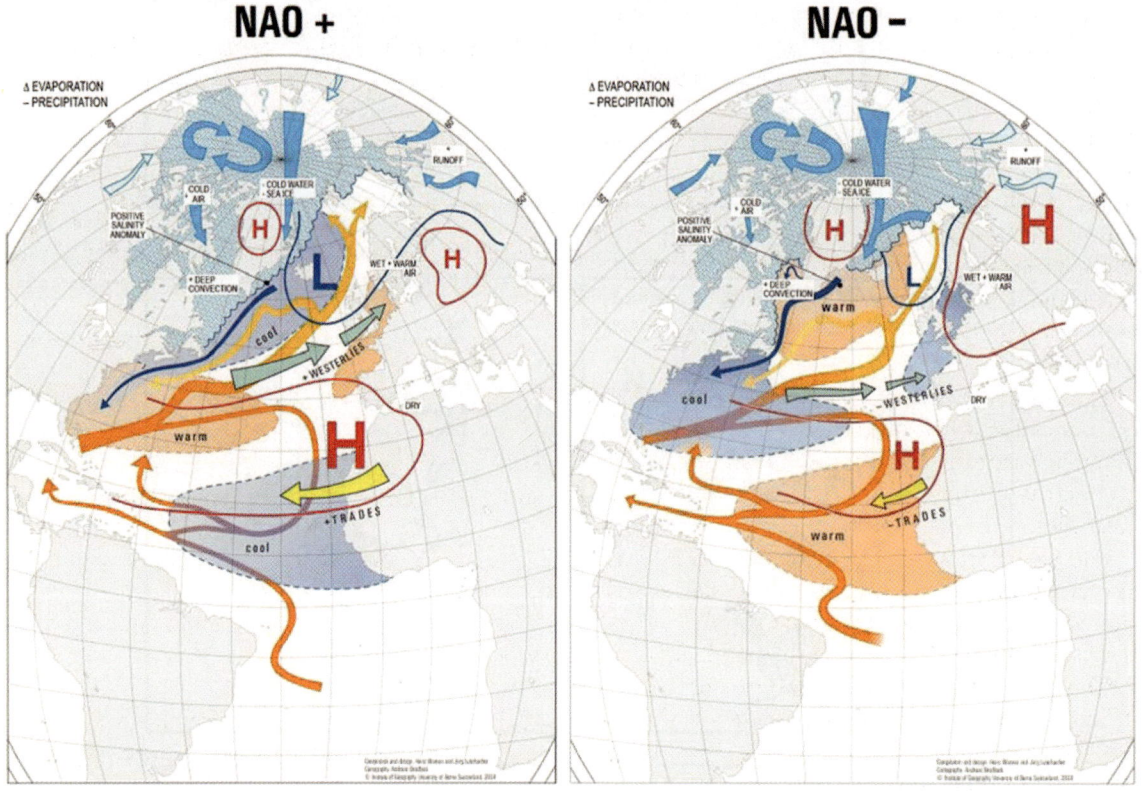

Fig. A.43. Weather conditions during positive (*left*) and negative (*right*) phase of the NAO (from Wanner et al. 2001)

The NAO can also be distinguished as an atmospheric teleconnection pattern. It is evident throughout the year in the NH, but its amplitude is largest during winter, when it accounts for about one-third of total SLP variance over the North Atlantic. A lot of different methods allow for identification of teleconnection patterns (Wallace and Gutzler 1981; Barnston and Livezey 1987). The NAO mode can be identified throughout the year, although its spatial pattern varies from season to season (Fig. A.41). It reveals the seasonal shift of the "centers of action", because the eigenvectors are constructed to explain maximum variance of pressure or geopotential height field.

Recently, another pattern, closely related to the NAO has been distinguished (Thompson and Wallace 1998, 2001). It is known as the Arctic Oscillation (AO) or Northern Hemisphere annular mode (NAM). AO is defined as a leading eigenvector of the monthly sea level pressure (SLP) field north of $20°$ N weighted by area (Fig. A.42). The spatial pattern of the AO in its positive phase has a strong low pressure center over the Arctic and a zonal high pressure band in the subtropics with two distinct centers, over the North Atlantic and the North Pacific. The spatial patterns of AO and NAO are very similar in the Atlantic sector. Some authors state that the NAO is a regional representation of the AO related with a more global pattern in the NH extratropics (Delworth and Dixon 2000). Others suggest that the NAO and AO represent the same phenomenon, for which different descriptions of dynamical processes are used (Wallace 2000).

The positive and negative phases of the NAO are associated with different spatial distributions of temperature and precipitation anomalies, not only across Europe, but across the whole NH extratropics.

The well developed Icelandic Low results in a flow of warm and wet air into north-western Europe whereas cold and relatively dry air comes to eastern Greenland. In the positive phase of the NAO in winter, temperature is above normal in all of Europe except its southern part, much of northern Eurasia and the central and western United States (US). The below normal temperature occurs in the northeastern part of the North America, southern Europe and northern Africa and over the Northern Pacific (Fig. A.43).

The NAO exerts a strong influence on winter precipitation also. In its positive phase above normal precipitation occurs in northern Europe and the eastern US, whereas a water deficit occurs in southern Europe, northern Africa and the northeastern part of North America. The storm track across the North Atlantic is shifted north. During the negative NAO phase, the storms wander more southerly along the Mediterranean region, bringing above normal precipitation to the Mediterranean area and Black Sea, whereas northern Europe then has precipitation below the average.

The NAO exhibits considerable seasonal and interannual variability, with prolonged periods of domination of positive or negative phases exerting a strong influence on different components of the ocean–atmosphere–sea-ice system: the localisation of warm and cool pools in the North Atlantic; the intensity of the subtropical and subpolar gyres, the Gulf Stream, the Mid-Atlantic and West Norwegian currents; the formation of sea-ice in the north (also in the Baltic Sea); runoff from big Siberian and Canadian rivers and the freshwater balance of the polar basin and many others (Wanner et al. 2001).

A.8 Statistical Background: Testing for Trends and Change Points (Jumps)

Hans von Storch, Anders Omstedt

A fundamental conceptual problem with trends and change points is that the statistical expression "a significant trend" or "a significant change point" is understood by some not as formally defined but by the everyday language meaning of the words "change point" and "trend" (Annex 8.1). There are well established procedures in the statistical literature to determine whether a given limited time series contains such instationarities as change points and trends. These tests almost always assume serially independent data, which for most physical and ecological environmental variables is not fulfilled (Annex 8.2). Therefore, pre-whitening and Monte Carlo methods need to be employed (Annex 8.3).

A.8.1 A Trend or a Change Point – a Property of a Random Process Generating Limited Time Series

Two key concepts in the description of nonstationary time series are jumps, or change points, and trends. Intuitively, these terms are quite clear, since they have a meaning in every day language. For instance, in the American Heritage Dictionary, the word "trend" is explained as "the general direction in which something tends to move" and "a general tendency or inclination", while a "change point" is explained as "a point of discontinuity, change, or cessation".

The meanings of the everyday language expressions imply changes beyond the range of recorded experience, in particular into the future. A trend means that we see the system of interest to have undergone systematic changes in the recent, documented past, and it is assumed that this tendency will continue for some time into the foreseeable future. It is often a trend "towards" something, e.g., higher prices or warmer temperatures. Similarly, if a development has a change point, then the new state will continue into the future, at least for some time. That is, in this language, the two terms contain a prediction of the foreseeable future.

The statistical definition of these terms is different, and the continuous blending of the two often causes difficulties in discussions about changing conditions.

In statistical thinking, a time series $X(t)$ of length T has a **trend** if

$$X(t) = \alpha t + n(t) \quad \text{with} \quad t = 0, \ldots T$$

is a valid description of $X(t)$; here α is a free parameter, t usually the time, and $n(t)$ a stationary random variable[18]. "Stationary" means that the random process generating $n(t)$ has the same properties for all considered t. Obviously, $X(t)$ is not stationary if $\alpha \neq 0$. Note that nothing is assumed about a state of $X(t)$ at times t prior to 0 or after T.

The time series $X(t)$ has a change point at time $t^\star$ if

$$X(t) = \alpha_1 + n(t) \quad \text{for} \quad t < t^\star \quad \text{and}$$
$$X(t) = \alpha_2 + n(t) \quad \text{for} \quad t \geq t^\star$$

is a valid description[19]. Here, α_1 and α_2 are constants and $n(t)$ a stationary process. Again, nothing is implied for the state of X prior to 0 and after T.

The validity of the expressions for a trend or a change point is examined in the formalism of a

statistical test, which features the properties "no trend" or "no change point" as null hypothesis and the properties "non-zero trend" or "existence of a change point at $t^\star$ with $0 < t^\star < T$" as alternative hypothesis.

To do so, an arithmetic expression $S(X, T)$ of the $T + 1$ data points $X(0)$ to $X(T)$ named "test statistics" is derived, which results in large numbers if the alternative hypothesis prevails and small numbers if the null hypothesis is a consistent description[20]. Then, the distribution of S is derived for the population of cases which satisfy the null hypothesis. If S_{95} is the 95%-ile of the distribution of S[21] then the null hypothesis is rejected if $S(X, T) > S_{95}$.

Rejecting the null hypothesis means to accept the alternative hypothesis. Note that the alternative hypothesis is not necessarily the negation of the null hypothesis. The latter is

$$X(t) = n(t)$$

so that the rejection would be

$$X(t) \neq n(t)$$

or "$X(t)$ is not a stationary process", which is not equivalent to either

$$X(t) = \alpha t + n(t) \quad \text{or}$$
$$X(t) = \alpha_1 + n(t) \quad \text{for} \quad t < t^\star \quad \text{and}$$
$$X(t) = \alpha_2 + n(t) \quad \text{for} \quad t \geq t^\star$$

It needs other arguments, preferably physical or ecological ones, to conclude that these specific alternative hypotheses are a rational choice. Also, the definition of S should be geared towards large values, when the specific alternative hypothesis is valid.

Rejection of the null hypothesis or acceptance of the alternative hypothesis does not imply that the trend or the state after the change point will continue beyond T. Instead it means that we assign the process, which has generated the finite time series the property described by the alternative hypothesis. Thus, if we would generate another limited time series, this would also have a trend, or a

---

[18]This is a linear trend in the mean. Clearly, one may construct also trends in the dispersion (variance) and other statistical parameters. Also, one may assume different forms of the trend, such as a cyclic $X(t) = sin(\alpha t/P) + n(t)$ or any other form.

[19]In principle this may be seen also as a "trend" with a step-function as trend. One could also define change points in terms of variability and other statistical properties.

[20]More precisely, S should attain numbers in a certain numerical range, when the data are inconsistent with the null hypothesis, and another range if they are consistent. For the sake of simpler language we assume that the former range contains small numbers, and the later large numbers.

[21]Or any other high percentile, which is subjectively chosen as sufficient to consider the data $X(0) \ldots X(T)$ to be inconsistent with null hypothesis.

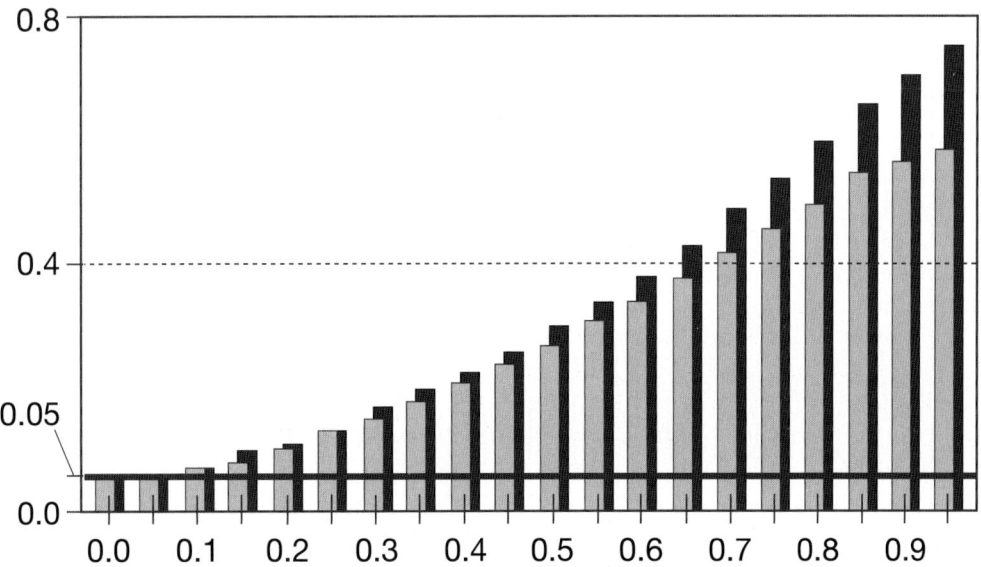

Fig. A.44. Rate of false rejections of the null hypothesis "Time series $X(1)\ldots X(T)$" contains no change point" with $T = 100$ (*light*) and $T = 500$ (*black*), when $X(t) = n(t)$ is generated by a red noise process with a memory term l given by the horizontal axis (adapted from Busuioc and von Storch 1996)

change point, in the limited time $[0,\,T]$. The random process, or the "dice" we are rolling, is generating a sequence of $T+1$ numbers $X(0)\ldots X(T)$. If we "roll the dice" again, then not the numbers $X(T+1)\ldots X(2T)$ are generated but a completely independent sequence of $T+1$ numbers.

A.8.2 Serial Dependency and its Effect on the Accuracy of Conventional Tests

There are a number of conventional test statistics, and their distributions, given in the statistical literature. For instance the Pettit-test (Pettit 1979) is often used in meteorological quarters, while the Mann-Kendall text (Mann 1945; Sneyers 1975) is popular for detecting trends. Other non-parametric trend tests are the Cox and Stuart test, the Daniel test and others (refer to e.g. Conover 1971). Thus, it seems that the detection of change points and trends should not pose a methodical challenge as standard routines can be used – it seems.

However, while this may be true in many applications, it is not true in most climatic applications. The reason is that these standard approaches assume that there is no serial dependence among the $n(t)$, i.e., that $n(t)$ and $n(t+\Delta)$ are independent. Because of memory in the physical (or ecological) processes, this condition is hardly fulfilled.

Instead, the lag correlation

$$c(\Delta) = \frac{1}{t-\Delta} \sum_{t=0}^{T-\Delta} X'(t)X'(t+\Delta)$$

is in most cases not zero, even if small[22] The violation of the condition of serial independence makes the test to reject a correct null hypothesis more frequently than formally stipulated by the adopted percentile S_{95}[23].

Monte Carlo experiments, in which serially correlated data without a trend or without a change point are examined with the Mann-Kendall test (Kulkarni and von Storch 1995) and with the Pettit-test (Busuioc and von Storch 1996), give an impression of how serious the problem is.

For instance, if the serial correlation is related to a short term "red" memory, i.e.,

$$n(t+1) = \lambda n(t) + m(t)$$

with a constant "memory term" λ and a stationary $m(t)$[24], the rate of false rejections, which was

---

[22]X' represents the normalized series of X, centered and rescaled to variance one.

[23]The test becomes "liberal" – and thus plainly false.

[24]This "red noise" is a so-called autoregressive process of first order; it is equivalent to a first order differential equation with a linear damping and a random forcing; see also von Storch and Zwiers, 2002. If $\lambda = 0$ then there is no serial correlation and the noise is called "white"; $m(t)$ is assumed to be white noise.

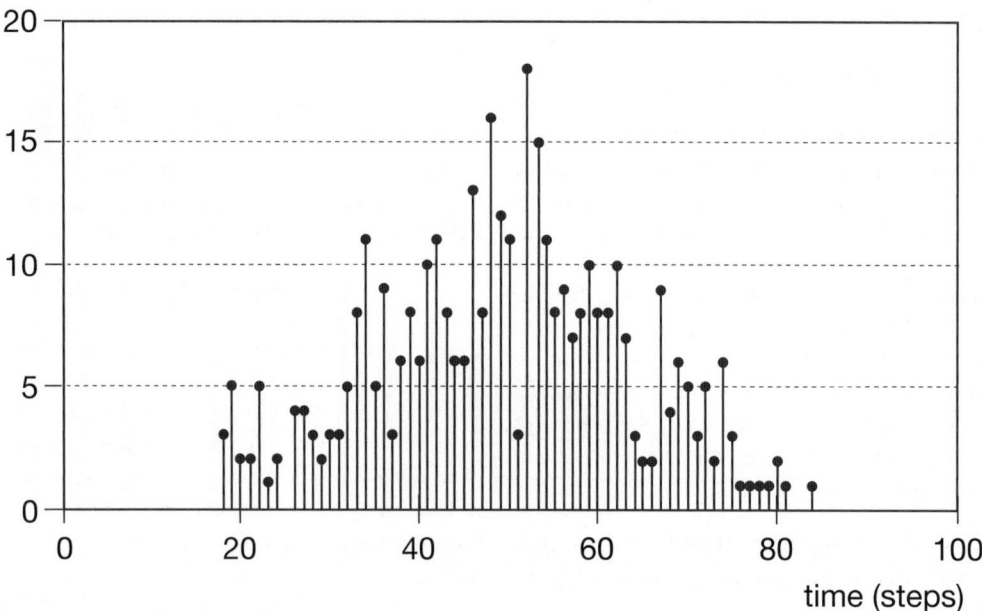

Fig. A.45. Frequency (in per mille) of false detections of a change point at the $t^\star$ values given at the horizontal axis. 1000 Monte Carlo cases have been evaluated, all generated by a linear trend overlaid with white noise ($\alpha = 0.005$; variance of noise $= 1$) (adapted from Busuioc and von Storch 1996)

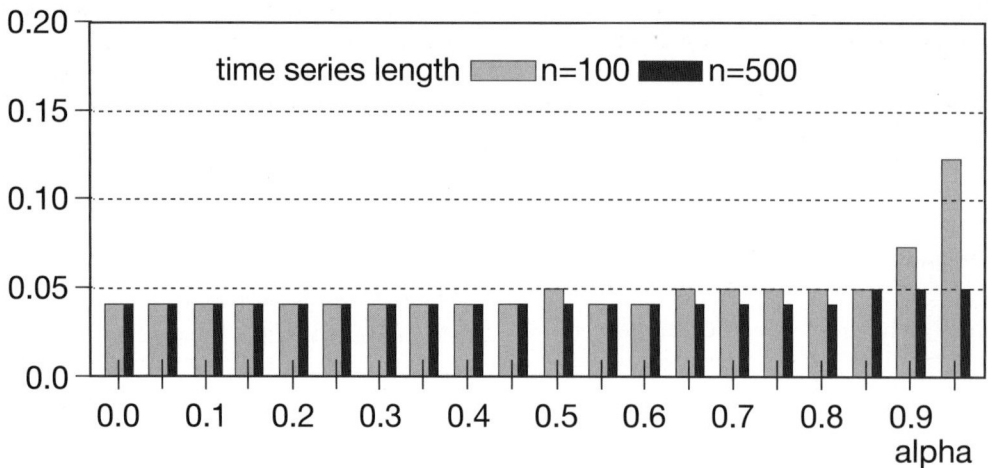

Fig. A.46. Rate of false rejections of the null hypothesis "Time series $X(1) \ldots X(T)$ contains no change point" with $T = 100$ (*light*) and $T = 500$ (*black*), when $X(t) = n(t)$ is generated by a red noise process with a memory term l given by the horizontal axis, when the Pettitt-test is applied to the pre-whitened, i.e., applied to $Y(t)$ instead of $X(t)$ (adapted from Busuioc and von Storch 1996)

stipulated to be 5%, of the Pettit-test dealing with the detection of change points, is markedly larger than 5% even for small $\lambda \geq 0.15$ (Fig. A.44).

For large λ, say 0.8, in more than 50% of applications of the Pettit test to a series without a change point and without other instationarities, the test indicates falsely that a change point is con-

tained in the data. A similar demonstration was provided by Kulkarni and von Storch (1995) for the case of the Mann-Kendall test of linear trends (not shown).

When the serial correlation is related to a linear trend, similarly false assessments of the alternative hypothesis happen. Figure A.45 shows the result for 1,000 cases of series with a trend given by

$$X(t) = \alpha t + n(t)$$

(Busuioc and von Storch, 1996). These series all do not fulfil the null hypothesis of stationarity, but they are all free of a change point. The time series have a length of $T = 100$, and the diagrams display how often a $t^\star$ is falsely associated with a change point. They all cluster in the middle; for instance the mid point $t^\star = 50$ is chosen in 1.5% of all 1000 cases, while $t^\star$'s smaller than 40 or larger than 60 are picked at a rate of on average 0.5%. $t^\star$'s smaller than 20 and larger than 80 never appear. In total, a false change point is identified in 38% of all cases.

A.8.3 Pre-whitening and Monte Carlo Approaches to Overcome the Serial Dependency Problem

At least two possibilities exist to overcome the problems related to serial dependence. One is pre-whitening and the other is Monte Carlo simulations.

"Pre-whitening" means to try to filter out the serial dependence. The detail of the filter depends on what is known, or assumed, about the serial dependence. For instance, if the serial dependence originates from a linear trend, then a difference filter, i.e.,

$$Y(t) = X(t+1) - X(t)$$

would be appropriate. If the serial dependence may be described by an auto-regressive process with memory λ, then a suitable filter is

$$Y(t) = X(t+1) - \lambda X(t)$$

Figure A.46 shows as an example the frequency of false rejections of the null hypothesis of no change point for different values of λ, when $Y(t)$ is tested instead of $X(t)$. Figure A.44 has demonstrated that with increasing λ's this rejection rate grows well above the stipulated level (of 5%). Except for very large values of λ, the test operates as required.

Another possibility is to construct a large ensemble of limited time series $X(0)\ldots X(T)$ with the same statistical properties of the series to be tested – without the specific property the test is dealing with, i.e., without a change point or without a trend. Then, by construction each member of the ensemble fulfils the null hypothesis. Then, for each member the test statistic S is determined, and finally by polling all S-values, an empirical distribution of the test statistic is derived.

A.9 References

Aarup T (2002) Transparency of the North Sea and Baltic Sea – a Secchi Depth data mining study. Oceanologia 44:323–337

Ackefors H (1969) Seasonal and vertical distribution of the zooplankton in the Askö area (northern Baltic proper) in relation to hydrographical conditions. Oikos 20:480–492

Ærtebjerg G, Andersen JH, Hansen OS (2003) Nutrients and eutrophication in Danish Marine Waters. A challenge for science and management. Ministry of Environment, National Environmental Research Institute, Denmark

Agnew MD, Palutikof JP (2001) Impacts of climate on the demand for tourism. In: Matzarakis A, de Freitas CR (eds) Proceedings of the First International Workshop on Climate, Tourism and Recreation. International Society of Biometeorology, Commission on Climate, Tourism and Recreation. Porto Carras, Halkidiki, Greece. WP4:1–10

Alenius P, Myrberg K, Nekrasov A (1998) The physical oceanography of the Gulf of Finland: A review. Boreal Env Res 3:97–125

Andrejev O, Myrberg K, Alenius P, Lundberg PA (2004a) Mean circulation and water exchange in the Gulf of Finland – A study based on three-dimensional modelling. Boreal Env Res 9:1–16

Andrejev O, Myrberg K, Lundberg PA (2004b) Age and renewal time of water masses in a semi-enclosed basin application to the Gulf of Finland. Tellus A 56:548–558

Andrén E, Andrén T, Sohlenius G (2000) The Holocene history of the southwestern Baltic Sea as reflected in a sediment core from the Bornholm Basin. Boreas 29:233–250

Andrén T (2003a) Baltiska Issjön – eller hur det började (The Baltic Ice Lake – or how it began). Havsutsikt 1/2003:4–5 (in Swedish)

Andrén T (2003b) Yoldiahavet – en viktig parentes (The Yoldia Sea – an important period). Havsutsikt 2/2003:6–7 (in Swedish)

Andrén T (2003c) Ancylussjön – fortfarande ett mysterium (The Ancylus Lake – still a mystery). Havsutsikt 3/2003:8–9 (in Swedish)

Andrén T (2004) Littorinahavet – en salt historia (The Littorina Sea – A salt history). Havsutsikt 1/2004:8–9

Andrén T, Björck J, Johnsen S (1999) Correlation of Swedish glacial varves with the Greenland (GRIP) oxygen isotope record. J Quat Sci 14:361–371

Aquaflow (2002) Miljøeffekter af havbrug i Finland (Environmental effects of aquacultures in Finland). Aquaflow ref: TL2002–088 (in Danish)

Arpe K, Hagemann S, Jacob D, Roeckner E (2005) The realisms of the ECHAM5 models to simulate the hydrological cycle in the Arctic and north-European area. Nordic Hydrology 36:349–367

Azam F, Fenchel T, Field JG, Gray J, Meyer-Reil LA, Thingstad F (1983) The ecological role of water-column microbes in the sea. Mar Ecol Prog Ser 10:257–263

Badeck FW, Bondeau A, Böttcher K, Doktor D, Lucht W, Schaber J, Sitch S (2004) Response of spring phenology to climate change. New Phytologist 162:295–309

Barnston AG, Livezey RE (1987) Classification seasonality and persistence of low frequency atmospheric circulation patterns. Mon Wea Rev 115:1083–1126

Bartnicki J, Gusev A, Lukewille A (2001) Atmospheric Supply of Nitrogen, Lead, Cadmium, Mercury and Lindane to the Baltic Sea in 1998. Joint EMEP centres report for HELCOM 2001

Bartnicki J, Gusev A, Barret K, Simpson K (2002) Atmospheric Supply of Nitrogen, Lead, Cadmium, Mercury and Lindane to the Baltic Sea in the period 1996 – 2000. Joint EMEP centres report for HELCOM 2002

Bartnicki J, Gusev A, Berg T, Fagerli H (2003) Atmospheric Supply of Nitrogen, Lead, Cadmium, Mercury and Lindane to the Baltic Sea in 2001. EMEP/MSC-W note 3/2003

Bartnicki J, Gusev A, Berg T, Fagerli H (2004) Atmospheric Supply of Nitrogen. Lead. Cadmium. Mercury and Lindane to the Baltic Sea in 2002. EMEP/MSC-W technical report 3/2004

Bartnicki J, Gusev A, Berg T, Fagerli H (2005) Atmospheric Supply of Nitrogen, Lead, Cadmium, Mercury and Lindane to the Baltic Sea in 2003. EMEP/MSC-W note 3/2005

Bartnicki J, Gusev A, Aas W, Fagerli H (2006) Atmospheric Supply of Nitrogen, Lead, Cadmium, Mercury and Dioxines/Furanes to the Baltic Sea in 2004. EMEP/MSC-W TECHNICAL REPORT 3/2006 OSLO, September 2006

Bazzaz FA (1990) The response of natural ecosystems to the rising global CO_2 levels. Ann Rev Ecol Systemat 21:167–196

Bell GD (2007) Climate Diagnostics Bulletin. Monthly Publication by U.S. Dept of Commerce, NOAA Climate Prediction Center

Bennike O, Jensen JB, Lemke W (1998) Fauna and flora in submarine early Holocene lake-marl deposits from the southwestern Baltic Sea. The Holocene 8:353–358

Bennike O, Jensen JB, Lemke W, Kuijpers A, Lomholt S (2004) Late- and postglacial history of the Great Belt, Denmark. Boreas 33:18–33

Benthien B (1996) Die Bäderlandschaft der südlichen Ostseeküste – ein Teil der zirkumbaltischen Erholungszone (Seaside resorts at the southern Baltic Sea coast: Part of the circum-baltic recreational zone). In: Greifswalder Beiträge zur Freizeit- und Tourismusforschung 7:7–13 (in German)

Bergh J, Linder S, Lundmark T, Elfving B (1999) The effect of water and nutrient availability on the productivity of Norway spruce in northern and southern Sweden. Forest Ecol Manag 119:51–62

Berglund BE, Sandgren P, Barnekow L, Hannon G, Jiang H, Skog G, Yu S (2005) Early Holocene history of the Baltic Sea as reflected in coastal sediments in Blekinge, southeastern Sweden. Quat Int 130:111–139

Bergström S, Carlsson B (1994) River runoff to the Baltic Sea: 1950–1990. Ambio 23:280–287

Bigano A, Goria A, Hamilton J, Tol R (2005): The Effect of Climate Change and Extreme Weather Events on Tourism. Nota Di Lavoro 30.2005. The Fondazione Eni Enrico Mattei Note di Lavoro Series, www.feem.it/Feem/Pub/Publications/WPapers/default.htm

Björck S (1995) A review of the history of the Baltic Sea, 130–80 ka BP. Quat Int 27:19–40

Björck S, Kromer B, Johnsen S, Bennike O, Hammarlund D, Lemdahl G, Possnert G, Rasmussen TL, Wohlfarth B, Hammer CU, Spurk M (1996) Synchronized terrestrial-atmospheric deglacial records around the North Atlantic. Science 274:1155–1160

Bogdanov VI, Medvedev M Yu, Solodov VA, Trapeznikov I, Yu A, Troshkov GA, Trubitsina AA, (2000) Mean monthly series of sea level observations (1777–1993) at the Kronstadt gauge. Reports of the Finnish Geodetic Institute 2000:1

Bradshaw RHW, Holmqvist BH, Cowling SA, Sykes MT (2000) The effects of climate change on the distribution and management of *Picea abies* in southern Scandinavia. Can J Forest Res 30:1992–1998

Breitzmann, KH (ed) (2004) Tourismus und Auslandstourismus im Ostseeraum (Tourism and tourism abroad in the Baltic Sea area). Selbstverlag des Ostseeinstituts, Beiträge und Informationen, Heft 14. Rostock (in German)

Breivik K, Wania F (2002a) Evaluating a model of the historical behavior of two hexachlorocyclohexanes in the Baltic Sea environment. Env Sci Technol 36:1014–1023

Breivik K, Wania F (2002b) Mass budgets pathways and equilibrium states of two hexachlorocyclohexanes in the Baltic Sea environment. Env Sci Technol 36:1024–1032

Brooks KM, Mahnken CVW (2003) Interactions of Atlantic salmon in the Pacific northwest environment II. Organic wastes. Fish Res 62:255–293

Busuioc A, von Storch H (1996) Changes in the winter precipitation in Romania and its relation to the large-scale circulation. Tellus 48A:538–552

Busuioc A, Chen D, Hellström C (2001) Temporal and spatial variability of precipitation in Sweden and its link with the large-scale atmospheric circulation. Tellus 53A:348–367

Cannell MGR (1989) Chilling thermal time and the date of flowering of trees. In: Wright C (ed) The Manipulation of Fruiting. Butterworths, London, pp. 99–113

Chen D, Hellström C (1999) The influence of the North Atlantic Oscillation on the regional temperature variability in Sweden: Spatial and temporal variations. Tellus 51A:505–516

Christensen PB, Rysgaard S, Sloth NP, Dalsgaard T, Schwærter S (2000) Sediment mineralization, nutrient fluxes, denitrification and dissimilatory nitrate reduction to ammonium in an estuarine fjord with sea cage trout farms. Aquat Microb Ecol 21:73–84

Conover WJ (1971) Practical nonparametric statistics. Wiley & Sons

Dahl E (1990) Probable effects of climatic change due to the greenhouse effect on plant productivity and survival in North Europe. In: Holten JI, Paulsen G, Oechel WC (eds) Effects of Climate Change on Terrestrial Ecosystems, Norwegian Institute for Nature Research, Trondheim, Norway, pp 81–83

Dahlström B (1995) Snow cover. In: Raab B, Vedin H (eds) Climate, lakes and rivers. National Atlas of Sweden, pp. 91–97

Davis HC (1976) A lateral boundary formulation for multi-level prediction models. Q J Roy Met Soc 102:405–418

de Pauw N, Jaspers E, Ackefors H, Wilkins N (1989) Aquaculture – A Biotechnology in Progress. Proc Europ Aquacult Soc, Bredene, Belgium, pp. 70–75

Dedkova I, Erdman L, Grigoryan S, Galperin M (1993) Assessment of airborne sulphur and nitrogen pollution of the Baltic Sea from European countries for 1987–1991. EMEP/MSC-E Moscow

Defant F (1972) Klima und Wetter der Ostsee (Climate and weather of the Baltic Sea). Kieler Meeresforsch 28:1–130 (in German)

Deleersnijder E, Campin JM, Delhez E (2001) The concept of age in marine modelling: I Theory and preliminary results. J Mar Sys 28:229–267

Delworth TL, Dixon KW (2000) Implications of the recent trend in the Arctic/North Atlantic Oscillation for the North Atlantic thermohaline circulation. J Clim 13:3721–3727

Dempster T, Sanchez-Jerez P, Bayle-Sempere JT, Gimenez-Casalduero F, Valle C (2002) Attraction of wild fish to sea-cage fish farms in the south-western Mediterranean Sea: Spatial and short-term temporal variability. Mar Ecol Progr Ser 242:237–252

Derwent R, Collins W, Johnson C, Stevenson D (2002) Global ozone concentrations and regional air quality. Env Sci Technol 36:379A-382A

Dewar RC, Belinda EM, Mcmurtrie RE (1999) Acclimation of the respiration/photosynthesis ratio to temperature: Insights from a model. Glob Change Biol 5:615–622

Döös K, Meier HEM, Döscher R (2004) The Baltic haline conveyor belt or the overturning circulation and mixing in the Baltic. Ambio 33:261–266

Duke RA, Liss PS, Merrill JT, Buat-Menard P, Hicks BB, Miller JM, Prospero JM, Arimoto R, Church TM, Ellis W, Galloway JM, Hansen L, Jickells TD, Knap AH, Reinhardt KH, Schneider B, Soudine A, Tokos JJ, Tsunogai S, Wollast R, Zhou M (1989) The input of atmospheric trace species to the world ocean. Rep Stud GESAMP 38

ECE (1993) The Environment in Europe and North America. Annotated Statistics

Edler L (1979) Phytoplankton succession in the Baltic Sea. Acta Bot Fennici 110:75–78

Eilola K, Stigebrandt A (1998) Spreading of juvenile freshwater in the Baltic proper. J Geophys Res 103,C12:27795–27807

Ekman M (1988) The world's longest continued series of sea level observations. Pure Appl Geophys 127:73–77

Ekman M (1996) A consistent map of the postglacial uplift of Fennoscandia. Terra Nova 8:158–165

Elken J (1994) Numerical study of fronts between the Baltic sub-basins. In: Proceedings of 19th Conference of the Baltic Oceanographers, Sopot 1:438–446

Elken J (1996) Deep water overflow circulation and vertical exchange in the Baltic proper. Estonian Marine Institute Report Series 6

Elken J, Pajuste M, Kõuts T (1988) On intrusive lenses and their role in mixing in the Baltic deep layers. Proceedings of the 16th Conference of the Baltic Oceanographers, Kiel, pp. 367–376

Elken J, Talpsepp L, Kõuts T, Pajuste M (1994) The role of mesoscale eddies and saline stratification in the generation of spring bloom heterogeneity in the southeastern Gotland Basin: An example from PEX '86. In: Dybern BI (ed) ICES Cooperative Research Report 201: Patchiness in the Baltic Sea, pp. 40–48

Elken J, Raudsepp U, Lips U (2003) On the estuarine transport reversal in deep layers of the Gulf of Finland. J Sea Res 49:267–274

Elmgren R (2001) Understanding human impact on the Baltic ecosystem: Changing views in recent decades. Ambio 30:222–231

EMEP (2000) Transboundary acidification and eutrophication in Europe. EMEP Summary report CCC and MSC-W, Oslo

EMEP (2003) Transboundary acidification eutrophication and ground level ozone in Europe. EMEP Status report 2003, Oslo

EMEP (2004) Transboundary level of acidification eutrophication and ground-level ozone in Europe. Joint MSC-W CCC CIAM ICP-M&M and CCE report, Oslo

EMEP (2006) Transboundary acidification eutrophication and ground level ozone in Europe from 1990 to 2004 in support for the review of the Gothenburg Protocol EMEP/MSC-W ETC/ACC ICP-Forests CCE ICP M&M report EMEP status report, 2006

Emery WJ, Thomson RE (1997) Data analysis methods in physical oceanography. Pergamon Press

Engström-Öst J, Koski M, Schmidt K, Viitasalo M, Jónasdóttir SH, Kokkonen M, Repka S, Sivonen K (2002) Effects of toxic cyanobacteria on a plankton assemblage: Community development during decay of *Nodularia spumigena*. Mar Ecol Prog Ser 232:1–14

Falarz M (2001) Zmienność wieloletnia wyst powania pokrywy śnieżnej w polskich Tatrach (Long term variability of snow cover in the Polish Tatra mountains). Folia Geogr Ser Geogr Phys 31–32:101–123 (in Polish)

Falarz M (2004) Variability and trends in the duration and depth of snow cover in Poland in the 20th century. Int J Climatol 24:1713–1727

FAO (2005) Fisheries global information system, www.fao.org

Feistel R, Nausch G, Matthäus W, Hagen E (2003) Temporal and spatial evolution of the Baltic deep water renewal in spring 2003. Oceanologia 45:623–642

Fennel W, Seifert T (1995) Kelvin wave controlled upwelling in the western Baltic. J Mar Syst 6:289–300

Fennel W, Sturm M (1992) Dynamics of the western Baltic. J Mar Syst 3:183–205

Fennel W, Seifert T, Kayser B (1991) Rossby radii and phase speeds in the Baltic Sea. Continent Shelf Res 11:23–36

Fischer H, Matthäus W (1996) The importance of the Drogden Sill in the Sound for major Baltic inflows. J Mar Syst 9:137–157

Fredén C (1986) Quaternary marine shell deposits in the region of Uddevalla and Lake Vänern. Sveriges Geologiska Undersökning. Rapporter och Meddelanden 46

Freeman M, Morén AS, Strömmer M, Linder S (2005) Climate change impacts on forests in Europe: Biological impact mechanisms. In: Kellomäki S, Leinonen S (eds) Management of European forests under changing climatic conditions. University of Joensuu, Faculty of Forestry Research, Notes 163: 46:115

GESAMP (2001) Planning and management for sustainable coastal aquaculture development. GESAMP Reports and Studies 68

Golenko NN, Beszczynska-Möller A, Piechura J, Walczowski W, Ameryk A (1999) Some results of research on internal waves in the Stolpe Sill area. Oceanologia 41:537–551

Gorokhova E, Fagerberg T, Hansson S (2004) Predation by herring (*Clupea harengus*) and sprat (*Sprattus sprattus*) on *Cercopagis pengoi* in a western Baltic Sea bay. ICES J Mar Sci 61:959–965

Grace J, Berninger F, Nagy L (2002) Impacts of climate change on the tree line. Ann Bot 90:537–544

Grasshoff K (1975) The hydrochemistry of landlocked basins and fjords. In: Riley JP, Skirrow G (eds) Chemical Oceanography (2nd ed.), pp. 455–597. Academic Press, London New York San Fransisco

Gustafsson BG (2001) Quantification of water salt oxygen and nutrient exchange of the Baltic Sea from observations in the Arkona Basin. Cont Shelf Res 21:1485–1500

Gustafsson BG, Andersson HC (2001) Modelling the exchange of the Baltic Sea from the meridional atmospheric pressure difference across the North Sea. J Geophys Res 106,C9:19731–19744

Gyllenhammar A (2004) Predictive modelling of aquatic ecosystems at different scales using mass-balances and GIS. PhD thesis, Uppsala Univ

Gyllenhammar A, Håkanson L (2005) Environmental consequence analyses of fish farms emissions related to different scales and exemplified by data from the Baltic – A review. Mar Env Res 60: 211–243

Hagen E, Feistel R (2004) Observations of low-frequency current fluctuations in deep water of the Eastern Gotland Basin/Baltic Sea. J Geophys Res 109 C03044 doi:101029/2003JC002017

Håkanson L (2005) Changes to lake ecosystem structure resulting from fish cage farm emissions. Lakes and Reservoirs: Research and Management 10:71–80

Håkanson L, Boulion V (2002) The lake foodweb – modelling predation and abiotic/biotic interactions. Backhuys Publishers, Leiden

Håkanson L, Ervik A, Mäkinen T, Möller B (1988) Basic concepts concerning assessments of environmental effects of marine fish farms. Nordic Council of Ministers, NORD88:90, Copenhagen

Håkanson L, Gyllenhammar A, Brolin A (2004) A dynamic model to predict sedimentation and suspended particulate matter in coastal areas. Ecol Model 175:353–384

Håkansson B, Broman B, Dahlin H (1993) The flow of water and salt in the Sound during the Baltic major inflow event in January 1993. ICES CM 1993/C:57

Håkansson B, Alenius P, Brydsten L (1996) The physical environment in the Gulf of Bothnia. Ambio Special Report No 8:5–12

Hall POJ, Anderson LG, Holby O, Kollberg S, Samuelsson M (1990) Chemical fluxes and mass balances in a marine fish cage farm. I. Carbon. Mar Ecol Progr Ser 61:61–73

Hall POJ, Holby O, Kollberg S, Samuelsson MO (1992) Chemical fluxes and mass balances in a marine fish cage farm. 4. Nitrogen. Mar Ecol Progr Ser 89:81–91

Hamilton J, Lau M (2004) The role of climate information in tourist destination choice decision-making. Centre for Marine and Climate Research, Hamburg University, www.uni-hamburg.de/Wiss/FB/15/Sustainability/climinfo.pdf

Hansson S, Larsson U, Johansson S (1990) Selective predation by herring and mysids and zooplankton community structure in a Baltic Sea coastal area. J Plank Res 12:1099–1116

Hass H, van Loon M, Kessler K, Stern R, Matthijsen J, Sauter F, Zlatev Z, Langner J, Foltescu V, Schaap M (2003) Aerosol modelling: Results and intercomparison from European regional-scale modelling systems. EUROTRAC-2 special report, EUROTRAC International Scientific Secretariat, GSF – National Research Center for Environment and Health. Munich, Germany

Havbrugsudvalget (2003) Bilag 4–11 til udvalgets rapport (Supplement 4–11 to Commission Report). Ministeriet for Fødevarer Landbrug og Fiskeri, (in Danish)

HELCOM (1989) Deposition of airborne pollution to the Baltic Sea area 1983–1985 and 1986. Baltic Sea Env Proc 32

HELCOM (1991) Airborne pollution load to the Baltic Sea 1986–1990. Baltic Sea Env Proc 39

HELCOM (1993) The Baltic Sea Joint Comprehensive Environmental Action Programme. Helsinki, 1993. Baltic Sea Env Proc 48:2–20

HELCOM (1996) Third Periodic Assessment of the State of the Marine Environment of the Baltic Sea 1989–1993. Background document. Baltic Sea Env Proc 64B

HELCOM (1997) Airborne pollution load to the Baltic Sea 1991–1995. Baltic Sea Env Proc 69

HELCOM (2002) Environment of the Baltic Sea Area 1994–1998. Baltic Sea Env Proc 82B

HELCOM (2003a) The Baltic Marine Environment 1999–2002. Baltic Sea Env Proc 87

HELCOM (2003b) HELCOM Report on Illegal Discharges Observed During Aerial Surveillance in 2003, www.helcom.fi/stc/files/shipping/spills2003.pdf

HELCOM (2005) www.helcom.fi/stc/files/shipping/Overview%20of%20ships%20traffic.pdf

Hertel O, Ambelas SC, Brandt J, Christensen JH, Frohn M, Frydendall J (2003) Operational mapping of atmospheric nitrogen deposition to the Baltic Sea. Atmos Chem Phys 3:2083–2099

Holmer M, Kristensen E (1992) Impact of marine fish cage farming on sediment metabolism and sulfate reduction of underlying sediments. Mar Ecol Progr Ser 80:191–201

Hongisto M, Joffre S (2005) Meteorological and climatological factors affecting transport and deposition of nitrogen compounds over the Baltic Sea. Boreal Env Res 10:1–17

Hongisto M, Sofiev M, Joffre S (2003) Hilatar a limited area simulation model of acid contaminants: II Model verification and long-tern simulation results. Atmos Env 37:1549–1560

Honkanen T, Helminen H (2000) Impacts of fish farming on eutrophication: Comparisons among different characteristics of ecosystem. Intern Rev Hydrobiol 85:673–686

Houmark-Nielsen M, Kjær K (2003) Southwest Scandinavia 40–15 kyr BP: Paleogeography and environmental change. J Quat Sci 18:1–18

Hungate BA, Dukes JS, Shaw MR, Luo Y, Field CB (2003) Nitrogen and climate change. Science 302:1512–1513

Huntley B (1991) How plants respond to climate change: Migration rates individualism and the consequences for plant communities. Ann Bot 67:15–22

Hurrell JW (1995) Decadal trends in the North Atlantic Oscillation: Regional temperatures and precipitation. Science 269:676–679

Ikävalko J, Thomsen HA (1997) The Baltic Sea ice biota (March 1994): A study of the protistan community. Europ J Protistol 33:229–243

ICES (2005) ICES Advice Vol. 8, www.ices.dk/products/icesadvice2005.asp

IOC (1997) Global sea level observing system (GLOSS) Implementation Plan – 1997. Intergovernmental Oceanographic Commission Technical Series, No 50, UNESCO

Isemer HJ, Rozwadowska A (1999) Solar radiation fluxes at the surface of the Baltic Proper. Part 2: Uncertainties and comparison with simple bulk parameterisations. Oceanologia 41:147–185

Jaagus J (1996) Spatial and temporal variability of snow cover duration in Estonia. In: Punning JM (ed) Estonia Geographical Studies. Estonian Academy Publishers, Tallinn, pp 43–59

Jaagus J (1997) The impact of climate change on the snow cover pattern in Estonia. Climatic Change 36:65–77

Jacobeit J, Jönsson P, Bärring L, Beck C, Ekström M (2001) Zonal indices for Europe 1780–1995 and running correlations with temperature. Climatic Change 48:219–241

Jäderholm C, Steingrube W (1996) Das finnische Mökkiwesen – ein landestypischer Lebensstil im Umbruch? (The Finnish "Mökkis" – a change in typically Finnish life-style?). Erdkunde 50:138–148 (in German)

Jakobsen F (1995) The major inflow to the Baltic Sea during January 1993. J Mar Syst 6:227–240

Jakobsen F (1996) The dense water exchange of the Bornholm Basin in the Baltic Sea. Dt Hydr Z 48,2: 133–145

Jakobsen F, Trebuchet C (2000) Observations of the transport through the Belt Sea and an investigation of momentum balance. Continent Shelf Res 20:293–311

Jensen JB, Bennike O, Witkowski A, Lemke W, Kuijpers A (1999) Early Holocene history of the southwestern Baltic Sea: The Ancylus Lake stage. Boreas 28:437–453

Jensen JB, Kuijpers A, Bennike O, Lemke W (2002) Balkat – The Baltic Sea without frontiers. Geologi nyt fra GEUS 4/2002

Johansson M, Gorokhova E, Larsson U (2004) Annual variability in ciliate community structure potential prey and predators in the open northern Baltic Sea proper. J Plank Res 26:67–80

Jonasson S, Michelsen A, Schmidt IK (1999) Coupling of nutrient cycling and carbon dynamics in the Arctic integration of soil microbial and plant processes. Appl Soil Ecol 11:135–146

Jones C, Ullerstig A (2002) The representation of precipitation in the RCA2 model. Rossby Centre Atmosphere Model Version 2. SWECLIM Newsletter 12, pp. 27–39

Jones PD, Moberg A (2003) Hemispheric and Large-Scale Surface Air Temperature Variations: An Extensive Revision and an Update to 2001. J Clim 16:206–223

Jones PD, Jónsson T, Wheeler D (1997) Extension to the North Atlantic Oscillation using early instrumental pressure observations from Gibraltar and south-west Iceland. Int J Climatol 17:1433–1450

Jönsson A, Broman B, Rahm L (2003) Variations in the Baltic Sea wave fields. Ocean Eng 30:107–126

Jutila E, Toivonen J (1985) Food consumption of salmon post-smolts (*Salmo salar* L) in the northern part of the Gulf of Bothnia. ICES CM 1985/M:21, p. 11

Kahru M, Horstmann U, Rud O (1994) Satellite detection of increased cyanobacteria blooms in the Baltic Sea: Natural fluctuation or ecosystem change. Ambio 23:469–472

Kallis A (1995) Estonia's place in the world of actinometry. In: Meteorology in Estonia in Johannes Letzmann's times and today. Estonian Academy Publishers, Tallinn, pp 95–100

Karlsson KG (1999) Satellite sensing techniques and applications for the purpose of BALTEX. Meteorol Z 9:111–116

Karlsson KG (2001) A NOAA AVHRR cloud climatology over Scandinavia covering the period 1991–2000. SMHI Reports Meteorology and Climatology 97

Kautsky U (1995) Ecosystem processes in coastal areas of the Baltic Sea. PhD thesis, Dept of Zoology, Stockholm University

Keevallik S, Russak V (2001) Changes in the amount of low clouds in Estonia (1955–1995). Int J Climatol 21:389–397

Keevallik S, Post P, Tuulik J (1999) European circulation patterns and meteorological situation in Estonia. Theor Appl Climatol 63:117–127

Kellomäki S, Leinonen S (2005) Management of European forests under changing climatic conditions. Final Report of the Project "Silvicultural Response Strategies to Climatic Change in Management of European Forests". University of Joensuu, Faculty of Forestry Research, Notes 163:1– 4–27

Kiørboe T (1993) Turbulence phytoplankton cell size and the structure of pelagic food webs. Adv Mar Biol 29:1–72

Kitaev L, Krueger O, Sherstyukov BG, Hobe H (2005a) Indications of influence of vegetation on the snow cover distribution. Russian Meteorology and Hydrology. Allerton Press Inc, New York, 7, pp. 61–69

Kitaev L, Forland E, Razuvaev V, Tveito OE, Krueger O (2005b) Distribution of snow cover over Northern Eurasia. Nordic Hydrology 36:311–319

Kitaev L, Razuvaev VN, Heino R, Forland E (2006) Duration of snow cover over Northern Europe. Russian Meteorology and Hydrology 3, Allerton Press Inc, New York, pp. 95–100

Kivi K (1986) Annual succession of pelagic protozoans and rotifers in the Tvärminne Storfjärden, SW coast of Finland. Ophelia Suppl 4:101–110

Kjær K, Houmark-Nielsen M, Richardt N (2003) Ice-flow patterns and dispersal of Erratics at the southwestern margin of the last Scandinavian ice sheet: Imprint after paleo ice-streams. Boreas 32: 130–148

Kononen K, Kuparinen J, Mäkelä K, Laanemets J, Pavelson J, Nômmann S (1996) Initiation of cyanobacterial blooms in a frontal region at the entrance to the Gulf of Finland, Baltic Sea. Limnol Oceanogr 41:98–112

Kononen K, Huttunen M, Hällfors S, Gentien P, Lunven M, Huttula T, Laanemets J, Lilover M, Pavelson J, Stips A (2003) Development of a deep chlorophyll maximum of Heterocapsa triquetra Ehrenb. at the entrance to the Gulf of Finland. Limnol Oceanogr 48:594–607

Körner C (1998) A re-assessment of high elevation treeline positions and their explanation. Oecologia 115:445–459

Kõuts T, Omstedt A (1993) Deep water exchange in the Baltic Proper. Tellus 45A:311–324

Kõuts T, Elken J, Lips U (1990) Late autumn intensification of deep thermohaline anomalies and formation of lenses in the Gotland Deep. Proceedings of the 17th Conference of the Baltic Oceanographers, Norrköping 1990, pp. 280–293

Kramer K (1995) Phenotypic plasticity of the phenology of seven European species in relation to climate warming. Plant Cell Env 18:93–104

Kulkarni A, von Storch H (1995) Monte Carlo experiments on the effect of serial correlation on the Mann-Kendall-test of trends. Met Zeitschrift 4:82–85

Kuparinen J, Leppänen JM, Sarvala J, Sundberg A, Virtanen A (1984) Production and utilization of organic matter in a Baltic ecosystem off Tvärminne, southwest coast of Finland. Rapp P v Réun. Cons. int. Explor. Mer. 193:180–192

Kuusisto E (1984) Snow accumulation and snowmelt in Finland. Publ of the Water Research Institute 55

Kuusisto E (1995) Hydrology and hydroenergetics of the Baltic Drainage. Proceedings of the First Study Conference of BALTEX, International BALTEX Secretariat Publication 3:18–27

Kuusisto E (2005) Snow as a geographic element. In: Seppälä M (ed) The Physical Geography of Fennoscandia. Oxford University Press, pp. 160–173

Lambeck K, Chappell J (2001) Sea level change through the last glacial cycle. Science 292:679–686

Larsson U, Hobro R, Wulff F (1986) Dynamics of a phytoplankton spring bloom in a coastal area of the northern Baltic Proper. Contributions from the Askö Laboratory 30:3–32

Lass HU, Mohrholz V (2003) On the dynamics and mixing of inflowing saltwater in the Arkona Sea. J Geophys Res 108,C2 3042 doi:101029/2002JC001465

Lass HU, Talpsepp L (1993) Observations of coastal jets in the Southern Baltic. Cont Shelf Res 13: 2–3:189–203

Lass HU, Prandke H, Liljebladh B (2003) Dissipation in the Baltic proper during winter stratification. J Geophys Res 108 No C6 3187 doi:101029/2002JC001401 2003

Laurila T, Jonson JE, Langner J, Sundet J, Tuovinen JP, Bergström R, Foltescu V, Tarvainen V, Isaksen ISA (2004) Ozone exposure scenarios in the Nordic countries during the 21st century. EMEP/MSC-W Technical Report 2/2004

Ledwith M (2002) Land cover classification using SPOT Vegetation 10-day composite images – Baltic Sea Catchment basin. GLC2000 Meeting Ispra, Italy, April 18–22.2002

Lee DS, Nemitz E, Fowler D, Kingdon RD (2001) Modelling atmospheric mercury transport and deposition across Europe and the UK. Atmos Env 35:5455–5466

Lehmann A, Hinrichsen HH (2000) On the wind driven and thermohaline circulation of the Baltic Sea. Phys Chem Earth,B 25:183–189

Lehmann A, Hinrichsen HH (2001) The importance of water storage variations for water balance studies of the Baltic Sea. Phys Chem Earth, B 26:383–389

Lehmann A, Hinrichsen HH (2002) Water heat and salt exchange between the deep basins of the Baltic Sea. Boreal Env Res 7:405–415

Lehmann A, Krauss W, Hinrichsen HH (2002) Effects of remote and local atmospheric forcing on circulation and upwelling in the Baltic Sea. Tellus 54A:299–316

Lehmann A, Lorenz P, Jacob D (2004) Modelling the exceptional Baltic Sea inflow events in 2002–2003. Geophys Res Lett, vol. 31, No. 21, L21308 101029/2004GL020830

Lehtonen KK (1997) Ecophysiology of two benthic amphipod species from the northern Baltic Sea. PhD thesis, University of Helsinki. Monogr Boreal Env Res No 7:1–33

Lemke W, Jensen JB, Bennike O, Witkowski A, Kuijpers A (1999) No indication of a deeply incised Dana River between Arkona Basin and Mecklenburg Bay. Baltica 12:66–70

Leppäkoski E, Gollasch S, Gruszka P, Ojaveer H, Olenin S, Panov V (2002) The Baltic – a sea of invaders. Can J Fish Aquat Sci 59:1175–1188

Lignell R, Heiskanen AS, Kuosa H, Gundersen K, Kuuppo-Leinikki P, Pajuniemi R, Uitto A (1993) Fate of a phytoplankton spring bloom: Sedimentation and carbon flow in the planktonic food web in the northern Baltic. Mar Ecol Prog Ser 94:239–252

Liljebladh B, Stigebrandt A (1996) Observations of deepwater flow into the Baltic Sea. J Geophys Res 101,C4:8895–8911

Lilover MJ, Lips U, Laanearu J, Liljebladh B (1998) Flow regime in the Irbe Strait. Aquat Sci 60: 253–265

Lindau R (2002) Energy and water balance of the Baltic Sea derived from merchant ship observations. Boreal Env Res 7,4:417–424

Lindfors V, Joffre SM, Damski J (1993) Meteorological variability of the wet and dry deposition of sulphur and nitrogen compounds over the Baltic Sea. Water Air Soil Pollut 66:1–28

Lips U, Lilover MJ, Raudsepp U, Talpsepp L (1995) Water renewal processes and related hydrographic structures in the Gulf of Riga. In: Hydrographic studies within the Gulf of Riga Project 1993–1994. Estonian Marine Institute Report Series No 1, pp. 1–34

Lise W, Tol RSJ (2005) Impact of climate on tourist demand. Climatic Change 55:429–449

Lloyd J, Farquhar GD (1996) The CO_2 dependence of photosynthesis plant growth responses to elevated atmospheric CO_2 concentrations and their interactions with soil nutrient status. I. General principles and forest ecosystems. Funct Ecol 10:4–32

Lohmann M (2003) Über die Rolle des Wetters bei Urlaubsreiseentscheidungen (How weather affects decisions for holiday destinations). Jahrbuch der Tourismuswirtschaft Schweiz. Institut für öffentliche Dienstleistungen und Tourismus, Universität St. Gallen

Lundqvist J, Wohlfarth B (2001) Timing and east-west correlation of south Swedish ice marginal lines during the Late Weichselian. Quat Sci Rev 20:1127–148

Mäkinen T (1991) Marine Aquaculture and Environment. Nordic Council of Minister, Nord, 1991, 22, pp. 9–23

Malanson GP, Cairns DM (1997) Effects of dispersal population delays and forest fragmentation on tree migration rates. Plant Ecol 131:67–79

Mann HB (1945) Non-parametric test against trend. Econometrica 13:245–259

Marmefelt E, Omstedt A (1993) Deep water properties in the Gulf of Bothnia. Cont Shelf Res 13: 169–187

Matuszko D (2003) Cloudiness changes in Cracow in the 20th century. Int J Climatol 23:975–984

Matthäus W (1984) Climatic and seasonal variability of oceanological parameters in the Baltic Sea. Beiträge zur Meereskunde 51:29–49

Matthäus W (1995) Natural variability and human impacts reflected in long-term changes in the Baltic Deep Water conditions – a brief review. Dt Hydr Z 47:47–65

Matthäus W, Franck H (1992) Characteristics of major Baltic inflows – a statistical analysis. Cont Shelf Res 12:1375–1400

Matthäus W, Lass HU (1995) The recent salt inflow into the Baltic Sea. J Phys Oceanogr 25:280–286

Mattsson J (1996) Some comments on the barotropic flow through the Danish Straits and the division of the flow between the Belt and the Öresund. Tellus 48:456–471

McGuire AD, Melillo JM, Joyce LA (1995) The role of nitrogen in the response of forest net primary production to elevated atmospheric carbon dioxide. Ann Rev Ecol Systemat 26:473–503

Meier HEM (2005) Modeling the age of Baltic Seawater masses: Quantification and steady state sensitivity experiments. J Geophys Res 110 C02006 doi:101029/2004JC002607

Meier HEM, Kauker F (2003) Modeling decadal variability of the Baltic Sea. Part 2: The role of freshwater inflow and large-scale atmospheric circulation for salinity. J Geophys Res 108 No C11 doi: 10 1029/2003JC001799 2003; 32–1–32–10

Meier HEM, Döscher R, Broman B, Piechura J (2004) The major Baltic inflow in January 2003 and preconditioning by smaller inflows in summer/autumn 2002: A model study. Oceanologia 46,4: 557–579

Melillo JM, Steudler PA, Aber JD, Newkirk K, Lux H, Bowles FP, Catricala C, Magill A, Ahrens T, Morrisseau S (2002) Soil warming and carbon-cycle feedbacks to the climate system. Science 298:2173–2176

Miętus M (1998) The Climate of the Baltic Sea Basin. Marine Meteorology and Related Oceanographic Activities. Report 41. WMO/TD-No. 933. World Meteorological Organization

Miguez-Macho G, Stenchikov GL, Robock A (2004) Spectral nudging to eliminate the effects of domain position and geometry in regional climate model simulations. J Geophys Res 109 D13104 doi:101029/2003JD004495

Möllmann C, Kornilovs G, Sidrevics L (2000) Long-term dynamics of main mesozooplankton species in the central Baltic Sea. J Plankton Res 22:2015–2038

Möllmann C, Kornilovs G, Fetter M, Köster FW (2004) Feeding ecology of central Baltic Sea herring and sprat. J Fish Biol 65:1563–1581

Munsterhjelm R (1997) The aquatic macrophyte vegetation of flads and gloes, S coast of Finland. Acta Bot Fenn 157:1–68

Munthe J, Wangberg I, Iverfeldt A, Lindquist O, Stromberg D, Sommar J, Gardfeldt K, Petersen G, Ebinghaus R, Prestbo E (2003) Distribution of atmopsheric mercury species in Northern Europe: Final results from the MOE project. Atmos Env 37–1:9–20

Murray NB, Cannell MGR, Smith I (1989) Date of budburst of fifteen tree species in Britain following climatic warming. J Appl Ecol 26:693–700

Myrberg K, Andrejev O (2003) Main upwelling regions in the Baltic Sea – a statistical analysis based on three-dimensional modelling. Boreal Env Res 8:97–112

Nakićenović N, Swart R (2000) Emissions Scenarios. A Special Report of Working Group III of the Intergovernmental Panel on Climate Change. Cambridge University Press

Nehring D, Matthäus W (1991) Current trends in hydrographic and chemical parameters and eutrophication in the Baltic Sea. Internationale Revue der gesamten Hydrobiologie 76:297–316

Nerheim S (2004) Shear-generating motions at various length scales and frequencies in the Baltic Sea – an attempt to narrow down the problem of horizontal dispersion. Oceanologia 46,4:477–503

Niemi Å (1975) Ecology of phytoplankton in the Tvärminne area SW coast of Finland. II. Primary production and environmental conditions in the archipelago and the sea zone. Acta Bot Fennica 105: 1–73

Nissling A, Westin L (1991) Egg buoyancy of Baltic cod (*Gadus morhua*) and its implications for cod stock fluctuations in the Baltic. Mar Biol 111:33–35

Norby RJ, DeLucia EH, Gielen B, Calfapietra C, Giardina CP, King JS, Ledford J, McCarthy HR, Moore DJP, Ceulemans R, De Angelis P, Finzi AC, Karnosky DF, Kubiske ME, Lukac M, Pregitzer KS, Scarascia-Mugnozza GE, Schlesinger WH, Oren R (2005) Forest response to elevated CO_2 is conserved across a broad range of productivity. Proceedings of the National Academy of Sciences, USA 102:18052–18056

Nordli Ø, Alexandersson H, Frich P, Førland E, Heino R, Jónsson T, Tveito OE (1997) The effect of radiation screens on Nordic time series of mean temperature. Int J Climatol 17:1667–1681

Nordvarg L, Håkanson L (2002) Predicting the environmental response of fish farming in coastal areas of the Åland archipelago (Baltic Sea) using management models for coastal water planning. Aquaculture 206:217–243

Novak B, Björck S (1998) Marine seismic studies in southern Kattegatt with special emphasis on longitudinal bars and their possible relationship to the drainage of the Ancylus Lake. GFF 120:293–302

Omstedt A, Axell LB (2003) Modeling the variations of salinity and temperature in the large Gulfs of the Baltic Sea. Continent Shelf Res 23:265–294

Omstedt A, Chen D (2001) Influence of atmospheric circulation on the maximum ice extent in the Baltic Sea. J Geophys Res 106,C3:4493–4500

Omstedt A, Nohr C (2004) Calculating the water and heat balances of the Baltic Sea using ocean modelling and available meteorological hydrological and ocean data. Tellus 56A:400–414

Omstedt A, Rutgersson A (2000) Closing the water and heat cycles of the Baltic Sea. Meteorol Z 9: 55–66

Omstedt A, Marmefelt E, Murthy CR (1993) Some flow charachteristics of the coastal boundary layer in the Bothnian Sea. Aqua Fennica 23,1:5–16

Omstedt A, Elken J, Lehmann A, Piechura J (2004) Knowledge of the Baltic Sea physics gained during the BALTEX and related programmes. Progr Oceanogr 63, Issues 1–2:1–28

Palz W, Greif J (1996) European Solar Radiation Atlas. Springer, Berlin Heidelberg New York

Pavelson J (1988) Nature and some characteristics of thermohaline fronts in the Baltic Proper. In: Proceedings of the 16th Conference of the Baltic Oceanographers, Kiel, pp 796–805

Pavelson J, Laanemets J, Kononen K, Nõmmann S (1997) Quasi-permanent density front at the entrance to the Gulf of Finland: Response to wind forcing. Cont Shelf Res 17,3:253–265

Peltonen H, Vinni M, Lappalainen A, Pönni J (2004) Spatial distribution patterns of herring (*Clupea harengus* L) sprat (*Sprattus sprattus* L) and the three-spined stickleback (*Gasterosteus aculeatus* L) in the Gulf of Finland, Baltic Sea. ICES J Mar Sci 61:966–971

Peltonen K (2002) Direct ground water inflow to the Baltic Sea. TemaNord 2002:503, Nordic Council of Ministers, Copenhagen, Denmark

Pershagen H (1981) Maxisnödjup i Sverige 1905–76. (Maximum snow cover in Sweden 1905–76) SMHI Reports Meteorology and Climatology, RMK 29 (in Swedish)

Petersen G, Krueger O (1993) Untersuchung und Bewertung des Schadstoffeintrags über die Atmosphäre im Rahmen von PARCOM (Nordsee) und HELCOM (Ostsee) – Teilvorhaben: Modellierung des großräumigen Transports von Spurenmetallen (Investigation and assessment of atmospheric deposition of pollutants within the framework of PARCOM (North Sea) und HELCOM (Baltic Sea) – Subproject: Modelling large scale transport of trace metals). GKSS Research Centre Geesthacht, Germany, GKSS 93/E/28 (in German)

Petersen G, Bloxam R, Wong S, Munthe J, Krüger O, Schmolke SR, Vinod Kumar A (2001) A comprehensive Eulerian modelling framework for airborne mercury species: Model development and applications in Europe. Atmos Env 35,17:3063–3074

Petersen JK, Loo LO (2004) Miljøkonsekvenser af dyrkning af blåmuslinger (Environmental consequences of common mussel cultivation), www.fvm.dk (in Danish)

Petersen KS, Rasmussen KL, Heinemeier J, Rud N (1992) Clams before Columbus. Nature 359:679

Pettit AN (1979) A non-parametric approach to the change point problem. App Statist 26–135

Piechura J, Beszczynska-Möller A (2004) Inflow waters in the deep regions of the southern Baltic sea-transport and transformations. Oceanologia 46,1:113–141

Piechura J, Walczowski W, Beszczynska-Möller A (1997) On the structure and dynamics of the water in the Slupsk Furrow. Oceanologia 39,1:35–54

Pizarro O, Shaffer G (1998) Wind-driven coastal-trapped waves off the Island of Gotland, Baltic Sea. J Phys Oceanogr 28,11:2117–2129

Poorter H, Navas ML (2003) Plant growth and competition at elevated CO2: On winners, losers, and functional groups. New Phytologist 157:175–198

Pussinen A, Meyer J, Zudin S, Lindner M (2005) European mitigation potential. In: Kellomäki S, Leinonen S (eds) Management of European Forests Under Changing Climatic Conditions. University of Joensuu, Faculty of Forestry Research. Notes 163:383–400

Pussinen A, Schelhaas MJ, Verkaik E, Heikkinen E, Liski J, Karjalainen T, Päivinen R, Nabuurs GJ (2001) Manual for the European Forest Information Scenario Model (EFISCEN 20). European Forest Institute, Joensuu, Finland, EFI, Internal Report 5:49

Raab B, Vedin H (1995) Climate, lakes, and rivers. National Atlas of Sweden

Räisänen J, Hansson U, Ullerstig A, Döscher R, Graham LP, Jones C, Meier M, Samuelsson P, Willén U (2003) GCM driven simulations of recent and future climate with the Rossby Centre coupled atmosphere – Baltic Sea regional climate model RCAO. SMHI Reports Meteorology and Climatology 101, SMHI, Norrköping, Sweden

Raudsepp U (1998) Current dynamics of estuarine circulation in the Lateral Boundary Layer. Estuar Coast Shelf Sci 47:715–730

Raudsepp U (2001) Interannual and seasonal temperature and salinity variations in the Gulf of Riga and corresponding saline water inflow from the Baltic Proper. Nordic Hydrology 32,2:135–160

Raudsepp U, Beletsky D, Schwab DJ (2003) Basin scale topographic waves in the Gulf of Riga. J Phys Oceanogr 33,5:1129–1140

Remane A (1934) Die Brackwasserfauna (Mit besonderer Berücksichtigung der Ostsee) (Brackish water fauna, with special emphasis on the Baltic Sea). Verhandlungen der Deutschen Zoologischen Gesellschaft 36:34–74 (in German)

Roads J, Raschke E, Rockel B (2004) BALTEX water and energy budgets in the NCEP/DOE reanalysis. II. Boreal Env Res 7,4:307–318

Robarts RD, Zohary T (1987) Temperature effects on photosynthetic capacity, respiration, and growth rates of bloom-forming cyanobacteria. N.Z. J Mar Freshwat Res 21:391–399

Rodhe H, Soederlund R, Ekstedt J (1980) Depositon of airborne pollutants on the Baltic Sea. Ambio 9: 168–173

Rodhe J (1998) The Baltic and the North Seas: A process-oriented review of the Physical Oceanography. In: Robinson A, Brink K (eds) The Sea, vol. 11. Wiley, New York, pp. 699–732

Roemer M, Beekmann M, Bergström R, Boersen G, Feldmann H, Flatøy F, Honore C, Langner J, Jonson JE, Matthijsen J, Memmesheimer M, Simpson D, Smeets P, Solberg S, Stern R, Stevenson D, Zandveld P, Zlatev Z (2003) Ozone trends according to ten dispersion models. EUROTRAC-2 special report, EUROTRAC International Scientific Secretariat, GSF – National Research Center for Environment and Health, Munich, Germany

Rogers JC (1984) The association between the North Atlantic Oscillation and the Southern Oscillation in the Northern Hemisphere. Mon Wea Rev 107:1999–2015

Rozwadowska A, Isemer HJ (1998) Solar radiation fluxes at the surface of the Baltic Proper. Part 1. Mean annual cycle and influencing factors. Oceanologia 40,4:307–330

Rubel F, Hantel M (1999) Correction of daily rain gauge measurements in the Baltic Sea drainage basin. Nordic Hydrology 30:191–208

Rubel F, Hantel M (2001) BALTEX 1/6-degree daily precipitation climatology 1996–1998. Met Atmos Phys 77:155–166

Rudstam LG, Hansson S (1990) On the ecology of *Mysis mixta* (Crustacea, Mysidacea) in a coastal area of the northern Baltic proper. Ann Zool Fennici 27:259–263

Ruprecht E, Kahl T (2003) Investigation of the atmospheric water budget of the BALTEX area using NCEP/NCAR reanalysis data. Tellus 55A:426–437

Rutgersson A, Bumke K, Clemens M, Foltescu V, Lindau R, Michelson D, Omstedt A (2001) Precipitation estimates over the Baltic Sea: Present state of the art. Nordic Hydrology 32:285–314

Rytkönen J, Siitonen L, Riipi T, Sassi J, Sukselainen J (2002) Statistical analyses of the Baltic maritime traffic. VTT Industrial Systems, Espoo. 108 s. + liitt. 44 s. VTT Research Report VAL34–012344

Sadowski M (1980) Rozkład przestrzenny zapasu wody w pokrywie śnieżnej w Polsce (Spatial distribution of water storage in the snow cover in Poland) Materiały Badawcze IMGW, ser Hydrologia i Oceanologia (in Polish)

Salemaa H, Tyystjärvi-Muuronen K, Aro E (1986) Life histories distribution and abundance of *Mysis mixta* and *Mysis relicta* in the northern Baltic Sea. Ophelia – Supplements 4:239–247

Sallnäs O (1990) A matrix model of the Swedish forest. Studia Forestalia Suecica 183:1– 23

Schelhaas MJ, Varis S, Schuck A, Nabuurs GJ (1999) EFISCEN's European Forest Resource Database European Forest Institute, Joensuu, Finland, www.efi.fi/projects/eefr

Schinke H, Matthäus W (1998) On the causes of major Baltic inflows – an analysis of long time series. Continent Shelf Res 18:67–97

Schneider B (1993) Untersuchung und Bewertung des Schadstoffeintrags über die Atmosphäre im Rahmen von PARCOM (Nordsee) und HELCOM (Ostsee) – Teilvorhaben: Messungen von Spurenmetallen (Investigation and assessment of atmospheric deposition of pollutants within the framework of PARCOM (North Sea) und HELCOM (Baltic Sea) – Subproject: Measurement of trace metals). GKSS Research Centre Geesthacht, Germany, GKSS 93/E/53 (in German)

Schneider B, Ceburnis D, Marks R, Munthe J, Petersen G, Sofiev M (2000) Atmospheric Pb and Cd input into the Baltic Sea: A new estimate based on measurements. Marine Chemistry 71:297–307

Schneider G (1990) Metabolism and standing stock of the winter mesozooplankton community in the Kiel Bight/western Baltic. Ophelia 32:237–247

Scott D, McBoyle G (2001) Using a 'tourism climate index' to examine the implications of climate change for climate as a tourism resource. In: Matzarakis A, de Freitas CR (eds) Proceedings of the First International Workshop on Climate Tourism and Recreation. International Society of Biometeorology, Commission on Climate Tourism and Recreation, Porto Carras, Halkidiki, Greece, WP4, pp. 1–10

Scott WB, Crossman EJ (1973) Freshwater fishes of Canada. Fish Res Board Can Bull 184:641–645

Segerstråle SG (1950) The amphipods on the coasts of Finland – some facts and problems. Soc Sci Fenn Comm Biol 10:1–26

Sellner KG, Olson MM, Olli K (1996) Copepod interactions with toxic and non-toxic cyanobacteria from the Gulf of Finland. Phycologia Suppl 35:177–182

Shaver GR, Canadell J, Chapin FS III, Gurevitch J, Harte J, Henry G, Ineson P, Jonasson S, Melillo J, Pitelka L, Rustad L (2000) Global warming and terrestrial ecosystems: A conceptual framework for analysis. BioScience 50:871–882

Sneyers R (1975) Sur l'analyse statistique des series d'observations (On the statistical analysis of observation series). Note technique No 143, WMO

Sofiev M, Grigoryan S (1996) Numerical modelling of hemispheric air transport of acid compounds. Comparison of three approaches. MSC-E Report 6/96, Moscow

Sofiev M, Maslyaev A, Gusev A (1996) Heavy metal model intercomparison. Methodology and results for Pb in 1990. MSC-E Report 2/96, Moscow

Sofiev M, Petersen G, Krueger O, Schneider B, Hongisto M, Jylha K (2001) Model simulations of the atmospheric trace metals concentrations and depositions over the Baltic Sea. Atmos Env 35,8:1395–1409

Sohlenius G, Emeis KC, Andrén E, Andrén T, Kohly A (2001) Development of anoxia during the Holocene fresh-brackish water transition in the Baltic Sea. Mar Geol 177:221–242

Sterr H, Ittekkot V, Klein RJT (1999) Weltmeere und Küsten im Wandel des Klimas (Oceans and coasts in a changing climate). PGM 143:24–31

Stettler J (1998) Natursport und Mobilität (Nature sports and mobility). In: Deutscher Sportbund (DSB) (ed): Sport und Mobilität. Dokumentation des 5. Symposiums zur ökologischen Zukunft des Sports im September 1997 in Bodenheim. Selbstverlag, Frankfurt

Stigebrandt A (1987) Computations of the flow of dense water into the Baltic Sea from hydrographical measurements in the Arkona Basin. Tellus 39A:170–177

Stigebrandt A (2001) Physical oceanography of the Baltic Sea. In: Wulff F, Rahm L, Larsson P (eds) A Systems Analysis of the Baltic Sea. Ecological Studies 148:19–68

Stigebrandt A, Gustafsson BG (2003) Response of Baltic Sea to climate change – Theory and observations. J Sea Res 49,4:243–256

Stigebrandt A, Lass HU, Liljebladh B, Alenius P, Piechura J, Hietala R, Beszczynska A (2002) DIAMIX – an experimental study of diapycnal deepwater mixing in the virtually tideless Baltic Sea. Boreal Env Res 7,4:363–369

Stipa T (2003) Baroclinic adjustment in the Finnish coastal current. Tellus 56A,1:79–87

Stipa T, Tamminen T, Seppälä J (1999) On the creation and maintenance of stratification in the Gulf of Riga. J Mar Syst 23:27–46

Strömgren M, Linder S (2002) Effects of nutrition and soil warming on stemwood production in a boreal Norway spruce stand. Global Change Biol 8:1195–1204

Sykes MT, Prentice IC (1995) Boreal forest futures: Modelling the controls on tree species range limits and transient responses to climate change. Water Air Soil Pollut 82:415–428

Sykes MT, Prentice IC, Cramer W (1996) A bioclimatic model for the potential distributions of north European tree species under present and future climates. J Biogeogr 23:203–233

Thompson DJW, Wallace JM (1998) The Arctic Oscillation signature in the wintertime geopotential height and temperature fields. Geophys Res Lett 25:1297–1300

Thompson DJW, Wallace JM (2000) Annular modes in the extratropical circulation. Part I: Month-to-month variability. J Clim 13:1000–1016

Thompson DJW, Wallace JM (2001) Regional Climate Impacts of the Northern Hemisphere Annular Mode. Science 293:85–89

Tooming H, Kadaja J (1999) Climate changes indicated by trends in snow cover duration and surface albedo in Estonia. Meteorol Z, NF 8:16–21

Tooming H, Kadaja J (2006) Handbook of Estonian snow cover. Estonian Meteorological and Hydrological Institute, Tallinn-Saku

Tooming H, Kadaja J (2000a) Snow cover and surface albedo in Estonia. Meteorol Z, NF 9,2:97–102

Tooming H, Kadaja J (2000b) Eesti lumikatte atlas (Snow cover atlas of Estonia). EMHI, Tallinn, 305 pp. (in Estonian)

Tveito OE, Førland EJ, Alexandersson H, Drebs A, Jonsson T, Vaarby-Laursen E (2001) Nordic climate maps. DNMI report 06/01 KLIMA

UBA (Umweltbundesamt) (2003) Mobilitätsstile in der Freizeit (Mobility styles during leisure time). Berichte 2/03, Berlin (in German)

Uitto A, Heiskanen AS, Lignell R, Autio R, Pajuniemi R (1997) Summer dynamics of the coastal planktonic food web in the northern Baltic Sea. Mar Ecol Prog Ser 151:27–41

UNESCO (1985) The International System of Units (SI) in Oceanography: Report of the IAPSO Working Group on Symbols, Units, and Nomenclature in Physical Oceanography (SUN) IAPSO Publication Scientifique No 32; UNESCO technical papers in marine science No 45

Uppala SM, Kållberg PW, Simmons AJ, Andrae U, da Costa Bechtold V, Fiorino M, Gibson JK, Haseler J, Hernandez A, Kelly GA, Li X, Onogi K, Saarinen S, Sokka N, Allan RP, Andersson E, Arpe K, Balmaseda MA, Beljaars ACM, van de Berg L, Bidlot J, Bormann N, Caires S, Chevallier F, Dethof A, Dragosavac M, Fisher M, Fuentes M, Hagemann S, Hólm E, Hoskins BJ, Isaksen L, Janssen PAEM, Jenne R, McNally AP, Mahfouf JF, Morcrette JJ, Rayner NA, Saunders RW, Simon P, Sterl A, Trenberth KE, Untch A, Vasiljevic D, Viterbo P, Woollen J (2005) The ERA-40 re-analysis. Q J Roy Met Soc 131:2961–3012doi:101256/qj04176

Urho L (2002) The importance of larvae and nursery areas for fish production. PhD thesis, University of Helsinki, pp. 118

van Loon H, Rogers JC (1978) The seesaw in winter temperatures between Greenland and northern Europe. Part I: General description. Mon Wea Rev 106:296–310

Viherluoto M, Kuosa H, Flinkman J, Viitasalo M (2000) Food utilisation of pelagic mysids *Mysis mixta* and *M. relicta* during their growing season in the northern Baltic Sea. Mar Biol 136:553–559

Viitasalo M, Katajisto T (1994) Mesozooplankton resting eggs in the Baltic Sea: Identification and vertical distribution in laminated and mixed sediments. Mar Biol 120:455–465

Viitasalo M, Vuorinen I, Saesmaa S (1995) Mesozooplankton dynamics in the northern Baltic Sea: implications of variations in hydrography and climate. J Plank Res 17:1857–1878

von Storch H, Zwiers FW (2002) Statistical Analysis in Climate Research. Cambridge, Cambridge University Press

von Storch H, Langenberg H, Feser F (2000) A spectral nudging technique for dynamical downscaling purposes. Mon Wea Rev 128:3664–3673

Waldron KM, Paegle J, Horel JD (1996) Sensitivity of a spectrally filtered and nudged limited-area model to outer model options. Mon Wea Rev 124:529–547

Walker GT (1924) Correlation in seasonal variation of weather, IX Mem. Ind Met Dept 25:275–332

Walker MD, Ingersoll RC, Webber PJ (1995) Effects of interannual climate variation on phenology and growth of two alpine forbs. Ecology 76:1067–1083

Wallace JM (2000) North Atlantic Oscillation /annular mode: Two paradigms – one phenomenon. Q J Roy Met Soc 126:791–805

Wallace JM, Gutzler DS (1981) Teleconnections in the geopotential height field during the Northern Hemisphere winter. Mon Wea Rev 109:784–812

Wallin M, Håkanson L, Persson J (1992) Belastningsmodeller för närsaltsutsäpp i kustvatten (Modelling nutrient discharge in coastal waters). Speciellt fiskodlingars miljöpåverkan – Nordiska ministerrådet, 1992:502 (in Swedish)

Wanner HS, Brönnimann S, Casty C, Gyalistras D, Luterbacher J, Schmutz C, Stephenson DB, Xoplaki E (2001) North Atlantic Oscillation – concepts and studies. Survey Geophys 22:321–381

Waring RH, Landsberg JJ, Williams M (1998) Net primary production of forests: A constant fraction of gross primary production? Tree Physiology 18:129–134

Werner I, Auel H (2004) Environmental conditions and overwintering strategies of planktonic metazoans in and below coastal fast ice in the Gulf of Finland (Baltic Sea). Sarsia 89:102–116

Winsor P, Rodhe J, Omstedt A (2001) Baltic Sea ocean climate: An analysis of 100 yr of hydrographic data with focus on the freshwater budget. Clim Res 18:5–15

Winsor P, Rodhe J, Omstedt A (2003) Erratum: Baltic Sea ocean climate: An analysis of 100 yr of hydrographic data with focus on the freshwater budget. Clim Res 18:5–15, 2001, Clim Res 25:183

Woodward FI (1987) Climate and Plant Distribution. Cambridge University Press, Cambridge

Xie P, Arkin P (1997) Global precipitation: A 17-year monthly analysis based on gauge observations satellite estimates and numerical model outputs. Bull Am Met Soc 78:2539–2558

Zhurbas V, Paka VT (1997) Mesoscale thermohaline variability in the Eastern Gotland Basin following the 1993 major Baltic inflow. J Geophys Res 102,C9:20917–20926

Zhurbas V, Paka VT (1999) What drives thermohaline intrusions in the Baltic Sea? J Mar Syst 21,1–4:229–241

Zhurbas V, Oh IS, Paka VT (2003) Generation of cyclonic eddies in the Eastern Gotland Basin of the Baltic Sea following dense water inflows: Numerical experiments. J Mar Syst 38:323–336

Zhurbas V, Stipa T, Mälkki P, Paka V, Golenko N, Hense I, Sklyarov V (2004) Generation of subsurface cyclonic eddies in the southeast Baltic Sea: Observations and numerical experiments. J Geophys Res 109 C05033 doi:101029/2003JC002074

Zlatev Z, Bergstrom R, Brandt J, Hongisto M, Johnson JE, Langner J, Sofiev M (2002) Studying sensitivity of air pollution levels caused by variations of different key parameters. TemaNord 2001:569. Nordic Council of Ministers, Copenhagen 2001

Acronyms and Abbreviations

If an acronym or abbreviation is used once or twice only locally, it is not necessarily included in this list.

3D	Three-dimensional
ACCELERATES	Assessing Climate Change Effects on Land Use and Ecosystems – From Regional Analyses to the European Scale, EU-funded project
ACIA	Arctic Climate Impact Assessment
AIS	Automatic Identification System
AO	Arctic Oscillation
AOGCM	Coupled Atmosphere-Ocean General Circulation Model
AOT-40	Ozone Index: Accumulated Ozone above the Threshold of 40 ppb
a.s.l	above sea level
ATEAM	Advanced Terrestrial Ecosystem Analysis and Modelling, EU-funded project
BACC	BALTEX Assessment of Climate Change for the Baltic Sea Basin
BALTEX	Baltic Sea Experiment
Baltic States	Estonia, Latvia and Lithuania
BASYS	Baltic Sea System Study, EU-funded project
BIL	Baltic Ice Lake
BP	Before Present
BSH	Bundesamt für Seeschifffahrt und Hydrography, Germany
C	Carbon
cal yr BP	Calibrated Years Before Present
CAVM	Circumpolar Arctic Vegetation Map
CC	Cloud Cover
CCEP	Climate Change and Energy Production
CCIRG	UK Climate Change Impacts Review Group
Cd	Cadmium
CDD	Consecutive Duration of Dry Days with Precipitation less than $1\,\mathrm{mm}$
CH_4	Methane
CMAP	Climate Prediction Centre Merged Analysis of Precipitation
CMIP2	Coupled Model Intercomparison Project, Phase 2
CO_2	Carbon Dioxide
$CO_3{}^{2-}$	Carbonate
CI	Confidence Interval
CPR	Continuous Plankton Recorder
CRU	Climate Research Unit at the University of East Anglia
CTD	Oceanographic device to measure Conductivity, Temperature, Depth
Cu	Copper
CWP	Clear-Water Phase

DDT	Dichloro-Diphenyl-Trichloroethane (a synthetic pesticide)
DIN	Dissolved Inorganic Nitrogen
DIP	Dissolved Inorganic Phosphorus
DJF	December–January–February
DOC	Dissolved Organic Carbon
DOM	Dissolved Organic Matter
DSi	Dissolved Silicate
DTR	Daily Temperature Range
DWD	Deutscher Wetterdienst (German Weather Service)
ECA	European Climate Assessment & Dataset, EU-funded project
ECHAM	Global climate model developed by Max-Plank-Institute Hamburg, Germany
ECE	Economic Commission for Europe
ECMWF	European Centre for Medium-Range Weather Forecasts
EMEP	European Monitoring and Evaluation Program
EMHI	Estonian Meteorological and Hydrological Institute
ENSEMBLES	Ensemble-based Predictions of Climate Changes and their Impacts, EU-funded project
ENSO	El Niño / Southern Oscillation
ERA-15	ECMWF Re-analysis Dataset 1978–1994
ERA-40	ECMWF Re-analysis Dataset 1958–2001
EU	European Union
FAD	First Arrival Date
FAO	Food and Agricultural Organisation (UN)
FIMR	Finnish Institute of Marine Research
GCM	Global Climate Model or General Circulation Model
GDR	German Democratic Republic
GESAMP	Joint Group of Experts on the Scientific Aspects of Marine Environmental Protection
GFDL	Geophysical Fluid Dynamics Laboratory
GHG	Greenhouse Gas
GIS	Geographical Information Systems
GISS	Goddard Institute for Space Studies
GPCP	Global Precipitation Climatology Project
GPP	Gross Primary Production
GPS	Global Positioning System
HCB	Hexa-Chloro-Benzene (a synthetic pesticide)
HCH	Beta-Benzene-Hexa-Chloride (a synthetic pesticide)
HCO_{3-}	Hydrogen Carbonate
HELCOM	Baltic Marine Environment Protection Commission
HIRHAM	Regional climate model, based on HIRLAM and ECHAM
HIRLAM	High Resolution Limited Area Model
Hg	Mercury
H_2S	Hydrogen Sulfide

IBFSC	International Baltic Sea Fisheries Commission
ICES	International Council for the Exploration of the Sea
IDAG	International ad-hoc Detection and Attribution Group
ILMAVA	Effect of Climate Change on Energy Resources in Finland, Finnish project
IMO	International Maritime Organization
INTAS	The International Association for the Promotion of Co-operation with Scientists from the New Independent States of the Former Soviet Union
IOC	Intergovernmental Oceanographic Commission
IOW	Baltic Sea Research Institute Warnemünde, Germany
IPCC	Intergovernmental Panel on Climate Change
IPG	International Phenological Gardens
JJA	June–July–August
K	Potassium
LAI	Leaf Area Index
MAM	March–April–Mai
MARPOL	International Convention for the Prevention of Pollution from Ships
MBI	Major Baltic Inflow
MIB	Maximum Annual Ice Extent in the Baltic Sea
MICE	Modelling the Impact of Climate Extremes, EU-funded project
MMT	Mean Migration Time
N	Nitrogen
N_2	Molecular Nitrogen
N_2O	Nitrous Oxide
NAM	Northern Hemisphere Annular Mode
NAO	North Atlantic Oscillation
NAOw	Winter North Atlantic Oscillation
NCAR	National Centre of Atmospheric Research
NCEP	National Centres for Environmental Prediction
NDVI	Normalized Differenced Vegetation Index
NGO	Non-Governmental Organisation
NH	Northern Hemisphere
NH_{4+}	Ammonium
NO_{2-}	Nitrite
NO_{3-}	Nitrate
NO_x	Nitrogen Oxides
NPP	Net Primary Production
O_2	Molecular Oxygen
O_3	Ozone
P	Phosphorus
PAH	Polycyclic Aromatic Hydrocarbons (organic pollutants)
PAX	Baltic Sea Patchiness Experiment
Pb	Lead
PBDE	Polybrominated Diphenyl Ethers (organic pollutants)

PCB	Polychlorinated Biphenyls (organic pollutants)
PEN	Potential Excess Nitrogen
POP	Persistent Organic Pollutant
POPCYCLING	Environmental Cycling of Persistent Organic Pollutants in the Baltic Region, EU-funded project
POSITIVE	Phenological Observations and Satellite Data: Trends in the Vegetation Cycle in Europe, EU-funded project
ppb	parts per billion
ppm	parts per million
PRUDENCE	Predictions of Regional Scenarios and Uncertainties for Defining European Climate Change Risks and Effects, EU-funded project
psu	Practical Salinity Unit
q850	Absolute Humidity at 850 hPa
RACCS	Regionalisation of Anthropogenic Climate Change Simulations, EU-funded project
RCA	Regional Climate Atmosphere Model
RCAO	Regional Climate Atmosphere Ocean Model
RCM	Regional Climate Model
RCO	Regional Climate Baltic Sea Model
RegClim	Regional Climate Development under Global Warming, Nordic project
S	Sulphur
Si	Silicate
SILMU	Finnish Research Programme on Climate Change
SilviStrat	Silvicultural Response Strategies to Climatic Change in Management of European Forests, EU-funded project
SLP	Sea Level Air Pressure
SMHI	Swedish Meteorological and Hydrological Institute
$SO_4{}^{2-}$	Sulphate
SON	September–October–November
SRES	Special Report on Emissions Scenarios
SST	Sea Surface Temperature
STARDEX	Statistical and Regional Dynamical Downscaling of Extremes for European Regions, EU-funded project
SW	South-West
SWECLIM	Swedish Regional Climate Modelling Programme
TAR	Third Assessment Report
TBT	Tri-Butyl-Tin (an organic pollutant)
TCI	Tourism Climate Index
THC	Thermohaline Circulation
TIN	Total Inorganic Nitrogen
Tn	Daily Minimum Air Temperature
TOC	Total Organic Carbon
Tx	Daily Maximum Air Temperature

UK	United Kingdom
UKMO	United Kingdom Meteorological Office
UN	United Nations
UNFCCC	United Nation's Framework Convention on Climate Change
UN-ECE	United Nations Economic Commission for Europe
UNEP	United Nations Environment Programme
UNESCO	United Nations Educational, Scientific and Cultural Organization
US	United States
USSR	Union of Soviet Socialist Republics
UV	Ultraviolet
VASClimO	Variability Analysis of Surface Climate Observations, German research project
WASA	Waves and Storms in the North Atlantic, EU-funded project
WMO	World Meteorological Organisation
WWF	World Wide Fund for Nature
YOY	Young of the Year
Zn	Zinc